NATURAL SCIENCES IN AMERICA

NATURAL SCIENCES IN AMERICA

Advisory Editor
KEIR B. STERLING

Editorial Board
EDWIN H. COLBERT
EDWARD GRUSON
ERNST MAYR
RICHARD G. VAN GELDER

QL
681
C85
1974
vol.1

DISCARDED

QL681 .C85 1974 v.1
Coues Elliott 1842
Key to North American birds.

COLORADO MOUNTAIN COLLEGE
LRC--WEST CAMPUS
Glenwood Springs, CO 81601

KEY
TO
NORTH AMERICAN BIRDS

By ELLIOTT COUES

Volume I

ARNO PRESS
A New York Times Company
New York, N. Y. • 1974

Reprint Edition 1974 by Arno Press Inc.

Reprinted from a copy in the Newark
  Public Library

NATURAL SCIENCES IN AMERICA
ISBN for complete set: 0-405-05700-8
See last pages of this volume for titles.

Manufactured in the United States of America

Library of Congress Cataloging in Publication Data

Coues, Elliott, 1842-1899.
    Key to North American birds.

    (Natural sciences in America)
    Reprint of the fifth ed. published in 1903 by
D. Estes, Boston.
    1. Birds--North America. 2. Birds--Collection
and preservation. I. Title. II. Series.
QL681.C85   1974   598.2'973   73-17816
ISBN 0-405-05732-6

KEY
TO
NORTH AMERICAN BIRDS.

Starlings
(Sturnus Vulgaris)

KEY
TO
NORTH AMERICAN BIRDS.

CONTAINING A CONCISE ACCOUNT OF EVERY SPECIES OF LIVING AND FOSSIL
BIRD AT PRESENT KNOWN FROM THE CONTINENT NORTH OF THE
MEXICAN AND UNITED STATES BOUNDARY, INCLUSIVE
OF GREENLAND AND LOWER CALIFORNIA,

WITH WHICH ARE INCORPORATED

GENERAL ORNITHOLOGY:
AN OUTLINE OF THE STRUCTURE AND CLASSIFICATION OF BIRDS;

AND

FIELD ORNITHOLOGY,
A MANUAL OF COLLECTING, PREPARING, AND PRESERVING BIRDS.

The Fifth Edition,
(ENTIRELY REVISED)

EXHIBITING THE NOMENCLATURE OF THE AMERICAN ORNITHOLOGISTS' UNION, AND INCLUDING
DESCRIPTIONS OF ADDITIONAL SPECIES

IN TWO VOLUMES.
VOLUME I.

By ELLIOTT COUES, A.M., M.D., PH.D.,

Late Captain and Assistant Surgeon U. S. Army and Secretary U. S. Geological Survey; Vice-President of the American
Ornithologists' Union, and Chairman of the Committee on the Classification and Nomenclature of North American Birds;
Foreign Member of the British Ornithologists' Union; Corresponding Member of the Zoölogical Society
of London; Member of the National Academy of Sciences, of the Faculty of the National
Medical College, of the Philosophical and Biological Societies of Washington.

PROFUSELY ILLUSTRATED.

BOSTON:
DANA ESTES AND COMPANY.
1903.

Entered according to Act of Congress, in the year 1872, by
F. W. PUTNAM AND ELLIOTT COUES,
In the Office of the Librarian of Congress at Washington.

Entered according to Act of Congress, in the year 1874, by
F. W. PUTNAM AND ELLIOTT COUES,
In the Office of the Librarian of Congress at Washington.

Copyright, 1882, 1884, and 1887,
BY ESTES AND LAURIAT.

Copyright, 1903,
BY DANA ESTES & CO.

UNIVERSITY PRESS:
JOHN WILSON AND SON, CAMBRIDGE.

To

SPENCER FULLERTON BAIRD,

NESTOR OF AMERICAN ORNITHOLOGISTS,

This Work,

BEARING TO OTHERS THE TORCH RECEIVED FROM HIM IN EARLIER DAYS,

Is Dedicated.

PUBLISHER'S PREFACE TO FIFTH REVISED EDITION.

THE present work constitutes the completion of Dr. Coues' life-long labors on behalf of the science of ornithology, too widely known and appreciated to require further mention here. In preparing it for publication the publishers have suffered extraordinary expense, difficulty, and delay by the loss of Dr. Coues' assistance in the proof-reading and illustrating of the book. The manuscript was finished but shortly before his death, and though fortunately complete in this form, was left in such shape as to present almost insuperable difficulties to the compositor or proof-reader, who lacked the author's direction and supervision.

The publishers have had the good fortune to secure the services of Mr. J. A. Farley, who has read the manuscript of the Systematic Synopsis, constituting Part Three or the body of the work, with the most painstaking care. To the scholarly zeal and conscientious spirit of fidelity and accuracy with which this ornithologist has carried out the task he set himself of presenting the fifth edition in exactly the form Dr. Coues would have wished, had he lived, the publishers and their readers owe an unlimited debt of gratitude. The result, though a posthumous book, is one which Dr. Coues would unquestionably have been proud to own as the crowning work of his life. As a scientific work, it is without doubt authoritative and definitive.

The science of ornithology has made vast strides since the publication of the fourth edition of this work, and the present issue has outgrown the limits of a single octavo volume. The following points briefly summarize the scope of the additions and changes from former editions:

1. Enlarged descriptions of species.

2. Accounts much fuller than in former editions, of the breeding habits of birds, particularly the detailed description of eggs.

3. The full collation in the *text* (not in an appendix, as in former editions) of the nomenclature of species in the Key, with the nomenclature and numeration of the American Ornithologists' Union Check-List.

4. The full synonymies and bibliographical references in the case of very many species — a new feature of the Key, and invaluable to students of all degrees of advancement. To the preparation of this important feature Dr. Coues brought his rare gifts as a bibliographer and nomenclator.

5. The previous very extensive series of illustrations has been largely increased by the addition of over two hundred new figures of species hitherto seldom figured, from life studies by Louis Agassiz Fuertes executed with a delicacy, beauty, and accuracy never before equalled.

6. The introductory (*i. e.* general) descriptions of ordinal, family, and other groups are much amplified over those in preceding editions of the Key, being of a broad scope which make plain the comparative relationships of North American families, genera, and species of birds, with extralimital forms (Old World and neotropical). This broad treatment makes the Key more than the purely faunal work its title would imply.

7. An appendix containing the additions to the American Ornithologists' Union-Check List of North American Birds and the changes in nomenclature not noted elsewhere which have been made since Dr. Coues' death.

DANA ESTES AND COMPANY.

BOSTON, October, 1903.

PREFACE TO THE FOURTH EDITION.

IN presenting a new edition of the KEY to those who are interested in North American Birds, the publishers desire the author to add a word by way of preface. But little need be said of a book which speaks for itself in passing through several editions to supply that demand for a standard textbook of ornithology which this work has itself done much to create, by stimulating and satisfying an interest in one of the most delightful departments of Natural History.

The part which the KEY has taken in the evolution of the subject since 1872 is sketched in the "Historical Preface" (pp. xxvi–xxx), first introduced in the Second Edition, 1884. Since the founding of the American Ornithologists' Union in 1883 the impetus then given to the study of birds has resulted in a momentum directly proportionate to the number of workers in this field and to the length of time these have been engaged. I could wish the fruits of such unparalleled activity were all sound and ripe, but they are not; growth has been forced to some extent in rival hot-houses, and the familiar parable of wheat and tares finds a fresh illustration. Too quick transition from an old to a new order of things in the technicalities of our subject has brought disorder, as usual. Till the pace slackens somewhat, so that we can see where we stand, I do not think it would be wise to recast the KEY.

Therefore, the only change in the present edition is the addition of a Second Appendix, beginning page 897.

E. C.

PREFACE TO THE THIRD EDITION.

THE second edition of the "Key," which appeared in May, 1884, has already been out of print for more than a year. Though aware of the continued demand for a standard work of reference, the author has been unable to meet it more promptly, having meanwhile accepted some other literary engagements which proved imperative in their demand upon his capacity for work. Slight as the requisite revision of this book has proven to be, it did not seem expedient to go to press again without recognizing the steps American Ornithology has taken during the past three years, though these may be called many rather than great ones. There is so little to change in the substance of the book that it has been thought decidedly best to reprint from the same plates, and put what new matter has come to hand in the form of an Appendix. However much there is that might have advantageously gone into the second edition, but did not, the author is satisfied with nearly everything that did go in, and quite ready to submit it all to the still further test of time. The transition from what some of his friends have called the "Couesian Period" may mean a change in form rather than in fact.

The naming of our birds, as an art distinguished from the science of knowing them, has lately been pitched in a key so high that the familiar notes of the former "Key" might jangle out of tune, or be lost entirely, were the attempt made to reset them just now. During the confusion unavoidably incident to such sweeping changes in nomenclature as we have recently made, it will be a decided benefit to the student, the sportsman, and the amateur, if not also to every working ornithologist, to be provided with a convenient means of comparing the older with the newer style of nomenclature we have adopted, until each one shall have grown accustomed to the change of spectacles. This accommodation is afforded by the present edition, which leaves the names and their numbers untouched in the body of the text, and then adjusts them to the new angle of vision in the Appendix, in parallel columns. Thus the new "Key" turns either way; or, to vary the metaphor, the renovated structure stands Janus-faced, looking both ways at once — backward upon its old self, of which it has no cause to be ashamed; forward upon another self, of which it has much reason to be proud.

The train of incidents which resulted in what may be called a nomenclatural explosion was fired at the founding of the American Ornithologists' Union at New York, in September, 1883. As one of three persons who brought that happy episode upon an unsuspecting bird-world, which nevertheless greeted their stroke with acclamation, the author must plead a modesty act in bar of trial of his pen on that particular count. But as the honor was his of presiding over the first Congress of the Union, whilst the ideas of its founders were shapen into a permanent and world-wide organization, so also it fell to his lot to appoint several committees for the despatch of business the Union at once took in hand; and of one of these he has to speak here.

This particular wheel within other wheels turned upon a resolution of the Union "that the Chairman appoint a committee of five, including himself, to whom shall be referred the question of a revision of the Classification and Nomenclature of the Birds of North America." Having accepted the situation, the author held with his esteemed colleagues many sessions of the Committee in Washington and New York, and in April, 1885, offered to the Union the result of much joint labor. The report of the Committee being accepted, it was ordered to be printed, and it appeared in 1886 in an octavo volume of 400 pages, entitled "The Code of Nomenclature and Check-list of North American Birds, adopted by the American Ornithologists' Union," etc.

The objects which we kept steadily in view were: first, to establish certain sound principles or canons of nomenclature applicable to zoölogy at large as well as to ornithology; and, secondly, to apply these rules consistently and effectually to the naming of North American birds. Others must be left to judge how well or ill these purposes may have been accomplished, but the simple fact is that no sooner had the book appeared than it became the standard and indeed the only recognized Nomenclator in American Ornithology. That which the Committee had stamped with the seal of the Union became the current coin of the realm, other than which our venerable fowl, The Auk, should know none.

In estimating the probable consequences for the long run, it is necessary to discriminate between any given ornithological fact and the handle we may agree to give that fact. The former is a natural fixity, the latter is a movable furniture; the former is subject to no authority we can set up, the latter is wholly arbitrary, determinable at our pleasure. Uniformity of nomenclature is so obvious and decided a practical convenience that even at the risk of seeming to laud work in which he had a hand, the author cannot too strongly urge compliance with the Union's code, and adherence to the set of names the Union has adopted. These may not be the best possible, but they are the best we have.

The author's insistence upon this point does not of course extend to any case where an error of ornithological fact may appear. That is an entirely different matter. Reserving to himself, as he certainly does, the right of individual judgment in every question of ornithological science, he is the last to persuade others to refrain from equal freedom of expert opinion. "So many men, so many minds," even when the number is only five; no individual opinion is necessarily reflected upon any point in the Code and Check-list; it is the collective voice of a majority of the Committee that is heard in every instance. The occasion for individual dissent on the part of any member of that body, as of any other writer upon the subject, arises when in his private capacity as an author he has, as it were, to pass upon and approve or disapprove any results of the labors of others. The Appendix to the present edition of the "Key" unavoidably brings up such an occasion. Yet that he may not even seem to reflect upon any of his co-workers, his criticism express or implied has been sedulously reduced to its lowest terms. It consists chiefly in declining to admit to the "Key" some forms that the Committee have deemed worthy of recognition by name. Indeed he has preferred to err, if at all, on the other side, desiring to give the user of this book the later results of the whole Committee.

Nevertheless he must here record an earnest protest, futile though it may be, against the fatal facility with which the system of trinomials lends itself to sad consequences in the hands of immature or inexperienced specialists. No allusion is here intended to anything that has been done, but he must reiterate what was said before (Key, p. xxvii) respecting what may be done hereafter if more judicious conservatism than we have enjoyed of late be not brought to bear down hard upon trifling incompetents. The "trinomial tool" is too sharp to be made a toy; and even if we do not cut our own fingers with it, we are likely to cut the throat of the whole system of naming we have reared with such care. Better throw the instrument away than use it to slice species so thin that it takes a microscope to perceive them. It may be assumed, as a safe rule of procedure, that it is useless to divide and subdivide beyond the fair average ability of ornithologists to recognize and verify the result. Named varieties of birds that require to be "compared with the types" by holding them up slantwise in a good strong light, — just as the ladies match crewels in the milliner's shop, — such often exist in the cabinets or in the books of their describers, but seldom in the woods and fields.

E. C.

SMITHSONIAN INSTITUTION,
   WASHINGTON, D.C., *April, 1887.*

HISTORICAL PREFACE.

WERE a modern Hesiod to essay — neither a cosmogony nor a theogony — but the genesis of even the least department of human knowledge, — were he to seek the beginnings of American Ornithology, he would find it only in Chaos. For from this sprang all things, great and small alike, to pass through Night and Nemesis to the light of days which first see orderly progress in the course of natural evolution, when is first established some sequence of events we recognize as causes and effects. Then there is system, and formal law; there science becomes possible; there its possible history begins.

Long was the time during which the birds of our country were known to its inhabitants, after the fashion of the people of those days, — known as things of which use could be made, and studied, too, that use might be made of them. But this period is prehistoric; no evidence remains, save in some quaint pictograph or rudely graven image. There followed a period — shorter by far than the former one, though it endures to-day — when the same birds awakened in other men an interest they could not excite in a savage breast, and the sense of beauty was felt. Use and Beauty! What may not spring from such divinely mated pair, when once they brood upon the human mind, like halcyons stilling troubled waters, sinking the instincts of the animal in the restful, satisfying reflections of the man?

The history of American Ornithology begins at the time when men first wrote upon American birds; for men write nothing without some reason, and to reason at all is the beginning of science, even as to reason aright is its end. The date no one can assign, unless it be arbitrarily; it was during the latter part of the sixteenth century, which, with the whole of the seventeenth, represents the formative or embryonic period during which were gathering about the germ the crude materials out of which an ornithology of North America was to be fashioned. As these accumulated and were assimilated, — as the writings multiplied and books bred books, "each after its kind," this special department of knowledge grew up, and its form changed with each new impress made upon its plastic organization.

Viewing in proper perspective these three centuries and more which our subject has seen — passing in retrospect the steps of its development — we find that it offers several phases, representing as many "epochs" or major divisions, of very unequal duration, and of scientific significance inversely proportionate to their respective lengths. All that went before 1700 constitutes the first of these, which may be termed the *Archaic* epoch. The eighteenth century witnessed an extraordinary event, the consequence of which to systematic zoölogy cannot be over-estimated; it occurred almost exactly in the middle of the century, which is thus sharply divided into a *Pre-Linnæan* epoch, before the institution of the binomial nomenclature, and a *Post-Linnæan* epoch, during which this technic of modern zoölogy was established, — each approximately of half a century's duration. In respect of our particular theme, the first quarter of the nineteenth century saw the "father of American ornithology," whose spirit pointed the crescent in the sky of the *Wilsonian* epoch. During the second quarter, these horns were filled with the genius of the *Audubonian* epoch. In the third, the plenteousness of a master mind has marked the *Bairdian* epoch.

Clearly as these six epochs may be recognized, there is of course no break between them; they not only meet, but merge in one another. The sharpest line is that which runs across Linnæus at 1758; but even that is only visible in historical perspective, while the assignation of the dates 1700 and 1800 is rather a chronological convenience than otherwise. Nothing absolutely marks the former; and Wilson was unseen till 1808.

The Archaic epoch stretches into the dim past with unshifting scene, even at the turning-point of the two centuries in which it lies. It is otherwise with the rest; their shapes have incessantly changed; and several have been the periods in each of them during which their course of development has been accelerated or retarded, or modified in some special feature. These changes have invariably coincided with — have in fact been induced by — the appearance of some great work; great, not necessarily in itself, but in its relation to the times, and thus in the consequences of the interaction between the times and the author who left the science other than he found it. The edifice as it stands to-day is the work of all, even of the humblest, builders; but its plan is that of the architects who have modelled its main features, and the changes they have successively wrought are the marks of progress. It is consequently possible, and it will be found convenient, to subdivide the epochs named (excepting the first) into lesser natural intervals of time, which may be called "periods," to each of which may attach the name of the architect whose design is expressed most clearly. I recognize fifteen such periods, of very unequal duration, to which specific dates may attach. Seven of these fall in the last century; eight in the three-quarters of the present century. We may pass them in brief review.

THE ARCHAIC EPOCH: TO 1700.

Mere mention or fragmentary notice of North American birds may be traced back to the middle of the sixteenth century; but, up to the eighteenth, no book entirely and exclusively devoted to the subject had appeared. The Turkey and the Humming-bird were among the earliest to appear in print; the latter forms the subject of the earliest paper I have found, exclusively and formally treating of any North American bird as such, and this was not until 1693, when Hamersly described the "American Tomineius," as it was called. One of the largest, as well as the smallest of our birds, — the turkey, early came in for a share of attention. The germs of the modern "faunal list," — that is to say, notes upon the birds of some particular region or locality, — appeared early in the seventeenth century, and continued throughout; but only as incidental and very slight features of books published by colonists, adventurers, and missionaries, in their several interests, — unless Hernandez's famous "Thesaurus" be brought into the present connection. Among such books containing bird-matter may be noted Smith's "Virginia," 1612; Hamor's "Virginia," 1615; Whitbourne's "Newfoundland," 1620; Higginson's "New England," 1630; Morton's "New English Canaan," 1632; Wood's "New England's Prospect," 1634; Sagard Theodat's "Voyage," 1632; Josselyn's "New England's Rarities," 1672; — and so on, with a few more, — sometimes mere paragraphs, sometimes a page or a formal chapter, — but scarcely anything to be now considered except in a spirit of curiosity.

THE PRE-LINNÆAN EPOCH: 1700–1758.

(1700–1730.)

The Lawsonian Period. — It may be a *lucus a non* to call this the "Lawsonian" period; but a name is needed for the portion of this epoch prior to Catesby, during which no other name is so prominent as that of John Lawson, Gentleman, Surveyor-General of North Carolina, whose "Description and Natural History" of that country contains one of the most considerable faunal lists of our birds which appeared before 1730, and went through many editions, — the last of these being published at Raleigh, in 1860. The several early editions devote some fifteen or twenty pages to birds, — an amount augmented considerably when Brickell appropriated the work in 1737. The Baron de la Hontan did similar service to Canadian birds in his "Voyages," 1793; but, on the whole, this period is scarcely more than archaic.

(1730–1748.)

The Catesbian Period. — This comprises the time when Mark Catesby's great work was appearing by instalments. "The Natural History of Carolina, Florida," etc., is the first really great work to come under our notice; its influence was immediate, and is even now felt. It is the "Audubon" of that time; a folio in two volumes, dating respectively 1731 and 1743, with an appendix, 1748; passing to a second edition in 1754, to a third in 1771, under the supervision of Edwards; reproduced in Germany, in "Seligmann's Sammlung," 1749-76. It was published in parts, the date of the first of which I believe to have been 1730, though it may have been a little earlier. Volume I, containing the birds, appears to have been issued in five parts, and was made up in 1731; it consists of a hundred colored plates of birds, with as many leaves of text; a few more birds are given in the appendix, raising the number to 113. These illustrations are recognizable almost without exception; most of the species are for the first time described and figured; they furnish the basis of many subsequently named in the Linnæan system; the work was eventually provided by Edwards with a Linnæan concordance or index; and altogether it is not easy to overestimate the significance of the Catesbian period, due to this one work; for no other book requires or indeed deserves to be mentioned in the same connection, though a few contributions, of somewhat "archaic" character, were made by various writers.

(1748-1758.)

The Edwardsian Period. — This bridges the interval between Catesby and the establishment of the binomial nomenclature, and finishes the Pre-Linnæan epoch. No great name of exclusive pertinence to North American ornithology appears in this decade. But the great naturalist whose name is inseparably associated with that of Catesby had begun in 1741 the "Natural History of Uncommon Birds," which he completed in four parts or volumes, in 1751, and in which the North American element is conspicuous. This work contains two hundred and ten colored plates, with accompanying text, forming a treatise which easily ranks among the half-dozen greatest works of the kind of the Pre-Linnæan epoch, and passed through several editions in different languages. Its impress upon American ornithology of the time is second only to that made by Catesby's, of which it was the natural sequence, if not consequence It bore similarly upon birds soon to be described in binomial terms, and was shortly followed by the not less famous "Gleanings of Natural History," 1758-64, a work of precisely the same character, and in fact a continuation of the former. Edwards also made some of our birds the subject of special papers before the Philosophical Society, as those of 1755 and 1758 upon the Ruffed Grouse and the Phalarope. It may be noted here that one of the few special papers upon any American bird which Linnæus published appeared in this period, he having in 1750 first described the Louisiana Nonpareil (*Passerina ciris*). This period also saw the publication of part of the original Swedish edition of Peter Kalm's "Travels," 1753-61, which went through numerous editions in different languages. Kalm was a correspondent of Linnæus; the genus of plants, *Kalmia*, commemorates his name; his work contains accounts of many of our birds, some of them the bases of Linnæan species; and he also published, in 1759, a special paper upon the Wild Pigeon. As in the Catesbian period, various lesser contributions were made, but none requiring comment. Thus Lawson, as representing the continuation of a preceding epoch, and the associated names of Catesby and Edwards in the present one, have carried us past the middle of the last century.

The Post-Linnæan Epoch: 1758–1800.

(1758–1766.)

The Linnæan Period. — An interregnum here, during which not a notable work or worker appears in North American ornithology itself. But events elsewhere occurred, the reflex action of which upon our theme is simply incalculable, fully requiring the recognition of this period. The dates, 1758–1766, are respectively those of the appearance of the tenth and of the twelfth edition of the "Systema Naturæ" of Linnæus. In the former the illustrious Swede first formally and consistently applied his system of nomenclature to all birds known to him; the latter is his completed system, as it finally left his hands; and from then to now, zoölogists and especially ornithologists have disputed whether 1758 or 1766 should be taken as the starting-point of zoölogical nomenclature. In ornithology, the matter is still at issue between the American and the British schools. However this may result, the fact remains that during this "Linnæan period," 1758 to 1766, we have the origin of all the tenable specific names of those of our birds which were known to Linnæus; the gathering up and methodical digestion and systematic arrangement of all that had gone before. Let this scant decade stand, — mute in America, but eloquent in Sweden, and since applauded to the echo of the world.

Nor is this all. The year 1760 saw the famous "Ornithologia" of Mathurin Jacques Brisson (born April 20, 1725 — died June 23, 1806), in six portly quartos with 261 folded plates, and elaborate descriptions in Latin and French of hundreds of birds, a fair proportion of which are North American. Many are described for the first time, though unfortunately not in the binomial nomenclature. The work holds permanent place; and most of the original descriptions of Brisson's are among the surest bases of Linnæan species.

(1766–1785.)

The Forsterian Period. — Nearly twenty years have now elapsed with so little incident that two brochures determine the complexion of this period. John Reinhold Forster was a learned and able man, whose connection with North American ornithology is interesting. In 1771 he published a tract, now very scarce and of no consequence whatever, entitled "A Catalogue of the Animals of North America." But it was the first attempt to do anything of the sort, — in short, the first thing of its kind. It gives 302 birds, neither described nor even named scientifically. But that was a large number of North American birds to even mention in those days, — more than Wilson gave in 1814. Forster followed up this exploit in 1772 with an interesting and valuable account of 58 birds from Hudson's Bay, occupying some fifty pages of the "Philosophical Transactions." Several of these birds were new to science, and were formally named, — such as our White-throated Sparrow, Black-poll Warbler, Hudsonian Titmouse, and Eskimo Curlew. Aside from its intrinsic merit, this paper is notable as the first formal treatise exclusively devoted to a collection of North American birds sent abroad. The period is otherwise marked by the publication in 1780 of Fabricius' "Fauna Grœnlandica," in which some 50 birds of Greenland receive attention; and especially by the appearance of a great statesman and one of the Presidents of the United States in the rôle of ornithologist, Thomas Jefferson's "Notes on the State of Virginia" having been first privately printed in Paris in 1782, though the authorized publication was not till 1787. It contains a list of 77 birds of Virginia, fortified with references to Catesby, Linnæus, and Brisson, as the author's authorities. There were many editions, one dating 1853.

The long publication in France of one of the monumental works on general ornithology coincides very nearly with this period. I refer of course to Buffon and his collaborators. The "Histoire Naturelle des Oiseaux," by Buffon and Montbeillard, dates in its original edition 1770–1783, being in nine quarto volumes with 264 plain plates. It forms a part of the grand set of volumes dating 1749–1804 in their original editions. With the nine bird-volumes are associated the magnificent series of colored plates known as the "Planches Enluminées," published in 42 fascicles from 1765 to 1781. The plates are 1008 in number, of which 973 represent birds.

(1785–1791.)

The Pennantian Period. — A great landmark — one of the most conspicuous of the last century — was set up with the appearance in 1785 of the second volume of Thomas Pennant's "Arctic Zoology." The whole work, in three quarto volumes with many plates, 1784–1787, was "designed as a sketch of the Zoölogy of North America." In this year, also, John Latham completed the third volume (or sixth part) of his "General Synopsis of Birds." These two great works have much in common, in so far as a more restricted treatise can be compared with a more comprehensive one; and in the history of our subject the names of Latham and Pennant are linked as closely as those of Catesby and Edwards. The parallel may be drawn still further; for neither Pennant nor Latham (up to the date in mention) used binomial names; their species had consequently no standing; but they furnished to Gmelin in 1788 the same bases of formally-named species of the thirteenth edition of the "Systema Naturæ," that Catesby and Edwards had afforded Linnæus in 1758 and 1766. Pennant treated upwards of 500 nominal species of North American Birds. The events at large of this brief but important period were the progress of Latham's Supplement to his Synopsis, the first volume of which appeared in 1787, though the second was not completed till 1801; the appearance in 1790 of Latham's "Index Ornithologicus," in which his birds receive Latin names in due form; and the publication in 1788 of the thirteenth edition of the "Systema Naturæ," as just said.

We are so accustomed to see "Linn." and "Gm." after the names of our longest-known birds that we almost unconsciously acquire the notion that Linnæus and Gmelin were great discoverers or describers of birds in those days. But the men who made North American ornithology what it was during the last century were Catesby, Edwards, Forster, Pennant, Latham, and Bartram. For "the illustrious Swede" was in this case little more than a methodical cataloguer, or systematic indexer; while his editor, Gmelin, was merely an industrious, indiscriminate compiler and transcriber. Neither of these men *discovered* anything to speak of in this connection.

(1791–1800.)

The Bartramian Period. — William Bartram's figure in the events we are sketching is a notable one, — rather more on account of his bearing upon Wilson's subsequent career than of his own actual achievements. Wilson is often called the "father of American ornithology;" if this designation be apt, then Bartram may be styled its godfather. Few are fully aware how much Wilson owed to Bartram, his "guide, philosopher, and friend," who published in 1791 his "Travels through North and South Carolina," containing much ornithological matter that was novel and valuable, including a formal catalogue of the birds of the Eastern United States, in which many species are named as new. I have always contended that those of his names which are identifiable are available, though Bartram frequently lapsed from strict binomial propriety; and the question furnishes a bone of contention to this day. Many birds which Wilson first fully described and figured were really named by Bartram, and several of the latter's designations were simply adopted by Wilson, who, in relation to Bartram, is as the broader and clearer stream to its principal tributary affluent. The notable "Travels," freighted with its unpretending yet almost portentous bird-matter, went through several editions and at least two translations; and I consider it the starting-point of a distinctively American school of ornithology.

We have seen, in several earlier periods, that men's names appear in pairs, if not also as mates. Thus, Catesby and Edwards; Linnæus and Gmelin; Pennant and Latham; and, perhaps, Buffon and Brisson. The Bartramian *alter ego* is not Wilson, but Barton, whose "Fragments of the Natural History of Pennsylvania," 1799, closed the period which Bartram had opened, and with it the century also. Benjamin Smith Barton's tract, a folio now very scarce, is doubly a "fragment," being at once a work never finished, and very imperfect as far as it went; but it is one of the most notable special treatises of the last century, and I think the first book published in this country that is entirely devoted to ornithology. But its author's laurels must rest mainly upon this count, for its influence or impression upon the course of events is scarcely to be recognized, — is incomparably less than that made by Bartram's "Travels," and by his mentorship of Wilson.

By the side of Bartram and Barton stand several lesser figures in the picture of this period. Jeremy Belknap treated the birds of New Hampshire in his "History" of that state (1792). Samuel Williams did like service for those of Vermont in his "History" (1794). Samuel Hearne, a pioneer ornithologist in the northerly parts of America, foreshadowed, as it were, the much later "Fauna Boreali-Americana" in the narrative of his journey from Hudson's Bay to the Northern Ocean — a stout quarto published in 1795. Here a chapter of fifty pages is devoted to about as many species of birds; and Hearne's observations have a value which "time, the destroyer," has not yet wholly effaced.

The Wilsonian Epoch: 1800–1824.

(1800–1808.)

The Vieillotian Period. — As we round the turn of the century a great work occupies the opening years, before the appearance of Wilson, — a work by a foreigner, a Frenchman, almost unknown to or ignored by his contemporaries in America, although he was already the author of several illustrated works on ornithology when, in 1807, his "Histoire Naturelle des Oiseaux de l'Amérique Septentrionale" was completed in two large folio volumes, containing more than a hundred engravings, with text relating to several hundred species of birds of North America and the West Indies; many of them figured for the first time, or entirely new to science. This work, bearing much the same relation to its times that Catesby's and Edwards' respectively did to theirs, is said to have been published in twenty-two parts of six plates each, probably during several years; but the date of its inception I have never been able to ascertain. However this may be, Vieillot alone and completely fills a period of eight years, during which no other notable or even mentionable treatise upon North American birds saw the light. Vieillot's case is an exceptional one. As the author of numerous splendidly illustrated works, all of which live; of a system of ornithology, most of the generic names contained in which are ingrained in the science; of very extensive encyclopædic work in which hundreds of species of birds receive new technical names: Vieillot has a fame which time rather brightens than obscures. Yet it is to be feared that the world was unkind during his lifetime. At Paris, he stood in the shadow of Cuvier's great name; Temminck assailed him from Holland; while, as to his work upon our birds, many years passed before it was appreciated or in any way adequately recognized. Thus, singularly, so great a work as the "Histoire Naturelle"—one absolutely characteristic of a period—had no appreciable effect upon the course of events till long after the times that saw its birth, when Cassin, Baird, and others brought Vieillot into proper perspective. There is so little trace of Vieillot during the Wilsonian and Audubonian epochs, that his "Birds of North America" may almost be said to have slept for half a century. But to-day, the solitary figure of the Vieillotian period stands out in bold relief.

(1808–1824.)

The Wilsonian Period.—The "Paisley weaver;" the "Scotch pedler;" the "melancholy poet-naturalist;" the "father of American ornithology,"—strange indeed are the guises of genius, yet stranger its disguises in the epithets by which we attempt to label and pigeon-hole that thing which has no name but its own, no place but its own. Alexander Wilson had genius, and not much of anything else—very little learning, scarcely any money, not many friends, and a paltry share of "the world's regard" while he lived. But genius brings a message which men must hear, and never tire of hearing; it is the word that comes when the passion that conceives is wedded with the patience that achieves. Wilson was a poet by nature, a naturalist by force of circumstances, an American ornithologist by mere accident,—that is, if anything can be accidental in the life of a man of genius. As a poet, he missed greatness by those limitations of passion which seem so sad and so unaccountable; as the naturalist, he achieved it by the patience that knew no limitation till death interposed. As between the man and his works, the very touchstone of genius is there; for the man was greater than all his works are. Genius may do that which satisfies all men, but never that which satisfies itself; for its inspiration is infinite and divine, its accomplishment finite and human. Such is the penalty of its possession.

Wilson made, of course, the epoch in which his work appeared, and I cannot restrict the Wilsonian period otherwise than by giving to Vieillot his own. The period of Wilson's actual authorship was brief; it began in September, 1808, when the first volume of the "American Ornithology" appeared, and was cut short by death before the work was finished. Wilson, having been born July 6, 1766, and come to America in 1794, died August 23, 1813, when his seventh volume was finished; the eighth and ninth being completed in 1814 by his friend and editor, George Ord. But from this time to 1824, when Bonaparte began to write, the reigning work was still Wilson's, nothing appearing during these years to alter the complexion of American ornithology appreciably. Wilson's name overshadows nearly the whole epoch, — not that others were not then great, but that he was so much greater. This author treated about 280 species, giving faithful descriptions of all, and colored illustrations of most of them. There are numerous editions of his work, of which the principal are Ord's, 1828-29, in three volumes; Jameson's, 1831, in four; Jardine's, 1832, in three; and Brewer's, 1840, in one; all of these, excepting of course the first one, containing Bonaparte's "American Ornithology" and other matter foreign to the original "Wilson." In 1814, just as "Wilson" was finished, appeared the history of the memorable expedition under Lewis and Clarke — an expedition which furnished some material to Wilson himself, as witness Lewis' Woodpecker, Clarke's Crow, and the "Louisiana" Tanager; and more to Ord, who contributed to the second edition of "Guthrie's Geography" an article upon ornithology. Ord's prominence in this science, however, rests mainly upon his connection with Wilson's work, as already noted. Near the close of the Wilsonian period, Thomas Say gave us important notices of Western birds, upon the basis of material acquired through Long's Expedition to the Rocky Mountains, the account of which appeared in 1823. In this work, Say described sundry species of birds new to science; but he was rather an entomologist than an ornithologist, and his imprint upon our subject is scarcely to be found outside the volume just named. A noted — some might say rather notorious — character appeared upon the scene during this period, in the person of C. S. Rafinesque, who seems to have been a genius, but one so awry that it is difficult to do aught else than misunderstand him, unless we confess that we scarcely understand him at all. In the elegant vernacular of the present day he would be called a crank; but I presume that term means that kind of genius which fails of interpretation; for an unsuccessful genius is a crank, and a successful crank is a genius. For the rest, the Wilsonian period was marked by great activity in Arctic exploration, in connection with the ornithological results of which appear prominently the names of William E. Leach and Edward Sabine.

As illustrating the relation between Wilson and Bartram, which I have already pointedly mentioned, I may quote a few lines from Ord's "Life of Wilson."[1]

---

[1] "His school-house and residence being but a short distance from Bartram's Botanic Garden, situated on the west bank of the Schuylkill: a sequestered spot, possessing attractions of no ordinary kind; an acquaintance was soon contracted with that venerable naturalist, Mr. William Bartram, which grew into an uncommon friendship, and continued without the least abatement until severed by death. Here it was that Wilson found himself translated, if we may so speak, into a new existence. He had long been a lover of the works of Nature, and had derived more happiness from the contemplation of her simple beauties, than from any other source of gratification. But he had hitherto been a mere novice; he was now about to receive instructions from one whom the experiences of a long life, spent in travel and rural retirement, had rendered qualified to teach. Mr. Bartram soon perceived the bent of his friend's mind, and its congeniality to his own; and took every pains to encourage him in a study, which, while it expands the faculties, and purifies the heart, insensibly leads to the contemplation of the glorious Author of Nature himself. From his youth Wilson had been an observer of the manners of birds; and since his arrival in America he had found them objects of uncommon interest; but he had not yet viewed them with the eye of a naturalist."

This was about 1800 — rather a little later. Wilson's "novitiate" was the Vieillotian period, almost exactly. Bartram survived till July 22, 1823, his eighty-fourth year; the date of his death thus coinciding very nearly with the close of the Wilsonian epoch and period.

The Audubonian Epoch: 1824–1853.

(1824–1831.)

The Bonapartian Period. — A princely person, destined to die one of the most famous of modern naturalists — Charles Lucien Bonaparte, early conceived and executed the plan of continuing Wilson's work in similar style, if not in the same spirit. He began by publishing a series of "Observations on the Nomenclature of Wilson's Ornithology," in the "Journal" of the Philadelphia Academy, 1824–25, republished in an octavo volume, 1826. This valuable critical commentary introduced a new feature, — decided changes in nomenclature resulting from the sifting and rectification of synonymy. It is here that questions of synonymy — to-day the bane and drudgery of the working naturalist — first acquire prominence in the history of our special subject. There had been very little of it before, and Wilson himself, the least "bookish" of men, gave it scarcely any attention. Bonaparte also in 1825 added several species to our fauna upon material collected in Florida by the now venerable Titian R. Peale, — whose honored name is thus the first of those of men still living to appear in these annals. Bonaparte's "American Ornithology," uniform with "Wilson," and generally incorporated therewith in subsequent editions, as a continuation of Wilson's work, was originally published in four large quarto volumes, running 1825–33. The year 1827, in the midst of this work of Bonaparte's, was a notable one in several particulars. Bonaparte himself was very busy, producing a "Catalogue of the Birds of the United States," which, with a "Supplement," raised the number of species to 366, and of genera to 83; nearly a hundred species having been thus become known to us since Ord laid aside the pen that Wilson had dropped. William Swainson the same year described a number of new Mexican species and genera, many of which come also into the "North American" fauna. But the most notable event of the year was the appearance of the first five parts of Audubon's elephant folio plates. In 1828–29, as may also be noted, Ord brought out his three-vol. 8vo edition of Wilson. In 1828, Bonaparte returned to the charge of systematically cataloguing the birds of North America, giving now 382 species; and about this time he also produced a comparative list of the birds of Rome and Philadelphia. His main work having been completed in 1833, as just said, Bonaparte continued his labors with a "Geographical and Comparative List of the Birds of Europe and North America," published in London in 1838. This brochure gives 503 European and 471 American species. The celebrated zoölogist wrote until 1857, but his connection with North American birds was only incidental after 1838. The period here assigned him, 1824–1831, may seem too short: but this was the opening of the Audubonian epoch — a period of brilliant inception, and one in which events that were soon to mature their splendid fruit came crowding fast; so that room must be made at once for others who were early in the present epoch.

(1831–1832.)

The Swainsonio-Richardsonian Period. — The "Fauna Boreali-Americana," the ornithological volume of which was published in 1831, made an impression so indelible that a period, albeit a brief one, must be put here. The technic of this celebrated treatise, more valuable for its descriptions of new species and genera than for its methods of classification, was by William Swainson, as were the elegant and accurate colored plates; the biographical matter, by Dr. (later Sir) John Richardson, increased our knowledge of the life-history of the northerly birds so largely, that it became a fountain of facts to be drawn upon by nearly every writer of prominence from that day to this. Each of the distinguished authors had previously appeared in connection with our birds, — Swainson as above said; Richardson in 1825, in the appendix to Captain Parry's "Journal." The influence of the work on the whole cannot be well overstated.

Two events, besides the appearance of the "Fauna," mark the year 1831. One of these is the publication of the first volume of Audubon's " Ornithological Biography," being the beginning of the text belonging to his great folio plates. The other is the completion of the bird-volumes of Peter Pallas' famous "Zoographia Rosso-Asiatica," one of the most important contributions ever made to our subject, treating so largely as it does of the birds of the region now called Alaska. The same year saw also the Jameson edition of "Wilson and Bonaparte."

(1832-1834.)

The Nuttallian Period.—Thomas Nuttall (born 1786—died 1859) was rather botanist than ornithologist; but the travels of this distinguished English-American naturalist made him the personal acquaintance of many of our birds, his love for which bore fruit in his " Manual of the Ornithology of the United States and Canada," of which the first volume appeared in 1832, the second in 1834. The work is notable as the first " handbook " of the subject; it possesses an agreeable flavor, and I think was the first formal treatise, excepting Wilson's, to pass to a second edition, as it did in 1840. Nuttall's name is permanent in our annals; and many years after he wrote, the honored title was chosen to be borne by the first distinctively ornithological association of this country, — the "Nuttall Ornithological Club," founded at Cambridge in 1873, and still flourishing.

(1834-1853.)

The Audubonian Period. — Meanwhile, the incomparable work of Audubon — "the greatest monument erected by art to nature" — was steadily progressing. The splendid genius of the man, surmounting every difficulty and discouragement of the author, had found and claimed its own. That which was always great had come to be known and named as such, victorious in its impetuous yet long-enduring battle with that curse of the world, — I mean the commonplace; the commonplace, with which genius never yet effected a compromise, since genius is necessarily a perpetual menace to mediocrity. Audubon and his work were one; he lived in his work, and in his work will live forever. When did Audubon die. We may read, indeed, " on Thursday morning, January 27th, 1851, when a deep pallor overspread his countenance. . . . Then, though he did not speak, his eyes, which had been so long nearly quenched, rekindled with their former lustre and beauty; his spirit seemed to be conscious that it was approaching the Spirit-land." And yet there are those who are wont to exclaim, "a soul! a soul! what is that?" Happy indeed are they who are conscious of its existence in themselves, and who can see it in others, every instant of time during their lives!

Audubon's first publication, perhaps, was in 1826, — an account of the Turkey-buzzard, in the "Edinburgh New Philosophical Journal," and some other minor notices came from his pen. But his energies were already focused on his life-work, with that intense and perfect absorption of self which only genius knows. The first volume of the magnificent folio plates, an hundred in number, appeared in 1827-30, in five parts; the second, in 1831-34, of the same number of plates; the third, in 1834-35, likewise of the same number of plates; the whole series of 4 volumes, 87 parts, 435 plates and 1065 figures of birds, being completed in June, 1839. Meanwhile, the text of the "Birds of America," entitled "Ornithological Biography," was steadily progressing, the first of these royal octavo volumes appearing in 1831, the fifth and last in 1839. In this latter year also appeared the "Synopsis of the Birds of North America," a single handy volume serving as a systematic index to the whole work. In 1840-44 appeared the standard octavo edition in seven volumes, with the plates reduced to octavo size and the text rearranged systematically; with a later and better nomenclature than that given in the "Ornithological Biography," and some other changes, including an appendix describing various new species procured during the author's journey to the upper Missouri in 1843. In the original elephant folios there were 435 plates; with the reduction in size the number was raised to 483, by the separation of various figures which had previously occupied the same plate; and to these 17 new ones were added, making 500 in all. The species of birds treated in the "Synopsis" are 491 in number; those in the work, as it finally left the illustrious author's hands, are 506 in number, nearly all of them splendidly figured in colors.

In estimating the influence of so grand an accomplishment as this, we must not leave Audubon "alone in his glory." Vivid and ardent was his genius; matchless he was both with pen and pencil in giving life and spirit to the beautiful objects he delineated with passionate love; but there was a strong and patient worker by his side, — William Macgillivray, the countryman of Wilson, destined to lend the sturdy Scotch fibre to an Audubonian epoch. The brilliant French-American naturalist was little of a "scientist." Of his work, the magical beauties of form and color and movement are all his; his page is redolent of Nature's fragrance: but Macgillivray's are the bone and sinew, the hidden anatomical parts beneath the lovely face, the nomenclature, the classification, — in a word, the technicalities of the science. Not that Macgillivray was only a closet-naturalist; he was a naturalist in the best sense — in every sense — of the word, and the "vital spark" is gleaming all through his works upon British birds, showing his intense and loyal love of Nature in all her moods. But his place in the Audubonian epoch in American ornithology is as has been said. The anatomical structure of American birds was first disclosed in any systematic manner, and to any considerable extent, by him. But only to-day, as it were, is this most important department of ornithology assuming its rightful place; and have we a modern Macgillivray to come?

The sensuous beauty with which Audubon endowed the object of his life was long in acquiring, with loss of no comeliness, the aspect more strict and severe of a later and maturer epoch. Audubon was practically accomplished in 1844, the year which saw his completed work; but I note no special or material change in the course of events, — no name of assured prominence, till 1853, when a new régime, that had meanwhile been insensibly established, may be considered to have closed the Andubonian epoch, — the Audubonian period thus extending through the nine years after 1844.

While Audubon was finishing, several mentionable events occurred. I have already spoken of Bonaparte's "List" of 1838, and of the 1840 edition of Nuttall's "Manual." Richardson in 1837 contributed to the Report of the Sixth Meeting of the British Association for the Advancement of Science an elaborate and important "Report on North American Zoölogy," relating in due part to birds. The distinguished Danish naturalist, Reinhardt, wrote a special treatise on Greenland Birds, 1838; W. B. O. Peabody one upon the birds of Massachusetts, 1839. The important Zoölogy of Captain Beechey's Voyage appeared in 1839, with the birds done by N. A. Vigors. Maximilian, Prince of Wied, published his "Reise in das Innere Nord-America" in 1839–41. Sixteen new species of birds from Texas were described and figured by J. P. Giraud in 1841, and the same author's useful "Birds of Long Island" was published in 1844. This year saw also the bird-volume of De Kay's "Zoölogy of New York." The Rev. J. H. Linsley furnished a notable catalogue of the birds of Connecticut in 1843. A name intimately associated with Audubon's is that of J. K. Townsend, whose fruitful travels in the West in company with Nuttall in 1834 resulted in adding to our list the many new species which were published by Townsend himself in 1837, and also utilized by Audubon. Townsend's "Narrative" of his journey appeared in 1839; and the same year saw the beginning of a large work which Townsend projected, an "Ornithology of the United States," which, however, progressed no further than one part or number, being killed by the octavo edition of Audubon. In 1837 I first find the name of a friend of Audubon which often appears in his work — that of Dr. Thomas Mayo Brewer, who wrote on the birds of Massachusetts in this year, and in 1840 brought out his useful and convenient duodecimo edition of "Wilson," in one volume. In 1844, Audubon's last effectual year, the brothers Wm. M. and S. F. Baird appear, with a list of the birds of Carlisle, Pennsylvania, having the year previously, in July, 1843, described two new species of flycatchers, in the first paper ever written by the one who was to make the succeeding epoch; and it is significant that the last bird in Audubon's work was named "*Emberiza bairdii.*"

Such were the aspects of the ornithological sky as the glorious Audubonian sun approached and passed the zenith; still more significant were the signs of the times as that orb neared its golden western horizon. In the interval between 1844 and 1853, Baird and Brewer continued; Cassin and Lawrence appeared in various papers; and round these names are grouped those of William Gambel, with new and interesting observations in the Southwest; of George A. McCall and S. W. Woodhouse, in the same connection; and of Holböll in respect of Greenland birds. The most important contributions were the several papers published by Gambel, in 1845 and subsequently, and Baird's Zoölogy of Stansbury's Expedition, 1852. But no period-marking, still less epoch-making, work accelerated the setting of the sun of Audubon.

The Bairdian Epoch: 1853–18 —.

(1853–1858.)

The Cassinian Period. — While much material was accumulating from the exploration of the great West, and the Bairdian period was rapidly nearing; while Brewer and

Lawrence were continuing their studies and writings, and many other names of lesser note were contributing their several shares to the whole result: the figure of John Cassin stands prominent. Cassin was born September 6, 1813, and passed from view in the Quaker City, January 10, 1869. Numerous valuable papers and several important works attest the assiduity and success with which he cultivated his favorite science to the end of his days. I think that his first paper was the description of a new hawk, *Cymindis wilsoni*, in 1847. Among his most important works are the Ornithology of the Wilkes Exploring Expedition; of the Perry Japan Expedition; and of the Gilliss Expedition to Chili. Aside from his strong coöperation with Baird in the great work to be presently noticed, Cassin's seal is set upon North American ornithology in the beautiful book begun in 1853 and finished in 1856, entitled "Illustrations of the Birds of California," etc., forming a large octavo volume, illustrated with fifty colored plates. His distinctive place in ornithology is this: he was the only ornithologist this country has ever produced who was as familiar with the birds of the Old World as with those of America. Enjoying the facilities of the then unrivalled collection of the Philadelphia Academy, his monographic studies were pushed into almost every group of birds of the world at large. He was patient and laborious in the technic of his art, and full of book-learning in the history of his subject; with the result, that the Cassinian period, largely by the work of Cassin himself, is marked by its "bookishness," by its breadth and scope in ornithology at large, and by the first decided change since Audubon in the aspect of the classification and nomenclature of the birds of our country. The Cassinian period marks the culmination of the changes that wrought the fall of the Audubonian sceptre in all that relates to the technicalities of the science, and consequently represents the beginning of a new epoch.

The peers of this period are only three, — Lawrence, Brewer, and Baird. The former of these, already an eminent ornithologist, continued his rapidly succeeding papers and was preparing his share of Baird's great work of 1858; though later his attention became so closely fixed upon the birds of Central and South America, that a "Lawrencian period" is to be found in the history of the ornithology of those countries rather than of our own. Dr. Brewer's various articles appeared, and in 1857 this author, so well known since Audubonian times, became the recognized leading oölogist of North America, through the publication of the first part of his "North American Oölogy" — a work unfortunately suspended at this point. Though thus fragmentary, this quarto volume stands as the first systematic treatise published in this country exclusively devoted to oölogy, and giving a considerable series of colored illustrations of eggs. But a larger measure of the world's regard became his much later, when, in 1874, appeared the great "History of North American Birds," in three quarto volumes, all the biographical matter of which was by him; and, even as I write, two more volumes are about to appear, in which he has like large share. Thus closely is the name of Brewer identified with the progress of the science for nearly half a century, — from 1837 at least, to 1884, some four years after his death, which occurred January 23, 1880. He was born in Boston, November 21, 1814.

Baird published little during the Cassinian period, being then intent upon the great work about to appear; but the number of workers in special fields attests the activity of the times. S. W. Woodhouse published his completed observations upon the birds of the Southwest in an illustrated octavo volume. Zadock Thompson's " Natural History of Vermont" (1853) paid attention to the birds of that state. Birds of Wisconsin were catalogued by P. R. Hoy; of Ohio, by M. C. Read and Robert Kennicott; of Illinois, by H. Pratten; of Indiana, by R. Haymond; of Massachusetts, by F. W. Putnam; and various other "faunal lists" and local annotations appeared, including President Jefferson's Virginian ornithology, three-quarters of a century out of date. Dr. T. C. Henry and Dr. A. L. Heermann wrote upon birds of the Southwest; Reinhardt continued observations on Greenland birds; Dr. Henry Bryant published some valuable papers. The since very eminent English ornithologist, Dr. P. L. Sclater, appeared during this period in the present connection. The series of Pacific Railroad Reports, which were to culminate, so far as ornithology is concerned, with the famous ninth volume, were in progress; the sixth volume, containing Dr. J. S. Newberry's valuable and interesting article upon the birds of California and Oregon, was published in 1857. Thus the Cassinian period, besides being marked as already said in its broader features, was notable in its details for the increase in the number of active workers, the extent and variety of their independent observations, and the consequent accumulation of materials ready to be worked into shape and system.

(1858–18—.)

The Bairdian Period. — The ninth volume of the "Pacific Railroad Reports" was an epoch-making work, bearing the same relation to the times that the respective works of Audubon and Wilson had sustained in former years. A great amount of material — not all of which is more than hinted at in the foregoing paragraph — was at the service of Professor Baird. In the hands of a less methodical, learned, and sagacious naturalist, — of one less capable of elaborating and systematizing, — the result would probably have been an ordinary official report upon the collections of birds secured during a few years by the naturalists of the several explorations and surveys for a railroad route from the Mississippi Valley to the Pacific Ocean. But having already transformed the eighth volume of the Reports from such a "public document" into a systematic treatise on North American Mammals, this author did the same for the birds of North America, with the coöperation of Cassin and Lawrence. This portly quarto volume, published in 1858, represents the most important and decided single step ever taken in North American ornithology in all that relates to the technicalities of the science. It effected a revolution — one already imminent in consequence of Cassin's studies — in classification and nomenclature, nearly all the names of our birds which had been in use in the Audubonian epoch being changed in accordance with more modern usages in generic and specific determinations. While the work contains no biographical matter, — nothing of the life-history of birds, it gives lucid and exact diagnoses of the species and genera known at the time, with copious synonymy and critical commentary. Various new genera are characterized, and many new species are described. The influence of the great work was immediate and widespread, and for many years the list of names of the 738 species contained in the work remained a standard of nomenclature from which few desired or indeed were in position to deviate. The value of the work was further enhanced in 1860 by its republication, identical in the text, but with the addition of an atlas of 100 colored plates. Many of these plates were the same as those which had appeared in other volumes of the Pacific Railroad Reports, notably the sixth and tenth and twelfth (the two latter volumes having appeared in 1859); others were those contained in the "Mexican Boundary Report" which had appeared under Professor Baird's editorship in 1859; about half of them were new.

I have spoken of the collaboration of Cassin and Lawrence in the production of this remarkable treatise. Considering it only as one of a series of reports upon the Pacific Railroad Surveys, I should bring into somewhat of association the names of those who contributed the ornithological portions of other volumes, as the fourth, sixth, tenth, and twelfth, — Dr. C. B. R. Kennerly, Dr. J. S. Newberry, Dr. A. L. Heermann, Dr. J. G. Cooper, and Dr. George Suckley. Nor should it be forgotten that numberless other collectors and contributors, whose specimens are catalogued throughout the volume, brought their hands to bear upon the erection of this grand monument.

But what of the genius of this work? — for I have not measured my words in speaking of Wilson and Audubon. Can any work be really great without that mysterious quality? Certainly not. This work is instinct with the genius of the times that saw its birth. This work is the spirit of an epoch embodied.

But here I must pause. My little sketch is brought upon the threshold of **contemporaneous** history, — to the beginning of the Bairdian period, of the close of which, as of the duration of the Bairdian epoch, it is not for me to speak. When the splendid achievements of American ornithologists during the past quarter of a century shall be seen in historical perspective; when the brilliant possibilities of our near future shall have become the realizations of a past; when the glowing names that went before shall have fired another generation with a noble zeal, a lofty purpose, and a generous emulation — then, perhaps, the thread here dropped may be recovered by another hand.

Yet a few words of Preface proper to the present work appear to be required. The original edition of the "Key" was published in October, 1872, in an issue of about 2,200 copies. It was not stereotyped, and has been for some years entirely out of print. It formed an imperial octavo of 361 pages, illustrated with 238 woodcuts in the text and 6 steel plates. It was designed as a manual or text-book of North American Ornithology. To meet this design, the Introduction consisted of a general account of the external characters of birds, an explanation of the technical terms used in describing them, and some exposition of the leading principles of classification and nomenclature. **An** artificial "key" or analysis of the genera, constructed upon a plan found practically useful in botany, but seldom applied to zoölogy, was introduced, to enable one who had some knowledge of the technical terms to refer a given specimen to its proper genus. Then, in the body of the work, each species was briefly described, with indication of its geographical distribution and references to several leading authorities. The families and orders of North American birds were also characterized, and a synopsis of the fossil birds was appended. The work introduced many decided changes in classification and nomenclature which the then state of the science seemed to require, and systematically recognized a large number of those subspecies or geographical races which are now indicated by the use of trinomial nomenclature, — a method now fully established and recognized as peculiar to the "American school." The central idea of the treatise was to enable one to identify and label his specimens, though he might have no other knowledge of ornithology than such as the book itself gave him. I have been given to understand that the work has answered its purpose, and has had a useful career; and I have long since been advised by my esteemed publishers that they were ready to issue a second edition, which I have only just now found time to complete.

The present edition of the "Key" is conceived in the same spirit as the former one, to fulfil precisely the same purpose. But it has been entirely rewritten, and is quite another work, though the old title is preserved. An author who practises his profession diligently for twenty years is apt to find fault with his first book, and seek to remedy its defects when opportunity offers. It has become quite clear to me, as it doubtless has to others, that the old "Key" no longer turns in the lock with ease and precision, — not that it has rusted from disuse, but that the more complicated mechanism of the lock requires its key to be refitted. During no previous period has our knowledge gone faster or farther or more surely than in the interval between the two editions of the "Key;" there are scores of active and enthusiastic workers where there was one before; scores of important treatises have appeared; the literature of the subject has been searched, sifted, and systematized; every corner of our country has been ransacked for birds, and the list of our species and subspecies has reached about 900 by the many late discoveries; active interest in this branch of science is no longer confined to professed ornithologists; the importance of avian anatomy is as fully recognized as is the beauty of the life-history of birds; a distinctively American school of ornithology has grown up, introducing radical changes in nomenclature and classification; a quarterly journal of ornithology has reached its ninth annual volume; an American Ornithologists' Union, the membership of which extends to every quarter of the globe, has been founded.

So rapid, indeed, has been the progress, and so radical the changes wrought during the last few years, that I doubt not this is the time to take our bearings anew and proceed with judicious conservatism. Neither do I doubt that just at this moment a new departure is imminent, hinging upon the establishment of the American Ornithologists' Union. It behooves us, therefore, to consider the question, not alone of where we stand to-day, but also, of whither we are tending; for we are certainly in a transition state, and not even the near future can as yet be accurately forecast. The pliability and elasticity of our trinomial system of nomenclature is very great; and the method lends itself so readily to the nicest discriminations of geographical races, — of the finest shades of variation in subspecific characters with climatic and other local conditions of environment, that our new toy may not impossibly prove a dangerous instrument, if it be not used with judgment and caution. We seem to be in danger of going too far, if not too fast, in this direction. It is not to cry "halt!" — for any advance is better than any standstill; but it is to urge prudence, caution, and circumspection, lest we be forced to recede ingloriously from an untenable position, — that these words are penned, with a serious sense of their necessity.

In the present unsettled and perplexing state of our nomenclature, when appeal to no "authority" or ultimate jurisdiction is possible, it is well to formulate and codify some canons of nomenclature by which to agree to abide. It is well to apply such canons rigidly, with thorough sifting of synonymy, no matter what precedents be disregarded, what innovations be caused. It is well to use trinomials for subspecific determinations. But it is not well to overdo the "variety business;" feather-splitting is no better than hair-splitting, and the liberties of the "American idea" must never degenerate into license. Our action in this regard must stop short of a point where an unfavorable reaction would be the inevitable result.

But I have digressed, in saying a warning word, from the point of the conclusion of this Preface, which is simply to describe the new edition of the "Key" with special reference to its difference from the former one. The classification and nomenclature are materially different, in consequence of the progress of our knowledge during the past twelve years. In 1873, a year after the old "Key" appeared, I published a "Check List," conformed exactly with the nomenclature of the "Key." In 1882, when I had recast the "Key," I published a second edition of the "Check List" in conformity with the new "Key." The present work, therefore, gives the same names, with scarcely any variance, though with a few additional ones; the new "Check List" and the new "Key" being practically one in all that pertains to nomenclature, and representing a particular phase of the subject. The numbering of the species, also, corresponds with that in the "Check List."

Part I. of the present work consists of my "Field Ornithology," originally published as a separate treatise in 1874, and now for the first time incorporated with the "Key." It is reprinted nearly verbatim, but with some little amplification towards its end, and the introduction of a few illustrations.

Part II. consists of the introductory matter of the old "Key," very greatly amplified. In its present shape it is a sort of "Closet Ornithology" as distinguished from a "Field Ornithology;" being a treatise on the classification and structure of birds, explaining and defining the technical terms used in ornithology, — in short, teaching the principles of the science and illustrating their application.

Part III., the main body of the work, describes all the species and subspecies of North American birds known to me, defines the genera, and characterizes the families and higher groups. The descriptions are much more elaborate than those of the old "Key," and I trust that such amplification has been made without loss of that sharpness of definition which was the aim of the first edition. I have kept steadily in view my main purpose — the ready identification of specimens. In many cases I have drawn upon my other works — such as the "Birds of the Colorado Valley," the "Birds of the Northwest," and several of my Monographs, — for available ready-made descriptions; but for the most part the matter of this kind is new. Scarcely any of this part of the old "Key" remains as it was. One improvement, I think, will be found in the removal of the unnecessary references to authorities which closed the descriptive paragraphs of the old "Key," and the utilization of the space thus gained by introducing terse biographical items, with special reference to nests and eggs, to song, flight, migrative and other habits; the technical descriptions of the species thus also epitomizing the life-history of the birds. Geographical distribution is also more fully treated, as its importance deserves. More attention has been paid to the description of the plumages of females and young birds. The specific names head their respective paragraphs, instead of tailing-off the same; they are also marked for accent, and their etymology is concisely stated, — though for this matter the student should continue to use the new "Check List."

As regards the artificial "key to the genera" of the old work, it has proven that too much was attempted in undertaking to carry the student at once to our refined modern genera. I have accordingly substituted artificial keys to the orders and families;

and throughout the work have analyzed species under their respective genera, these under their subfamilies or families, and these again under their orders.

Part IV. consists of a Synopsis of the Fossil birds of North America, corresponding to the appendix of the old "Key," but augmented by later discoveries. As before, this part of the work has been revised by Professor O. C. Marsh.

In the mechanical execution of the work, it has been my aim to compress the most matter into the least space and leave no waste paper, in order to keep the treatise within a single portable volume of convenient text-book size. I judge that there is nearly four times as much matter in the present volume as there was in the original edition, the page being much more closely printed, in a smaller type, and on thinner paper.

The old "Key" was insufficiently illustrated, and the average character of the cuts was not entirely satisfactory. The present edition more than doubles the number of illustrations. These are in part original, in part derived from various sources, all of which are duly accredited in the text. The basis of the series is of course the cuts of the former edition; but many of these have been discarded and replaced by better ones. About fifty of the most effective engravings were secured by my publishers from Brehm's "Thierleben;" nearly as many more are from Dixon's "Rural Bird Life," the American edition of which is owned by the same firm. A few have been copied from D. G. Elliot's "Birds of America," and a few others from the Proceedings of the Zoölogical Society of London. About fifty of the prettiest ones were drawn by Mr. Edwin Sheppard and engraved by Mr. H. H. Nichols, expressly for this edition. Another set — how many there are of them I do not know — are from my own drawings, and have mostly appeared in other of my publications. Several of Mr. R. Ridgway's drawings have been placed at my service, through his kind attentions, and with Professor Baird's permission. I am indebted to Dr. R. W. Shufeldt, U. S. A., for about thirty original anatomical drawings, as well as for the colored frontispiece. Mr. Henry W. Elliott has kindly put at my disposition several of his own artistic compositions, and I have received some very beautiful engravings with the compliments of the Century Company of New York.

It is always agreeable to pay one's respects when due, and acknowledge assistance and encouragement received in the preparation of one's books. Yet what an embarrassment is mine now! For there is no writer of repute on North American ornithology, and scarcely a leader of the science at large, who has not assisted in the making of the "Key;" and there is no reader of the work who has not encouraged its author to produce this new edition. I am trebly in debt, — to thousands whose names I know not; to hundreds I only know by name and fame; to scores of tried and trusted friends.

But let me say how much I am indebted to my compositors and proof-readers of the University Press at Cambridge for the skill with which they have turned copy into print, and to the proprietors of that justly-celebrated establishment for the pains they have taken in making the book an example of beautiful and accurate typography. Let me recognize here the liberality and generosity of my friend, Mr. Dana Estes, senior of the firm of Estes and Lauriat, in permitting me to make the book to suit myself, and in sparing no expense to which he might be put in consequence. Let me not forget that during its preparation, as for many years previously, I have enjoyed to the fullest extent the privileges of the Smithsonian Institution and the National Museum, through the courtesy of Professor Baird, my access to the great collection of birds being always facilitated by the attentions of Mr. Robert Ridgway, the Curator of Ornithology. And may that less tangible but not less real source of strength which inheres in the sympathetic and genial intercourse of a lifetime continue to be mine to draw upon, for all my works, from my warm friend, J. A. Allen, the first President of the American Ornithologists' Union.

"Prefaces," says some one, "ever were and still are but of two sorts; . . . still the author keeps to his old and wonted method of prefacing, when, at the beginning of his book he enters, either with a halter about his neck, submitting himself to his reader's mercy whether he shall be hanged, or no; or else in a huffing manner he appears with the halter in his hand, and threatens to hang his reader, if he gives him not his good word." But I wish neither to hang nor be hanged; I wish the work were better than it is, for my reader's sake; I wish the author were better than he is, for my own sake; and above all I wish that every author may rise superior to his best work, to the end that the man himself be judged above his largest achievements. It is well to do great things, but better still to be great.

E. C.

SMITHSONIAN INSTITUTION,
WASHINGTON, D. C., APRIL, 1884.

CONTENTS TO VOLUME I.

	PAGE
DEDICATION	i
PUBLISHER'S PREFACE	iii
PREFACE, FOURTH EDITION	v
PREFACE, THIRD EDITION	vii
HISTORICAL PREFACE	xi
CONTENTS	xxxi
IN MEMORIAM, ELLIOTT COUES	xxxv

PART I.

FIELD ORNITHOLOGY.

§ 1. Implements for collecting, and their use ... 1
2. Dogs ... 9
3. Various suggestions and directions for field-work ... 9
4. Hygiene of collectorship ... 19
5. Registration and labelling ... 21
6. Instruments, materials, and fixtures for preparing birdskins ... 25
7. How to make a birdskin ... 28
8. Miscellaneous particulars ... 45
9. Collection of nests and eggs ... 50
10. Care of a collection ... 54

PART II.

GENERAL ORNITHOLOGY.

§ 1. Definition of birds ... 59
2. Principles and practice of classification ... 65
3. Definitions and descriptions of the exterior parts of birds ... 81
    a. Of the feathers, or plumage ... 81
    b. The topography of birds ... 96
        1. Regions of the body ... 99
        2. Of the members; their parts and organs ... 105
            i. The bill ... 105
            ii. The wings ... 111
            iii. The tail ... 120
            iv. The feet ... 124

xxxii   CONTENTS.

	PAGE
§ 4. An introduction to the Anatomy of birds	139
a. Osteology: the osseous system, or skeleton	140
1. The spinal column	143
2. The thorax: ribs and sternum	148
3. The pectoral arch	151
4. The pelvic arch	153
5. The skull	155
b. Neurology: the nervous system; organs of special senses	180
c. Myology: the muscular system	198
d. Angeiology: the vascular or circulatory systems	201
e. Pneumatology: the respiratory system	205
f. Splanchnology: the digestive system	215
g. Oölogy: the uro-genital organs	221
5. Directions for using the artificial keys	233
ARTIFICIAL KEY TO THE ORDERS AND SUBORDERS	236
ARTIFICIAL KEY TO THE FAMILIES	237
TABULAR VIEW OF THE GROUPS HIGHER THAN GENERA	240

PART III.

SYSTEMATIC SYNOPSIS OF NORTH AMERICAN BIRDS.

Order PASSERES: Insessores, or Perchers Proper	244
Suborder ACROMYODI, POLYMODI, or OSCINES: Singing Birds	246
Family TURDIDÆ: Thrushes, etc.	247
Subfamily *Turdinæ*: Typical Thrushes	248
Subfamily *Myiadestinæ*: Fly-catching Thrushes; Solitaires	259
Family CINCLIDÆ: Dippers	260
Family SYLVIIDÆ: Old World Warblers, Kinglets, etc.	261
Subfamily *Sylviinæ*: Old World Warblers	261
Subfamily *Regulinæ*: Kinglets	262
Subfamily *Polyoptilinæ*: Gnat-catchers	264
Family CHAMÆIDÆ: Wren-tits	266
Family PARIDÆ: Titmice, or Chickadees	267
Subfamily *Parinæ*: True Titmice	267
Family SITTIDÆ: Nuthatches	276
Family CERTHIIDÆ: Creepers	278
Subfamily *Certhiinæ*: Typical Creepers	279
Family TROGLODYTIDÆ: Wrens; Thrashers, etc.	280
Subfamily *Miminæ*: Mockingbirds; Thrashers	281
Subfamily *Troglodytinæ*: Wrens	289
Family MOTACILLIDÆ: Wagtails and Pipits	300
Family MNIOTILTIDÆ: American Warblers	304
Family CŒREBIDÆ: Honey Creepers	346
Family TANAGRIDÆ: Tanagers	347
Family HIRUNDINIDÆ: Swallows	350

CONTENTS. xxxiii

	PAGE
Family AMPELIDÆ: Chatterers	357
Subfamily *Ampelinæ:* Waxwings	358
Subfamily *Ptilogonatinæ:* Flysnappers	360
Family VIREONIDÆ: Vireos, or Greenlets	361
Family LANIIDÆ: Shrikes	369
Subfamily *Laniinæ:* True Shrikes	369
Family FRINGILLIDÆ: Finches, etc.	373
Family ICTERIDÆ: American Starlings; Blackbirds, etc.	463
Subfamily *Agelæinæ:* Marsh Blackbirds	465
Subfamily *Sturnellinæ:* Meadow Starlings	471
Subfamily *Icterinæ:* American Orioles; Hang-nests	474
Subfamily *Quiscalinæ:* American Grackles	479
Family CORVIDÆ: Crows, Jays, Pies, etc.	484
Subfamily *Corvinæ:* Crows	485
Subfamily *Garrulinæ:* Jays and Pies	492
Family STURNIDÆ: Old World Starlings	502
Subfamily *Sturninæ:* Typical Starlings	502
Family ALAUDIDÆ: Larks	503
Suborder PASSERES MESSOMYODI, or CLAMATORES: Non-melodious or Songless Passeres	509
Family TYRANNIDÆ: American Flycatchers	510
Subfamily *Tyranninæ:* True Tyrant Flycatchers	510
Family COTINGIDÆ: Cotingas	534
Subfamily *Tityrinæ:* Tityrines	534

Sincerely Yours,
Elliott Coues.

IN MEMORIAM: ELLIOTT COUES.[1]
Born 9th September, 1842. — Died 25th December, 1899.

IN the life of every nation, society, or individual, no matter how peaceful, prosperous, or happy the record of the past may have been, no matter how encouraging and bright the future may be for further advancement, increased progress and greater achievements in the path that always leads onward and upward, toward the ultimate fulfilment of the highest destiny that may be attained, in the varying, shifting career that all must follow while accomplishing the pilgrimage of earth, yet in the experience of all, even amidst the rush of a restless activity, there comes a time to mourn. A time when the daily duties are temporarily neglected or wholly laid aside, when the engrossing pursuits that occupy the thoughts and call for the utmost energies of man's nature cease for the moment to interest the mind, when the smile vanishes and joyous laughter no longer cheers the heart, when the voice sinks to a whisper low and soft, as the sense of some irreparable loss comes with stunning force to overwhelm the soul. To this Society, to all its individual members, and to some of us in a peculiar and intimate relationship such a time has surely come, for as we are gathered here to-day, one engaging presence, one vitalizing force, one attractive personality, one brilliant mind is no longer in our midst, to grace, strengthen, and assist us in our deliberations, and in the accomplishment of duties that must be met. Who shall measure the extent of the loss sustained by various branches of scientific and historical research, by this and kindred societies, by those of us who have parted from an intimate friend and colleague of many vanished years, as well as the younger men just entering upon the scientific field, in the recent death of our former President and late colleague, Elliott Coues? No one occupied a more prominent position in our midst than he, and no one held it by a stronger claim, founded on exceptional ability, in brilliant work successfully accomplished.

On September 9th, 1842, in the town of Portsmouth, New Hampshire, Elliott Coues was born, and as soon as he could exhibit a preference for any object, his taste for ornithology was manifested, and even when only able to toddle about the nursery, a poster of one of the old-style menageries rendered him oblivious to all other attractions and no book nor story interested him unless animals were their subjects. So early did the tastes and preferences that were to be the chief controlling influences of his life declare themselves. When he was eleven years of age his father, Samuel Elliott Coues, removed to Washington, in which city our late colleague was destined to pass a large part of his life, and where some of his most

---

[1] An address delivered at the Eighteenth Congress of the American Ornithologists' Union, Cambridge, Mass., Nov. 13, 1900.

important works were to be written. For a time he attended Gonzaga College, a Jesuit Institution, and where, to one of his ardent temperament, the gorgeous ritual of the Romish church would be apt to make a deep impression; but his was to be an energetic life that demanded a wide field for its activity, and could not be pent amid cloistered shades or cathedral aisles. In his early days he was rather inclined to neglect the classics, replying once to a remonstrance of his father, "I only want just enough of these things to facilitate my other work," but later he appreciated the importance of a thorough knowledge of the ancient tongues and they had no more earnest advocate than himself. At the age of seventeen he entered Columbia College, now Columbian University, took his degree of A.B. in 1861, Honorary M.A. in 1862, became a Medical Cadet in 1862, M.D. in 1863 and Acting Assistant Surgeon, United States Army, in the same year, and Assistant Surgeon in 1864. When he passed his examination for the United States Army medical corps, he was obliged to tell them he was not of age, and he was appointed a volunteer surgeon for one year before he could receive his commission, and that year he passed at Mount Pleasant Hospital near Washington. For seventeen years he continued in the service of the United States, and was made a brevet Captain, resigning in 1881 in order to devote himself entirely to his scientific and literary pursuits.

During his army life he was stationed at various posts, mostly those situated in the western part of the United States, and he was also attached to some of the most important Government Surveys of the Territories and little known parts of our country, such as the one under the command of Dr. F. V. Hayden, and that of the Northern Boundary Commission which surveyed the forty-ninth parallel westward from the Lake of the Woods. In these great expeditions he served as surgeon and naturalist, and gained in the field that intimate knowledge of our birds and mammals which was to make him in the near future one of the most illustrious naturalists of our country and of our time. He had now become so absorbed in his scientific pursuits that the monotonous routine of an army post was most distasteful, and when he was detached from the surveying expeditions and ordered back to his first station at Fort Whipple, Arizona, he endeavored to obtain a different assignment, one more congenial to him and better adapted for his scientific work, and when this proved impossible he resigned from the army and took up his abode in Washington, where he resided until his death.

Although he was a writer on many and various subjects, his first scientific work was done in ornithology, and as early as 1861, when he was but nineteen years of age, he made his début as an author in a well-conceived and executed paper, that would have been highly creditable to a far more experienced hand, entitled "A Monograph of the Tringæ of North America." In his scientific studies Coues was fortunate in having for his mentor the late Professor Baird, and between them the strongest friendship existed and which only terminated with the death of the senior naturalist. From this period Coues's contributions to literary, scientific, and philosophic subjects never ceased, for his energies were unlimited and he became one of the most prolific writers of our day. In 1869 he was elected Professor of Zoölogy and Comparative Anatomy in Norwich University, Vermont, but the duties of army life prevented him from accepting this position; but after he retired from the service of the United States he accepted the chair of anatomy at the National Medical College in the medical department of Columbian University, Washington, where he lectured acceptably for ten years. He was also one of the contributors to the Century Dictionary, and had editorial charge of General Zoölogy, Biology, and Comparative Anatomy, and furnished some forty thousand words to this monumental work as his share of the enterprise; devoting to it the greater part of his labor for seven years. Another immense undertaking to which he devoted some years of painstaking work was a "Bibliography of Ornithology," certain instalments of which alone have been published, the greater portion still remaining in manuscript. He also began a "History of North American Mammals," but though considerable progress with it was accomplished nothing was ever published.

From 1861 to 1881 he completed three hundred works and papers, the major portion devoted to ornithology; and although he always kept up his interest in that science and was more or less an active contributor to it all his life, his later years were more particularly devoted to historical research. The titles to his scientific writings of all kinds, minor papers, reviews, and special works, number nearly one thousand, and he was the author or joint author of thirty-seven separate volumes. The work by which he will probably be best known and remembered, and which has had above all others the most important influence on ornithology in our own land, is his "Key to North American Birds," a work that in its conception and the masterly manner in which it is carried out in all its details stands as one of the best if not *the* best bird book ever written. His knowledge of North American mammals was as extensive and intimate as was that of our birds, and the "Fur Bearing Animals," published in 1877, as well as the Monographs on the Muridæ, Zapodidæ, Saccomyidæ, Haplodontia, and Geomyidæ in the "North American Rodentia," also issued in 1877, bear ample witness to this fact. It is impossible, however, in a comparatively brief address to enumerate the titles of his works, and to this audience they would seem like twice-told tales, for with the more important you are thoroughly familiar, and the minor ones are being constantly met with and referred to by you in the pursuit of your investigations.

We know what he has done in Natural Sciences, and although he rests from his labors, and the eloquent tongue is silent and the still more eloquent pen lies motionless, never more to perpetuate the virile thoughts that struggled for expression in the active mind, yet his works remain and speak with no uncertain tones for him. I would, however, pass from the consideration of him as an author and facile writer, and present him to you as the man, as he really was, for although many persons were acquainted with Coues few I believe really knew him. It is now nearly forty years ago, when on a visit to Professor Baird in Washington, one evening, in company with my old friend Dr. Gill, I first met Elliott Coues. He was then in his teens, a student of medicine, frank, simple, honest, and confiding, with a boy's generous impulses, and the glorious enthusiasm of the ornithologist manifest in speech and action. The friendship then formed continued without a break or a hasty word ever having been exchanged with tongue or pen throughout all the intervening years. And yet we thought very differently on many subjects;

but such was our confidence in each other's honest intention and unreserved frankness that we could, and did many times, argue on different sides, both orally and in writing, with an energetic earnestness that would have been highly dangerous to our continued friendship if we had not understood each other so well. And first among his most eminent characteristics was his love of truth, and he was constantly striving with all the force of his energetic nature to search it out and take its teaching to himself wherever he might find it, careless where it might lead him or what preconceived views or opinions it might overthrow or destroy. He believed with Carlyle that "there is no reliance for this world or any other but just the truth, there is no hope for the world but just so far as men find out and believe the truth and match their own lives to it." It was therefore in his search for truth and an attempt to apply the principles of physical science to psychical research that in 1880 he became affiliated with the Theosophical Society of India and was elected President of its American Board of Control, and was continued in that office for several years. He was much interested in the subject and investigated its principles and methods with his usual thoroughness, even visiting Europe in company with Madame Blavatsky and other prominent members of the sect, and his connection with this and kindred societies resulted in the production of several publications such as "Biogen" and the "Dæmon of Darwin." But the knowledge that he gained of this interesting but peculiar doctrine was not of that satisfying character as to cause him to hold fast to its tenets, nor to enable him to retain his respect for its leaders, and although he gives no reasons for the action, yet in the memorandum in which he records his election as President in 1885 and his re-election in the following year, with characteristic frankness he states that he was expelled from the Society in 1889. Those of us who have little sympathy with the claims asserted by the disciples of Theosophy cannot but regard his expulsion from the Society as having conferred a greater honor upon him than his election to the Presidency, and can easily imagine the action he may have taken in the Council to cause such a result after he finally satisfied himself that the doctrine could not substantiate its claims. He detested shams of all kinds and hurled the full force of his invective against those who had proved themselves unworthy or who strove to appear entitled to more than was their due.

As a critic in certain lines he was unrivalled and exhibited the highest practice of the art in his reviews, dwelling most upon what was meritorious in the treatment of the subject before him, for he believed true criticism was to seek that which was praiseworthy rather than something to condemn. But no one could be more caustic in his treatment, nor wield a sharper weapon, when he found that praise would be misapplied and it would be kinder to act as the skilful surgeon does, create wounds in order that the patient's recovery might be more sure and lasting. Rarely, however, for one who published so much, was he severe in his writings, though none had the power to be more so; but when, from whatever the cause that influenced him, he permitted himself to indulge in phrases that would be remembered and might possibly leave a sting, he set down "naught in malice," but employed a phraseology that he honestly believed was best suited to the case in hand, and after some such severe articles had been issued, he has spoken to me in the kindest way of the author of the work or act he had so criticised or condemned, apparently entirely unconscious that it could possibly affect any friendly relations or be the means of any estrangement. It was the sentiment advanced, or the conclusion reached, that was the object of his attack, not the individual who was the author. In all his critical reviews there is no thought of self, but only desire to do justice to his subject and to its author, and if anything could be charged against him on this point, it was an evident inclination always to find something to praise.

In his scientific writings he was always extremely lucid and conservative in his methods, and he had but little sympathy for the hair-splitting and microscopic variations in the appearance of animals that is the joy and delight of some naturalists in these later days. He was a scholar and knew his Greek and Latin; and with a scholar's instinct and abhorrence of incorrect phraseology, he strove with all his might to inculcate not only in his own scientific writings but in those of others the true principles of etymology and philology; and both by tongue and pen, in the keen analytical style of which he was an undisputed master, he strove with all the force of his energetic personality against the unfortunate and mistaken doctrine that the perpetuation of errors can ever be permissible, much less commendable. He possessed a command of language gained by few, and the beauty of his style and his felicity of expression has created numerous pen pictures of the habits and appearances of our wild creatures that have never been excelled by any writer, if indeed they have been equalled.

While a keen and just critic himself, he was very sensitive regarding the opinion of others towards his own productions, and sought the approbation of those who were bound closely to him either by earthly ties or an intimate friendship, or whose knowledge of the subject under consideration caused their opinion to be of special value. This extreme sensitiveness is best illustrated by an act committed in his youthful days, when after having labored for several years upon a work on Arizona, on reading his manuscript to one who, if not competent to judge of the importance of his labors, he had the right to expect would exhibit sympathy for his efforts, and who must at least have been impressed with its thoroughness and beauty of diction, yet was only able to consider its value as a commercial asset, and therefore commented upon it so unfavorably, and with such strength of expression, that, utterly disheartened at the want of appreciation for that which had been so long a labor of love and of which he was so proud of his ability to produce, on the impulse of the moment he cast the "copy" into the fire, where it was consumed, and then suffered a severe attack of illness in consequence of his loss by his hasty act.

Of a most affectionate disposition, he sought and enjoyed the society of his friends and those with sympathetic tastes; and although he possessed strong convictions and firm opinions, yet no one more readily yielded to the views of another whose opportunities to reach a correct decision had been greater than his own, and this was always effected with a courtesy that caused his friendly opponent to regret he could not himself yield and reverse their positions. He loved science and scientific work, and scorned to employ his talents and his knowledge merely for financial considerations; and although he could command large sums for his labor, he preferred to devote himself to pure science, which, if less remunerative pecuniarily, achieves a more lasting result and one of greater honor.

After all these years of scientific work, his thoughts and labors turned to a new channel, that of historical research, and the last eight or ten years of his life were devoted to editing the journals of the early explorers of our continent, and he made many long and wearisome journeys over the various routes taken by these hardy pioneers in order to familiarize himself with the country traversed and locate the many places mentioned, but which had no designation on any published map. His former army life and his great experience as a naturalist eminently fitted him for this task, and probably no one could have proved himself so competent to fulfil this duty. The first of these works was that of the Expedition of Lewis and Clarke, which appeared in 1893, followed in 1895 by the Expedition of Zebulon M. Pike. In 1897 came the Henry & Thompson Journals; in 1898 appeared the Fowler Journal and the Narrative of Charles Larpentuer, forty years a Fur Trader on the Upper Missouri; and during this year The Diary of Francisco Garces, on the trail of a Spanish Pioneer: in all, fifteen volumes. All of these books bear the impress of his most conscientious care and wonderful minuteness of annotation; and it is to Coues more than to any other that the original sources of the early explorations of the western portion of our country, beyond the Mississippi, are preserved.

It was during an arduous journey in New Mexico and Arizona in the summer of 1899, undertaken, as he wrote me, as a "still hunt for old Spanish MSS.," and to refresh his memory of the country described by Francisco Garces, and render still more effective his editing of the Diary in his possession, that Coues's splendid physique and robust health, that for so long seemed to defy fatigue and exposure, gave way, and he was brought to Santa Fé in a rather critical condition, where for a month he was very ill, but in September he came to Chicago. He seemed to be getting better, and at my last interview with him, during which his condition was freely discussed, although he fully appreciated the gravity of his case, yet he expressed the hope, and perhaps he thought it was clearly among the possibilities, that he might be present at the last meeting of this Society in Philadelphia. Regarding him, as I then did, as in a critical condition, I could not share this hope, although I encouraged him in his belief, or what seemed to be his belief, for Coues had been too long a skilled medical practitioner to try and deceive himself; but from his references to his attendant physician it was clearly apparent that he preferred to advance the opinion of his medical adviser, of whom he spoke in the highest terms, rather than any of his own. He was greatly changed in appearance, but the old fire and enthusiasm, that I had so often admired and not infrequently contended with in friendly conflict during so many years, was not a whit abated, and he spoke with all his old-time interest of the work he had himself in view and that of others. But the voice was feeble and the frame was weak, and he was filled with a restlessness that was foreign to him. But when I bade him an adieu, which was to be our last on earth, he was cheerful and spoke hopefully of meeting soon again. As you all know, his condition became more serious after he arrived at his home in Washington, and an expert examination at Johns Hopkins Hospital in Baltimore gave but little hope for the preservation of his life. During these last days I received a number of letters from him explaining frankly his condition and how few were his chances for life, and just before submitting to the operation came one virtually bidding me farewell and announcing the close of our correspondence, that had extended over many years. On the 6th of December the operation was performed, and for a short time there was a probability that his life would be prolonged; but it was not to be, for he had finished his work and he was to rest from his labors. Throughout his illness he exhibited the natural bravery of spirit habitual to him; not a murmur or complaint of the excessive and lasting pain, but gentle and courteously appreciative of every attention, and at the last overcoming for an instant the weakness that denoted the approach of that moment when his freed spirit should depart and soar above all earthly things, he raised himself in his bed, and with all the old-time vigor of voice exclaimed, "Welcome, oh, welcome, beloved death!" and sinking backwards on the pillow he was at rest. Nevermore shall you welcome to your midst this courteous gentleman, who was the considerate friend, the able counsellor, the chivalrous debater, the one most capable of leadership, yet always willing to yield to another, the trained scientist, the accomplished anatomist, the able naturalist, the conscientious historian. His was a life of intense activity, and that which his hand found to do he did with all his might; and of none can it be more appropriately said, "Nihil tetigit quod non ornavit."

Coues, as may be readily supposed, was the recipient of many scientific honors, and he was an Honorary or Active member of a very large number of societies, both in this country and in Europe, and at the time of his election to our National Academy he was, I believe, its youngest member. The list of scientific societies with which he was connected numbers between fifty and sixty, far too many for me to attempt to give their titles at this time, yet none of them was so distinguished but that it received as well as conferred an honor by having his name upon its rolls. As a naturalist Coues will always hold the highest rank in the estimation of all who are familiar with his works; and in that galaxy of eminent names which sheds so great a brilliancy on the scientific annals of our own land, none shall appear in the years to come more lustrous than that of our late distinguished colleague and friend. But the brilliant mind no longer teems with thoughts of earth, and the hand that executed its commands lies motionless, and we, who are drawing near to that shining portal through which he has so lately passed, and from whose farther side no steps are ever retraced by any one of mortal birth, may never look upon his like again, whose pen was the "pen of a ready writer," fit instrument to convey and render permanent the eloquence of thought, beauty of diction, and facility of exprestion of Nature's illustrious disciple and interpreter.

D. G. ELLIOT, F. R. S. E., ETC.

PART I.

FIELD ORNITHOLOGY:

BEING A

MANUAL OF INSTRUCTION FOR COLLECTING, PREPARING, AND PRESERVING BIRDS.

FIELD ORNITHOLOGY must lead the way to Systematic and Descriptive Ornithology. The study of Birds in the field is an indispensable prerequisite to their study in the library and the museum. Directions for observing and collecting birds, for preparing and preserving them as objects of natural history, will greatly help the student on his way to become a successful Ornithologist, if he will faithfully and intelligently observe them. It is believed that the practical Instructions which the author has to give will, if followed out, enable any one who has the least taste or aptitude for such pursuits to become proficient in the necessary qualifications of the good working ornithologist. These instructions are derived from the writer's own experience, reaching in time over twenty years, and extending in area over large portions of North America. Having made in the field the personal acquaintance of most species of North American birds, and having shot and skinned with his own hands several thousand specimens, he may reasonably venture to speak with confidence, if not also with authority, respecting methods of study and manipulation. Feeling so much at home in the field, with his gun for destroying birds, and his instruments for preserving their skins, he wishes to put the most inexperienced student equally at ease; and therefore beg to lay formality aside, that he may address the reader familiarly, as if chatting with a friend on a subject of mutual interest.

§ 1.—IMPLEMENTS FOR COLLECTING, AND THEIR USE.

The Double-barrelled Shot Gun is your main reliance. Under some circumstances you may trap or snare birds, catch them with bird-lime, or use other devices; but such cases are exceptions to the rule that you will shoot birds, and for this purpose no weapon compares with the one just mentioned. The soul of good advice respecting the selection of a gun is, *Get the best one you can afford to buy;* go the full length of your purse in the matters of material and workmanship. To say nothing of the prime requisite, safety, or of the next most desirable quality, efficiency, the durability of a high-priced gun makes it cheapest in the end.

Style of finish is obviously of little consequence, except as an index of other qualities; for inferior guns rarely, if ever, display the exquisite appointments that mark a first-rate arm. There is really so little choice among good guns that nothing need be said on this score; you cannot miss it if you pay enough to any reputable maker or reliable dealer. But collecting is a specialty, and some guns are better adapted than others to your particular purpose, which is the destruction, as a rule, of small birds, at moderate range, with the least possible injury to their plumage. Probably three-fourths or more of the birds of a miscellaneous collection average under the size of a pigeon, and were shot within thirty yards. A *heavy* gun is therefore unnecessary, in fact ineligible, the extra weight being useless. You will find a gun of 7½ to 8 pounds weight most suitable. For similar reasons the *bore* should be small; I prefer 14 gauge, and should not think of going over 12. To judge from the best sporting authorities, *length of barrel* is of less consequence than many suppose; for myself, I incline to a rather long barrel, — one nearer 33 than 28 inches, — believing that such a barrel *may* throw shot better; but I am not sure that this is even the rule, while it is well known that several circumstances of loading, besides some almost inappreciable differences in the way barrels are bored, will cause guns apparently exactly alike to throw shot differently. Length and crook of *stock* should of course be adapted to your figure, — a gun may be made to fit you, as well as a coat. For wild-fowl shooting, and on some other special occasions, a heavier and altogether more powerful gun will be preferable.

Breech-Loader *vs.* **Muzzle-Loader**, a case long argued, may be considered settled in favor of the former. Provided the mechanism and workmanship of the breech be what they should, there are no valid objections to offset obvious advantages, some of which are these: ease and rapidity of loading, and consequently delivery of shots in quick succession; facility of cleaning; compactness and portability of ammunition; readiness with which different-sized shot may be used. This last is highly important to the collector, who never knows the moment he may wish to fire at a very different bird from such as he has already loaded for. The muzzle-loader must always contain the fine shot with which nine-tenths of your specimens will be secured; if in both barrels, you cannot deal with a hawk or other large bird with reasonable prospects of success; if in only one barrel, the other being more heavily charged, you are crippled to the extent of exactly one-half of your resources for ordinary shooting. Whereas, with the breech-loader you will habitually use mustard-seed in both barrels, and yet can slip in a different shell in time to seize most opportunities requiring large shot. This consideration alone should decide the case. But, moreover, the time spent in the field in loading an ordinary gun is no small item; while cartridges may be charged in your leisure at home. This should become the natural occupation of your spare moments. No time is really *gained;* you simply change to advantage the time consumed. Metal shells, charged with loose ammunition, and susceptible of being reloaded many times, may be used instead of any special fixed ammunition which, once exhausted in a distant place (and circumstances may upset the best calculations on that score), leaves the gun useless. On charging the shells mark the number of the shot used on the outside wad; or better, use colored wads, say plain white for dust shot, and red, blue, and green for certain other sizes. If going far away, take as many shells as you think can possibly be wanted — and a few more.

Experience, however, will soon teach you to prefer paper cartridges for breech-loaders. They may of course be loaded according to circumstances, with the same facility as metal shells, and even reloaded if desired. It is a good deal of trouble to take care of metal shells, to prevent loss, keep them clean, and avoid bending or indenting; while there is often a practical difficulty in recapping — at least with the common styles that take a special primer. Those fitted with a screw top holding a nipple for ordinary caps are expensive. Paper cartridges come already capped, so that this bother is avoided, as it is not ordinarily worth while to reload them. They are made of different colors, distinguishing various sizes of shot used without employ of colored wads otherwise required. They may be taken into the field empty and loaded on occasion to suit; but it is better to pay a trifle extra to have them loaded at the shop. In such case, about four-fifths of the stock should contain mustard-seed, nearly all the rest about No. 7, a very few being reserved for about No. 4. Cost of ammunition is hardly appreciably increased; its weight is put in the most conveniently portable shape; the whole apparatus for carrying it, and loading the shells, is dispensed with; much time is saved, the entire drudgery (excepting gun-cleaning) of collecting being avoided. I was prepared in this way during the summer of 1873 for the heaviest work I ever succeeded in accomplishing during the same length of time. In June, when birds were plentiful, I easily averaged fifteen skins a day, and occasionally made twice as many. As items serving to base calculations, I may mention that in four months I used about two thousand cartridges, loaded, at $42 per M., with seven-eighths of an ounce of shot and two and three-fourths drachms of powder; only about three hundred were charged with shot larger than mustard-seed. In estimating the size of a collection that may result from use of a given number of cartridges, it may not be safe for even a good shot to count on much more than half as many specimens as cartridges. The number is practically reduced by the following steps: — Cartridges lost or damaged, or originally defective; shots missed; birds killed or wounded, not recovered; specimens secured unfit for preservation, or not preserved for any reason; specimens accidentally spoilt in stuffing, or subsequently damaged so as to be not worth keeping; and finally, use of cartridges to supply the table.

Other Weapons, etc. — An ordinary *single-barrel* gun will of course answer; but is a sorry makeshift, for it is sometimes so poorly constructed as to be unsafe, and can at best be only just half as effective. This remark does not apply to any of the fine single-barrelled breech-loaders now made. You will find them very effective weapons, and they are not at all expensive. An arm now much used by collectors is a kind of breech-loading pistol, with or without a skeleton gun-stock to screw into the handle, and taking a particular style of metal cartridge, charged with a few grains of powder, or with nothing but the fulminate. They are very light, very cheap, safe and easy to work, and astonishingly effective up to twenty or thirty yards; making probably the best "second choice" after the matchless double-barrelled breech-loader itself. The *cane-gun* should be mentioned in this connection. It is a single-barrel, lacquered to look like a stick, with a brass stopper at the muzzle to imitate a ferule, countersunk hammer and trigger, and either a simple curved handle, or a light gunstock-shaped piece that screws in. The affair is easily mistaken for a cane. Some have acquired considerable dexterity in its use; my own experience with it is very limited and unsatisfactory; the handle always hit me in the face, and I generally missed my bird. It has only two recommendations. If you approve of shooting on Sunday and yet scruple to shock popular prejudice, you can slip out of town unsuspected. If you are shooting where the law forbids destruction of small birds, — a wise and good law that you may sometimes be inclined to defy, — artfully careless handling of the deceitful implement may prevent arrest and fine. A *blow-gun* is sometimes used. It is a long slender tube of wood, metal, or glass, through which clay-balls, tiny arrows, etc., are projected by force of the breath. It must be quite an art to use such a weapon successfully, and its employment is necessarily exceptional. Some uncivilized tribes are said to possess marvellous skill in the use of long bamboo blow-guns; and such people are often valuable employés of the collector. I have had no experience with the noiseless *air-gun*, which is, in effect, a modified blow-gun, compressed air being the explosive power. Nor can I say much of various methods of *trapping* birds that may be practised. On these points I must leave you to your own devices, with the remark that horse-hair *snares*, set over a nest, are often of great service in securing the parent of eggs that might otherwise remain *unidentified*. I have no practical knowledge of *bird-lime* ; I believe it is seldom used in this country. A method of *netting* birds alive, which I have tried, is both easy and successful. A net of fine green silk, some 8 or 10 feet square, is stretched perpendicularly across a narrow part of one of the tiny brooks, overgrown with briers and shrubbery, that intersect many of our meadows. Retreating to a distance, the collector beats along the shrubbery making all the noise he can, urging on the little birds till they reach the almost invisible net and become entangled in trying to fly through. I have in this manner taken a dozen sparrows and the like at one "drive." But the gun can rarely be laid aside for this or any similar device.

Ammunition.— The best *powder* is that combining strength and cleanliness in the highest compatible degree. In some brands too much of the latter is sacrificed to the former. Other things being equal, a rather coarse powder is preferable, since its slower action tends to throw shot closer. Some numbers are said to be "too quick" for fine breech-loaders. Inexperienced sportsmen and collectors almost invariably use too coarse *shot*. When unnecessarily large, two evils result: the number of pellets in a load is decreased, the chances of killing being correspondingly lessened; and the plumage is unnecessarily injured, either by direct mutilation, or by subsequent bleeding through large holes. As already hinted, shot cannot be too fine for your routine collecting. Use "mustard-seed," or "dust-shot," as it is variously called; it is smaller than any of the sizes usually numbered. As the very finest can only be procured in cities, provide yourself liberally on leaving any centre of civilization for even a country village, to say nothing of remote regions. A small bird that would have been torn to pieces by a few large pellets, may be riddled with mustard-seed and yet be preservable ; moreover, there is, as a rule, little or no bleeding from such minute holes, which close up by the elasticity of the tissues involved. It is astonishing what large birds may be brought down with the tiny pellets. I have killed hawks with such shot, knocked over a wood ibis at forty yards and once shot a wolf dead with No. 10, though I am bound to say the animal was within a few feet of me. After dust-shot, and the nearest number or two, No. 8 or 7 will be found most useful. Waterfowl, thick-skinned sea-birds, like loons, cormorants, and pelicans, and a few of the largest land birds, require heavier shot. I have had no experience with the substitution of fine gravel or sand, much less water, as a projectile; besides shot I never fired anything at a bird except my ramrod, on one or two occasions, when I never afterwards saw either the bird or the stick. The comparatively trivial matter of *caps* will repay attention. Breech-loaders not discharged with a pin take a particular style of short cap called a "primer;" for other guns the *best* water-proof lined caps will prevent annoyance and disappointment in wet weather, and may save you an eye, for they only *split* when exploded ; whereas, the flimsy cheap ones — that "G D" trash, for instance, sold in the corner grocery at ten cents a hundred — usually fly to pieces. Cut felt *wads* are the only suitable article. Ely's "chemically prepared" wadding is the best. It is well, when using plain wads, occasionally to drive a greased one through the barrel. Since you may sometimes run out of wads through an unexpected contingency, always keep a wad-cutter to fit your gun. You can make serviceable wads of pasteboard, but they are inferior to felt. Cut them on the flat sawn end of a stick of firewood : the side of a plank does not do very well. Use a wooden mallet, instead of a hammer or hatchet, and so save your cutter. Soft paper is next best after wads; I have never used rags, cotton or tow, fearing these tinder-like substances might leave a spark in the barrels. Crumbled leaves or grass will answer at a pinch. I have occasionally, in a desperate hurry, loaded and *killed* without any wadding.

Other Equipments.— (*a.*) *For the Gun.* A gun-case will come cheap in the end, especially if you travel much. The usual box, divided into compartments, and well lined, is the best, though the full length leather or india-rubber cloth case answers very well. The box should contain a small kit of tools, such as mainspring-vice, nipple-wrench, screw-driver, etc. A stout hard-wood cleaning rod, with wormer, will be required. It is always safe to have parts of the gun-lock, especially mainspring, in duplicate. For muzzle-loaders extra nipples and extra ramrod heads and tips often come into use. For breech-loaders the apparatus for charging the shells is so useful as to be practically indispensable. (*b.*) *For ammunition.* Metal shells or paper cartridges may be carried loose in the large lower coat pocket, or in a leather satchel. There is said to be a chance of explosion by some unlucky blow, when they are so carried, but I never knew of an instance. Another way is to fix them separately in a row in snug loops of soft leather sewn continuously along a stout waist-belt; or in several such horizontal rows on a square piece of thick leather, to be slung by a strap over the shoulder. But better than anything else is a stout linen *vest,* similarly furnished with loops holding each a cartridge; this distributes the weight so perfectly, that the usual " forty rounds " may be carried without feeling it. The appliances for loose ammunition are almost endlessly varied, so every one may consult his taste or convenience. But now that everybody uses the breech-loader, shot-pouches and powder-flasks are among the things that were. (*c.*) *For specimens.* You must always carry *paper* in which to wrap up your specimens, as more particularly directed beyond. Nothing is better for this purpose than writing-paper; "rejected" or otherwise useless MSS. may thus be utilized. The ordinary game bag, with leather back and network front, answers very well; but a light basket, fitting the body, such as is used by fishermen, is the best thing to carry specimens in. Avoid putting specimens into *pockets,* unless you have your coat-tail largely excavated: crowding them into a close pocket, where they press each other, and receive warmth from the person, will injure them. It is always well to take a little cotton into the field, to plug up shot-holes, mouth, nostrils, or vent, immediately, if required. (*d.*) *For Yourself.* The indications to be fulfilled in your clothing are these: Adaptability to the weather; and since a shooting-coat is not conveniently changed, while an overcoat is ordinarily ineligible, the requirement is best met by different underclothes. Easy fit, allowing perfect freedom of muscular action, especially of the arms. Strength of fabric, to resist briers and stand wear; velveteen and corduroy are excellent materials. Subdued color, to render you as inconspicuous as possible, and to show dirt the least. Multiplicity of pockets — a perfect shooting-coat is an ingenious system of hanging pouches about the person. Broad-soled, low-heeled boots or shoes, giving a firm tread even when wet. Close-fitting cap with prominent visor, or low soft felt hat, rather broad brimmed. Let india-rubber goods alone; the field is no place for a sweat-bath.

Qualifications for Success. — With the outfit just indicated you command all the required appliances that you can *buy,* and the rest lies with yourself. Success hangs upon your own exertions; upon your energy, industry, and perseverance; your knowledge and skill; your zeal and enthusiasm, in collecting birds, much as in other affairs of life. But that your efforts — maiden attempts they must once have been if they be not such now — may be directed to best advantage, further instructions may not be unacceptable.

To Carry a Gun without peril to human life or limb is the *a b c* of its use. "There's death in the pot." Such constant care is required to avoid accidents that no man can give it by continual voluntary efforts: safe carriage of the gun must become an unconscious habit, fixed as the movements of an automaton. The golden rule and whole secret is: *the muzzle must never sweep the horizon ;* accidental discharge should send the shot into the ground before your feet, or away up in the air. There are several safe and easy ways of holding a piece: they will be employed by turns to relieve particular muscles when fatigued. 1. Hold it in the hollow of the arm (preferably the left, as you can recover to aim in less time than from the right), across the front of your person, the hand on the grip, the muzzle elevated about 45°. 2. Hang it by the trigger guard hitched over the forearm brought round to the breast, the stock passing behind the upper arm, the muzzle pointing to the ground a pace or so in front of you. 3. Shoulder it, the hand on the grip or heel-plate, the muzzle pointing upward at least 45°. 4. Shoulder it reversed, the hand grasping the barrels about their middle, the muzzle pointing forward and downward: this is perfectly admissible, but is the most awkward position of all to recover from. *Always carry a loaded gun at half-cock*, unless you are about to shoot. Most good guns are now fitted with rebounding locks, an arrangement by which the hammer is thrown back to half-cock as soon as the blow is delivered on the pin. This admirable device is a great safe-guard, and is particularly eligible for breech-loaders, as the barrels may be unlocked and relocked without touching the hammers. Unless the lock fail, accidental discharge is impossible, except under these circumstances: *a*, a direct blow on the nipple or pin; *b*, catching of both hammer and trigger simultaneously, drawing back of the former and its release whilst the trigger is still held,—the chances against which are simply incalculable. Full-cock, ticklish as it seems, is safer than no-cock, when a tap on the hammer or even the heel-plate, or a slight catch and release of the hammer, may cause discharge. Never let the muzzle of a loaded gun point toward your own person for a single instant. Get your gun over fences, or into boats or carriages, before you get over or in yourself, or at any rate no later. Remove caps or cartridges on entering a house. Never aim a gun, loaded or not, at any object, unless you mean to press the trigger. Never put a loaded gun away long enough to forget whether it is loaded or not; never leave a loaded gun to be found by others under circumstances reasonably presupposing it to be unloaded. Never put a gun where it can be knocked down by a dog or a child. Never imagine that there can be any excuse for *leaving* a breech-loader loaded under any circumstances. Never forget that the idiots who kill people because they " did n't know it was loaded," are perennial. Never forget that though a gunning accident may be sometimes interpreted (from a certain standpoint) as a " dispensation of Providence," such dispensations happen oftenest to the careless.

To Clean a Gun properly requires some knowledge, more good temper, and most " elbow-grease;" it is dirty, disagreeable, inevitable work, which laziness, business, tiredness, indifference, and good taste will by turns tempt you to shirk. After a hunt you are tired, have your clothes to change, a meal to eat, a lot of birds to skin, a journal to write up. If you " sub-let " the contract the chances are it is but half fulfilled; serve yourself, if you want to be well served. If you cannot find time for a regular cleaning, an intolerably foul gun may be made to do another day's work by swabbing for a few moments with a wet (not dripping) rag, and then with an oiled one. For the full wash use cold water first; it loosens dirt better than hot water. Set the barrels in a pail of water; wrap the end of the cleaning rod with tow or cloth, and pump away till your arms ache. Change the rag or tow, and the water too, till they both stay clean for all the swabbing you can do. Fill the barrels with boiling water till they are well heated; pour it out, wipe as dry as possible inside and out, and set them by a fire. Finish with a *light* oiling, inside and out; touch up all the metal about the stock, and polish the wood-work. Do not remove the locks oftener than is necessary; every time they are taken out, something of the exquisite fitting that marks a good gun may be lost; as long as they work smoothly take it for granted they are all right. The same direction applies to nipples. To keep a gun well, under long disuse, it should have had a particularly thorough cleaning; the chambers should be packed with greasy tow; greased wads may be rammed at intervals along the barrels; or the barrels may be filled with melted tallow. Neat's-foot is recommended as the best easily procured oil; porpoise-oil which is, I believe, used by watch-makers, is the very best; the oil made for use on sewing-machines is excellent; " olive " oil (made of lard) for table use answers the purpose. The quality of any oil may be improved by putting in it a few tacks, or scraps of zinc, — the oil expends its rusty capacity in oxidizing the metal. Inferior oils get "sticky." One of the best preventives of rust is mercurial ("blue") ointment: it may be freely used. Kerosene will remove rust; but use it sparingly for it "eats" sound metal too.

To Load a Gun effectively requires something more than knowledge of the facts that the powder should go in before the shot, and that each should have a wad a-top. Probably the most nearly universal fault is use of too much shot for the amount of powder; and the next, too much of both. The rule is *bulk for bulk* of powder and shot. If not exactly this, then rather less shot than powder. It is absurd to suppose, as some persons who ought to know better do, that the more shot in a gun the greater the chances of killing. The projectile force of a charge cannot possibly be greater than the *vis inertiæ* of the gun as held by the shooter. The explosion is manifested in all directions, and blows the shot one way simply and only because it has no other escape. If the resistance in front of the powder were greater than elsewhere, the shot would not budge, but the gun would fly backward, or burst. This always reminds me of Lord Dundreary's famous conundrum — Why does a dog wag his tail? Because he is bigger than his tail; otherwise the tail would wag *him*. A gun shoots shot because the gun is the heavier; otherwise the shot would shoot the gun. Every unnecessary pellet is a pellet against you, not against the game. The experienced sportsman uses about one-third less shot than the tyro, with proportionally better result, other things being equal. As to powder, moreover, a gun can only burn just so much, and every grain blown out unburnt is wasted if nothing more. No express directions for absolute weight or measures of either powder or shot can be given; in fact, different guns take as their most effective charge such a variable amount of ammunition, that one of the first things you have to learn about your own arm is, its normal charge-gauge. Find out, by assiduous target practice, what absolute amounts (and to a slight degree, what relative proportion) of powder and shot are required to shoot the furthest and distribute the pellets most evenly. This practice, furthermore, will acquaint you with the gun's capacities in every respect. You should learn exactly what it will and what it will not do, so as to feel perfect confidence in your arm within a certain range, and to waste no shots in attempting miracles. Immoderate recoil is a pretty sure sign that the gun was overloaded, or otherwise wrongly charged; and all force of recoil is subtracted from the impulse of the shot. It is useless to ram powder very hard; two or three smart taps of the rod will suffice, and more will not increase the explosive force. On the shot the wad should simply be pressed close enough to fix the pellets immovably. All these directions apply to the charging of metal or paper cartridges as well as to loading by the muzzle. The latter operation is so rarely required, now that guns of every grade break at the breach, that advice on this score may seem quite anachronistic; nevertheless, I let what I said in the original edition stand. When about to recharge one barrel see that the hammer of the other stands at half-cock. Do not drop the ramrod into the other barrel, for a stray shot might impact between the swell of the head and the gun and make it difficult to withdraw the rod. During the whole operation keep the muzzle as far from your person as you conveniently can. Never force home a wad with the flat of your hand over the end of the rod, but hold the rod between your fingers and thumb; in case of premature explosion, it will make just the difference of lacerated finger tips, or a blown-up hand. Never look into a loaded gun-barrel; you might as wisely put your head into a lion's mouth to see what the animal had for dinner. After a miss-fire hold the gun up a few moments and be slow to reload; the fire sometimes "hangs" for several seconds. Finally, let me strongly impress upon you the expediency of *light loading* in your routine collecting. Three-fourths of your shots need not bring into action the gun's full powers of execution. You will shoot more birds under than over 30 yards; not a few you must secure, if at all, at 10 or 15 yards; and your object is always to kill them with the least possible damage to the plumage. I have, on particular occasions, loaded even down to ½oz. of shot and 1½dr. of powder. There is astonishing force compressed in a few grains of powder; an astonishing number of pellets in the smallest load of mustard-seed. If you *can* load so nicely as to just drive the shot into a bird and not through it and out again, do so, and save half the holes in the skin.

To Shoot successfully is an art which may be acquired by practice, and can be learned only in the school of experience. No general directions will make you a good shot, any more than a proficient in music or painting. To tell you that in order to hit a bird you must point the gun at it and press the trigger, is like saying that to play on the fiddle you must shove the bow across the strings with one hand while you finger them with the other; in either case the result is the same, a noise — *vox et præterea nihil* — but neither music nor game. Nor is it possible for every one to become an artist in gunnery; a "crack shot," like a poet, is born, not made. For myself I make no pretensions to genius in that direction; for although I generally make fair bags, and have destroyed many thousand birds in my time, this is rather owing to some familiarity I have gained with the habits of birds, and a certain knack, acquired by long practice, of picking them out of trees and bushes, than to skilful shooting from the sportsman's standpoint; in fact, if I cut down two or three birds on the wing without a miss I am working quite up to my average in that line. But any one not a purblind "butter fingers," can become a reasonably fair shot by practice, and do good collecting. It is not so hard, after all, to sight a gun correctly on an immovable object, and collecting differs from sporting proper in this, that comparatively few birds are shot on the wing. But I do not mean to imply that it requires less skill to collect successfully than to secure game; on the contrary, it is finer shooting, I think, to drop a warbler skipping about a tree-top than to stop a quail at full speed; while hitting a sparrow that springs from the grass at one's feet to flicker in sight a few seconds and disappear is the most difficult of all shooting. Besides, a crack shot, as understood, aims unconsciously, with mechanical accuracy and certitude of hitting; he simply wills, and the trained muscles obey without his superintendence, just as the fingers form letters with the pen in writing; whereas the collector must usually supervise his muscles all through the act and see that they mind. In spite of the proportion of snap shots of all sorts you will have to take, your collecting shots, as a rule, are made with deliberate aim. There is much the same difference, on the whole, between the sportsman's work and the collector's, that there is between shot-gun and rifle practice, collecting being comparable to the latter. It is generally understood that the acme of skill with the two weapons is an incompatibility; and, certainly, the best shot is not always the best collector, even supposing the two to be on a par in their knowledge of birds' haunts and habits. Still a hopelessly poor shot can only attain fair results by extraordinary diligence and perseverance. Certain principles of shooting may perhaps be reduced to words. Aim deliberately directly at an immovable object at fair range. Hold over a motionless object when far off, as the trajectory of the shot curves downward. Hold a little to one side of a stationary object when very near, preferring rather to take the chances of missing it with the peripheral pellets, than of hopelessly mutilating it with the main body of the charge. Fire at the first fair aim, without trying to improve what is good enough already. Never "pull" the trigger, but *press* it. Bear the shock of discharge without flinching. In shooting on the wing, fire the instant the but of the gun taps your shoulder; you will miss at first, but by and by the birds will begin to drop, and you will have laid the foundation of good shooting, the knack of "covering" a bird unconsciously. The habit of "poking" after a bird on the wing is an almost incurable vice, and may keep you a poor shot all your life. (The collector's frequent necessity of poking after little birds in the bush is just what so often hinders him from acquiring brilliant execution.) Aim *ahead* of a flying bird — the calculation to be made varies, according to the distance of the object, its velocity, its course and the wind, from a few inches to several feet; practice will finally render it intuitive.

§ 2. — DOGS.

A Good Dog is one of the most faithful, respectful, affectionate and sensible of brutes; deference to such rare qualities demands a chapter, however brief. A trained dog is the indispensable servant of the sportsman in his pursuit of most kinds of game; but I trust I am guilty of no discourtesy to the noble animal, when I say that he is a luxury rather than a necessity to the collector — a pleasant companion, who knows almost everything except how to talk, who converses with his eyes and ears and tail, shares comforts and discomforts with equal alacrity, and occasionally makes himself useful. So far as a collector's work tallies with that of a sportsman, the dog is equally useful to both; but finding and telling of *game* aside, your dog's services are restricted to companionship and retrieving. He may, indeed, flush many sorts of birds for you; but he does it, if at all, at random, while capering about; for the brute intellect is limited after all, and cannot comprehend a naturalist. The best trained setter or pointer that ever marked a quail could not be made to understand what you are about, and it would ruin him for sporting purposes if he did. Take a well-bred dog out with you, and the chances are he will soon trot home in disgust at your performances with jack-sparrows and tomtits. It implies such a lowering and perversion of a good dog's instincts to make him really a useful servant of yours, that I am half inclined to say nothing about retrieving, and tell you to make a companion of your dog, or let him alone. I was followed for several years by "the best dog I ever saw" (every one's gun, dog, and child is the best ever seen), and a first-rate retriever; yet I always preferred, when practicable, to pick up my own birds, rather than let a delicate plumage into a dog's mouth, and scolded away the poor brute so often, that she very properly returned the compliment, in the end, by retrieving just when she felt like it. However, we remained the best of friends. Any good setter, pointer, or spaniel, and some kinds of curs, may be trained to retrieve. The great point is to teach them not to "mouth" a bird; it may be accomplished by sticking pins in the ball with which their early lessons are taught. Such dogs are particularly useful in bringing birds out of the water, and in searching for them when lost. One point in training should never be neglected: teach a dog what "to heel" means, and make him obey this command. A riotous brute is simply unendurable under any circumstances.

§ 3. — VARIOUS SUGGESTIONS AND DIRECTIONS FOR FIELD-WORK.

To be a Good Collector, and nothing more, is a small affair; great skill may be acquired in the art, without a single quality commanding respect. One of the most vulgar, brutal, and ignorant men I ever knew was a sharp collector and an excellent taxidermist. Collecting stands much in the same relation to ornithology that the useful and indispensable office of an apothecary bears to the duties of a physician. A field-naturalist is always more or less of a collector; the latter is sometimes found to know almost nothing of natural history worth knowing. The true ornithologist goes out to study birds alive and destroys some of them simply because that is the only way of learning their structure and technical characters. There is much more about a bird than can be discovered in its dead body, — how much more, then, than can be found out from its stuffed skin! In my humble opinion the man who only gathers birds, as a miser money, to swell his cabinet, and that other man who gloats, as miser-like, over the same hoard, both work on a plane far beneath where the enlightened naturalist stands. One looks at Nature, and never knows that she is beautiful; the other knows she is beautiful, as even a corpse may be; the naturalist catches her sentient expression, and knows how beautiful she is! I would have you to know and love her; for fairer mistress never swayed the heart of man. Aim high! — press on, and leave the half-way house of mere collectorship far behind in your pursuit of a delightful study, nor fancy the closet its goal.

Birds may be sought anywhere, at any time; they should be sought everywhere, at all times. Some come about your doorstep to tell their stories unasked. Others spring up before you as you stroll in the field, like the flowers that enticed the feet of Proserpine. Birds flit by as you measure the tired roadside, lending a tithe of their life to quicken your dusty steps. They disport overhead at hide-and-seek with the foliage as you loiter in the shade of the forest, and their music now answers the sigh of the tree-tops, now ripples an echo to the voice of the brook. But you will not always so pluck a thornless rose. Birds hedge themselves about with a bristling girdle of brier and bramble you cannot break; they build their tiny castles in the air surrounded by impassable moats, and the drawbridges are never down. They crown the mountain-top you may lose your breath to climb; they sprinkle the desert where your parched lips may find no cooling draught; they fleck the snow-wreath when the nipping blast may make you turn your back; they breathe unharmed the pestilent vapors of the swamp that mean disease, if not death, for you; they outride the storm at sea that sends strong men to their last account. Where now will you look for birds?

And yet, as skilled labor is always most productive, so expert search yields more than random or blundering pursuit. *Imprimis;* The more varied the face of a country, the more varied its birds. A place all plain, all marsh, all woodland, yields its particular set of birds, perhaps in profusion: but the kinds will be limited in number. It is of first importance to remember this, when you are so fortunate as to have choice of a collecting-ground; and it will guide your steps aright in a day's walk anywhere, for it will make you leave covert for open, wet for dry, high for low and back again. Well-watered country is more fruitful of bird-life than desert or even prairie; warm regions are more productive than cold ones. As a rule, variety and abundance of birds are in direct ratio to diversity and luxuriance of vegetation. Your most valuable as well as largest bags may be made in the regions most favored botanically, up to the point where exuberance of plant-growth mechanically opposes your operations.

Search for particular Birds can only be well directed, of course, by a knowledge of their special haunts and habits, and is one of the mysteries of wood-craft only solved by long experience and close observation. Here is where the true naturalist bears himself with conscious pride and strength, winning laurels that become him, and do honor to his calling. Where to find *game* ("game" is anything that vulgar people do not ridicule you for shooting) of all the kinds we have in this country has been so often and so minutely detailed in sporting-works that it need not be here enlarged upon, especially since, being the best known, it is the least valuable of ornithological material. Most *large* or otherwise conspicuous birds have very special haunts that may be soon learned; and as a rule such rank next after game in ornithological disesteem. Birds of prey are an exception to these statements; they range everywhere, and most of them are worth securing. Hawks will unwittingly fly in your way oftener than they will allow you to approach them when perched: be ready for them. Owls will be startled out of their retreats in thick bushes, dense foliage, and hollow trees, in the daytime; if hunting them at night, good aim in the dark may be taken by rubbing a wet lucifer match on the sight of the gun, causing a momentary glimmer. Large and small waders are to be found by any water's edge, in open marshes, and often on dry plains; the herons more particularly in heavy bogs and dense swamps. Under cover, waders are oftenest approached by stealth; in the open, by strategy; but most of the smaller kinds require the exercise of no special precautions. Swimming birds, aside from water-fowl (as the "game" kinds are called), are generally shot from a boat, as they fly past; but at their breeding places many kinds that congregate in vast numbers are more readily reached. There is a knack of shooting loons and grebes on the water; if they are to be reached at all by the shot it will be by aiming not directly at them but at the water just in front of them. They do not go under just where they float, but kick up behind like a jumping-jack and plunge *forward*. Rails and several kinds of sparrows are confined to reedy marshes. But why prolong such desultory remarks? Little can be said to the point without at least a miniature treatise on ornithology; and I have not yet even alluded to the diversified host of small insectivorous and granivorous birds that fill our woods and fields. The very existence of most of these is unknown to all but the initiated; yet they include the treasures of the ornithologist. Some are plain and humble, others are among the most beautiful objects in nature; but most agree in being *small*, and therefore liable to be overlooked. The sum of my advice about them must be brief. Get over as much ground, both wooded and open, as you can thoroughly examine in a day's tramp, and go out as many days as you can. It is not always necessary, however, to keep on the tramp, especially during the migration of the restless insectivorous species. One may often shoot for hours without moving more than a few yards, by selecting a favorable locality and allowing the birds to come to him as they pass in varied troops through the low woodlands or swampy thickets. Keep your eyes and ears wide open. Look out for every rustling leaf and swaying twig and bending blade of grass. Hearken to every note, however faint; when there is no sound, listen for a chirp. Habitually move as noiselessly as possible. Keep your gun *always* ready. Improve every opportunity of studying a bird you do not wish to destroy; you may often make observations more valuable than the specimen. Let this be the rule with all birds you recognize. But I fear I must tell you to shoot an unknown bird on sight; it may give you the slip in a moment and a prize may be lost. One of the most fascinating things about field-work is its delightful uncertainty: you never know what's in store for you as you start out; you never can tell what will happen next; surprises are always in order, and excitement is continually whetted on the chances of the varied chase.

For myself, the time is past, happily or not, when every bird was an agreeable surprise, for dewdrops do not last all day; but I have never yet walked in the woods without learning something pleasant that I did not know before. I should consider a bird new to science ample reward for a month's steady work; *one* bird new to a locality would repay a week's search; a day is happily spent that shows me any bird that I never saw alive before. How then can you, with so much before you, keep out of the woods another minute?

All Times are good times to go a-shooting; but some are better than others. (*a.*) *Time of year.* In all temperate latitudes, spring and fall — periods of migration with most birds — are the most profitable seasons for collecting. Not only are birds then most numerous, both as species and as individuals, and most active, so as to be the more readily found, but they include a far larger proportion of rare and valuable kinds. In every locality in this country the periodical visitants outnumber the permanent residents; in most regions the number of regular migrants, that simply pass through in the spring and fall, equals or exceeds that of either of the sets of species that come from the south in spring to breed during the summer, or from the north to spend the winter. Far north, of course, on or near the limit of the vernal migration, where there are few if any migrants *passing through*, and where the winter birds are extremely few, nearly all the bird fauna is composed of "summer visitants;" far south, in this country, the reverse is somewhat the case, though with many qualifications. Between these extremes, what is conventionally known as "a season" means the period of the vernal or autumnal migration. For example, the body of birds present in the District of Columbia (where I collected for several years) in the two months from April 20th to May 20th, and from September 10th to October 10th, is undoubtedly greater, as far as individuals are concerned, than the total number found there at all other seasons of the year together. As for species, the number of migrants about equals that of summer visitants; the permanent residents equal the winter residents, both these being fewer than either of the first mentioned sets; while the irregular visitors, or stragglers, that complete the bird fauna, are about, or rather less than one-half as many as the species of either of the other categories. About Washington, therefore, I would readily undertake to secure a greater *variety* of birds in the nine weeks above specified than in all the rest of the year; for in that time would be found, not only all the permanent residents, but nearly all the migrants, and almost all the summer visitants; while the number of individual birds that might be taken exceeds, by quite as much, the number of those procurable in the same length of time at any other season. *Mutatis mutandis*, it is the same everywhere in this country. Look out then, for "the season;" work all through it at a rate you could not possibly sustain the year around; and make hay while the sun shines. (*b.*) *Time of day.* Early in the morning and late in the afternoon are the best times for birds. There is a mysterious something in these diurnal crises that sets bird-life astir, over and above what is explainable by the simple fact that they are the transition periods from repose to activity, or the reverse. Subtle meteorological changes occur; various delicate instruments used in physicists' researches are sometimes inexplicably disturbed; diseases have often their turning point for better or worse; people are apt to be born or die; and the susceptible organisms of birds manifest various excitements. Whatever the operative influence, the fact is, birds are particularly lively at such hours. In the dark, they rest — most of them do; at noonday, again, they are comparatively still; between these times they are passing to or from their feeding grounds or roosting places; they are foraging for food, they are singing; at any rate, they are in motion. Many migratory birds (among them warblers, etc.) perform their journeys by night; just at daybreak they may be seen to descend from the upper regions, rest a while, and then move about briskly, singing and searching for food. Their meal taken, they recuperate by resting till towards evening; feed again and are off for the night. If you have had some experience, don't you remember what a fine spurt you made early that morning? — how many unexpected shots offered as you trudged home belated that evening? Now I am no fowl, and have no desire to adopt the habits of the hen-yard; I have my opinion of those who like the world before it is aired; I think it served the worm right for getting up, when caught by the early bird; nevertheless I go shooting betimes in the morning, and would walk all night to find a rare bird at daylight. (*c.*) *Weather.* It rarely occurs in this country that either heat or cold is unendurably severe; but extremes of temperature are unfavorable, for two reasons: they both occasion great personal discomfort; and in one extreme only a few hardy birds will be found, while in the other most birds are languid, disposed to seek shelter, and therefore less likely to be found. A still, cloudy day of moderate temperature offers as a rule the best chance; among other reasons, there is no sun to blind the eyes, as always occurs on a bright day in one direction, particularly when the sun is low. While a bright day has its good influence in setting many birds astir, some others are most easily approached in heavy or falling weather. Some kinds are more likely to be secured during a light snowfall, or after a storm. Singular as it may seem, a thoroughly wet day offers some peculiar inducements to the collector. I cannot well specify them, but I heartily indorse a remark John Cassin once made to me: — " I like," said he, " to go shooting in the rain sometimes; there are some curious things to be learned about birds when the trees are dripping, things too that have not yet found their way into the books."

How many Birds of the Same Kind do you want? — *All you can get* — with some reasonable limitations; say fifty or a hundred of any but the most abundant and widely diffused species. You may often be provoked with your friend for speaking of some bird he shot, but did not bring you, because, he says, "Why, you've got one like that!" Birdskins are capital; capital unemployed may be useless, but can never be worthless. Birdskins are a medium of exchange among ornithologists the world over; they represent value, — money value and scientific value. If you have more of one kind than you can use, exchange with some one for species you lack; both parties to the transaction are equally benefited. Let me bring this matter under several heads. (*a.*) Your own "series" of skins of any species is incomplete until it contains at least one example of each sex, of every normal state of plumage, and every normal transition stage of plumage, and further illustrates at least the principal abnormal variations in size, form, and color to which the species may be subject; I will even add that every different faunal area the bird is known to inhabit should be represented by a specimen, particularly if there be anything exceptional in the geographical distribution of the species. Any additional specimens to all such are your *only* "duplicates," properly speaking. (*b.*) Birds vary so much in their size, form, and coloring, that a "specific character" can only be precisely determined from examination of a large number of specimens, shot at different times, in different places; still less can the "limits of variation" in these respects be settled without ample materials. (*c.*) The *rarity* of any bird is necessarily an arbitrary and fluctuating consideration, because in the nature of the case there can be no natural unit of comparison, nor standard of appreciation. It may be said, in general terms, no bird is actually "rare." With a few possible exceptions, as in the cases of birds occupying extraordinarily limited areas, like some of the birds of paradise, or about to become extinct, like the pied duck, enough birds of all kinds exist to overstock every public and private collection in the world, without sensible diminution of their numbers. "Rarity" or the reverse is only predicable upon the accidental (so to speak) circumstances that throw, or tend to throw, specimens into naturalists' hands. *Accessibility* is the variable element in every case. The fulmar petrel is said (on what authority I know not) to exceed any other bird in its aggregate of individuals; how do the skins of that bird you have handled compare in number with specimens you have seen of the "rare" warbler of your own vicinity? All birds are common somewhere at some season; the point is, have collectors been there at the time? Moreover, even the arbitrary appreciation of "rarity" is fluctuating, and may change at any time; long sought and highly prized birds are liable to appear suddenly in great numbers in places that knew them not before; a single heavy "invoice" of a bird from some distant or little-explored region may at once stock the market, and depreciate the current value of the species to almost nothing. For example, Baird's bunting and Sprague's lark remained for thirty years among our special desiderata, only one specimen of the former and two or three of the latter being known. Yet they are two of the most abundant birds of Dakota, where in 1873 I took as many of both as I desired; and specimens enough have lately been secured to stock all the leading museums of this country and Europe. (*d.*) Some practical deductions are to be made from these premises. Your object is to make yourself acquainted with all the birds of your vicinity, and to preserve a complete suite of specimens of every species. Begin by shooting every bird you can, coupling this sad destruction, however, with the closest observations upon habits. You will very soon fill your series of a few kinds, that you find almost everywhere, almost daily. Then if you are in a region the ornithology of which is well known to the profession, at once stop killing these common birds — they are in every collection. You should not, as a rule, destroy any more robins, bluebirds, song-sparrows, and the like, than you want for yourself. Keep an eye on them, studying them always, but turn your actual pursuit into other channels, until in this way, gradually eliminating the undesirables, you exhaust the bird fauna as far as possible (you will not *quite* exhaust it — at least for many years). But if you are in a new or little-known locality, I had almost said the very reverse course is the best. The chances are that the most abundant and characteristic birds are "rare" in collections. Many a bird's range is quite restricted: you may happen to be just at its metropolis; seize the opportunity, and get good store, — yes, up to fifty or a hundred; all you can spare will be thankfully received by those who have none. Quite as likely, birds that are scarce just where you happen to be, are so only because you are on the edge of their habitat, and are plentiful in more accessible regions. But, rare or not, it is always a point to determine the exact geographical distribution of a species; and this is fixed best by having specimens to tell each its own tale, from as many different and widely separated localities as possible. This alone warrants procuring one or more specimens in every locality; the commonest bird acquires a certain value if it be captured away from its ordinary range. An Eastern bluebird (*Sialia sialis*) shot in California might be considered more valuable than the "rarest" bird of that State, and would certainly be worth a hundred Massachusetts skins; a varied thrush (*Turdus nævius*) killed in Massachusetts is worth a like number from Oregon. But let all your justifiable destruction of birds be tempered with mercy; your humanity will be continually shocked with the havoc you work, and should never permit you to take life wantonly. Never shoot a bird you do not fully intend to preserve, or to utilize in some proper way. Bird-life is too beautiful a thing to destroy to no purpose; too sacred a thing, like all life, to be sacrificed, unless the tribute is hallowed by worthiness of motive. " Not a sparrow falleth to the ground without His notice."

I should not neglect to speak particularly of the care to be taken to secure full suites of *females*. Most miscellaneous collections contain four or more males to every female, — a disproportion that should be as far reduced as possible. The occasion of the disparity is obvious : females are usually more shy and retiring in disposition, and consequently less frequently noticed, while their smaller size and plainer plumage, as a rule, further favor their eluding observation. The difference in coloring is greatest among those groups where the males are most richly clad, and the shyness of the mother birds is most marked during the breeding season, just when the males, full of song, and in their nuptial attire, become most conspicuous. It is often worth while to neglect the gay Benedicts, to trace out and secure the plainer but not less interesting females. This pursuit, moreover, often leads to discovery of the nests and eggs, — an important consideration. Although both sexes are generally found together when breeding, and mixing indiscriminately at other seasons, they often go in separate flocks, and often migrate independently of each other; in this case the males usually in advance. Towards the end of the passage of some warblers, for instance, we may get almost nothing but females, all our specimens of a few days before having been males. The notable exceptions to the rule of smaller size of the female are among rapacious birds and many waders, though in these last the disparity is not so marked. I only recall one instance, among American birds, of the female being more richly colored than the male — the phalaropes. When the sexes are notably different in adult life, the *young* of both sexes usually resemble the adult female, the young males gradually assuming their distinctive characters. When the adults of both sexes are alike, the young commonly differ from them.

In the same connection I wish to urge a point, the importance of which is often overlooked; it is our practical interpretation of the adage, " a bird in the hand is worth two in the bush." Always keep the first specimen you secure of a species till you get another; no matter how common the species, how poor the specimen, or how certain you may feel of getting other better ones, *keep it*. Your most reasonable calculations may come to naught, from a variety of circumstances, and *any* specimen is better than no specimen, on general principles. And in general, do not, if you can help it, discard any specimen *in the field*. No tyro can tell what will prove valuable and what not; while even the expert may regret to find that a point comes up which a specimen he injudiciously discarded might have determined. Let a collection be " weeded out," if at all, only after deliberate and mature examination, when the scientific results it affords have been elaborated by a competent ornithologist; and even then, the refuse (with certain limitations) had better be put where it will do *some* good, than be destroyed utterly. For instance, I myself once valued, and used, some Smithsonian "sweepings"; and I know very well what to do with specimens, *now*, to which I would not give house-room in my own cabinet. If forced to reduce bulk, owing to limited facilities for transportation in the field (as too often happens), throw away according to *size*, other things being equal. Given only so many cubic inches or feet, eliminate the few *large* birds which take up the space that would contain fifty or a hundred different little ones. If you have a fine large bald eagle or pelican, for instance, throw it away first, and follow it with your ducks, geese, etc. In this way, the bulk of a large miscellaneous collection may be reduced one half, perhaps, with very little depreciation of its actual value. The same principle may be extended to other collections in natural history (excepting fossils, which are always weighty, if not also bulky); very few bird-skins, indeed, being as valuable contributions to science as, for example, a vial of miscellaneous insects that occupies no more room may prove to be.

What is "A Good Day's Work?" — Fifty birds shot, their skins preserved, and observations recorded, is a *very* good day's work; it is sharp practice, even when birds are plentiful. I never knew a person to *average* anywhere near it; even during the "season" such work cannot possibly be sustained. You may, of course, by a murderous discharge into a flock, as of blackbirds or reedbirds, get a hundred or more in a moment; but I refer to collecting a fair variety of birds. You will do very well if you *average* a dozen a day during the seasons. I doubt whether any collector ever averaged as many the year around; it would be over four thousand specimens annually. The greatest number I ever procured *and prepared* in one day was forty, and I have not often gone over twenty. Even when collecting regularly and assiduously, I am satisfied to average a dozen a day during the migrations, and one-third or one-fourth as many the rest of the year. Probably this implies the shooting of about one in five not skinned for various reasons, as mutilation, decay, or want of time.

Approaching Birds. — There is little if any trouble in getting near enough to shoot most birds. With notable exceptions, they are harder to see when near enough, or to hit when seen; particularly small birds that are almost incessantly in motion. As a rule — and a curious one it is — difficulty of approach is in direct ratio to the *size* of the bird; it is perhaps because large conspicuous birds are objects of more general pursuit than the little ones you ordinarily search for. The qualities that birds possess for self-preservation may be called *wariness* in large birds, *shyness* in small ones. The former make off knowingly from a suspicious object; the latter fly from anything that is strange to them, be it dangerous or not. This is strikingly illustrated in the behavior of small birds in the wilderness, as contrasted with their actions about towns; singular as it may seem, they are more timid under the former circumstances than when grown accustomed to the presence of man. It is just the reverse with a hawk or raven, for instance; in populous districts they spend much of their time in trying to save their skins, while in a new country they have not learned, like Indians, that a white man is "mighty uncertain." In stealing on a shy bird, you will of course take advantage of any cover that may offer, as inequalities of the ground, thick bushes, the trunks of trees; and it is often worth while to make a considerable détour to secure unobserved approach. I think that birds are more likely, as a rule, to be frightened away by the movements of the collector, than by his simple presence, however near, and that they are more afraid of noise than of mere motion. Crackling of twigs and rustling of leaves are sharp sounds, though not loud ones; you may have sometimes been surprised to find how distinctly you could hear the movements of a horse or cow in underbrush at some distance. Birds have sharp ears for such sounds. Form a habit of stealthy movement; *it tells*, in the long run, in comparison with lumbering tread. There are no special precautions to be taken in shooting through high open forest; you have only to saunter along with your eyes in the tree-tops. It is ordinarily the easiest and on the whole the most renumerative path of the collector. In traversing fields and meadows move briskly, your principal object being to flush birds out of the grass; and as most of your shots will be snap ones, keep in readiness for instant action. Excellent and varied shooting is to be had along the hedge rows, and in the rank herbage that fringes fences. It is best to keep at a little distance, yet near enough to arouse all the birds as you pass: you may catch them on wing, or pick them off just as they settle after a short flight. In this shooting, two persons, one on each side, can together do more than twice as much work as one. Thickets and tangled undergrowth are favorite resorts of many birds; but when very close, or, as often happens, over miry ground, they are hard places to shoot in. As you come thrashing through the brush, the little inhabitants are scared into deeper recesses; but if you keep still a few minutes in some favorable spot, they are reassured, and will often come back to take a peep at you. A good deal of standing still will repay you at such times; needless to add, you cannot be too lightly loaded for such shooting, when birds are mostly out of sight if a dozen yards off. When yourself concealed in a thicket, and no birds appear, you can often call numbers about you by a simple artifice. Apply the back of your hand to your slightly parted lips, and suck in air; it makes a nondescript "screeping" noise, variable in intonation at your whim, and some of the sounds resemble the cries of a wounded bird, or a young one in distress. It wakes up the whole neighborhood, and sometimes puts certain birds almost beside themselves, particularly in the breeding season. Torturing a wounded bird to make it scream in agony accomplishes the same result, but of course is only permissible under great exigency. In penetrating swamps and marshes, the best advice I can give you is to tell you to get along the best way you can. Shooting on perfectly open ground offers much the same case; you must be left to your own devices. I will say, however, you can ride on horseback, or even in a buggy, nearer birds than they will allow you to walk up to them. Sportsmen take advantage of this to get within a shot of the upland plover, usually a very wary bird in populous districts; I have driven right into a flock of wild geese; in California they often train a bullock to graze gradually up to geese, the gunner being hidden by its body. There is one trick worth knowing; it is not to let a bird that has seen you know by your action that you have seen *it*, but to keep on unconcernedly, gradually sidling nearer. I have secured many hawks in this way, when the bird would have flown off at the first step of direct approach. Numberless other little arts will come to you as your wood-craft matures.

Recovering Birds. — It is not always that you secure the birds you kill; you may not be able to find them, or you may see them lying, perhaps but a few feet off, in a spot practically inaccessible. Under such circumstances a retriever does excellent service, as already hinted; he is equally useful when a bird properly "marked down" is not found there, having fluttered or run away and hidden elsewhere. The most difficult of all places to find birds is among reeds, the eternal sameness of which makes it almost impossible to rediscover a spot whence the eye has once wandered, while the peculiar growth allows birds to slip far down out of sight. In rank grass or weeds, when you have walked up with your eye fixed on the spot where the bird seemed to fall, yet failed to discover it, drop your cap or handkerchief for a mark, and hunt around it as a centre, in enlarging circles. In thickets, make a "bee line" for the spot, if possible keeping your eye on the spray from which the bird fell, and not forgetting where you stood on firing; you may require to come back to the spot and take a new departure. You will not seldom see a bird just shot at fly off as if unharmed, when really it will drop dead in a few moments. In all cases therefore when the bird does not drop at the shot, follow it with your eyes as far as you can; if you see it finally drop, or even flutter languidly downward, mark it on the principles just mentioned, and go in search. Make every endeavor to secure wounded birds, on the score of humanity; they should not be left to pine away and die in lingering misery if it can possibly be avoided.

Killing Wounded Birds. — You will often recover winged birds, as full of life as before the bone was broken; and others too grievously hurt to fly, yet far from death. Your object is to kill them as quickly and as painlessly as possible, without injuring the plumage. This is to be accomplished, with all small birds, by suffocation. The respiration and circulation of birds is very active, and most of them die in a few moments if the lungs are so compressed that they cannot breathe. Squeeze the bird tightly across the chest, under the wings, thumb on one side, middle finger on the other, forefinger pressed in the hollow at the root of the neck, between the forks of the merrythought. Press firmly, hard enough to fix the chest immovably and compress the lungs, but not to break in the ribs. The bird will make vigorous but ineffectual efforts to breathe, when the muscles will contract spasmodically; but in a moment more, the system relaxes with a painful shiver, light fades from the eyes, and the lids close. I assure you, it will make you wince the first few times; you had better habitually hold the poor creature behind you. You can tell by its limp feel and motionlessness when it is dead, without watching the sad struggle. Large birds obviously cannot be dealt with in this way; I would as soon attempt to throttle a dog as a loon, for instance, upon which all the pressure you can give makes no sensible impression. A winged hawk, again, will throw itself on its back as you come up, and show such good fight with beak and talons, that you may be quite severely scratched in the encounter: meanwhile the struggling bird may be bespattering its plumage with blood. In such a case — in any case of a large bird making decided resistance — I think it best to step back a few paces and settle the matter with a light charge of mustard-seed. Any large bird once secured may be speedily dispatched by stabbing to the heart with some slender instrument thrust in under the wing — care must be taken too about the bleeding; or, it may be instantly killed by piercing the brain with a knife introduced into the mouth and driven upward and obliquely backward from the palate. The latter method is preferable as it leaves no outward sign and causes no bleeding to speak of. With your thumb, you may indent the back part of a bird's skull so as to compress the cerebellum; if you can get deep enough in, without materially disordering the plumage, or breaking the skin, the method is unobjectionable.

Handling Bleeding Birds. — Bleeding depends altogether upon the part or organ wounded; but other things being equal, violence of the hæmorrhage is usually in direct proportion to the size of the shot-hole; when mustard-seed is used it is ordinarily very trifling, if it occur at all. Blood flows oftener from the orifice of *exit* of a shot, than from the wound of entrance, for the latter is usually plugged with a little wad of feathers driven in. Bleeding from the mouth or nostrils is the rule when the lungs are wounded. When it occcurs, hold up the bird by the feet, and let it drip; a general squeeze of the body in that position will facilitate the drainage. In general, hold a bird so that a bleeding place is most dependent; then, pressure about the part will help the flow. A "gob" of blood, which is simply a forming clot, on the plumage may often be dexterously flipped almost clean away with a snap of the finger. It is first-rate practice to take cotton and forceps into the field to plug up shot-holes, and stop the mouth and nostrils and vent on the spot. I follow the custom of the books in recommending this, but I will confess I have rarely done it myself, and I suspect that only a few of our most leisurely and elegant collectors do so habitually. Shot-holes may be found by gently raising the feathers, or blowing them aside; you can of course get only a tiny plug into the wound itself, but it should be one end of a sizable pledget, the rest lying fluffy among the feathers. In stopping the mouth or vent, ram the fluff of cotton, entirely inside. You cannot conveniently stop up the nostrils of small birds separately; but take a light cylinder of cotton, lay it transversely across the base of the upper mandible, closely covering the nostrils, and confine it there by tucking each end tightly into the corner of the mouth. In default of such nice fixing as this, a pinch of dry loam pressed on a bleeding spot will plaster itself there and stop further mischief. Never try to *wipe off* fresh blood that has already wetted the plumage; you will only make matters worse. Let it dry on, and then — but the treatment of blood-stains, and other soilings of plumage, is given beyond.

Carrying Birds Home Safe. — Suppose you have secured a fine specimen, very likely without a soiled or ruffled feather; your next care will be to keep it so till you are ready to skin it. But if you pocket or bag it directly, it will be a sorry-looking object before you get home. Each specimen must be separately cared for, by wrapping in stout paper; writing paper is as good as any, if not the best. It will repay you to prepare a stock of paper before starting out; your most convenient sizes are those of a half-sheet of note, of letter, and of cap respectively. Either take these, or fold and cut newspaper to correspond; besides, it is always well to have a *whole* newspaper or two for large birds. Plenty of paper will go in the breast pockets of the shooting-coat. Make a " cornucopia," — the simplest thing in the world, but, like tying a particular knot, hard to explain. Setting the wings closely, adjusting disturbed feathers, and seeing that the bill points straight forward, thrust the bird head first into one of these paper cones, till it will go no further, being bound by the bulge of the breast. Let the cone be large enough for the open end to fold over or pinch together entirely beyond the tail. Be particular not to crumple or bend the tail-feathers. Lay the paper cases in the game bag or great pocket so that they very nearly run parallel and lie horizontal; they will carry better than if thrown in at random. Avoid overcrowding the packages, as far as is reasonably practicable; moderate pressure will do no harm, as a rule, but if great it may make birds bleed afresh, or cause the fluids of a wounded intestine to ooze out and soak the plumage of the belly, — a very bad accident indeed. For similar obvious reasons, do not put a large heavy bird on top of a lot of little ones; I would sooner sling a hawk or heron over my shoulder, or carry it by hand. If it goes in the bag, see that it gets to the bottom. Avoid putting birds in pockets that are close about your person; they are almost always unduly pressed, and may gain just enough additional warmth from your body to make them begin to decompose before you can get at skinning them. Handle birds no more than is necessary, especially white-plumaged ones; ten to one your hands are powder-begrimed: and besides, even the warmth and moisture of your palms may tend to injure a delicate feathering. Ordinarily pick up a bird by the feet or bill; as you need both hands to make the cornucopia, let the specimen dangle by the toes from your teeth while you are so employed. In catching at a wounded bird, aim to cover it entirely with your hand; but whatever you do, never seize it by the tail, which then will often be left in your hands for your pains. Never grasp wing-tips or tail-feathers; these large flat quills would get a peculiar crimping all along the webs, very difficult to efface. Finally, I would add there is a certain knack or art in manipulating, either of a dead bird or a birdskin, by which you may handle it with seeming carelessness and perfect impunity; whilst the most gingerly fingering of an inexperienced person will leave its rude trace. You will naturally acquire the correct touch; but it can be neither taught nor described.

A Special Case. — While the ordinary run of land birds will be brought home in good order by the foregoing method, some require special precautions. I refer to sea birds, such as gulls, terns, petrels, etc., shot from a boat. In the first place, the plumage of most of them is, in part at least, white and of exquisite purity. Then, fish-eating birds usually vomit and purge when shot. They are necessarily fished all dripping from the water. They are too large for pocketing. If you put them on the thwarts or elsewhere about the boat, they usually fall off, or are knocked off, into the bilge water; if you stow them in the cubby-hole, they will assuredly soil by mutual pressure, or by rolling about. It will repay you to pick them from the water by the bill, and shake off all the water you can; hold them up, or let some one do it, till they are tolerably dry; plug the mouth, nostrils, and vent, if not also shot-holes; wrap each one separately in a *cloth* (*not* paper) or a mass of tow, and pack steadily in a covered box or basket taken on board for this purpose. With such precautions as these birds most liable to be soiled reach the skinning table in perfect order; and your care will afterward transform them into specimens without spot or blemish.

§ 4. — HYGIENE OF COLLECTORSHIP.

It is Unnecessary to speak of the Healthfulness of a pursuit that, like the collector's occupation, demands regular bodily exercise, and at the same time stimulates the mind by supplying an object, thus calling the whole system into exhilarating action. Yet collecting has its perils, not to be overlooked if we would adequately guard against them, as fortunately we may, in most cases, by simple precautions. The dangers of taxidermy itself are elsewhere noticed; but, besides these, the collector is exposed to vicissitudes of the weather, may endure great fatigue, may breathe miasm, and may be mechanically injured.

Accidents from the Gun have been already treated; a few special rules will render others little liable to occur. The secret of safe *climbing* is never to relax one hold until another is secured; it is in spirit equally applicable to scrambling over rocks, a particularly difficult thing to do safely with a loaded gun. Test rotten, slippery, or otherwise suspicious holds before trusting them. In lifting the body up anywhere, keep the mouth shut, breathe through the nostrils, and go slowly. In *swimming*, waste no strength unnecessarily in trying to stem a current; yield partly, and land obliquely lower down; if exhausted, float; the slightest motion of the hands will ordinarily keep the face above water; and in any event keep your wits collected. In fording deeply, a heavy stone will strengthen your position. Never sail a boat experimentally; if you are no sailor, take one with you or stay on land. In crossing a high, narrow footpath, never look lower than your feet; the muscles will work true if not confused with faltering instructions from a giddy brain. On soft ground, see what, if anything, has preceded you; large hoof-marks generally mean that the way is safe; if none are found, inquire for yourself before going on. Quicksand is the most treacherous, because far more dangerous than it looks; but I have seen a mule's ears finally disappear in genuine mud. Cattle paths, however erratic, commonly prove the surest way out of a difficult place, whether of uncertain footing or dense undergrowth.

Miasm. — Unguarded exposure in malarious regions usually entails sickness, often preventable, however, by due precautions. It is worth knowing, in the first place, that miasmatic poison is most powerful between sunset and sunrise; more exactly, from the damp of the evening until night vapors are dissipated; we may be out in the daytime with comparative impunity, where to pass a night would be almost certain disease. If forced to camp out, seek the highest and dryest spot, put a good fire on the swamp side, and also, if possible, let trees intervene. Never go out on an empty stomach; just a cup of coffee and a crust may make a decided difference. Meet the earliest unfavorable symptoms with quinine; I should rather say, if unacclimated, anticipate them with this invaluable agent. Endeavor to maintain high health of all functions by the natural means of regularity and temperance in diet, exercise, and repose.

"Taking Cold." — This vague "household word" indicates one or more of a long varied train of unpleasant affections, nearly always traceable to one or the other of only two causes: *sudden change* of temperature, and *unequal distribution* of temperature. No extremes of heat or cold can alone effect this result; persons frozen to death do not "take cold" during the process. But if a part of the body be rapidly cooled, as by evaporation from a wet article of clothing, or by sitting in a draught of air, the rest of the body remaining at an ordinary temperature; or if the temperature of the whole be suddenly changed by going out into the cold, or, especially, by coming into a warm room, there is much liability of trouble. There is an old saying, —

> " When the air comes through a hole
> Say your prayers to save your soul; "

and I should think almost any one could get a "cold" with a spoonful of water on the wrist held to a key-hole. Singular as it may seem, sudden warming when cold is more dangerous than the reverse; every one has noticed how soon the handkerchief is required on entering a heated room on a cold day. Frost-bite is an extreme illustration of this. As the Irishman said on picking himself up, it was not the fall, but stopping so quickly that hurt him; it is not the lowering of the temperature to the freezing point, but its subsequent elevation, that devitalizes the tissue. This is why rubbing with snow, or bathing in cold water, is required to restore safely a frozen part; the arrested circulation must be very gradually re-established, or inflammation, perhaps mortification, ensues. General precautions against taking cold are almost self-evident, in this light. There is ordinarily little if any danger to be apprehended from wet clothes, so long as exercise is kept up; for the "glow" about compensates for the extra cooling by evaporation. Nor is a complete drenching more likely to be injurious than wetting of one part. But never sit still wet; and in changing rub the body dry. There is a general tendency, springing from fatigue, indolence, or indifference, to neglect damp feet; that is to say, to dry them by the fire; but this process is tedious and uncertain. I would say especially, off with the muddy boots and sodden socks at once; dry stockings and slippers, after a hunt, may make just the difference of your being able to go out again or never. Take care never to check perspiration; during this process, the body is in a somewhat critical condition, and sudden arrest of the function may result disastrously, even fatally. One part of the business of perspiration is to equalize bodily temperature, and it must not be interfered with. The secret of much that might be said about *bathing* when heated, lies here. A person overheated, panting it may be, with throbbing temples and a *dry* skin, is in danger partly because the natural cooling by evaporation from the skin is denied, and this condition is sometimes not far from a "sunstroke." Under these circumstances, a person of fairly good constitution may plunge into the water with impunity, even with benefit. But if the body be already cooling by sweating, rapid abstraction of heat from the surface may cause internal congestion, never unattended with danger. Drinking ice-water offers a somewhat parallel case; even on stooping to drink at the brook, when flushed with heat, it is well to bathe the face and hands first, and to taste the water before a full draught. It is a well-known excellent rule, not to bathe immediately after a full meal; because during digestion the organs concerned are comparatively engorged, and any sudden disturbance of the circulation may be disastrous. The imperative necessity of resisting drowsiness under extreme cold requires no comment. In walking under a hot sun, the head may be sensibly protected by green leaves or grass in the hat; they may be advantageously moistened, but not enough to drip about the ears. Under such circumstances the slightest giddiness, dimness of sight, or confusion of ideas, should be taken as a warning of possible sunstroke, instantly demanding rest and shelter.

Hunger and Fatigue are more closely related than they might seem to be; one is a sign that the fuel is out, and the other asks for it. Extreme fatigue, indeed, destroys appetite; this simply means, temporary incapacity for digestion. But even far short of this, food is more easily digested and better relished after a little preparation of the furnace. On coming home tired, it is much better to make a leisurely and reasonably nice toilet than to eat at once, or to lie still thinking how tired you are; after a change and a wash you will feel like a "new man," and go to table in capital state. Whatever dietetic irregularities a high state of civilization may demand or render practicable, a normally healthy person is inconvenienced almost as soon as his regular meal-time passes without food; a few can work comfortably or profitably fasting over six or eight hours. Eat before starting; if for a day's tramp, take a lunch; the most frugal meal will appease if it do not satisfy hunger, and so postpone its urgency. As a small scrap of practical wisdom, I would add, keep the remnants of the lunch, if there are any; for you cannot always be sure of getting in to supper.

Stimulation. — When cold, fatigued, depressed in mind, and on other occasions, you may feel inclined to resort to artificial stimulus. Respecting this many-sided theme I have a few words to offer of direct bearing on the collector's case. It should be clearly understood in the first place that a stimulant confers no strength whatever; it simply calls the powers that be into increased action at their own expense. Seeking real strength in stimulus is as wise as an attempt to lift yourself up by the boot-straps. You may gather yourself to leap the ditch and you clear it; but no such muscular energy can be sustained; exhaustion speedily renders further expenditure impossible. But now suppose a very powerful mental impression be made, say the circumstance of a succession of ditches in front, and a mad dog behind; if the stimulus of terror be sufficiently strong, you may leap on till you drop senseless. Alcoholic stimulus is a parallel case, and is not seldom pushed to the same extreme. Under its influence you never can tell when you *are* tired; the expenditure goes on, indeed, with unnatural rapidity, only it is not felt at the time; but the upshot is you have all the original fatigue to endure and to recover from, *plus* the fatigue resulting from over-excitation of the system. Taken as a fortification against cold, alcohol is as unsatisfactory as a remedy for fatigue. Insensibility to cold does not imply protection. The fact is the exposure is greater than before; the circulation and respiration being hurried, the waste is greater, and as sound fuel cannot be immediately supplied, the temperature of the body is soon lowered. The transient warmth and glow over, the system has both cold *and* depression to endure; there is no use in borrowing from yourself and fancying you are richer. Secondly, the value of any stimulus (except in a few exigencies of disease or injury) is in proportion, not to the intensity, but to the equableness and durability of its effect. This is one reason why tea, coffee, and articles of corresponding qualities, are preferable to alcoholic drinks; they work so smoothly that their effect is often unnoticed, and they "stay by" well; the friction of alcohol is tremendous in comparison. A glass of grog may help a veteran over the fence, but no one, young or old, can shoot all day on liquor. I have had so much experience in the use of tobacco as a mild stimulant that I am probably no impartial judge of its merits: I will simply say I do not use it in the field, because it indisposes to muscular activity, and favors reflection when observation is required; and because temporary abstinence provokes the morbid appetite and renders the weed more grateful afterwards. Thirdly, undue excitation of any physical function is followed by corresponding depression, on the simple principle that action and reaction are equal; and the balance of health turns too easily to be wilfully disturbed. Stimulation is a draft upon vital capital, when interest alone should suffice; it may be needed at times to bridge a chasm, but habitual living beyond vital income infallibly entails bankruptcy in health. The use of alcohol in health seems practically restricted to purposes of sensuous gratification on the part of those prepared to pay a round price for this luxury. The three golden rules here are, — never drink before breakfast, never drink alone, and never drink bad liquor; their observance may make even the abuse of alcohol tolerable. Serious objections for a naturalist, at least, are that science, viewed through a glass, seems distant and uncertain, while the joys of rum are immediate and unquestionable; and that intemperance, being an attempt to defy certain physical laws, is therefore eminently unscientific.

§ 5 — REGISTRATION AND LABELLING.

A mere Outline of a Field Naturalist's Duties would be inexcusably incomplete without mention of these important matters; and, because so much of the business of collecting *must* be left to be acquired in the school of experience, I am the more anxious to give explicit directions whenever, as in this instance, it is possible to do so.

Record your Observations Daily. — In one sense the specimens themselves are your record, — *prima facie* evidence of your industry and ability; and if labelled, as I shall presently advise, they tell no small part of the whole story. But this is not enough; indeed, I am not sure that an ably conducted ornithological journal is not the better half of your operations. Under your editorship of labelling, specimens tell what they know about themselves; but you can tell much more yourself. Let us look at a day's work: You have shot and skinned so many birds and laid them away labelled. You have made observations about them before shooting, and have observed a number of birds that you did not shoot. You have items of haunts and habits, abundance or scarcity; of manners and actions under special circumstances, as of pairing, nesting, laying, rearing young, feeding, migrating, and what not; various notes of birds are still ringing in your ears; and finally, you may have noted the *absence* of species you saw a while before, or had expected to occur in your vicinity. Meteorological and topographical items, especially when travelling, are often of great assistance in explaining the occurrences and actions of birds. Now *you* know these things, but very likely no one else does; and you know them *at the time*, but you will not recollect a tithe of them in a few weeks or months, to say nothing of years. Don't trust your memory: it will trip you up; what is clear now will grow obscure; what is found will be lost. Write down everything while it is fresh in your mind; write it out in full: time so spent now will be time saved in the end, when you offer your researches to the discriminating public. Don't be satisfied with a dry-as-dust item; clothe a skeleton fact, and breathe life into it with thoughts that glow; let the paper smell of the woods. There's a pulse in a new fact; catch the rhythm before it dies. Keep off the quicksands of mere memorandum — that means something "to be remembered," which is just what you cannot do. Shun abbreviations; such keys rust with disuse, and may fail in after times to unlock the secret that should have been laid bare in the beginning. Use no signs intelligible only to yourself: your note-books may come to be overhauled by others whom you would not wish to disappoint. Be sparing of sentiment, a delicate thing, easily degraded to drivel: crude enthusiasm always hacks instead of hewing. Beware of literary infelicities: "the written word remains," it may be, after you have passed away; put down nothing for your friend's blush, or your enemy's sneer; write as if a stranger were looking over your shoulder.

Ornithological Book-keeping may be left to your discretion and good taste in the details of execution. Each may consult his preferences for rulings, headings, and blank forms of all sorts, as well as particular modes of entry. But my experience has been that the entries it is advisable to make are too multifarious to be accommodated by the most ingenious formal ruling; unless, indeed, you make the conventional heading "Remarks" disproportionately wide, and commit to it everything not otherwise provided for. My preference is decidedly for a plain page. I use a strongly bound blank book, cap size, containing at least six or eight quires of *good* smooth paper; but smaller may be needed for travelling, even down to a pocket note-book. I would not advise a multiplicity of books, splitting up your record into different departments: let it be journal and register of specimens combined. (The registry of *your own collecting* has nothing to do with the register of your *cabinet of birds*, which is sure to include a proportion of specimens from other sources, received in exchange, donated, or purchased. I speak of this beyond.) I have found it convenient to commence a day's record with a register of the specimens secured, each entry consisting of a duplicate of the bird's label (see beyond), accompanied by any further remarks I have to offer respecting the particular specimens; then to go on with the full of my day's observations, as suggested in the last paragraph. You thus have a "register of collections" in chronological order, told off with an unbroken series of numbers, checked with the routine label-items, and continually interspersed with the balance of your ornithological studies. Since your private field-number is sometimes an indispensable clew to the authentication of a specimen after it has left your own hands, *never duplicate it*. If you are collecting other objects of natural history besides birds, still have but one series of numbers; duly enter your mammal, or mineral, or whatever it is, in its place, with the number under which it happens to fall. Be scrupulously accurate with these and all other *figures*, as of dates and measurements. Always use black ink; the "fancy" writing-fluids, even the useful carmine, fade sooner than black, while lead-pencilling is never safe.

Labelling. — This should *never* be neglected. It is enough to make a sensitive ornithologist shiver to see a specimen without that indispensable appendage — a label. I am sorry to observe that the routine labelling of most collections is far from being satisfactory. A well-appointed label is something more than a slip of paper with the bird's name on it, and is still defective, if, as is too often the case, only the locality and collector are added. A complete label records the following particulars: 1. *Title* of the survey, voyage, exploration, or other expedition (if any), during which the specimen was collected. 2. *Name* of the person in charge of the same (and it may be remarked that the less he really cares about birds, and the less he actually interests himself to procure them, the more particular he will be about this). 3. *Title* of the institution or association (if any) under the auspices or patronage of which the specimen was procured, or for which it is designed. 4. *Name of collector;* partly to give credit where it is due, but principally to fix responsibility, and authenticate the rest of the items. 5. *Collector's number*, referring to his note-book, as just explained; if the specimen afterwards forms part of a general collection it usually acquires another number by new registry; the collector's then becoming the "original," as distinguished from the "current," number. 6. *Locality*, perhaps the most important of all the items. A specimen of unknown or even uncertain origin is worthless or nearly so; while lamentable confusion has only too often arisen in ornithological writings from vague or erroneous indications of locality: I should say that a specimen "not authentic" in this particular had better have its *supposed* origin erased and be let alone. Nor will it do to say simply, for instance, "North America" or even "United States." The general geographical distribution of birds being according to recognized faunal areas, ornithologists generally know already the quarter of the globe from which any bird comes; the locality of particular specimens, therefore, should be fixed down to the very spot. If this be obscure add the name of the nearest place to be found on a fairly good map, giving distance and direction. 7. *Date of collection*, — day of the month, and year. Among other reasons for this may be mentioned the fact that it is often important to know what season a particular plumage indicates. 8. *Sex*, and if possible also *age*, of the specimen, — an item that bespeaks its own importance. Ornithologists of all countries are agreed upon certain signs to indicate sex. These are: ♂ for *male*, ♀ for *female*, — the symbols respectively of Mars and Venus. Immaturity is often denoted by the sign $_o$; thus, ♂$_o$, young male. Or, we may write ♀ *ad.*, ♀ *yg.*, for adult female, young female, respectively. It is preferable, however, to use the language of science, not our vernacular, and say ♂ *juv. (juvenis,* young). "*Nupt.*" signifies breeding plumage; "*hornot.*" means a bird of the year. 9. *Measurements* of length, and of extent of wings; the former can only be obtained approximately, and the latter not at all, from a prepared specimen. 10. *Color* of the eyes, and of the bill, feet, or other naked or soft parts, the tints of which may change in drying. 11. *Miscellaneous particulars*, such as contents of stomach, special circumstances of capture, vernacular name, etc. 12. *Scientific name of the bird.* This is really the least important item of all, though generally thought to take precedence. But a bird labels itself, so to speak; and nature's label may be deciphered at any time. In fact, I would enjoin upon the collector *not* to write out the supposed name of the bird in the field, unless the species is so well known as to be absolutely unquestionable. Proper identification, in any case to which the slightest doubt may attach, can only be made after critical study in the closet with ample facilities for examination and comparison. The first eight items, and the twelfth, usually constitute the face of a label; the rest are commonly written on the back. Labels should be of light cardboard, or very stiff writing paper; they may be dressed attractively, as fancy suggests; the general items of a large number of specimens are best printed; the special ones must of course be written. Shape is immaterial; small "cards" or "tickets" are preferred by some, and certainly look very well when neatly appointed; but I think, on the whole, that a shape answering the idea of a "slip" rather than a "ticket" is most eligible. A slip about three inches long and two thirds of an inch wide will do very well for anything, from a hawk to a humming-bird. Something like the "shipping tag" used by merchants is excellent, particularly for larger objects. It seems most natural to attach the string to the *left-hand* end. The slip should be tied so as to swing just clear of the bird's legs, but *not* loose enough to dangle several inches, for in that case the labels are continually tangling with each other when the birds are laid away in drawers. The following diagrams show the face and back of the last label I happened to write before these lines were originally penned; they represent the size and shape that I find most convenient for general purposes; while the "legend" illustrates every one of the twelve items above specified.

Smithsonian   Explorations in Dakota.   Dr. Elliott Coues, U. S. A.   Institution.

No. 2655. Buteo borealis (Gm.) V. ♀ juv.

Fort Randall, Missouri River.   Oct. 29, 1872.

Obverse.

23.00 × 53.00 × 17.50. — Eyes yellowish-gray; bill horn-blue, darker at tip; cere wax-yellow; tarsi dull yellowish; claws bluish-black. Stomach contained portions of a rabbit; also, a large tapeworm.

Reverse.

Directions for Measurement may be inserted here, as this matter pertains rightfully to the recording of specimens. The following instructions apply not only to length and extent, but to the principal other dimensions, which may be taken at any time. For large birds, a tape-line showing inches and fourths will do; for smaller ones, a foot-rule graduated for inches and eighths, or better, decimals to hundredths, must be used; and for all nice measurements the dividers are indispensable. "*Length:*" Distance between the tip of the bill and end of the longest tail-feather. Lay the bird on its back on the ruler on a table; take hold of the bill with one hand and of both legs with the other; pull with reasonable force to get the curve all out of the neck; hold the bird thus with the tip of the bill flush with one end of the ruler, and see where the end of the tail points. Put the tape-line in place of the ruler, in the same way, for larger birds. "*Extent:*" Distance between the tips of the outspread wings. They must be *fully* outstretched, with the bird on its back, crosswise on the ruler, its bill pointing to your breast. Take hold of right and left metacarpus with the thumb and forefinger of your left and right hand respectively, *stretch* with reasonable force, getting one wing-tip flush with one end of the ruler, and see how much the other wing-tip reaches. With large birds pull away as hard as you please, and use the table, floor, or side of the room; mark the points and apply tape-line. "*Length of wing:*" Distance from the carpal angle formed at the bend of the wing to the end of the longest primary. Get it with compasses for small birds. In birds with a convex wing, do not lay the tape-line over the curve, but under the wing in a straight line. This measurement is the one called, for short, "the wing." "*Length of tail:*" Distance from the roots of the rectrices to the end of the longest one. Feel for the pope's nose; in either a fresh or dried specimen there is more or less of a palpable lump into which the tail-feathers stick. Guess as near as you can to the middle of this lump; place the end of the ruler opposite this point, and see where the tip of the longest tail-feather comes. *" Length of bill :"* Some take the curve of the upper mandible; others the side of the upper mandible from the feathers; others the gape, etc. I take the *chord of the culmen.* Place one foot of the dividers on the culmen just where the feathers end; no matter whether the culmen runs up on the forehead, or the frontal feathers run out on the culmen, and no matter whether the culmen is straight or curved. Then with me the *length of the bill* is the shortest distance from the point just indicated to the tip of the upper mandible; measure it with the dividers. In a straight bill of course it is the length of the culmen itself; in a curved bill, however, it is quite another thing. *" Length of tarsus :"* Distance between the joint of the tarsus with the leg above, and that with the first phalanx of the middle toe below. Measure it *always* with dividers, and in *front* of the leg. *" Length of toes :"* Distance in a straight line along the upper surface of a toe from the point last indicated to the root of the claw on top. Length of toe is to be taken *without* the claw, unless otherwise specified. *" Length of the claws :"* Distance in a *straight line* from the point last indicated to the tip of the claw. *" Length of head "* is often a convenient dimension for comparison with the bill. Set one foot of the dividers over the base of the culmen (determined as above) and allow the other to slip snugly down over the arch of the occiput.

§ 6. — INSTRUMENTS, MATERIALS, AND FIXTURES FOR PREPARING BIRDSKINS.

Instruments. — The only indispensable instrument is a pair of scissors *or* a knife; although practically you want both of these, a pair of spring forceps, and a knitting-needle, or some similar wooden or ivory object, yet I have made hundreds of birdskins consecutively without touching another tool. *" Persicos odi, puer, apparatus!"* I always mistrust the emphasis of a collector who makes a flourish of instruments. You might be surprised to see what a meagre, shabby-looking kit our best taxidermists work with. Stick to your scissors, knife, forceps, and needle. But you may as well buy, at the outset, a common dissecting-case, just what medical students begin business with; it is very cheap, and if there are some unnecessary things in it, it makes a nice little box in which to keep your tools. The case contains, among other things, several scalpels, just the knives you want; a " cartilage-knife," which is nothing but a stout scalpel, suitable for large birds; the best kind of scissors for your purpose, with short blades and long handles — if " kneed" at the hinge so much the better; spring forceps, the very thing; a blow-pipe, useful in many ways and answering well for a knitting-needle; and some little steel-hooks, chained together, which you may want to use. But you will also require, for large birds, a very heavy pair of scissors, or small shears, short-bladed and long-handled, and a stout pair of bone-nippers. Have some pins and needles; surgical needles, which cut instead of punching, are the best. Get a hone or strop, if you wish, and a feather duster. Use of scissors requires no comment, and I would urge their habitual employ instead of the knife-blade; I do nine-tenths of my cutting with scissors, and find it much the easiest. A double-lever is twice as effective as a single one, and besides, you gain in cutting soft, yielding substances by opposing two blades. Moreover, scalpels need constant sharpening; mine are generally too dull to cut much with, and I suppose I am like other people — while scissors stay sharp enough. The flat, thin ivory or ebony *handle* of the scalpel is about as useful as the blade. Finger-nails, which were made before scalpels, are a mighty help. Forceps are almost indispensable for seizing and holding parts too small or too remote to be grasped by the fingers. The knitting-needle is wanted for a specific purpose noted beyond. The shears or nippers are only needed for what the ordinary scissors are too weak to do. Our instruments, you see now, are " a short horse soon curried."

Materials. — (*a.*) *For stuffing.* "What do you stuff 'em with?" is usually the first question of idle curiosity about taxidermy, as if that were the great point; whereas, the stuffing is so small a matter that I generally reply, "anything, except brickbats!" But if stuffing birds were the final cause of *Cotton*, that admirable substance could not be more perfectly adapted than it is to the purpose. Ordinary raw cotton-batting or wadding is what you want. When I can get it I never think of using anything else for small birds. I would use it for all birds were expense no object. Here *tow* comes in; there is a fine, clean, bleached article of tow prepared for surgical dressings; this is the best, but any will do. Some say chop your tow fine; this is harmless, but unnecessary. A crumpled newspaper, wrapped with tow, is first-rate for a large bird. Failing cotton or tow, any *soft, light, dry, vegetable substance* may be made to answer, — rags, paper, crumbled leaves, fine dried grass, soft fibrous inner bark, etc.; the down of certain plants, as thistle and silkweed, makes an exquisite filling for small birds. But I will qualify my remark about brickbats by saying: *never put hair, wool, feathers, or any other* ANIMAL *substance in a birdskin;* far better leave it empty: for, as we shall see in the sequel, bugs come fast enough, without being invited into a snug nest. (*b.*) *For preserving.* ARSENIC, — not the pure metal properly so called, but arsenic of the shops, or arsenious acid, — is the great preservative. Use dry powdered arsenic, plenty of it, and nothing else. There is no substitute for arsenic worthy of the name, and no preparation of arsenic so good as the simple substance. Various kinds of "arsenical soap" were and may still be in vogue; it is a nasty greasy substance, not fit to handle; and although efficacious enough, there is a very serious hygienic objection to its use.[1] Arsenic, I need not say, is a violent irritant poison, and must therefore be duly *guarded*, but may be used with perfect impunity. It is a very *heavy* substance, not appreciably volatile at ordinary temperatures, and therefore not liable, as some suppose, to be breathed, to any perceptible, much less injurious, extent. It will not even at once enter the pores of healthy unbroken skin; so it is no matter if it gets on the fingers. The exceedingly minute quantity that may be supposed to find its way into the system in the course of time is believed by many competent physicians to be rather beneficial as a tonic. I will not commit myself to this; for, though I have never felt better than when working daily with arsenic, I do not know how much my health was improved by the out-door exercise always taken at the same time. The simple precautions are, not to let it lie too long in contact with the skin, nor get into an abrasion, nor under the nails. It will convert a scratch or cut into a festering sore of some little severity; while if lodged under the nails it soon shows itself by soreness, increased by pressure; a white speck appears, then a tiny abscess forms, discharges and gets well in a few days. Your precautions really respect other persons more than yourself; the receptacle should be conspicuously labelled "POISON!" Arsenic is a good friend of ours; besides preserving our birds, it keeps busybodies and meddlesome folks away from the scene of operations, by raising a wholesome suspicion of the taxidermist's surroundings. It may be kept in the tin pots in which it is usually sold; but some shallower, *broader* receptacle is more convenient. A little drawer say 6 × 6 inches, and an inch deep, to slip under the edge of the table, or a similar compartment in a large drawer, will be found handy. A salt-spoon, or little wooden shovel whittled like one, is nice to use it with, though in effect, I always shovel it up with the handle of a scalpel. As stated, there is no substitute for arsenic;

---

[1] "Strange as it may appear to some, I would say avoid especially all the so-called arsenical soaps; they are at best but filthy preparations; besides, it is a fact to which I can bear painful testimony that they are, especially when applied to a greasy skin, poisonous in the extreme. I have been so badly poisoned, while working upon the skins of some fat water birds that had been prepared with arsenical soap, as to be made seriously ill, the poison having worked into the system through some small wounds or scratches on my hand. Had pure arsenic been used in preparing the skins, the effect would not have been *as bad*, although grease and arsenic are generally a blood-poison in *some* degree; but when combined with 'soap' the effect, at least as far as my experience goes, is much more injurious." (MAYNARD, *Guide*, p. 12.) In endorsing this, I would add that the combination is the more poisonous, in all probability, simply because the soap, being detersive, mechanically facilitates the entrance of the poison, without, however, chemically increasing its virulence.

but at a pinch you can make temporary shift with the following, among other articles: — table salt, or saltpetre, or charcoal strewn plentifully; strong solution of corrosive sublimate, brushed over the skin inside; creosote; impure carbolic acid; these last two are quite efficacious, but they smell horribly for an indefinite period. A bird threatening to decompose before you can get at it to skin, may be saved for a while by squirting weak carbolic acid or creosote down the throat and up the fundament; or by disembowelling, and filling the cavity with powdered charcoal. (*c.*) *For cleansing. Gypsum* is an almost indispensable material for cleansing soiled plumage. "Gypsum" is properly native hydrated sulphate of lime; the article referred to is "plaster of Paris" or gypsum heated up to 260° F. (by which the water of crystallization is driven off) and then finely pulverized. When mixed with water it soon solidifies, the original hydrate being again formed. The mode of using it is indicated beyond. It is most conveniently kept in a shallow tray, say a foot square, and an inch or two deep, which had better, furthermore, slide under the table as a drawer; or form a compartment of a larger drawer. *Keep gypsum and arsenic in different-looking receptacles*, not so much to keep from poisoning yourself, as to keep from *not* poisoning a birdskin. They look much alike, and skinning becomes such a mechanical process that you may get hold of the wrong article when your thoughts are wandering in the woods. Gypsum, like arsenic, has no worthy rival in its own field; some substitutes, in the order of their applicability, are: — corn-meal, probably the best thing after gypsum; calcined magnesia (very good, but too light — it floats in the air, and makes you cough); bicarbonate of magnesia; powdered chalk ("prepared chalk," *creta præparata* of the drug shops, is the best kind); fine wood-ashes; clean dry loam. No article, however powdery when dry, that contains a glutinous principle, as for instance gum-arabic or flour, is admissible. (*d.*) *For wrapping*, you want a thin, pliable, strong paper; water-closet paper is the very best; newspaper is pretty good. For making the cones or cylinders in which birdskins may be set to dry, a stiffer article is required; writing paper answers perfectly.

Naturalists habitually carry a Pocket Lens, much as other people do a watch. You will find a magnifying glass very convenient in your search for the sexual organs of small birds when obscure, as they frequently are, out of the breeding season; in picking lice from plumage, to send to your entomological friend, who will very likely pronounce them to be of a "new species;" and for other purposes.

Fixtures. When travelling, your fixtures must ordinarily be limited to a collecting-chest; you will have to skin birds on the top of this, on the tail-board of a wagon, or on your lap, as the case may be. The chest should be very substantial — iron-bound is best; strong as to hinges and lock — and have handles. A good size is $30 \times 18 \times 18$ inches. Let it be fitted with a set of trays; the bottom one say four inches deep; the rest shallower; the top one very shallow, and divided into compartments for your tools and materials, unless you fix these on the under side of the lid. Start out with all the trays full of cotton or tow. At home, have a room to yourself, if possible; taxidermy makes a mess to which your wife may object, and arsenic must not come in the way of children. At any rate have your own table. I prefer plain deal that may be *scrubbed* when required; great cleanliness is indispensable, especially when doing much work in hot weather, for the place soon smells sour if neglected. I use no special receptacle for offal, for this only makes another article to be cleaned; lay down a piece of paper for the refuse, and throw the whole away. A perfectly smooth surface is desirable. I generally have a large pane of window-glass on the table before me. It will really be found advantageous to have a scale of inches scratched on the edge of the table; only a small part of it need be fractionally subdivided; this replaces the foot-rule and tape-line, just as the tacks of a dry-goods counter answer for the yardstick. You will find it worth while to rig some sort of a derrick arrangement, which you can readily devise, on one end of the table, to hitch your hook to, if you hang your birds up to skin them; they should swing clear of everything. The table should have a large general drawer, with a little drawer for gypsum and arsenic already mentioned, unless these be kept elsewhere. Stuffing may be kept in a box under the table, and make a nice footstool; or in a bag slung to the table leg.

Query: Have you cleansed the bird's plumage? Have you plugged the mouth, nostrils, and vent? Have you measured the specimen and noted the color of the eyes, bill, and feet, and prepared the labels, and made the entry in the register? Have you got all your apparatus within arm's length? Then we are ready to proceed.

§ 7.—HOW TO MAKE A BIRDSKIN.
a. THE REGULAR PROCESS.

Lay the Bird on its Back, the bill pointing to your right [1] elbow. Take the scalpel like a pen, with edge of blade uppermost, and run a straight furrow through the feathers along the middle line of the belly, from end of the breast-bone to the vent. Part the feathers completely, and keep them parted.[2] Observe a strip of skin either perfectly naked, or only covered with short down; this is the line for incision. Take scissors, stick in the pointed blade just over the end of the breast-bone, cut in a straight line thence to and *into* the vent; cut extremely shallow.[3]

Take the forceps in your left hand, and scalpel in your right, both held pen-wise, and with the forceps seize and lift up one of the edges of the cut skin, gently pressing away the belly-walls with the scalpel-point; no cutting is required; the skin may be peeled off without trouble. Skin away till you meet an obstacle; it is the thigh. Lay down the instruments; with your left hand take hold of the leg outside at the shank; put your right forefinger under the raised flap of skin, and feel a bump; it is the *knee;* push up the leg till this bump comes into view; hold it so. Take the scissors in your right hand; tuck one blade under the concavity of the knee, and sever the joint at a stroke; then the thigh is left with the rest of the body, while the rest of the leg is dissevered and hangs only by skin. Push the leg further up till it has slipped out of its sheath of skin, like a finger out of a glove, down to the heel-joint. You have now to clear off the flesh and leave the bone there; you may scrape till this is done, but there is a better way. Stick the *closed* points of the scissors in among the muscles just below the head of the bone, then separate the blades just wide enough to grasp the bone; snip off its head; draw the head to one side; all the muscles follow, being there attached; strip them *downward* from the bone; the bone is left naked, with the muscle hanging by a bundle of tendons ("leaders") at its foot; sever these tendons collectively at a stroke. This whole performance will occupy about three seconds, after practice; and you may soon discover you can nick off the head of the bone of a small bird with the thumb-nail. Draw the leg bone back into its sheath, and leave it. Repeat all the foregoing steps on the other side of the bird. If you are bothered by the skin-flaps settling against the belly-walls, insert a fluff of cotton.

---

[1] Reverse this and following directions for *position*, if you are left-handed.

[2] The motion is exactly like stroking the right and left sides of a moustache apart; you would never dress the hairs smoothly away from the middle line, by poking from ends to root; nor will the feathers stay aside, unless stroked away from base to tips.

[3] The skin over the belly is thin as tissue paper in a small bird; the chances are you will at first cut the walls of the belly too, opening the cavity; this is no great matter, for a pledget of cotton will keep the bowels in; nevertheless, try to divide skin only. Reason for cutting *into* vent: this orifice makes a nice natural termination of the incision, buttonhole-wise, and may keep the end of the cut from tearing around the root of the tail. Reason for beginning to cut *over* the edge of the sternum: the muscular walls of the belly are very thin, and stick so close to the skin that you may be in danger of attempting to remove them with the skin, instead of removing the skin from them; whereas, you cannot remove anything but skin from over the breast bone, so you have a guide at the start. You can tell skin from belly-wall, by its livid, translucent whitishness instead of redness.

HOW TO MAKE A BIRDSKIN.   29

Keep the feathers out of the wound; cotton and the moustache movement will do it. Next you must sever the tail from the body, leaving a small "pope's-nose" for the feathers to stay stuck into. Put the bird in the hollow of your lightly closed left hand, tail upward, belly toward you; or, if too large for this, stand it on its breast on the table in similar position. Throw your left forefinger across the front of the tail, pressing a little backward; take the scissors, cut the end of the lower bowel free first, then peck away at bone and muscle with cautious snips, till the tail-stump is dissevered from the rump, and the tail hangs only by skin. You will soon learn to do it all at one stroke; but you cannot be too careful at first; you are cutting right down on to the skin over the top of the pope's-nose, and if you divide this, the bird will part company with its tail altogether. Now you have the rump-stump protruding naked; the legs dangling on either side; the tail hanging loose over the bird's back between them. Lay down scissors, take up forceps [1] in your left hand; with them seize and hold the stump of the rump; and with point or handle of scalpel in the other hand, with finger-tips, or with thumb-nail (best), gently press down on and peel away skin.[2] No cutting will be required (usually) till you come to the wings: the skin peels off (usually) as easily as an orange-rind; as fast as it is loosened, *evert* it; that is make it continually turn itself more and more completely inside out. Work thus till you are stopped by the obtruding wings.[3] You have to sever the wing from the body at the shoulder, just as you did the leg at the knee, and leave it hanging by skin alone. Take your scissors,[4] as soon as the upper arm is exposed, and cut through flesh and bone alike at one stroke, a little below (outside of) the shoulder-joint. Do the same with the other wing. As soon as the wings are severed the body has been skinned to the root of the neck; the process becomes very easy; the neck almost slips out of its sheath of itself; and if you have properly attended to keeping the feathers out of the wound and to continual eversion of the skin, you now find you have a naked body connected dumb-bell-wise by a naked neck to a cap of reversed skin into which the head has disappeared, from the inside of which the legs and wings dangle, and around the edges of which is a row of plumage and a tail.[5] Here comes up an important consideration: the skin, plumage, legs, wings, and tail together *weigh* something, — enough to *stretch* [6] unduly the skin of the neck, from the small cylinder of which they are now suspended; the whole mass must be *supported*. For small birds, gather it in the hollow of your left hand, letting the body swing over the back of your hand out of the

---

[1] Or at this stage you may instead stick a hook into a firm part of the rump, and hang up the bird about the level of your breast; you thus have both hands free to work with. This is advisable with all birds too large to be readily taken in hand, and will help you, *at first*, with any bird. But there is really no use of it with a small bird, and you may as well learn the best way of working at first as afterward.

[2] The idea of the whole movement is exactly like ungloving your hand from the wrist, by turning the glove inside out to the very finger tips. Some people say, *pull* off the skin; I say *never pull* a bird's skin under any circumstances: *push it off*, always operating at lines of contact of skin with body, never upon areas of skins already detached.

[3] The elbows will get in your way before you reach the point of attack, namely, the shoulder, unless the wings were completely relaxed (as was essential, indeed, if you measured alar expanse correctly). Think what a difference it would make, were you skinning a man through a slit in the belly, whether his arms were stretched above his head, or pinned against his ribs. It is just the same with a bird. When properly relaxed the wings are readily pressed away toward the bird's head, so that the shoulders are encountered before the elbows.

[4] Shears will be required to crash through a *large* arm-bone. Or, you may with the scalpel unjoint the shoulder. The joint will be found higher up and deeper among the breast muscles than you might suppose, unless you are used to carving fowls at table. With a small bird, you may snap the bone with the thumb-nail and tear asunder the muscles in an instant.

[5] You find that the little straight cut you made along the belly has somehow become a hole larger than the greatest girth of the bird; be undismayed; it is all right.

[6] If you have up to this point properly *pushed* off the skin instead of *pulling* it, there is as yet probably no stretching of any consequence; but, in skinning the head, which comes next, it is almost impossible for a beginner to avoid stretching to an extent involving great damage to the good looks of a skin. Try your utmost, by delicacy of manipulation at the lines of contact of skin with flesh, and only there, to prevent *lengthwise* stretching. Crosswise distension is of no consequence; in fact more or less of it is usually required to skin the head, and it tends to counteract the ill effect of undue elongation.

way; for large ones, rest the affair on the table or your lap. To skin the head, secure the body in the position just indicated, by confining the neck between your left thumb and forefinger; bring the right fingers and thumb to a cone over the head, and draw it out with gentle force; or, holding the head itself between the left thumb and forefinger, insert the handle of the scalpel between the skin and skull, and pry a little, to enlarge the neck-cylinder of skin enough to let the head pass. It will generally[1] slip out of its hood very readily, as far as its greatest diameter;[2] there it sticks, being in fact pinned by the *ears*. Still holding the bird as before, with the point of the scalpel handled like a nut-picker, or with your thumb-nail, detach the delicate membrane that lines the ear-opening; do the same for the other ear. The skull is then shelled out to the *eyes*, and will skin no further of its own accord, being again attached by a membrane, around the border of the eye-socket. Holding the scalpel as before, run its edge around an arc (a semicircle is enough to let you into the orbit) of the circumference, dissevering the membrane from the bone. Reverse the scalpel, and scoop out the eyeball with the end of the handle; you bring out the eye betwixt the ball of your thumb and the handle of the instrument, tearing apart the optic nerve and the conjunctival tissue, but taking care not to open the eyeball[3] or lacerate the eyelids. Do the same with the other eye. The head is then skinned far enough; there is no use of getting *quite* to the base of the bill. You have now to get rid of the brain and flesh of the nape and jaws,[4] and leave most of the skull in; the cranial dome makes the only perfect "stuffing" for the skin of the head. This is all done at once by only four particular cuts. Hold the head between your left thumb and fingers, the bill pointing towards you, the bird's palate facing you; you observe a space bounded behind by the base of the skull where the neck joins, in front by the floor of the mouth, on either side by the prongs of the under jaw,—these last especially prominent. Take the scissors; stick one blade just inside one branch of the lower jaw, thence into the eye-socket which lies below (the head being upside down), thence into the brain-box; make a cut parallel with the jaw, just inside of it, bringing the upper scissor blade perpendicularly downward, crashing through the skull just inside of the angle of the jaw. Duplicate this cut on the other side. Connect the anterior ends of these cuts by a transverse one across the floor and roof of the mouth. Connect the posterior ends of the side cuts by one across the back of the skull near its base,—just where the nape-muscle ceases to override the cranium. You have enclosed and cut out a squarish-shaped mass of bone and muscle, and, on gently pulling the neck (to which of course it remains attached), the whole affair comes out, bringing the brain with it, but leaving the entire roof of the skull supported on a scaffolding of jaw-bone. It only remains to skin the wings. Seize the arm-stump with fingers or forceps; the upper arm is readily drawn from its sheath as far as the elbow; but the wing must be skinned to the wrist (carpus—"bend of the wing"); yet it will not come out so easily, because the secondary quills grow to one of the fore-arm bones (the ulna), pinning down the skin the whole way along a series of points. To break up these connections, hold the upper arm firmly with the left thumb and forefinger, the convexity of the elbow looking towards you; press the right thumb-nail closely against the back edge of the ulna, and strip downward, scraping the bone with the nail the whole way. If you only hit the line of adhesions, there is no trouble at all about this. Now you want to

---

[1] The special case of head too large for the calibre of the neck is treated beyond.

[2] And you will at once find a great apparent increase of amount of free skin in your hand, owing to release and extension of all that was before shortened in length by circular distension, in enlargement of the neck-cylinder.

[3] An eyeball is much larger than it looks from the outside; if you stick the instrument straight into the socket, you may punch a hole in the ball and let out the water; a very disagreeable complication. Insinuate the knife-handle close to the rim of the socket, and hug the wall of the cavity throughout.

[4] You may of course at this stage cut off the neck at the nape, punch a hole in the base of the skull, dig out the brains, and scrape away at the jaw-muscles till you are satisfied or tired; an unnecessary job, during which the skin may have become dry and shrivelled and hard to turn right side out. The operation described in the text may require ten seconds, perhaps.

leave in one of the two fore-arm bones, to preserve sufficiently the shape of the limb, but to remove the other, with the upper-arm bone and all the flesh. It is done in a moment: stick the point of the scissors between the heads of the two fore-arm bones, and cut the hinder one (ulna) away from the elbow; then the other fore-arm bone (radius), bearing on its near end the elbow and the whole upper arm, is to be stripped away from the ulna, taking with it the flesh of the fore-arm, and to be cut off at its far end close to the wrist-joint, one stroke severing the bone and all the tendons that pass over the wrist to the hand; then the ulna, bare of flesh, is alone left in, attached at the wrist. Draw gently on the wing from the outside till it slips into the natural position whence you everted it. Do the same for the other wing. This finishes the skinning process. The skin is now to be turned right side out. Begin any way you please, till you see the point of the bill reappearing among the feathers; seize it with fingers or forceps, as convenient, and use it for gentle traction. But by no means pull it out by holding on to the rear end of the skin — that would infallibly stretch the skin. Holding the bill, make a cylinder of your left hand and coax the skin backward with a sort of milking motion. It will come easily enough, until the final stage of getting the head back into its skull-cap; this may require some little dexterity; but you cannot fail to get the head in, if you remember what you did to get it out. When this is fairly accomplished, you for the first time have the pleasure of seeing something that looks like a birdskin. Your next[1] care is to apply arsenic. Lay the skin on its back, the opening toward you and wide spread, so the interior is in view. Run the scalpel-handle into the neck to dilate that cylinder until you can see the skull; find your way to the orifices of the legs and wings; expose the pope's-nose; thus you have not only the general skin surface, but all the points where some traces of flesh were left, fairly in view. Shovel in arsenic; dump some down the neck, making sure it reaches and plentifully besprinkles the whole skull; drop a little in each wing hole and leg hole; leave a small pile at the root of the tail; strew some more over the skin at large. The simple rule is, put in as much arsenic as will *stick* anywhere. Then close the opening, and shake up the skin; move the head about by the bill; rustle the wings and move the legs; this distributes the poison thoroughly. If you have got in more than is necessary, as you may judge by seeing it piled up dry, anywhere, hold the skin with the opening downward over the poison-drawer, and give it a flip and let the superfluous powder fall out. Now for the "make up," upon which the beauty of the preparation depends. First get the empty skin into good shape. Let it lie on its back; draw it straight out to its natural length. See that the skin of the head fits snugly; that the eyes, ears, and jaws are in place. Expand the wings to make sure that the bone is in place, and fold them so that the quills override each other naturally; set the tail-feathers shinglewise also; draw down the legs and leave them straddling wide apart. Give the plumage a preliminary dressing; if the skin is free from kinks and creases, the feathers come naturally into place; particular ones that may be awry should be set right, as may be generally done by stroking, or by lifting them free repeatedly, and letting them fall; if any (through carelessness) remain turned into the opening, they should be carefully picked out. Remove all traces of gypsum or arsenic with the feather duster. The stuffing is to be put in through the opening in the belly; the art is to get in just enough, in the right places. It would never do to push in pellets of cotton, as you would stuff a pillow-case, till the skin is filled up; no subsequent skill in setting could remove the distortion that would result. It takes just *four*[2] pieces of stuffing — one for each eye, one for the neck, and one for the body;

---

[1] Some direct the poisoning to be done while the skin is still wrong side out; and it may be very thoroughly effected at that stage. I wait, because the arsenic generally strews over the table in the operation of reversing the skin, if you use as much as I think advisable; and it is better to have a cavity to put it *into* than a surface to strew it *on*.

[2] For any ordinary bird up to the size of a crow. It is often directed that the leg-bones and wing-bones be wrapped with cotton or tow. I should not think of putting anything around the wing-bones of any bird up to the size of an eagle, swan, or pelican. Examination of a skinned wing will show how extremely compact it is, except

while it requires rather less than half as much stuffing as an inexperienced person might suppose. Take a shred of cotton that will make a tight ball as large as the bird's eye; stick it on the end of your knitting-needle, and by twirling the needle whilst the cotton is confined in your finger tips, you make a neat ball. Introduce this through the belly-opening, into the eye-socket; if you have cut away skull enough, as already directed, it will go right in; disengage the needle with a reverse twirl, and withdraw it. Take hold of the bill with one hand, and with the forceps in the other, dress the eyelids neatly and naturally over the elastic substance within. Repeat for the other eye. Take next a shred of cotton that will roll into a firm cylinder rather less than the size of the bird's neck. Roll it on the needle much as you did the eye-ball, introduce it in the same way, and ram it firmly into the base of the skull; disengage the needle by twirling it the other way, and withdraw it, taking care not to dislodge the cotton neck. If now you peep into the skin you will see the end of this artificial neck; push it up against the skin of the breast, — it must not lie down on the back between the shoulders.[1] The body-wad comes next; you want to imitate the size and shape of the bird's trunk. Take a mass of cotton you think will be enough, and take about *half* of this; that will be plenty (cotton is very elastic). It should make a tolerably firm ball, rather egg-shaped, swelling at the breast, smaller behind. If you simply squeeze up the cotton, it will not stay compressed; it requires a motion something like that which bakers employ to knead dough into the shape of a loaf. Keep tucking over the borders of the cotton till the desired shape and firmness are attained. Insert the ball between the blades of the forceps in such way that the instrument confines the folded-over edges, and with a wriggling motion insinuate it aright into the body. Before relaxing the forceps, put your thumb and forefinger in the bird's armpits, and pinch the shoulders together till they almost touch; this is to make sure that there is no stuffing between the shoulders, — the whole mass lying breastwards. Loosen the forceps and withdraw them. If the ball is rightly made and tucked in, the elasticity of the cotton will chiefly expend itself in puffing out the breast, which is just what is wanted. Be careful not to push the body *too far in*; if it impacts against the skin of the neck, this will infallibly stretch, driving the shoulders apart, and no art will remedy the unsightly gape resulting. You see I dwell on this matter of the shoulders; the whole knack of stuffing correctly focuses just over the shoulders. If you find you have made the body too large, pull it out and make a smaller one; if it fits nicely about the shoulders, but is too long to go in, or too puffy over the belly, let it stay, and pick away shreds at the open end till the redundancy is remedied. Your bird is now stuffed. Close the opening by bringing the edges of the original cut together. There is no use of sewing[2] up the cut, for a small bird; if the stuffing is correct, the feathers will hide the opening; and if they do not, it is no matter. You are not making an object for a show case, but for a naturalist's

just at the shoulder. What you remove will never make any difference from the outside, while you would almost inevitably get in too much, not of the right shape, and make an awkward bulging no art would remedy; I say, then, leave the wings of all but the largest birds *empty*, and put in very little under any circumstances. As for legs, the whole host of small perching birds need no wrapping whatever; depend upon it you will make a nicer skin without wrapping. But large birds and those with very muscular or otherwise prominent legs must have the removal of flesh compensated for. I treat of these cases beyond.

[1] Although a bird's neck is really, of course, in direct continuation of the back-bone, yet the natural sigmoid curve of the neck is such that it virtually takes departure rather from the breast, its lower curve being received between the prongs of the merrythought. This is what we must imitate instead of the true anatomy. If you let the end of the neck lie between the shoulders, it will infallibly press them apart, so that the interscapular plumage cannot shingle over the scapular feathers as it should, and a gaping place, showing down or even naked skin, will result. Likewise if the neck be made *too large* (the chances are that way, at first), the same result follows. These seemingly trifling points are very important indeed; I never made a decent birdskin till I learned to get the neck small enough and to shove the end of it against the breast.

[2] But sew it up, if you please, though you may be perhaps giving the man who subsequently mounts the bird the trouble of ripping out the stitches. Stitches, however, will not come amiss with a *large* bird. I generally, in such cases, *pin* the edges of the cut in one or more places.

cabinet. Supposing you to have been so far successful, little remains to be done; the skin already looks very much like a dead bird; you have only to give the finishing touches, and "set" it. Fixing the wings nicely is a great point. Fold each wing closely; see that the carpal bend is well defined, that the coverts show their several oblique rows perfectly, that all the quills override each other like shingles. Tuck the folded wings close up to the body — rather on the bird's back than along its sides; see that the wing tips meet over the tail (*under the tail as the bird lies on its back*); let the carpal angle nestle in the plumage; have the shoulders close together, so that the interscapulars shingle over the scapulars. If the wing be pressed in *too* tightly, the scapulars will rise up on end; there must be neither furrow nor ridge about the insertion of the wings; everything must lie perfectly smooth. At this stage of the process, I generally lift up the skin gingerly, and let it slip head first through one hand after the other, pressing here or there to correct a deformity, or uniformly to make the whole skin compact. The wings set, next bring the legs together, so that the bones within the skin lie parallel with each other; bend the heel-joint a little, to let the tarsi *cross* each other about their middle; lay them sidewise on the tail, so that the naturally flexed toes lie flat, all the claws mutually facing each other. See that the neck is perfectly straight, and, if anything, shortened rather than outstretched; have the crown of the head flat on the table, the bill pointing straight forward,[1] the mandibles shut tightly.[2] Never attempt any "fancy" attitudes with a birdskin; the simpler and more compactly it is made up the better.[3] Finally, I say, hang over your bird (if you have time); dress better the feathers that were well dressed before; perfect every curve; finish caressingly, and put it away tenderly, as you hope to be shriven yourself when the time comes.

There are several ways of laying a birdskin. A common, easy, and slovenly way is to thrust it head first into a paper *cone;* but it makes a hollow-chested, pot-bellied object, unpleasant to see, and renders your nice work on the make-up futile. A paper *cylinder*, corresponding in calibre to the greatest girth of the birdskin, binds the wings well, and makes a good ordinary specimen, — perhaps better than the average. Remarking that there are some detestable practices, such as hanging up a bird by a string through the nose (methods only to be mentioned to be condemned), I will tell you the easiest and best way, by which the most elegant and tasteful results are almost necessarily secured. The skins are simply laid away in cotton, just as they come from your hands. Take a considerable wad of cotton, make a "bed" of it, lay the specimen in, and tuck it up nicely around the edges. In effect, I generally take a thin sheet of cotton wadding, the sizing of which confers some textile consistency, and wrap the bird completely but lightly in it. By loosening or tightening a trifle here or there, laying down a "pillow" or other special slight pressure, the most delicate contour-lines may be preserved with perfect fidelity. Unnecessary pother is sometimes made about *drying*

---

[1] Exceptions. Woodpeckers, ducks, and some other birds treated of beyond, are best set with the head flat on one side, the bill pointing obliquely to the right or left; owls, with the bill pointing straight up in the air as the bird lies on its back.

[2] If the mandibles gape, run a thread through the nostrils and tie it tightly under the bill. Or, since this injures the nostrils (and we frequently want to examine their structure) stick a pin in under the bill close to the gonys, driving it obliquely into the palate. Sometimes the skin of the throat looks sunken betwixt the sides of the jaw. A shred of cotton introduced with forceps through the mouth will obviate this.

[3] Don't cock up the head, trying to impart a knowing air — it cannot be done, and only makes the poor bird look ridiculous. Don't lay the skin on one side, with the legs in perching position, and don't spread the wings — the bird will never perch nor fly again, and the suggestion is unartistic because incongruous. The only permissible departure from the rule of severe simplicity is when some special ornament, as a fine crest, may be naturally displayed, or some hidden markings are desired to be brought out, or a shape of tail or wing to be perpetuated; but in all such cases the "flowery" inclination should be sparingly and judiciously indulged. It is, however, frequently desirable to give some special set to *hide a defect*, as loss of plumage, etc.; this may often be accomplished very cunningly, with excellent result. No rules for this can be laid down, since the details vary in every case; but in general the weak spot may be hidden by contracting the skin of the place, and then setting the bird in an attitude that naturally corresponds, thus making a virtue of necessity.

skins; the fact being that under ordinary circumstances they could not be kept from drying perfectly; and they dry in exactly the shape they are set, if not accidentally pressed upon. At sea, however, or during unusually protracted wet weather, they of course dry slowly, and may require some attention to prevent mildew or souring, especially in the cases of very large, thick-skinned, or greasy specimens. Thorough poisoning, and drying by a fire, or placing in the sun, will always answer. Very close packing retards drying. When travelling, or operating under other circumstances requiring economy of space, you must not expect to turn out your collection in elegant order. Perfection of contour-lines can only be secured by putting each specimen away by itself; undue pressure is always liable to produce unhappily *outré* configuration of a skin. Trays in a packing box are of great service in limiting possibilities of pressure; they should be shallow; one four inches deep will take a well stuffed henhawk, for example, or accommodate from three to six sparrows a-top of one another. It is well to sort out your specimens somewhat according to size, to keep heavy ones off little ones; though the chinks around the former may usually be economized with advantage by packing in the less valuable or the less neatly prepared of the latter. When limited to a travelling chest, I generally pass in the skins as fast as made, packing them "solid" in one sense, yet hunting up a nice resting-place for each. If each rests in its own cotton coffin, it is astonishing how close they may be laid without harm, and how many will go in a given space; a tray 30 × 18 × 4 inches will easily hold three hundred and fifty birds six inches long. As a tray fills up, the drier ones first put in may be submitted to more pressure. A skin originally dried in good shape may subsequently be pressed perfectly flat without material injury; the only thing to avoid being *contortion*. The whole knack of packing birds corresponds to that of filling a trunk *solidly* full of clothes, as may easily be done without damage to an immaculate shirt-front. Finally, I would say, never put away a bird unlabelled, not even for an hour; you may forget it or die. Never tie a label to a bird's bill, wing, or tail; tie it securely to *both* legs where they cross, and it will be just half as liable to become detached as if tied to one leg only. Never paste a label, or even a number, on a bird's plumage. Never put in glass eyes before mounting. Never paint or varnish a bird's bill or feet. Never replace missing plumage of one bird with the feathers of another — no, not even if the birds came out of the same nest.

b. SPECIAL PROCESSES; COMPLICATIONS AND ACCIDENTS.

The Foregoing Method of procedure is a routine practice applicable to three-fourths if not nine-tenths of the "general run" of birds. But there are several cases requiring a modification of this programme; while several circumstances may tend to embarrass your operations. The principal special conditions may therefore be separately treated to your advantage.

Size. — Other things being equal, a large bird is more difficult to prepare than a small one. In one case, you only need a certain delicacy of touch, easily acquired and soon becoming mechanical; in the other, demand on your strength may be made, till your muscles ache. It takes longer, too;[1] I could put away a dozen sparrows in the time I should spend over an eagle; and I would rather undertake a hundred humming-birds than one ostrich. For

---

[1] The reader may be curious to know something of the statistics on this score — how long it ought to take him to prepare an ordinary skin. He can scarcely imagine, from his first tedious operations, how expert he may become, not only in beauty of result, but in rapidity of execution. I have seen taxidermists make good small skins at the rate of ten an hour; but this is extraordinary. The quickest work I ever did myself was eight an hour, or an average of seven and a half minutes apiece, and fairly good skins. But I picked my birds, all small ones, well shot, labelled, measured, and plugged beforehand, so that the rate of work was exceptional, besides including only the actual manipulations from first cut to laying away. No one *averages* eight birds an hour, even excluding the necessary preliminaries of cleansing, plugging, etc. Four birds an hour, everything included, is good work. A very eminent ornithologist of this country, and an expert taxidermist, once laid a whimsical wager, that he would skin and stuff a bird before a certain friend of his could pick all the feathers off a specimen of the same kind. I forget the time, but he won, and his friend ate crow, literally, that night.

"large" birds, say anything from a hen-hawk upward, various special manipulations I have directed may be foregone, while however you observe their general drift and intent. You may open the bird as directed, or, turning it tail to you, cut with a knife.[1] Forceps are rarely required; there is not much that is too small to be taken in hand. As soon as the tail is divided, hang up the bird by the rump, so you will have both hands free. Let it swing clear of the wall or table, at any height most convenient. The steel hooks of a dissecting case are not always large enough; use a stout fish-hook with the barb filed off. Work with your nails, assisted by the scalpel if necessary. I know of no bird, and I think there is none, in this country at least, the skin of which is so intimately adherent by fibrous or muscular tissue as to require actual dissecting throughout; a pelican comes, perhaps, as near this as any; but in many cases the knife may be constantly employed with advantage. Use it with long clean sweeping strokes, hugging the skin rather than the body. The knee and shoulder commonly require disarticulation, unless you use bone-nippers or strong shears; the four cuts of the skull may presuppose a very able-bodied instrument, even a chisel. The wings will give you the most trouble, and they require a special process; for you cannot readily break up the adhesions of the secondary quills to the ulna, nor is it desirable that very large feathers should be deprived of this natural support. Hammer or nip off the great head of the upper arm-bone, just below the insertion of the breast muscles; clean the rest of that bone and leave it in. Tie a string around it (what sailors call "two half hitches" gives a secure hold on the bony cylinder), and tie it to the other humerus, inside the skin, so that the two bones shall be rather less than their natural distance apart. After the skin is brought right side out, attack the wings thus: Spread the wing under side uppermost, and secure it on the table by driving a tack or brad through the wrist-joint; this fixes the far end, while the weight of the skin steadies the other. Raise a whole layer of the under wing-coverts, and make a cut in the skin thus exposed, from elbow to wrist, in the middle line between the two forearm bones. Raise the flaps of skin and all the muscle is laid bare; it is to be removed. This is best done by lifting each muscle from its bed separately, slipping the handle of the scalpel under the individual bellies; there is little if any bony attachment except at each end, and this is readily severed. Strew in arsenic; a little cotton may be used to fill the bed of muscle removed from a *very* large bird; bring the flaps of skin together, and smooth down the coverts; you need not be particular to sew up the cut, for the coverts will hide the opening; in fact, the operation does not show at all after the make-up. Stuffing of large birds is not commonly done with only the four pieces already directed. The eyeballs, and usually the neck-cylinder, go in as before; the body may be filled any way you please, provided you do not put in too much stuffing nor get any between the shoulders. All large birds had better have the leg-bones wrapped to nearly natural size. Observe that the leg-muscles do not form a cylinder, but a cone; let the wrapping taper naturally from top to bottom. Attention to this point is necessary for all large or medium-sized birds with naturally prominent legs. The large finely feathered legs of a hawk, for example, ought to be well displayed; with these birds, and also with rails, etc., moreover, imitate the bulge of the thigh with a special wad laid inside the skin. Large birds commonly require also a special wad introduced by the mouth, to make the swell of the throat; this wad should be rather fluffy than firm. As a rule, do not fill out

---

[1] Certain among larger birds are often opened elsewhere than along the belly, with what advantage I cannot say from my own experience. Various water birds, such as loons, grebes, auks, gulls, and ducks (in fact any swimming bird with dense under plumage) may be opened along the side by a cut under the wings from the shoulder over the hip to the rump; the cut is completely hidden by the make-up, and the plumage is never ruffled. But I see no necessity for this; for, as a rule, the belly opening can, if desired, be completely effaced with due care, though a very greasy bird with white under plumage generally stains where opened, in spite of every precaution. Such birds as loons, grebes, cormorants, and penguins are often opened by a cut across the fundament from one leg to the other; their conformation in fact suggests and favors this operation. I have often seen water birds slit down the back; but I consider it very poor practice.

large birds to their natural dimensions; they take up too much room. Let the head, neck, and legs be accurately prepared, but leave the main cavity one-third if not one-half empty; no more is required than will fairly smooth out creases in the skin. Reduce bulk rather by flattening out than by general compression. Use tow instead of cotton; and if at all short of tow, economize with paper, hay, etc., at least for the deeper portions of the main stuffing. Large birds may be "set" in a great quantity of tow; wrapped in paper, much like any other parcel; or simply left to dry on the table, the wings being only supported by cushioning or other suitable means.

Shape. — Some special configurations have been noticed in the last paragraph, prematurely perhaps, but leading directly up to further considerations respecting *shape* of certain birds as a modifying element in the process of preparation. As for skinning, there is one extremely important matter. . Most ducks, many woodpeckers, flamingoes, and doubtless some others with which I am not familiar, cannot be skinned in the usual way, because the head is too large for the calibre of the neck and cannot be drawn through. In such cases, skin as usual to the base of the skull, cut off the head there (inside the skin of course), and operate upon it, after turning the skin right side out, as follows: Part the feathers carefully in a straight line down the back of the skull, make a cut through the skin, just long enough to permit the head to pass, draw out the skull through this opening, and dress it as already directed. Return it, draw the edges of the cut nicely together, and sew up the opening with a great many fine stitches. Simple as it may appear, this process is often embarrassing, for the cut has an unhappy tendency to wander about the neck, enlarging itself even under the most careful manipulation; while the feathers of the parts are usually so short, that it is difficult to efface all traces of the operation. I consider it very disagreeable; but for ducks I know of no alternative. I have however found out a way to avoid it with woodpeckers, excepting the very largest; it is this: Before skinning, part the eyelids, and plunge the scalpel right into the eyeballs; seize the cut edge of the ball with the forceps, and pull the eye right out. It may be dexterously done without spilling the eye-water on the plumage; but, for fear of this, previously put a little pile of plaster on the spot. Throw arsenic into the socket, and then fill it with cotton poked in between the lids. The eyes are thus disposed of. Then, in skinning, when you come to the head, dissever it from the neck and work the skull as far out as you can; it may be sufficiently exposed, in all cases, for you to gouge out the base of the skull with the scissors, and get at the brain to remove it. Apply an extra large dose of arsenic, and you will never hear from what jaw-muscle has been left in. In all these cases, as already remarked, the head is preferably set lying on one side, with the bill pointing obliquely to the right or left. Certain birds require a special mode of *setting*; these are, birds with very long legs or neck, or both, as swans, geese, pelicans, cormorants, snakebirds, loons, and especially cranes, herons, ibises, and flamingoes. Long legs should be doubled completely on themselves by bending at the heel-joint, and either tucked under the wings, or laid on the under surface; the chief point is to see that the toes lie flat, so that the claws do not stick up, to catch in things or get broken off. A long neck should be carefully folded; not at a sharp angle with a crease in the skin, but with a short curve, and brought round either to the side of the bird or on its breast, as may seem most convenient. The object is to make a "bale" of the skin as nearly as may be, and when it is properly effected it is surprising what little space a crane, for instance, occupies. But it is rarely, if ever, admissible to bend a tail back on the body, however inconveniently long it may be. Special dilations of skin, like the pouch of a pelican, or the air sacs of a prairie hen, may be moderately displayed.

Thin Skin. — Loose Plumage. — It is astonishing how much resistance is offered by the thin skin of the smallest bird. Though no thicker than tissue paper, it is not very liable to tear if deftly handled; yet a rent once started often enlarges to an embarrassing extent if the skin be stretched in the least. Accidental rents and enlargements of shot-holes should be neatly sewn up, if occurring in an exposed place; but in most cases the plumage may be set to hide the openings. The trogons are said to have remarkably thin and delicate skin; I have never handled one in the flesh. Among our birds, the cardinal grosbeak and the species of *Caprimulgidæ* have, I think, about the tenderest skins. The obvious indication in all such cases is simply a little extra delicacy of manipulation. In skinning most birds, you should not loose more than a feather or two, excepting those loosened by the shot. Pigeons are peculiar, among our birds, for the very loose insertion of their plumage; you will have to be particularly careful with them, and in spite of all your precautions a good many feathers will probably drop. As stripping down the secondary quills from the forearm, in the manner already indicated, will almost invariably set these feathers free from the skin, I recommend you not to attempt it, but to dress the wings as prescribed for large birds.

Fatness. — Fat is a substance abhorred of all dissectors; always in the way, embarrassing operations and obscuring observations; while it is seldom worth examination after its structure has once been ascertained. It is particularly obnoxious to the taxidermist, since it is liable to soil the plumage during skinning, and also to soak into the feathers afterwards; and greasy birdskins are never pleasing objects. A few birds never seem to have any fat; some, like petrels, are always oily; at times, especially in the indolent autumn season, when birds have little to do but feed, the great majority acquire an *embonpoint* doubtless to their own satisfaction, but to the taxidermist's discomfort. In all such cases gypsum should be lavishly employed. Strew plaster plentifully, from the first cut all through the operation; dip your fingers in it frequently, as well as your instruments. The invaluable absorbent will deal with most of the "running" fat. When the skin is completely reversed, remove as much of the solid fat as possible; it is generally found occupying the areolar tissue of particular definite tracts, and most of it may usually be peeled or flaked off in considerable masses. Since the soft and oozy state of most birds' fat at ordinary temperatures may be much improved by cold, it will repay you to leave your birds on ice for a while before skinning, if you have the means and time to do so; the fat will become quite firm. There is a device for preventing or at any rate lessening the soiling of the plumage so apt to occur along the line of your incision; it is invaluable in all cases of white plumage. Take a strip of cloth of greater width than the length of the feathers, long enough to go up one side of the cut and down the other. Sew this closely to the skin all around the cut, and it will form an apron to guard the plumage. You will too frequently find that a bird, prepared without soiling and laid away apparently safe, afterwards grows greasy; if the plumage is white, it soon becomes worse than ever by showing dust that the grease catches. Perhaps the majority of such birds in our museums show the dirty streak along the belly. The reason is, that the grease has oozed out along the cut, or wherever else the skin has been broken, and infiltrated the plumage, being drawn up apparently by capillary attraction, just as a lampwick "sucks up" oil. Sometimes, without obviously soiling the plumage, the grease will run along the thread that ties the label, and make a uniformly transparent piece of "oil-paper." I have no remedy to offer for this gradual infiltration of the plumage. It will not wash out, even with soap and water. Possibly careful and persistent treatment with an ether might be effective, but I am not prepared to say it would be. Removal of all fat that can be got off during skinning, with a liberal use of plaster, will in a measure prevent a difficulty that remains incurable.

Bloodstains, etc. — In the nature of the case, this complication is of continual occurrence; fortunately it is easier dealt with than greasiness. Much may be done in the field to prevent bloodying of the plumage, as already said. A little blood does not show much on a dark plumage; but it is of course conspicuous on light or white feathers. Dried blood may often be scraped off, in imitation of the natural process by which a bird cleanses its plumage with the bill; or be pulverized by gently twiddling the feathers between the fingers, and then blown off. But feathers may by due care be *washed* almost as readily as clothing; and we must ordinarily resort to this to remove all traces of blood, especially from white surfaces. If properly dried they do not show the operation. With a soft rag or pledget of cotton dipped in warm water bathe the place assiduously, pressing down pretty hard, only taking care to stroke the feathers the right way, so as not to crumple them, until the red color disappears; then you have simply a wet place to deal with. Press gypsum on the spot; it will cake; flake it off and apply more, till it will no longer stick. Then raise the feathers on a knife-blade and sprinkle gypsum in among them; pat it down and shake it up, wrestling with the spot till the moisture is entirely absorbed. Two other fluids of the body will give you occasional annoyance, — the juices of the alimentary canal and the eye-water. Escape of the former by mouth, nostrils, or vent is preventable by plugging these orifices, and its occurrence is inexcusable. But shot often lacerates the gullet, crop, and bowels, and though nothing may flow at the time, subsequent jolting or pressure in the game-bag causes the escape of fluids: a seemingly safe specimen may be unwrapped to show the whole belly-plumage a sodden brown mass. Such accidents should be treated precisely like bloodstains; but it is to be remarked that these stains are not seldom indelible, traces usually persisting in white plumage at least in spite of our best endeavors. Eye-water, insignificant as it may appear, is often a great annoyance. This liquor is slightly glairy, or rather glassy, and puts a sort of sizing on the plumage difficult to efface; the more so since the soiling necessarily occurs in a conspicuous place, where the plumage is too scanty and delicate to bear much handling. It frequently happens that a lacerated eyeball, by the elasticity of the coats, or adhesion of the lids, retains its fluid till this is pressed out in manipulating the parts; and recollecting how the head lies buried in plumage at that stage of the process, it will be seen that not only the head, but much of the neck and even the breast may become wetted. If the parts are extensively soaked, the specimen is almost irreparably damaged, if not ruined. Plaster will absorb the moisture, but much of the sizing may be retained on the plumage; therefore, though the place seems simply wet, it should be thoroughly washed with water before the gypsum is applied. I always endeavor to prevent the accident; if I notice a lacerated eyeball, I extract it before skinning, in the manner described for woodpeckers. Miscellaneous stains, from the juices of plants, etc., may be received; all such are treated on general principles. Blood on the beak and feet of rapacious birds, mud on the bill and legs of waders, etc., etc., may be washed off without the slightest difficulty. A land bird that has fallen in the water should be recovered as soon as possible, picked up *by the bill*, and shaken; most of the water will run off, unless the plumage is completely soaked. It should be allowed to dry just as it is, without touching the plumage, before being wrapped and bagged. If a bird fall in soft mud, the dirt should be scraped or snapped off as far as this can be done without plastering the feathers down, and the rest allowed to dry; it may afterward be rubbed fine and dusted off, when no harm will ensue, except to white feathers which may require washing.

Mutilation. — You will often be troubled, early in your practice, with broken legs and wings, and various lacerations; but the injury must be very severe (such as the carrying away of a limb, or blowing off the whole top of a head) that cannot be in great measure remedied by care and skill. Suppose a little bird, shot through the neck or small of the back, comes apart while being skinned; you have only to remove the hinder portion, be that much or little, and go on with the rest as if it were the whole. If the leg bone of a small bird be broken near the heel, let it come away altogether; it will make little if any difference. In case of the same accident to a large bird that ought to have the legs wrapped, whittle out a peg and stick it in the hollow stump of the bone; if there is no stump left, file a piece of stout wire to a point and stick it into the heel joint. If the forearm bone that you usually leave in a small bird is broken, remove it and leave the other in; if both are broken, do not clean the wings so thoroughly that they become detached; an extra pinch of arsenic will condone the omission. In a large bird, if both bones of the forearm are broken, splint them with a bit of wood laid in between, so that one end hitches at the elbow, the other at the wrist. A humerus may be replaced like a leg bone, but this is rarely required. If the skull be smashed, save the pieces, and leave them if you can; if not, imitate the arch of the head with a firm cotton-ball. A broken tarsus is readily splinted with a pin thrust up through the sole of the foot: if too large for this, use a pointed piece of wire. There is no mending a bill when part of it is shot away; for I think the replacing of part by putty, stucco, etc., inadmissible; but if it be only fractured, the pieces may usually be retained in place by winding with thread, or with a touch of glue or mucilage. It is singular, by the way, what unsightliness results from a very trifling injury to the bill; much, I suppose, as a boil on a person's nose is peculiarly deplorable. I have already hinted how artfully various weak places in a skin, due to mutilation or loss of plumage, may be hidden.

Decomposition. — It might seem unnecessary to speak of what may be *smelled out* so readily as animal putrescence; but there are some useful points to be learned in this connection, besides the important sanitary precautions that are to be deduced. Immediately after death the various fluids of the body begin to "settle" (so to speak), and shortly after the muscular system as a rule becomes fixed in what is technically called *rigor mortis*. This stiffening usually occurs as the animal heat dies away; but its onset, and especially its duration, is very variable, according to circumstances, such as cause of death; although in most cases of sudden violent death of an animal in previous good health, it seems to depend chiefly upon temperature, being transient and imperfect, or altogether wanting, in hot weather. As it passes off, the whole system relaxes, and the body soon becomes as "limp" as at the moment of death. This is the period immediately preceding decomposition; in fact, it may be considered as the stage of incipient putridity; it is very brief in warm weather, and it should be seized as the last opportunity of preparing a bird without inconvenience and even danger. If not skinned at once, putrescence becomes established; it is indicated by the effluvium (at the outset "sour," but rapidly acquiring a variety of disgusting odors); by the distension of the abdomen with gaseous products of decomposition; by the loosening of the cuticle, and consequently of the feathers; and by other signs. If you part the feathers of a bad-smelling bird's belly to find the skin swollen and livid or greenish, while the feathers come off at a touch, the bird is too far gone to be recovered without trouble and risk that no ordinary specimen warrants. It is a singular fact that this early putrescence is more poisonous than utter rottenness; as physicians are aware, a post-mortem examination at this stage, or even before it, involves more risk than their ordinary dissecting-room experience. It seems that both natural and pathological poisons lose their early virulence by resolution into other products of decay. The obvious deduction from all this is to skin your birds soon enough. Some say they are best skinned perfectly fresh, but I see no reason for this; when I have time to choose, I take the period of rigidity as being preferable on the whole; for the fluids have then "settled," and the limbs are readily relaxed by manipulation. If you have a large bag to dispose of, and are pressed for time, set them in the coolest place you can find, preferably on ice; a slight lowering of temperature may make a decided difference. Disembowelling, which may be accomplished in a moment, will materially retard decomposition. Injections of creosote or dilute carbolic acid will arrest decay for a time, for an indefinitely long period if a large quantity of these antiseptics be employed. When it becomes desirable (it can never be *necessary*) to skin a putrescent bird, great care must be exercised not only to accomplish the operation, but to avoid danger. I must not, however, unconsciously lead you to exaggerate the risk, and will add that I think it often overrated. I have probably skinned birds as " gamey " as any one has, and repeatedly, without being conscious of any ill effects. I am sure that no poison, ordinarily generated by decomposition of a body *healthy* at death, can compare in virulence with that commonly resulting after death by many diseases. I also believe that the gaseous products, however offensive to the smell, are innocuous as a rule. The danger practically narrows down to the absorption of fluids through an abraded surface; the poison is rarely taken in by natural pores of healthy skin, if it remain in contact but a short time. Cuts and scratches may be closed with a film of collodion, or covered with isinglass or court plaster, or protected by rubber cots on the fingers. The hands should, of course, be washed with particular care immediately after the operation, and the nails scrupulously dressed. Having never been poisoned (to my knowledge), I cannot give the symptoms from personal experience; but I will quote from Mr. Maynard:

" In a few days numerous pimples, which are exceedingly painful, appear upon the skin of the face and other parts of the person and, upon those parts where there is chafing or rubbing, become large and deep sores. There is a general languor and, if badly poisoned, complete prostration results; the slightest scratch becomes a festering sore. Once poisoned in this manner (and I speak from experience), one is never afterward able to skin any animal that has become in the least putrid, without experiencing some of the symptoms above described. Even birds that you handled before with impunity, you cannot now skin without great care. The best remedy in this case is, as the Hibernian would say, not to get poisoned, ... bathe the parts frequently in cold water; and, if chafed, sprinkle the parts after bathing, with wheat flour. These remedies, if persisted in, will effect a cure, if not too bad; then, medical advice should be procured without delay."[1]

How to mount Birds. — As some may not improbably procure this volume with a reasonable expectation of being taught to *mount* birds, I append the required instructions, although the work only professes to treat of the preparation of skins for the cabinet. As a rule, the purposes of science are best subserved by *not* mounting specimens; for display. the only end attained, is not required. I would strongly advise you not to mount your rarer or otherwise particularly valuable specimens; select for this purpose nice, pretty birds of no special scientific value. The principal objections to mounted birds are, that they take up altogether too much room, require special arrangements for keeping and transportation, and cannot be handled for study with impunity. Some might suppose that a mounted bird would give a better idea of its figure and general aspect than a skin; but this is only true to a limited extent. Faultless mounting is an art really difficult, acquired by few; the average work done in this line shows something of caricature, ludicrous or repulsive, as the case may be. To copy nature faithfully by taxidermy requires not only long and close study, but an artistic sense; and this last is a rare gift. Unless you have at least the germs of the faculty in your composition, your taxidermal success will be incommensurate with the time and trouble you bestow. My own taxidermal art is of a low order, decidedly not above average; although I have mounted a great many birds that would compare very favorably with ordinary museum work, few of them have entirely answered my ideas. A live bird is to me such a beautiful object that the slightest taxidermal flaw in the effort to represent it is painfully offensive; perhaps this makes me place the standard of excellence too high for practical purposes. I like a good honest birdskin that does not pretend to be anything else; it is far preferable to the

---

[1] Avoid all mechanical irritation of the inflamed parts; touch the parts that have ulcerated with a stick of lunar caustic; take a dose of salts; use syrup of the iodide of iron, or tincture of the chloride of iron, say thirty drops of either, in a wineglass of water, thrice daily; rest at first, exercise gradually as you can bear it; and skin no birds till you have completely recovered.

ordinary taxidermal abortions of the show-cases. But if, after the warnings that I mean to convey in this paragraph, you still wish to try your hand in the higher department of taxidermy, I will explain the whole process as far as manipulation goes; the art you must discover in yourself.

The operation of skinning is precisely the same as that already given in detail; then, instead of stuffing the skin as directed above, to lie on its back in a drawer, you have to stuff it so that it will stand up on its feet and look as much like a live bird as possible. To this end a few additional implements and materials are required. These are: *a*, annealed wire of various numbers; it may be iron or brass, but must be perfectly annealed, so as to retain no elasticity or "spring;" *b*, several files of different sizes; *c*, some slender, straight, brad awls; *d*, cutting pliers; *e*, setting needles, merely sewing or darning needles stuck in a light wooden handle, for dressing individual feathers; *f*, plenty of pins (the long, slender insect pins used by entomologists are the best) and sewing thread; *g*, an assortment of glass eyes. (The fixtures and decorations are noticed, beyond, as occasion for their use arises.)

There are two principal methods of mounting, which may be respectively styled *soft* stuffing and *hard* stuffing. In the former, a wire framework, consisting of a single anterior piece passing in the middle line of the body up through the neck and out at top of the head, is immovably joined behind with two pieces, one passing through each leg; around this naked forked frame soft stuffing is introduced, bit by bit, till the proper contour of the skin is secured. I have seen very pretty work of this kind, particularly on small birds; but I consider it much more difficult to secure satisfactory results in this way than by hard stuffing, and I shall therefore confine attention to the latter. This method is applicable to all birds, is readily practised, facilitates setting of the wings, arranging of the plumage, and giving of any desired attitude. In hard stuffing, you make a firm ball of tow rolled upon a wire of the size and shape of the bird's body and neck together; you introduce this whole, afterwards running in the leg wires and clinching them immovably in the mass of tow.

Having your empty skin in good shape, as already described; cut three pieces of wire of the right[1] size; one piece somewhat longer than the whole bird, the other pieces two or three times as long as the whole leg of the bird. File one end of each piece to a fine sharp point; try to secure a three-edged cutting point like that of a surgical needle, rather than the smooth punching point of a sewing-needle, as the former perforates more readily. Have these wires perfectly straight.[2] Bend a small portion of the unfiled end of the longer wire irregularly upon itself, as a convenient nucleus for the ball of tow.[3] Take fine clean tow, in loose dossils, and wrap it round and round the wire nucleus, till you make a firm ball, of the size and shape of the bird's body and neck. Study the contour of the skinned body: notice the swelling breast-muscles, the arch of the lower back, the hollow between the furcula into which the neck, when naturally curved, sinks. Everything depends upon correct shaping of the artificial body; if it be misshapen, no art can properly adjust the skin over it. Firmness of the tow ball and accurate contour may both be secured by wrapping the mass with sewing thread, loosening here, tightening there, till the shape is satisfactory. Be particular to secure a *smooth* superficies; the skin in drying will shrink close to the stuffing, disclosing its irregularities, if there be any, by the maladjustment of the plumage that will ensue. Observe especially that the neck, though the direct continuation of the backbone, dips at its lower end into the hollow of the merry-thought, and so virtually begins there instead of directly between the shoulders.

---

[1] The right size is the smallest that will support the whole weight of the stuffing and skin without bending, when a piece is introduced into each leg. If using too thick wire, you may have trouble in thrusting it through the legs, or may burst the tarsal envelope.

[2] If accidentally kinky, the finer sizes of wire may be readily straightened by drawing strongly upon them so as to stretch them a little. Heavier wire must be hammered out straight.

[3] Cotton will not do at all; it is too soft and elastic, and moreover will not allow of the leg wires being thrust into it and there clinched.

The three mistakes most likely to be made by a beginner are, getting the body altogether too large, not firm enough, and irregular. When properly made, it will closely resemble the bird's body and neck, with an inch or several inches of sharp-pointed wire protruding from the anterior extremity of the neck of tow. You have now to introduce the whole affair into the skin. With the birdskin on its back, the tail pointing to your right elbow, and the abdominal opening as wide as possible, hold the tow body in position relative to the skin; enter the wire, pass it up through the neck, bring the sharp point exactly against the middle of the skull, pierce skull and skin, causing the wire to protrude some distance from the middle of the crown. Then by gentle means insinuate the body, partly pushing it in, partly drawing the skin over it, till it rests in its proper position. This is just like drawing on a tight kid glove, and no more difficult. See that the body is *completely* encased; you must be able to close the abdominal aperture entirely. You have next to wire the legs. Enter the sharp point of one of the leg-wires already prepared, exactly at the centre of the sole of the foot, thrusting it up inside the tarsal envelope the whole length of the "shank," thence across the heel joint [1] and up along the next bone of the leg, still inside the skin. The point of the wire will then be seen within the skin, and may be seized and drawn a little further through, and you will have passed a wire entirely out of sight all the way along the leg. The end of the wire is next to be fixed immovably in the tow ball. Thrust it in at the point where the knee, in life, rests against the side of the body.[2] Bring the point to view, bend it over and reinsert it till it sticks fast. There are no special directions to be given here; fasten the wire in any way that effectually prevents "wabbling." You may find it convenient to wire both legs before fastening either, and then clinch them by twisting the two ends together. But remember that the leg-wires may be fixed respecting each other, yet permit a see-saw motion of the body upon them. This must not be; the body and legs must be fixed upon a jointless frame. Having secured the legs, close the abdominal opening nicely, either by sewing or pinning; you may stick pins in anywhere, as freely as in a pin-cushion; the feathers hide their heads. Stick a pin through the pope's nose to fix the tail in place.

All this while the bird has been lying on its back, the neck stretched straight in continuation of the body, wired stiffly, the legs straddling wide apart, straight and stiff, the wings lying loosely, half-spread. Now bring the legs together, parallel with each other, and make the sharp bend at the heel joint that will bring the feet naturally under the belly (over it, as the bird lies on its back). Pick up the bird by the wires that project from the soles and set it on its stand, by running the wires through holes bored the proper distance apart, and then securing the ends by twisting. The temporary stand that you use for this purpose should have a heavy or otherwise firm support, so as not easily to overturn during the subsequent manipulations. At this stage the bird is a sorry-looking object; but if you have stuffed correctly and wired securely, it will soon improve. Begin by making it *stand* properly. The common fault here is placing the tarsi too nearly perpendicular. Perching birds, constituting the majority, habitually stand with the tarsi more nearly *horizontal* than perpendicular, and generally keep the tarsi parallel with each other. Wading and most walking birds stand with the legs more nearly upright and straight. Many swimming birds straddle a little; others rarely if ever. See that the toes clasp the perch naturally, or are properly spread on the flat surface. Cause the flank feathers to be correctly adjusted over the tibiæ (and here I will remark that with most birds little, if any, of the tibiæ shows in life), the heel joint barely, if at all, projecting

---

[1] There is occasionally difficulty in getting the wire across this joint, from the point sticking into the enlarged end of the shin-bone. In such case, take stout pliers and pinch the joint till the bone is smashed to fragments. The wire will then pass and the comminution will not show. If there is any trouble in passing the wire through the tarsus, bore a hole for it with a brad awl.

[2] This point is further forward and more belly-ward than you might suppose. Observe the skinned body again, and see where the lower end of the thigh lies. If you insert the wire too far back, you cannot by any possibility balance the bird naturally on its perch; it will look in imminent danger of toppling over.

from the general plumage. It is a common fault of stuffing not to draw the legs closely enough to the body. Above all, look out for the centre of gravity; though you have really fastened the bird to its perch, you must not let it look as if it would fall off if the wires slipped; it must appear to rest there of its own accord. Next, give the head and neck a preliminary setting, according to the attitude you have determined upon. This will bring the plumage about the shoulders in proper position for the setting of the wings, to which you may at once attend. If the body be correctly fashioned and the skin of the shoulders duly adjusted over it, the wings will fold into place without the slightest difficulty. All that I have said before about setting the wings in a skin applies here as well; but in this case they will not *stay* in place, since they fall by their own weight. They must be pinned up. Holding the wing in place, thrust a pin steadily through near the wrist joint, into the tow body. Sometimes another pin is required to support the weight of the primaries; it may be stuck into the flank of the bird, the outer quill feather resting directly upon it. With large birds a sharp pointed wire must replace the pin. When properly set, the wing-tips will fall together or symmetrically opposite each other, the quills and coverts will be smoothly imbricated, the scapular series of feathers will lie close, and no bare space will show in front of the shoulder. Much depends upon the *final* adjustment of the head. The commonest mistake is getting it too far away from the body. In the ordinary attitudes of most birds little neck shows, the head appearing nestled upon the shoulders. If the neck appears too long, it is not to be contracted by pushing the head directly down upon it, but by making an S curve of the neck. No precise directions can be given for the set of the head, but you may be assured it is a delicate, difficult matter; the slightest turn of the bill one way or another may alter the whole expression of the bird. You will of course have determined beforehand upon your attitude, upon what you wish the bird to appear to be doing; then, let your meaning be pointed by the bird's bill.

On the general subject of striking an attitude, and giving expression to a stuffed bird, little can be said to good purpose. If you are to become proficient in this art, it will come from your own study of birds in the field, your own good taste and appreciation of bird life. The manual processes are easily described and practised; it is easy to grind paint, I suppose, but not so to be an artist. I shall therefore only follow the above account of the general processes with some special practical points. After "attitudinizing" to your satisfaction, or to the best of your ability, the plumage is to be carefully "dressed." Feathers awry may be set in place with a light spring forceps, or needles fixed in a handle, one by one if necessary. When no individual feather seems out of place, it often occurs that the general plumage has a loose, slovenly aspect. This is readily corrected by wrapping with fine thread. Stick a pin into the middle of the back, another into the breast, and perhaps others, elsewhere. Fasten the end of a spool of sewing cotton to one of the pins, and carry it to another, winding the thread about among the pins, till the whole surface is covered with an irregular network. Tighten to reduce an undue prominence, loosen over a depression; but let the wrapping as a whole be light, firm, and even. This procedure, nicely executed, will give a smoothness to the plumage not otherwise attainable, and may be made to produce the most exquisite curves, particularly about the head, neck, and breast. The thread should be left on till the bird is perfectly dry; it may then be unwound or cut off, and the pins withdrawn. When a particular patch of skin is out of place, it may often be pulled into position and pinned there. You need not be afraid of sticking pins in anywhere: they may be buried in the plumage and left there, or withdrawn when the skin is dry. In addition to the main stuffing, a little is often required in particular places. As for the legs, they should be filled out in all such cases as I indicated earlier in this section; small birds require no such stuffing. It is necessary to fill out the eyes so that the lids rest naturally; it may be done as heretofore directed, or by putting in pledgets of cotton from the outside. A little nice stuffing is generally required about the upper throat. To stuff a bird with spread wings requires a special process, in most cases. The wings are to be wired, exactly as directed for the legs; they may then be placed in any shape. But with most small birds, and those with short wings, simple pinning in the half-spread position indicating fluttering will suffice; it is readily accomplished with a long, slender insect pin. I have already spoken of fixing the tail by pinning or wiring the pope's nose to the tow body; it may be thus fixed at any desired elevation or depression. There are two ways of spreading the tail. One is to run a pointed wire through the quills, near their base, where the wire will be hidden by the coverts; each feather may be set at any required distance from the next by sliding it along this wire. This method is applicable to large birds; for small ones the tail may be fixed with the desired spread by enclosing it near its base in a split match, or two slips of card-board, with the ends tied together. This holds the feathers until they dry in position, when it is to be taken off. Crests may be raised, spread, and displayed on similar principles. A small crest, like that of a cardinal or cherry bird, for instance, may be held up till it dries in position by sticking in behind it a pin with a little ball of cotton on its head. It is sometimes necessary to make a bird's toes grasp a support by tying them down to it till they dry. The toes of waders that do not lie evenly on the surface of the stand may be tacked down with small brads. The bill may be pinned open or shut, as desired, by the method already given. Never paint or varnish a bird's bill or feet.

Substitution of an artificial eye for the natural one is essential for the good looks of a specimen. Glass eyes, of all sizes and colors, may be purchased at a moderate cost. The pupil is always black; the iris varies. You will, of course, secure the proper color if it is known, but if not, put in a dark brown or black eye. It is well understood that this means nothing; it is purely conventional. Yellow is probably the next most common color; then come red, white, blue, and green, perhaps approximately in this order of frequency. But do not use these striking colors at hap-hazard; sacrificing truth, perhaps, to looks. Eyes are generally inserted after the specimen is dry. Remove a portion of the cotton from the orbit, and moisten the lids till they are perfectly pliable; fix the eye in with putty or wet plaster of Paris, making sure that the lids are naturally adjusted over it. It goes in obliquely, like a button through a button-hole. Much art may be displayed in this little matter, making a bird look this way or that, to carry out the general "expression."

On finishing a specimen, set it away to dry; the time required varies, of course, with the weather, the size of the bird, its fatness, etc. The more slowly it dries the better; there is less risk of the skin shrinking irregularly. You will often find that a specimen set away with smooth plumage and satisfactory curves dries more or less out of shape, perhaps with the feathers raised in places. I know of no remedy; it may, in a measure, be prevented by scrupulous care in making the body smooth and firm, and in securing slow, equable drying. When perfectly dry remove the wrapping, pull out the superfluous pins or wires, nip off the others so short that the ends are concealed, and insert the eyes. The specimen is then ready to be transferred to its permanent stand.

Fixtures for the display of the object of course vary interminably. We will take the simplest case, of a large collection of mounted birds for public exhibition. In this instance, uniformity and simplicity are desiderata. "Spread eagle" styles of mounting, artificial rocks and flowers, etc., are entirely out of place in a collection of any scientific pretensions, or designed for popular instruction. Besides, they take up too much room. Artistic grouping of an extensive collection is usually out of the question; and when this is unattainable, half-way efforts in that direction should be abandoned in favor of severe simplicity. Birds look best on the whole in uniform rows, assorted according to size, as far as a natural classification allows. They are best set on the plainest stands, with circular base and a short cylindrical crossbar on a lightly turned upright. The stands should be painted dead-white, and be no larger than is necessary for secure support; a neat stiff paper label may be attached. A small collection of birds, as an ornament to a private residence, offers a different case; here, variety of attitude and appropriate imitation of the birds' natural surroundings are to be secured. A miniature tree, on which a number of birds may be placed, is readily made. Take stout wire, and by bending it, and attaching other pieces, get the framework of the tree of the desired size, shape, and number of perches. Wrap it closely with tow to a proper calibre, remembering that the two forks of a stem must be together only about as large as the stem itself. Gather a basket full of lichens and tree moss; reduce them to coarse powder by rubbing with the hands; besmear the whole tree with mucilage or thin glue, and sift the lichen powder on it till the tow is completely hidden. This produces a very natural effect, which may be heightened by separately affixing larger scraps of lichen, or little bunches of moss; artificial leaves and flowers may be added at your taste. The groundwork may be similarly prepared with a bit of board, made adhesive and bestrewn with the same substance; grasses and moss may be added. If a flat surface is not desired, soak stout pasteboard till it can be moulded in various irregular elevations and depressions; lay it over the board and decorate it in the same way. Rocks may be thus nicely imitated, with the addition of powdered glass of various colors. Such a lot of birds is generally enclosed in a cylindrical glass case with arched top. As it stands on a table to be viewed from different points, it must be presentable on all sides. A niche in parlor or study is often fitted with a wall-case, which, when artistically arranged, has a very pleasing effect. As such cases may be of considerable size, there is opportunity for the display of great taste in grouping. A place is not to be found for a bird, but a bird for the place, — waders and swimmers below on the ground, perchers on projecting rests above. The surroundings may be prepared by the methods just indicated. One point deserves attention here; since the birds are only viewed from the front, they may have a "show-side" to which everything else may be sacrificed. Birds are represented flying in such cases more readily than under other circumstances, supported on a concealed wire inserted in the back of the case. I have seen some very successful attempts to represent a bird swimming, the duck being let down part way through an oval hole in a plate of thick glass, underneath which were fixed stuffed fishes, shells, and seaweed. It is hardly necessary to add that in all ornamental collections, labels or other scientific machinery must be rigorously suppressed.

Transportation of mounted birds offers obvious difficulty. Unless very small, they are best secured immovably inside a box by screwing the foot of the stands to the bottom and sides, so that they stay in place without touching each other. Or, they may be carefully packed in cotton, with or without removal of the stands. Their preservation from accidental injury depends upon the same care that is bestowed upon ordinary fragile ornaments of the parlor. The ravages of insects are to be prevented upon the principles to be hereafter given in treating of the preservation of birdskins.

§ 8. — MISCELLANEOUS PARTICULARS.

Determination of Sex. — This is an important matter, which must never be neglected. For although many birds show unequivocal sexual distinctions of size, shape, and color, like those of the barnyard cock and hen for instance, yet the outward characteristics are more frequently obscure, if not altogether inappreciable, on examination of the skin alone. Young birds, moreover, are usually indistinguishable as to sex, although the adults of the same species may be easily recognized. The rule results, that the sexual organs should be examined as the only infallible indices. The essential organs of masculinity are the *testicles*; similarly, the *ovaries* contain the essence of the female nature. However similar the accessory sexual structures may be, the testicles and ovaries are always distinct. The male organs of birds never leave the cavity of the belly to fill an external bag of skin (*scrotum*) as they do among mammalia; they remain within the abdomen, and lie in the same position as the ovaries of the female. Both these organs are situated in the belly opposite what corresponds to the "small of the back," bound closely to the spine, resting on the front of the kidneys near their fore end. The testicles are a pair of subspherical or rather ellipsoidal bodies, usually of the same size, shape, and color, and are commonly of a dull opaque whitish tint. They always lie close together. A remarkable fact connected with them is, that they are not always of the same size in the same bird, being subject to periodical enlargement during the breeding season, and corresponding atrophy at other seasons. Thus the testicles of a house sparrow, no bigger than a pin's head in winter, swell to the size of peas in April. The ovary (for although this organ is paired originally, only one is usually functionally developed in birds) will be recognized as a flattish mass of irregular contour, and usually whitish color; when inactive, it simply appears of finely granular structure which may require a hand lens to be made out; when producing eggs, its appearance is unmistakable. Both testis and ovary may further be recognized by a thread leading to the end of the lower bowel, — in one case the sperm-duct, in the other the oviduct; the latter is usually much the more conspicuous, as it at times transmits the perfect egg. There is no difficulty in reaching the site of these organs. Lay the bird on the left side, its belly toward you: cut with the scissors through the belly-walls diagonally from anus to the root of the last rib, or further, snipping across a few of the lower ribs, if these continue far down, as they do in a loon for instance. Press the whole mass of intestines aside collectively, and you at once see to the small of the back. There you observe the kidneys, — large, lobular, dark reddish masses moulded into the concavity of the sacrum (or back middle bone of the pelvis); and on their surface, towards their fore end, lie testes or ovary, as just described. The only precaution required is, not to mistake for testicles a pair of small bodies capping the kidneys. These are the *adrenals* or " supra-renal capsules," — organs whose function is unknown, but with which at any rate we have nothing to do in this connection. They occur in both sexes, and if the testicles are not immediately seen, or the ovary not at once recognized, they might easily be mistaken for testicles. Observe, that instead of lying in front, they *cap* the kidneys; that they are usually yellowish instead of opaque whitish; and that they have not the firm, smooth, regular sphericity of the testicles. The testes, however, vary more in shape and color than might be expected, being sometimes rather oblong or linear, and sometimes grayish or livid bluish, or reddish. There is occasionally but one. The sex determined, use the sign ♂ or ♀ to designate it, as already explained. In the very rare cases of impotence or sterility among birds, of course no organs will be observed; but I should dislike to become responsible for such labelling without very careful examination. The organs of a small bird out of the breeding season are never conspicuous, but may always be found on close scrutiny, unless the parts are disintegrated by a shot.

Recognition of Age is a matter of ornithological experience requiring in many or most cases great familiarity with birds for its even approximate accomplishment. There are, however, some unmistakable signs of immaturity, even after a bird has become full-feathered, that persist for at least one season. These are, in the first place, a peculiar soft fluffy " feel" of the plumage; the feathers lack a certain smoothness, density, and stiffening which they subsequently acquire. Secondly, the bill and feet are softer than those of the adults; the corners of the mouth are puffy and flabby, the edges and point of the bill are dull, and the scales, etc., of the legs are not sharply cut. Thirdly, the flesh itself is tender and pale colored. These are some of the points common to all birds, and are independent of the special markings that belong to the youth of particular species. Some birds are actually larger for a while after leaving the nest, than in after years when the frame seems to shrink somewhat in acquiring the compactness of senility. On the other hand, the various members, especially the bill and feet, are proportionally smaller at first. Newly growing quills are usually recognized on sight, the barrel being dark colored and full of liquid, while the vanes are incomplete. In studying, for example, the shape of a wing or tail, there is always reason to suspect that the natural proportions are not yet presented, unless the quill is dry, colorless, and empty, or only occupied with shrunken white pith.

Examination of the Stomach frequently leads to interesting observations, and is always worth while. In the first place, we learn most unquestionably the nature of the bird's food, which is a highly important item in its natural history. Secondly, we often secure valuable specimens in other departments of zoölogy, particularly entomology. Birds consume incalculable numbers of insects, the harder kinds of which, such as beetles, are not seldom found intact in their stomachs; and a due percentage of these represent rare and curious species. The gizzards of birds of prey, in particular, should always be inspected, in search of the small mammals, etc., they devour; and even if the creatures are unfit for preservation, we at least learn of their occurrence, perhaps unknown before in a particular region. Mollusk-feeding and fish-eating birds yield their share of specimens. The alimentary canal is often the seat of parasites of various kinds, interesting to the helminthologist; other species are to be found under the skin, in the body of muscle, in the brain, etc. Most birds are also infested with external parasites of many kinds, so various that almost every leading species has its own sort of louse, tick, etc. Since these creatures are only at home with a *live* host, they will be found crawling on the surface of the plumage, preparing for departure, as soon as the body cools after death. There is thus much to learn of a bird aside from what the prepared specimen teaches, and moreover apart from regular anatomical investigations. Whenever practicable, brief items should be recorded on the label, as already mentioned.

Restoration of Poor Skins. — If your cabinet be a "general" one, comprising specimens from various sources, you will frequently happen to receive skins so badly prepared as to be unpleasant objects, besides failing to show their specific characters. There is of course no supplying of missing parts or plumage; but if the defect be simply deformity, this may usually be in a measure remedied. The point is simply to *relax* the skin, and then proceed as if it were freshly removed from the bird; it is what bird-stuffers constantly do in mounting birds from prepared skins. The relaxation is effected by moisture alone. Remove the stuffing; fill the interior with cotton or tow saturated with water, yet not dripping; put pads of the same under the wings; wrap the bill and feet, and set the specimen in a damp, cool place. Small birds soften very readily and completely; the process may be facilitated by persistent manipulation. This is the usual method, but there is another, more thorough and more effective; it is exposure to a vapor-bath. The appointments of the kitchen stove furnish all the apparatus required for an extempore "steamer;" the regular fixture is a tin vessel much like a washboiler, with closed lid, false bottom, and stopcock at lower edge. On the false bottom is placed a heavy layer of gypsum, completely saturated with water; the birds are laid on a perforated tray above it; and a gentle heat is maintained over a stove. The vapor penetrates every part of the skin, and completely relaxes it, without actually wetting the feathers. The time required varies greatly of course; observation is the best guide. The chief precaution is not to let the thing get too hot. Professor Baird has remarked that crumpled or bent feathers may have much of their original elasticity restored by dipping in hot water. Immersion for a few seconds suffices, when the feathers will be observed to straighten out. Shaking off superfluous water, they may be simply left to dry, or they may be dried with plaster. The method is chiefly applicable to the large feathers of the wings and tail. Soiled plumage of dried skins may be treated exactly as in the case of fresh skins.

Mummification. — As before mentioned, decay may be arrested by injections of carbolic acid and other antiseptics; if the tissues be sufficiently permeated with these substances, the body will keep indefinitely; it dries and hardens, becoming, in short, a "mummy." Injection should be done by the mouth and vent, be thorough, and be repeated several times as the fluid dries in. It is an improvement on this to disembowel and fill the belly with saturated tow or cotton. Due care should be taken not to soil the feathers in any case, nor should the carbolic solution come in contact with the hands, for it is a powerful irritant poison. I mention the process chiefly to condemn it as an atrocious one; I cannot imagine what circumstances would recommend it, while only an extreme emergency could justify it. It is further objectionable because it appears to lend a dingy hue to some plumages, and to dull most of them perceptibly. Birds prepared — rather unprepared — in this way, may be relaxed by the method just described, and then skinned; but the operation is rather difficult.

Wet Preparations. — By this term is technically understood an object immersed in some preservative fluid. It is highly desirable to obtain more information of birds than their stuffed skins can ever furnish, and their structure cannot be always examined by dissection on the spot. In fact, a certain small proportion of the birds of any protracted or otherwise " heavy " collecting may be preferably and very profitably preserved in this way. Specimens in too poor plumage to be worth skinning may be thus utilized; so may the *bodies* of skinned birds, which, although necessarily defective, retain all the viscera, and also afford osteological material. Alcohol is the liquid usually employed, and, of all the various articles recommended, seems to answer best on the whole. I have used a very weak solution of chloride of zinc with excellent results; it should not be strong enough to show the slightest turbidity. As glass bottles are liable to break when travelling, do not fit corners, and offer practical annoyance about corkage, rectangular metal cans, preferably of copper, with screw-lid opening, are advisable. They are to be set in small, strong, wooden boxes, made to leave a little room for the lid wrench, muslin bags for doing up separate parcels, parchment for labels, etc. Unoccupied space in the cans should be filled with tow or a similar substance, to prevent the specimens from swashing about. Labelling should be on parchment; the writing should be perfectly dry before immersion; india-ink is the best. Skinned bodies should be numbered to correspond with the dried skin from which taken; otherwise they may not be identifiable. Large birds thrown in unskinned should have the belly opened, to let in the alcohol freely. Birds may be skinned, after being in alcohol, by simply drying them: they often make fair specimens. They are best withdrawn by the bill, that the "swash" of the alcohol at the moment of emersion may set the plumage all one way, and hung up to dry untouched. Watery moisture that may remain after evaporation of the alcohol may be dried with plaster.

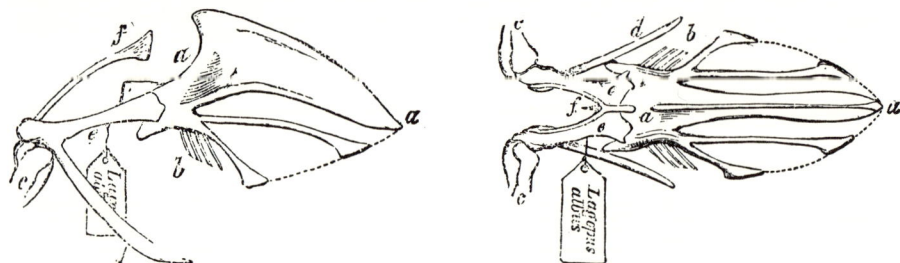

FIGS 1, 2. — Views of sternum and pectoral arch of the ptarmigan, *Lagopus albus*, reduced; after A. Newton. 1, lateral view, with the bones upside down; 2, viewed from below. *a*, sternum or breast-bone, showing two long slender lateral processes; *b*, ends of sternal ribs; *c*, ends of humerus, or upper arm-bone, near the shoulder-joint; *d*, scapula, or shoulder-blade; *e*, coracoid; *f*, merry-thought, or furculum (clavicles).

Osteological and other Preparations (figs. 1–3). — While complete skeletonizing of a bird is a special art of some difficulty, and one that does not fall within the scope of this treatise, I may mention two bony preparations very readily made, and susceptible of rendering

MISCELLANEOUS PARTICULARS.

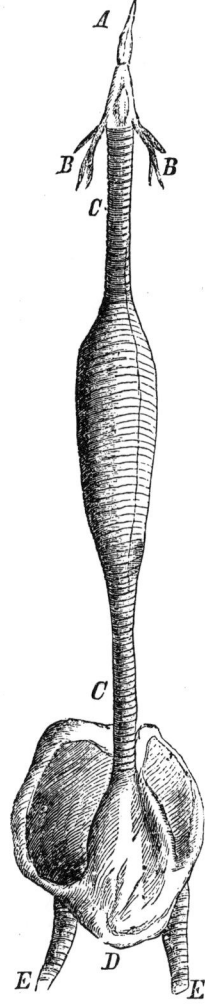

FIG. 3. — Trachea or windpipe of the male red-breasted merganser, *Mergus serrator*, about ½ nat. size, viewed from above (behind); after Newton. A, tongue; B B, its attachments; C C, windpipe, dilated in the middle and swelling below into a bony box, D; E E, bronchial tubes, going to lungs.

ornithology essential service. I refer to the skull, and to the breast-bone with its principal attachments. These parts of the skeleton are, as a rule, so highly characteristic that they afford in most cases invaluable zoölogical items. To save a skull is of course to sacrifice a skin, to all intents; but you often have mutilated or decayed specimens that are very profitably utilized in this way. The breast-bone (figs. 1, 2, *a*) excepting when mutilated, is always preservable with the skin, and for "choice" invoices may form its natural accompaniment. You want to remove along with it the *coracoids* (the stout bones connecting the breast-bone with the shoulders, figs. 1, 2, *e*), the merry-thought (figs. 1, 2, *f*) intervening between these bones, and the shoulder-blades (figs. 1, 2, *d*), all without detachment from each other, for these bones collectively constitute the "shoulder-girdle," or *scapular arch*. Slice off the large breast muscles close to the bone, and divide their insertions into the wing-bones (*c*); scrape or cut away the muscles that tie the shoulder-blades to the chest; snip off the ribs (figs. 1, 2, *b*) close to the side of the breast-bone; sever a tough membrane usually found between the prongs of the wish-bone; then, by taking hold of the *shoulders* (figs. 1, 2, at *c*), you can lift out the whole affair, dividing some slight connections underneath the bone and behind it. The following points require attention: the breast-bone often has long slender processes behind and on the sides (the common fowl and the ptarmigan are extreme illustrations of this, as shown in the figures), liable to be cut by mistake for ribs, or to be snapped; the shoulder-blades usually taper to a point, easily broken off; the merry-thought is sometimes very delicate or defective. When travelling, it is generally not advisable to make perfect preparations of either skull or sternum; they are best dried with only superfluous flesh removed, and besprinkled with arsenic. The skull, if perfectly cleaned, is particularly liable to lose the odd-shaped, pronged bones that hinge the jaw, and the freely movable pair that push on the palate from behind. Great care should be exercised respecting the identification of these bones, particularly the sternum, which should invariably bear the number of the specimen to which it belongs; the label should be tied to the coracoid bone. A skull is more likely to be able to speak for itself, and, besides, is not usually accompanied by a skin; nevertheless, any record tending to facilitate its recognition should be duly entered on the register. There are methods, with which I am not familiar, of making elegant bony preparations. You may secure very good results by simply boiling the bones; or, what is perhaps better, macerating them in water till the flesh is completely rotted away, and then bleaching them in the sun. A little potassa or soda hastens the process. With breast-bones, if you can stop the process just when the flesh is completely dissolved but the tougher ligaments remain, you secure a "natural" preparation, as it is called; if the ligaments go too, the associate parts of a large specimen may be wired together, those of a small one glued. I think it best, with skulls, to clean them entirely of ligament as well as muscle; for the underneath parts are usually those conveying the most desirable information, and they should not be in the slightest degree obscured. Since in such case the anvil-shaped bones, the palatal cylinders already mentioned, and sometimes other portions come apart, the whole are best kept in a suitable box. I prefer to see a skull with the sheath of the beak removed, though in some cases, particularly of hard-billed birds, it may profitably be left on. The completed preparations should be fully labelled by writing on the bone, in preference to an accompanying or attached paper slip, which *may* be lost. Some object to this, as others do to writing on eggs, that it "defaces" the specimen; but I confess I see in dry bones no beauty but that of utility.

"In many families of birds, as the ducks (*Anatidæ*), the *trachea* or windpipe of the male affords valuable means of distinguishing between the different natural groups, or even species, chiefly by the form of the bony labyrinth, or *bulla ossea*, situated at or just above the divarication of the bronchial tubes. A little trouble will enable the collector in all cases to preserve this organ perfectly, as represented in the annexed engraving (fig. 3). Before proceeding to skin the specimen, a narrow-bladed knife should be introduced into its mouth and by taking hold of the tongue (*A*) by the fingers or forceps, the muscles (*B B*) by which it is attached to the lower jaw should be severed as far as they can be reached, care being of course taken not to puncture the windpipe (*C C*); and later in the operation of skinning, when dividing the body from the neck or head, not to cut into or through it. This done, the windpipe can be easily withdrawn entire and separated from the neck, and then the sternal apparatus being removed as before described, its course must be traced to where, after branching off in a fork (*D*), the bronchial tubes (*E E*) join the lungs. At these latter points it is to be cut off. Then rinsing it in cold water, and leaving it to dry partially, it may, while yet pliant, be either wrapped round the sternum, or coiled up and labelled separately." — (*A. Newton.*)

§ 9. — COLLECTION OF NESTS AND EGGS.

Ornithology and Oölogy are twin studies, or rather one includes the other. A collection of nests and eggs is indispensable for any thorough study of birds; and many persons find peculiar pleasure in forming one. Some, however, shrink from "robbing birds' nests" as something particularly cruel; a sentiment springing, no doubt, from the sympathy and deference that the tender office of maternity inspires; but with all proper respect for the humane emotion, it may be said simply, that birds'-nesting is not nearly so cruel as bird-shooting. What I said in a former section, in endeavoring to guide search for birds, applies in substance to hunting for their nests; the essential difference is, that the latter are of course stationary objects, and consequently more liable to be overlooked, other things being equal, than birds themselves. Most birds nest on trees or bushes; many on the ground and on rocks; others in hollows. Some build elegant, elaborate structures, endlessly varied in details of form and material; others make no nest whatever. In this country, egging is chiefly practicable in May and during the summer; but some species, particularly birds of prey, begin to lay in January, while, on our southern border at least, the season of reproduction is protracted through September; so there is really a long period for search. Particular nests, of course, like the birds that build them, can only be found through ornithological knowledge; but general search is usually rewarded with a varied assortment. The best clew to a hidden nest is the actions of the parents; patient watchfulness is commonly successful in tracing the bird's home. As the science of oölogy has not progressed to the point of determining from the nests and eggs to what bird they belong, in even a majority of cases, the utmost care in authentication is indispensable. To be worth anything, not to be worse than worthless in fact, an egg must be identified beyond question; must be not only unsuspected, but above suspicion. A shade of suspicion is often attached to dealers' eggs; not necessarily implying bad faith or even negligence on the dealers' part, but from the nature of the case. It is often extremely difficult to make an unquestionable determination, as for instance when numbers of birds of similar habits are breeding close together; or even impossible, as in case the parent eludes observation. Sometimes the most acute observer may be mistaken, circumstances appearing to prove a parentage when such is not the fact. It is in general advisable to secure the parent with the eggs: if shot or snared on the nest, the identification is simply unquestionable. If you do not yourself know the species, it then becomes *necessary* to secure the specimen, and retain it with the eggs. It is not required to make a perfect preparation; the head, or better, the head and a wing, will answer the purpose. When egging in downright earnest, a pair of climbing irons, a coil of $\frac{3}{8}$ inch rope, and a tin collecting box filled with cotton, become practically indispensable; these are the only field implements required in addition to those already specified.

Preparing Eggs. For blowing eggs, a set of special tools is needed. These are "egg-drills," — steel implements with a sharp-pointed conical head of rasping surface, and a slender shaft; several such, of different sizes, are needed; also, blow-pipes of different sizes, a delicate thin pair of scissors, light spring forceps, some little hooks, and a small syringe. They are inexpensive, and may be had of any dealer in naturalists' supplies. (See figs. 4–7.) Eggs

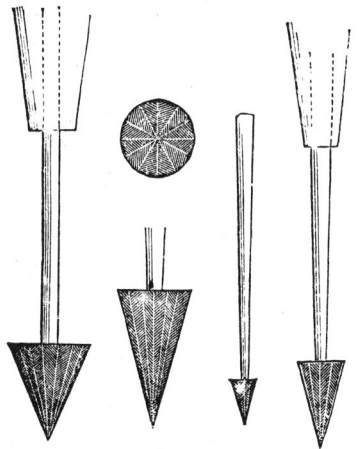

FIG. 4. — Egg-drills, different sizes, nat. size; after Newton.

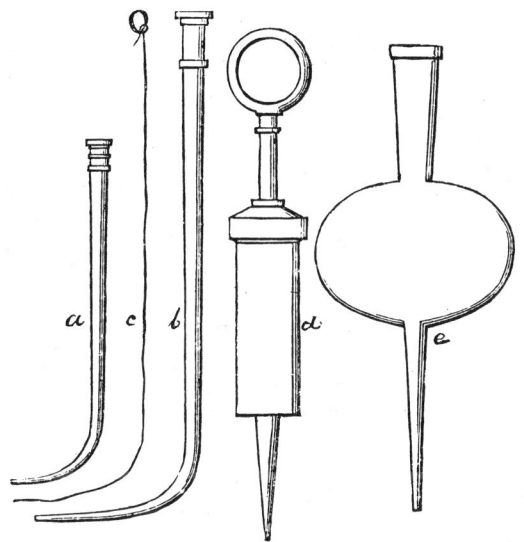

FIG. 5. — Instruments for blowing eggs; after Newton. *a, b,* blow-pipes, $\frac{1}{2}$ nat. size; *c*, wire for cleansing them; *d*, syringe, $\frac{1}{2}$ nat. size (the ring of the handle must be large enough to insert the thumb); *e*, bulbous insufflator, for sucking eggs.

should never be blown in the old way of making a hole at each end; nor are two holes anywhere usually required. Opening should be effected on one side, preferably that showing least conspicuous or characteristic markings. If two are made, they should be rather near together; on the same side at any rate. But one is generally sufficient, as the fluid contents can escape around the blow-pipe. Holding the egg gently but steadily in the fingers,[1] apply the point of

---

[1] The usual method of emptying eggs through one small hole is doubtless supposed to be a very modern trick; but it dates back at least to 1828, when M. Danger proposed "a new method of preparing and preserving eggs for the cabinet," which is practically the one now followed, though he used a three-edged needle to prick the hole, instead of our modern drill, and did not appear to know some of our ways of managing the embryo. I make this reference to his article to call attention to one of the tools he recommends, which I think would prove useful, as being better than the fingers for holding an egg during drilling and blowing. The simple instrument will be understood from a glance at the figure given in the Nuttall Bulletin, iii, 1878, p. 191. The oval rings are covered with light fabric, like mosquito-netting or muslin, and do not touch the egg, which is held lightly but securely in the netting. The cost would be trifling, and danger might be avoided by Danger's method.

the drill perpendicularly to the surface, unless it be preferred to prick with a needle first. A twirling motion of the instrument gradually enlarges the opening by filing away the shell, and so bores a smooth-edged circular hole. This should be no larger than is required to insert the blow-pipe loosely, with room for the contents to escape around it. Nor is it always necessary to *insert* the pipe; a fine stream of water may be easily injected by holding the instrument close to the egg, but not quite touching. The blowing should be continuous and equable, rather than forcible; a strong puff easily bursts a delicate egg. Be sure that all the contents are removed; then rinse the interior thoroughly with clean water, either by taking a mouthful and sending it through a blow-pipe, or with the syringe. Blowing eggs is a rather fatiguing process, more so than it might seem; the cheek muscles soon tire, and the operator actually becomes "blown" himself before long. The operation had better be done over a basin of water, both to receive the contents, and to catch the egg if it slip from the fingers. The membrane lining the shell should be removed if possible. It may be seized by the edge around the hole, with the forceps, and drawn out, or picked out with a bent pin. But this is scarcely to be accomplished in the case of fresh eggs, when the membrane may be simply pared smoothly around the edge

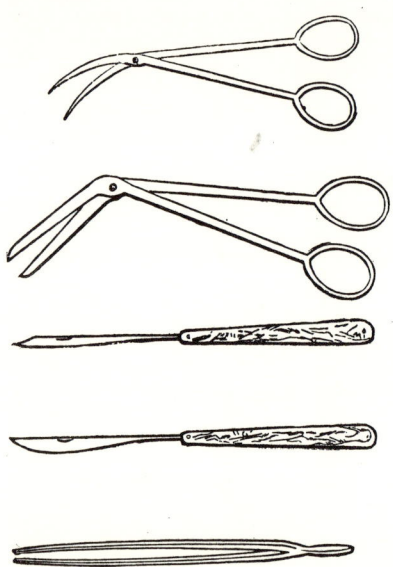

FIG. 6. — Scissors, knives, and forceps, $\frac{1}{2}$ nat. size; after Newton.

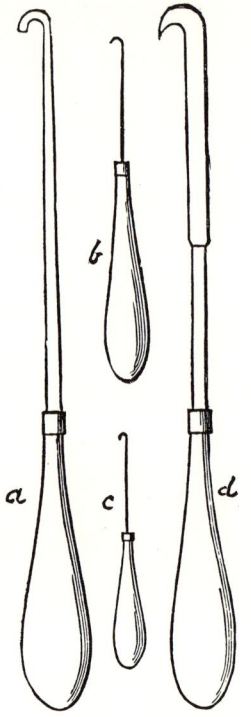

FIG. 7. — Hooks for extracting embryos, nat. size; after Newton. *a*, *b*, *c*, plain hooks; *d*, bill-hook, having cutting edge along the concavity.

of the hole. Eggs that have been incubated of course offer difficulty, in proportion to the size of the embryo. The hole may be drilled, as before, but it must be larger; and as the drill is apt to split a shell after it has bored beyond a certain size of hole, it is often well to prick, with a fine needle, a circular series of minute holes almost touching, and then remove the enclosed circle of shell. This must be very carefully done, or the needle will indent or crack the shell, which, it must be remembered, grows more brittle towards the time of hatching. Well-formed embryos cannot be got bodily through any *hole* that can be made in an egg; they must be extracted piecemeal. They may be cut to pieces with the slender scissors introduced through the hole, and the fragments be picked out with the forceps, hooked out, or blown out. No embryo should be forced through a hole too small; there is every probability that the shell will burst at the critical moment. Addled eggs, the contents of which are thickened or hardened, offer some difficulty, to overcome which persistent syringing and repeated rinsing are required; or it may be necessary to fill them with water, and set them away for such length of time that the contents dissolve by maceration; carbonate of soda is said to hasten the solution; the process may be repeated as often as may be necessary. In no event must any of the animal contents be suffered to remain in the shell. When emptied and rinsed, eggs should be gently wiped dry, and set hole downward on blotting-paper to drain.[1] Broken eggs may be neatly mended, sometimes with a film of collodion, or a bit of tissue paper and paste, or the edges may be simply stuck together with any adhesive substance. Even when fragmentary a rare egg is worth preserving. Eggs should ordinarily be left empty; indeed, the only case in which any filling is admissible is that of a defective specimen to which some slight solidity can be imparted with cotton. It is unnecessary even to close up the hole. It is best, on all accounts, to keep eggs in *sets*, a " set" being the natural clutch, or whatever less number was taken from a nest. The most scrupulous attention must be paid to accurate, complete, and permanent labelling. So important is this, that the undeniable defacing of a specimen, by writing on it, is no offset to the advantages accruing from such fixity of record. It is practically impossible to attach a label, as is done with a bird-skin, and a loose label is always in danger of being lost or displaced. Write on the shell, then, as many items as possible; if done neatly, on the side in which the hole was bored, at least one good " show side " remains. An egg should always bear the same number as the parent, in the collector's record. In a general collection, where separate ornithological and oölogical registers are kept, identification of egg with parent is nevertheless readily secured, by making one the numerator the other the denominator of a fraction, to be simply inverted in its respective application. Thus, bird No. 456, and egg No. 123, are identified by making the former $\frac{456}{123}$ the latter $\frac{123}{456}$. All the eggs of a clutch should have the same number. If the shell be large enough, the name of the species should be written on it; if too small, it should be accompanied by a label, and may have the name indicated by a number referring to a certain catalogue. According to my " Check List," for example, "No. 1" would indicate *Turdus migratorius*. The date of collection is a highly desirable item; it may be abbreviated thus; 3 | 6 | 82 means June 3, 1882. It is well to have the egg authenticated by the collector's initials at least. Since " sets " of eggs may be broken up for distributions to other cabinets, yet permanent indication of the size of the clutch be wanted, it is well to have some method. A good one is to write the number of the clutch on each egg composing it, giving each egg of the set, moreover, its individual number. Supposing for example the clutch No. $\frac{123}{456}$ contained five eggs; one of them would be $\frac{123}{456}$ | 5 | 1: the next $\frac{123}{456}$ | 5 | 2, and so on. But it should be remembered that all such arbitrary memoranda must be systematic, and be accompanied by a key. Eggs may be kept in cabinets of shallow drawers in little pasteboard trays, each holding a set, and containing a paper label on which various items that cannot be traced on the shell are written in full.

FIG. 8. — Nat. size.

[1] *Reinforcing the Eggshell before Blowing.* — Fig. 8 " shows a piece of paper, a number of which, when gummed on to an egg, one over the other, *and left to dry*, strengthen the shell in such a manner that the instruments above described can be introduced through the aperture in the middle and worked to the best advantage, and thus a fully formed embryo may be cut up, and the pieces extracted through a very moderately sized hole; the number of thicknesses required depends, of course, greatly upon the size of the egg, the length of time it has been incubated, and the stoutness of the shell and the paper. Five or six is the least number that it is safe to use. Each piece should be left to dry before the next is gummed on. The slits in the margin cause them to set pretty smoothly, which will be found very desirable; the aperture in the middle of each may be cut out first, or the whole series of layers may be drilled through when the hole is made in the egg. For convenience' sake, the papers may be prepared already gummed, and moistened when put on (in the same way that adhesive postage labels are used). Doubtless, patches of linen or cotton cloth would answer equally well. When the operation is over, a slight application of water (especially if warm) through the syringe will loosen them so that they can be easily removed, and they can be separated from one another, and dried to serve another time. The size represented in the sketch is that suitable for an egg of moderate dimension, such as that of a common fowl. The most effectual way of adopting this method of emptying eggs is by using *very many layers of thin paper and plenty of thick gum*, but this is, of course, the most tedious. Nevertheless, it is quite worth the trouble in the case of really rare specimens, and they will be none the worse for operating upon from the delay of a few days caused by waiting for the gum to dry and harden. The naturalist to whom this method first occurred has found it answer remarkably well in every case that it has been used, from the egg of an eagle to that of a humming-bird, and among English oölogists it has been generally adopted." (*A. Newton*, in Smiths. Misc. Coll. 139, 1860.)

Such trays should all be of the same depth,—half an inch is a convenient depth for general purposes; and of assorted sizes, say from one inch by one and one-half inches up to three by six inches; it is convenient to have the dimensions regularly graduated by a constant factor of, say half an inch, so that the little boxes may be set side by side, either lengthwise or crosswise, without interference. Eggs may also be kept safely, advantageously, and with attractive effect, in the nests themselves, in which a fluff of cotton may be placed to steady them. When not too bulky, too loosely constructed, or of material unsuitable for preservation, nests should always be collected.[1] Those that are very closely attached to twigs should not be torn off. Nests threatening to come to pieces, or too frail to be handled without injury, may be secured by sewing through and through with fine thread: indeed, this is an advisable precaution in most cases. Packing eggs for transportation requires much care, but the precautions to be taken are obvious. I will only remark that there is no safer way than to leave them in their own nests, each wrapped in cotton, with which the whole cavity is to be lightly filled; the nests themselves being packed close enough to be perfectly steady.

§ 10.—CARE OF A COLLECTION.

Well Preserved Specimens will last "forever and a day," so far as natural decay is concerned. I have handled birds in good state, shot back in the twenties, and have no doubt that some eighteenth century preparations are still extant. The precautions against defilement, mutilation, or other mechanical injury, are self-evident, and may be dismissed with the remark, that *white* plumages, especially if at all greasy, require the most care to guard against soiling. We have, however, to fight for our possessions against a host of enemies, individually despicable but collectively formidable,— foes so determined that untiring vigilance is required to ward off their attacks even temporarily, whilst in the end they prove invincible. It may be said that to be eaten up by insects is the natural end of all bird-skins not sooner destroyed.

[1] "*A Plea for the Study of Nests*," made by Mr. Ernest Ingersoll in his excellent "Birds'-Nesting," suits me so well that I will transcribe it. "Whether or not it is worth while to collect nests—for there are many persons who never do so—is, it seems to me, only a question of room in the cabinet. As a scientific study there is far more advantage to be obtained from a series of nests than from a series of eggs. The nest is something with which the will and energies of the bird are concerned. It expresses the character of the workman; is to a certain extent an index of its rank among birds,— for in general those of the highest organization are the best architects,— and give us a glimpse of the bird's mind and power to understand and adapt itself to changed conditions of life. Over the shape and ornamentation of an egg the bird has no control, being no more able to govern the matter than it can the growth of its beak. There is as much difference to me, in the interest inspired, between the nest and the egg of a bird, as between its brain and its skull,— using the word brain to mean the seat of intellect. The nest is always more or less the result of conscious planning and intelligent work, even though it does follow a hereditary habit in its style; while the egg is an automatic production varying, if at all, only as the whole organization of the bird undergoes change. Don't neglect the nests then. In them more than anywhere else lies the key to the mind and thoughts of a bird,— the spirit which inhabits that beautiful frame and bubbles out of that golden mouth. And is it not this inner life,— this human significance in bird nature,— this *soul* of ornithology, that we are all aiming to discover? Nests are beautiful, too. What can surpass the delicacy of the humming-bird's home glued to the surface of a mossy branch or nestling in the warped point of a pendent leaf; the vireo's silken hammock; the oriole's gracefully swaying purse; the blackbird's model basket in the flags; the snug little caves of the marsh wrens; the hermitage-huts of the shy wagtails and ground-warblers, the stout fortresses of the sociable swallows! Moreover, there is much that is highly interesting which remains to be learned about nests, and which can only be known by paying close attention to these artistic masterpieces of animal art. We want to know by what sort of skill the many nests are woven together that we find it so hard even to disentangle; we want to know how long they are in being built; whether there is any particular choice in respect to location; whether it be a rule, as is supposed, that the female bird is the architect, to the exclusion of her mate's efforts further than his supplying a part of the materials. Many such points remain to be cleared up. Then there is the question of variation, and its extent in the architect of the same species in different quarters of its ranging area. How far is this carried, and how many varieties can be recorded from a single district, where the same list of materials is open to all the birds equally? Variation shows individual opinion or taste among the builders as to the suitability of this or that sort of timber or furniture for their dwellings, and observations upon it thus increase our acquaintance with the scope of ideas and habits characteristic of each species of bird."

Insect Pests (Figs. 9, 10, 11, 12) with which we have to contend belong principally to the two families *Tineidæ* and *Dermestidæ* — the former are moths, the latter beetles. The moths are of species identical with, and allied to, the common clothes moth, *Tinea flavifrontella*, the carpet moth, *T. tapetzella*, etc., — small species observed flying about our apartments and museums, in May and during the summer. The beetles are several rather small thick-set species, principally of the genera *Dermestes* and *Anthrenus*. I am able to figure species of these genera, with their larval stages, and of two other genera, *Ptinus* and *Sitodrepa*, through the attentions of Prof. C. V. Riley, the eminent entomologist. The larvæ ("caterpillars" of the moths, and "grubs" of the beetles) appear to be the chief agents of the destruction. The presence of the mature insects is usually readily detected; on disturbing an infested suite of specimens the moths

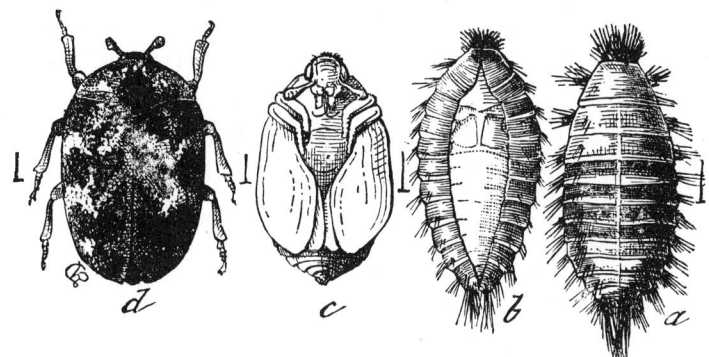

FIG. 9. — *Anthrenus scrofulariæ*, enlarged; the short line shows nat. size. *a*, *b*, larvæ; *c*, pupa; *d*, imago.

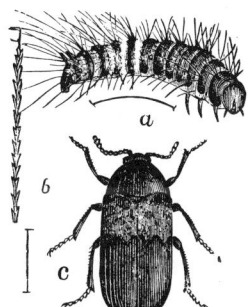

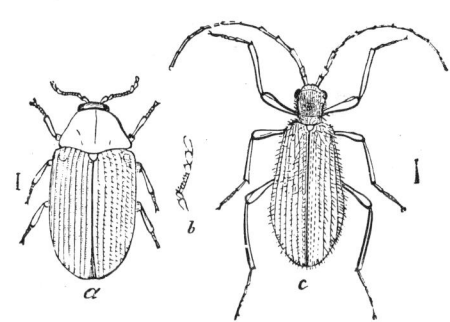

FIG. 10. — *Dermestes lardarius*, enlarged. *a*, larva; *b*, an enlarged hair; *c*, imago.

FIG. 11. — *Sitodrepa panicea*, enlarged. *a*, imago; *b*, its antenna, more enlarged.

FIG. 12. — *Ptinus brunneus*.

flutter about, and the beetles crawl as fast as they can into shelter, or simulate death. The insidious larvæ, however, are not so easily observed, burrowing as they do among the feathers, or in the interior of a skin; whilst the minute eggs are commonly altogether overlooked. But the "bugs" are not long at work without leaving their unmistakable traces. Shreds of feathers float off when a specimen is handled, or fly out on flipping the skin with the fingers, and in bad cases even whole bundles of plumes come away at a touch. Sometimes, leaving the plumage intact, bugs eat away the horny covering of the bill and feet, making a peculiarly unhappy and irreparable mutilation. I suppose this piece of work is done by a particular insect, but if so I do not know what one. It would appear that when the bugs effect lodgment in any one skin, they usually finish it before attacking another, unless they are in great force. We may consequently, by prompt removal of an infested specimen, save further depredations;

nevertheless, the rest become "suspicious," and the whole drawer or box should be quarantined, if not submitted to any of the processes described beyond. Our lines of defence are several. We may mechanically oppose entrance of the enemy; we may meet him with abhorrent odors that drive him off, sicken or kill him, and finally we may cook him to death. I will notice these methods successively, taking occasion to describe a *cabinet* under head of the first.

Cases for Storage or Transportation should be rather small, for several reasons. They are easier to handle and pack. There are fewer birds pressing each other. Particular specimens are more readily reached. Bugs must effect just so many more separate entrances to infest the whole. Small lids are more readily fitted tight. For the ordinary run of small birds I should not desire a box over 18×18×18, and should prefer a smaller one; for large birds, a box just long enough for the biggest specimen, and of other proportions to correspond fairly, is most eligible. Whatever the dimensions, a proper box presupposes perfect jointing; but if any suspicion be entertained on this score, stout paper should be pasted along all the edges, both inside and out. We have practically to do with the lid only. If the lot is likely to remain long untouched, the cover may be screwed very close and the crack pasted like the others. Under other and usual circumstances the lid may be provided with a metal boss fitting a groove lined with india rubber or filled with wax. An excellent case may be made of tin with the lid secured in this manner, and further fortified with a wooden casing. Birdskins entirely free from insects or their eggs, encased in some such secure manner, will remain intact indefinitely; but there is misery in store if any bugs or nits be put away with them.

Cabinets. — As a matter of fact, most collections are kept readily accessible for examination, display, or other immediate use, and this precludes any disposition of them in "hermetical" cases. The most we can do is to secure tight fitting of movable woodwork. The "cabinet" is most eligible for private collections. This is, in effect, simply a bureau, or chest of drawers, protected with folding doors, or a front that may be detached, either of plain wood or sashing for panes of glass. It is simply astonishing how many birdskins of average size can be accommodated in a cabinet that makes no inconvenient piece of furniture for an ordinary room. A cabinet may of course be of any desired size, shape, and style. In general it will be better to put money into excellence of fitting rather than elegance of finish; the handsomest front does not compensate for a crack in the back or for a drawer that hitches. There should not be the slightest flaw in the exterior, and doors should fit so tightly that a puff of air may be felt on closing them. The greatest desideratum of the interior work, next after close fitting yet smooth running of the drawers, is economy of space. This is secured by making the drawers as thin as is consistent with stability; by having them slide by a boss at each end fitting a groove in the side wall, instead of resting on horizontal partitions; and by hinged countersunk handles instead of knobs. I do not recommend, except for a suite of the smallest birds, a multiplicity of shallow drawers, accommodating each one layer of specimens; it is better to have fewer deeper drawers, into which light shallow movable trays are fitted. These trays never need be of stuff over one-eighth or one-fourth of an inch thick, and may have bottoms of stiff pasteboard glued or tacked on. They may vary from one-half inch to two inches in depth, but this dimension should always be some factor of the depth of the drawer, so that a certain number of trays may exactly fill it. They should be just as long as one transverse dimension of the drawer, and rather narrow, so that two or more are set side by side. Finally, though they may be of different depths, they should be of the same length and breadth, so as to be interchangeable. They may simply rest on top of each other, or slide on separate projections inside the drawer. Such trays are extremely handy for holding particular sets of specimens, to be carried to the study table without disturbing the rest of the collection.

If a collection be so extensive that any particular specimen may not be readily hunted up, it will be found convenient to have the drawers themselves labelled with the name of the group within. A collection should always be methodically arranged — preferably according to some approved or supposed natural classification of birds; this is also the readiest mode, since, with some conspicuous exceptions, birds of the same natural group are approximately of the same *size*. If I were desired to suggest proportions for a private cabinet of most general eligibility, I should say four feet high, by three feet wide, by two feet deep, in the clear; this makes a portly yet not unwieldy looking object. It is wide enough for folding-doors, to be secured by bolts at top and bottom, and lock; not so high that the top drawer is not readily inspected; and of proportionate depth. Such a case will take seven drawers six inches deep either of the full width, or in two series with a median partition; these drawers will hold anything up to an eagle or crane. A part of them at least should have a full complement of such trays as I have described, — say three or four tiers of the shallower trays, three trays to a tier, each about two feet long by about a foot wide; and one or two tiers of deeper trays.

To Destroy Bugs. — In our present case prevention is not the best remedy, simply because it is not always practicable; in spite of all mechanical precautions the bugs will get in. We have, therefore, to see what will destroy them, or at least stop their ravages. It is a general rule that any pungent aromatic odor is obnoxious to them, and that any very light powdery substance restrains their movements by getting into the joints and breathing pores. Both these qualities are secured in the ordinary "insect powder," to be had of any leading druggist. It should be lavishly strewn on and among the skins, and laid in the corners of the drawers and trays. Thus employed it proves highly effective, and is on the whole the most eligible substance to use when a collection is constantly handled. Camphor is a valuable agent. Small fragments may be strewn about the drawers, or a lump pinned in mosquito netting in a corner. Benzine is also very useful. A small saucer full may be kept evaporating, or the liquid may be sprinkled — even poured — directly over the skins; it is very volatile and leaves little or no stain. It is, however, obviously ineligible when a collection is in constant use. My friend Mr. Allen informs me he has used sulphide of carbon with great success. The objection to this agent is, that it is a stinking poison; should be used in the open air, to escape the ineffably disgusting and deleterious odors, and its employ is properly restricted to cases for storage. When the bill or feet show they are attacked, further depredation may be prevented by pencilling with a strong solution of corrosive sublimate; a weaker solution, one that leaves no white film, on drying, on a black feather, may even be brushed over the whole plumage. Mr. Ridgway tells me that oil of bitter almonds is equally efficacious. But remember that these poisons must be used with care. Specimens may be buried in coarse refuse tobacco leaves. One or another of these lines of defence will commonly prove successful in destroying or driving off mature insects, and even in stopping the ravages of the larvæ; but I doubt that any such means will kill the "nits." With these we must deal otherwise; and their destruction no less that that of their parents is assured, if we subject them to a high temperature. Baking bird-skins is really the only process that can make us feel perfectly safe. Infected specimens, along with suspected ones, should be subjected to a dry heat, from $212°$ F. up to any degree short of singeing the plumage. This is readily done by putting the birds in a wooden tray in any oven — they must however be watched, unless you have special contrivances for regulating the temperature. How long a time is required is probably not ascertained with precision; it will be well to bake for several hours. When the beetles and larvæ are found completely parched, it may be confidently believed that the unseen eggs are out of the hatching way forever.

Two Items. — One is, that arsenic helps to keep out the bugs, besides preventing decay — a fact that should never be forgotten, and that should give sharper edge to my advice respecting lavish use of the substance at the outset. If it be true, as some state, that bugs can eat arsenic without dying, it is also true that they do not relish it; and in entering a case of skins they will burrow by preference in those holding the least of it. This fact is continually exhibited in large collections, where if two birds be side by side, one being duly arsenicized and the other not so, one will be taken and the other left. My second item, with its proper deduction, will form, I think, a fitting conclusion to this treatise. It is a fact in the natural history of these our pests, that they are fond of peace and quiet,—they do not like to be disturbed at their meals. So they rarely effect permanent lodgment in a collection that is constantly handled, though the doors stand open for hours daily. As a consequence, the degree of our diligence in *studying* birdskins is likely to become the measure of our success in preserving them. I once read a work, by an eminent and learned divine, on the "Moral Uses of Dark Things," under which head the author included everything from earthquakes to mosquitoes. If there be a moral use in the "dark thing" that museum pests certainly are to us, we have it here. The very bugs urge on our work.

Fig. 13.— WILSON'S SCHOOL-HOUSE, NEAR GRAY'S FERRY, PHILADELPHIA. From a drawing by M. S. Weaver, Oct. 22, 1841, received by Elliott Coues, February, 1879, from Malvina Lawson, daughter of Alexander Lawson, Wilson's engraver. See article in the "Penn Monthly," June, 1879, p. 443. The drawing was first engraved on wood, and published, by Thomas Meehan, in the "Gardener's Monthly," August, 1880, p. 248. The present impression is from an electrotype of that wood-cut. The size of the original is 5.10 × 3.95 inches. This reminder of early days of "Field Ornithology" in America may be further attested by the signature of

PART II.

GENERAL ORNITHOLOGY:

AN OUTLINE OF THE

STRUCTURE AND CLASSIFICATION OF BIRDS.

§ 1.—DEFINITION OF BIRDS.

GENERAL ORNITHOLOGY, like Field Ornithology, is a subject with which the student must have some acquaintance, if he would hope to derive either pleasure or profit from the Birds of North America. For any intelligent understanding of this subject, he must become reasonably familiar with the technical terms used in describing and classifying birds, and learn at least enough of the structure of these creatures to appreciate the characters upon which all description and classification is based. Extensive and varied and accurate as may be his random perception of objects of natural history, his knowledge is not scientific, but only empirical, until reflection comes to aid observation, and conceptions of the significance of what he knows are formed by logical processes in the mind. For

Science (Lat. *scire*, to know) is knowledge set in order — knowledge disposed after the rational method that best shows, or tends to show best, the true relations of observed facts. Sound scientific facts are the natural basis of all philosophic truth, and the safest stepping-stones to religious faith — to that wisdom which comes only of knowing the relation which material entities bear to spiritual realities. The orderly knowledge of any particular class of facts — the methodical disposition of observations upon any particular set of objects — constitutes a Special Science. Thus

Ornithology (Gr. ὄρνιθος, *ornithos*, of a bird; λόγος, *logos*, a discourse) is the Science of Birds. Ornithology consists in the rational arrangement and exposition of all that is known of birds, and the logical inference of much that is not known. Ornithology treats of the physical structure, physiological functions, and mental attributes of birds; of their habits and manners; of their geographical distribution and geological succession; of their probable ancestry; of their every relation to one another and to all other animals, including man. The first business of Ornithology is to define its ground — to answer the question,

What is a Bird? — There is every reason to believe that a Bird is a greatly modified Reptile, being the offspring by direct descent of some reptilian progenitor; and there is no reason to suppose that any bird ever had any other origin than by due process of hatching out of an egg laid by its mother after fecundation by its father — just what we believe to have been the invariable method during the period of the world known to human history. There is no reason to believe that any bird was ever originally created and endowed with the characters it now possesses; but that every bird now living is the naturally modified lineal descendant of parents that were less and less like itself, and more and more like certain reptiles, the further removed they were in the line of avian ancestry from such birds as are now living. This is the Darwinian logic of observed facts, upon which the modern Theory of Evolution is based, in opposition to the tradition of the special creation of every species of animal; which latter has no scientific basis whatever, and is consequently accepted as true by few thoughtful persons who are capable of forming independent judgments. Accordingly,

Birds and Reptiles — even those of the present geologic epoch — share so many and such important structural characters, that we unite the two classes, *Aves* and *Reptilia*, in one primary group of *Vertebrata*, or animals with a backbone. This group is called *Sauropsida*, or reptiliform; it is contrasted, on the one hand, with *Ichthyopsida*, or fish-like vertebrates, including Batrachians as well as Fishes; and, on the other, with *Mammalia*, the province of *Vertebrata* which includes Man and all other animals that suckle their young. We find that

Sauropsida (Gr. σαῦρος, *sauros*, a reptile; ὄψις, *opsis*, appearance), or lizard-like Vertebrates, agree with one another, and differ from other animals, in the following important combination of characters, substantially as laid down by Professor Huxley — some of the characters being shared by *Ichthyopsida*, and some by *Mammalia*, but the sum of the characters being distinctive of *Sauropsida*: They are all oviparous (laying eggs hatched outside the body of the parent), or ovoviviparous (laying eggs hatched inside the body of the parent), being never viviparous (bringing forth alive young nourished before birth by the blood of the mother). The embryo develops those fœtal organs called *amnion* and *allantois*, and is nourished before hatching by a great quantity of yolk in the egg. There are no mammary glands to furnish the young with milk after birth. The generative, urinary, and digestive organs come together behind in a common receptacle, the *cloaca*, or sewer, and their products are discharged by a single orifice. The kidneys of the early embryo, called *Wolffian bodies*, are soon replaced functionally by permanent kidneys, and structurally by the *testes* of the male and the *ovaries* of the female. The cavity of the *abdomen*, or belly, is not separated from that of the *thorax*, or chest, by a complete muscular partition, or *diaphragm*. The two lateral hemispheres of the brain are not connected by a transverse commissure, or *corpus callosum*. Air is always breathed by true lungs, never by gills. The blood, which may be cold or hot, has red oval nucleated corpuscles; the heart has either three or four separate chambers — four in birds, in which the circulation of hot blood is completely double, *i. e.*, in the lungs and right side of the heart, in the body at large and left side of the heart. The *aortic* arches are several; or if but one, as in adult birds, it is the right, not the left as in Mammals. The *centra*, or bodies, of the vertebræ are ossified, but have no terminal *epiphyses*. The skull hinges upon the backbone by a single median protuberance, or *condyle*, and the basioccipital part bearing the condyle is completely ossified. The lower jaw, or *mandible*, consists of several separate pieces, the articular one of which hinges upon a movable *quadrate* bone; and there are other peculiarities in the formation of the skull. The ankle-joint is situated, not, as in Mammals, between the *tarsal* bones and those of the leg, but between two rows of tarsal bones. The skin is usually covered with outgrowths, in the form of scales or feathers. — Different as are any living members of the class of Birds from any known Reptiles, the characters of the two groups converge in geologic history so closely, that the presence of *feathers* in the avian class, and their absence from the reptilian, is one of the most positive differences. The oldest known birds are from the Jurassic rocks of Europe and North America. These birds had teeth, and various other strong peculiarities of structure, which no living members of the class have retained.

AVES, or the Class of Birds, may be distinguished from other *Sauropsida* by the following sum of characters: The body is covered with feathers, a kind of skin-outgrowth no other animals possess. The blood is hot; circulation is completely double; the heart is perfectly four-chambered; there is but one (the right) aortic arch, and only one pulmonary artery springs from the heart; the aortic and the pulmonary artery have each three semilunar valves. The lungs are fixed and moulded to the cavity of the thorax, and some of the air-passages run through them to admit air to other parts of the body, as under the skin and in various bones. Reproduction is oviparous; the eggs are very large, in consequence of the copious yolk and white; have a hard chalky shell, and are hatched outside the body of the parent. There are always four limbs, of which the fore or pectoral pair are strongly distinguished from the hind or pelvic pair by being modified into *wings*, fitted for flying, if at all, by means of feathers — not of skin as in the cases of such mammals, reptiles, and fishes as can fly. The terminal part of the limb is compressed and reduced, bearing never more than three digits, only two of which ever have claws, and no claws being the rule. There are not more than two separate *carpals*, or wrist-bones, in adult recent birds (with very rare exceptions); nor any distinct interclavicular bone. The clavicles are complete (with rare exceptions), and coalesce to form a " wish-bone " or " merry-thought." The *sternum*, or breast-bone, is large, usually *carinate*, or keeled, and the ribs are attached to its sides only; it is developed from two to five or more centres of ossification. The *sacral* vertebræ proper have no expanded ribs abutting against the *ilia*; the ilia, or haunch-bones, are greatly prolonged forward; the socket for the head of the *femur*, or thigh-bone, is a ring, not a cup; the *ischia* and *pubes* are prolonged backward in parallel directions, and neither of these bones ever unites with its fellow in a ventral symphysis (except in *Struthio* and *Rhea*). The *fibula*, or outer bone of the leg, is incomplete below, taking no part in the ankle-joint. The *astragalus*, or upper bone of the tarsus, unites with the *tibia*, or inner bone of the leg, leaving the ankle-joint between itself and other tarsal bones, the lower of which latter similarly unites with the bones of the instep, or *metatarsus*. There are never more than four metatarsal bones, and the same number of digits; the first or inner metatarsal bone is usually free, and incomplete above; the other three anchylose (fuse) together, and with distal tarsal bones, as already said, to form a compound tarso-metatarsus. Recent birds, at any rate, have a certain saddle-shape of the ends of the bodies of some vertebræ. Such birds have also no teeth and no fleshy lips; the jaws are covered with horny or leathery integument, as the feet are also, when not feathered.

The Position of the Class Aves among other Vertebrates is definite. Birds come in the scale of development next below the Class *Mammalia*, and no close links between Birds and Mammals are known; the most bird-like known mammal, the duck-billed platypus of Australia (*Ornithorhynchus paradoxus*), being several steps beyond any known bird. Birds are the higher one of the two classes of *Sauropsida* — the lower class, *Reptilia*, connecting with the Batrachians (frogs, toads, newts, etc.) and so with the Fishes, *Ichthyopsida*. In this Vertebrate series, Birds constitute what is called a *highly specialized* group; that is to say, a very particular off-shoot, or, more literally, a side-issue, of the Vertebrate genealogical tree, which in the present geological era has become developed into very numerous (about 11,500) species, closely agreeing with one another in the sum of their physical characters. In comparison with other classes of Vertebrates, all birds are much alike; there is a less degree of difference among them than that among the members of any other classes of Vertebrates. Their likeness to each other is strong, and their difference from any other Vertebrates is peculiar; this makes them the "highly specialized" class they are recognized to be. The structural difference between a Humming-bird and an Ostrich, for example, is not greater in degree than that subsisting between the members of some of the orders of Reptiles; whence some hold, with reason, that Birds should not form a class *Aves*, but an order, or at most a sub-class, of *Sauropsida*, and thus be compared not with a class *Reptilia* collectively, but with other Sauropsidan orders, such as *Chelonia* (turtles), *Sauria* (lizards), *Ophidia* (serpents), etc. The practical convenience of starting with a "class" *Aves*, however, is so great, that such classificatory value will probably long continue to be ascribed, as heretofore, to Birds collectively. I have spoken of Birds as a particular "side-issue" or lateral branch of the Vertebrate "tree of life"; hence it is not to be supposed that they are in the direct line of genealogical descent. Though they stand as a group next below Mammals in the scale of evolution, it does not follow that Mammals were developed from any such creature as a Bird has come to be, nor that Birds have been evolved from any such Reptiles as those of the present day. It is one of the popular misunderstandings of the Theory of Evolution, to imagine that all the lower forms of animals are in the genetic line of development of the higher forms; that man, for example, was once a gorilla or a chimpanzee — actually such an ape. The theory simply requires all forms of life to be developed from *some* antecedent form, presumably, and in most cases certainly, lower in the scale of organization. Thus man and the gorilla are both descendants of some common progenitor, more or less unlike either of these existing creatures. All Mammals are similarly the modified descendants of some more primitive stock, from which stock sprang also all *Sauropsida*, mediately or immediately; therefore, a Mammal is not a modified Bird, though higher in the scale; and, though a Bird is a modified Reptile, it is not a modification of any such snake or lizard as now exists. The most bird-like reptiles known are not the Pterodactyls, or Flying Reptiles (*Pterosauria*), as might be supposed; but of that remarkable order, the *Ornithoscelida*, comprising the Dinosaurians, which "present a large series of modifications intermediate in structure between existing *Reptilia* and *Aves*," and are therefore inferentially in the direct ancestral line of modern Birds.

Fig. 14. — Oldest known ornithological treatise, illustrating also the art of lithography in the Jurassic period, engraved by *Archæopteryx lithographica*. From the original slab in the British Museum; after A. Newton, *Ency. Brit.*

Geologic Succession of Birds. — Birds have been traced back in geologic time to Cretaceous and Jurassic epochs of the Mesozoic or Mid-Life period of the world's history. The earliest *ornithichnites* — the fossils so called because supposed to indicate the presence of Birds by their foot-prints — were discovered about the year 1835 in the Triassic formation in Connecticut. But the creatures which made these tracks are now believed to have been Dinosaurian Reptiles. The oldest *ornitholite*, or fossil certainly known to be that of a true Bird, is the famous *Archæopteryx*, found by Andreas Wagner in 1861 in the Oölitic slate of Solenhofen in Bavaria. This has a long lizard-like tail of 20 vertebræ, from each of which springs a well-developed *feather* on each side; feathers of the wings are also well preserved; bones of the hand are not fused together, as they are in recent Birds; and the jaws bear true teeth. This Bird has served as the basis of one of the primary divisions of the class *Aves*; though it has many reptilian char-

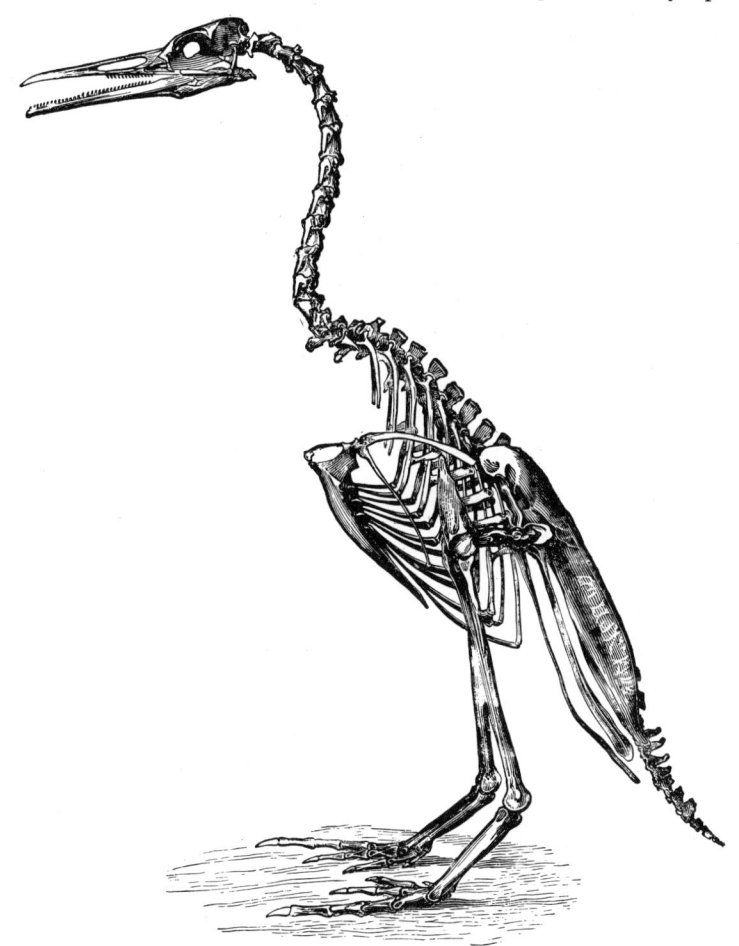

FIG. 15. — Restoration of *Hesperornis regalis*. After Marsh.

acters, it is a true Bird. A Bird (*Laopteryx priscus*) believed to be also of Jurassic age was discovered in 1881 in North America. The great gap between these ancient Avians and latter-day birds has been to some extent bridged by the discovery in 1870-72 of Birds from Cretaceous formations of North America; such genera as *Ichthyornis* and *Hesperornis* forming types of two remarkable groups, *Odontotormæ* and *Odontolcæ*, or Birds with teeth in sockets, and Birds with teeth in grooves. In both the tail is short, as in ordinary birds. In *Ichthyornis*, though the wings are well developed, with fused metacarpals, and the sternum is keeled, the vertebræ present the primitive character of being biconcave. In *Hesperornis* the vertebræ are saddle-shaped, as usual, but the sternum is flat, as in existing Ostriches, and the wings are rudimentary, wanting metacarpals. Some 20 species of several other genera of American Cretaceous Birds have been described. Remains of Birds multiply in the next period, the Tertiary. Those of the Eocene or early Tertiary are largely and longest known from discoveries made in the Paris Basin, among them *Gastornis parisiensis*, as large as an Ostrich; some of these belong to extinct genera, others to genera which still flourish; none are known to have true teeth, or otherwise to be as primitive as the reptile-like forms of the Cretaceous.

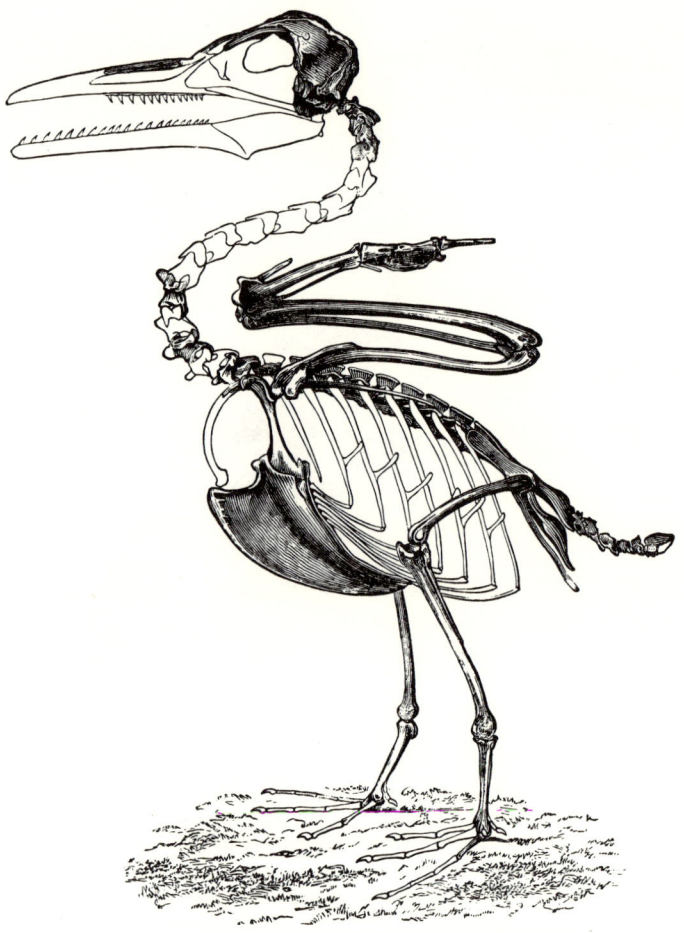

FIG. 16.— Restoration of *Ichthyornis victor*. After Marsh.

The Miocene or Middle Tertiary has proven specially rich in remains of Birds, including some of extinct genera, but in largest proportion referable to modern types. Later Tertiary (Pliocene and Post-pliocene) birds almost all belong to living genera, and some are apparently of living species. Extinct birds coeval with man, their bones bearing his marks, are found in various caves. Subfossil birds' bones occur in shell-heaps (kitchenmiddens) and elsewhere, of course contemporaneous with man, and some of them scarcely prehistoric. One of the oldest of these is the gigantic *Æpyornis maximus* of Madagascar, of which we have not only the bones, but the egg. The immense Moas, or *Dinornithes* of New Zealand, were among the later of these to die, portions of skin, feathers, etc., having been found. With Moa-remains are found those of *Harpagornis*, a raptorial bird large enough to have preyed upon Moas. Finally, various birds have been exterminated in historic times, some of them within the lifetime of persons now living. The Dodo of Mauritius, *Didus ineptus*, is the most celebrated one of these, of the living of which we have documentary evidence down to 1681; the Solitaire of Rodriguez, *Pezophaps solitarius*, the Géant, *Leguatia gigantea*, and several others of the same Mascarene group of islands, are in similar case. The Great Auk, *Plautus impennis*, is supposed to have become extinct in 1844; a Parrot, *Nestor productus*, was last known to be living in 1851; various Parrots, Rails, and other birds have likewise disappeared within a very few years. At least two North American birds, Pallas' Cormorant, *Phalacrocoræ perspicillatus*, and the Labrador Duck, *Camptolæmus labradorius*, are lately deceased. (See NEWTON, *Ency. Brit.*, 9th ed., art. Birds.)

FIG. 17. — Restoration of *Leguatia gigantea*. From Packard, after Schlegel.

§ 2. — PRINCIPLES AND PRACTICE OF CLASSIFICATION.

Having seen what a Bird is, and how it is distinguished from other animals, our next business is to inquire how birds are related to and distinguished from one another, as the basis of

Classification: a prime object of ornithology, without the attainment of which birds, however pleasing they are to the senses, do not satisfy the mind, which always strives to make orderly disposition of its knowledge, and so discover the reciprocal relations and interdependencies of the things it knows. Classification presupposes that there do exist such relations, according to which we may arrange objects in a manner which facilitates their comprehension, by bringing together what is like, and separating what is unlike; and that such relations are the results of evolutionary law. It is, therefore,

Taxonomy (Gr. τάξις, *taxis*, arrangement, and νόμος, *nomos*, law), or the rational, *lawful* disposition of observed facts. Just as taxidermy is the art of fixing a bird's skin in a natural manner, so taxonomy is the science of arranging birds in the most natural manner — in the way that brings out most clearly their natural affinities, and so shows them in their proper relations to each other. This is the greatest possible help to the memory in its attempt to retain its hold upon great numbers of facts. But taxonomy, which involves consideration of the greatest problems of ornithology, as of every other branch of biology (*biology* being the science of life and living things in general), is beset with gravest difficulties, springing from our defective knowledge. We could only perfect our taxonomy by having before us a specimen of every kind of bird that exists, or ever existed; and by thoroughly understanding how each is related to and differs from every other one. This is obviously impossible; in point of fact, we do not know all the birds now living, and only a small number of extinct birds have come to light; so that many of the most important links in the chain of evidence are missing, and many more cannot be satisfactorily joined together. With these springs of ignorance and sources of error must be reckoned also the risk of going wrong through natural fallibility of the human mind. The result is, that "natural classification," like the elixir of life or the philosopher's stone, is a goal still distant; and as a matter of fact, the present state of the ornithological system is far from being satisfactory. It is obvious that birds, or any other objects, may be "classified" in numberless ways — in as many ways as are afforded by all their qualities and relations, — to suit particular purposes, or to satisfy particular bents of mind. Hence have arisen, in the history of the science, very many different schemes of classification; in fact, nearly every leader of ornithology has proposed his own "system," and enjoyed a more or less respectable and influential following. Systems have been based upon this or that set of characters, and erected from this or that preconception in the mind of the systematist. Down to quite recent days, modifications of the external parts of birds, particularly of the bill, feet, wings, and tail, were almost exclusively employed for purposes of classification; and the mental point of view was, that each species of bird was a separate creation, and as much of a fixture in Nature's museum as any specimen in the naturalist's cabinet. Crops of classifications have been sown in the fruitful soil of such blind error, but no lasting harvest has been reaped. The confusion thus engendered has brought about the inevitable reaction; and the newest fashion is decidedly the opposite extreme — that of counting external features of little consequence in comparison with anatomical characters. Much ingenuity has been wasted in arguing the superiority of each of these characters for the purposes of classification; as if a natural classification should not be based upon *all* points of structure! as if internal and external characters were not reciprocal and mutually exponent! But the genius of modern taxonomy seems to be so certainly right — to be tending so surely, even if slowly, toward the desired consummation, that all differences of opinion, we may hope, will be settled, and defect of knowledge, not perversity of mind, be the only obstacle left in the way of success. The taxonomic goal is not now to find a way in which birds may be most *conveniently* arranged, described, and catalogued; but to discover their *pedigree*, and thus construct their family tree. Such a genealogical table, or *phylum* (Gr. φῦλον, *phulon*, tribe, race, stock), is rightly considered the only taxonomy worthy the name — the only true or *natural* classification. In attempting this end, we proceed upon the belief that, as explained above, all birds, like all other animals and plants, are related to each other *genetically*, as offspring are to parents; and that to discover their genetic relationships is to bring out their true affinities — in other words, to reconstruct the actual taxonomy of Nature. In this view, there can be but one "natural" classification, to the perfecting of which all increase in our knowledge of the *structure* of birds infallibly tends. The classification now used is the result of our best endeavors to accomplish this purpose, and represents what approach we have made to this end. It is based upon principles of Evolution which most naturalists are satisfied have been demonstrated. It is necessarily a

Morphological Classification — that is, one based solely upon consideration of structure or form (μορφέ, *morphē*, form); and for the following reasons: Every offspring tends to take on precisely the structure or form of its parents, as its natural physical heritage; and the principle involved, or the *law of heredity*, would, if nothing interfered, keep the descendants perfectly true to the physical characters of their progenitors; they would "breed true" and be exactly alike. But counter influences are incessantly operative, in consequence of varying conditions of environment; plasticity of organization of all creatures rendering them more or less susceptible of modification by such means, they become *unlike* their ancestors in various ways and to different degrees. On a large scale is thus accomplished, by *natural selection* and other natural agencies, just what man does in a small way in producing and maintaining different breeds of domestic animals. Amidst such shifting scenes, degrees of likeness or unlikeness of physical structure indicate with exactitude nearness or remoteness of organisms in kinship. Morphological characters are therefore the surest guides we can have to the blood-relationships we desire to establish; and such relationships *are* the "natural affinities" which classification aims to discover and formulate. As already said, taxonomy consists in tracing pedigrees and constructing the *phylum;* it is like tracing any leaf or twig of a tree to its branchlet, this to its bough, this to its trunk or main stem. The student will readily perceive, from what has been said, the impossibility of *naturally* arranging any considerable number of birds in any *linear series* of groups, one after the other. To do so is a mechanical necessity of book-making, where groups have to succeed one another, on page after page. *Some* groups will follow naturally; others will not; no connected chain is possible, because no such single continuous series exists in nature. In cataloguing, or otherwise arranging a series of birds for description, we simply begin with the highest groups, and make our juxtapositions as well as we can, in order to have the fewest breaks in the series.

Morphology being the only safe clue to natural affinities, and the key to all rational classification, the student cannot too carefully consider what is meant by this term, or too sedulously guard against misinterpreting morphological characters, and so turning the key the wrong way. The chief difficulty he will encounter comes from *physiological adaptations of structure;* and this is something that must be understood. The expression means that birds, or any animals, widely different in their morphological characters, may have certain parts of their organization modified in the same way, thus bringing about a seemingly close resemblance between organisms not nearly related to each other. For example: a Phalarope, a Coot, and a Grebe, all have *lobate* feet — that is, their feet are fitted for swimming in the same way, namely, by development of flaps or lobes on the toes. A striking but superficial and therefore unimportant resemblance in a certain particular exists between these birds, on the strength of which they used to be classed together in a group called *Pinnatipedes*, or "fin-footed" birds. But, on sufficient examination, these three birds are found to be very unlike in other respects; the sum of their unlikenesses requires us to separate them quite widely in any natural system. The group *Pinnatipedes* is therefore unnatural, and the appearance of affinity is proven to be deceptive. Such resemblance in the condition of the feet is simply functional, or physiological, and is not correspondent with structural or morphological relationships. The relation between these three birds is *analogical;* it is an inexact superficial resemblance between things profoundly unlike, and therefore having little *homological* or exact relationship. *Analogy* is the apparent resemblance between things really unlike — as the wing of a bird and the wing of a butterfly, as the lungs of a bird and the gills of a fish. *Homology* is the real resemblance or true relation between things, however different they may appear to be — as the wing of a bird and the foreleg of a horse, the lungs of a bird and the swim-bladder of a fish. Analogy commonly rests upon mere functional, *i. e.* physiological, modifications; homology is grounded upon structural, *i. e.* morphological, identity or unity. Analogy is the correlative of physiology, homology of morphology; but the two may be coincident, as when identical structures are used for the same purposes and are therefore physiologically identical. Physiological diversity of structure is incessant, and continually interferes with morphological identity of structure, to obscure or obliterate the indications of affinity the latter would otherwise express clearly. It is obvious that birds might be classified physiologically, according to their adaptive modifications or analogical resemblances, just as readily as upon any other basis: for example, into those that perch, those that walk, those that swim, etc.; in fact, most early classifications rested upon such considerations. It is also evident, that when functional modifications happen to be *coincident* with structural affinities — as when the turning of the lower larynx into a music-box coincides with a certain type of structure — such modifications are of the greatest possible service in classification. But since all sound taxonomy rests on morphology, on real structural affinity, we must be on our guard against those physiological "appearances" which are proverbially "deceptive." I trust I make the principle clear to the student. Its practical application is another matter, only to be learned in the school of experience. This question of

Homology or Analogy may be thus summed : Birds are *homologically related*, or naturally allied or affined, according to the sum of like structural characters employed for similar purposes; they are *analogically related* according to the sum of unlike characters employed for similar purposes. A Loon and a Cormorant, for instance, are closely affined, because they are both fitted in the same way for the pursuit of their prey by flying under water. A Dipper (family *Cinclidæ*) and a Loon (family *Gaviidæ*) are analogous, in so far as both are fitted to pursue their prey by flying under water; but they stand near opposite extremes of the ornithological system; they have little affinity beyond their common birdhood; very different structure being modified to attain the same end. A Crow has vocal organs almost identical in structure with those of a Nightingale, and the organization of the two birds is in other respects very similar; their affinity or homology is therefore close, though the Crow is a hoarse croaker, the Nightingale an impassioned musician.

The Reason why Morphological Classification is so important as to require adoption has been clearly stated by Huxley, whose words I cannot do better than quote in this connection. Speaking of animals, not as physiological apparatuses merely; not as related to other forms of life and to climatic conditions; not as successive tenants of the earth; but as fabrics, each of which is built upon a certain plan, he continues : —

"It is possible and conceivable that every animal should have been constructed upon a plan of its own, having no resemblance whatever to the plan of any other animal. For any reason we can discover to the contrary, that combination of natural forces which we term Life might have resulted from, or been manifested by, a series of infinitely diverse structures ; nor would anything in the nature of the case lead us to suspect a community of organization between animals so different in habit and in appearance as a porpoise and a gazelle, an eagle and a crocodile, or a butterfly and a lobster. Had animals been thus independently organized, each working out its life by a mechanism peculiar to itself, such a classification as that now under contemplation would be obviously impossible ; a morphological or structural classification plainly implying morphological or structural resemblances in the things classified.

"As a matter of fact, however, no such mutual independence of animal forms exists in nature. On the contrary, the members of the animal kingdom, from the highest to the lowest, are marvellously connected. Every animal has something in common with all its fellows; much, with many of them; more, with a few; and usually, so much with several, that it differs but little from them.

"Now, a morphological classification is a statement of these gradations of likeness which are observable in animal structures, and its objects and uses are manifold. In the first place, it strives to throw our knowledge of the facts which underlie, and are the cause of, the similarities discerned, into the fewest possible general propositions, subordinated to one another, according to their greater or less degree of generality; and in this way it answers the purpose of a *memoria technica*, without which the mind would be incompetent to grasp and retain the multifarious details of anatomical science.

"But there is a second and even more important aspect of morphological classification. Every group in that classification is such in virtue of certain structural characters, which are not only common to the members of the group, but distinguish it from all others; and the statement of these constitutes the definition of the group.

"Thus, among animals with vertebræ, the class *Mammalia* is definable as those which have two occipital condyles, with a well ossified basi-occipital; which have each ramus of the mandible composed of a single piece of bone and articulated with the squamosal element of the skull; and which possess mammæ and non-nucleated red blood-corpuscles.

"But this statement of the characters of the class *Mammalia* is something more than an arbitrary definition. It does not merely mean that naturalists agree to call such and such animals *Mammalia*: but it expresses, firstly, a generalization based upon, and constantly verified by, very wide experience; and, secondly, a belief arising out of that generalization. The generalization is that, in nature, the structures mentioned are always found associated together; the belief is that they always have been, and always will be, found so associated. In other words, the definition of the class *Mammalia* is a statement of a law of correlation, or coexistence, of animal structures, from which the most important conclusions are deducible." (Introd. to Classif. of Animals, 8vo, London, 1869, pp. 2, 3.)

But broad as such laws of correlation of structure are, and important as are the conclusions deducible, we must guard against presuming upon infallibility either of the data or of the deduction, as the author just quoted goes on to show. Such caution is specially required where there is no obvious reason for the particular combination that may be found to exist. In the case of the ostrich-like birds (*Ratitæ*), for example, we can understand how a flat, unkeeled breast-bone, a particular arrangement of shoulder-bones, and a rudimentary state of wing-bones, are found in combination, because all these modifications of structure are evidently related to loss of power of flight; and, in fact, no exception is known to the generalization, that such conditions of sternal, coraco-scapular, and humeral bones always coexist. But in all known struthious (ratite) birds, this state of the bones in mention coexists with a peculiar modification of bones of the palate, and no necessary connection between these two sets of diverse characters is conceivable. Now, if we only knew struthious birds, and found the combination in mention to hold with them all, we should doubtless declare our belief, that any bird having such palatal characters would also be found to possess such imperfect wing-apparatus. But this would be going too far; for we know that Tinamous (*Dromæognathæ*) have such a palate, yet have a keeled sternum and functionally developed wings. To take another case, derived from consideration of a large number of existing birds: it is an observed fact, that a particular arrangement of plates upon the back of the tarsus, a peculiar modification of the lower larynx or voice organ, and an undeveloped or abortive condition of the first large feather on the hand, are found associated in a vast series of birds, constituting the group of *Passeres* called *Oscines*. What possible connection there can be between these three separate and apparently independent modifications we cannot even surmise; but that they have some natural and necessary connection we cannot doubt, and that the connection is causal, not fortuitous, is a logical inference from the observed fact, that birds which present this particular combination are also closely related in other structural characters — that is, that they have all been subjected to operative influences which have conspired to produce the modifications observed. Given, then, a bird with a known oscine larynx, but unknown as to its feet and wings, it would be a reasonable inference that these members, when discovered, would present the characters observed to occur in like cases. But the first Lark (*Alaudidæ*) examined would show this inference to be fallible; for the tarsus of such a bird is differently disposed, though a lark has an elaborate singing apparatus, and only nine instead of ten developed primaries. Once more: the development of a keeled sternum, a peculiar saddle-shape of certain vertebræ, and lack of true teeth, are characters coexisting in all the higher birds; and, as far as these birds are concerned, we have no hint that such a combination is ever broken. In fact, however, the singular Cretaceous *Ichthyornis* shows us a pattern of bird in which a well-keeled sternum and perfectly formed wing coexist with teeth in reptile-like jaws and with fish-like biconcave vertebræ. What we learn from this case indeed breaks down one of the most precise definitions we might have made (and indeed did make) respecting birds at large; but in its failure we are taught how great is the modification of geologically recent birds from their primitive generalized ancestry; we learn something likewise of the steps of such modification, and of the length of time required for the process. It is the history of attempts to frame definitions of groups in zoölogy, that they are all liable to be negatived by new discoveries, and therefore to be broken down and require remodelling as our knowledge increases. It is to be readily perceived that the ability to draw distinctions and make definitions of groups is as much the *gauge of our ignorance* as the test of our knowledge; for all groups, like all species, come to be such by modification so gradual, so slight in each successive increment of difference, that, if all the steps of the process were before our eyes, we should be able to limit no groups whatever in a positive, unqualified manner. All would merge insensibly into one another, be inseparably linked in as many series as there have been actual lines of evolutionary progress, and finally converge to the one or few starting points of organized beings.

Practically, however, the case is quite the reverse — happily for the comfort of the working naturalist, however sadly the philosopher may deplore the ignorance implied. Degrees of likeness and unlikeness do exist, which when rightly interpreted enable us to mark off groups of all grades with much facility and precision, and thus erect a morphological classification which recognizes and defines such degrees, and explains them upon the principles of Evolution. The way in which the principles of such classification are to be practically applied gives occasion for some further remarks upon

Zoölogical Characters. — A "character," in zoölogical language, is any point of structure which may be perceived and described for the purpose of comparing or contrasting animals with one another. Thus, conditions of sternum, palate, tarsus, larynx, as noted in preceding paragraphs, are each of them "characters" which may be used in describing individual birds, or in framing definitions of groups of birds. Morphological characters, with which the classification we have adopted alone concerns itself, may be derived from the structure of a bird considered in any of its relations, or as affected by any of the conditions to which it is subjected. Thus *embryological* characters are those afforded by the bird during the progress of its development in the egg, from the almost structureless germ to the fully formed chick. Such characters of the embryo in its successive stages are of the utmost significance; for it is a fact, that the germ of each of the higher organisms goes through a series of developmental changes which, at each succeeding step in the unfolding of its appropriate plan of structure, causes it to resemble the adult state of animals lower than itself in the scale of organization. In fine, the history of the evolution of every *individual* bird epitomizes the history of those changes which birds collectively have undergone in becoming what they are by modified descent from lower organisms. Such transitory stages of any embryo, therefore, give us glimpses of those revolutionary processes which have affected the group to which it belongs. Any bird, for example, when a germ, is at first on the plane of organization of the very lowest known creatures — it is one of the *Protozoa*. As its germ develops, and its structure becomes more complicated by the formation of parts and organs successively differentiated and specialized, it rises higher and higher in the scale of being. At a certain stage very early reached (for the steps by which it becomes like any *in*vertebrate are very speedily passed over), it resembles a fish in possessing gill-like slits, several aortic arches, no true kidneys, no amnion, etc. Further advanced, losing its gills, gaining kidneys and amnion, etc., it rises to the dignity of a reptile, and at this stage it is more like a reptile than like a bird; having, for example, a number of separate bones of the wrist and ankle, no feathers, etc. The assumption of its own appropriate characters, *i. e.*, those by which it passes from a reptilian creature to become a bird, is always the last stage. We can thus actually see, inside any egg-shell, exactly those progressive steps of development of the individual bird which we believe to have been taken on a grand scale in nature for the evolution of the class *Aves* from lower forms of life; and the lesson learned is fraught with significance. It is nothing less than the demonstration in *ontogeny* (genesis of the individual) of that *phylogeny* (genesis of the phylum) by which groups of creatures come to be. The interior of any adult bird, again, furnishes us with all kinds of ordinary *anatomical* characters, derived from the way we perceive the different organs and systems of organs to be fashioned in themselves, and arranged with reference to one another. The finishing of the outward parts of a bird gives us the ordinary *external* characters, in the way in which the skin and its appendages are modified to form the covering of the bill and feet, and to fashion all kinds of feathers. Birds being of opposite sexes, and such difference being not only indicated in the essential sexual organs, but usually also in modifications in size or shape of the body or quality of the plumage and other outgrowths, a set of *sexual* characters are at our service. Birds are also sensibly modified in their outward details of feathering by times of the year when the plumage is changed, and this renders appreciation of *seasonal* characters possible. All such circumstances, and others that could be mentioned, such as effects of climate, of domestication, etc., in so far as they affect the *structure* of birds, conspire to produce zoölogical "characters," as these are above defined. Such characters, according as they result from more or less profound impressions made upon the organism, are of more or less "value" in taxonomy; being of all grades, from the trivial ones that serve to distinguish the nearest related species or varieties, to the fundamental ones that serve to mark off primary divisions. Thus the "character" of possessing a backbone is common to all animals of an immense series, called *Vertebrata*. The "character" of feathers is common to all the class *Aves;* of toothless jaws to all modern birds; of a keeled sternum to all the sub-class *Carinatæ;* of feet fitted for perching to all *Passeres;* of a musical apparatus to all *Oscines;* of nine primaries to all *Fringillidæ;* of crossed mandibles to all of the genus *Loxia;* of white bands on the wings to all of the species *Loxia leucoptera.* There is thus seen a sliding scale or valuation of characters, from those involving the most profound or *primitive* modifications of structure to those resting upon the most superficial or *ultimate* impressions. It will also be obvious, that every ulterior modification presupposes inclusion of all the prior ones; for a White-winged Crossbill, to be itself, must be a loxian, fringilline, oscine, passerine, carinate, modern, avian, vertebrated animal. The more characters, of all grades, that birds share in common, the more closely are they related, and conversely. Obviously, possession of more or fewer characters in common results in

Degrees of Likeness. — Were all birds alike, or did they all differ by the same characters to the same degree, no classification would be possible. It is a matter of fact, that they do exhibit all degrees of likeness possible within limits of their Avian nature; it is a matter of belief, that these degrees are the necessary result of Evolution, — of descent with modification from a common ancestry; and that, being dependent upon that process, they are capable of explaining it if rightly interpreted. For example: Two White-winged Crossbills, hatched in the same nest, scarcely differ perceptibly (except in sexual characters) from each other and from the pair that laid the eggs. We call them "specifically" identical; and the sum of the differences by which they are distinguished from any other kinds of Crossbills is their "specific character." All the individual Crossbills which exhibit this particular sum constitute a "species." In this case, the genetic relationship of offspring and parent is unquestionable; it is an observed fact. Now turn to the extremely opposite case. The difference between our Crossbills and the Jurassic *Archæopteryx* is the greatest known to subsist between any two birds whatsoever. But *Archæopteryx* and *Loxia* are also separated by an immense interval of *time*, and presumably by correspondingly enormous differences in *conditions of environment* — in their physical surroundings. It is a logical inference that these two things — difference in physical structure, and difference in physical environment — are in some way correlated and coördinated. If we presume, upon the theory of evolution, that despite the great difference, a Crossbill is genetically related to some such bird as an *Archæopteryx*, as truly as it is to its actual parents, only much more remotely, and that the difference is due to modifications impressed upon its stock in the course of time, conformably with changing conditions of environment, we shall have a better explanation of the difference than any other as yet offered — an explanation, moreover, which is corroborated by all the related facts we know, and with which no known facts are irreconcilable. But to correctly gauge and formulate the degrees of likeness or unlikeness between any two birds is to correctly "classify" them; and if these degrees rest, as we believe they do, upon nearness or remoteness of genetic relationship, classification upon such basis becomes the truest attainable formulation of "natural affinities." It is the province of morphological classification to search out those natural affinities which the structure of birds indicates, and express them by dividing birds into groups, and subdividing these into other groups, of greater or lesser "value," or grade, according to the more or fewer characters shared in common — that is, according to degrees of likeness — that is, again, according to genealogical relationship or consanguinity.

Zoölogical Groups. — To carry any scheme of classification into practical effect, naturalists have found it necessary to invent and apply a system of grouping objects whereby the like may come together and be separated from the unlike. They have also found it expedient to give names to all these groups, of whatever grade, such as *class, order, family, genus, species;* and to stamp each such group with the *value* of its grade, or its relative rank in the scale, so that it may become currency among naturalists. The student must observe, in the first place, that the value of each such coinage is wholly arbitrary, until sanctioned and fixed by common consent. The term "class," for example, simply indicates that naturalists agree to use that word to designate a conventional group of a particular grade or value. Indispensable as is some such acceptable medium of exchange of ideas among naturalists, their groups are *not* fixed, have *no* natural value, and in fact have no actual existence in the treasury of Nature. It cannot be too strongly impressed upon the student that Nature makes no bounds — *Natura non facit saltus;* there are no such abrupt transitions in the unfolding of Nature's plan, no such breaks in the chain of being, as he would be led to suppose by our method of defining and naming groups. He must consider the words "class," "order," etc., as wholly arbitrary terms, invented and designed to express our ideas of the *relations* which subsist between any animals or sets of animals. Thus, for example, by the term "Class of Birds" we signify simply the kind and degree of likeness which all birds share, such being also the kind and degree of their unlikeness from any other animals; the word "class" being simply the name or handle of the generalization we make respecting their relations with one another and with other animals; it represents an abstract idea, is the expression of a relation. True, all birds *embody* the idea; but "class" is nevertheless an abstraction. Now, as intimated earlier in this essay, definition of the idea we attach to the term — limitation of the class *Aves* — depends entirely upon how much we know of the relation intended to be expressed. It so happens, that no animals are known which cannot be decided to belong, or not to belong, to the conventional Class of Birds, because we have found it convenient and expedient to consider the presence of feathers a fair criterion, or necessary qualification. But what if an animal be discovered the covering of whose body is half-way between the scales of a lizard and the plumes of a bird, and whose structure is otherwise as equivocal? This may happen any day. A feather is certainly a modified scale; a feather has doubtless been developed out of a scale. In the case supposed, we should have to modify our definition of the "Class of Birds"; that is, change our ideas upon the subject, and alter the boundary-line we established between the classes of birds and reptiles; whereas, were a "class" something naturally definite, independent, and fixed, all that we could learn about it would only tend to establish it more surely. The same obscurity and uncertainty of definition attaches to groups of every grade — from the Animal "Kingdom" itself, which cannot be cut clear of the Vegetable "Kingdom" — down through classes, orders, families, genera, species, and varieties — yes, to the *individual* itself which, however unmistakable among higher organisms, cannot always be predicated of the lowermost forms of Life. Such divisions, of whatever grade, as we are able to establish for the purposes of classification, depend entirely upon the breaks and defects in our knowledge. There is no such thing as drawing "hard and fast" lines anywhere, for none such exist in Nature.

Taxonomic Equivalence of Groups. — But, however arbitrary they may be, or however obscure or fluctuating may be their boundaries, groups we must have in zoölogy, and groups of different grades, to express different degrees of likeness of the objects examined, and so to "classify" them. It is a great convenience, moreover, to have a recognized sliding-scale of valuation of groups from the highest to the lowest, and an accepted valuation. Just as in a thermometric scale, there are "degrees" designated as those of the boiling-point of water, the heat of the blood, the freezing of water, of mercury, etc.; so there are certain degrees of likeness conventionally designated as those of *class, order, family, genus,* and *species;* always accepted in the order here given, from higher to lower groups. (There are various others, and especially a number of intermediate groups, generally distinguished by the prefix *sub-*, as *subfamily;* but those here given are generally adopted by English-speaking naturalists, and suffice to illustrate the point I wish to make.) It may sound like a truism to say, that groups of the same grade bearing the same name, whatever that may be, must be of the same value,

— must be based upon and distinguished by characters of equal or equivalent importance. *Equivalence of groups* is necessary to the stability and harmony of any classificatory system. It will not do to frame an order upon one set of characters here, and there a family upon a similar set of characters; but order must differ from order, and family from family, by an equal or corresponding amount of difference. Let a group called a family differ as much from the other families in its own order as it does from some other order, and by this very circumstance it is not a family but an order itself. It seems a very simple proposition, but it is too often ignored, and always with practical ill result. Two points should be remembered here: First, that absolute size or numerical bulk of a group has nothing to do with its taxonomic value: one order may contain a thousand species, and another be represented by a single species, without having its ordinal valuation affected thereby. Secondly, any given character may assume different importance, or be of different value, in its application to different groups. Thus, the number of developed primaries, whether nine or ten, is a family character almost throughout *Oscines;* but in one oscine family (*Vireonidæ*) it has scarcely generic value. It is difficult, however, to determine such a point as this without long experience. Nor is it possible, in fact, to make our groups correspond in value with entire exactitude. The most we can hope for is a reasonable approximation. As in the thermometric simile above given, "blood heat" and other points fluctuate, so does order not always correspond with order, nor family with family, in actual significance. What degree of difference shall be "ordinal"? What shall be a difference of "family"? What shall be "generic" and what "specific" differences? Such questions are more easily asked than answered. They demand critical consideration.

Valuation of Characters. — In a general way, of course, the greater the difference between any two objects, the more "important" or "fundamental" are the characters by which they are distinguished. But what makes a character "important" or the reverse? Obviously, what it signifies represents its importance. We are classifying morphologically, and upon the theory of Evolution; and in such a system a character is important or the reverse, simply as an exponent of the principles, or an illustration of the facts, of evolutionary processes of Nature, according to the unfolding of whose plans of animal fabrics the whole structure of living beings has been built up. Why is possession of a back-bone such a "fundamental" character that it is used to establish one of the primary branches of the animal kingdom? It is not because so many millions of creatures possess it, but because it was introduced so early in the evolutionary process, and because its introduction led to the most profound modification of the whole structure of the animals which became possessed of a vertebral column. Why is possession by a bird of biconcave vertebræ so significant? Not because all modern birds have saddle-shaped vertebræ, but because to have biconcave vertebræ is to be fish-like in that respect. Why is presence or absence of teeth so important? Not that teeth served those old birds better than a horny beak serves modern ones, but because teeth are a reptilian character. Obviously, to be fish-like or reptile-like is to be by so much unbird-like; the degree of difference thus indicated is enormous; and a character that indicates such degree of difference is proportionally "important" or "fundamental." By knowledge of facts like these, and by the same process of reasoning, a naturalist of tact, sagacity, and experience is able to put a pretty fair valuation upon any given character; he acquires the faculty of perceiving its significance, and according to what it signifies does it possess for him its taxonomic importance. As a matter of fact, it seems that characters of all sorts are to be estimated *chronologically.* For, if animals have come to be what they are by any process that took time to be accomplished, characters earliest established are likely to be the most fundamental ones, upon the introduction of which the most important train of consequences ensued. Feathers, for example, as *Archæopteryx* teaches us, were in full bloom in the Jurassic period, and they are still the most characteristic possession of birds: all birds have them; no other animals have them; they are a class character. If they had been taken on quite recently, we may infer that many creatures otherwise entirely avian might not possess them, and they would have in classification less significance than that now rightly attributed to them. On the other hand, we cannot suppose that the finishing touches, by which, in the presence of white bands on the wings of *Loxia leucoptera*, and their absence in *Loxia curvirostra*, these two "species" are distinguished, were not very lately given to these birds. It is a very late step in the process, and correspondingly insignificant; it is of that value or importance which we call "specific." The same method of reasoning is available for determining the value of any character whatever, and so of estimating the grade of the group which we establish upon such character. As a rule, therefore, the length of time a character has been in existence, and its taxonomic value, are correlated, and each is the exponent of the other.

"**Types of Structure.**" — In no department of natural history has the late revolution in biological thought been more effective than in remodeling, presumably for the better, the ideas underlying classification. In earlier days, when "species" were supposed to be independent creations, it was natural and almost inevitable to regard them as fixed facts in nature. A species was as actual and tangible as an individual, and the notion was, that, given any two specimens, it should be perfectly possible to decide whether they were of the same or different species, according to whether or not they answered the "specific characters" laid down for them. The same fancy vitiated all ideas upon the subject of genera, families, and higher groups. A "genus" was to be discovered in nature, just like a species; to be named and defined. Then species that answered the definition were "typical;" those that did not do so well were "sub-typical;" those that did worse, were "aberrant." A good deal was said of "types of structure," much as if living creatures were originally run into moulds, like casting type-metal, to receive some indelible stamp; while — to carry out my simile — it was supposed that by looking at some particular aspect of such an animal, as at the face of a printer's type, it could be determined in what box in the case the creature should be put; the boxes themselves being supposed to be arranged by Nature in some particular way to make them fit perfectly alongside each other by threes or fives, or in stars and circles, or what not. How much ingenuity was wasted in striving to put together such a Chinese puzzle as these fancies made of Nature's processes and results, I need not say; suffice it, that such views have become extinct, by the method of natural selection, and others, apparently better fitted to survive, are now in the struggle for existence. Rightly appreciated, however, the expression which heads this paragraph is a proper one. There are numberless "types of structure." It is perfectly proper to speak of the "vertebrate type," meaning thereby the whole plan of organization of any vertebrate, if we clearly understand that such a type is not an independent or original model conformably with which all back-boned animals were separately created, but that it is one modification of some more general plan of organization, the unfolding of which may or did result in other besides vertebrated animals; and that the successive modifications of the vertebrate plan resulted in other forms, equally to be regarded as "types," as the reptilian, the avian, the mammalian. Upon this understanding, a group of any grade in the animal kingdom is a "type of structure," of more general or more special significance, presumably according to the longer or shorter time it has been in existence. An individual specimen is "typical" of a species, a species is "typical" of a genus, etc., if it has not had time enough to be modified away from the characters which such species or genus expresses. Any set of individuals, that is, any progeny, which become modified to a degree from their progenitors, introduce a new type; and continually increasing modification makes such a type specific, generic, and so on, in succession of time. There must have been a time, for example, when the Avian and Reptilian "types" began to diverge from each other, or, rather, to branch apart from their common ancestry. In the initial step of their divergence, when their respective types were beginning to be formed, the difference must have been infinitesimal. A little further along, the increment of difference became, let us say, equivalent to that which serves to distinguish two species. Wider and wider divergence increased the difference till genera, families, orders, and finally the classes of *Reptilia* and *Aves*, became established. In one sense, therefore, — and it is the usual sense of the term, — the "type" of a bird is that one which is *furthest removed from* the reptilian type, — which is most highly *specialized* by differentiation to the last degree from the characters of its primitive ancestors. One of the *Oscines*, as a Thrush or Sparrow, would answer to such a type, having lost the low, primitive, *generalized* structure of its early progenitors, and acquired very special characters of its own, representing the extreme modification which the stock whence it sprung has undergone. In a broader sense, however, the type of a bird is simply the stock from which it originated; and in such sense the highest birds are the least typical, being the furthest removed and the most modified derivatives of such stock, the characters of which are consequently remodeled and obscured to the last degree. Two opposite ideas have evidently been confused in the use of the word "Type." They may be distinguished by inventing the word *teleotype* (Gr. τέλεος, *teleos*, final, *i. e.*, accomplished or determined) in the usual sense of the word *type*, and using the word we already possess, *prototype* (Gr. πρῶτος, *protos*, first, leading, determining), in the broader sense of the earlier plan whence any teleotype has been derived by modification. Thus, *Archæopteryx* is prototypic of modern birds, all of which latter are teleotypic of their ancestors. It may be further observed that any form which is teleotypic in its own group, is prototypic of those derived from it. Thus, the *Archæopteryx*, so prototypic of modern birds, was a very highly specialized teleotype of its own ancestry. A little reflection will also make it clear that the same principle of antitypes (opposed types) is applicable to any of our *groups* in zoölogy. *Any group is teleotypic of the next greater group of which it is a member; prototypic of the next lesser one.* Any species is teleotypic of its genus; any genus, of its family; any family, of its order; and conversely; that is to say, any species represents one of the ulterior modifications of the plan of its genus. The Class of Birds, for example, is one of the several teleotypes of Vertebrata, *i. e.*, of the vertebrate plan of structure; representing, as it does, one of several ways in which the vertebrate prototype is accomplished. Conversely, the Class of Birds is prototypical of its several orders, representing the plan which these orders severally unfold in different ways. And so on, throughout any series of animals, backwards and forwards in the process of their evolution; any given form being teleotypic of its predecessors, prototypic of its successors. All existing forms are necessarily teleotypic — only prototypic for the future. Prototype, in the sense here conveyed, indicates what is often expressed by the word *archetype*. But the latter, as I understand its use by Owen and others, signifies an ideal plan never actually realized; the "archetype of the vertebrate skeleton," for example, being something no vertebrate ever possessed, but a theoretical model — a generalization from all known skeletons. The correspondence of my use of "prototypic" with a common employ of "archetypic," and of "teleotypic" as including both "attypic" and "etypic," is noted below.[1]

The actual and visible genetic relationships of living forms being practically restricted to individuals of the same species — parents and offspring " specifically " identical — it would seem at first sight that species must be the modified descendants of their respective genera, in order

[1] "*Archetypical* characters are those which a group derives from its progenitor, and with which it commences, but which in much modified descendants are lost; such, for example, is the dental formula of the Educabilia (M ⅜ PM ⅘ C ⅓ 1 ⅜ × 2), — a formula, as shown by Owen, very prevalent among early members of the group, but generally departed from more or less in those of the existing faunas. *Attypical* characters are those to the acquisition of which, as a matter of fact, we find that forms, in their journey to a specialized condition, tend . . . *Etypical* characters are exceptional ones, and which are exhibited by an eccentric offshoot from the common stock of a group." (Gill, Pr. Am. Assoc. Adv. Sci. xx, 1873, p. 293.) To illustrate in birds: A generalized lizard-like type of sternum is *archetypic* of any bird's sternum. The sternum of the lizard-like animals whence birds actually descended is *prototypic;* the keeled sternum of a carinate bird is *attypical* in most birds, *etypical* in the peculiar state in which it is found in *Stringops;* but equally *teleotypic* in both instances.

to be teleotypic of any such next higher group. But nothing descends from a genus, or any other group; everything descends from individuals; a "genus," like any other group, is an abstract statement of a relation, not a begetter of anything. To illustrate: the "genus *Turdus*" is represented by many species: if these species be rightly allocated in the genus, they are all the modified descendants of a form which was, before they severally branched off, a specific form; the "genus *Turdus*" in the abstract is simply that form; and that form is prototypic of its derivatives. In the concrete, as represented by its teleotypes, the genus *Turdus* sums the modifications which these have collectively undergone, without specifying the particular modifications of any of them; it expresses the way in which they are all like one another, and in which they are all unlike the representatives of any other genus. Thus what is above advanced is seen to hold, though genera and all other groups are actual descendants of individuals specifically identical.

Generalized and Specialized Forms. — Taking any one group of animals — say the genus *Turdus*, of numerous species — and considering it apart from any other group, we perceive that it represents a certain assemblage of characters peculiar to itself, aside from those more fundamental ones it includes of its family, order, etc. Its particular characters we call "generic." Among the numerous teleotypic forms it includes, there is a wide range of specific variation, within the limits of generic relationship. Some of its species are modified further away than some others are from the generic standard or type to which all conform more or less perfectly. The former, having more peculiarities of their own, are said to be the most *specialized*; the latter, having fewer peculiarities, are the least specialized. Those that are the least specialized are obviously the most *generalized*; and this means, that we believe them to be nearest to the stock whence all have together descended with modification. The application of this illustration to great groups shows us the principle upon which any form is said to be *generalized* or *specialized*. *Ichthyornis*, with its fish-like vertebræ, reptile-like teeth, bird-like sternum and shoulder-girdle, is a very generalized form. A Thrush is the opposite extreme of a highly specialized form. The two are also separated by an enormous interval of time: one being very old, the other quite new; a chronological sequence is here perceived. Since the evolutionary processes concerned in the modification on the whole represent progress from simplicity to complexity of organization, and therefore ascent in the scale of organization, a *generalized* type, an *ancient* type, and a *simple* type are on the whole synonymous, and to be contrasted with *specialized*, *recent*, and *complex* types. They therefore respectively correspond to

"Low" and "High" in the Scale of Organization. — All existing birds are very closely related, notwithstanding the great numerical preponderance of the class in the present geological epoch. This outbreak, as it were, of birds upon the modern scene, is like the nearly simultaneous bursting into bloom of a mass of flowers at the end of one branch of the Sauropsidan stem. All modern birds, in fact, are strongly specialized forms, so much so that it is difficult to predicate "high" or "low" within such a narrow scale. The great group *Passeres*, for example, comprehending a majority of all known birds, is scarcely more different from other birds than are the families of reptiles from each other, and among *Passeres* we have little to go upon in deciding "high" or "low" beyond the musical ability of *Oscines*. It is hard to see much difference in actual complexity of organization between birds regarded as lowest, as an Ostrich or a Penguin, and those conceded to be highest, as a Swallow or Sparrow. Nevertheless, in a larger perspective, as between a fish, a reptile, and a bird, the student will readily perceive the bearing of the ideas attached to the terms "low" and "high" in the scale of organization. Creatures rise in the scale by a number of correlated modifications and in the course of time (for it takes *time* to evolve a class of birds from sauropsidan stock as really as it does to develop the germ of an egg into the body of a chick). Progressive differentiation and specialization of structure and function in due course elaborates diversity from sameness, complexity from simplicity, the "high" special from the "low" general plan of organization; the culmination in man of the vertebrate type, first faintly foreshadowed in the embryonic Ascidian. No one should venture to foretell the result of infinitesimal increments in elevation of structure and function, nor presume to limit the infinite possibilities of evolutionary processes, either in this actual world or in the foretold next one.

As to "evidences of design" in the plan of organized beings, it may be said simply that every creature is perfectly "designed" or fitted for its appropriate activities, and perfectly adapted to its conditions of environment. In fact, it must be so fitted and adapted, or it would perish. Whether it so determines itself, or is so determined, is a teleological question. The truth remains that every creature is perfect in its own way. A worm is as perfectly fitted to be a worm, as is a bird to be a bird; in fact, were it not, it would either turn into something else, or cease to be. A spade is as perfect an organization *of the spade kind*, as is a steam-engine of that kind of an organization; though the difference in complexity of structure and functional capacity, like that between the lowly organized ascidian generality and the highly organized avian speciality, is enormous.

One word more: The class of mammals is highest in the scale of organization. The class of birds is next highest. But it does not follow, from this relation sustained by *Mammalia* and *Aves* collectively, that every mammal must be more highly organized than every bird. It is difficult to say how a mole or a mouse is a more elaborate or more capable creature than a canary-bird, physically or mentally. The relative rank of two groups is determined by balancing the aggregate of their structural characters. In large series, the average of development, not the extremes either way, is taken into account; so that the lowest members of a higher group may be below the highest members of the next lower group. The common phrase, "below par," or "above par," is most applicable to such cases.

Machinery of Classification. — The inexperienced student may be glad to be given some explanation of the way in which the taxonomic principles we have discussed are applied, and carried into practical effect in classifying birds. Our machinery for that purpose is our inheritance from those naturalists who held very different views from those which touch the evolutionary key-note of modern classification. It is clumsy, and does not work well as a means of expressing the relations we now believe to be sustained by all organisms toward one another; but it is the best we have. Systematic zoölogy, or the practice of classification, has failed to keep pace with the principles of the science; we are greatly in need of some new and sharper "tools of thought," which shall do for zoölogy what the system of symbols and formulæ has done for chemistry. *We want some symbolic formulation of our knowledge.* The invention of a practicable scheme of classification and nomenclature, which should enable us to formulate what we mean by *Merula migratoria*, as a chemist symbolizes by SO_4H_2 what he understands hydrated sulphuric acid to be, would be an inestimable boon to working naturalists. The mapping out of groups with connecting lines to indicate their genetic relations, in the form of a "phylum," is a common practice; but that, like any other pictorial representation of a "family tree," is not the graphic symbolization required. We already have a mother of the required invention in the necessity of the case, and may hope that the father will not be long in coming.

Under the present system, Birds are called a "Class" of Vertebrates, and are subdivided into "orders," "families," "genera," "species" and "varieties," as already sufficiently indicated. Groups intermediate to any of these may be recognized; and if so, are usually distinguished by the prefix *sub-*. Many other terms are in occasional use, as "tribe," "race," "series," "cohort," "super-family;" but those first mentioned are the best established ones among English-speaking naturalists. Their sequence is fixed, as above, from higher to lower, in relative rank.¹ With the exceptions to be presently noted, the names of any groups are arbitrary, at the will of the person who founds and designates them. The framer of a genus, or the describer of a species, calls it what he pleases, and the name he gives holds, subject to certain statutory regulations which naturalists generally agree to abide by. The exceptions are the names of families and subfamilies, the former commonly being made to end in -*idæ*, the latter in -*inæ*: family *Turdidæ*; subfamily *Turdinæ*. This is a great convenience, since we always know the rank intended to be noted by these forms. The names of groups higher than species are almost invariably single words; as, order *Passeres*; but sometimes, especially in cases of intermediate groups, two words are used, one qualifying the other; as, suborder *Passeres Acromyodi*, or Oscine Passeres. A generic or subgeneric name is always a single word; these, and the names of all higher groups, invariably begin with a capital letter.

Until quite recently, the scientific name of any individual bird almost invariably consisted of two terms, generic and specific — the name of the genus, followed by the name of the species; as, *Merula migratoria*, for the Robin. This is the "binomial nomenclature" (badly so called, for "binominal" would be better), introduced by Linnæus in the middle of the last century. It was a great improvement upon the former method of giving either single arbitrary names to birds, often a mere Latin translation of their vernacular nickname, or long descriptive names of several words; probably no other single improvement in a method of nomenclature ever did so much to make the technique of nomenclature systematic. To couple the two terms at all was a great thing, the convenience of which we who never felt its want can hardly appreciate. To follow the generic by the specific term was itself of the same advantage that it is to have the Smiths and Browns of a directory entered under S and B, instead of by Johns and Jameses; besides according with the genius of the Romance languages, which commonly put the adjective after the noun. A Frenchman, for example, would say, *Bec-croisé aux ailes blanches de l'Amérique septentrionale*, or "Bill-crossed to the wings white of the America north," where we should say, "North American White-winged Cross-bill," and Linnæus would have written *Loxia leucoptera*. The binomial scheme worked so well that it came to have the authority and force of a statute, which few subsequent naturalists have been inclined, and fewer have ventured, to violate; while it became an *ex post facto* law to prior naturalists, ruling them out of court altogether, as far as the legitimacy of any of the names they had bestowed was concerned. It necessarily rested, however, or at any rate proceeded upon, the false idea of a species as a fixity. Linnæus himself experienced the inadequacy of his system to deal binomially with those lesser groups than species, commonly called "varieties," now better designated as "conspecies" or "subspecies"; and he often used a third word, separated however from the binomial name by intervention of the sign "var." or some other symbol. Thus, if he had supposed an American Crossbill to be a variety of a European *Loxia leucoptera*, he might have called it *Loxia leucoptera*, a, *americana*. Many years ago I urged the necessity of recognizing by name a great number of forms of our birds intermediate between nominal species, and connected by links so perfect, that our handling of "species" required thorough reconsideration. The dilemma arose, through our very intimate knowledge of the climatic and geographical variation of "species," either to discard a great number that had been described, and so ignore all the ultimate modifications of our bird-forms; or else to recognize as good species the same large number of forms that we knew shaded into each so completely that no specific character could be assigned. In the original edition of the present work (1872), I compromised the matter by reducing to the rank of varieties the nominal species that were known or believed

---

¹ The expression "higher group," in the sense of relative rank in the taxonomic scale, will of course be distinguished from the same expression when applied to the relative rank in the scale of organization of the objects classified. An order of birds is a "higher group" than a family of birds, in the former sense, but no higher than an order of worms, in the latter sense

to intergrade; and the original edition of my Check List (1873) distinguished such by the sign "var."-intervening between the specific and the subspecific name. I subsequently determined to do away with the superfluous term "var.," and in the next edition of the Check List (1882) and Key (1884) adopted a purely trinomial system of naming the equivocal forms as subspecies; as, *Loxia curvirostra americana*. This method was found to work so well, that it was immediately adopted and officially formulated in the Code of Nomenclature (1886) of the American Ornithologists' Union, and thus came into universal use in this country. Trinomialism is confidently commended as a boon to our brethren over the sea, who perceive its usefulness, yet continue to handle it gingerly, Linnæus being still something of a fetich on the more conservative side of the water. It is the most distinctive feature of what English ornithologists call "the American school."

The Student cannot be too well assured, that no such things as species, in the old sense of the word, exist in nature, any more than have genera or families an actual existence. Indeed they cannot be, if there is any truth in the principles discussed in our earlier paragraphs. Species are simply ulterior modifications, which once were, if they be not still, inseparably linked together; and their nominal recognition is a pure convention, like that of a genus. More practically hinges upon the way we regard them than turns upon our establishment of higher groups, simply because upon the way we decide in this case depends the scientific *labeling of specimens*. If we are speaking of a Robin, we do not ordinarily concern ourselves with the family or order it belongs to, but we do require a technical name for constant use. That name is compounded of its genus, species, and variety. No infallible rule can be laid down for determining what shall be held to be a species, what a conspecies, subspecies, or variety. It is a matter of tact and experience, like appreciation of the value of any other group in zoölogy. There is, however, a convention upon the subject, which the present workers in ornithology in this country find available; at any rate, we have no better rule to go by. We treat as "specific" any form, however little different from the next, that we do not know or believe to intergrade with that next one — between which and the next one no intermediate equivocal specimens are forthcoming, and none, consequently, are supposed to exist. This is to imply that the differentiation is accomplished, the links are lost, and the characters actually become "specific." We treat as "subspecific" of each other any forms, however different in their extreme manifestation, which we know to intergrade, having the intermediate specimens before us, or which we believe with any good reason do intergrade. If the links still exist, the differentiation is still incomplete, and the characters are not specific, but only subspecific, in the literal sense of these terms. In the latter case, the oldest approved name is retained as the specific one, and to it is appended the subspecific designation: as *Merula migratoria propinqua*. The specific and subspecific names are preferably written with a small initial letter, even when derived from the name of a person or place.

One other term than those just considered sometimes forms part of a bird's scientific name: this is the *subgenus*. When introduced, it always follows the generic term, in parentheses; thus, *Turdus (Hylocichla) mustelinus*. This is cumbrous, especially when there are already three terms, and is little used in this country. I discarded it altogether in 1884, and so did the American Ornithologists' Union in 1886. There is no real difference between a subgenus and a genus, and modern genera have so multiplied that one can easily find a single name for any generic refinement he may wish to indulge.

It has always been customary to write after a bird's name the name of the original describer of the species, as the authority or voucher for the validity of the species named. But as genera multiplied, it was often found necessary to change the generic name, the species being placed in another genus than that to which its original namer had referred it. Then the name of the person who originated the new combination was commonly suffixed, presumably as authority for the validity of the classification implied. As this was to ignore the proprietorship of the original describer, it became customary to retain such describer's name in parentheses and add that of the classifier; thus, *Turdus migratorius* Linnæus; *Planesticus migratorius* (Linn.) Bonaparte. The practice still prevails; it is no more objectionable than any other harmless exhibition of human vanity. The student will find it carefully carried out in my Check List of 1873 and 1882, and entirely discarded in the Second and subsequent editions of the present work.

It would take me too far to go fully into the rules of nomenclature: some few points may be noted. A proper sense of justice to the describers of new genera, species, and subspecies, prompts us to preserve inviolate the names they see fit to bestow, with certain salutary provisions. Hence arises the "law of priority." The *first* name given during or since 1758 is to be retained and used, if it can be identified with reasonable certitude, — that is, if we think we know what the giver meant by it. But it is to be discarded, and the next name in priority of time substituted, if it is "glaringly false or of express absurdity,"—as calling an American bird "*cafer*," or a black one "*albus*." No generic name can be duplicated in zoölogy, and one once void for any reason cannot be revived and used in any connection. The same specific name cannot be used twice in the same genus.

In my judgment, the best set of rules for naming objects of natural history ever devised is the Code of Nomenclature promulgated by the American Ornithologists' Union in 1886. Its canons are applicable not only to ornithology, but also to all other branches of zoölogy. They have acquired the force of statutory regulations in this country, and the student who would be more than an amateur must learn them. He will also do well to obey them until he becomes a professional ornithologist and can afford to express opinions of his own. For myself I subscribe to the Code in its entirety, with two exceptions. I will never obey a canon which would oblige me to use a "glaringly false" name, for falsity is foreign to science. Nor shall I ever have anything but contempt for Canon XL., which would make me misspell a name for no other reason than that it was misspelled in the beginning; for that would be a matter "of express absurdity, and therefore contemptible." The committee who devised this Code were: Elliott Coues, Chairman; J. A. Allen, Robert Ridgway, William Brewster, and H. W. Henshaw.

The Actual Classification of Birds has undergone radical modification of late years, though the same machinery is employed for its expression. This is as would be expected, seeing how profoundly the theory of Evolution has affected our principles of classification, how completely the morphological has replaced other systems, and how steadily our knowledge of the structure of birds, and their chronological relations, has progressed. Nevertheless, the ornithological system is still in a transition state, and the classification implied by my arrangement of North American birds in the present work must be regarded as tentative and provisional. In the original edition of the Key the classification was vitiated at the outset by physiological considerations,[1] and in some other respects was open to decided improvement, as I trust the present edition shows. The table given on a succeeding page will afford the student a *coup d'œil* of the groups, from subclass to subfamily, which I have been led to adopt; it represents, as far as it goes, a classification of birds at large. The principal groups, higher than families, which are absent from the North American Fauna, are: the whole of the *Ratitæ*, or Struthious birds; the *Dromæognathæ*, embracing the South American Tinamous; the *Sphenisci*, Penguins of the Southern Hemisphere, and several small superfamily groups belonging in the vicinity of the Columbine, Gallinaceous, Lemicoline, and Anserine birds.

As to the primary divisions of *Aves*, it seems certain that these must be made with special

---

[1] In primarily dividing birds into *Aves aereæ*, *Aves terrestres*, and *Aves aquaticæ*, after Lilljeborg, I should do myself the justice to say, however, that the fact that these divisions did not rest upon morphological characters of any consequence was expressly stated (pp. 8 and 276 of the orig. ed.).

reference to the extraordinary form from the Jurassic and to the radical difference between Ratite and Carinate Birds. The subclass *Carinatæ*, which includes all other existing birds, seems not to be primarily divisible into a few orders, such as were in vogue not many years ago; but to be split directly into a large number — perhaps about twenty — groups of approximately equivalent value, to be conventionally designated as orders, if we take *Carinatæ* as a subclass of the class *Aves*. *Passeres* seems to be one of the most firmly established of these "ordinal" groups; but neither *Passeres* nor any other leading group of birds has any such taxonomic grade as the groups of the same name have in other branches of zoölogy. "*Picariæ*" is one of the most unsatisfactory, and I have no doubt it will be abolished. The arrangement offered on a subsequent page has perhaps some claims to consideration.

With this glance at some taxonomic principles and practices, I pass to an outline of the structure of birds, some knowledge of which is indispensable to any appreciation of ornithological definitions and descriptions. It is necessary to be brief, and I shall confine myself mainly to consideration of those points, and explanation of those technical terms, which the student needs to understand in order to use the present volume easily and successfully. Here I will insert a tabular illustration of a sequence of zoölogical groups, from highest to lowest, under which a bird may fall: —

Kingdom, *Animalia*: Animals.
  Branch, *Vertebrata*. Back-boned Animals.
    Province, *Sauropsida*: Lizard-like Vertebrates.
      Class, *Aves*: Birds.
        Subclass, *Carinatæ*: Birds with keeled breast-bone.
          Order, *Passeres*: Perching Birds.
            Suborder, *Oscines*: Singing Birds.
              Family, *Turdidæ*: Thrush-like Birds.
                Subfamily, *Turdinæ*: True Thrushes.
                  Genus, *Turdus*: Typical Thrushes.
                    Subgenus, *Hylocichla*: Wood Thrushes.
                      Species, *ustulatus*: Olive-backed Thrush.
                        Subspecies, *aliciæ*: Alice's Thrush.

§ 3. — DEFINITIONS AND DESCRIPTIONS OF THE EXTERIOR PARTS OF BIRDS.

a. OF THE FEATHERS, OR PLUMAGE.

Feathers are possessed only by birds, and all birds possess them. Feathers are therefore diagnostic of the class *Aves*. Feathers are modified scales; like scales, hair, horns, claws, etc., they are outgrowths of the integument, or skin covering the body, and therefore belong to the class of *epidermic* (Gr. ἐπί, *epi*, upon; δέρμα, *derma*, skin), or *exoskeletal* (Gr. ἐξ, *ex*, out; σκελετόν, *skeleton*, dried; in the sense of "outer skeleton") structures. The horny coverings of beak and feet are of the same class, but very differently developed. The development of feathers is a complicated process, and the result is correspondingly complex. Besides being the most highly developed or specialized, wonderfully beautiful and perfect kind of tegumentary outgrowth — besides fulfilling in a singular manner the function of covering and protecting the body — feathers have their particular *locomotory* office: that of accomplishing the act of flying in a manner peculiar to birds. For all vertebrates, excepting birds, that progress through the air — the flying fish (*Exocœtus*) with its enlarged pectoral fins; the flying reptile (*Draco* or *Pterodactylus*) with its skinny parachute; the flying mammal (bat) with its great webbed fingers — accomplish aërial locomotion by means of tegumentary *expansions*. Birds alone fly with tegumentary *outgrowths*, or appendages. These peculiar structures are very light, weighing little in proportion to their bulk ; and some kinds of feathers are very strong and elastic, easier to bend than to break — in fact, the horny part of a feather is a very tough substance. Feathers make extremely poor conductors of heat, and consequently a warm covering of the body.

All a bird's feathers, of whatever kind, collectively constitute its *ptilosis* (Gr. πτίλον, *ptilon*, a feather) or PLUMAGE (Lat. *pluma*, a plume or feather). In many cases the first plumage of the nestling or newly-hatched bird is a short-lived set of feathers so different from the crop next grown and longer worn as to give rise to the technical distinction between

Neossoptiles and Teleoptiles (Gr. νεοσσός, *neossos*, a young bird, chick, or fledgling ; τέλεος, *teleos*, finished, final, or mature ; and πτίλον, *ptilon*, a feather). A neossoptile is an unfinished feather which precedes a final feather, is borne upon the latter for a while, and then drops off. All birds do not have neossoptiles, and these temporary feathers form but a sparse and scanty covering of some birds which possess them. Such is the case with those birds which are commonly said to hatch naked, and which stay in the nest until they are fully fledged; though even in these instances the few straggling hair-like feathers which may be first observed are neossoptiles. In the highly exceptional case of the Mound-birds (*Megapodidæ*) neossoptiles are shed before the chick is hatched, so that the apparently first but actually second set of feathers are teleoptiles. Neossoptiles are copious enough to form the complete downy covering of those young birds which hatch clothed and are able to run about or swim almost immediately, as in the cases of the chicks, ducklings, or goslings of the poultry-yard, the unfledged young of plovers, snipes, and many others; such a covering is also speedily acquired by various birds which hatch naked or nearly so, yet remain long in the nest, as the squabs of pigeons, and the nestlings of herons, gulls, and most other water-birds. The generalization may be made, that neossoptiles are most copious and conspicuous in the lower orders of birds, as the walkers, waders and swimmers, least so in the higher Passerine and Picarian orders. This distinction agrees very well with what are explained beyond as *altricial* or *psilopædic* birds on the one hand, and *præcocial* or *ptilopædic* birds on the other hand ; less exactly, with birds called *nidicolous* and *nidifugous*, or those which remain some time in the nest and those which can leave it at once. Neossoptiles are always weak, fluffy, hairy or downy feathers — in fact, they form the first "downy plumage" of any bird which possesses such a covering. Their character will be better understood by the student after he has read what is said beyond of the structure of feathers. The distinction between neossoptiles and teleoptiles is not that the former are downy, for many of the latter are equally downy ; but that neossoptiles are shed or moulted, from the ends of the teleoptiles upon which they are borne, not from the skin itself. In fine, a neossoptile is simply the temporary, deciduous, terminal portion of an ungrown teleoptile, though it may be the only kind of a feather the young bird possesses. The whole plumage of every adult bird consists of teleoptiles, whose several kinds are described beyond.

Development of Feathers. In a manner analogous to that of hair, a feather grows in a little pit or pouch formed by an inversion of the dermal or true-skin layer of the integument as well as of the epidermal or scarf-skin layer. This pit is the *feather-follicle;* it supports the base of a little conical pimple, the *feather-papilla,* upon which the future feather is to be moulded. The outermost layer of epidermal cells is called the *epitrichium ;* the subjacent layers form the *Malpighian stratum,* which enters into the structure both of the follicle and of the feather itself. The cells of this stratum, rapidly multiplying and growing downward into the pit, separate into two sets, one of which lines the whole wall of the follicle, while the other covers a mass of cells which have meanwhile shot up in the centre of the follicle from the underlying dermal layer of integument. This central mass is the pulp which is to nourish the rapidly growing feather; it becomes a soft spongy network and furnishes the blood supply, but is not otherwise transformed into the substance of the feather. For the latter is entirely epidermal, being built up from the cells of that portion of the Malpighian stratum which covers the central pulp. This portion subdivides into three layers. The outermost layer sprouts out of the skin in the form of a horny cylindric sheath, and is the well-known object we call a "pin-feather." The thick intermediate layer makes most of the feather itself, set free when

Fig. 18. — **Symmetrical Figures from Forming Feathers;** *a*, dove; *b*, turkey. — "In the summer of 1869, whilst examining the feather capsule of a nestling dove, the microscopic slide was suddenly covered with a multitude of exquisite forms. . . . The next day my German farmer climbed to the dove's nest and procured a few more pin-feathers. Some of these were cut into fine shreds, rubbed in a drop of water, and placed under the microscope. In a short period the figures of yesterday were again before me. From the cut surfaces of the portions of the pin-feathers I had placed under the lens, granules appeared to stream forth like blood, covering the microscopic slide in countless numbers. Mingled with these were numerous larger cells of a globular or oval form, having a transparent centre. These and the granules gave to the water a slightly glutinous consistency. As the fluids on the glass dried, lines at different angles shot across the slide, looking much as though an unseen camel's hair pencil had been swiftly drawn in opposite directions, sometimes at right angles, but frequently at angles more acute. Probably at the moment of transition from a fluid to a solid condition, the transparent nucleated cells assumed the form of a square, a lozenge, a starry hexagon, a cross, or any other beautiful figure which could be formed of the parts which suddenly appeared in the spherical cells, these parts seeming at first, in some instances at least, to consist of minute triangles. At the same moment the little granules moved to order, and there before the astonished gaze were diamonds such as Aladdin might have envied, in form as varied, but far more symmetrical, than the frost-work on a window pane of a winter's morning." (Grace Anna Lewis, Am. Nat., v, 1871, p. 675.)

the sheath that contains it peels off. The innermost layer simply sheathes the pulp, and is finally transformed into the pith which may be observed inside the hollow quill as a set of little caps or thimbles. Such development of feathers as is here briefly sketched holds good both for neossoptiles and for teleoptiles, the formation of the latter simply completing the process begun with the former. When the final feather has completed its growth the activity of the follicle ceases as long as the feather stays in place; but when the feather drops, as it always does when it is worn out, the follicle renews its function and grows another feather in the same manner as before, except that this teleoptile is never preceded by a neossoptile. The steps of the process by which a feather expands into its complex figure from such a simple *matrix* or mould of form is thus graphically illustrated by Huxley:

"The integument of birds is always provided with horny appendages, which result from the conversion into horn of the cells of the outer layer of the epidermis. But the majority of these appendages, which are termed 'feathers,' do not take the form of mere plates developed upon the surface of the skin, but are evolved within sacs from the surfaces of conical papillæ of the dermis. The external surface of the dermal papilla, whence a feather is to be developed, is provided upon its dorsal [upper] surface with a median groove, which becomes shallower towards the apex of the papilla. From this median groove lateral furrows proceed at an open angle, and passing round upon the under surface of the papilla, become shallower, until, in the middle line, opposite the dorsal median groove, they become obsolete. Minor grooves run at right angles to the lateral furrows. Hence the surface of the papilla has the character of a kind of mould, and if it were repeatedly dipped in such a substance as a solution of gelatine, and withdrawn to cool until its whole surface was covered with an even coat of that substance, it is clear that the gelatinous coat would be thickest at the basal or anterior end of the median groove, at the median ends of the lateral furrows, and at those ends of the minor grooves which open into them; while it would be very thin at the apices of the median and lateral grooves, and between the ends of the minor grooves. If, therefore, the hollow cone of gelatine, removed from its mould, were stretched from within; or if its thinnest parts became weak by drying: it would tend to give way, along the inferior median line, opposite the rod-like cast of the dorsal median groove and between the ends of the casts of the lateral furrows, as well as between each of the minor grooves, and the hollow cone would expand into a flat feather-like structure with a median shaft, as a 'vane' formed of 'barbs' and 'barbules.' In point of fact, in the development of a feather such a cast of the dermal papilla is formed, though not in gelatine, but in the horny epidermic layer developed upon the mould, and, as this is thrust outward, it opens out in the manner just described. After a certain period of growth the papilla of the feather ceases to be grooved, and a continuous horny cylinder is formed, which constitutes the 'quill.'" (Introd. Classif. Anim., p. 71.)

Structure of Feathers. — A perfect feather, possessing all the structures a teleoptile can have developed, consists of the following named parts: (1) a main stem, shaft, or *scape* in two portions, *calamus* and *rhachis;* (2) a supplementary stem, aftershaft, or *hyporhachis;* (3) each stem bearing on each side a web, vane, or *vexillum;* (4) each web composed of a series of barbs or *rami;* (5) each barb bearing on each side a series of barbules or *radii;* (6) most barbules bearing a set of barbicels or *cilia;* (7) some barbicels forming hooklets or *hamuli.* Exactly how these several parts or structures combine to compose the feather is next to be shown.

(1) The main stem, shaft, or *scape* (Lat. *scapus*, a stalk) is usually divided into two well distinguished parts, *calamus* and *rhachis*. (*a*) The calamus (Lat. a reed) is the part next the body of the bird inserted by one end into the skin, and at the other end supporting the rest of the feather. This is the tube, barrel, or "quill" proper; a hard, horny, hollow, semitransparent cylinder, bearing no webs, and containing on the interior a little loose dry pith in the form of a series of caps or thimbles, sometimes called the "soul." These are the remains of the innermost layer of the inner division of the Malpighian stratum. One end of this quill tapers to its insertion, and is marked by the trace of what was an opening when the feather was growing; this is the *umbilicus inferior*. The other end of the calamus passes directly into the rhachis, at a point marked by a little pit, the *umbilicus superior*, on the under side of the feather (nearest the bird's body). The *rhachis* (Gr. ῥάχις, *rhachis*, a spine or ridge) is the direct continuation of the calamus to the tip of the feather, but differs in character, being a four-sided prism, squarish in cross section, tapering gradually to a fine point: it is less horny than the barrel, very elastic, opaque, and solidly pithy; it alone bears the vexilla, serving as a midrib between the two vanes for their whole extent. The rhachis is usually grooved length-

wise on its under side, this groove being best marked on the large feathers of the wings and tail; and it is commonly much longer than the calamus.

(2) The aftershaft or *hyporhachis* (Gr. ὑπό, *hupo*, and ῥάχις), when well developed is like a duplicate of the main feather, from the under side of the stem of which it springs, at junction of calamus with rhachis, close by the umbilicus superior. It is generally very small in comparison with the main part of the feather, though quite as large in a few birds, as Cassowaries, Emeus, and Moas. This counterpart or "counterfeit" is not developed in all groups of birds, nor on all feathers of any bird; its presence or absence, whether by non-acquisition or subsequent reduction, thus becomes a classificatory character of some importance. It is never well developed, but generally minute or wanting on the large strong wing- and tail-feathers; is best marked as an appendage of small contour feathers, and especially down feathers. The aftershaft may bear vanes, and generally does; but the barbs and barbules are never connected by barbicels or hooklets (as presently to be described for ordinary feathers), and therefore this supplementary feather is of a fluffy or downy texture, not close-webbed. The appearance of double feathers in the Emeu and some other ratite birds results from the equal size of the aftershaft and main shaft; the former is well developed though inconspicuous in Parrots, Gulls, Herons, and most raptorial birds; it is small and very weak in the great Passerine series, in most waders, and many Gallinæ birds; still smaller in or absent from the Duck tribe, in Totipalmate birds, in some Picarians, in Owls, Pigeons, and the Ostriches and Kiwis. More detailed notice of presence or absence of aftershafts will be found under heads of the groups of birds treated in this work.

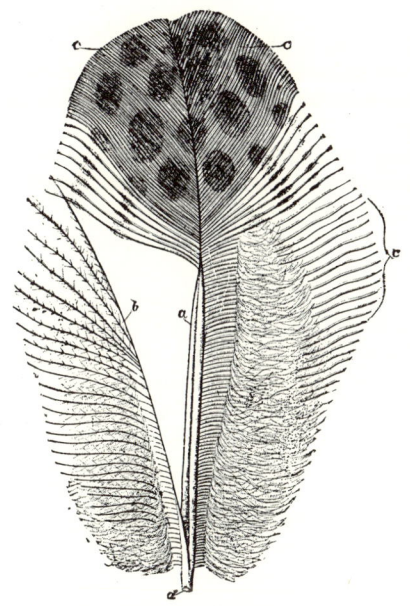

FIG. 19. — A partly pennaceous, partly plumulaceous feather, from Argus pheasant; after Nitzsch. *ad*, main stem; *d*, calamus; *a*, rhachis; *c, c, c*, vanes, cut away on left side in order not to interfere with *b*, the after-shaft, the whole of the right vane of which is likewise cut away.

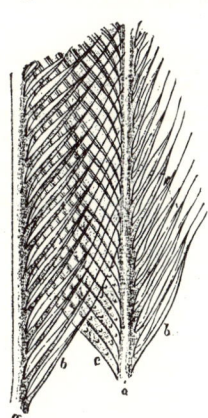

FIG. 20. — Two barbs, *a, a*, of a vane, bearing anterior, *b, b*, and posterior, *c*, barbules; enlarged; after Nitzsch.

(3) Each web, vane, or *vexillum* (Lat. *vexillum*, a standard; pl. *vexilla*) consists of a series of parallel, mutually appressed, flat, narrowly linear or lance-linear laminæ or plates, each one of which is set by its end obliquely on the rhachis (or on the hyporhachis, as the case may be), diverging at a varying open angle, and ending in a free point. Every such narrow flat plate constitutes a

(4) *Barb* or *ramus* of the vane (Lat. *barba*, a beard; *ramus*, a branch; pl. *rami*). The barbs may be likened to the blades of a pocket knife, with the sharp edge turned toward the under side of the feather, the blunt back of the blade turned to the other side. Barbs of the outer webs of many feathers are deeper, stronger, and shorter than those of inner webs, and commonly set on the rhachis at a more acute angle; this difference is best marked on large feathers of the wings and tail. The number of barbs to a vane is very variable; there may be several hundred. Now, if these barbs simply lay alongside one another, like leaves of a book,

without any means of holding together, the feather would have no texture or consistency — there would be no true web; therefore they are connected by means of

(5) **Barbules** or *radii* (Lat. *radius*, a ray; pl. *radii*). Just as the rhachis bears its two series of barbs, so does each barb bear two series of processes or plates of the second order. These *barbules*, as they are called (Lat., dimin. of *barba*), or *radii*, are to the barbs exactly what the barbs are to the main shaft, and are similarly given off from both sides of the thick upper border of a barb; they make the vane truly a *web* — that is, they so connect the barbs together that some slight force is required to pull them apart. Barbules are variously shaped, but generally flat sidewise, to pack together closely, with an upper and under edge at base, rapidly tapering to a slender thready end; and are long enough for each one to reach obliquely over several barbules of the next barb. Their number on most feathers is very great; a feather with a few hundred barbs to each web may have several hundred thousand barbules. All the structures thus far described may be seen by the naked eye or with a simple pocket lens; but a microscope is required to make out the minute structures by means of which the barbules confer consistency on the barbs. These are the

(6), (7) **Barbicels** or *cilia* (barbicel, another dimin. of Lat. *barba;* and *cilium*, an eyelash; pl. *cilia*), and **hooklets** or *hamuli* (Lat. *hamulus*, a little hook, dimin. of *hamus*, a hook; pl. *hamuli*). Both of these minute structures are simply a sort of fringe to a barbule, as if the end and part of the lower edge of the barbule were frayed out, and only differ from each other in that barbicels are plain hair-like processes, while hamuli are hooked at the end; they are not found on all feathers, nor on all parts of any feathers. There are countless millions of barbicels and hamuli on the main feathers of every bird which has smooth webby surface plumage and well-formed wings and tail; but their absence characterizes all neossoptiles, all supplementary feathers, and all the downy or hairy under plumages to be presently noticed. Barbicels occur on both anterior and posterior rows of barbules, though rarely on the latter; hooklets are confined to anterior series of barbules, which, as we have seen, overlie the posterior rows, forming a diagonal mesh-work. The purpose of this beautiful structure is evident; barbules are interlocked, and the whole made a web; for each hooklet of one barbule catches hold of a barbule from the next barb in front, any barbule thus holding on to as many barbules of the next barb as it has hooklets; while, to facilitate this interlocking, barbules have a thickened or folded-over upper edge of the right size for hooklets to grasp. The arrangement is shown in fig. 22, where a, a, a, a, are four barbs in transverse section, viewed from the cut surfaces, with their anterior, b, b, b, b, and posterior, c, c, c, c, barbules, the former bearing the hooklets which catch over the edge of the latter.

Types of Feathery Structure. — But all feathers do not answer the above complete description. The aftershaft may be wanting, as we have seen. Hooklets may not be developed, as frequently happens. Barbicels may be few or entirely lacking. Barbules may be similarly deficient, or so defective as to be only recognized by their position and relations. Even barbs may be few or lacking on one side of the shaft, or on both sides, as in certain bristly or hair-like styles of feathers. Finally the main stem may be a mere filament, without obvious distinction of calamus and rhachis. Consideration of these and other modifications of feather-structure has led to recognition of *three* types or plans : 1. The perfectly feathery, *plumous*, or *pennaceous* (Lat. *pluma*, a plume, or *penna*, a feather fit for writing with; fig. 23), as above described. 2. The downy or *plumulaceous* (Lat. *plumula*, a little plume, a down-feather), when the stem is short and weak, with soft rhachis and barbs, long slender thready barbules, little knots in place of barbicels, no hooklets, and consequently no smooth webbing. 3. The hairy, bristly, or *filoplumaceous* (Lat. *filum*, a thread), with a very long, slender stem, rudimentary or very small vanes composed of fine cylindrical barbs and barbules, if any, and no barbicels, knots, or hooklets. There is no abrupt definition between these types of structure;

in fact, the same feather may be constructed on more than one of these plans, in different parts of its length, as in fig. 19, partly pennaceous, partly plumulaceous. All feathers are built upon one or another, or some combination, or modification, of these types; and, in all their endless diversity, may be reduced to four or five.

Different Kinds of Feathers. — 1. *Contour-feathers, pennæ* or *plumæ* proper, have a perfect stem composed of calamus and rhachis, with smooth-webbed vanes of pennaceous structure, at least in part, usually plumulaceous toward the base. These form the great bulk of surface-plumage exposed to light; their beautiful tints give a bird its colors; they are the most modified in detail of all, from the fish-like scales of a penguin's wings to the glittering jewels of a Humming-bird, and all the endless array of tufts, crests, ruffs, and other ornaments of the feathered tribes; even the imperfect bristle-like feathers above mentioned may belong among them. The most conspicuous contour-feathers are the large ones of the wings and tail; these are also the most perfect feathers, except for lack of an aftershaft. Some contour-feathers are of fluffy texture, assume singular shapes, and grow to great lengths; such are commonly ornamental, and may be confined to the nuptial plumage, or characteristic of the male sex, as the aigrettes of many Herons, and the plumas of Paradise birds. Such feathers may not only lack all the minute structures above described, but even have the webs decomposed, owing to fewness of barbules or of barbs themselves. It would take me too far afield to go fully into their numerous variations. Contour-feathers are usually individually moved by subcutaneous muscles, of which there may be several to one feather, passing to be attached to the sheath of the tube, inside the skin, in which the stem is inserted. These muscles may be plainly seen under the skin of a goose, and every one has observed their operation when a hen shakes herself after a sand-bath, or erects her top-knot, or any other bird ruffles up its plumage. 2. *Down-feathers, plumulæ*, are characterized by a downy structure throughout. They more or less completely invest the body, but are almost always hidden beneath contour-feathers, like padding about the bases of the latter; occasionally they come to light, as in the fleecy ruff about the Condor's neck, and then usually replace contour-feathers; they have an aftershaft, or none; sometimes no rhachis at all, the barbs then being sessile in a tuft at the end of the calamus. They often stand in a regular quincunx (:·:) between four contour-feathers. All neossoptiles are of downy structure, though they belong to a different category of feathers, as we have seen, and we are now talking only of teleoptiles. Down-feathers, as a rule, are more copious in water birds than in land birds; swan's-down and eider-down are fine examples, and may be used by both birds and people to warm their respective nests. 3. *Semiplumes, semiplumæ*, may be said to unite the characters of the last two, possessing the pennaceous stem of the former, and the plumulaceous vanes of the latter; they are with or without aftershaft. They stand among pennæ, as plumulæ do, about the edges of patches of the former, or in parcels by themselves,

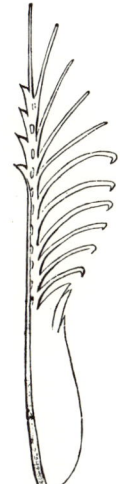

FIG 21. — A single barbule, bearing barbicels and hooklets; magnified; after Nitzsch.

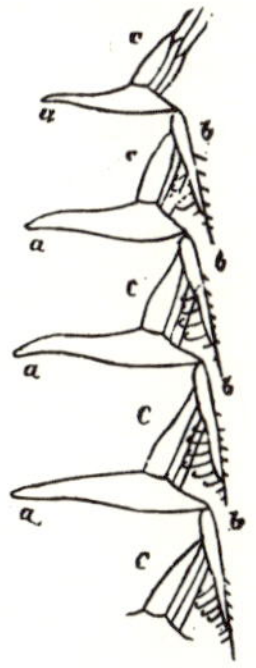

FIG. 22. — Four barbs in cross section, *a, a, a, a,* bearing anterior, *b, b, b, b,* and posterior *c, c, c, c,* barbules, the former bearing hooklets which catch over the latter; magnified; after Nitzsch.

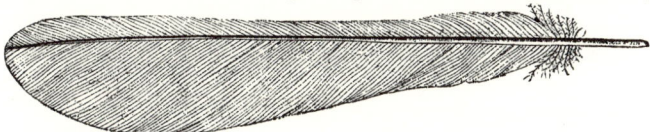

FIG. 23. — A feather from the tail of a Kingbird. *Tyrannus tyrannus,* almost entirely pennaceous; no aftershaft. From nature, by Coues.

but are always covered by contour-feathers. As implied in the name, the alleged distinctions of semiplumes are equivocal, and not easy to verify in all cases. Better marked are — 4. *Filoplumes, filoplumæ*, or thread-feathers, which have an extremely slender, almost invisible stem, not well distinguished into calamus and rhachis, and usually no vane, unless a terminal tuft of barbs be held for such. Long as they are, they are usually hidden by contour-feathers, close to which they stand as accessories, one or more seeming to issue out of the very sacs in which larger feathers are implanted. Sometimes they come to the surface, as the hairs on the neck of birds of the genus *Criniger*, so named from this fact; and some think the thready white plumes on the neck and flanks of Cormorants in nuptial plumage are filoplumaceous. Typical filoplumes are the nearest approach to *hairs* that birds have; they are very well shown on domestic poultry, being what a good cook finds it necessary to singe off after plucking a fowl for the table. 5. Certain down-feathers are remarkable for continuing to grow indefinitely, and with this unlimited growth is associated a continual breaking down of the ends of the barbs. Such plumulæ, from being always dusted over with dry, scurfy exfoliation, are called *powder-down;* they may be entitled to rank as a fifth kind of teleoptiles, which I have named *pulviplumes*. They occur in the Hawk, Parrot, and Gallinaceous tribes; also in certain Picarian birds (*Leptosomus* and *Podargus*); and especially in Herons and their allies. They are always present in the latter, where they may be readily seen as at least two large patches of greasy or dusty, whitish feathers, matted over the hips and on the breast. Pulviplumes are said to be luminous at times with a sort of phosphorescence; but what good it does a bird to wear such fungus-like puff-balls is unknown.

Colors of Feathers, in almost endless diversity of shade, hue, or tint, are reducible to three categories (see Newton's Dict., p. 95). 1. *Chemical, absorptive,* or *pigmentary* colors, due to the deposition in the caratine or substance of the feather of certain pigments, either in the form of fixed granules, or diffuse solution. Such colors are unvarying in any light in which they may be viewed. Some kinds of pigment have been distinguished by name as follows: *Zoomelanin*, or black; *zoonerythrin*, or red; and *zooxanthin*, or yellow; the names being derived from Greek words meaning "animal," and "black," "red," "yellow." To these add *turacin*, the particular red pigment of birds of the genus *Turacus*, family *Musophagidæ;* and *turacoverdin*, the green pigment of the same birds; the red color is due to copper and the green to iron. Browns are due to varying mixtures of red and black pigments. White is no color, but results from the molecular structure of the feather, in the absence of pigment. Gloss, of whatever color, is due to smooth polish of the surface of a feather. 2. What have been called *objective structural* colors result from surface-conditions of the feather in connection with underlying pigments. All blues, most greens, and some yellows belong in this category, as no blue pigment is known, and under the microscope these colors are always seen to depend upon the structure that overlies pigment of a different color. For example, the color basis of a blue feather may be a brownish or blackish pigment, and the blue only show as a condition of the surface of the barbs and barbules. 3. *Subjective structural, prismatic* or so called *metallic* colors constitute iridescence, or the glittering scintillation of those feathers which change rainbow-like according to the position in which they are viewed by the eye with regard to light, *i. e.*, to angle of incidence of light-rays. Iridescence is thus wholly due to superficial texture of a feather, without regard to the subjacent dark or black pigment. Prismatic hues are mostly confined to exposed surfaces of feathers, and to barbules which lack barbicels, and also have a particular disposition. Iridescence is to be distinguished from mere sheen, gloss, or "bloom" of a feather; it is carried to its pitch of perfection in Humming-birds, though many other groups of birds also exhibit this optical phenomenon.

Whatever be the coloration normal to any bird, that is its *chrosis* (Gr. χρῶσις, *chrosis*, coloring). But any bird may exhibit abnormal color or lack of color, either as a pathological

condition, or as due to particular diet, or to direct artificial tincture; this has been called *heterochrosis* (Gr. ἕτερος, *heteros*, other, and χρῶσις). The principal abnormal conditions are:
1. *Albinism*, in which the bird is white, wholly or in part; a "white Blackbird" is no misnomer, and white Crows or Ravens are well known. This is the commonest affection of the kinds now under consideration; any bird may, and many birds do, become entirely pure white, from failure of pigment; such are called *albinos*. 2. *Melanism*, or abnormal blackness, from excess of dark pigment. It is much less common than albinism, but by no means rare. 3. *Xanthism* (Gr. ξανθὸς, *xanthos*, yellow), or yellowness, as when a red, orange, or green bird turns out more or less yellow. 4. *Erythrism* (Gr. ἐρυθρὸς, *eruthros*, red), or redness. Both the last two cases are somewhat special ones, considered as abnormities. Feeding upon cayenne pepper may produce erythrism; in Brazil, where counterfeit species of *Chrysotis*, a genus of Parrots, are fashionable, they "are produced by the rubbing in of the cutaneous secretion of a toad, *Bufo tructorius*, into the budding feathers of the head, which then turn out yellow instead of green" (Newton, Dict., p. 99). It should be noted that all these heterochroses are abnormal; normal changes of plumage with age or season, and normal differences of plumage, are treated beyond. Neither dechromatism nor aptosochromatism is here in question.

Feather Oil Gland. — Birds do not perspire, and cutaneous glands, corresponding to the sweat-glands and sebaceous follicles so common in *Mammalia*, are hardly known among them. But their "oil-can" is a kind of sebaceous follicle, which may be noticed here in connection with other tegumentary appendages. This is a two-lobed or rather heart-shaped gland, saddled upon the "pope's nose," at the root of the tail, and hence sometimes called the *uropygial* (Lat. *uropygium*, rump), or rump gland; is also known as the *elæodochon* (Gr. ἐλαιοδόχος, *elaiodochos*, containing oil). It is composed of numerous slender tubes or follicles which secrete a greasy fluid, the ducts of which, uniting successively in larger tubes, finally open by one or more pores, commonly upon a little nipple-like elevation. Birds press out a drop of oil with the beak and dress the feathers with it, in the well-known operation called "preening." The gland is present in most birds; it is large and always present in aquatic birds, which have need of waterproof plumage; smaller in land-birds, as a rule, and wanting in some. The presence or absence of this singular structure, and whether or not it is surmounted by a particular circlet of feathers, distinguishes certain groups of birds, and has become much used in classification, as it was supposed to be related in some occult manner to the cœca of the intestine.

Pterylography. — Feathered Tracts and Unfeathered Spaces. — Excepting certain birds having obviously naked spaces, as about the head or feet, all would be taken to be fully feathered. So they are all *covered with feathers*, but it does not follow that feathers are everywhere implanted upon the skin. On the contrary, a uniform and continuous *pterylosis* is the rarest of all kinds of feathering; though such occurs, almost or quite perfectly, among certain birds, as Ostriches and their allies Penguins, and Toucans. If we compare a bird's skin to a well-kept park, part woodland, part lawn, then where feathers grow is the woodland, where they do not grow is the lawn. The former places are called *tracts* or *pterylæ* (Gr. πτερόν, *pteron*, a plume, and ὕλη, *hule*, woods; literally, "feather-forests"); the latter, *spaces*, or *apteria* (Gr. α privative, and πτερόν); they reciprocally distinguish certain definite areas. Not only are *pterylæ* and *apteria* thus definite, but their size, form, and arrangement mark whole families and even orders of birds, so that *pterylosis* becomes available, and is indeed found to be important, for purposes of classification. *Pterylography*, or the description of this matter, was first (1833) made a special study by the celebrated Nitzsch, who laid down the general plan of pterylosis which obtains in the great majority of birds, as follows: 1. The spinal or dorsal tract (*pteryla spinalis*; fig. 24, 1), running along the middle of a bird above from nape of neck to tail; subject to great variation in width, to dilation and contraction, to forking, to sending out branches, to interruption, to enclosing an apterium, etc. 2. Humeral or arm tracts (*pt. humerales*; Lat. *humerus*, shoulder, or upper arm-bone; fig. 24, 2), always present, one on each wing; they are narrow bands, running from the shoulder obliquely backward upon the upper arm-bone, parallel with the shoulder-blade. 3. Femoral or thigh tracts (*pt. femorales*; Lat. *femur*, thigh; fig. 24, 3); a similar oblique band upon the outside of each thigh, subject to great variation. 4. The ventral tract (*pt. ventralis*; Lat. *venter*, belly; fig. 24, 8), which forms most of the plumage on the under part of a bird, commencing at or near the throat, and continuing to the vent; like the dorsal tract, it is very variable, is broad or narrow, branched, etc., though always consisting of right and left halves, with a median apterium; thus, Nitzsch enumerates *seventeen* distinct modifications, and there are others. The foregoing are mostly isolated tracts, that is, bands nearly surrounded by complementary apteria; the following are, in general, continuously and uniformly feathered, and thus practically equivalent to the part of the body they represent: Thus, 5, the head tract (*pt. capitalis*; Lat. *caput*,

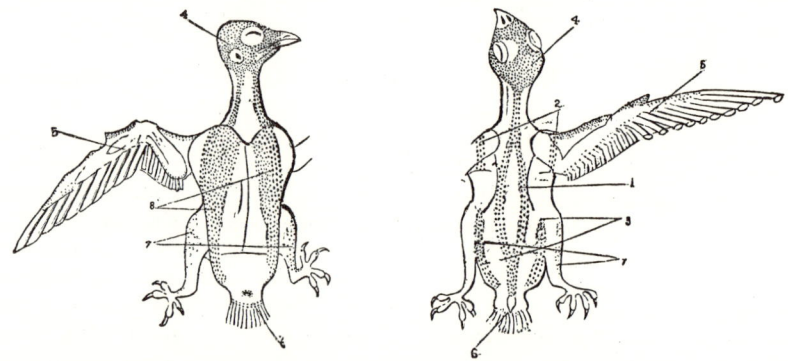

FIG. 24. — Pterylosis of *Micropus apus*, drawn by Coues after Nitzsch; right hand upper, left hand lower, surface. 1. spinal tract; 2. humeral; 3. femoral; 4. capital; 5. alar; 6. caudal; 7. crural; 8. ventral.

capitis, head; fig. 24, 4) clothes the head, and generally runs into the beginning of both dorsal and ventral tracts. 6. There may be a recognizable neck tract (*pt. colli*; Lat. *collum*, neck), and in some cases, as Herons, what Nitzsch called *pt. colli laterales*. 7. The alar or wing tract (*pt. alaris*; Lat. *ala*, wing; fig. 24, 5) represents all feathers that grow upon the wing, excepting those of the humeral tract. 8. The caudal or tail tract (*pt. caudalis*; Lat. *cauda*, tail; fig. 24, 6) includes the tail-feathers proper and their coverts, and usually receives the termination of dorsal, ventral, and femoral tracts. 9. The crural or leg tract (*pt. cruralis*; Lat. *crus*, *cruris*, leg; figs. 24, 7) clothes the legs as far as these are feathered, which is generally to the heel, always below the knee, and sometimes to the toes or even the claws. 10. The uropygial or rump tract (*pt. uropygii*) is confined to the elæodochon, which may be uniformly feathered, or naked except for a peculiar circlet of feathers which surmounts it. I need not enumerate the *apteria*, as these are merely the complements of the pterylæ. The highly important special "flight-feathers" of the wings and "rudder-feathers" of the tail are to be examined beyond, in describing those members for purposes of classification.

Endysis and Ecdysis. — Putting on and off Plumage. — Newly hatched birds, as already said, are partially or entirely covered for some time with a kind of down neossoptiles, entirely different from such teleoptiles as they ultimately acquire. The relation between these two sets of feathers has already been fully shown. Down is scanty, leaving much or all of the body naked, in most *altricial* birds, or such as are reared by the parents in the nest (Lat. *altrix*, female nourisher); but thick and puffy in some *Altrices*, and in all *Præcoces* (Lat. *præcox*, precocious), which run about at birth. Since many birds which require to be reared in the nest also hatch clothed, or very speedily become downy, a more exact distinction may be drawn by using the terms *ptilopædic* and *psilopædic* (Gr. πτίλον, *ptilon*, a feather; ψιλός, *psilos*, bare; and παῖς, *pais*, a child) respectively for those birds which hatch feathered or naked; a chicken and a canary-bird are familiar examples. Ptilopædic birds are also called dasypædic (Gr. δασύς, *dasus*, hairy), which means the same thing. It is the rule, that the higher birds are born helpless and naked or nearly so, requiring to be reared in the nest till their true feathers grow; the reverse with lower birds, as the walking, wading, and swimming tribes; and a primary division of birds has even been proposed upon this physiological distinction. It offers, however, too many exceptions; thus, no birds are more naked and helpless at birth than young Cormorants. Probably all præcocial birds are also ptilopædic and all psilopædic birds altricial; but the converse is far from holding good, many Altrices, as Hawks and Owls, being also ptilopædic. In other words, psilopædic birds are always altricial, but ptilopædic birds may be either altricial or præcocial. In view of this distinction which does not always distinguish, it has been proposed to drop the terms, and substitute *nidicolous* (Lat. *nidus*, nest, and *colere*, to inhabit) for those birds which stay some time in the nest, those which leave it at once being termed *nidifugous* (Lat. *nidus*, nest, and *fugere*, to flee). Thus, all *Altrices* are *Nidicolæ*, and all *Præcoces* are *Nidifugæ*; in either case without prejudice to the question whether the former are psilopædic or ptilopædic.

In any case, true feathers are soon gained, in some days or weeks, those of wings and tail being usually the first to sprout. The acquisition of plumage is called *endysis* (ἔνδυσις, *endusis*, putting on). The renewal of plumage is a process familiar to all, in its generalities, under the term "moult," or *ecdysis* (Gr. ἔκδυσις, *ekdusis*, putting off), though the details of the process have been worked out satisfactorily for comparatively few species of birds, and we have to be cautious in making statements concerning this subject; for unexpected exceptions may be unprovided for unless our language is guarded. Feathers are of such rapid growth, and make such a drain upon vital energies, that we easily understand how critical are periods of moult. The first plumage is usually worn but a short time; then another more or less complete change commonly occurs. The moult is annual, as a rule; and in many cases more than one moult is required before a mature bird attains the perfection of its feathering. It is well known how different many birds are the first year in their coloration from that afterward acquired; sometimes changes progress for several years; and some birds appear to have a period of senile decline. All such changes are necessarily connected, if not with actual moult, as is the rule, then at any rate with wear and tear and repair of the plumage. The first plumage having been gained, under whatever conditions peculiar to the species, it is the general rule, that birds are thereafter subject to *single*, or *annual*, moult; possibly there is no exception to the rule that a healthy adult bird renews its plumage at least once a year. This change commonly occurs when the duties of incubation are concluded, and the well-worn plumage most needs renewal, as happens in late summer and early autumn months in our latitudes; though some of our birds, as Swallows and Hawks, may put off the process till winter. Many birds, however, moult twice a year, the additional moult usually occurring in spring, when a fresh nuptial suit is acquired; in such cases, the moult is said to be *double*, or *semi-annual*. Such additional moult is generally incomplete; that is, all the feathers are not shed and renewed, but more or fewer new ones are gained, with more or less loss of old ones, if any. The most striking ornaments donned for the breeding season, as the elegant plumes of many Herons, are usually worn but a brief time, being doffed in advance of the general fall moult; and males of very many birds which put on special nuptial ornaments make room for these by doffing feathers from the parts where the new ones grow. A few birds, as Ptarmigan (*Lagopus*), regularly have a third or *triple* moult, " shedding their feathers as usual by annual moult in summer, then soon changing by another (partial) moult to pure white for the winter, then in spring moulting again more or less to assume their wedding dress." As a rule, feathers are moulted so gradually, particularly those of wings and tail, and so simultaneously upon right and left sides of the body, that birds are at no time deprived of power of flight; moreover, the first *flight-feathers* acquired by young birds are usually kept till the next season. But those that fly very early, before they are half grown, as so many gallinaceous birds do, include their first weak wing-feathers in the general moult which occurs to young and old in autumn. The Duck family (*Anatidæ*) and some others offer the remarkable case, that they drop their wing-quills so nearly all at once as to be for some time deprived of power of flight; and on regaining them the males acquire a *postnuptial* plumage very different from that gay attire they last wore — a dingy dress like that of the female. Numberless other birds, like our Tanagers, the males of which are brilliant in breeding dress, moult into a postnuptial plumage in which they resemble their homely mates. It is difficult to lay down any rules of moulting for particular groups of birds, since very closely related species may differ greatly in respect to their changes of plumage, and the subject has not yet received the attention its interest and importance should claim for it.

The physiological processes involved in endysis and ecdysis are analogous to those concerned in shedding of hair of mammals, casting of cuticle of reptiles; for hair, cuticle, and feather are alike cuticular or epidermal structures, as we have seen (p. 81). Therefore it need surprise no one to learn that feathers are not the only tegumentary appendages subject to moult. Some birds shed portions of the horny covering of the bill, feet, claws, even eyelids. Thus in the Grouse family (*Tetraonidæ*) the greatly overgrown claws of some species in the winter season are reduced in size by moult or by mechanical wearing away (as is also the case with some Lemmings among mammals); and some Grouse develop along the sides of the toes a fringe of horny process which is regularly shed and renewed. The bill of Redpolls of the genus *Ægiothus* enlarges in summer, bulging out into a redundant growth of horn, which in winter is mechanically worn down till the bill resumes its usual acutely conical shape. Our White Pelican regularly sheds a curious horny outgrowth of the upper mandible. But the most remarkable known cases of such ecdysis of horn are found in various species of the Auk family (*Alcidæ*); for a full account of which the reader is referred to my article in the body of this book, where that family is treated at length.

Aptosochromatism. — It is certain that many birds change the colors of their plumage without losing or gaining any feathers, and thus independently of moult. This is what I call *aptosochromatism* (Gr. a, privative; πτῶσις, *ptosis*, a falling off or away; χρῶμα, *chroma*, color, complexion). Though I coined this word many years ago, and some of the facts to which it applies were known long before my time, certain writers have had the hardihood to deny the facts and decry the term. It is asserted by such persons that a feather becomes a dry dead appendage as soon as it attains its growth; which is not true. A feather, like a hair or a claw, retains vitality for a time after it ceases to grow; and does not die until it is ready to be cast like a foreign body. True, there is no blood circulation after the pin-feather stage is past, and the "soul" of the feather has turned to dry pith; but a certain degree of vascularity persists for some time thereafter, maintaining vital connection with the body, and permitting certain molecular changes of pigmentation in the substance of the feather. The full-grown hairs of a mammal long retain a sort of circulation which in some cases is capable of altering their color; witness the bleaching of black or brown human hair in a few hours under some strong mental emotions of grief or terror. Feathers are in precisely the same case. Nay, more; a different degree of vitality can easily be shown to persist in different parts of the same feather. Thus, the primaries of many Gulls acquire definite white tips; and these wear away sooner than the black portion of the feathers, so that the white spots are lost as neatly as if they had been snipped away with scissors. It seems to be a rule, that heavy pigmentation tends to make a feather more durable than it would otherwise be. Again, the identical primary of a Gull acquires after it is full grown a greater extent of white or pearly web than it had before, by actual absorption or decomposition of black pigment in a portion of the web, the bleaching being thus progressive. Such changes could not go on in a dead feather; they are physiological processes, or at any rate chemical processes, in living tissue, — not merely mechanical alterations due to wear and tear of dead substance; and they affect coloration of plumage far more profoundly than is commonly recognized, as well as in a far greater number of cases than have been ascribed, as they should be, to aptosochromatism. It has been proven in the case of the red and gray phases of our Screech Owl (*Megascops asio*) that aptosochromatism is effected by actual alteration of pigmentation, without any loss of old or gain of new feathers. Erythrism and melanism, and numerous other alterations of color, may be undergone by birds without any moult; such change may be brought about by a particular diet, and certainly this could not occur if a grown feather were a dead feather, lacking all vascular connection with the fluids or humors of the body. Such physiological or chemical processes as are concerned in depigmentation and repigmentation of grown yet living feathers may be likened to the changes undergone by chlorophyll in the leaves of plants which change from green to yellow or scarlet while they still live, and do not lose vascular connection with the stem till they turn brown, wither, and drop. Persons who pluck live geese understand this matter better than some ornithologists do; they resort to this cruel process because they can get a better price for feathers torn from the living body of the poor bird, than for the identical feathers taken from the same goose dead, because the former are more elastic and more durable.

Aptosochromatism is thus primarily a physiological and chemical fact. But it extends to and is directly connected with a certain mechanical process by which plumage may be profoundly affected in coloration without loss or gain of any feathers. Now, if the student will refer back to what I have said regarding color, he will recall the facts, that pigmentary colors are often dependent upon the *texture* of feathers for their optical effect. For example, there are no blue pigments, but plenty of birds are blue by objective structural coloration; and any alteration in texture or structure of a feather is liable to produce a change of color. In fact, this sort of aptosochromatism is very common; it consists in shedding *certain parts* of a feather which have less vitality than the rest, and therefore break off and drop away before the whole feather dies and follows suit. Not only barbicels and barbules may be thus moulted, without visibly altering the shape of the feather, though very likely with some change of objective structural coloration, and in cases of iridescence with entire change of subjective structural coloration; but also some of the barbs themselves may fall away from the rest, with great change in the figure of the webs, and consequently great alteration in color of plumage if, as is usually the case, the lost portion of the webs be differently pigmented from the part that remains. Few ornithologists seem to be aware of the prevalence of this sort of aptosochromatism as a factor in modifying or entirely changing the coloration of birds. The male of our Bobolink, for example, acquires his faultless black plumage by shedding the long yellowish tips of the feathers which just before had veiled those portions of his wedding suit. The Snow Bunting, which has no spring moult, passes to the pure black and white nuptial plumage by dropping the brown edges and ends of black centred feathers; in this case, so much of each feather is lost that the shape changes from a broadly rounded to a sharply pointed contour. In the related genus of Longspurs I have found that certain uniformly glossy black areas result in like manner from loss of deciduous gray or brown portions of the webs. I have above spoken of this kind of aptosochromatism as "a certain mechanical process." It is mechanical in the sense of breakage and loss of parts of a feather, but this is not due to actual abrasion or wear and tear, and would not occur if there were no physiological process concerned; for if the parts in question were not devitalized they would not be deciduous, nor drop before the whole feather was ready to fall. While it is not probable, as some have claimed, that a worn feather can mend its ragged edges by a new growth of barbs, barbules, or barbicels, it is certain that a fresh feather retains for a while after it is full grown those molecular movements in its substance which may result in deposition of additional pigment, and in absorption or decomposition of pigment already laid down; so that some colored areas may be extended or restricted, and also change color to some degree, during the lifetime of an individual feather — that is, without moult — that is, in a word, aptosochromatism. Once more; if we turn from consideration of color-change in the webs of feathers to such as may be readily observed in their *shafts*, we find the same thing again. The surface of a rhachis is smooth, firm, and solidly horny, quite like the corneous covering of the bill. Now the beak of some birds, as of the genus *Leucosticte* for example, is black in summer and yellow in winter, and this is aptosochromatism, for nobody imagines that the horny sheath of the bill is shed in this genus; it is an actual alteration in color from black to yellow and back again. The same thing occurs, for instance, in the shaft of a Gull's primary, which alters from blackish to yellowish or white in a certain portion of its extent corresponding to the gradual extension of white areas in the adjoining portions of each web of the same feather, and has nothing whatever to do with the moult of that feather. The notorious inconstancy of coloration of what are called the "soft parts" of most water birds is another case in point. Such as these are "softer" than feathers, indeed, but horny epidermis is only "soft" in comparison with harder horn, not to the degree of what is commonly called vascularity, for it has no blood vessels. I adduce these facts to bring all the epidermal structures of birds into proper correlation, showing that feathers do not differ from beaks or claws so much as some have assumed in the degree of that kind of vascularity which they retain for a while after they have ceased to grow, and that in the interval between maturity and moult they may continue subject to color-changes (*a*) by pigmentary vicissitudes, (*b*) by structural modifications; both of which modes of alteration in coloration come under the head of aptosochromatism, or change of plumage without loss or gain of any feathers.

Plumage-changes with Sex, Age, and Season. — Aside from any consideration of the way in which plumage changes, whether by moult or otherwise, the fact remains that most birds of the same species differ more or less from one another according to certain circumstances. The dissimilarity is not only in coloration, though this is the usual and most pronounced difference, but also in the degree of development of plumes, — their size, form, and texture. Since young birds are those which have not come to sexual vigor; since breeding recurs at regular periods of adult life, annually or oftener; and since males and females usually differ in plumage, — nearly all the various dresses worn by different individuals of the same species are correlated with conditions of the reproductive system. As the internal generative organs represent of course the essential or *primary* sexual characters, all those of plumage just indicated may be properly classed as *secondary sexual characters*. These are of great importance, not only in practical ornithology, but as the basis of some of the soundest views that have been advanced respecting the evolution of specific characters in this class of animals. The generalizations may be made: that when the sexes are strikingly different in plumage, the young at first resemble the female; when the adults are alike, the young are different from either; when seasonal changes are great, the young resemble the fall plumage of the parents; and, further, that when the adults of two related species of the same genus are nearly alike, the young are usually intermediate, their specific characters not being fully developed. Specific characters are often to be found only in the male, the females of two related species being scarcely distinguishable, though the males may be told apart at a glance. Extraordinary developments of feathers, as to size, shape, and color, are often confined to one sex, usually the male. The more richly, extensively, or peculiarly the male is adorned, the simpler the female in comparison, as the

Peacock and Peahen. The Wise Man of Late has formulated several categories of secondary sexual characters, giving the following rules or classes of cases: "1. When the adult male is more beautiful or conspicuous than the adult female, the young of both sexes in their first plumage closely resemble the adult female, as with the common Fowl and Peacock; or, as occasionally occurs, they resemble her much more closely than they do the adult male. 2. When the adult female is more conspicuous than the adult male, as sometimes though rarely occurs [chiefly with certain birds of prey and snipe-like birds], the young of both sexes in their first plumage resemble the adult male. 3. When the adult male resembles the adult female, the young of both sexes have a peculiar first plumage of their own, as with the Robin [usual]. 4. When the adult male resembles the adult female, the young of both sexes in their first plumage resemble the adults [unusual]. 5. When the adults of both sexes have a distinct winter and summer plumage, whether or not the male differs from the female, the young resemble the adults of both sexes in their winter dress, or much more rarely in their summer dress, or they resemble the females alone. Or the young may have an intermediate character; or again they may differ greatly from the adults in both their seasonal plumages. 6. In some few cases the young in their first plumage differ from each other according to sex; the young males resembling more or less closely the adult males, and the young females more or less closely the adult females." — (Darwin, Desc. of Man, ed. 1881, p. 466.)

Summary of Secondary Sexual Characters of Birds. — The temptation to give the conclusion of the whole matter in Darwin's own words, summary of his views of Sexual Selection as so important a factor in Natural Selection, need not be resisted. I therefore quote again from the work last cited, pp. 496–499.

"Most male birds are highly pugnacious during the breeding season, and some possess weapons adapted for fighting with their rivals. But the most pugnacious and the best armed males rarely or never depend for success solely upon their power to drive away or kill their rivals, but have special means for charming the female. With some it is the power of song, or of giving forth strange cries, or instrumental music, and the males in consequence differ in their vocal organs, or in the structure of certain feathers. From the curiously diversified means for producing various sounds, we gain a high idea of the importance of this means of courtship. Many birds endeavor to charm the female by love-dances or antics, performed on the ground or in the air, and sometimes at prepared places. But ornaments of many kinds, the most brilliant tints, combs, and wattles, beautiful plumes, elongated feathers, top-knots, and so forth, are by far the commonest means. In some cases mere novelty appears to have acted as a charm. The ornaments of the males must be highly important to them, for they have been acquired in not a few cases at the cost of increased danger from enemies, and even at some loss of power in fighting with their rivals. The males of very many species do not assume their ornamental dress until they arrive at maturity, or they assume it only during the breeding season, or the tints then become more vivid. Certain ornamental appendages become enlarged, turgid, and brightly colored during the act of courtship. The males display their charms with elaborate care and to the best effect; and this is done in the presence of the females. The courtship is sometimes a prolonged affair, and many males and females congregate at an appointed place. To suppose that the females do not appreciate the beauty of the males, is to admit that their splendid decorations, all their pomp and display, are useless; and this is incredible. Birds have fine powers of discrimination, and in some few cases it can be shown that they have a taste for the beautiful. The females, moreover, are known occasionally to exhibit a marked preference or antipathy for certain individual males.

"If it be admitted that the females prefer, or are unconsciously excited by the more beautiful males, then the males would slowly but surely be rendered more and more attractive through sexual selection. That it is this sex which has been chiefly modified, we may infer from the fact that, in almost every genus where the sexes differ, the males differ much more from one another than do the females; this is well shown in certain closely-allied representative species, in which the females can hardly be distinguished, whilst the males are quite distinct. Birds in a state of nature offer individual differences which would amply suffice for the work of sexual selection; but we have seen that they occasionally present more strongly-marked variations which recur so frequently that they would immediately be fixed, if they served to allure the female. The laws of variation must determine the nature of the initial changes, and will have largely influenced the final result. The gradations, which may be observed between the males of allied species, indicate the nature of the steps through which they have passed. They explain also in the most interesting manner how certain characters have originated, such as the indented ocelli on the tail-feathers of the peacock and the ball and socket ocelli on the wing-feathers of the Argus pheasant. It is evident that the brilliant colors, top-knots, fine plumes, &c., of many male birds cannot have been acquired as a protection; indeed, they sometimes lead to danger. That they are not due to the direct and definite action of the conditions of life, we may feel assured, because the females have been exposed to the same conditions, and yet often differ from the males to an extreme degree. Although it is probable that changed conditions acting during a lengthened period have in some cases produced a definite effect on both sexes, or sometimes on one sex alone, the more important result will have been an increased tendency to vary or to present more strongly-marked individual differences; and such differences will have afforded an excellent ground-work for the action of sexual selection.

"The laws of inheritance, irrespectively of selection, appear to have determined whether the characters acquired by the males for the sake of ornament, for producing various sounds, and for fighting together, have been transmitted to the males alone or to both sexes, either permanently, or periodically during certain seasons of the year. Why various characters should have been transmitted sometimes in one way and sometimes in another, is not in most cases known; but the period of variability seems often to have been the determining cause. When the two sexes have inherited all characters in common, they necessarily resemble each other; but as the successive variations may be differently transmitted, every possible gradation may be found, even within the same genus, from the closest similarity to the widest dissimilarity between the sexes. With many closely-allied species, following nearly the same habits of life, the males have come to differ from each other chiefly through the action of sexual selection; whilst the females have come to differ chiefly from partaking more or less of the characters thus acquired by the males. The effects, moreover, of the definite action of the conditions of life, will not have been masked in the females, as in the males, by the accumulation through sexual selection of strongly-pronounced colors and other ornaments. The individuals of both sexes, however affected, will have been kept at each successive period nearly uniform by the free intercrossing of many individuals.

"With species, in which the sexes differ in color, it is possible or probable that some of the successive variations often tended to be transmitted equally to both sexes; but that when this occurred the females were prevented from acquiring the bright colors of the males, by the destruction which they suffered during incubation. There is no evidence that it is possible by natural selection to convert one form of transmission into another. But there would not be the least difficulty in rendering a female dull-colored, the male being still kept bright-colored, by the selection by successive variations, which were from the first limited in their transmission to the same sex. Whether the females of many species have actually been thus modified, must at present remain doubtful. When, through the law of the equal transmission of characters to both sexes, the females were rendered as conspicuously colored as the males, their instincts appear often to have been modified so that they were led to build domed or concealed nests.

"In one small and curious class of cases the characters and habits of the two sexes have been completely transposed, for the females are larger, stronger, more vociferous and brighter colored than the males. They have, also, become so quarrelsome that they often fight together for the possession of the males, like the males of other pugnacious species for the possession of the females. If, as seems probable, such females habitually drive away their rivals, and by the display of their bright colors or other charms endeavor to attract the males, we can understand how it is that they have gradually been rendered, by sexual selection and sexually-limited transmission, more beautiful than the males — the latter being left unmodified or only slightly modified.

"Whenever the law of inheritance at corresponding ages prevails, but not that of sexually-limited transmission, then if the parents vary late in life — and we know that this constantly occurs with our poultry, and occasionally with other birds — the young will be left unaffected, whilst the adults of both sexes will be modified. If both these laws of inheritance prevail and either sex varies late in life, that sex alone will be modified, the other sex and the young being unaffected. When variations in brightness or in other conspicuous characters occur early in life, as no doubt often happens, they will not be acted on through sexual selection until the period of reproduction arrives; consequently if dangerous to the young, they will be eliminated through natural selection. Thus we can understand how it is that variations arising late in life have so often been preserved for the ornamentation of the males; the females and the young being left almost unaffected, and therefore like each other. With species having a distinct summer and winter plumage, the males of which either resemble or differ from the females during both seasons or during the summer alone, the degrees and kinds of resemblance between the young and the old are exceedingly complex; and this complexity apparently depends on characters, first acquired by the males, being transmitted in various ways, as limited by age, sex, and season.

"As the young of so many species have been but little modified in color and other ornaments, we are enabled to form some judgment with respect to the plumage of their early progenitors; and we may infer that the beauty of our existing species, if we look to the whole class, has been largely increased since that period, of which the plumage gives us an indistinct record. Many birds, especially those which live much on the ground, have undoubtedly been obscurely colored for the sake of protection. In some instances the upper exposed surface of the plumage has been thus colored in both sexes, whilst the lower surface in the males alone has been variously ornamented through sexual selection. Finally, from the facts given in these four chapters [pp. 358-499 of the work in citation], we may conclude that weapons for battle, organs for producing sound, ornaments of many kinds, bright and conspicuous colors, have generally been acquired by the males through variation and sexual selection, and have been transmitted in various ways according to the several laws of inheritance — the female and the young being left comparatively but little modified."

b. TOPOGRAPHY OF BIRDS.

The Contour of a Bird with the feathers on is spindle-shaped, or *fusiform* (Lat. *fusus*, a spindle), tapering at both ends; it represents two cones joined base to base at the middle or greatest girth of the body, tapering in front to the tip of the bill, behind to the end of the tail. The obvious design is easiest cleavage of air in front, and least drag or wash behind, in the act of flying. This shape is largely produced by the lay of the plumage; a naked bird presents several prominences and depressions, this irregular contour being reducible, in general terms, to two spindles or double cones. The head tapers to a point in front, at the tip of the bill, and contracts behind toward the middle of the neck, in consequence of diminution in bulk of the muscles by which it is slung on the neck; which last is somewhat contracted or hour-glass shaped near the middle, swelling where it is slung to the body. The body is largest in front and tapers to the tail.

The Centre of Gravity is admirably preserved beneath the centre of the body, and opposite the points where it is supported by the wings. The enormous breast-muscles of a bird are among its heaviest parts, sometimes weighing, to speak roundly, as much as one-sixth of the whole bird. Now these are they that effect all the movements of the wings at the shoulder-joints, lifting as well as lowering the wings. Did these pectoral muscles pull straight, the lifters would have to be *above* the shoulder-joint; but they all lie below it, and the lifters accomplish their office by running through pulleys to change the line of their traction. They work like men hoisting sails from the deck of a vessel; and thus, like a ship's cargo, a bird's chief weight is kept below the centre of motion. Top-heaviness is further obviated by the way in which birds with a long heavy neck and head draw these parts in upon the breast, and extend the legs behind, as is well shown by the attitude of a heron flying. The nice adjustment of balance by the variable extension of the head and feet is exactly like that produced in weighing by shifting a weight along the arm of a steel-yard; and together with the slinging of the chief weight under the wings instead of over or even between them, enables a bird to easily keep right side up in flight.

The Exterior of a Bird is divided for purposes of description into *seven* parts:— 1. Head (Lat. *caput*); 2. Neck (Lat. *collum*); 3. Body proper, or trunk (Lat. *truncus*); 4. Bill or beak (Lat. *rostrum*); 5. Wings (Lat. pl. *alæ*); 6. Tail (Lat. *cauda*); 7. Feet (Lat. pl. *pedes*). Of these, 1, 2, 3, head, neck, and trunk, are collectively termed *body* (Lat. *corpus*), in distinction from 4, 5, 6, 7, which are *members* (Lat. *membra*). Wings and feet are of course double or paired parts. The bill is strictly but a part of the head; but its manifold uses as an organ of prehension make it functionally a hand, and therefore one of the "members."

The Head has the general shape of a four-sided pyramid; of which the base is applied to the end of the neck, therefore not appearing from the exterior, and the apex of which is frustrated at the base of the bill. The uppermost side is more or less convex or vaulted, sloping in every direction; the under side is flattish and horizontal; the lateral surfaces are flattish and vertical; all similarly taper forward. The departures from any such typical shape are endless in degree and variable in kind, giving rise to numerous general descriptive terms, such as "head flattened," "head globular," but not susceptible of exact definition. The head is moulded, of course, upon the skull, corresponding in a general way to the brain-cavity of the cranium proper, both in size and shape; but it differs in several particulars. In the first place, there is the scaffolding of the jaws; secondly, large excavations to receive the eye-balls, and smaller ones for the ear-parts; thirdly, muscular and sometimes glandular masses overlying the bone; and lastly, in some birds, large hollow spaces in bone between the inner and outer tables or plates of the cranial walls. Each side of the head presents two openings for *eye* (Lat. *oculus*) and *ear* (Lat. *auris*), the position of which is variable, both absolutely and in relation to each other. But in the vast majority of birds, the eye is strictly lateral in situation, and near the middle of the side of the head; while the ear is behind and a little below the eye, near the articulation of the lower jaw. But the shape of the skull of Owls is such, that the eyes are directed forward, and such birds are said to have "eyes anterior." Owls also have enormous outer ears, in some cases provided with a movable flap or conch, closing upon the opening like the lid of a box; and in many cases their ear-parts, and some of the cranium itself, is unsymmetrical. In most birds the ear-opening is quite small, and only covered by modified feathers. In Woodcock and Snipe, owing to the way the brain-box is tilted up, the ears are below and not behind the eyes. The *mouth* (Lat. *os*, gen. *oris*) is always a fissure across the front of the head. The cleavage varies, both in extent and direction; the latter is usually horizontal, or nearly so, but may trend much downward; the former varies from a minimum, in which the cleft does not reach back of the horny part of the bill, as in a snipe, to the maximum seen in fissure-billed birds like Swifts and Goatsuckers, which gape almost from ear to ear. There are no other openings in the head proper, for the nostrils are always in the bill.

The Neck, in effect, is a simple cylinder, rendered somewhat hour-glass shaped, as above said. It consists of a movable chain of bones, or *cervical vertebræ* (Lat. *cervix*, neck ; *verto*, I turn), enveloped in muscle, along which in front lie the gullet (Lat. *œsophagus*) and windpipe (Lat. *trachea*), with associate blood-vessels, nerves, etc. Its length is very variable, as is the number of its bones, the latter ranging from 8 to about 26. Bearing as it does the head, with the bill, which serves as a hand, the neck is extremely flexible, to permit necessarily varied movements of this handy member. Its least length may be that which allows the point of a bird's beak to reach the oil-gland on the rump ; its greatest length sometimes exceeds that of the body and tail together, as in the case of a Swan, Crane, or Heron. The length is usually in direct proportion to that of the legs, in obvious design of allowing the beak to touch the ground easily to pick up food. The neck is habitually carried in a double curve, like an open S or italic *f*, the lower belly of the curve, convex forward, fitting in between the forks of the merry-thought (Lat. *furculum*), the upper curve, convex forward, holding the head horizontal at the same time. This "sigmoid flexure" (*sigma*, Greek S), highly characteristic of a bird's neck, is produced by saddle-shaping of the articular surfaces of nearly all its bones. The mechanical arrangement is such, that the sigma may be easily bent till the upper end (head) rests on the lower convexity, or as easily straightened to a right line ; but little if any further deviation in opposite curvature is permitted. As a generalization, the neck may be called relatively longest in wading birds, as Herons, Cranes, Ibises, etc.; shortest in perching birds, as the great majority of small *Insessores;* intermediate in swimming birds. But many swimmers, as Swans and Cormorants, have extremely long necks ; and some waders, as Plovers, have very short ones. A long neck is a rarity among higher birds (above *Gallinæ*), in most of which the head seems to nestle upon the shoulders. The longer the neck, the more sinuous and flexible is it likely to be. Anatomically, the neck ends in front at the articulation of the *atlas* (first cervical vertebra) with the skull, and behind at the first vertebra which bears free jointed ribs reaching the sternum. The shape of

The Body proper, or Trunk, is obviously referable to that of an egg ; it is *ovate* (Lat. *ovum*, an egg ; whence *oval*, the plane figure represented by the middle lengthwise section of an egg ; *ovate* or *ovoid*, the solid figure). The swelling of the breast represents the greatest diameter of the egg, usually near the larger end. But an ovoid is never perfectly expressed, and departures from such figure are numberless. In general, perching birds have the body nearly of ovate shape ; among waders, the figure is usually *compressed*, or flattened vertically, as is well seen in Herons, and still better in Rails, where the lateral narrowing is at an extreme ; among swimmers, the body is always more or less *depressed*, or flattened horizontally, and especially underneath, that the birds may rest on water with more stability, as well shown by a Duck or Diver. Anatomically the body begins with the foremost one of the *dorsal vertebræ*, or those that bear true ribs ; laterally, it ceases quite definitely at the shoulder-joints, the whole fore limb being outside the general content of the trunk ; behind, in mid-line, it includes everything, only the tail-*feathers* themselves being beyond it; behind and laterally, it includes more or less of the legs, for these are generally buried in common integument of the body nearly or quite to the knee-joint, sometimes to the heel-joint; though in anatomical strictness the trunk is limited by the hip-joint. The rib-bearing extent of the back-bone, ribs themselves, and the greatly enlarged breast-bone (Lat. *sternum*) compose the cavity of the chest (Lat. *thorax*). Upon this bony box, which contains the heart and lungs and some other viscera, are saddled on each side the bones of the *shoulder-girdle* or *scapular-arch*, namely, the shoulder-blades (Lat. *scapula*), the *coracoids*, and the collar-bones (Lat. *clavicula*), all three of which come together at the shoulder-joint. The thoracic cavity is not separated by any partition or *diaphragm* from that of the belly (Lat. *abdomen*), which with the *pelvis*, or basin, contains the digestive, urinary, and genital organs. The pelvis is composed, in dorsal mid-line, of so many vertebræ (*dorso-lumbar, sacral* proper, and *urosacral*, as become immovably joined to one another, and laterally of the confluent haunch-bones. The numerous *anchylosed* (or confluent) vertebræ compose the *sacrum*. The haunch-bones or *ossa innominata* consist on each side of three bones, *ilium, ischium,* and *pubis*, in adult life more or less perfectly anchylosed. Where they all three come together on each side is the hip-joint or *coxa*. The remaining bones, usually included among those of the body proper, are the *coccygeal* or caudal vertebræ. (For anatomical detail see beyond, under *Osteology*, etc.)

Topography of the Body. — Besides being thus divided into head, neck, trunk, and members, the exterior of the body is further subdivided or mapped out into *regions* for purposes of description. It is necessary for the student to become familiar with the "topography" of a bird, as this kind of mapping out may be called, for names of regions or outer areas are incessantly used in ordinary descriptive ornithology. Many more names have been applied than are in common use; I shall try to define and explain all those which are usually employed, beginning with the parts of the *body*, and ending with those of the *members*.

1. REGIONS OF THE BODY.

Upper and Under Parts. — Draw a line from corner of mouth along side of head and neck to and through shoulder-joint and thence along side of body to root of tail; all above this line, including upper surfaces of wings and tail, are *upper parts;* all below it, including under surfaces of wings and tail, are *under parts;* for which the short words "above" and "below" often stand. The distinction is arbitrary, but so convenient as to be practically indispensable. It will be seen how an otherwise lengthy description, enumerating parts that lie over or under the "lateral line," can be put in so few words as, for example, "above, green; below, yellow." Many birds' colors have some such simple general distribution. These parts are also *dorsal* (Lat. *dorsum*, back) and *ventral* (Lat. *venter*, belly) surfaces or aspects. Upper parts of the body proper, or trunk, have also received the general name of *notæum* (Gr. νῶτος, *notos*, back); under parts, similarly restricted, that of *gastræum* (Gr. γαστήρ, *gaster*, belly). but these terms are not much used. These two are *never naked*, while both head and neck may be variously bare of feathers. The only exception is the transient condition of certain birds during incubation, when, like the Eider Duck, they pull off feathers to furnish the nest, or when the plumage, as usually happens, wears off. The gastræum is rarely ornamented with feathers different in texture or structure from those of the plumage at large; but such a case is furnished by Lewis' woodpecker (*Asyndesmus torquatus*), and much more notable cases are those of certain Birds of Paradise, Storks, etc. The notæum, on the contrary, is often the seat of extraordinary development of feathers, either in size, shape, or texture, or all three of these qualities; as the singularly elegant dorsal plumes of many Herons. Individual feathers of the notæum are mostly pennaceous, straight, lanceolate; and as a whole lie smoothly shingled or *imbricated*. The ventral feathers are usually more largely plumulaceous, and less flat and imbricated, but even more compact — that is, thicker — than those of the upper parts; especially among water birds, where they are more or less curly, and very thick-set. There are subdivisions of the

Notæum. — Beginning where neck ends, and ending where tail-coverts begin (see fig. 25, 12), this part of a bird is subdivided into *back* (Lat. *dorsum;* fig. 25, 11) and *rump* (Lat. *uropygium;* fig. 25, 13). These are in direct continuation of each other, and their limits are not precisely defined; feathers of both grow on the *pteryla dorsalis.* In general, we should call the anterior two-thirds or three-fourths of notæum "back," and the rest "rump." With the former are generally included the scapular or shoulder-feathers, *scapulars* or *scapularies;* these are they that grow on the *pterylæ humerales.* The region of notæum they represent is called *scapulare* (Lat. *scapula*, shoulder-blade), and that part of notæum strictly between them is called *interscapulare* (fig. 25, 10); it is often marked, as in the Chipping Sparrow, with streaks or some other distinctive coloration. A part of dorsum, lying between interscapulare and uropygium, is sometimes recognized as "lower back" (Lat. *tergum*); but this distinction is not practically useful. To uropygium probably also belong feathers of the *pterylæ femorales*, or at

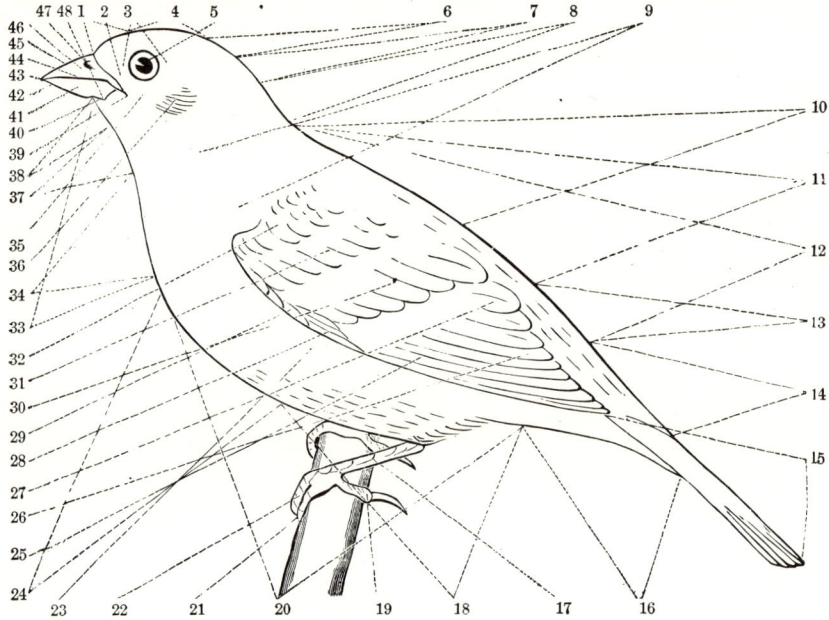

FIG. 25. — Topography of a Bird. 1, forehead (*frons*). 2, lore (*lorum*). 3, circumocular region. 4, crown (*vertex*). 5, eye. 6, hind head (*occiput*). 7, nape (*nucha*). 8, hind neck (*cervix*). 9, side of neck. 10, interscapular region. 11, back proper (*dorsum*), including 10. 12, *notæum*, or upper part of body proper, including 10, 11, and 13. 13, rump (*uropygium*). 14, upper tail-coverts (*tectrices superiores*). 15, tail (*cauda*). 16, under tail-coverts (*crissum* or *tectrices inferiores*). 17, tarsus. 18, abdomen. 19, hind toe (*hallux*). 20, *gastræum*, including 18 and 24. 21, outer or fourth toe. 22, middle or third toe. 23, side of body. 24, breast (*pectus*). 25, primaries. 26, secondaries. 27, so-called tertiaries; nos. 25, 26, 27 are flight-feathers or *remiges*. 28, primary coverts. 29, *alula*, or bastard wing. 30, greater coverts (*tectrices majores*). 31, median coverts (*tectrices medianæ*). 32, lesser coverts (*tectrices minores*). 33, "throat," including 34, 37, 38. 34, *jugulum*, or lower throat. 35, malar region. 36, auriculars. 37, *gula*, or middle throat. 38. *mentum*, or chin. 39, angle of commissure, or corner of mouth. 40, ramus of under mandible. 41, side of under mandible. 42, *gonys*. 43, *apex*, or tip of bill. 44, *tomia*, or cutting edges of bill. 45, *culmen*, or ridge of upper mandible, corresponding to gonys. 46, side of upper mandible. 47, nostril (*naris*). 48 passes across bill a little in front of its *base*.

any rate these are commonly included with rump in descriptions; but they more properly represent *flanks* (Lat. *ilia*, or *hypochondria*) — that is, sides of rump. They are sometimes the seat of largely developed or otherwise peculiarly modified feathers, as the snowy flank-plumes of the White-bellied Swift (*Aëronautes saxatilis*) or Violet-green Swallow (*Tachycineta thalassina*), which meet over the rump. The whole of notæum, taken together with upper surfaces of wings, is called the *mantle* (Lat. *stragulum*, a cloak); often a convenient term, as in describing Gulls and Terns, for example. In like manner, the

Gastræum is subdivided into regions, called *breast* (Lat. *pectus;* fig. 25, 24), *belly* (Lat. *abdomen;* fig. 25, 18), and *sides of body* (Lat. *pleura;* fig. 25, 23). The "sides" or pleuræ belong as much to dorsal as to ventral aspects of a bird's body; but in consequence of the underneath-freighted shape, the line we drew passes so high up along them, that they are almost entirely given over to gastræum. The *breast* begins over the merry-thought where jugulum (see beyond) ends; on either hand, it slopes up to "sides"; behind, its extension is indefinite. It should properly reach as far as the breast-bone does, to the limit of the thorax; but in many birds this would leave almost nothing for abdomen, and the limit would fluctuate with almost every family of birds, the sternum being so variable in length. Practically, therefore, without reference to the breast-bone, "breast" or *pectus* is restricted to the *swelling* anterior part of gastræum, which we call belly or *abdomen* as soon as it begins to straighten out and flatten. Abdomen, like pectus, rounds up on either hand into sides; behind, it ends in a transverse line passing across the anus. It has been unnecessarily divided into *epigastrium* or "pit of the stomach," and *venter* or lower belly; but these terms are rarely used. (*Crissum* is a frequent name of some indefinite region immediately about the vent; sometimes meaning flanks, sometimes vent-feathers or under tail-coverts proper; I refer to it again in connection with these last.) Though these boundaries seem fluctuating and not perfectly satisfactory, a little practice will enable the student to appreciate their proper use in descriptions, and to employ them himself with sufficient accuracy. The adjectival terms are respectively *pectoral, abdominal,* and *lateral.* The anterior continuation of the trunk, or the

Neck (Lat. *collum*) is likewise subdivided into regions. Its lateral aspects, except in birds that have lateral neck-tracts of feathers, are formed by the meeting over its sides of feathers that grow on dorsal and ventral pterylæ, the skin being usually not planted with feathers on its sides. Partly on this account, perhaps, a distinct region is not often named; we say simply "sides of neck," or "neck laterally" (*parauchenia,* fig. 25, 9). The neck behind, or its dorsal (upper) aspect, is divided into two portions: a lower, "hind neck" proper, or "scruff of neck" (Lat. *cervix;* fig. 25, 8), next to back; and an upper, or "nape of neck" (Lat. *nucha;* fig. 25, 7), adjoining hind head. These are otherwise respectively known as the *cervical* and *nuchal region;* and, in speaking of both together, we usually say "neck behind." The front of the neck has been needlessly subdivided, and these subregions vary with almost every writer. It suffices to call it *throat* (Lat. *gula,* fig. 25, 37, or *jugulum,* 34); remembering that the *jugular* portion is lowermost, vanishing in breast, and the *gula* uppermost, running into chin along under surface of head. *Guttur* is a term sometimes used to include gula and jugulum together: it is equivalent to "throat," as just defined; the adjective is *guttural.* Though generally covered with feathers, the neck is frequently naked in part. When naked behind, it is usually cervix that is bare, as so characteristically occurs in Herons, from interruption of forward extension of pteryla spinalis. Nucha is seldom if ever naked, except as an extension of general bald-headedness. Gula is similarly naked from above downward, as conspicuously illustrated in the order *Steganopodes,* comprising Pelicans, Cormorants, etc., which have a bare gular pouch; and as seen in many Vultures, whose baldness extends over nucha and gula, and even all around the neck, as in the Condor, whose nakedness ends with so singular a collar of close-set, downy feathers. The lower throat or jugulum becomes naked in a few birds, in which a distended crop or craw protrudes, pushing apart feathers of two branches of pteryla ventralis as these ascend the neck. The rule is, that the neck is not the seat of enlarged or otherwise highly developed feathers, which might restrict the requisite freedom of its motion; but there are some signal exceptions, among which may be instanced the Grouse family. The Ruffed Grouse has a singular umbrella-like tuft on each side of the neck: the Pinnated Grouse has still more curious winglets in the same situation, covering bare distensible skin: the Sharp-tailed Grouse is in somewhat similar but less pronounced case; while the Cock-of-the-plains has some extraordinary jugular developments of feathers in connection with his subcutaneous tympanum. Cervix proper almost never has modified feathers, but often a transverse coloration different from that of the rest of the upper parts; when conspicuous, this is called "cervical collar," to distinguish it from guttural or jugular "collars" or rings of color. Nucha is frequently similarly marked with a "nuchal band;" often special developments there take the form of *lengthening* of feathers, and we have a "nuchal crest." More particularly in birds of much variegated colors, guttur and jugulum are marked *lengthwise* with stripes and streaks, of which those on the sides are apt to be different from those along the middle line in front. Jugulum occasionally has lengthened feathers, as in many Herons. Higher up, the neck in front may have variously lengthened or otherwise modified feathers. Conspicuous among these are the *ruffs* or tippets of some birds, especially of the Grebe family (*Podicipedidæ*), and of the male ruff (*Pavoncella pugnax*). But these, and a few other modifications of feathers of upper neck, are more conveniently considered with those of the

Head. — Though smaller than any of the areas already considered, the head has been more minutely mapped out, and much detail is required by the number and importance of its recognizable parts or regions. Without intending to mention all that have been named, I describe all needed to be known for any practical purposes.

"Top of head" is a collective term for all the upper surface, from base of bill to nape, and laterally about to level of upper border of eyes; this is *pileum* or "cap" (fig. 25, 1, 4, 6): it is divided into three portions. The *forehead*, frontal region, or simply "the front" (Lat. *frons*; fig. 25, 1), includes all that slopes upward from bill, — generally to about opposite anterior border of eyes. *Middle head* or crown (Lat. *corona*, or *vertex*; fig. 25, 1), includes top of head proper, or highest part, from rise of forehead to fall of hind-head toward nucha. This slope is *hind-head* (Lat. *occiput*; fig. 25, 6). The lateral border of all three constitutes the superciliary line, that is, line over eye (Lat. *super*, over; *cilia*, little hairs, especially of the brows). "Crown" means the same thing as pileum. The adjectives of the several words are *frontal, coronal* or *vertical,* and *occipital* (pileum has none in use, coronal being said instead).

"Side of head" is a general term defining itself; it presents for consideration several regions. The *orbital* or *circumorbital* region, or simply *orbit* (Lat. *orbis*, an orb, here meaning socket of eyeball; fig. 25, 3), is a small space forming a ring around eye. It includes eye, and especially eyelids (Lat. *palpebræ*). The points where these meet, in front and behind, respectively, are *anterior canthus* and *posterior canthus* (Gr. κανθός, *kanthos*, Lat. *canthus*, a tire). The orbital region is subdivided into *supra-orbital, infra-orbital, ante-orbital,* and *post-orbital,* according as its upper, under, front, or back portion is desired to be specially designated. The situation of the orbit varies much in different groups of birds; it is generally midway, as said above, but may be higher or lower, jammed on toward bill, or pushed far up and back, as strikingly shown in Woodcock. In Owls, the orbital region is exaggerated into a great disc of radiating feathers, conferring a peculiar physiognomy. The *aural* or *auricular* (Lat. *auris*, or *auriculum,* ear; fig. 25, 36) region lies about the external opening of the ear, or *meatus auditorius;* its position varies in heads of different shapes, but it nearly always lies behind and a little below eye. Wherever located, it may be recognized at a glance, by a peculiar texture of feathers (the *auriculars*) which overlie the meatus. Doubtless to offer least obstacle to sound, these are a parcel of loose-webbed little plumes, which may be collectively raised and turned forward, exposing orifice of ear; they are extremely large in those Owls which have complicated external ear-parts, and in such they form a portion of the great facial disc. The term "temporal region" or "temple" is not often used in ornithology, not being well distinguished from a post-orbital space between eye and ear, and having nothing special about it. At lowermost back corner of side of head, generally just behind and below ear, may be seen or felt a hard protuberance; this is the sharpest corner-stone of the head, being the place where the lower jaw hinges upon the skull. This is called "angle of jaw"; it is a good landmark, which must by no means be confused with "angle of mouth," where horny parts of the beak come together. The *lore* (Lat. *lorum*, a strap, or bridle; hence, place where the cheek-strap passes; fig. 25, 2) includes pretty much all the space between eye and side of base of upper mandible; a considerable part of it is simply ante-orbital. Thus we say of a Hawk, "lores bristly"; and examination of a bird of that kind will show how large a space is covered by the term. Lore, however, should properly be restricted to a narrow line between eye and bill in direction of nostrils. It is excellently shown in Herons and Grebes, where "naked lores" is a distinctive character. The lore is frequently the seat of specially modified or specially colored feathers. The rest of side of head, including space between angle of jaw and bill, has the name of *cheek* (Lat. *gena*; fig. 25, 35). It is bounded above by loral, infra-orbital, and auricular regions; below, by a line along lower edge of bony prong of under mandible. It is cleft in front for a varying distance by backward extension of gape of mouth; above this gape is more properly *gena*, or *malar region* (Lat. *mala*, upper jaw) in strictness; below it is *jaw* (*maxilla*), or rather "side of jaw." The lower edge of jaw definitely separates side of head from "under surface" of head, which is a space bounded behind by an imaginary line drawn straight across from one angle of jaw to the other, and running forward to a point between forks of under mandible. As already hinted, "throat" (*gula*; fig. 25, 37) extends upward and forward into this space without obvious dividing line; it runs into *chin* (Lat. *mentum*; fig. 25, 38), which is the (varying in extent) anterior part of under surface of head. Anteriorly, mentum may be marked off, opposite the point where feathers end on side of lower jaw, from a feathery space (when any) *between* branches of upper mandible itself; this space is called *interramal* (Lat. *inter*, between; *ramus*, fork).

The head is often striped lengthwise with different colors, apt to take definite position; these lines have received special names. *Median vertical line* is one along middle of pileum, from base of bill to nucha; *lateral vertical lines* bound it on either side. *Superciliary line* has already been noticed; below it runs the *lateral line*; that part of it before eye, is loral or ante-orbital; behind eye, post-orbital; when these are continuous through eye, they form a *trans-ocular* (Lat. *trans*, across; *oculus*, eye) *line*; below this is *malar line*, or cheek-stripe (Lat. *frenum*, a bridle); below this, on under jaw, *maxillary* or *submaxillary line*; in the middle below, *mental* or *gular lines*.

No other part of the body has so variable a ptilosis as the head. In most birds it is wholly and densely feathered; but it ranges from this condition to one wholly naked; though such nakedness means only absence of perfect contour feathers, for most birds with unfeathered heads have a hair-like growth of filoplumes. Our examples of naked-headed birds are Turkeys, Vultures, Cranes, and some of Ibises. Associated with more or less complete baldness, is frequent presence of various fleshy outgrowths, as *combs, wattles, caruncles* (warty excrescences), *lobes*, and *flaps* of all sorts, even to enumerate which would exceed our limits. The parts of the barnyard cock exemplify the whole; among North American birds they are very rare, being almost confined to Turkeys. Sometimes *horny plates* take the place of feathers on part of the head; as the frontal shields of Coots and Gallinules. A common form of head-nakedness marks one whole order of birds, *Steganopodes*, which have mentum and more or less of gula naked, and transformed into a sort of pouch, extremely developed in Pelicans, and well seen in Cormorants. The next commonest is definite bareness of lores, as in all Herons and Grebes; in the former including the whole circum-orbital region. A little orbital space is bare in many birds, as vulturine Hawks and some Pigeons; species of Grouse have a bare warty supra-orbital space.

Among water-birds particularly, more or less of the interramal space is almost always unfeathered; the nakedness always proceeds from before backward. With the rare exceptions of a narrow frontal line, and a little space about angle of mouth, no other special parts of the head than those above given are naked in any North American bird, unless associated with general baldness.

The opposite condition, that of redundant feathering, gives rise to all the various *crests* (Lat., pl. *cristæ*) that form such striking ornaments of many birds. Crests proper belong to top of head, but may be also held to include those growths on its side; these together being called crests in distinction to the ruffs, ruffles, beard, etc., of gula or mentum. Crests may be divided into two kinds: 1, where feathers are simply lengthened or otherwise enlarged; and 2, where texture, and sometimes even structure, is altered. Nearly all birds possess the power of moving and elevating the feathers on the head, simulating a slight crest in moments of excitement. The general form of a crest is a full, soft elongation of coronal feathers collectively; when perfect, such a crest is *globular*, as in the genus *Pyrocephalus;* generally, however, feathers lengthen on occiput more than on vertex or front, and this gives us the simplest and commonest form. Such crests, when more particularly occipital, are usually connected with lengthening of nuchal feathers, and are likely to be of a thin, pointed shape, as well shown in the Kingfisher. Coronal or vertical crests proper are apt to be different rather in coloration than in much elongation of feathers; they are perfectly illustrated in the Kingbird, and other species of the genus *Tyrannus*. Frontal crests are the most elegant of all; they generally rise as a pyramid from the forehead, as excellently shown in the Bluejay, Cardinal, Tufted Titmouse, and others. All the foregoing crests are generally single, but sometimes double; as shown in the two lateral occipital tufts of "horned" Larks, in all tufted or "horned" Owls, and in some Cormorants. Lateral crests are, of course, always double, one on each side of the head; they are of various shapes, but need not be particularized here, since they mostly belong to the second class of crests — those consisting of texturally modified feathers. It is a general, though not exclusive, character of these last that they are *temporary;* while the other kind is only changed with the general moult, these are assumed for a short period only, the breeding season; and they are often distinctive of *sex*. Occurring on top of head, they furnish remarkable ornaments of birds. I need only instance the elegant helmet-like plumes of Partridges of the genus *Lophortyx;* the graceful flowing train of *Oreortyx picta;* the similar plumes of Night and other Herons. Most Cormorants and some Auks possess lateral plumes of similar description; these, and those of Herons, are usually *deciduous;* while those of the Partridges above mentioned last as long as the general plumage. In many birds, especially Grebes, these lateral plumes are associated or coalesce with *ruffs*, which are singular lengthening and modifying of feathers of auriculars, genæ and gula; and are almost always temporary. *Beards*, or special lengthening of mental feathers alone, are comparatively rare; we have no good example among our birds, but a European vulture, *Gypaëtus barbatus*, is one. The feathers sometimes become *scaly* (*squamous*), forming, for instance, the exquisite gorgelets or frontlets of Hummingbirds. They are often *bristly* (*setaceous*), as about the lores of nearly all Hawks, the forehead of the Dabchicks, Meadow-larks, etc. A particular set of bristles, which grow in single series along the gape or *rictus* of many birds, are called *rictal bristles* or *vibrissæ*. These are more or less developed in nearly all small insectivorous birds; they are large, stiff, and highly characteristic of the family *Tyrannidæ*, or Tyrant Flycatchers; while in some of Goatsuckers (*Caprimulgidæ*) they are prodigiously long, and in one species of that family (*Antrostomus carolinensis*) have lateral filaments. While usually all unlengthened head-feathers point backward, they are sometimes *erect*, forming a velvety pile, or they may radiate from a given point, as from the eye in most Owls, where they form a *disc*.

In the foregoing paragraph I mention only a few styles of crests, chiefly needed to be known in the study of our native birds; there are many others, with endless modifications, among exotic birds; to these, however, I cannot even allude by name. Peculiarities of nasal feathers, and others around base of bill, are noticed below. Forms of crests are illustrated by many figures given *passim* in the present work.

2. OF THE MEMBERS: THEIR PARTS AND ORGANS.

I. THE BILL.

The Bill (Lat. *rostrum*) is hand and mouth in one: the instrument of *prehension*. As hand, it takes, holds, and carries food or other substances, and in many instances, *feels;* as mouth, it tears, cuts, or crushes, according to the nature of the substances taken; assuming functions of both lips and teeth, neither of which do any recent birds possess. An organ thus essential to a prime function of birds, one directly related to their various modes of life, is of much consequence in a taxonomic point of view; yet its structural modifications are so various and so variously interrelated, that it is more important in framing genera than families or orders; more *constant* characters must be employed for higher groups. The general shape of the bill is referable to the *cone*. This shape combines great strength with great delicacy; the end is fine to apprehend the smallest objects, while the base is stout to manipulate the largest. But in no bird is the cone expressed with entire precision; and, in most, the departure from this figure is great. The bill always consists of two, Upper and Under or Lower

Mandibles (fig. 26), which lie, as their names indicate, above and below, and are separated by a horizontal fissure — the mouth. Each mandible consists of certain projecting skull-bones, sheathed with more or less *horny* integument in place of true skin. The framework of the Upper Mandible is (chiefly) a bone called *intermaxillary*, or better, *premaxillary*. In general, this is a three-pronged or tripodal bone running to a point in front, with the uppermost prong, or foot, implanted upon the middle of the forehead, and the other two, lower and horizontal, running into the sides of the skull in front. The basis of the Under Mandible is a compound bone called *inferior maxillary* or *inframaxillary;* it is U- or V-shaped, with a point or convexity in front, and prongs running to either side of base of skull behind, to be there movably hinged. These two bones, with certain accessory ones of the upper mandible, as *palate* bones, etc., together with the horny investment, constitute the Jaws. *Both* jaws, in birds, are *movable;* the under, by the joint just mentioned; the upper, either by a joint at, or by elasticity of bones of, the forehead; and by a singular muscular and bony apparatus in the palate, further notice of which is given beyond, under head of Anatomy (Osteology). Motion of the upper mandible is freest in Parrots, where both fronto-maxillary and palato-maxillary sutures exist. When closed, the jaws meet and fit along

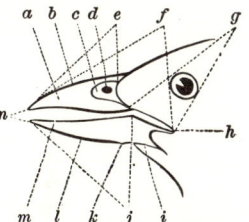

FIG. 26. — Parts of a Bill. *a*, side of upper mandible; *b*, culmen; *c*, nasal fossa; *d*, nostril; *e* (see below); *f*, gape, or whole commissural line; *g*, rictus; *h*, commissural point or angle of the mouth; *i*, ramus of under jaw; *j*, tomia of under mandible (the reference lines *e* should have been drawn to indicate the corresponding tomia of upper mandible): *k*, angle of gonys; *l*, gonys; *m*, side of under mandible; *n*, tips of mandibles.

their apposed edges or surfaces, in the same manner and for the same purposes as lips and teeth of man or other vertebrates. All bills, thus similarly constituted, have been divided into

Four Classes, representing as many ways in which the two mandibles close upon each other at the end: 1. *Epignathous* (Gr. ἐπί, *epi*, upon, γνάθος, *gnathos*, jaw) way, plan, or type, in which the upper mandible is longer than the under, and its tip is evidently bent down over the tip of the lower. 2. *Hypognathous* (Gr. ὑπό, *hupo*, under), in which the lower mandible is longer than the other. 3. *Paragnathous* (Gr. παρά, *para*, at or by), in which both are of about equal length, and neither is evidently bent over the other. 4. *Metagnathous* (Gr.

μετά, *meta*, with, beside, etc.), in which the points of the mandibles cross each other. The second and fourth of these forms are extremely rare; they are exemplified, respectively, by Skimmers and Crossbills (genera *Rhynchops* and *Loxia*). The first is common, occurring throughout Birds of Prey and Parrots, and among Petrels, Gulls, etc. The great majority of birds exhibit the third; and there is such evident gradation of paragnathism into epignathism, that it is necessary to restrict the latter to cases of its complete development, exhibited in the intermaxillary bone divested of its horny sheath, which often, as among Flycatchers, etc., forms a little overhanging point, but does not constitute epignathism. These four classes, though always determinable, and convenient in descriptions, are purely arbitrary — that is, they by no means correspond to any four large groups of birds; but, on the contrary, usually only mark families and subdivisions of families; and the four types may be seen in closely related genera. The general shape of the bill has also furnished

Other Classes, for many years used as a large basis for ornithological classification, even for establishment of orders; but which progress of the science has shown to be merely as convenient as, and only less arbitrary than, the foregoing. The principal of these are represented by the following types: A, among land birds. 1. *Fissirostral* (Lat. *fissus*, cleft, and *rostrum*), or cleft, in which the bill is small, *short*, and with a very large gap running down the side of the head; as in the Swallow, Chimney-swift, Whippoorwill. 2. *Tenuirostral* (Lat. *tenuis*, slender), or slender, in which the bill is slim, *long*, and with a short cleft; as in the Humming-bird, Creeper, Nuthatch. 3. *Dentirostral* (Lat. *dens*, a tooth), or toothed, in which, with a various general shape, there is present a nick, tooth, or evident lobe in the apposed edges of one or both mandibles near the end; as in the Shrike, Vireo, and some Wrens, Thrushes, Warblers. 4. *Conirostral* (Lat. *conus*, a cone), or conical, sufficiently defined by its name, and illustrated by the Finch family and some allied ones. — B, among water birds. 5. *Longirostral* (Lat. *longus*, long), or long, an aquatic style of the tenuirostral, best exhibited in the Snipe family. 6. *Pressirostral* (Lat. *pressus*, pressed), or compact, illustrated by Plovers, etc., and quite likely analogous to the conirostral. 7. *Cultrirostral* (Lat. *culter*, a knife), cutting, perhaps analogous to the dentirostral, exemplified by Herons. 8. *Lamellirostral* (Lat. *lamella*, a little plate), or lamellate, in which the bill is furnished with series of little laminæ along the apposed edges of both mandibles, as in Swans, Geese, Ducks, Mergansers, Flamingoes, and certain Petrels. None of these terms is now used to indicate a natural group, nor have we such absurdities as "orders" *Fissirostres, Tenuirostres*, etc. Swallows, for instance, and Swifts are equally fissirostral, though only distantly related to each other; a Swift is closely related to a Hummingbird, though the latter is extremely tenuirostral; and birds of contiguous genera may be dentirostral or not. The terms are nevertheless convenient to use in descriptions. Some similar terms, expressing special modifications, as *unguirostral* (Lat. *unguis*, a hook), *acutirostral* (Lat. *acutus*, sharp), etc., are also employed.

Other Forms. — A bill is called *long*, when notably longer than head proper; *short*, when notably shorter; *medium*, in neither of these conditions. It is *compressed*, when higher than wide, at base at least, and generally for some portion of its length; *depressed*, when wider than high; *terete* (Lat. *teres*, cylindric), under neither of these conditions. It is *recurved*, when curved upward; *decurved*, when curved downward; *bent*, when the variation in any direction is at an angle, as in Flamingoes and the Wry-billed Plover; *straight*, when not out of line with axis of head. A bill is *obtuse* (said chiefly of the paragnathous sort) when it rapidly comes to an end that therefore is not fine, or when the end is knobby; it is *acute*, when it runs to a sharp point; *acuminate*, when equally sharp and slenderer; *attenuate*, when still slenderer; *subulate* (awl-shaped), when slenderer still; *acicular* (needle-shaped), when slenderest possible, as in some Hummingbirds and Phalaropes. A bill is *arched, vaulted, turgid*, *tumid, inflated,* etc., when its outlines, both crosswise and lengthwise, are notably more or less convex; and *contracted,* when some, or the principal, outlines are concave (said chiefly of depressions about base of upper mandible, or of concavity along sides of both mandibles). A bill is *hamulate* (Lat. *hamus,* a hook), or *unguiculate* (Lat. *unguis,* a claw), when strongly epignathous, as in rapacious birds, whose upper mandible is like the talon of a carnivorous beast; it is *dentate,* when toothed, as in a Falcon; if there are a number of similar "teeth," it is *serrate* (Lat. *serra,* a saw), like a saw, or *denticulate,* as in Motmots, Trogons, some Hummingbirds, etc.; it is *cultrate* (knife-like), when extremely compressed and sharp-edged, as in an Auk or Skimmer; if much curved as well as cultrate, it is *falcate* (Lat. *falx,* a reaping-hook; scythe-shaped); and each mandible may be oppositely falcate, as in a Crossbill, constituting metagnathism. A *gibbous* bill is one which has a pronounced hump or knob, as that of some Swans and Scoters; and some bills are *appendaged* with various leathery or skinny lobes or flaps. A bill much flattened and widened at end (rare) is *spatulate* (Lat. *spatula,* a spoon); examples: Spoonbill, Shoveler Duck, and the extraordinary little Sandpiper whose technical name is *Eurynorhynchus pygmæus.* One is called *lamellate,* when it has a series of plates or processes just inside the edges of the mandibles, as in all Ducks, etc., furnishing a sifter or strainer of water — just what is effected in the whale by the "bone" in its mouth. The commonest shape of a bill is *conical,* as in any Finch, Bunting, or Warbler; probably the next commonest is that called by some ornithologists *grypaniform,* such as is exhibited by any Thrush or Warbler — the grypaniform being a mild case of epignathism, usually associated with weak toothing or nicking. Finally, the far end of the bill, of whatever shape, is called the *tip* or *apex* (fig. 26, *n*); the near end, joined to the rest of the skull, the *base;* the rest is the *continuity.*

Particular shapes of bills are almost endlessly varied, and cannot be given; the student who uses this book to the end will find many of them described, and "there are others." One of the most curious cases is that of the New Zealand Huia, *Heteralocha acutirostris,* in which shape of bill is a sexual character; for in the ♂ the bill is comparatively short and straight, but in the ♀ it is about twice as long and curved almost in the arc of a circle.

Covering of the Bill. — (*a*) In a great majority of birds, including nearly all **perchers**, many walkers, and some waders and swimmers, the sheathing of both mandibles is wholly **hard, horny,** or **corneous** (Lat. *cornu,* a horn); it is integument modified much as in case of nails or claws of beasts, by thickening and hardening of outer layers of malpighian cells. In nearly all waders and most swimmers, the sheath becomes softer, and wholly or partly of a dense, leathery texture. But many swimmers furnish bills as hard-covered as any, while some perchers have the integument partly quite soft, so that no unexceptional rule can be laid down; moreover, gradations from one extreme to the other are insensible. Probably the softest bill is found in *Scolopacidæ,* where it is skinny throughout, and in typical Snipes and Woodcocks vascular and nervous at tip, becoming a true organ of touch, used to feel for worms out of sight in the mud. In all the Duck order the bill is likewise soft; but there it always ends in a hard, horny *unguis* or "nail," more or less distinct; and such a horny claw also occurs in other water birds with softish bills, as Pelicans. An interesting modification occurs in the Pigeon order (*Columbæ*); these birds have the bill hard or hardish at tip and through most of continuity, but toward and at base of upper mandible the sheath changes to a soft, tumid, skinny texture, overarching the nostrils; and the case is much the same with most Plovers. But the most important feature in this connection is afforded by Parrots and all Birds of Prey — one so remarkable that it has received a distinct name: CERE (or *ceroma*). The cere (Lat. *cera,* wax; because it looks waxy) is a dense membrane saddled on the upper mandible at base, so different from the rest of the bill, that it might be questioned whether it does not more properly belong to head than to bill, were it not that the nostrils open in it. A cere is often densely feathered, as in the Carolina paroquet, in the bill proper of which no nostrils are seen, these being hidden in the feathered cere, which, therefore, might easily be mistaken for the bird's forehead. A sort of false cere occurs in some water birds, as Jaegers or Skuas gulls (genera *Stercorarius* and *Megalestris*). The tumid nasal skin of Pigeons is sometimes called a cere; but the term had better be restricted to the birds first above named. The under mandible probably never presents softening except as a part of general skinniness of bill; it may have a nail at the end, as it does in the Duck family (*Anatidæ*). (*b*) The covering is either *entire* or *pieced*. In most birds it is entire — that is, the sheath of either mandible may be pulled off whole, like the finger of a glove. But in many birds it is divided into parts by various lines of slight connection, and then comes off in pieces; as is the case with some water birds, particularly Petrels, where the divisions are regular, and the pieces have received distinctive names. Thus the pieces which I named in 1866 for the Albatross are: *culminicorn*, along ridge of bill; *latericorn*, along each side of upper mandible; *unguicorn*, on the hook of the bill; *naricorn*, encasing each nostril; *ramicorn*, along each side of under mandible; *inferior unguicorn*, at tip of under mandible, and *interramicorn*, between the two lower edges of the inferior unguicorn. Many Auks (*Alcidæ*) also have the covering of the bill in particular pieces, and it is an extraordinary fact that such parts are of a secondary sexual character, being assumed at the breeding season and afterward *moulted* like feathers. Such condition of the sheath, or of its special developments, is called *caducous* or *deciduous*. The entire covering of both jaws together is called *rhamphotheca* (Gr. ῥάμφος, *hramphos*, beak; θήκη, *theke*, sheath); of the upper alone, *rhinotheca* (Gr. ῥίς, *hris*, nose); of the under, *gnathotheca* (Gr. γνάθος, *gnathos*, jaw); but these terms are not much used, nor are *dertrotheca* (Gr. δέρτρον, *dertron*, hook) and *myxotheca* (Gr. μύξα, Lat. *myxa*) for the superior and inferior unguicorns, respectively. (*c*) The covering is otherwise variously marked; sometimes so strongly that similar features are impressed upon the bones beneath. The most frequent marks are various *ridges* (Lat. pl. *carinæ*, keels) of all lengths and degrees of expression, straight or curved, vertical, oblique, horizontal, lengthwise, or transverse; a bill so marked is said to be *striate* (Lat. *stria*, a streak) or *carinate*; when numerous and irregular, the ridges are called *rugæ* (Lat. *ruga*, a wrinkle), and a bill is said to be *corrugated* or *rugose*. When the elevations are in points or spots instead of lines, they are called *puncta* (Lat. *punctum*, a point); a bill so furnished is *punctate*, but the last word is oftener employed to designate the presence of little *pits* or depressions, as in the dried bill of a Snipe toward the end. Larger softish, irregular knobs or elevations pass under the general name of *warts* or *papillæ*, and a bill so marked is *papillose;* when the processes are very large and soft, a bill is said to be *carunculate* (Lat. *caro*, flesh, diminutive *carunculus*, little bit of flesh). Various linear *depressions*, often but not always associated with carinæ, are grooves or *sulci* (Lat. *sulcus*, a furrow), and the bill is then called *sulcate*. Sulci, like carinæ, are of all shapes, sizes, and positions; when very large and definite, they are sometimes called *canaliculi*, or channels. The various knobs, "horns," and large special features of bill cannot be here particularized. Any of the foregoing features may occur on both mandibles, and they are exclusive of that special mark of the upper, the *nasal fossa* in which the nostrils open, and which is considered below. We have still to notice special parts of either mandible; and will begin with the simplest, the

Under Mandible (*mandibula*, or *maxilla inferior*). — In most birds this is a little shorter and narrower and not nearly so deep as the upper mandible; sometimes quite as large, or even larger. The upper edge, double (*i. e.*, there is an edge on both sides), is called the mandibular *tomium*, or in the plural, *tomia* (Gr. τέμνειν, *temnein*, to cut; fig. 26, *j*); this is received against, and usually a little within, the corresponding edge of the upper mandible. The prongs already mentioned are *mandibular rami* (pl. of Lat. *ramus*, a branch; fig. 26, *i*); these meet at some point in front, either at a short angle (like >) or with a rounded joining (like ⊐); in either case this is called *angulus menti* or *mental angle*. At their point of union there is a prominence, more or less marked (fig. 26, *k*) ; this is the GONYS (corrupted from Gr. γόνυ, *gonu*, a knee; hence, any similar protuberance). That is to say, this point is gonys proper (sometimes called *angle of the gonys* or *gonydeal angle*) ; but the term *gonys* is extended to apply to the whole line of union of rami, from gonys proper to tip of under mandible ; and in descriptions it means, then, *under outline of bill* for a corresponding distance (fig. 26, *l*). This important term is constantly used in describing birds.* Gonys is to under mandible what a keel is to a boat; it is the opposite of ridge or *culmen* of upper mandible. It varies greatly in length. Ordinarily it forms one-half to three-fourths of the under outline. Sometimes, as in conirostral birds, a Sparrow for example, it represents nearly all this outline ; while in a few birds it makes the whole, and in some, as the Puffin, is actually longer than the lower mandible proper, because it extends backward in a point. Other birds may have almost no gonys ; as a Pelican, where the rami only meet at the extreme tip, or the whole Duck family, where there is hardly more. As the student must see, length of gonys is simply a matter of how extensive is fusion of rami, and that, similarly, their mode of fusion, as in a sharp ridge, a flat surface, a straight line, a curve, etc., results in corresponding modifications of its special shape. The *interramal space* is complementary to length of gonys ; sometimes it runs to tip of bill, as in a Pelican, sometimes there is next to none, as in a Puffin ; while its width depends upon degree of divergence, and straightness or curvature, of the rami. This space may be occupied by naked skin of the floor of the mouth, or partly or completely feathered. The surface between tomium and lower edge of rami and gonys together is *side of under mandible* (fig. 26, *m*). Each mandibular ramus is sometimes called *gnathidium ;* and that portion of the rami which corresponds to length of gonys is known as *myxa.* The most important feature of the

Upper Mandible is the *culmen* (Lat. for top of anything; fig. 26, *b*). The culmen is to the upper mandible what the ridge is to the roof of a house ; it is the upper profile of the bill — *highest middle lengthwise line of bill;* it begins where feathers end on the forehead, and extends to tip of upper mandible. According to shape of bill it may be straight, convex, concave, or even somewhat ∽ -shaped ; or double-convex, as in the Tufted Puffin : but in most cases it is convex, with increasing convexity toward the tip. Sometimes it rises up into a thin elevated crest, as in the genus *Crotophaga,* and in Puffins (*Fratercula*), when the upper mandible is said to be *keeled*, and the culmen itself to be *cultrate ;* sometimes it is a furrow instead of a ridge, as toward the end of a Snipe's bill ; but generally it is simply the uppermost line of union of the gently convex and sloping *sides of upper mandible* (fig. 26, *a*). In a great many birds, especially those with depressed bill, as all Ducks, there is really no culmen ; then the *median lengthwise* line of surface of upper mandible takes place and name of culmen. The culmen generally stops about opposite the proper base of the bill ; then the feathers sweep across its end, and downward across the sides of the upper mandible, usually also obliquely backward. Variations in both directions are frequent; feathers may run out in a point on culmen, shortening the latter, or a culmen may run up the forehead, parting feathers ; either in a point, as in Rails and Gallinaceous birds, or as a broad plate of horn, as in Coots and Gallinules. A culminal point between feathers of the forehead forms an *angulus frontalis* or *frontal angle ;* and the same terms are used for extension of feathers in a point on the culmen. The lower edge (double) of the upper mandible is the *maxillary tomium,* as far backward as it is hard and horny. The most conspicuous feature of the upper mandible in most birds is the

Nasal Fossa (Lat. *fossa,* a ditch), or *nasal groove* (fig. 26, *c*), in which each nostril opens. The upper prong of the intermaxillary bone is usually separated some way from each lateral

---

* The word *gonys* originated with Illiger in 1811. It is a mistake for *genys* (Gr. γένυς, *genus*), meaning lower jaw or chin. But it is firmly established in ornithology, and supplied with a fictitious etymology to suit, as in my text. (See, for example, Sundevall, Tentamen, or the Century Dictionary.) The adjective *gonydeal* is a monstrous abortion of a word, but in good current usage.

prong; the skinny or horny sheath that stretches betwixt them is usually sunken below the general level of the bill, especially in those birds whose prongs are long or widely separated; this "ditch" is what we are about. It is called *fossa* when short and wide, with varying depth; *sulcus* or groove when long and narrow; the former is well illustrated in Gallinaceous birds; the latter in nearly all wading birds and many swimmers. When the intermaxillary prongs are soldered throughout, or are very short and close together, there is no (or no evident) nasal depression, the nostrils then opening flush with the general surface. The

Nostrils or **Nares** (Lat. pl. of *naris*, fig. 26, *d*), two in number, vary in *position* as follows: they are *lateral*, when on sides of upper mandible (almost always); *culminal*, when together on the ridge (rare); *superior* or *inferior* when evidently above or below midway betwixt culmen and tomia; *basal*, when at base of upper mandible; *sub-basal* when near it (usual); *median* when at or near middle of upper mandible (frequent, as in Cranes, Geese, etc.); *terminal* when beyond this (very rare; *no* birds have nostrils at end of bill, except the Kiwis, *Apterygidæ*). Nostrils are *pervious*, when open, as in nearly all birds; *impervious*, when not visibly open, as among Cormorants and other birds of the same order; *perforate*, when there is no *septum* (partition) between them, so that you can look through them from one side of the bill to the other, as in the Turkey-buzzard, Crane, etc.; *imperforate*, when partitioned off from each other, as in most birds; but different ornithologists use these terms interchangeably, saying *nares perviæ* of nostrils which communicate with each other, and *nares imperviæ* of nostrils shut off from each other by an internasal septum. Principal *shapes* of nostrils may be thus exhibited: — a line, *linear* nostrils; a line variously enlarged at either end, *clavate, club-shaped, oblong, ovate* nostrils; a line, enlarged in the middle, *oval* or *elliptic* nostrils; this passing insensibly into a circle, *round* or *circular* nostrils; and more or less linear nostrils may be either longitudinal, as in most birds, or oblique, as in a few; almost never directly transverse (up and down). Rounded nostrils may have a raised border or rim; when this is prolonged they become *tubular*, as in the Goatsucker family and all Petrels. Usually, nostrils are defined entirely by the substance surrounding them; as a cere, in Hawks, Owls, Parrots; softish skin, in a Pigeon, Plover, or Snipe, and much swollen in the first named of these birds; or horn, in most birds; but often their contour is partly formed by a special development, somewhat distinct either in form or texture, called the *nasal scale*, or *operculum*. Generally, it forms a sort of overhanging arch or portico, as well shown in Gallinaceous birds, among Wrens, etc. A curious case of this is seen in the European Wryneck (*Iynx torquilla*), where the scale floors instead of roofing the nostrils. In the singular Kagu (*Rhinochetus jubatus*), the operculum forms a large movable scroll, apparently capable of closing the aperture. The nostrils also vary in being *feathered* or *naked*, the nasal fossa being a place where frontal feathers are apt to run out in paired points (called *antiæ*), embracing a small portion of the culmen (called *mesorhinium*). Such extension of feathers may completely fill and hide the fossa, as in Grouse and Ptarmigan; but it oftener runs for a varying distance toward, or *above* and beyond, the nostrils, as in Hummingbirds; sometimes similarly below them, as in a Chimney-swift; and the nostrils may be densely feathered when there is no evident fossa, as in an Auk. When thus feathered in varying degree, they are still open to view; another condition is, their being covered and hidden by modified feathers not growing on the bill itself, but on the forehead. These are usually bristly (*setaceous*), and form two tufts, close-pressed and directed forward, as is perfectly shown in a Crow; or, the feathers may be less modified in texture, and form either two *tufts*, one over each nostril, or a single *ruff*, embracing the whole base of the upper mandible, as in Nuthatches, Titmice, Red-polls, Snow Buntings, and many other northern *Fringillidæ*. Bristles or feathers thus growing forward are called *retrorse* (Lat. *retrorsum*, backward; here used in the sense of *in an opposite direction from* the lay of the general plumage; but they should properly be called *antrorse*, *i. e.*, forward). Nostrils, whether culminal or lateral, are, like eyes and ears, *always two* in number, though they may be united in one tube, as in Petrels.

The Gape. — It remains to consider what results from relations of mandibles to each other. When a bill is opened, there is a cleft or fissure between upper and under mandibles; this is the *gape* or *rictus* (Lat. *rictus*, mouth in the act of grinning). Though thus really meaning the open *space* between mandibles, *gape* generally signifies the *line of their closure*. *Commissure* (Lat. *committere*, to put or join together) properly means the point where the gape ends behind — that is, *angle of mouth*, *angulus oris*, where apposed edges of mandibles join each other; but, like *gape*, it is extended to the whole line of closure, from commissural point to tip of bill. So we say, "commissure straight," or "commissure curved"; also, "commissural edge" of either mandible (equivalent to "tomial edge"), in distinction from culmen or gonys. But it would be well to have more precision in this matter. Let, then, *tomia* (fig. 26, *j*) be the true cutting edges of either mandible from tip to base of bill proper; let *rictus* (fig. 26, *g*) be their edges thence to the *commissural point* (fig. 26, *h*) where they join when the bill is open; *commissural line* (fig. 26, *f*) to include both when the bill is closed. The gape is *straight*, when rictus and tomia are both straight and lie in the same line; *curved*, *sinuate*, when they lie in the same curved or waved line; *angulated*, when they are straight, or nearly so, but do not lie in the same line, and therefore meet at an angle. (An important distinction: see under family *Fringillidæ* in the Synopsis.)

The "Egg Tooth." — Finally, it is to be observed that unhatched birds are provided with a tool for working their way into the world by chipping the eggshell. This interesting instrument is a small sharp knob or boss at the tip of the upper mandible, such as also exists in some reptiles; it may readily be observed in a newly hatched chick of domestic fowl. It consists of a deposit of hard calcareous matter in the middle layers of epidermis, not connected with the underlying bone, but breaking through the epidermal layers to come in contact with the eggshell that is to be chipped at one point and thus cracked open. Soon after hatching, the calcareous substance of this curious little drill is cast off, and the layers of epidermis through which the point of the drill projected cease to be distinguishable from the rest of the horny covering of the bill.

II. THE WINGS.

Definition. — Pair of anterior or *pectoral* limbs organized for flight by means of epidermal outgrowths (feathers). Used for this purpose by birds in general; but by Ostriches and their allies only as outriggers to aid running; by Penguins as fins for swimming under water; used also in the latter capacity by some birds that fly well, as Divers, Cormorants, Dippers. Wanting in no recent birds, but imperfect in all *Ratitæ*, among which the wings are greatly reduced in the Emeu, Cassowary, and Apteryx, while in Moas (*Dinornithidæ*), as in the Cretaceous *Hesperornis*, only a rudimentary humerus is known. To understand their structure we must notice particularly

The Bony Framework (figs. 27, 28, 29). — The skeleton of a bird's wing is built upon a plan common to the fore or pectoral limb of most vertebrates, so that its bones and joints may readily be compared and identified with those of any lizard or mammal, including man. But the member is highly specialized; being fitted for accomplishing flight, not only by development of feathers, but also by modifications in the bones themselves. The axes of the bones have a special direction with reference to each other and to the axes of the body; the movements of the joints are peculiar in some respects; and the end of the wing, from the wrist outward, is peculiarly constructed, by loss of some of the digits that five-fingered animals possess,

112   GENERAL ORNITHOLOGY.

and by compression of those that are left, as more particularly said beyond. The wing proper begins at the shoulder-joint, where it hinges freely in a shallow socket formed conjointly by the shoulder-blade or *scapula*, and by the *coracoid*; these two bones, with the clavicles, collar-bones or merry-thought (*furculum*) forming the *shoulder-girdle*, or *pectoral arch* (figs. 56, 59).

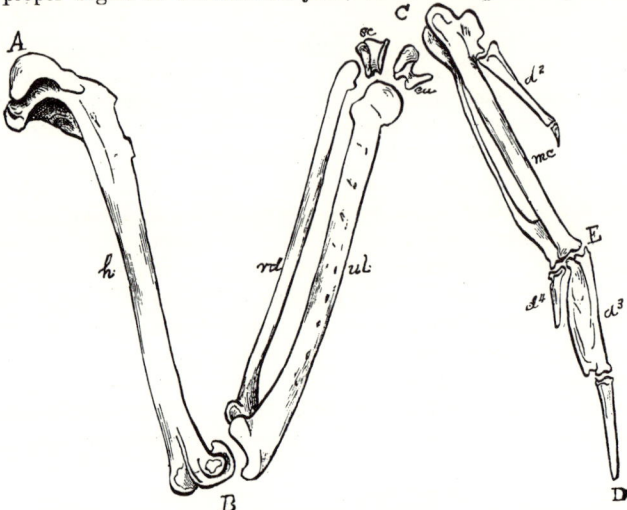

FIG. 27. — Bones of right wing of a duck, *Clangula islandica*, from above, ¾ nat. size. (Dr. R. W. Shufeldt, U.S.A.) A, shoulder, *omos*; B, elbow, *ancon*; C, wrist, *carpus*; D, end of principal finger; E, end of hand proper, *metacarpus*. AB, upper arm, *brachium*; BC, fore-arm, *antibrachium*; CD, whole hand or pinion, *manus*; composed of CE, hand proper or *metacarpus*, excepting d^2; ED, or $d^2\ d^3$, d^4, fingers, digits, *digiti*. h, *humerus*; rd, *radius*; ul, *ulna*; sc, outer carpal, *scapholunare* or *radiale*; cu, inner carpal, *cuneiforme* or *ulnare*; these two composing wrist or *carpus*. mc, the compound hand-bone, or *metacarpus*, composed of three metacarpal bones, bearing as many digits — the outer digit seated upon a protuberance at the head of the metacarpal, the other two situated at the end of the bone. d^2, the outer or radial digit, commonly called the thumb or *pollex*, composed of two *phalanges*; d^3, the middle digit, of two phalanges; d^4, the inner or ulnar digit, of one phalanx d^2 is the seat of the feathers of the *bastard wing* or *alula*. D to C (whole pinion), seat of the flight-feathers called *primaries*; C to B (fore-arm), seat of the *secondaries*; at B and above it in direction of A, seat of the *tertiaries* proper; below A, in direction of B, seat of *scapularies* (upon pteryla humeralis), often called tertiaries. The wing shown half-spread: complete extension would bring $ABCD$ into a right line; in complete folding C goes to A, and D to B; all these motions *nearly* in the plane of the paper. The elbow-joint and wrist are such perfect hinges. that, in opening or closing the wing, C cannot sink below the paper, nor D fly up above the paper, as would otherwise be the effect of the pressure of the air upon the flight-feathers. Observe also: rd and ul are two rods connecting B and C; the construction of their jointing at B and C, and with each other, is such, that they can *slide lengthwise* a little upon each other. Now when the point C, revolving about B, approaches A in the arc of a circle, rd pushes on sc, while ul pulls back cu; the motion is transmitted to D, and makes this point approach B. Conversely, in opening the wing, rd pulls back sc, and ul pushes on cu, making D recede from B. In other words. the angle ABC cannot be increased or diminished without similarly increasing or diminishing the angle BCD; so that no part of the wing can be opened or shut without automatically opening or shutting the rest, — an interesting mechanism by which muscular power is correlated and economized. This latter mechanism is further illustrated in fig. 28, where rc and uc show respectively the size, shape and position of the radial condyle and ulnar condyle of the humerus. It is evident that in the flexed state of the elbow, as shown in the middle figure. the radius, rd, is so pushed upon that its end projects beyond ul, the ulna; while in the opposite condition of extension, shown in the lower figure, rd is pulled back to a corresponding extent.

The wing ordinarily consists, in adult life, of ten or eleven actually separate bones; in embryos (see fig. 29) there are indications of several more at the wrist (*carpus*), which speedily lose their identity by fusing together and with bones of the hand (*metacarpus*). Aside from these, there is often an accessory ossicle at the shoulder-joint (fig. 56, *ohs*), sometimes one at the wrist-joint, occasionally an extra bone at the end of the principal finger. Among *Ratitæ*, the carpal bones are reduced to one in a Cassowary, to none in an Emeu and a Kiwi; all of which birds have but a single digit. The *Archæopteryx* had the most bones of any known bird, with three separate metacarpals, three free digits, and altogether nine phalanges. The normal or usual number of wing-bones is shown in fig. 27, taken from a duck (*Clangula islandica*), in which there are eleven.

The upper arm-bone, h, reaching from shoulder A to elbow B, is the *humerus*, which alone forms the first segment of the wing. In the closed wing, the humerus lies nearly in the position of the same bone in man when the elbow is against the body; in extension of the wing, the elbow is borne away from the body, as when we raise the arm, but carry it neither forward nor backward. A peculiarity of the bird's humerus is, that it is rotated on its axis through about the quadrant of a circle, so that what is the front of the human bone is the outer aspect in a bird. The humerus is a cylindric bone, straightish or somewhat italic ƒ-shaped, with a globular head to fit the socket of the shoulder, a strong pectoral ridge for insertion of breast muscles, and at the lower end two *condyles* (fig. 28, *rc*, *uc*), or surfaces for articulation with a pair of succeeding bones. The second segment is the fore-arm, *cubit* or *antibrachium*, extending from elbow to wrist, B to C, fig. 27; this has two parallel bones of about equal lengths. These are *ulna*, *ul*, and *radius*, *rd*; the ulna, inner and posterior, the larger of the two, bears quills of the secondary series; the radius is slenderer, outer, and anterior. The enlarged upper end of the ulna is called *olecranon*, or "head of the elbow." The third segment of the wing is the pinion, hand, or *manus*, to be considered in its three successive portions: wrist or *carpus*; hand proper or *metacarpus*; and fingers or *digits*: in all, C to D in fig. 27. In adult

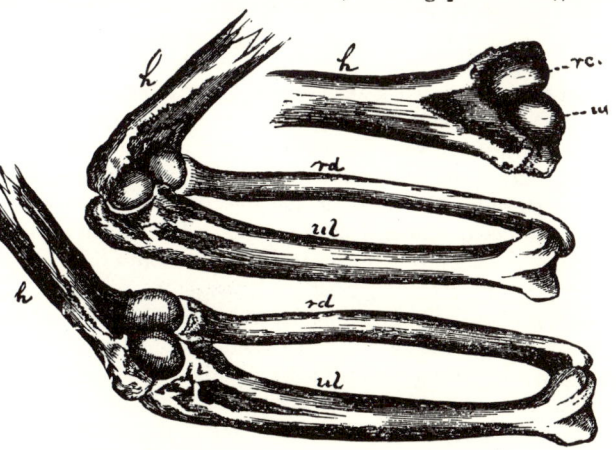

FIG. 28.—Mechanism of elbow-joint. (See explanation of fig. 27.)

life, the carpus almost always consists of two small knobby carpal bones, extremely irregular in shape, called *scapholunar*, *sc*, and *cuneiform*, *cu*; or *radiale* and *ulnare*, because one of them is at the end of the radius, and the other at the end of the ulna. In embryos, several more cartilaginous or gristly nodules are demonstrable; their number varies in different birds. The theory is, that birds' ancestors had the following number of carpals: three in a proximal or first row, named *radiale*, *intermedium*, and *ulnare*; one median, called *centrale*; and five in a distal row, being one for each of the five ancestral digits (though no more than three have ever been demonstrated). It is believed with reason that the actual radiale consists of an ancestral radiale fused with an intermedium; that the actual ulnare consists of an ancestral ulnare fused with a centrale; and it is certain that, whatever number of distal carpals can be demonstrated in any case, they all fuse with the metacarpal bones. Thus a bird's carpals are reduced to the two abovesaid, and one of these disappears in some ratite birds. The hand proper or metacarpus, C to E (exclusive of *d* 2), in all recent adult birds, consists of a single metacarpal bone; but this is a

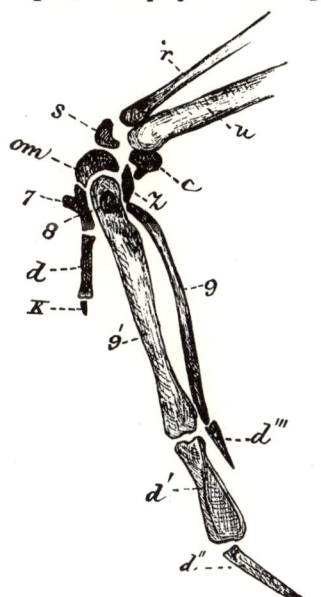

FIG. 29, from a *young* grouse (*Centrocercus urophasianus*, six months old), is designed to show the composition of the carpus and metacarpus before the elements of these bones fuse together: *r*, radius; *u*, ulna; *s*, scapholunar or radiale; *c*, cuneiform or ulnare; *om*, a carpal bone believed to be os magnum, later fusing with the metacarpus; *z*, a carpal bone, supposed to be unciform, later fusing with metacarpus; 8, an unidentified fifth carpal bone, which may be called *pentosteon*, later fusing with the metacarpus; 7, radial or outer metacarpal bone, bearing the pollex or outer digit, consisting of two phalanges, *d* and *k*; 9', principal (median) metacarpal bone, bearing the middle finger, consisting of the two phalanges, *d'*, *d''*; 9, inner or ulnar metacarpal, bearing a digit of one phalanx, *d'''*. The pieces marked *om*, *z*, 7, 8, 9. all fuse with 9'. (From nature by Dr. R. W. Shufeldt, U. S. A.)

compound bone; for, besides including one or more carpal bones in itself, as already shown, it consists of three metacarpal bones fused in one, corresponding to the three fingers or digits which nearly all birds possess. In fact, it is three metacarpals in one, plus certain carpals; its full name would therefore be *carpo-metacarpus*. Much the greater part of this composite bone corresponds to a bird's middle finger; a small, short part, only at the base and on the radial side, corresponds to the outer finger, as seen in the figure above the bone marked $d\,2$; while that part corresponding to the inner finger is slender, nearly as long as the rest of the bone, and often fused therewith only at its two ends, leaving between itself and the main metacarpal an open space, as seen opposite the letters *mc* in the figure. The metacarpus thus compounded articulates at the wrist with both the free carpals; it bears the digits, almost invariably three in number, with which the wing is finished off; they are marked $d\,2$, $d\,3$, $d\,4$ in the figure. They are the *radial*, *median*, and *ulnar* digits. The median digit, $d\,3$, extending from E to D in the figure, is much the largest of the three, and forms the main continuation of the hand; it ordinarily consists of two jointed *phalanges*, or bones placed one after the other, but may have a third *phalanx*; the first or proximal phalanx is much larger than the other one or two. The inner or ulnar digit, $d\,4$, is borne upon the distal end of the metacarpal bone, alongside the first phalanx of the middle digit; it ordinarily consists of a single small phalanx, but sometimes there is another (the *Archæopteryx* had four); it enjoys little if any freedom of motion, and occasionally fuses with the first phalanx of the middle finger. The outer or radial digit, $d\,2$, is borne upon the projection near the *base* of the metacarpus, alongside which it lies, away from the other two fingers; it ordinarily consists of two phalanges, of which the terminal one is small, and often wanting; it enjoys considerable motion, being quite freely articulated with the metacarpus, except in Penguins. No bird has, and none is known to have had, more than these three digits; and in the Cassowary, Emeu, and Kiwi there is only one, the inner and outer being lost or reduced to mere traces. Such is the compactness and consolidation of a bird's hand that all the fingers act almost like a single stout tapering digit, only the outer one being capable of much individual action; though in the Archæopteryx the three metacarpals were free bones like the digits, and the whole hand more like that of a lizard. A bird's three digits are supposed by some to correspond to the thumb and fore and middle fingers of our hands; in this view, the radial digit is called *pollex*, which means thumb; and the next one, *index* or forefinger. But I agree with others who consider that birds have lost the first and fifth digits of the ancestral five-fingered, consequently the three they retain correspond to our fore, middle, and ring fingers, or our 2d, 3d, and 4th digits, and so I have marked them $d\,2$, $d\,3$, $d\,4$, in the figure.

The resemblance of a bird's digits to those of a lizard or mammal is increased by the *claws* (Lat. *ungues*) which some birds possess. The *Archæopteryx* had claws upon all three of its finger-tips. In recent birds, claws are found on the ends of the radial and middle fingers, especially the former; and in some embryos, as of the Ostrich, there is said to be a rudimentary claw on the ulnar digit. The adult Ostriches of the genera *Struthio* and *Rhea* have claws on the radial and middle digits, and so do some *Anatidæ*, and various Birds of Prey, the Cassowary, Emeu, and Kiwi have a claw on the middle digit; one on the radial digit is well shown by the Turkey-buzzard and other *Cathartidæ*, various Anserine and Gallinaceous birds, some Birds of Prey; and such a claw has occasionally been found on an oscine bird. The occurrence of claws is more or less irregular, and probably more frequent than is yet known.

The Mechanism of these Bones is admirable. The shoulder-joint is free, much like our own, permitting the humerus to swing all about; though the principal motions are to and from the side of the body (*adduction* and *abduction*), and up and down in a vertical plane. The elbow-joint is a very strict hinge, permitting motion in one plane, nearly that of the wing itself. The finger-bones have little individual motion, as we have already seen. The construction of the wrist-joint is quite peculiar. In the first place the two bones of the forearm are so fixed in relation to each other, that the radius cannot roll over the ulna, like ours. If you stretch your arm upon the table, you can, without moving the elbow, turn the hand over so that either the palm or the knuckles are downward. This is a rotary motion of the bones of the forearm, called *pronation* and *supination ;* the prone when the palm touches the table, supine when the knuckles are downward. This rotation is absent from the bird's arm ; if it could occur, the action of the air upon the pinion-feathers would throw them all " at sea " during the strokes of the wing, rendering flight difficult or impossible. The hingeing of the hand upon the wrist is such, also, that the hand does not move up and down, as ours can, in a plane perpendicular to the surface of the wing, but in the same plane as that surface. The motion is that which would take place in our hand if we could bring the little finger and its border of the hand so far around as to touch the corresponding border of the forearm. It is a motion of adduction, not of flexion, and its opposite, abduction, not extension, by which a wing is folded and spread. Such *abduction* is the way in which the hand is " extended " upon the wrist-joint, increasing and completing the unfolding of the wing that begins by the true extension of the forearm upon the elbow and abduction of the upper arm from the body. In a word, a wing is spread by the motion of abduction at the shoulder and wrist, of extension at the elbow ; it is closed by adduction at the shoulder and wrist, and flexion at the elbow. The numerous muscles which unfold or straighten out the wing are called *extensors ;* those that bend or close it are *flexors.* Extensors lie upon the back of the upper arm, and the front of the forearm and hand, their " leaders" or tendons passing over the *convexities* of the elbow and of the wrist. The flexors occupy the opposite sides of the limb, with tendons in the concavities of the joints. The most powerful muscles of the wings are the great *pectoral* or breast muscles, acting upon the upper end of the humerus ; there are several of them, exerted in throwing out the arm from the body, and in giving both the up and down wing-strokes. Tendons are generally strong inelastic cords ; but there is an interesting arrangement of an elastic cord in a bird's wing. In fig. 27, $A B C$ is a deep angle formed by the naked bones, but none such is visible from the exterior, because the space is filled by a fold of skin passing from C to near A. But C approaches and recedes from A as the wing is folded or unfolded, and a cord long enough to reach $A-C$ would be slack in the folded wing, did not its elasticity enable it to contract and stretch, keeping the anterior border of the wing straight and smooth. (For another automatic mechanism, see explanation of fig. 28.)

The point C is a highly important landmark in practical ornithology ; it represents, in any folded wing, a very prominent point, the distance from which to the tip of the longest flight-feather is a special measurement known as that of " the wing." It is the convexity of the carpus, commonly called the " carpal angle," or " bend of the wing." Having thus glanced at the bony structure and mechanism of the wing, we are ready to examine the

Feathers of the Wing (fig. 30). — How important these are will be evident from the consideration that they are the bird's chief *organs of locomotion ;* for without them the wing would be useless for flight. We also remember that such means of locomotion is the great specialty of birds. Wing-feathers are those which grow upon the *pteryla alaris*. They are of two main sorts : the *flight-feathers* proper, or long stiff quills, collectively called *remiges* (Lat. *remex,* pl. *remiges,* rowers) ; and the smaller, weaker feathers overlying them, and hence called *coverts,* or *tectrices* (Lat. *tectrix,* pl. *tectrices,* coverers). To these may be added as a third distinct group the *bastard quills,* which constitute the

Alula, or Ala Spuria (Lat. *alula,* little wing, diminutive of *ala,* wing *; spuria,* spurious, bastard). The "little wing" is simply the small parcel of feathers which grow upon the "thumb" (see fig. 27, $d\,2$; 29, d and $k\,;$ 30, al). Highly significant as these may be in a morphological point of view, as representing what this part of the wing may have been in early times, they are so much reduced in modern birds as to be of little account in practical ornithology. In fact, the unpractised student may fail to recognize them at first. They form a small packet on the fore outer border of the pinion near the carpal angle, and lie smoothly upon the upper surface of the wing, strengthening and finishing off what would be otherwise a weak spot in the contour of the wing-border. It is quite easy, on recognizing them, to lift them collectively a little away from the other feathers, owing to the slight mobility the thumb possesses. In fact, they are sometimes quite obtrusive, when faulty taxidermy has discomposed them. They are not often conspicuously modified either in size or color. In a few birds (*e.g.*, *Cathartes*), a *claw* will be found at the end of the joint which bears them. The student must be careful to discriminate between the use of the word *spurious* in the present connection and its application to a rudimentary condition of the first *remex* (see p. 119). The

Wing-Coverts overlie the bases of the large quills on both the upper and under surfaces of the wing. They are therefore conveniently divided into an *upper* set (*tectrices superiores*) and an under set (*tect. inferiores*). The former are so much more conspicuous than the latter that they are always understood when "upper" is not specified. The latter are sometimes collectively called "the lining of the wings." Coverts include all the *small* feathers of the wings excepting the bastard quills; they extend a varying distance along the bases of the flight-feathers. The ordinary disposition and division of the upper coverts is as follows: One set, rather long and stiffish, grow upon the pinion, and are close-pressed upon the bases of the outer nine or ten remiges, covering these large feathers about as

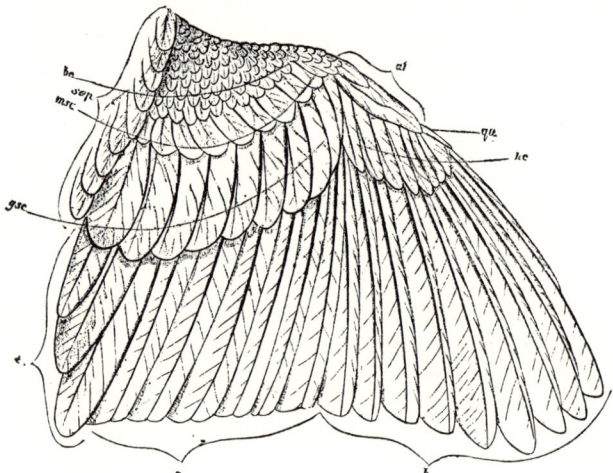

FIG. 30. — Feathers of a sparrow's wing; nat. size. (For explanation see text.)

far as their structure is plumulaceous. These are the *upper* PRIMARY *coverts*, or coverts of the primaries (fig. 30, *pc*); they are ordinarily the least conspicuous of any. All the rest of the upper coverts are SECONDARY; they spring mostly from the forearm. These are considered in three groups or *rows*. The *greater upper secondary coverts*, called simply the "greater coverts" (*tectrices majores*, fig. 30, *gsc*,) are the first, outermost, longest row, reaching nearest the tips of the flight-feathers; they overlie the bases of nearly all the remiges, excepting the first nine or ten. The *median upper secondary coverts*, shortly known as the "middle coverts" (*tectrices mediæ*), are a next row, shorter and therefore less exposed, but still quite evidently forming a special series (fig. 30, *msc*). It is a common feature of these median coverts that they shingle over each other contrary-wise to the way the greater coverts are imbricated, the outer vane of one being under the inner vane of the next outer one. All the rest of the upper secondary coverts, forming several indistinguishable rows, pass under the general name of *lesser coverts* (*tectrices minores*; fig. 30, *bc*). The greater coverts furnish an excellent zoölogical character; for in no *Passeres* are they more than half as long as the remiges they cover, while the reverse is the case in most birds of lower orders. Woodpeckers, however, though non-passerine, have quite short coverts. The *under coverts* have the same general arrangement as the upper; but they are more alike and less distinctly disposed in rows or series; so that for practical purposes they pass under the general name of under wing-coverts, or *lining of the wing*. Since, when the wing is particularly marked on the under side, it is the coverts and not the remiges that are highly or variously colored, the common expression "wing below," or "under surface of the wing," refers to the coverts more particularly. We should distinguish, however, from the under coverts in general, the *axillars*, or axillary feathers (Lat. *axilla*, the arm-pit). These are the innermost feathers lining the wings, lying close to the body; almost always longer, stiffer, narrower, or otherwise peculiarly modified. In ducks, for example, and many of the waders, as snipe and plover, they are remarkably well developed. The color of the axillaries is the principal distinction between some species of plovers. The

Remiges, or Flight-Feathers (fig. 30, *b*, *s*, and *t*), give the wing its general character, mainly determining both its size and its shape; they represent most of its surface and of its inner and outer borders, and all of its posterior outline, forming a great expansion of which the bony and fleshy framework is insignificant in comparison. The shape of the wing is indeed primarily affected by the relative lengths of its bony segments, the upper arm being, in a humming-bird, for example, very short in comparison with the terminal portion of the limb, and in an albatross again, both upper and forearm being greatly lengthened; still in any case it is the flight-feathers that mainly determine the contour of the wing, by their absolute degree of development, their lengths proportionately to one another, and their individual shapes. They collectively form a thin, elastic, flattened surface for striking the air, quite firm along the front border where the bone and muscle lie, thence growing more mobile and resilient toward the posterior border and along the outer edge. Such surface may be quite flat, as in such birds as cut the air with long, pointed wings, like oar-blades; but it is generally a little concave underneath and correspondingly convex above; such arching or vaulting of the wing-surface being usually associated with a short, broad, rounded wing, as in the gallinaceous tribe, and being least in birds which have the thinnest and sharpest wings. Corresponding differences in the mode of flight result. The short, rounded wing confers a powerful though labored flight for short distances, usually accompanied by a whirring noise resulting from the rapidity of the wing-beats; birds that fly thus are almost always thickset and heavy. The long, pointed wing gives a noiseless, airy, skimming flight, indefinitely prolonged, and accomplished with more deliberate wing-beats; birds of this style of wing are generally trim and elegant. These, of course, are merely generalizations of the extremes of modes of flight, mixed and gradated in every degree in actual bird-life. Thus the humming-bird, which has sharp, thin wings, whirs them fastest of all birds, — so rapidly that the eye cannot follow the strokes, merely perceiving a haze about the bird while the ear hears the buzzing. The combination of acuteness and concavo-convexity is a remarkably strong one, conferring a rapid, vigorous, whistling flight, as that of a duck or pigeon, or the splendid hurtling of a falcon. An ample wing, as one both long and broad without being pointed is called, is well displayed by such birds as herons, ibises, and cranes; the flight may be strong and sustained, but is rather slow and heavy. The longest-winged birds are found among the swimmers, particularly the pelagic family of the petrels, and some of the whole-webbed order, as pelicans, particularly the frigate-pelican. The last named, *Tachypetes aquilus*, has perhaps the longest wings for its bulk of body of any bird whatever, as well as the shortest feet. The American vultures are likewise of great alar expanse in proportion to their weight. The shortest wings, among birds possessing perfect remiges, occur among the lower swimmers, as auks and divers, and among some of the Gallinæ. The great auk is, or was, perhaps the only flightless bird with well-formed flight-feathers, only too small to subserve their usual purpose; though certain South American ducks are said to be in similar predicament. In the penguins, the whole wing-structure is degraded, and the remiges abort in scale-like feathers, the wings being reduced to fins both in form and function. The whole of the existing *Ratitæ* have rudimentary or very imperfect wings, as was the case with the Cretaceous *Hesperornis*; but the contemporary of the latter, *Icthyornis*, and the still more ancient *Archæopteryx*, appear both to have had excellent ones.

The disposition of the remiges in their mutual relations is very noteworthy. They have a rigid hollow barrel of great resistant powers, considering the amount of substance, — just like the cylindrical stem of the cereal plant; a stout, solid, highly elastic shaft; the outer web narrower than the inner, with its barbs set at a more acute angle upon the shaft. Any one of these stiffer outer vanes *overlies* the broader and more yielding inner vane of the next outer feather, which, on receiving the impact of air from below, resists as it were with the strength of a second shaft superimposed. Though the "way of an eagle in the air" was a mystery to the wise man of old, the mechanics of ordinary flight are now better understood. But the sailing of some birds for an indefinite length of time, up as well as down, without visible motion of the wings, and without reference to the wind, remains an enigma. The flight of the albatross and turkey vulture, I venture to affirm, is not yet explained. The riddle of The Wing will be read when we know how the archsaurian escaped from ilus to æther.

The *number* of true remiges ranges from about sixteen, as in a humming-bird, to upwards of fifty, as in the albatross. Their *shape* is quite uniform, minor details aside. They are the stiffest, strongest, most perfectly *pennaceous* of feathers, without evident hyporhachis, if any. They are generally *lanceolate*, that is, tapering regularly and gradually to an obtuse point, though not infrequently more parallel-sided, especially those of the secondary and tertiary series. Either or both webs may be incised toward the end; that is, more or less abruptly narrowed; this is called *emargination* (see fig. 343); their ends may be transversely or obliquely truncate, or nicked in various ways. In a few birds, apparently for purposes of sexual ornamentation, they are developed in bizarre shapes of beauty, with evident decrease of utility as flight-feathers. Those of the ostrich and penguin tribes share the peculiarities of the general plumage of these extraordinary birds. Remiges are divided into three classes or series, according to where they grow upon the limb, whether upon the hand, the fore-arm, or the upper arm. In this distinction is involved one of the most important considerations of practical ornithology, of which the student must make himself master. The three classes of quill-feathers are: 1. the *primaries*; 2. the *secondaries*; 3. the *tertiaries*.

The Primaries (Fig. 30, *b*) are those remiges which grow upon the pinion, or hand- and finger-bones collectively (fig. 27, *C* to *D*). Whatever the total number of the remiges may be, *in nearly all birds with true remiges the Primaries are either* NINE *or* TEN *in number*. The humming-bird with sixteen remiges, the albatross with fifty or more, each have ten primaries. The grebes and a few other birds are said to have eleven primaries: if this be so, it is at any rate highly exceptional. No instance of a higher number than this is known to me. Again, it is only among the highest *Passeres* that the number nine is found, the *Oscines* having indifferently nine or ten. In a good many *Oscines*, rated as nine-primaried, there are actually ten, though the outermost is so rudimentary, and even out of alignment with the developed primaries, that it is not counted as one of them. Among *Oscines*, just this difference of one evident and unquestionable primary more or less forms one of the best distinctions between the families of that suborder. So the tenth feather in a bird's wing, counting from the outside, becomes a crucial test in many cases; for, if it be last primary, the bird is one thing; if it be first secondary, the bird is another. In such cases the necessity, therefore, of determining exactly which it is becomes evident. Of course it is always possible to settle the question by striking at the roots of the remiges and seeing how many are seated on the pinion; but this generally involves some defacing of the specimen, and there is usually an easier way of determining. Hold the wing half-spread: then, in most *Oscines*, the primaries come sloping down on one side, and the secondaries similarly on the other, to form where they meet a reëntrant angle in the general contour of the posterior border of the wing; the feather that occupies this notch is the one we are after, and unluckily it is sometimes last primary, sometimes first secondary. But observe that primaries are so to speak, *self-asserting, emphatic, italicized*, remiges, stiff, strong, and obstinate; while secondaries are *retiring, whispering, in brevier*, limber, weak, and yielding. Their different character is almost always shown by *something* in their shape or texture which the student will soon learn to recognize, though it cannot well be described. Let him examine fig. 30, where *b* marks the nine primaries of a sparrow's wing, and *s* indicates the secondaries; he will see a difference at once. The primaries express themselves, though with diminishing emphasis, to the last one; then the secondaries begin to tell a different tale. Among North American birds the only ones with NINE primaries are the families *Motacillidæ, Vireonidæ, Coerebidæ, Sylvicolidæ, Hirundinidæ, Tanagridæ, Fringillidæ, Icteridæ*, part of *Vireonidæ*, and the genus *Ampelis*. The condition of the *first* primary, whether *spurious* or not, is often of great help in this determination. The first primary is called "spurious" when it is very short — say one third, or less, as long as the second, or longest, primary. Among *Passeres*, a spurious first primary only occurs in certain ten-primaried *Oscines*: whence it is evident, that to find such short first primary is equivalent to determining the presence of ten primaries, though not to find it does not prove there are only nine; the count should be made in all cases in which the outer primary is more than one-third as long as the next. The difference between nine primaries, and ten with the first spurious, is excellently illustrated among the species of *Vireo*. Any thrush, nuthatch, titmouse, or creeper shows a spurious primary to advantage, — large enough not to be overlooked, small enough not to be mistaken.

The Secondaries (Fig. 30, *s*) are those remiges which are seated on the fore-arm (fig. 27, *B* to *C*). They vary in number from six to forty or more. They have the peculiarity of being attached to one of the bones of the fore-arm, the *ulna*. If an ulna be examined closely, there will be seen a row of little points showing the attachment; such are indicated in fig. 27, along *ul*, and in fig. 31. The secondaries present no points necessary to dwell upon here, after what has been said of the primaries.

FIG. 31. — Ulna of *Colaptes mexicanus*, showing points of attachment of the secondaries. (Dr. R. W. Shufeldt, U. S. A.)

They are enormously developed in the Argus pheasant, and have curious shapes in some other exotic birds. They are often long enough to cover the primaries completely when the wing is closed, as in grebes; on the other hand, they are extremely short in the swifts and humming-birds.

The Tertiaries (Fig. 30, *t*) are properly the remiges which grow upon the upper arm, *humerus*. But such feathers are not very evident in most birds, and the two or three innermost secondaries, growing upon the very elbow, and commonly different from the rest in form or color, pass under the name of "tertiaries." Again, in some cases, scapular feathers (fig. 30, *scp*,) are called tertiaries, especially when long or otherwise conspicuous. But there is an evident and proper distinction. Scapulars belong to the *pteryla humeralis* (see p. 90); while tertiaries, whether seated on the elbow or higher up the arm, are the innermost remiges of the *pteryla alaris*. These inner remiges are often shortly called *tertials*; though the longer name is more correct, besides being conformable with the names of the other two series of remiges. Tertiaries often afford good characters for description, in peculiarities of their size, shape, or color. Thus it is very common among *Fringillidæ* for these feathers to be parti-colored differently from the other remiges. In many birds they are long and "flowing"; as in the families *Motacillidæ* and *Alaudidæ*, where they reach about to the end of the primaries when the wing is closed. Their development is similar in many *Scolopacidæ*. In such cases, the feather-border of the wing pronounces the letter **W** quite strongly, — outer lower angle at point of primaries; middle upper angle at reëntrance between primaries and secondaries; inner lower angle at point of tertiaries.

The "point of the wing" is at the tip of the longest primary. It is best expressed when the first primary is longest. Sometimes the end is so much rounded off, that the midmost primary may be the longest one, the others being graduated on both sides of this projecting point. In speaking of the relative lengths of remiges, we always mean the way in which their tips fall together, not the actual total lengths of the feathers. Thus a second primary, whose tip falls opposite the tip of the first one, is said to be of equal length, though it may actually be longer, being seated higher up on the pinion. The development of the primaries also furnishes one of the most important measurements of birds: for the expression "length of wing," or simply "the wing," means the distance from the "bend of the wing," or carpal angle, to the end of the longest primary. The integument of the wing does not very often develop anything but feathers. Occasionally

Claws and Spurs are found upon the pinion. Claws have been already noticed (p. 114). They are properly so called, being horny growths comparable in every way to those upon the ends of the toes, like the claws of beasts, or human nails. A *spur* (Lat. *calcar*), however, is something different, though of the same horny texture, since it does not terminate a digital phalanx, but is off-set from the side of the hand. It is exactly like the spur on the leg of a fowl, which obviously is not a claw. The spur-winged goose (*Plectropterus*), pigeon (*Didunculus*), plovers (*Chettusia*, etc.), and the doubly-spurred screamer (*Palamedea*), afford examples of such outgrowths, of which the Jaçanas (*Parra*) furnish the only, though a very well-marked, illustration among North American birds. (See fig. 53 *ter*.)

III. THE TAIL.

Its Bony Basis. — Time was when birds flew about with long, lizard-like, bony and fleshy tails, having the feathers inserted in a row on either side like the hairs of a squirrel's. But we have changed all that *distichous* arrangement since when the *Archæopteryx* was steered with such a rudder through the scenes of its Jurassic life. Now the true separate *coccygeal* bones are few, generally about nine in number, and so short and stunted that they do not project beyond the general plumage, — in fact scarcely beyond the border of the pelvis. Anteriorly, within the bony basin of the pelvis, there are several vertebræ, which, fusing together and with the true *sacrum*, are termed *urosacral* or false tail-bones. To these succeed the true caudal vertebræ, movable upon each other and upon the urosacrum. The last one of these, abruptly larger than the rest, and of peculiar shape, bears all the large tail-feathers, which radiate from it like the blades of a fan. The true caudal vertebræ collectively form the *coccyx* (Gr. κόκκυξ, *kokkux*, a cuckoo; from fancied resemblance of the human tail-bones to a cuckoo's bill); the enlarged terminal one is the *vomer* (Lat. *vomer*, a plough-share, from its shape; not to be confused with a bone of the skull of same name) or *pygostyle* (Gr. πυγή, *puge*, rump, and στῦλος, *stulos*, a stake, pale). The pygostyle, however, is a compound bone, consisting of several stunted coccygeal vertebræ fused in one. The bones are moved by appropriate muscles, and upon the surface is seated the elæodochon (p. 89). The whole bony and muscular affair is familiar to every one as the "pope's nose" of the Christmas turkey; it is a bird's real tail, of which the feathers are merely appendages. In descriptive ornithology, however, the anatomical parts are ignored, the word "tail" having reference solely to the feathers. These, like those of the wings, are of two sorts: the coverts or *tectrices*, and the rudders or *rectrices* (Lat. *rectrix*, pl. *rectrices*, a ruler, guider; because they seem to steer the bird's flight); corresponding exactly to the coverts and remiges of the wings. The

Tail-Coverts are the numerous comparatively small and weak feathers which overlie and underlie the rectrices, covering their bases and extending a variable distance toward their ends, contributing to the firmness and symmetry of the tail. They pass smoothly out from the body, by gradual lengthening, there being seldom, if ever, any obvious outward distinction between them and feathers of the rump and belly; but they belong to the *pteryla caudalis* (p. 90). The natural division of the coverts is into an *upper* and *under* set (*tectrices superiores, tectrices inferiores*). The inferior coverts are the best distinguished from the general plumage, the anus generally dividing off these "vent-feathers," as they are sometimes called. It is to the bundle of under tail-coverts, behind the vent, that the term *crissum* is most properly applied. Neither set is ever entirely wanting; but one or the other, particularly the upper one, may be very short, as in a cormorant, or duck of the genus *Erismatura*, exposing the quills almost to their bases. While the upper coverts are usually shorter and fewer than the under ones, reaching less than half-way to the end of the tail, they sometimes take on extraordinary development and form the bird's chiefest ornament. The gorgeous, iridescent, argus-eyed train of the peacock consists of enormous tectrices, not rectrices; the elegant plumes of the paradise trogon, *Pharomacrus mocinno*, several times longer than the bird itself, are likewise coverts. Occasionally, a pair of coverts lengthens and stiffens, and then resembles true tail-feathers; as in the Ptarmigan (*Lagopus*). The crissal feathers are more uniform in development; they ordinarily form a compact, definite bundle, as well shown in a duck, where they reach about to the end of the tail. In some of the storks, they become plumes of considerable pretensions; and in the wonderful humming-bird, *Loddigesia mirabilis*, the middle pair stiffens to resemble rectrices and projects far beyond the true tail. The

Rectrices, Rudders, or true tail-feathers, like the remiges or rowers, are usually stiff, well-pronounced feathers, pennaceous to the very base of the vexilla, without after-shafts, as a rule, and with the outer web narrower than the other in most cases. They are always *in pairs;* that is, there is an equal number of feathers on the right and left half of the tail; and their number, consequently, is an even one. The exceptions to this rule are so few and irregular, and then only among birds with the higher numbers of rectrices, that such are probably to be regarded as mere anomalies, from accidental arrest of a feather. They are imbricated over each other in this wise: the central pair are highest, lying with *both* their webs over the next feather on either side, the inner web of one of these middle feathers indifferently underlying or overlying that of the other; all thus successively overlying the next outer one so that they would form a pyramid were they thick instead of being so flat. The arrangement is perceived at once in the accompanying diagram; where it will be seen, also, that *spreading* the tail is the divergence of *a* from *b*, while closing the tail is bringing *a* and *b* together under *c*. The motion is effected by certain muscles that draw on either side upon the bases of the quills collectively; they are the same that pull the whole tail to one side or the other, acting like the tiller-ropes of a boat's rudder. The *general*

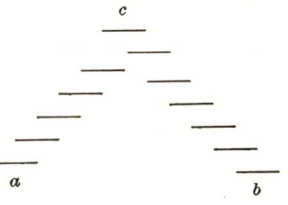

Shape of a Rectrix is shown in fig. 23. Such a feather is ordinarily straight, somewhat clubbed or oblong, widening a little, regularly and gradually toward the tip, where it is gently rounded off. But the departures from such shape, or any that could be assumed as a standard, are numberless, and in some cases extreme. In fact, none of a bird's feathers are more variable than those of the tail; it is impossible to specify all the shapes they assume. While most are straight, some are curved — and the curvature may be to or from the middle line of the body, in the horizontal plane, or up and down, in the vertical plane. Some shapes have received particular names. A rectrix broad to the very tip, and there cut squarely off, is said to be *truncate;* one such cut obliquely off is *incised,* especially when, as often happens, the outline of the cut-off is concave. A *linear* rectrix is very narrow, with parallel sides; a *lanceolate* one is broader at the base, thence tapering regularly and gradually to the tip. A notably pointed rectrix is said to be *acute;* when the pointing is produced by abrupt centraction near the tip, as in most woodpeckers, the feather is *acuminate.* A very long and slender, more or less linear feather is called *filamentous,* as the lateral pair of a barn swallow or most sea swallows. The vanes sometimes enlarge abruptly at the end, forming a spoon-shaped or *spatulate* feather;

Fig. 32. — The Lyre-bird of Australia, *Menura superba,* to show the unique *lyrate* shape of the tail. (From Amer. Nat.)

or such a spoon may result from narrowing of the vanes near the end, or their entire absence, as in the "racket" of a saw-bill (*Momotus*). The vanes are sometimes wavy as if crimped; our *Plotus* is a fine example of this. Sometimes the vanes are entirely loosened, the barbs being remote from each other, as in the exotic genus *Stipiturus,* and some parts of the wonderful caudal appendage of the male lyre-bird (*Menura superba*). When the rhachis projects beyond the vanes, the feather is *spinose,* or better, *mucronate* (Lat. *mucro,* a pricker), as excellently shown in the chimney-swift, *Chætura* (fig. 375). A pair of feathers abruptly extending far beyond the others are called *long-exserted,* after the analogous use of the term in botany. Tail-feathers also differ much in their consistency, from the softest and weakest, not well distinguished from coverts, to such stiff and rugged props as the woodpeckers possess. They are downy and very rudimentary in a few birds, notably all the grebes, *Podicipedidæ,* which are commonly said to have no tail. The tinamous of South America (*Dromæognathæ*) are also very closely docked. The

Typical Number of Rectrices is *twelve.* This holds in the great majority of birds. It is so uniform throughout the great group Oscines, that the rare exceptions seem perfectly anomalous (ten in *Edoliidæ* or *Dicruridæ*). In the other group of *Passeres* (*Clamatores*) it is usually twelve, sometimes ten. Ten is the rule among *Picariæ,* though many have twelve, a very few only eight, as in the genus *Crotophaga.* The whole of the woodpeckers (*Picidæ*) have *apparently ten;* but really *twelve*, of which the outer one on each side is spurious, very small, and hidden between the bases of the second and third feathers. Birds of prey (*Raptores*) have about twelve. In pigeons the rule is twelve or fourteen, as in all our genera; but sixteen are found in some and twenty in one case. In birds below these, the number increases directly; there are often or usually more than twelve in the grouse, and there may be sixteen, eighteen, or twenty, as among our own genera of *Tetraonidæ*. Wading birds, often having but twelve, furnish instances of as many as twenty. Those swimming birds with large well-formed tails, as the *Longipennes*, and some *Anatidæ*, have the fewest, as twelve, sometimes fourteen, rarely sixteen; those with short soft tails have the most, as sixteen to twenty-four (forty in some domestic pigeons). Among the penguins there are thirty-two or more. The *Archæopteryx* appears to have had forty, — a pair to each free caudal vertebra; and this may be considered the prototypic relation between the bones and feathers of the tail. The

Typical Shape of the Tail, as a whole, is the *fan*. The modifications of form, however, which are greater and more varied than those of the wing, are susceptible of better definition, and many of them have received special names. Taking the simplest case, where the rectrices are all of the same length, we have what is called the *even, square,* or *truncate* tail. The other forms depart from this mainly by shortening or lengthening of certain feathers. A tail nearly or quite even may have the two central feathers long-exserted, as seen in the jaegers (*Stercorarius*), and tropic-birds (*Phaëthon*). The most frequent departure from the even shape results from gradual shortening of successive rectrices from the middle to the outer ones. This is called, in general, *gradation* or *graduation* (Lat. *gradus*, a step); such shortening may be to any degree. More precisely, graduation means shortening of each successive feather to the same extent, — say, each half an inch shorter than the next; but such exactitude is not often expressed. When the feathers shorten by more and more, we have the true *rounded* tail, probably the commonest form among birds; thus, the gradation between the middle and next pair may be just appreciable, and then increase regularly to an inch between the next and the lateral feather. The opposite gradation, by less and less shortening, gives the wedge-shaped or *cuneate* (Lat. *cuneus*, a wedge) tail; it is well shown by the magpie (*Pica*) in which, as in many other birds, the middle feathers would be called long-exserted were the rest all as short as the outer one is. A cuneate tail, especially if the feathers be narrow and lanceolate, is also called *acute*, or pointed, as in the sprig-tailed duck (*Dafila*) or sharp-tailed grouse (*Pediœcetes*). The generic opposite of the gradated is the *forked* tail; in which the lateral feathers successively increase in length from the middle to the outermost pair. The least appreciable forking is called *emargination*, and a tail thus shaped is said to be *emarginate;* when it is better marked, as, for instance, an inch of forking in a tail six inches long, the tail is truly *forked* or *furcate* (Lat. *furca*, a fork). But the degrees of furcation, like those of gradation, are so insensibly varied, that qualified expressions are usual; as, "slightly forked," "deeply forked." Deep furcation is usually accompanied by more or less narrowing or filamentous elongation of the lateral pair of rectrices, as in the barn swallows (*Hirundo*) and most of the sea-swallows (*Sterna*). An advisable term to express such an extreme furcation is *forficate* (Lat. *forfex*, scissors), when the depth of the fork is at least equal to the length of the shortest feathers; it occurs among our birds in those last named, in the species of the flycatcher genus *Milvulus*, and elsewhere. *Double-forked* and *double-rounded* tails are not uncommon; they result from combination of both opposite gradations, in this way: The middle feathers being of a certain length, the next two or three pairs progressively increasing in length, and the rest successively decreasing, the tail is evidently forked centrally, rounded externally, which is the double-rounded form, each half of the tail being rounded; it is shown in the genera *Myiadestes* and *Anous*. Now if with middle feathers as before, the next pair or two decrease in length, and then the rest increase to the outermost, we have the double-forked, a common style among sandpipers, as if each half of the tail were forked. But in such case, the forking is slight, merely emargination, being little more than protrusion of the middle pair of feathers in an otherwise lightly forked tail; and in the double-rounded form the gradation is seldom if ever great.

I should also allude to shapes of tail resulting from the relative positions of the feathers. Prominent among these is the *complicate* or *folded* tail of the barn-yard fowl, and others of the *Phasianidæ*,—a very familiar but not common form. It is only retained while the tail is closed and cocked up,—for when it is lowered and spread in flight it flattens out. The opposite disposition of the feathers is seen to some extent in our crow blackbirds (*Quiscalus*), where the lateral feathers slant upward from the lowermost central pair, like the sides of a boat from its keel; this is the *scaphoid* (Gr. σκάφη, a boat) or *carinate* (Lat. *carina*, a keel) tail. Our "boat-tailed" grackle has been so named on this account. One of the most beautiful and wonderful of all the shapes of the tail is illustrated by the male of the lyre-bird (*Menura superba*, fig. 32), in which the feathers are anomalous both in shape and in texture, and the resulting form of the whole is unique. Various shapes, which the student will readily name from the foregoing paragraphs, are illustrated in many other figures of this work. It should be remembered that, to determine the shape, the tail should be *nearly* closed; for spreading will obviously make a square tail round, an emarginate one square, etc. I append a diagram of the principal forms (fig. 33).

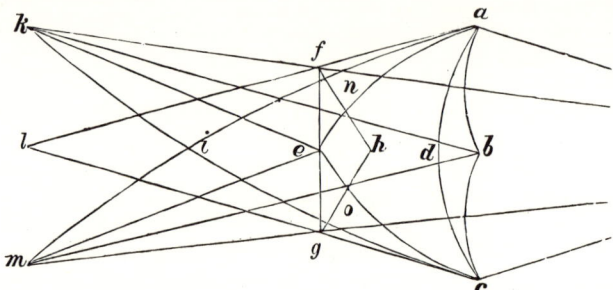

FIG. 33.—Diagram of shapes of tail. *adc*, rounded; *aec*, gradate; *aic*, cuneate-gradate; *alc*, cuneate; *abc*, double-rounded; *feg*, square; *fhg*, emarginate; *fneog*, double-emarginate; *kim*, forked; *kem*, deeply forked; *kbm*, forficate.

IV. THE FEET.

The Hind Limbs, in all birds, are organized for progression—all can walk, run, or hop on land, though the power to do so is very slight in some of the lower swimming birds, as loons and grebes, and certain of the lower perching birds, as hummers, swifts, goatsuckers, and kingfishers. They are specially fitted for perching on trees, bushes, and other supports requiring to be grasped, in the great majority of birds, as throughout the *Passeres*, *Picariæ*, *Accipitres*, *Columbæ*, and, in fact, many water-birds; there being few forms, mainly found among three-toed birds, or those in which the hind toe is short, weak, and elevated, in which the extremity of the limb has not decided grasping power. The limb becomes a paddle for swimming either on or in the water in many cases. In not a few, as parrots and birds of prey, the foot is serviceable as a hand. Those kinds of birds which live in trees and bushes habitually progress, even when on level ground, in a series of hops, or rather leaps, both feet being moved together: in all the lower birds, however, the feet move one after the other, as in ordinary walking or running. The modifications of the hind limb are more numerous, more diverse, and more important in their bearing on classification than those of either bill, wing, or tail; their study is consequently a matter of special interest.

Their Bony Framework (fig. 34).—Beginning at the hip-joint, and ending at the extremities of the several toes, the skeleton of the hind limb consists in the vast majority of adult birds of *twenty* bones. This is the typical and nearly the average number; birds scarcely ever have more, and the principal lessenings of the number result from the absence of one or two toes, or a slight reduction in the number of the joints of some toes, or absence of the knee-cap. Of the normal twenty, fourteen are bones of the toes; one is an incomplete bone connecting the hind toe with the foot; one is the knee-cap, and four are the principal bones of the thigh (1), leg (2), and foot (1). The first or uppermost is the thigh-bone or *femur* (Lat. *femur*; adjective, *femoral*), *fm*, from hip to knee, *A* to *B* in the figure. It is a rather short, quite stout, cylindrical bone, enlarging above and below. Above it has a globular head, *a*, standing off obliquely from the shaft, received in the *acetabulum* (Lat. *acetabulum*, a kind of receptacle) or socket of the hip, and a prominent shoulder or *trochanter*, which abuts against the brim of the acetabulum. Below, it expands into two *condyles* (Gr. κόνδυλος, a knob), for articulation with both the bones it meets at the knee. It is the same bone as the femur of a quadruped or of man, and corresponds to the *humerus* of the wing. In the knee-joint, many or most birds have a small ossicle, and a few have two such bony nodules, not shown in the figure, but nearly in the position of the letter *B*: the knee-pan or knee-cap, *patella* (Lat. *patella*). The thigh is the first *segment* of the limb; the next segment is the leg proper, or *crus* (Lat. *crus*, the shin; adjective, *crural*), *B* to *C* in the figure, or from knee to heel. This segment is occupied by two bones, the *tibia* (Lat. *tibia*, a tube, trumpet), *tb*, and *fibula* (Lat. *fibula*, a splint, clasp), *fi*. Of these the tibia is the principal, larger, inner

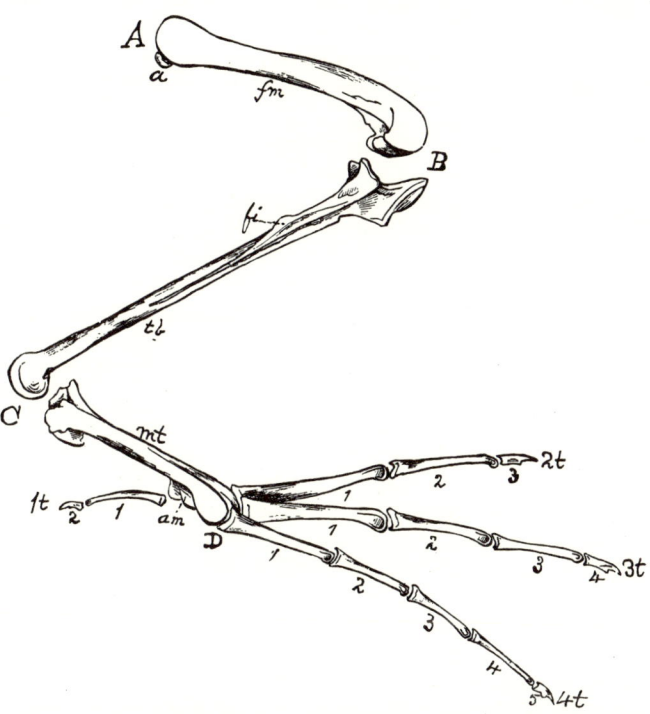

FIG. 34. — Bones of a bird's hind limb: from a duck, *Clangula islandica*, ¾ nat. size; Dr. R. W. Shufeldt, U. S. A. *A*, hip: *B*, knee: *C*, heel or ankle-joint; *D*, bases of toes. *A* to *B*, thigh or "second joint"; *B* to *C*, crus, leg proper, "drumstick," often wrongly called "thigh"; *C* to *D*, metatarsus, foot proper, corresponding to our instep, or foot from ankle to bases of toes; in descriptive ornithology *the tarsus;* often called "shank." From *D* outward are the toes or digits. *fm*, femur; *tb*, tibia, principal (inner) bone of leg; *fi*, fibula, lesser (outer) bone of leg; *mt*, principal metatarsal bone, consisting chiefly of three fused metatarsal bones; *am*, accessory metatarsal, bearing 1*t*, first or hind toe, with two joints; 2*t*, second toe, with three joints; 3*t*, third toe, with four joints; 4*t*, fourth toe, with five joints. At *C* there are in the embryo some small tarsal bones, not shown in the figure, uniting in part with the tibia, which is therefore a *tibio-tarsus*, in part with the metatarsus, which is therefore a *tarso-metatarsus;* the ankle-joint being therefore between two rows of tarsal bones, not, as it appears to be, directly between tibia and metatarsus.

bone, running quite to the heel; the fibula is smaller, and (with rare exceptions, as in some of the penguins) only runs part way down the outside of the tibia as a slender pointed spike, close pressed against or even partly fused with the shaft of the tibia. Above, at the knee, both bones articulate with the femur; the tibia with both the femoral condyles, the fibula only with the outer condyle. Above, the tibia has an irregularly expanded head or *cnemial* process (Gr.

κνήμη, *kneme*, same as Lat. *crus*), which in some birds, as loons, runs high up in front above the knee-joint. Below, the tibia alone forms the ankle-joint, *C*, by articulating with the next bone. For this purpose it ends in an enlarged *trochlear* (Gr. τροχαλία), or pulley-like surface, presenting a little forward as well as downward, above which, in many birds, there is a little bony bridge beneath which tendons passing to the foot are confined. This finishes the leg, consisting of thigh, *A B*, and leg proper, *B C*, bringing us to the ankle-joint at the heel, *C*.

Now a bird's legs, unlike ours, are not separate from the body from the hip downward; but, for a variable distance, are enclosed within the general integument of the body. The freedom of the limb is greatest among the high perching birds, and especially the *Raptores*, which use the feet like hands, and least among the lowest swimmers. The range of variation, from greatest freedom to most extensive enclosure of the limb, is from a little above *B* nearly to *C*, as in the case of a loon, grebe, or penguin. In no bird is the knee, *B*, seen outside the general contour of the plumage: it must be looked or felt for among the feathers, and in most prepared skins will not be found at all, the femur having been removed. It is a point of little practical consequence, though bearing upon the generalization just made. The first *joint*, or bending of the limb, that appears beyond a bird's plumage is the *heel*, *C*; and this is what, in loose popular parlance, is called "knee," upon the same erroneous notions that make people call the wrist of a horse's fore-leg "knee." People also call a bird's *crus* or leg proper, *B* to *C*, the "thigh," and disregard the true thigh altogether. This confusion is inexcusable; any one, even without the slightest anatomical knowledge, can tell knee from heel at a glance, whatever their respective positions relative to the body. *Knee* is at junction of thigh and leg proper; it always *bends forward*; *heel* is at junction of leg with foot, and always bends backward. This is as true of a bird, which is *digitigrade*, that is, walks on its toes with its heels in the air, as it is of a man, who is *plantigrade*, that is, walks on the whole sole of the foot, with the heel down to the ground. In a carver's language, the thigh is the "second joint" (from below); the leg is the "drumstick"; the rest of a fowl's hind limb does not usually come to table, having no flesh upon it.

Before proceeding to the next segment of the limb, I must dwell upon the ankle-joint, situated at the heel, — the point *C*, — corresponding to the carpal angle or bend of the wing, *C*, in fig. 27. There we found, in adult birds, two small carpal bones, or bones of the wrist proper; and noted the presence in the embryo of several other carpals (fig. 29), which early fuse with the metacarpus. Just so in the ankle, there are in embryonic life several *tarsal* bones, or bones of the *tarsus* (Lat. *tarsus*, the ankle); *all* of which, however, soon disappear, so that there appears to be no tarsus, or collection of little bones between the tibia and the next segment of the limb, the *metatarsus*. An upper tarsal bone, or series of tarsal bones, fuses with the lower end of the tibia, making this leg-bone really a *tibio-tarsus*; and similarly, a lower bone or set of bones fuses with the upper end of the metatarsus, making this bone a *tarso-metatarsus*. So there are left no free bones in the ankle-joint, which thus appears to be immediately between the leg-bone and the principal foot-bone; but which is nevertheless really between two series of tarsal bones, the identity of which has been lost.[1]

---

[1] The exact homologues of a bird's vanishing tarsal bones are still questioned. Gegenbaur showed the so-called epiphysis or shoe of bone at the foot of the tibia, and the similar cap of bone on the head of the principal metatarsal bone, to be true tarsal elements. Morse went further, showing the tibial epiphysis, or upper tarsal bone of Gegenbaur to be really two bones, which he held to correspond with the tibiale and fibulare, or *astragalus* and *calcaneum* of mammals; these subsequently combining to form the single upper tarsal bone of Gegenbaur, and finally becoming anchylosed with the tibia to form the bitrochlear condylar surface so characteristic of the tibia of *Aves*. The distal tarsal ossicle he believed to be the *centrale* of reptiles. Wyman discovered the so-called "process of the astragalus" to have a distinct ossification, and Morse interpreted it as the *intermedium* of reptiles. Later views, however, as of Huxley and Parker, limit the tibial epiphysis to the *astragalus* alone of mammals. If these opinions be correct, other tarsal elements (more than one) are to be looked for in the epiphysis of the metatarsus. Whatever the final determination of these obscure points may be, it is certain that, as said in the text above, the lower end of a bird's tibia and the upper end of a bird's metatarsus include true tarsal elements, just as the upper

The next segment of the limb, *C* to *D*, or the foot proper, is represented by the principal *metatarsal* bone, *mt*. This corresponds to the human *instep* or arch of the foot, nearly from the ankle-joint quite to the roots of the toes. The metatarsal bone, like the metacarpal of the hand, which it represents in the foot, is a compound one. Besides including the evanescent tarsal element or elements already specified, it consists of *three* metatarsal bones consolidated in one, just as the metacarpal is tripartite. Among recent birds, the three are partly distinct only in the penguins; but in all, excepting ostriches, the original distinction is indicated by three prongs or stumps at the lower end of the bone, forming as many articular surfaces for the three anterior toes. The other toe most birds possess, the hind toe, is hinged upon the metatarsus in a different way, by means of a small separate metatarsal bone, quite imperfect; this is the *accessory metatarsal, am*. It is situated near the lower end toward the inner side of the principal metatarsal bone, and is of various shapes and sizes; it has no true jointing with the latter, but is simply pressed close upon it, much as the fibula is applied to the tibia, or partly soldered with it. Above, it is defective; below, it bears a good facet for articulation with the hind toe. ☞ In spite of anatomical proprieties, the metatarsal part of a bird's foot — from heel to base of toes — from *C* to *D*, is in ordinary descriptive ornithology *invariably* called "*The Tarsus*"; a wrong name, but one so firmly established that it would be finical and futile to attempt to substitute the correct name. In the ordinary attitude of most birds, it is held more or less upright, and seems to be rather "leg" than a part of the "foot." It is vulgarly called "the shank." These points must be ingrained in the student's mind to prevent confusion. (See fig. 112 *bis*, p. 235.)

The *digits* of the foot, or *toes*, upon which alone most birds walk or perch, consist of certain numbers of small bones placed end to end, all jointed upon one another, and the basal or proximate ones of each toe separately jointed either with the principal or the accessory metatarsal bone. Like those of the fingers, these bones are called *phalanges* (Lat. *phalanx*, a rank or series) or *internodes* (because coming between any two joints or nodes of the toes). The furthermost one of each almost invariably bears a nail or claw (*unguis*). The phalanges are of various relative lengths, and of a variable number in the same or different toes. But all these points, being matters of descriptive ornithology rather than of anatomy proper, are fully treated beyond, as is also the special horny or leathery covering of the feet usually existing from the point *C* outward. We may here glance at the

Mechanism of these Bones. — The hip is a ball-and-socket joint, permitting round-about as well as fore-and-aft movements of the whole limb, though more restricted than the shoulder-joint. The knee is usually a strict *ginglymus* (Gr. γίγγλυμος, *gigglumos*, hinge) or hinge-joint, allowing only backward and forward motion; and so constructed that the forward movement of the leg is never carried beyond a right line with the femur, while the backward is so extensive that the leg may be quite doubled under the thigh. In some birds there is a slight rotatory motion at the knee, very evident in certain swimmers, by which the foot is thrown outward, so that the broad webbed toes may not "interfere." The heel or ankle-joint is a strict hinge; its bendings are just the reverse of those of the knee; for the foot cannot pass back of a right line with the leg, but can come forward till the toes nearly touch the front of the knee. In some birds the details of structure are such that, with the assistance of certain muscles, the foot is *locked* upon the leg when completely straightened out, so firmly that some little muscular effort is required to overcome the obstacle; birds with this arrangement sleep securely standing on one leg, which is the design of the mechanism. The jointing of the toes with the prongs of the metatarsus is peculiar; for the articular surfaces are so disposed in a certain obliquity, that when

---

end of the metacarpus includes carpal elements; and that a bird's ankle-joint is *not* tibio-tarsal or between leg-bone and foot-bones, as in mammals, but between proximal and distal series of tarsal bones, and therefore *medio*-tarsal, as in reptiles.

the toes are brought forwards, at right angles or thereabouts with the foot, they spread apart from each other automatically in the action, and the diverging toes of the foot thus opened are pressed upon the ground or against the water. When the toes are bent around in the opposite direction, they automatically come together and lie in a bundle more or less parallel with one another, besides being each bent or flexed at their several nodes. The mechanism is best marked in the swimmers, which, for advantageous use of their webbed toes, must present a broad surface to the water in giving the backward stroke, and bring the foot forward with the toes closed, presenting only an edge to the water, — all on the principle of the feathering of oars in rowing. It is carried to an extreme in a loon, where, when the foot is closed, the digit marked 2*t* in the figure lies below and behind 3*t*. It is probably least marked in birds of prey, which give the clutch with their talons spread. The jointings of the individual phalanges of the toes upon one another are simple hinges, permitting motion of extension to a right line or a little beyond in some cases, with very free flexion in the opposite direction. On the whole, the mechanics of a bird's foot are less peculiar than those of the wing, and quite those of the limbs of a quadruped.

In ordinary hopping, walking, and running, and in perching as well, only the toes rest upon or grasp the support, from *D* to beyond *C* being more or less vertically over *D*. Such resting of the toes is complete for 2 *t*, 3 *t*, 4 *t* in the figure, or for all the anterior toes; but for the hind toe it varies according to the length and position of that digit, from complete incumbency, like that of the front toes, to mere touching of the tip of that toe, or not even this: the hind toe is then sure to be functionless. But many of the lower birds, such as loons and grebes, cannot stand at all upright on their toes, resting with the heel touching the ground; and in many such cases the tail furnishes additional support, making a tripod with the feet, as in the kangaroo. Such birds might be called *plantigrade* (Lat. *planta*, the sole; *gradus*, a step) in strict anatomical conformity with the quadrupeds so designated. The others are all *digitigrade*, standing or walking on their toes alone. But no birds progress on the ends of their toes, or toe-nails, as hoofed quadrupeds do. A bird's ordinary walking or running is the same as ours, so far as the ordinary mechanics of the motions are concerned; but its so-called "hopping" is really leaping, both legs moving at once. Most birds, down to *Columbæ*, leap when on the ground, a mode of progression characteristic of the higher orders; but many of the more terrestrial *Passeres* and *Accipitres* progress by ordinary walking when on the ground, as is invariably the case with parrots, pigeons, gallinaceous birds, and all waders and swimmers.

The student need scarcely be reassured that, whatever their modifications, their relative development, motions, and postures, the several segments of both fore and hind limbs of any vertebrate, quadruped or biped, feathered or featherless, are fixed in one morphologically identical series, thus: 1, shoulder or hip-joint; 2, upper arm or thigh, humerus or femur; 3, elbow or knee-joint; 4, fore-arm or leg proper, radius and ulna or tibia and fibula; 5, wrist, bend of wing, carpus, or heel, ankle, tarsus; 6, hand proper, metacarpus, or foot proper, metatarsus; 7, digits with their phalanges, of hand or foot, fingers or toes. 2, first segment; 4, second segment; 5, third segment (not separate in foot of bird); 6 and 7, fourth segment, in the wing called manus or pinion, in the leg, pes. Observe the improper naming of parts, in the case of the hind limb, whereby 1, 2, 3, are not generally counted; 4 is called "thigh"; 5 is called "knee"; 6 is called "leg" or "shank"; 7 is called "foot." Observe also that in descriptive ornithology 6 is "*the tarsus*."

The Plumage of the Leg and Foot varies within wide limits. In general, the leg is feathered to the heel, *C*, and the rest of the limb is bare of feathers. The thigh is *always* feathered, as part of the body plumage (*pteryla femoralis*). The crus or leg proper (thigh of vulgar language, *B* to *C*) is feathered in nearly all the higher birds, and in swimming birds without exception; in the loons, the feathering even extends on the heel-joint. It is among the walking and especially the wading birds that the crus is most extensively denuded; it may be naked half-way up to the knee. A few waders, — among ours, chiefly in the snipe family, — have the crus apparently clothed to the heel-joint; but this is due, in most if not all cases, to the length of the feathers, for probably in none of them does the pteryla cruralis itself extend to the joint. Crural feathers are nearly always short and inconspicuous; but sometimes long and flowing, as in the "flags" of most hawks, and in our tree-cuckoos. The *tarsus* (I now and hereafter use the term in its ordinary acceptation — *C* to *D* in fig. 34; *trs* in fig. 36) in the vast majority of birds is entirely naked, being provided with a horny or leathery sheath of integument like that covering the bill. Such is its condition in the *Passeres* and *Picariæ* (with few exceptions, as among swifts and goatsuckers); in the waders without exception, and in nearly all swimmers (the frigate-bird, *Tachypetes*, has a slight feathering). The *Raptores* and *Gallinæ* furnish the most feathered tarsi. Thus, feathered tarsi is the rule among owls (*Striges*); frequent, either partial or complete, in hawks and eagles, as in *Aquila*, *Archibuteo*, *Falco*, *Buteo*, etc. All our grouse, and perhaps all true grouse, have the tarsus more or less feathered (fig. 35). The *toes* themselves are feathered in a few birds, as several of the owls, and all the ptarmigans (*Lagopus*). Partial feathering of the tarsus is often continued downward, to the toes or upon them, by sparse modified feathers in the form of bristles; as is well shown in the barn-owl (fig. 47). When incomplete, the feathering is generally wanting behind and below, and it is almost invariably continuous above with the crural plumage. But in that spirit of perversity in which birds delight to prove every rule we establish by furnishing exceptions, the tarsus is sometimes partly feathered discontinuously. A curious example of this is afforded by the bank-swallow, *Cotile riparia*, with its little tuft of feathers at the base of the hind toe; and some varieties of the barn-yard fowl sprout monstrous leggings of feathers from the side of the tarsus.

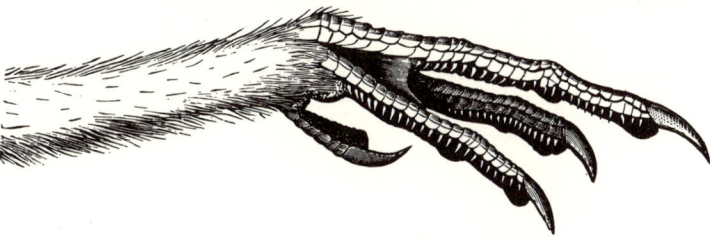

FIG. 35. — Feathered tarsus of a grouse, *Cupidonia cupido*. Nat. size.

The Length of Leg, relatively to the size of the bird, is extremely variable; a thrush or sparrow probably represents about average proportions of the limb. The shortest-legged bird known is probably the frigate-pelican, *Tachypetes;* which, though a yard long more or less, has a tibia not half as long as the skull, and a tarsus under an inch. The leg is very short in many Picarian birds, as hummers, swifts, goatsuckers, kingfishers, trogons, etc., in many of which it scarcely serves at all for progression. Among *Passeres*, the swallows resemble swifts in shortness of their hind limbs. It is pretty short likewise in many zygodactyle, yoke-toed or scansorial birds, as woodpeckers, cuckoos, and parrots. In most swimming birds the limb may also be called short, especially in its femoral and tarsal segments; while the broad-webbed toes are comparatively longer. The leg lengthens in the lower perching birds, as many hawks and some of the terrestrial pigeons; it is still longer among walkers proper, such as the gallinaceous birds, and reaches its maximum among the waders, especially the larger ones, such as cranes, herons, ibises, storks, and flamingoes; among all of which it is correlated with extension of the neck. Probably the longest-legged of all birds for its size is the stilt (*Himantopus*). Taking the tarsus alone as an index of length of the whole limb, this is in the frigate under one-thirty-sixth of the bird's length; a flamingo, four feet long, has a tarsus a foot long: a stilt, fourteen inches long, one of four inches; so that the maximum and minimum lengths of tarsus are nearly thirty and under three per cent. of a bird's whole length.

The Horny Integument of the Foot requires particular attention. That part of the limb which is devoid of feathers is covered, like the bill, by a hardened, thickened, modified integument, varying in texture from horny to leathery. This sheath is called the *podotheca* (Gr. πούς, ποδός, *pous, podos*, foot, and θήκη, *theke*, sheath). It is more corneous in land birds, and in water birds more leathery; this general distinction has but few exceptions. The perfectly horny envelope is tight, and immovably fixed or nearly so, while the skinny styles of sheath are looser, and may usually be slipped about a little. The integument may differ on different parts of the same leg, and in fact generally does so to some extent. Unlike the sheath of the bill, the podotheca is never simple and continuous, being divided and subdivided in various ways. The lower part of the crus, when naked, and the tarsus and toes, always have their integument cut up into scales, plates, tubercles, and other special formations, which have received particular names. The manner and character of such divisions are often of the utmost consequence in classification, especially among the higher birds, since they are quite significant of genera, families, and even some larger groups.

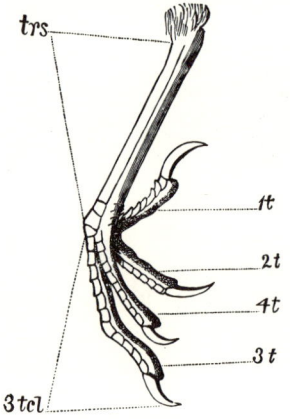
Fig. 36.—Booted laminiplantar tarsus of a robin. Nat. size.

Fig. 37.—Scutellate laminiplantar tarsus of a cat-bird. Nat. size.

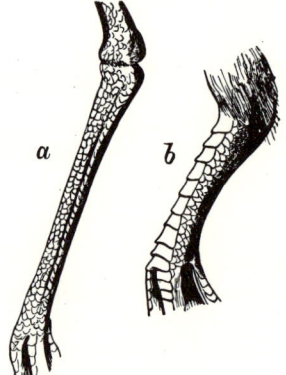
Fig. 38.—*a*. Reticulate tarsus of a plover. Nat. size. *b*. Scutellate and reticulate tarsus of a pigeon. Nat. size.

The commonest division of the podotheca is into *scales* or *scutella* (Lat. *scutellum*, a little shield; pl. *scutella*, not *scutellæ* as often written); figs. 37, and 38, *b*. These are generally of large comparative size, arranged in definite vertical series up and down the tarsus and along the toes, and apt to be somewhat imbricated, or fixed shingle-wise, the lower edge of one overlapping the upper edge of the next. The great majority of birds have such scutella. They oftenest occur on the front of the tarsus (or *acrotarsium*, corresponding to our "instep"), and almost invariably on the tops of the toes (collectively called *acropodium*); frequently also on the sides and back of the tarsus or *planta*; not so often on the crus, and rarely if ever on the sides and under surfaces of the toes. A tarsus so disposed as to its podotheca is said to be *scutellate*,—scutellate before (fig. 37), or behind, or both, as the case may be. The term is equally applicable to the acropodium, but is not so often used because scutellation of the upper sides of the toes is so universal as to be taken for granted unless the contrary condition is expressly said. The most notorious case of the *Oscine podotheca* (figs. 36, 37), characterizing that great group of birds, is given beyond (next paragraph).

Plates, or *reticulations* (Lat. *reticulum*, a web; fig. 38, *a*) result from the cutting up of the envelope in various ways by cross lines. Plates are of various shapes and sizes, and grade usually into true scutella, from which however they are generally distinguished by being smaller, or of irregular contour, or not in definite rows, or lacking the appearance of imbrication; but there is no positive distinction. They are oftenest *hexagonal* (six-sided), a form best adapted to close packing, as shown very perfectly in the cells of the honey-bee's comb; but they may have fewer sides, or be *polygonal* (many-sided), or even circular; when crowded in one direction and loosened in another the shape tends to be oval or even linear. A leg so furnished is said to be *reticulate*: the reticulation may be entire, or be associated with scutellation, as often happens (fig. 38, *b*). A particular case of reticulation is called *granulation* (Lat. *granum*, a grain): when the plates become elevated into little tubercles, roughened or not. Such a leg is said to be *granular, granulated*, or *rugose*: it is well shown by parrots, and the fish-hawk (*Pandion*). When the harder sorts of scales or plates are roughened without obvious elevation, the leg is said to be *scabrous* or *scarious* (Lat. *scabrum*, a scab). But *scabrous* is also said of the under surfaces of the toes, when these develop special *pads*, or wart-like bulbs (called *tylari*): as is well shown in the sharp-shinned and many other hawks. The softer sorts of legs, and especially the webs of swimming birds, are often marked crosswise or *cancellated* with a lattice work of lines, these however not being strong enough to produce plates; it is more like the lines seen on our palms and finger-tips. The plates of a part of the leg occasionally develop into actual *serrations;* as witnessed along the hinder edge of a grebe's tarsus. When an unfeathered tarsus shows *no* divisions of the podotheca in front (along the acrotarsium), or only two or three scales close by the toes, it is said to be *booted* or *greaved;* and such a podotheca is *holothecal* (Gr. ὅλος, *holos*, whole, entire, and θήκη; fig. 36). The generic opposite is *schizothecal* (Gr. σχίζω, I cleave), whether by scutellation or reticulation or in any other way the integument may be cut up. A booted or holothecal tarsus chiefly occurs in the higher *Oscines*, and is supposed by many, particularly German ornithologists, to indicate the highest type of bird structure. It is, however, found in a few water birds, as Wilson's stormy petrel and other species of *Oceanites*. It is not a common modification. Exceptions aside, it only occurs in connection with an equally particular condition of the *sides and back* of the tarsus, or *planta*. In almost all *Oscine Passeres* (*Alaudidæ* are an exception), which constitute the great bulk of the large order *Passeres*, the planta is covered with one pair of plates or *laminæ*, one on each side, meeting behind in a sharp ridge; a condition called *laminiplantar*, in distinction from the opposite, *scutelliplantar*, state of the parts. A holothecal podotheca only occurs in connection with the laminiplantar condition, the combination resulting in the perfect "boot." Among North American birds, the genus *Oceanites* aside, it is exhibited by the following genera, and by these only: *Turdus, Cinclus, Saxicola, Sialia, Regulus, Cyanecula, Phylloscopus, Chamæa, Myiadestes;* and even birds of these genera, when *young*, show scutella which disappear with age by progressive fusion of the acrotarsial podotheca. (Compare figs. 36, 37.)

The Crus, when bare of feathers below, may, like the tarsus, be scutellate or reticulate before or behind, or both; such divisions of the crural integument being commonly seen in long-legged wading birds. Or, again, this integument may be loose, softish, and movable, not obviously divided, and passing directly into ordinary skin.

The Tarsus, in general, may be called subcylindrical: it is often quite circular in cross-section; generally thicker from before backward, and only rarely wider from one side to the other than in the opposite direction; but such a shape as this last is exhibited by the penguins. When the transverse thinness is noticeable, the tarsus is said to be *compressed;* and such compression is very great in a loon, in which the tarsus is almost like a knife blade. Quite cylindrical tarsi occur chiefly when there are similar scales or plates before and behind, as happens in the larks (*Alaudidæ*); they are rare among land birds, common among waders. Those swimming birds with a very thin skinny podotheca are apt to show traces of the four-sidedness of the metatarsal bone. The tarsus in the vast majority of land birds is seen on close inspection to be somewhat ovate or drop-shaped on cross-section, — gently rounded in front, more compressed laterally, and sharp-ridged behind. This results from the *laminiplantation* described above, and is equally well exhibited by most passerine birds, whether they have booted or anteriorly scutellate tarsi. The line of union of anterior scutella with posterolateral plates on the sides of the tarsus is generally in a straight vertical line, — either a mere line of flush union, or a ridge, or oftener a groove (well seen in the crows), which may or may not be filled in with a few small narrow plates. In the Clamatorial *Passeres*, represented by our flycatchers, the tarsus is enveloped in a scroll-like podotheca of irregularly arranged plates, the edges of the scroll meeting along the inner side of the tarsus. But the full consideration of special states of the tarsal envelope, however important and interesting, would be part of a systematic treatise on ornithology, rather than of an outline sketch like this.

The Number of Toes (individually, *digiti*; collectively, *podium*) is *four*: there are *never* more. There are *two* in the ostrich alone, in which both inner and hind toe are wanting.

FIG. 39. — Tridactyle foot of sanderling, *Calidris arenaria*; nat. size.

There are three in all the other struthious birds (*Rheidæ, Casuariidæ*), excepting *Apteryx*, which has four. There are likewise three, the hind toe being suppressed, in the tinamine genera *Calodromas* and *Tinamotis* (*Dromæognathæ*); throughout the auk family (*Alcidæ*); in the petrel genus *Pelecanoïdes*; apparently in the albatrosses (*Diomedeinæ*); usually in the gull genus *Rissa*; in the flamingo genus *Phœnicoparra*; throughout the bustard family (*Otididæ*), and among various related forms, as *Œdicnemus, Esacus, Cursorius*; in the plovers (*Charadriidæ*), excepting *Squatarola*; and in the bush-quails (*Turnicidæ*), excepting *Pedionomus*. In higher birds, three toes are a rare anomaly, only known to occur in three genera of woodpeckers (*Picoïdes, Sasia*, and *Tiga*), and in one galbuline genus (*Jacamaralcyon*), by loss of the hind toe; in two genera of kingfishers (*Ceyx* and *Alcyone*), by suppression of the inner front toe; and in the passerine genus *Cholornis*, by defect of the outer front toe. North American three-toed birds are these only: the woodpeckers of the genus *Picoïdes*; all auks (*Alcidæ*), and albatrosses (*Diomedeinæ*; in these, however, there is a rudiment of the hind toe); all plovers (*Charadriidæ*, excepting one, *Squatarola*); the oyster-catchers (*Hæmatopus*); the sanderling (*Calidris*, fig. 39); the stilt (*Himantopus*). Birds with two toes are said to be *didactyle*; with three, *tridactyle*; with four, *tetradactyle*. In the vast majority of cases, birds have *three* toes in front and *one* behind. Occasionally, either the hind toe, or the outermost front toe, is *versatile*, that is, susceptible of being turned either way. Such is the condition of the outer front toe in most owls (*Striges*) and in the fish-hawk (*Pandion*). We have no case of true versatility of the hind toe among North American birds; but several cases of its stationary somewhat lateral position, as in goatsuckers (*Caprimulgidæ*), some of the swifts (*Cypselidæ*), the loons (*Colymbidæ*), and all the totipalmate swimmers (*Steganopodes*). Nor have we any example of that rarest of all conditions (seen in some *Cypselidæ*, and the African *Coliidæ*) in which *all four* toes are turned forward. The arrangement of toes *in pairs*, two before and two behind, is quite common, being the characteristic of scansorial birds and some others, as all the parrots and woodpeckers, cuckoos, trogons, etc. Such arrangement is called *zygodactyle* or *zygodactylous* (Gr. ζυγόν, *zugon*, a yoke; δάκτυλος, *daktulos*, a digit); and birds exhibiting it are said to be *yoke-toed* (fig. 45). In all yoke-toed birds, excepting the trogons, it is the *outer* anterior toe which is reversed; in trogons, the

EXTERNAL PARTS OF BIRDS.—THE FEET. 133

inner one. In nearly every three-toed bird, all three toes are anterior; our single exception is in the genus *Picoïdes*, where the true hind toe is wanting, the outer anterior one being reversed as usual in zygodactyles. No bird has more toes behind than in front. Birds' toes, and their respective joints, are

Numbered, in a certain definite order, as follows (see figs. 34, 36): hind toe = *first* toe, $1\,t$; inner anterior toe = *second* toe, $2\,t$; middle anterior toe = *third* toe, $3\,t$; outer anterior toe = *fourth* toe, $4\,t$. Such identification of $1\,t$, $2\,t$, $3\,t$, $4\,t$ applies to the ordinary case of three toes in front and one behind. But, obviously, it holds good for any other arrangement of the toes, if we only know which one is changed in position,—a thing always easy to learn, as we shall see at once. In birds with the hind toe reversed, leaving all four in front, the same order is evident, though then $1\,t$ is the inner anterior, $2\,t$ the next, etc.; for it always happens, when a hind toe turns forward, that it turns on the *inner* side of the foot. Similarly, in yoke-toed birds (excepting *Trogonidæ*), it is the *outer* anterior which is turned backward, as above said; then, evidently, inner hind toe = $1\,t$; inner front toe = $2\,t$; outer front toe = $3\,t$; outer hind toe = $4\,t$. In *Trogonidæ*, with *inner* front toe reversed, the correction of the formula is easily made. Moreover, when the number of toes decreases from four to three or two, the digits are almost always reduced in the same order: thus, in three-toed birds, $1\,t$ is the missing one; in the two-toed ostrich, $1\,t$ and $2\,t$ are gone. The only known exceptions to this generalization are afforded by two exotic genera of kingfishers, *Ceyx* and *Alcyone*, in which $2\,t$ is defective; and by the anomalous passerine *Cholornis* of China, in which $4\,t$ is in like case. The rule is proven by the

Number of Phalanges, or joints, of the digits. The constancy of the joints in birds' toes is remarkable,—it is one of the strongest expressions of the highly monomorphic character of *Aves*. In *all* birds, excepting *Procellariidæ*, $1\,t$ when present has *two* joints (not counting, of course, the accessory metatarsal). In *all* birds, $2\,t$ when present has *three* joints. In *nearly all* birds, $3\,t$ has *four* joints. In *nearly all* birds, $4\,t$ has *five* joints. Thus, any digit has one more joint than the number of itself. The exceptions to this regularity consist in the lessening of the number of joints of $1\,t$ or $3\,t$ by *one*, and of $4\,t$ by *one* or *two*. So when the joints do not run 2, 3, 4, 5, for toes 1 to 4, they run either, 1, 3, 4, 5, or 2, 3, 4, 4, or 2, 3, 3, 3. (These statements do not regard the anomalous cases of *Ceyx*, *Alcyone*, and *Cholornis*—see above.) This variability is nearly confined to certain Picarian birds: our examples of it are in certain genera of *Cypselinæ*, fig. 40, where the ratio is 2, 3, 3, 3, of *Caprimulginæ*, fig. 41, where it is 2, 3, 4, 4; and the petrel family, with 1, 3, 4, 5. Such admirable conservatism enables us to tell what toes are missing in any case, or what ones are out of the regular position. Thus, in *Picoïdes*, the hind toe, apparently $1\,t$, is known to be $4\,t$, because it is five-jointed; in a trogon, the inner hind toe is $2\,t$, being three-jointed; in the ostrich, with only two toes, $3\,t$ and $4\,t$ are seen to be preserved, because they are respectively four- and five-jointed. (See fig. 34, where the digits and their phalanges are numbered.) Besides this interesting numerical ratio, the phalanges have other inter-relations of some consequence in classification, resulting from their comparative lengths. In some families of birds, one or more of the *basal* or proximal phalanges (those next to the foot—opposed to *distal*, or those at the ends of the digits) of the front toes are extremely short, being mere nodules of bone (fig. 40); in other and more frequent cases, they are the longest of all, as in figs. 34, 41. On the whole, they generally decrease in length from proximal to distal extremity, and the last one of any toe is quite small, serving merely

FIG. 40.—Phalanges of Cypseline foot, 2, 3, 3, 3.

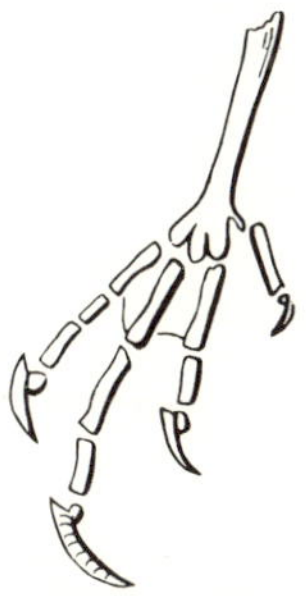

FIG. 41.—Phalanges of Caprimulgine foot, 2, 3, 4, 4.

as a core to the claw. The difference in the lengths of the several phalanges, like that of the digits themselves, makes the toes more efficient in grasping, since they thereby clasp more perfectly upon an irregular object. The design and the principle are the same as seen in the human hand, in which model instrument the digits and their joints are all of different lengths.

The Position of the Digits, other than in respect to their *direction*, is important. In *all* birds the front toes are inserted on the metatarsus on the same level, or so nearly in one horizontal plane that the difference is not notable. The same may be said of the hind toes when they are a pair, as in zygodactyle birds. But the *hind toe*, or *hallux*, as it is often called, when present and single, varies remarkably in position with reference to the front toes; and this matter requires special notice, as it is important in classification. The insertion of this digit varies, from the very bottom of the tarsus (*metatarsus*), where it is on a level with the front toes, to some distance up the bone. When the hallux is flush with the bases of the other toes, so that its whole length is on the ground, it is said to be *incumbent*. When just so much raised that its tip only touches the ground, it is called *insistent*. When inserted so high up that it does not reach the ground, it is termed *remote* (*amotus*) or *elevated*. But as the precise position varies insensibly, so that the foregoing distinctions are not readily perceived, it is practically best to recognize only two of these three conditions, saying simply " hind toe elevated," when it is inserted fairly above the rest, and " hind toe not elevated," when its insertion is flush with that of the other toes. In round terms : it is characteristic of all *insessorial* (Lat. *insedo*, I sit upon) or perching birds to have the hind toe DOWN ; of all other birds to have it UP (when present). The exceptions to the first of these statements are extremely rare ; among North American birds they are chiefly furnished by certain genera of *Caprimulgidæ*, perhaps also of *Cypselidæ*, and of *Cathartidæ*. But among other *Raptores* besides *Cathartidæ*, especially certain owls (*Striges*), and in some of the pigeons (*Columbidæ*), the hind toe is not quite down, or is decidedly uplifted (as in *Starnœnas*, for example). It is elevated in all our rasorial birds (*Gallinæ*); elevated in all our waders excepting the herons and some of their allies (*Herodiones*), though not very markedly so in the rail family (*Rallidæ*). It is elevated in *all* swimming birds, whether lobe-footed or completely or partly web-footed, but in the totipalmate order (*Steganopodes*), where the hallux is lateral in position and webbed with the inner toe, the elevation is slight. Now since, curiously enough, the only ones of our insessorial genera (see above) that have the hind toe up, have also little webs between the front toes — since some *Raptores* are our only other insessorial birds with any such true webbing — since herons and some of their allies are our only birds with such webbing that have the hallux down — the following rule is perhaps infallible for North American birds : *Consider the hind toe* UP *in any bird with any true webbing or lobing of the front toes*, excepting herons and some of their allies and some birds of prey. The converse also holds almost as well ; for our only birds with fully cleft anterior toes and hind toe up, are the rails and gallinules (*Rallidæ*), the black-bellied plover (*Squatarola helvetica*), our only four-toed plover, the turn-stone (*Strepsilas interpres*), the American woodcock (*Philohela minor*), the European woodcock (*Scolopax rusticula*), Wilson's snipe (*Gallinago wilsoni*), and most of the sandpipers (*Scolopacidæ*). If the sense of this paragraph is taken in, the student who wishes to use my artificial " key " will seldom be puzzled to know whether to take the toe up or down.

The Hallux has other Notable Characters. — It is *free* and *simple*, in the vast majority of birds : in all *insessorial* birds, nearly all *cursorial* (Lat. *cursor*, a courser), and most *natatorial* (Lat. *natator*, a swimmer) forms. Its length, claw included, may equal or surpass that of the longest anterior toe ; and generally exceeds that of one or two of these. It is never so long as when *incumbent ;* when thus down on a level with the rest it also acquires its greatest mobility and functional efficiency. In most *Passeres* it is virtually provided with a special muscle for independent movement, so that it may be perfectly apposable to the other toes collectively, just as our thumb may be brought against the tip of any finger. In general, it shortens as it rises on the metatarsus; and probably in no bird in which it is truly elevated is it as long as the shortest anterior toe. It is short, barely touching the ground, in most wading birds; shorter still in some swimmers, as the gulls, where it is probably functionless; it is incomplete in one genus of gulls (*Rissa*), where it bears no perfect claw; it has only one phalanx and is represented only by a short immovable claw in the petrels (*Procellariidæ*); it disappears in the birds named in the last paragraph but two above, and in some others. It is never actually soldered with any other toe, for any noticeable distance; but it is webbed to the base of the inner toe in the loons (*Colymbus*), and to the whole length of the toe in all the *Steganopodes* (fig. 52). It may also be independently webbed; that is, be provided with a separate flap or lobe of free membrane. This lobation of the hallux is seen in all our sea-ducks and mergansers (*Fuligulinæ* and *Merginæ*), and in all the truly lobe-footed birds, as coots (*Fulica*), grebes (*Podicipedidæ*) and phalaropes (*Phalaropodidæ*). The modes of union of the anterior toes with one another may be finally considered under the head of the

Three leading Modifications of the Avian Foot. — Birds' feet are modelled, on the whole, upon one or another of three plans, furnishing as many *types of structure;* which types, though they run into one another, and each is variously modified, may readily be appreciated. These plans are the perching or *insessorial*, the walking or wading, *cursorial* or *grallatorial*, and the swimming or *natatorial* — in fact, so well distinguished are they, that carinate birds have even been primarily divided into groups corresponding to these three evidences of physiological adaptation of the structure of the Avian *pes*. Independently of the number and position of the digits, the plans are pretty well indicated by the method of union of the toes, or their entire lack of union. 1. *The insessorial type.* (*a*) In order to make a foot the most of *a hand*, that is, to fit it best for that grasping function which the perching of birds upon trees and bushes requires, it is requisite that the digits should be as free and movable as possible, and that the hind one should be perfectly apposable to the others. Compare the human hand, for example, with the foot, and observe the perfection secured by the perfect freedom of the fingers and especially the appositeness of the thumb. In the most accomplished insessorial foot, the front toes are *cleft to the base*, or only coherent to a very slight extent; the hind toe is completely incumbent, and as long and flexible as the rest. Our thrushes (*Turdidæ*) probably show as complete cleavage as is ever seen, practically as much as that of the human fingers; the cleft between the inner and middle toe being to the very base, while the outer is only joined to the middle for about the length of its own basal joint. This is the typical *passerine* foot (figs. 36, 37, 42, 43). There may be somewhat more cohesion of the toes at base, as in the wrens, titmice, creepers, vireos, etc., without, however, obscuring the true passerine character. As regards this matter, the point is, that when the toes are united at all, it is by their actual *cohesion* there, not by movable webbing. Besides the typical passerine, there are several other modifications of the insessorial foot. (*b*) Thus a kingfisher shows what is called a *syndactyle* or *syngnesious* (Gr. σύν, *sun*, together; γνήσιος, *gnesios*, relating to way of birth) foot (fig. 44), where the outer and middle toes cohere for most of their extent and have a broad sole in common. It is a degradation of the insessorial foot, and not a common

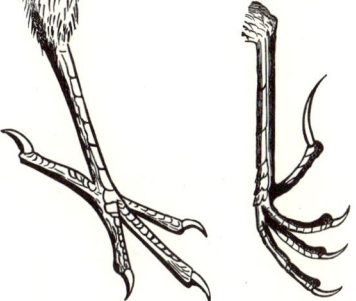

FIGS. 42, 43. — Typical passerine feet. (The right-hand fig, is *Plectrophanes lapponicus*, nat. size.)

one either; seen in those perching birds which scarcely use their feet for progression, but simply for sitting motionless. (*c*) The *zygodactyle* or yoke-toed modification has been sufficiently noted (fig. 45). It was formerly made much of, as a *scansorial* or *climbing* type of foot, and an absurd " order " of birds has been called *Scansores*. But many of the zygodactyle birds do not climb, as the cuckoos; while the most nimble and adroit of climbers, such as the nuthatches and creepers, retain a typically passerine foot. The "scansorial" is simply one modification of the insessorial plan, and has little classificatory significance, — no more than that attaching to the particular condition of the insessorial foot (*d*) which results from elevation or versatility of the hind toe, as in some *Cypselidæ* and *Caprimulgidæ*. This is an abnormality which has received no special name; it is generally associated with some little webbing of the anterior toes at base, which is a departure from the true insessorial plan, or with abnormal reduction of the phalanges of the third and fourth toes, as explained above (figs. 40, 41). (*e*) The *raptorial* is another modification of the insessorial foot. It is advantageous to a bird of prey to be able to spread the toes as widely as possible, that the talons may seize the prey like a set of grappling irons; and accordingly the toes are widely divergent from each other, the outer one in the owls and a few hawks being quite versatile. In a foot of raptorial character, the toes are cleft profoundly, or, if united at base, it is by movable webbing; the claws are immensely developed, and the under-surfaces of the toes are scabrous or bulbous for greater security of the object grasped. Any hawk or owl or old-world vulture exhibits the raptorial insessorial foot (figs. 46, 47).
2. *The cursorial* or *grallatorial type*. The gist of this plan lies in the decrease or entire loss of the grasping function, and in the elevation, reduction in length, or loss of the hind toe; the foot is a good foot, but nothing of a hand. The columbine birds, which are partly terrestrial, partly arboreal,

FIG. 45. — Zygodactyle foot of a woodpecker, *Ceophlœus pileatus*, nat. size.

FIG. 44. — Syndactyle foot of kingfisher, nat. size.

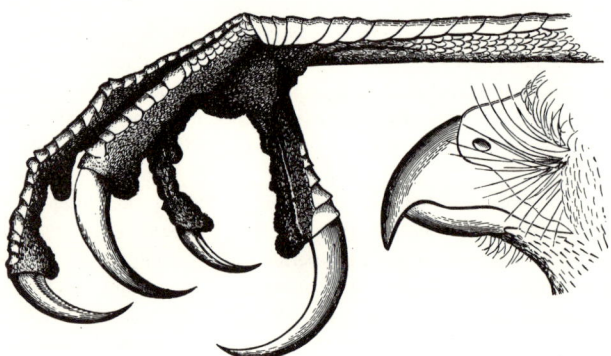

FIG. 46. — Raptorial foot of a hawk, *Accipiter cooperi*, nat. size.

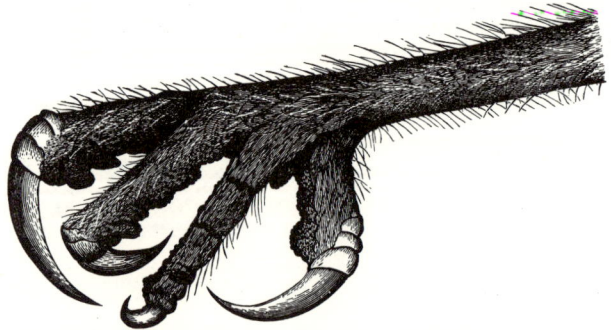

FIG. 47. — Raptorial foot of an owl, *Aluco flammeus*, nat. size.

exhibit the transition from the perching to the gradient foot, in some reduction of the hind toe, which is nevertheless in most cases still on the same level as the rest (fig. 38, *b*). In the gallinaceous or rasorial (Lat. *rasor*, a scraper) birds, which are essentially terrestrial, and noted for their habit of scratching the ground for food, the hind toe is decidedly elevated and shortened in almost all of the families (fig. 35). Such reduction and uplifting of the hallux is carried to an extreme in most of the waders, or *grallatores*, in many of which this toe disappears (figs. 38, *a*, 39). It is scarcely practicable to recognize special modifications of such gradient or grallatorial feet, since they merge insensibly into one another. The herons, which are the most arboricole of the waders, exhibit a reversion to the insessorial type, in the length and incumbency of the hallux. The mode of union of the front toes of the walkers and waders is somewhat characteristic. The toes are either cleft quite to the base, or there joined by small webs; probably never actually coherent. Such basal webbing of the toes is called *semipalmation* ("half-webbing"). It is actually the same thing that occurs in many birds of prey, in most gallinaceous birds, etc.; the term is mostly restricted, in descriptive ornithology, to those *wading* birds, or *grallatores*, in which it occurs. Such basal webs generally run out to the end of the first, or along part of the second, phalanx of the toes; usually farther between the outer and middle than between the middle and inner toes. Such a foot is well illustrated by the semipalmated plover (*Ægialites semipalmatus*), semipalmated sandpiper (*Ereunetes pusillus*, fig. 48), and willet (*Symphemia semipalmata*, fig. 49). In a few wading birds, as the avocet and flamingo, the webs extend to the ends of the toes. This introduces us at once to the *third* main modification of the foot, 3. *The natatorial type.* Here the foot is transformed into a swimming implement, usually with much if not entire abrogation of its function as foot or hand. Swimming birds with few exceptions are notoriously bad walkers, and few of them are perchers. The swimming type is presented under two principal modifications: — (*a.*) In the *palmate* or ordinary webbed foot, all the front toes are united by ample webs (fig. 50). The palmation is usually complete, extending to the ends of the toes; but one or both webs may be so deeply *incised*, that is, cut away, that the palmation is practically reduced to semipalmation, as in terns of the genus *Hydrochelidon* (fig. 51). The *totipalmate* is a special case of palmation, in which all four toes are webbed; this characterizes the whole order *Steganopodes* (fig. 52). (*b.*) In the *lobate* foot, a paddle results not from connecting webs, but from a series of *lobes* or flaps along the sides of the individual toes; as in the coots, grebes, phalaropes, and sun-birds (*Heliornithidæ*). Lobation is usually associated with semipalmation, as is well seen in the grebes (*Podicipedidæ*). In the snipe-like phalaropes (*Phalaropodidæ*), lobation is present as a modification of a foot otherwise quite cursorial. The most emphatic cases of lobation are those in which each joint of the toes has its own flap, with a free convex border; the membranes as a whole therefore present a scolloped outline (figs. 53, 53 *bis*). Such lobes are merely a development of certain *marginal fringes* or processes exhibited by many non-lobate or non-palmate birds. Thus, if the foot of some of the gallinules be examined in a fresh state, the toes will be seen to

FIG. 48. — Semipalmation in *Ereunetes;* nat. size.

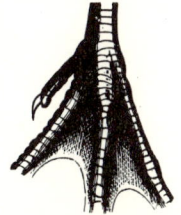

FIG. 49. — Semipalmated bases of toes of *Symphemia;* nat. size.

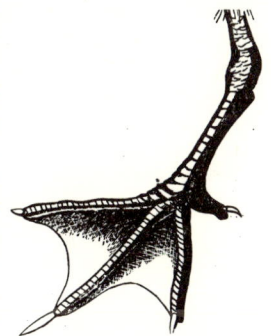

FIG. 50. — Palmate foot of a tern, *Sterna forsteri;* nat size.

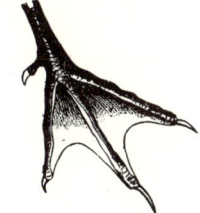

FIG. 51. — Incised palmation of *Hydrochelidon lariformis;* nat. size.

have a narrow membranous margin running the whole length. The same thing is evident in a great many waders, and on the free borders of the inner and outer toes of web-footed birds.

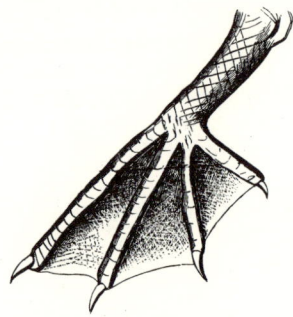

FIG. 52. — Totipalmate foot of a pelican; reduced.

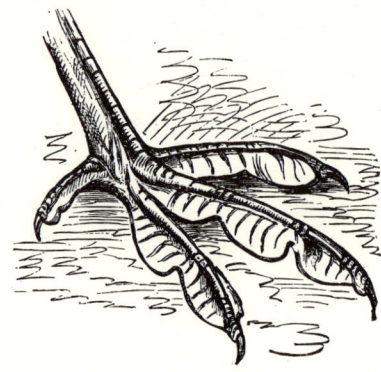

FIG. 53. — Lobate foot of a coot; reduced.

In the grouse family (*Tetraonidæ*), marginal fringes are very conspicuous; there being a great development of hard horny substance, fringed into a series of sharp teeth or *pectinations* (fig. 35). These formations appear to be deciduous, that is, to fall off periodically, like parts of the claws of some quadrupeds (lemmings).

Claws and Spurs. — With rare anomalous exceptions, as in the case of an imperfect hind toe, every digit terminates in a complete *claw*. The general shape is remarkably constant in the class; variations being rather in degree than in kind. A cat's claw is about the usual shape: it is *compressed, arched, acute*. The great talons of a bird of prey are only an enlargement of the typical shape; and, in fact, they are scarcely longer, more curved, or more acute than those of a delicate canary bird; they are simply stouter. The claws of scansorial birds are very acute and much curved, as well as quite large. The under surface of the claw is generally excavated, so that the transverse section, as well as the lengthwise outline below, is concave, and the under surface is bounded on either side by a sharp edge. One of these edges, particularly the inner edge of the middle claw, is expanded or dilated in a great many birds; in some it becomes a perfect *comb*, having a regular series of teeth.

FIG. 53 *bis*. — Lobate foot of phalarope, *Lobipes hyperboreus*; nat. size.

This *pectination* (Lat. *pecten*, a comb), as it is called, only occurs on the inner edge of the middle claw. It is beautifully shown by all the true herons (*Ardeidæ*); by the whip-poor-wills and night-hawks (*Caprimulgidæ*, fig. 41); by the frigate pelican (*Tachypetes*); and imperfectly by the barn owl (*Aluco flammeus*). It is supposed to be used for freeing parts of the plumage that cannot be reached by the bill from parasites; but this is very questionable, seeing that some of the shortest-legged birds, which cannot possibly reach much of the plumage with the comb, possess that instrument. Claws are more *obtuse* among the lower birds than in the insessorial and scansorial groups, as the columbine and gallinaceous (*rasorial*) orders, and most natatorial families. Obtuseness is generally associated with flatness or depression; for in proportion as a claw becomes less acute, so does it lose its arcuation, as a rule. This is well illustrated by Wilson's petrel (*Oceanites oceanicus*), as compared with others of the same family. Such condition is carried to an extreme in the grebes (*Podicipedidæ*), the claws of which birds resemble human finger-nails. Otherwise, deviations from curvature, without loss of acuteness, are chiefly exhibited by the *hind* claw of many terrestrial *Passeres*, as in the whole family *Alaudidæ* (larks), and some of the finches (*Fringillidæ*), as the species of "long-spur" (*Centrophanes*). But all the claws are straight, sharp, and prodigiously long, in birds of the genus *Parra* (fig. 53 *ter*); these jaçanás being enabled to run lightly over the floating leaves of aquatic plants by so much increase in the spread of their toes that they do not "slump in." Claws are also variously *carinate* or ridged, *sulcate* or grooved. In a few cases they are rounded underneath, so as to be nearly circular in cross-section, as is the case with those of the fish-hawl (*Pandion*). They are always horny (*corneous*). They take name from and are reckoned by their respective digits: thus, 1 *cl.* = claw of 1 *t*; 2 *cl.* = claw of 2 *t*, etc.

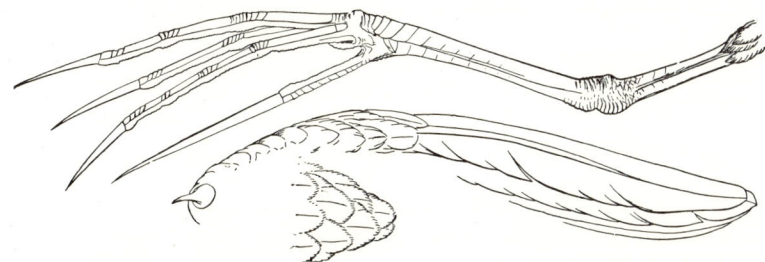

Fig. 53 *ter*. — Foot of *Jacana* (*Asarcia*) *spinosa*, nat. size, showing the long, straight claws. (From Pr. U. S. Nat. Mus. The *spurred* wing of the same bird is also shown. See p. 120.)

Spurs (Lat. *calcar*, a spur) are developed on the metatarsal bones of a few birds. They are of the nature of claws, being hard, horny modifications of the epiderm: but they have nothing to do with the digits. They possess a bony core upon which they are supported, like the horns of cattle. Such growths chiefly occur in gallinaceous birds: the spurs of the domestic fowl are a familiar case. Sometimes there are a pair of such weapons on each foot, as in the *Pavo bicalcaratus*. The only instance of their occurrence among indigenous birds of North America is offered by the wild turkey (*Meleagris gallopavo*). Metatarsal spurs are characteristic of the male sex; they are offensive weapons, and belong to the class of "secondary sexual characters" (p. 95). (For wing-spurs, as shown in fig. 53 *ter*, see p. 120.)

§ 4. — AN INTRODUCTION TO THE ANATOMY OF BIRDS.

Anatomical Structure now affords ornithologists many and the most important of the characters used in classification. In fact, few if any of the groups above genera can be securely established without consideration of internal parts and organs, as well of exterior modifications of structure. Therefore, the student who really "means business" must be on speaking terms at least with avian anatomy. For example, none could in the least intelligently understand a wing or a leg without knowing the bony framework of those members. Yet, for me to adequately set this matter forth would be to occupy this whole volume with anatomy; whereas, I can only devote a few pages to the entire subject. In such embarrassment, which attends any attempt to treat a great theme in a short way that shall not also be a small way, attention must be mainly confined to those points which bear most directly upon systematic ornithology as distinguished from pure anatomy, in order to bring forward the structures which are more particularly concerned in the classification of birds. I wish to give a fair account of the skeleton, as osteological characters are of the utmost importance for the determination of natural affinities; and to continue with some notice of prominent features of the muscular, vascular, respiratory, digestive, urogenital, and nervous systems, and organs of the special senses, as the eye and ear. The tegumentary system has already been treated at some length (pp. 81–96); so has the osseous system, so far as the bones of the limbs are concerned (pp. 111–115, 124–128, 133). What further I shall have to say is designed merely as an introduction to the rudiments of avian anatomy, and is supposed to be addressed to beginners only.

a. Osteology: The Osseous System, or Skeleton.

Osteology (Gr. ὀστέον, *osteon*, a bone; λόγος, *logos*, a word) is a scientific description of bone in general and of bones in particular. Bone consists of an animal basis or matrix (Lat. *matrix*, a mould) hardened by deposit of earthy salts, chiefly phosphate of lime. Bone is either preformed in the gristly substance called *cartilage* (Lat. *cartilago*, gristle), and results from the substitution of the peculiar osseous tissue for the cartilaginous tissue, or it is formed directly in ordinary connective tissue, such as that of most membranes or any ligaments of the body. Bone tissue presents a peculiar microscopic structure, in which it differs from teeth, as it does also in not being developed from mucous membrane; the substance is called *ostein*, as distinguished from *dentine*. Though very dense and hard, bone has a copious blood-supply, and is therefore very *vascular ;* the nutrient fluid penetrates every part in a system of vessels called *Haversian canals*. In the natural state bone is covered with a tough membrane called *periosteum* (Gr. περί, *peri*, around, and ὀστέον), which is to bone what bark is to a tree. The bones collectively constitute the *osseous system*, otherwise known as the *skeleton* (Gr. σκελετόν, dried, as bones usually are when studied). The skeleton is divided into the *endoskeleton* (Gr. ἔνδον, *endon*, within), consisting of the bones inside the body; and the *exoskeleton* (Gr. ἐξ, *ex*, out of), or those upon the surface of the body, of which birds have none. Certain bones developed apart from the systematic endoskeleton, in fibrous tissue, are called *scleroskeletal* (Gr. σκληρός, *scleros*, hard), as the ossified tendons or leaders of a turkey's leg, the ring of ossicles in a bird's eye (an *ossicle* is any small bone). *Sesamoid* (Gr. σησαμη, *sesame*, a kind of pea) bones, so often found in the ligaments and tendons about joints, are probably best considered scleroskeletal. The endoskeleton is divided into bones of the *axial skeleton*, so called because they lie in the axis of the body, as those of the skull, backbone, chest, pelvis, and shoulder-girdle; and of the *appendicular skeleton*, including bones of the limbs, considered as diverging appendages of the trunk. The skeleton is jointed; bones join either by immovable *suture*, or by movable *articulation* (Lat. *articulus*, a joint, dimin. of *artus*, a limb). In free articulations, the opposing surfaces are generally smooth, and lubricated with a fluid called *synovia*. Progressive ossification often causes bones originally distinct to *coössify*, that is, to fuse together; this is termed *ankylosis* or *anchylosis ;* bones so melted together are said to be *ankylosed* or *anchylosed* (Gr. ἀγκύλωσις or ἀγχύλωσις, the stiffening of joints in a bent position). Thus all the bones of a bird's brain-box are anchylosed together, though the box at first consists of many distinct ones; and the determination of such osseous elements or integers in compounded bones is a very important matter, as a clue to their morphological composition. The names of most individual bones, chiefly derived from the old anatomists, are arbitrary and have little scientific signification; many are fanciful and misleading; bones named since anatomy passed from the empiric stage, when it was little more than the art of dissecting and describing, however, have as a rule better naming. The shaft of a long bone is its *continuity :* the enlargements usually found at its extremities are called *condyles* (Gr. κόνδυλος, *kondulos*, a lump, knot, as of the knuckles). Points where ossification commences in cartilage or membrane, are *ossific centres*, or *osteoses ;* valuable clues, usually, to the elements of compound bones. But ossification of individual simple bones may begin in more than one spot, and the several osteoses afterward grow together. This is especially the case with the *ends* of bones, which often make much progress in ossification before they unite with the shaft or main part; such caps of bone, as long as they are disunited, are called *epiphyses* (Gr. ἐπί, *epi*, upon; φύσις, *phusis*, growth). Protrusive parts of bones have the general name of *processes*, or *apophyses* (Gr. ἀπό, *apo*, away from, and φύσις); such have generally no ossific centres, being mere outgrowths. But many parts of a vertebra, which are called "apophyses," have independent ossific centres. The progress of ossification is usually rapid and effectual.

THE ANATOMY OF BIRDS. — OSTEOLOGY.

The skeleton of birds is noted for the number and extent of its anchyloses, a great tendency to coössification and condensation of bone-tissue resulting from the energy of the vital activities in this hot-blooded, quick-breathing class of creatures. Birds', bones are remarkably hard and compact. When growing, they are solid and marrowy, but in after life more or fewer of them become hollow and are filled with air. This *pneumaticity* (Gr. πνευματικός, *pneumatikos*, windy) is highly characteristic of the avian skeleton. Air penetrates the skull-bones from the nose and ear-passages, and may permeate all of them. It gains access to the bones of the trunk and limbs by means of air-tubes and air-sacs which connect with the air-passages in the lungs; such sacs, sometimes of great extent, are also found in many places in the interior of the body, beneath the skin, etc.; sometimes the whole subcutaneous tissue is pneumatic. The extent to which the skeleton is aërated is very variable. In many birds only the skull, in a few the entire skeleton, is in such condition; ordinarily the greater part of the skull, and the lesser part of the trunk and limbs, is pneumatized. The passage of air in some cases is so free, as into the arm-bone for example, that a bird with the windpipe stopped can breathe

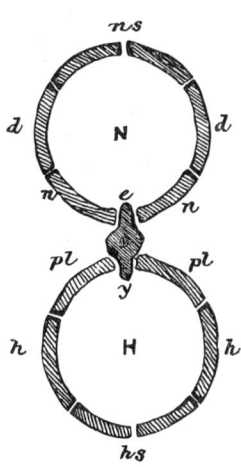

FIG. 54. — Ideal plan of the double-ringed body of a vertebrate. *N*, neural canal; *H*, hæmal canal; the body separating them is the centrum of any vertebra, bearing *e*, an epapophysis, and *y*, a hypapophysis; *n, n*, neurapophyses; *d, d*, diapophyses; *ns*, bifid neural spine; *pl, pl*, pleurapophyses; *h, h*, hæmapophyses; *hs*, bifid hæmal spine. Drawn by Dr. R. W. Shufeldt, U. S. A., after Owen.

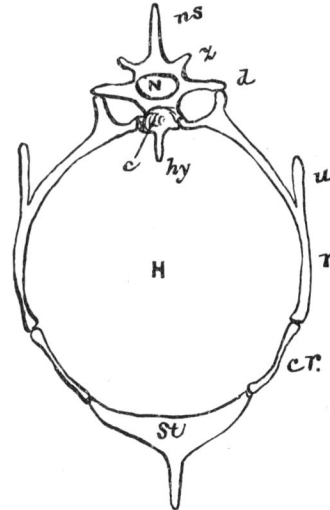

FIG 55. — Actual section of the body in the thoracic region of a bird. *N*, neural canal; *H*, hæmal canal; *c*, centrum of a dorsal vertebra; *hy*, hypapophysis; *d*, diapophysis; *z*, zygapophysis; *ns*, neural spine; *r*, pleurapophysis, or vertebral part of a free rib, bearing *u*, uncinate process or epipleura; *cr*, hæmapophysis or sternal part of the same; *st*, section of sternum or breast-bone (hæmal spine). Designed by Dr. R. W. Shufeldt, U. S. A.

for an indefinite period through a hole in the humerus. Pneumaticity is not directly nor necessarily related to power of flight; some birds which do not fly at all are more pneumatic than some of the most buoyant. (On the general pneumaticity of the body see beyond under head of the respiratory system.)

The Axial Skeleton (figs. 54, 55, 56) of a bird or any *vertebrated* animal, that is, one having a back-bone, exhibits in cross-section two rings or hoops, one above and the other below a central point, like the upper and lower loops of a figure **8**. The upper ring is the *neural arch* (Gr. νεῦρον, *neuron*, a nerve), so called because such a cylinder encloses a section of the cerebro-spinal axis, or principal nervous system of a vertebrate (brain and spinal cord, whence arise all the nerves of the body, excepting those of the sympathetic nervous system). The lower ring is the *hæmal arch* (Gr. αἷμα, *haima*, blood), which similarly contains a section of the principal blood-vessels and viscera. Fig. 55 shows such a section, made across the *thoracic* or chest-region of the trunk. Here the upper ring (neural) is contracted, only surrounding the slender spinal cord, while the lower ring is expanded to enclose the heart and

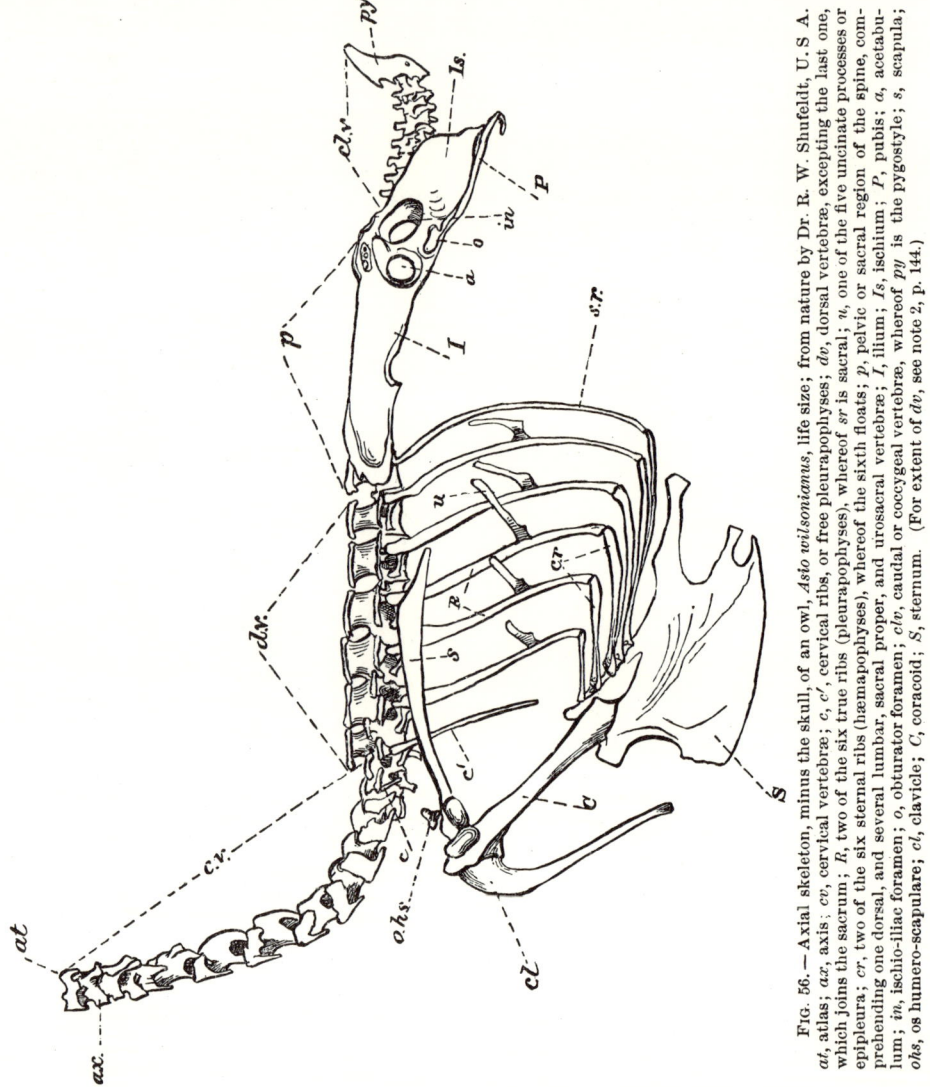

FIG. 56.—Axial skeleton, minus the skull, of an owl, *Asio wilsonianus*, life size; from nature by Dr. R. W. Shufeldt, U.S.A. *at*, atlas; *ax*, axis; *cv*, cervical vertebræ; *c*, *c'*, cervical ribs, or free pleurapophyses; *dv*, dorsal vertebræ, excepting the last one, which joins the sacrum; *R*, two of the six true ribs (pleurapophyses), whereof *sr* is sacral; *u*, one of the five uncinate processes or epipleura; *cr*, two of the six sternal ribs (hæmapophyses), whereof the sixth floats; *p*, pelvic or sacral region of the spine, comprehending one dorsal, and several lumbar, sacral proper, and urosacral vertebræ; *I*, ilium; *Is*, ischium; *P*, pubis; *a*, acetabulum; *in*, ischio-iliac foramen; *o*, obturator foramen; *clv*, caudal or coccygeal vertebræ, whereof *py* is the pygostyle; *s*, scapula; *ohs*, os humero-scapulare; *cl*, clavicle; *C*, coracoid; *S*, sternum. (For extent of *dv*, see note 2, p. 144.)

lungs. Such a section, made in the region of the skull, would show the reverse; the upper ring greatly inflated to contain the brain, the lower contracted and otherwise greatly modified into bones of the jaws. Thus the trunk of a vertebrate is a double-barrelled tube; one tube above for the nervous system, the other below for the viscera at large; the partition between the two being a jointed chain of solid bones from one end of the body to the other. These solid bones are the *centrums* or *bodies of vertebræ*, in the trunk; and in the head certain bones which in some respects correspond with the centrums of vertebræ. The entire chain or series of vertebræ composes the back-bone or *spinal column;* with its connections (thorax and pelvis) and anterior continuation (skull) it is the *axial skeleton.* The skull is considered by some competent anatomists to consist of modified vertebræ. The skull-bones have certainly the position and relations of parts of vertebræ; to a certain extent they resemble vertebræ, as in being divisible into several segments, like as many vertebral segments; they are also directly in the axis of the body, enclosing a part of the cerebro-spinal nervous system above, and portions of the visceral systems below. But supposed strict morphological correspondence of cranial bones with vertebræ is not supported by their mode of development, and is now generally denied, the relation being considered rather analogical and physiological than homological and morphological.

1. THE SPINAL COLUMN.

A **Vertebra** (so called from the flexibility of the chain of vertebræ; Lat. *verto*, I turn) consists of a solid body or *centrum*, and more or fewer processes or apophyses, some of which have separate ossific centres. Plate-like processes which arch upward from either side of a centrum to enclose the neural canal are the *neural arches* or *neurapophyses* (fig. 54, *n, n*); at their union in the middle line above they commonly send up a process called the *neural spine* (*ns*). Transverse processes from the sides of the neural arch are *diapophyses* (Gr. διά, *dia*, across) (figs. 54, 55, *d, d*). Oblique processes from the sides of the same arches, serving to lock them together, are *zygapophyses* (Gr. ζυγον, *zugon*, a yoke; fig. 55, *z*); there are two on each side; one anterior, on the front border of an arch, a *pre-zygapophysis;* one posterior, on the hind border, a *post-zygapophysis.* From the under-side of a centrum, in the middle line, there is often a *hypapophysis* (Gr. ὑπό, *hupo*, under; fig. 55, *hy*). These several processes, with some others not necessary to mention here, make with the centrum *a vertebra* in strictness; that is, when existing at all, they are completely consolidated with one another and with the centrum into one bone. But certain important elements of a vertebra, developed from independent ossific centres, may or may not anchylose therewith, in different regions of the same spinal column. These are the *pleurapophyses* (Gr. πλευρόν, *pleuron*, a rib; fig. 54 *pl*; fig. 55, *r*). Any *rib* is in fact the pleurapophysial element of a vertebra; it may be, and in most regions of the spinal column it is, quite small when existing at all, and anchylosed with the vertebra to which it belongs, as an integral portion thereof. Only in the lower region of the neck, and throughout the thoracic region, such pleurapophyses elongate, and are movably articulated with their respective vertebræ; they then become the "ribs" of ordinary language. Moreover, the true thoracic ribs of birds are jointed near the middle, each thus consisting of two pieces; the upper piece is pleurapophysis proper: the lower is called a *hæmapophysis* (fig. 54, *h*; fig. 55, *cr*); it corresponds to a "costal cartilage" of human anatomy. Once again: since the *sternum* (breast-bone) is theoretically, and doubtless archetypically, a solidified set of those parts of the vertebral segments which complete the hæmal arches below, each segment of a sternum to which a hæmapophysis is articulated is called a *hæmal spine*, being compared to a neural spine above. Aside from any consideration of the ribs proper and sternum, or free pleurapophyses, hæmapophyses, and hæmal spines, any "vertebra" of ordinary language is the compound bone which consists of centrum and neur-, di-, pre- and post-zyg-, pleur-, hyp- and other -apophyses, if any, and neural spine; the latter being often called the "spinous process."

The **Vertebræ** join one another, forming a continuous chain. Their centra are placed end to end, one after another; their neural arches are also locked together by the zygapophyses, when such articular processes are developed. Zygapophyses bear upon their free ends smooth articular facets, the faces of which are mostly horizontal; those of the pre-zygapophyses looking downward, and overriding the reversed faces of the post-zygapophyses. The mode of jointing of the centra of such vertebræ as are freely movable upon each other is highly characteristic of birds, in so far as the *shapes* of the articular ends of the vertebral centra are concerned. In anatomy at large, a vertebral centrum which is cupped or hollowed at both ends, is of course bi-concave. Such a vertebra is called *amphicœlous* (Gr. ἀμφί, *amphi*, on both sides; κοῖλος, *koilos*, hollowed); this is the rule in fishes, and obtained in some extinct Cretaceous birds, as *Ichthyornis;* it is unknown in recent birds.[1] A centrum cupped in front only is *procœlous;* one cupped only behind is *opisthocœlous* (Gr. ὄπισθε, *opisthe*, behind). Such structure necessarily results in a ball-and-socket jointing of vertebræ. In those vertebræ of birds in which this arrangement obtains, it is always the *posterior* face of a centrum which is cupped, the anterior one being balled; such vertebræ are therefore opisthocœlous. But in the freest vertebral articulation of birds, that existing in the region of the neck, another modification occurs. Both ends of each vertebra are *saddle-shaped ; i. e.*, concave in one direction, convex in the other; a condition which may be called *heterocœlous* (Gr. ἕτερος, *heteros*, contrary). The concavo-convexity of any one vertebra fits the reciprocal concavo-convexity of the next. *Anterior* faces of heterocœlous vertebræ are concave crosswise, up-and-down convex; *posterior* faces are the reverse; consequently, such vertebræ are procœlous in horizontal section, but in vertical section opisthocœlous. The various physical characters of vertebræ in different regions of the body, and their connections with and relations to other parts of the body, have caused their division into several sets, as cervical, dorsal, etc., which are best considered separately.

Cervical Vertebræ (fig. 56, *cv*) are those of the *neck :* all those in front of the thorax or chest, which do not bear free pleurapophyses in adult life, or the free pleurapophyses of which, if any, are not in two-jointed pieces and do not reach the breast-bone; *i. e.*, have no hæmapophyses. It is advisable, in birds, to draw this line between cervical and succeeding vertebræ, no other being equally practicable ; for, on the one hand, one, two or more of the cervicals (recognizable as such by their general conformation and free articulation) may have long free ribs, movably articulated; and all the cervicals, excepting usually the first, or first and second, have short pleurapophyses, anchylosed in adult life, but free in the embryo; while, on the other hand, a vertebra, apparently dorsal by its configuration and even its anchylosis with the dorsal series, may be entirely cervical in its pleurapophysial character.[2] Thus, in fig. 56, of an owl's trunk, the bone which is apparently first dorsal, and is so marked (*dv*), bears a free styliform "riblet" an inch long (*c'*), only it is not jointed, and does not reach the sternum ; while the next to the last cervical has a minute but still free rib (*c*). In a raven's neck before me, the last cervical rib is about two inches long, articulating by well-defined head and shoulder to body and lateral process of the vertebra; the penultimate rib is about half an inch long, with one articulation to the lateral process; while the next anterior vertebra (third from the last) has a minute ossicle, as a free "riblet." The rule is *two* such free pleurapophyses or cervical ribs of any considerable length : sometimes one; rarely three; in the cassowary four. Rudimentary pleurapophyses may usually be traced up to the second cervical vertebra, as slender

---

[1] Except to this statement, however, the oddly-massed pygostyle, which, in birds where a terminal disc develops inferiorly, may be distinctly cupped at both ends, as it is in a raven for example.

[2] The case is very puzzling; the more so because, viewing the whole series of birds, the ambiguous "cervico-dorsal," or two such equivocal vertebræ, may lean in different cases in opposite directions when the whole sum of characters is taken into account. Therefore it may be best, as already said, to make the possession of a jointed sternum-reaching rib the criterion of the *first* dorsal vertebra, even though an antecedent one may have the physical characters of a dorsal, and be anchylosed with the dorsal series. This is the view taken by Huxley, who says: "The first dorsal vertebra is defined as such by the union of its ribs with the sternum by means of a sternal rib." (Anat. Vert. Anim., 1872, p. 237.) Owen appears to regard as dorsal any of the vertebræ in question which bear free ribs. The actual uncertainty in the case, and the discrepant reckoning by different authors, prevents us from making a satisfactory count of the numbers of the two series of vertebræ in any given case. Thus, fig 56, as marked by Dr. Shufeldt, shows *six* dorsals (*dv*), to which is to be added the one under *p*, bearing the rib *sr;* and from which is to be subtracted the anterior one, bearing the rib *c'*, which is to be regarded as cervical, though its physical characters are evidently those of the dorsal series.

stylets or riblets, completely anchylosed with the neural arches in adult life, and lying parallel with the long axes of the bones. The anchylosis of pleuropophyses distinguishes most cervical vertebræ in another way: for from it results, on each side of the neural arch, a *foramen* (Lat. *foramen*, a hole, pl. *foramina*), through which blood-vessels (vertebral artery and vein) pass to and from the skull. The series of these foramina is called the *vertebrarterial canal;* none such exist in those posterior cervical vertebræ which bear free ribs; thus, in the raven the canal begins abruptly at the *fourth* from the last cervical. But, as in *Rhea* for instance (and doubtless in many other cases), the vertebrarterial canal shades visibly into the series of foramina formed by the spaces between the head and shoulder of any rib and the side of the vertebra to which it is attached; such being, as I suppose, the true morphology of the canal. The cervical is the most *flexible* region of a bird's spine; the articular ends of the vertebral bodies are the most completely saddle-shaped (heterocœlous); the zygapophyses are large and flaring, overriding each other extensively; the largest processes are at the fore ends of the bones; the appositions of the central and zygapophysial articular surfaces are collectively such, that the column tends to bend in an S-shape or sigmoid curve. The vertebral bodies are more or less contracted in the middle, or somewhat hour-glass-shaped; on several lower cervicals, hypapophyses are likely to be well developed; as are neural spines toward both the beginning and end of the series. The vertebræ on the whole are large; their neural canal is also of ample calibre. The first two cervicals are so peculiarly modified for the articulation of the skull as to have received special names. The *first* one, fig. 56, *at*, the *atlas* (so called because it bears up the head, as the giant Atlas was fabled to support the firmament), is a simple ring, apparently without a centrum. The lower part of the ring is deeply cupped to receive the condyle of the occiput into ball-and-socket joint. The *second* cervical is the *axis*, *ax*, which subserves rotary movements of the skull. It has a peculiar tooth-like *odontoid* (Gr. ὀδούς, ὀδόντος, *odous, odontos*, tooth; εἶδος, *eidos*, form) process, borne upon the anterior end of its body, fitting into the lower part of the atlantal ring; about which pivot the atlas, bearing the head, revolves like a wheel upon an eccentric axis. The cervicals of birds vary greatly in number; according to Huxley there are never fewer than eight, and there may be as many as twenty-three; Stejneger gives twenty-four for some of the swans. Twelve to fourteen may be about an average number.

Thoracic or Dorsal Vertebræ (fig. 56, *dv*) extend from the cervical to or into the pelvic region of the spine. In most animals, and in ordinary anatomical language, a "dorsal" is one which bears a distinct free rib, and is therefore truly thoracic, since "ribs" are the sidewalls of the chest. But in birds, as we have seen, certain cervicals have distinct elongate ribs; and, as will be seen soon, long jointed pleurapophyses are usually found in that region commonly called "sacral." The first dorsal, in birds, is arbitrarily considered to be that one which bears the first rib which is jointed, and which reaches the sternum by its lower (hæmapophysial) half. Five or six vertebræ of birds commonly answer this description; though the last one which bears a long free jointed rib (which may or may not reach the sternum) is commonly anchylosed with the sacrum, as *sr*. So few as only *three* hæmapophysis-bearing ribs may reach the sternum. There may also be a long free-jointed rib which "floats" at *both* ends; *i. e.*, is articulated neither with the sternum nor with the vertebra to which it belongs as in the loon, for example. As the dorsal series thus shades insensibly behind into another series, the lumbar (which has no free, nor any *distinct* ribs,—ribs that one would not hesitate to call such), it is best to consider as dorsal or thoracic all those vertebræ, succeeding the last cervical (which is to be determined as explained in the last paragraph), which have *distinct jointed ribs*, whatever the connection or disconnection of such pleurapophyses at either end. On this understanding, one, sometimes two or even three "dorsal" vertebræ anchylose with the pelvic region of the spine. Fixity of the dorsal region being of advantage to flight, these vertebræ are very tightly locked together; not only by the close apposition or **even**

anchylosis of their bodies and processes, but also, in many cases, by ossifications of the tendons of muscles of the back, and coössifications of these with the vertebræ, like a set of splints, till the consolidation of the thoracic is only surpassed by that of the pelvic region of the spine. Dorsal vertebræ also usually differ a good deal from most cervicals in having shorter bodies, laterally compressed, producing a ridge which runs along their middle line below; in lacking a vertebrarterial canal; in having on each side two articular facets, — one on the body and the other on the transverse process, for the head and shoulder of a rib. They are further distinguished, usually, by having large spinous processes, in the form of high, long, thin, squarish plates, often or usually anchylosed together. Their transverse processes are also very prominent laterally, thin and horizontal, and often anchylosed. More or fewer dorsals may bear large hypapophyses; which, as in the loon, may bifurcate at their ends into two flaring plates. Such processes continue a similar series from the neck, and are in relation to the advantageous action of the muscles (*rectus colli anticus* and *longus colli*) by which the neck is made to straighten out from the lower curve of its sigmoid flexure.

The "**Sacrum**" **of a Bird** (figs. 57, and 60) is commonly considered to be that large solid mass of numerous anchylosed vertebræ in the region of the pelvis, covered in by, and fused more or less completely with, the principal bones of the pelvis, or haunch-bones (*ilia*). But in this consolidation of an extremely variable number (averaging perhaps twelve, but running up to at least twenty, eleven to thirteen being usual) of bones are included vertebræ which in other animals belong to several different sets — dorsal, lumbar, sacral proper, and coccygeal or caudal. We have just seen that one or two, even three, vertebræ, which are dorsal according to the definition agreed upon, may enter into the composition of the "sacrum," being firmly anchylosed therewith, and their long ribs issuing out from underneath the ilia, as shown in fig. 56, *sr*. Next comes one bone, or a series of several (two to five or more) bones, anchylosed together by their bodies and spinous processes, and also anchylosed with the ilia by means of stout lateral bars of bone sent transversely outward on either side from their respective centra to abut against the ilia. These cross-bars correspond in general form and position with the transverse process of the last true rib-bearing dorsal, — that process against which the shoulder of any developed rib abuts; they are variously considered to be, to represent, or to include rudimentary ribs; and such difference of view may be warranted by the state of the parts in different birds. However this may be, the bones just described are *lumbar* vertebræ (Lat. *lumbus*, the loin; where such vertebræ are situated in man and other mammals); which certainly possess abortive ribs in some cases. On successive lumbars the cross-bars, whatever their nature, commonly slip lower and lower downward (belly-ward) on the vertebral bodies, till the last ones are quite down to the level of the ventral aspect of the centrum; these are also commonly the stoutest, most directly transverse, and most nearly horizontal of the series of processes, abutting against the ilia a little in advance of the socket of the thigh bone. This ends a series of consolidated "sacral" vertebræ which are termed collectively "dorso-lumbar,"

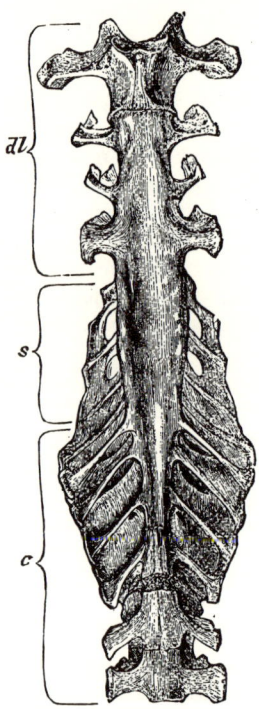

Fig. 57. — The "sacrum" of a young fowl, seen from below, nat. size; after Parker. *dl*, dorsolumbar series, whereof the first is dorsal proper, the next three are lumbar; *s*, the sacral series proper, or true sacrum, consisting of five vertebræ; *c*, the urosacral series, being those caudal vertebræ, six in number, which anchylose with one another and with the sacrum.

— all of them anterior to the true *sacrum* of a bird. The *sacrum proper* (fig. 57, *s*) consists of those few vertebræ — three, four, or five — from foramina between which issue the spinal nerves that form the net-work called the *sacral plexus*. These true sacral vertebræ are ribless, and may be recognized, in a general way, by the absence of anything like the cross-bars above described, issuing from the vertebral centra; though their neural arches send off some small bars or plates to fuse with the ilia. These sacrals proper are at or near the middle of the whole sacral mass. After these come a large number — from five to ten or more — of vertebræ which, from their following the true sacrals, though consolidated therewith and with one another, are considered to belong to what would be the caudal region of other animals, and are hence called "tail-sacrals," *uro-sacrals* (Gr. οὐρα, tail, fig. 57, *c*.) These continue to send off a series of little plate-like processes from their neural arches, just as the true sacrals do; but, in addition to these, processes are given off from the bodies of the uro-sacrals, corresponding in position and relation to those which proceed from the bodies of the lumbars, and being apparently of the same morphological character (pleurapophysial). These "riblets" are, however, quite slender, and also oblique in two directions; for instead of being transverse and nearly horizontal, they trend very obliquely backward and upward; they also shorten consecutively from before backward. The cross-bars of the latter uro-sacrals, however, are stouter and altogether more like those of a lumbar vertebra. The appearances described are those seen from below, or on the ventral aspect. Above, on the back of the pelvis, the line of confluent spinous processes of the dorso-lumbars is commonly distinct, separated a little from the flaring lips of the ilia. Such distinct formation may continue throughout the sacral and uro-sacral regions; oftener, however, the line of spinous process sinks, flattens, and widens into a horizontal plate which becomes perfectly confluent with the ilia along the posterior portion of their extent; such smooth, somewhat lozenge-shaped surface being quite continuous with the superficies of the pelvis, but perforated with more or fewer pairs of intervertebral foramina. — Such is the general character of a bird's complex sacrum; the description is taken chiefly from a raven (*Corvus corax*); the figure from the common fowl, after Parker. The kidneys are moulded into the recesses between the sacral and uro-sacral vertebræ and in the concavity of the ilia. The general shape of a "sacrum," viewed from below, is fusiform, broadest across the sacral bodies proper or just in front of them, tapering toward either end; the face of the sacrum is also flattest about the middle, more or less ridged before and behind from compression of the vertebral bodies. It has little if any lengthwise curvature, and that chiefly in the uro-sacral region, where the concavity is downward. The total number of bones may be less than twelve, or more than twenty. The extensive anchyloses in this region of the spine are in evident adaptation to bipedal locomotion, which requires fixity hereabouts, that the trunk may not bend upon the fulcrum represented by a line drawn through the hip-joints, which are situated about opposite the middle of the sacral mass, as shown by the arrow, *ac*, in fig. 60. (The word "sacrum," a "sacred thing," curious in this application, is very ancient in human anatomy, commemorating some superstitious or ritualistic notion, respecting this part of the body.)

The Coccygeal, or Caudal Vertebræ (fig. 56, *clv*) proper, terminate the spinal column. They are called "coccygeal," from the fancied resemblance of the human tail-bones collectively to the beak of a cuckoo (Gr. κόκκυξ, *kokkux*). The caudals are all the *free* bones situated behind the anchylosed uro-sacrals. The series commonly begins opposite the point where the pelvic bones end; it consists of a variable number of bones, from the twenty long slender ones which the *Archæopteryx* possessed, down to seven or fewer separate ones. The usual number is eight without the pygostyle. They are stunted, degraded vertebræ, whose chief office is to support the tail-feathers: for the leash of nerves which emerge from the spinal canal to form the sacral plexus by so much diminish the spinal cord that a mere thread is left to penetrate the tail, though the neural arches of all the coccygeals be still pervious. All may be freely movable, as in the American Ostrich (*Rhea*); but in almost all birds only the anterior ones are distinct and vertebra-like, the rest, to a variable number, being abortive, and melted into that extraordinary affair called the "ploughshare" or *pygostyle* (Gr. πυγή, *puge*, the rump; στῦλος, a post), which may consist of no fewer than ten such metamorphosed tail-bones. It has usually a shape suggesting the share of a plough (see fig. 56, *py*), but is too variable to be concisely described. The pygostyle supports the tail-feathers; and as these are morphologically one pair to each rectrix-bearing vertebra, the number of tail-feathers may be primarily equal to the number of vertebræ which fuse in the pygostyle. Thus the swan is said to have ten vertebræ in this mass; our wild swan (*Cygnus columbianus*) has twenty tail-feathers. In this view, six should be the usual composition of the share-bone. A bird's tail is really more extensive and lizard-like than commonly supposed; thus the swan, besides its ten in the pygostyle, has seven free caudals, and ten uro-sacrals — twenty-seven post-sacral vertebræ in all (Huxley). In the raven, the free caudals are six, exclusive of the pygostyle. These all have large flaring transverse processes and moderate spinous processes, and the latter ones are also provided with hypapophyses, some of which are bifurcate. The pygostyle in many birds expands below into a large circular or polygonal disc.

2. THE THORAX: RIBS AND STERNUM.

The Thorax (Gr. θώραξ, a coat of mail; in anat., the chest; adj. *thoracic;* see fig. 56) is the bony box formed by the ribs on each side, the breast-bone below, and the back-bone above. In birds, it is very extensive, including most or all of the abdominal as well as the thoracic viscera, and its cavity is not partitioned off from that of the belly by a completed *diaphragm,* though a rudimentary structure of that kind is found in the class. The thorax is usually soldered behind to the pelvis by union of one or more pairs of ribs with the ilia; in front it always and entirely bears the *pectoral arch* (see p. 151). The thorax is very movable in birds, by reason of the great length and jointedness of the ribs.

The Ribs (Lat. *costa,* a rib; pl. *costæ;* adj. *costal;* see fig. 56, *c, c', R, cr, sr, u*), as said above, are the pleurapophysial elements of vertebræ, which remain small and anchylosed, or become long and free. In the latter state only are they "ribs" in ordinary language. The one or more cervical ribs, however elongated, and the abortive lumbar and uro-sacral ribs, are to be excluded from the present description, and have been already considered. *True ribs* are those which belong to the dorsal vertebræ proper, and are jointed in themselves; that is, have articulated *hæmapophyses* (see p. 143), by which they may or do articulate with the sternum. Such true ribs are *fixed,* when they reach from back-bone to breast-bone; *floating,* when either or neither of these connections is made. Usually the last rib, though bearing a perfect hæmapophysis, does not reach the sternum; in the loon, for example, the last rib floats at *both* ends, having connection neither with vertebra nor sternum; and the two next ribs float at their sternal ends. The perfected ribs are few, — five or six is a usual number, though nine are hæmapophysis-bearing in the loon. The last rib at least is usually "sacral;" *i. e.,* belongs to a dorsal vertebra which is anchylosed with the "sacral" mass; and two or even, as in the loon, three ribs may likewise issue out from under cover of the ilia. These "sacral ribs" are furthermore distinguished by being devoid of the *epipleural* or *uncinate processes* (Lat. *uncus,* a hook; fig. 56, *u*) with which other true ribs are furnished, forming a series of splint-bones proceeding obliquely from one rib to shingle over the next succeeding one, and thus increase the stability of the thoracic side-walls. Such splints may be either articulated or anchylosed with their respective ribs; they have independent ossific centres. The upper (pleurapophysial) part of a rib, or "vertebral rib," when perfected, articulates with the side of the body of a vertebra by its head or *capitulum* (Lat. dimin. of *caput*, head), and also with the lateral process of the same vertebra by its shoulder or *tuberculum* (Lat. dimin. of *tuber*, a swelling). In well-marked cases, the head and shoulder are quite far apart, the rib seeming prolonged above; either of these vertebral connections may be disestablished, the other remaining, or both may be lost. The lower (hæmapophysial) part of a rib, or "sternal rib," articulates with the side of the sternum by a simple enlargement; the ends of those sternal ribs which thus join the sternum tend to cluster closely together at a part of the breast-bone called its *costal process* (fig. 58); those which do not make the sternal connection are simply bundled together. Commonly five or six, sometimes four, rarely only three ribs reach the sternum. The ribs are ordinarily as slender and strict as those shown in fig. 56; but in *Apteryx*, for example, their pleurapophysial parts are expansive and plate-like. They lengthen rapidly from before backward, both in their vertebral and their sternal moieties; these parts meet at angles of decreasing acuteness from before backward; but these angles, as those of the ribs both with vertebræ and sternum, incessantly increase and diminish in the respiratory movements of the chest; all being in expiration more acute, and more obtuse in inspiration.

The Avian Sternum (Gr. στέρνον, *sternon*, the breast; fig. 56, S) is highly specialized; its extensive development is peculiar to the class of Birds, and its modifications are of more importance in classification than those of any other single bone. Thereupon it becomes an interesting object. Theoretically it is a collection of hæmal spines of vertebræ. Though such morphological character is appreciable in those animals which have a long jointed sternum, the segments of which, answering to pairs of ribs, develop from separate centres, there is little or nothing in the development or physical characters of the avian sternum to favor this view. The great bone floors the chest and more or less of the belly, and furnishes the main *point d'appui* of both the bony and muscular apparatus of flight, receiving important bones of the scapular arch and giving origin to the immense pectoral muscles. (See also fig. 58.)

Birds offer *two* leading types of sternal structure, the *ratite* and the *carinate*, or the "raft-like" and the "boat-like", according as the bone is flat or keeled (Lat. *ratis*, a raft; adj. *ratite*; in an arbitrary nom. pl., *Ratitæ*, a name of one of the leading divisions of birds: Lat. *carina*, a keel; adj. *carinate*: nom. pl. *Carinatæ*, name of another such division). 1. In all struthious birds, comprehending the ostrich and its allies (and also in the Cretaceous *Hesperornis*), the sternum is a flattish, or rather concavo-convex, buckler-like bone, of somewhat squarish or rhomboidal shape, developed from a single pair of lateral centres of ossification, — a "flat boat," without any keel, built with reference to an important modification of the shoulder-girdle, and a reduced or rudimentary condition of the wings, which are unfit for flight. 2. In all flying birds, and some which from other than any fault of the sternum do not fly, — comprising all remaining recent birds, or *Carinatæ*, and also the Cretaceous *Ichthyornis*, — the sternum is keeled and develops from a median centre of ossification as well as from lateral paired centres; usually two of these, making five in all. In a few *Carinatæ* the keel is rudimentary, as the flightless ground parrot of New Zealand, *Stringops habroptilus;* or otherwise anomalous, as in the extraordinary *Opisthocomus cristatus*, where it is cut away in front, and in the rail-like *Notornis*, where the sternum is extremely like a lizard's. In general, the development of the *keel* is an index of wing-power, whether for flying or swimming, or both; the effectiveness of the pectoral muscles being rather in proportion to depth of keel than to extent of the sides of the "boat-bone;" thus, the keel is enormous in swifts (*Cypselidæ*) and humming-birds (*Trochilidæ*).

The carinate sternum normally develops from five centres, having consequently as many separate pieces in early life. Two of these are lateral and in pairs; the third is median and single. The median ossification, which includes the keel, is the *lophosteon*(Gr. λόφος, *lophos*, a crest; ὀστέον, *osteon*, a bone). The anterior lateral piece, that with which the ribs, or some of them, articulate, is the *pleurosteon* (Gr. πλευρόν, *pleuron*, a rib); in adult life this becomes the *costal process*, so prominent in *Passeres* (fig. 58). The posterior lateral piece is the *metosteon* (Gr. μετά, *meta*, after). From the latter are derived the pair, or two pairs, of lateral processes which the posterior border of the sternum has in so many birds. In fine, the extent of ossification of the lophosteon and metostea, and the mode of their coösification, determines all those various shapes of the posterior border of the sternum which, being commonly characteristic of genera and higher groups, are described for purposes of classification. Thus, if the lophosteon and the metostea are completely ossified and to the same extent behind, the posterior border of the sternum will be transverse, and perfectly bony. Such a sternum is said to be *entire*. If the lophosteon is longer than the lateral pieces, the sternum will have a central pointed or rounded projection; when such a formation is called the middle *xiphoid* process (Gr. ξίφος, *xiphos*, a sword: εἶδος, *eidos*, form). The projection of the metostea, not infrequent, similarly gives a pair of external lateral xiphoid processes. But such processes oftener result merely from defects of coösification between the elements of the sternum. Thus, there is often a deep notch in the posterior border of the sternum between the lophosteon and the metosteon of each side; the sternum is then said to be *single-notched* or *single-emarginate* (one pair of notches, one on each side; fig. 58). This conformation prevails throughout the great group *Passeres*, possibly without exception; it is therefore highly characteristic of that order, though a great many other birds also have it. In the natural state, the notch is filled in with membrane. Such a notch may also be converted into a "fontanelle" or *fenestra* (Lat. *fenestra*, a window), which is simply a hole in the bone, the metostea having grown to the lophosteon at their extremities, but left an opening between. Such a sternum is called *fenestrate*, more exactly *uni-fenestrate* (Lat. *unus*, one; one window on each side). Now, the parts remaining as before, let either each half of the lophosteon, or each metosteon, be notched or fenestrate; obviously then, such a sternum is *double-notched* or *bi-fenestrate*, having four notches, or holes, two on each side, — two notches, or two holes; or notched *and* fenestrate, having a notch and a hole on each side. The latter is very frequent: when occurring, the hole is generally nearest the middle line, the notch exterior. Irregularity of ossification, converting a hole into a notch, and conversely, may in any case result in lack of symmetry; but this is a mere individual peculiarity. When there are two notches on each side, as in fig. 56, the sternum has evidently a median and two lateral backward extensions, which are then called respectively the *middle, internal lateral*, and *external lateral* xiphoid processes. Notching of the lophosteon in the middle line, at least to any extent, must be very rare, if indeed it ever occurs. The extreme case of emargination of the sternum is afforded by the *Gallinæ*, and is highly characteristic of that group. Here the lophosteon is extremely narrow, and fissured deeply away from the metostea, which latter are deeply forked; the arrangement giving rise to two very long slender lateral processes on each side (figs. 1 and 2, p. 48). The sternum of the tinamou, a dromæognathous bird, is still more deeply emarginated, but the extremely long and slender lateral processes, which enclose an oval contour, are simple, not forked.

In a very few birds there are centres of ossification additional to those above described. In *Turnix*, there are said by Parker to be a pair of centres between the pleurostea, which he names *coracostea*, because related to the part of the sternum with which the coracoids (see p. 146) unite. The same authority describes for *Dicholophus* a posterior median cartilaginous flap having a separate centre, named *urosteon* (Gr. οὖρα, *oura*, tail). In various birds the sternum is eked out in the middle line behind by cartilage which has no ossification.

The sternum, especially of the higher birds, develops in the middle line in front a beak-like process called the rostrum or *manubrium* (Lat. *manubrium*, a handle); its size and shape vary; it is well-marked in Passerine birds (fig. 58); and may be bifurcate at the end and run down the front of the keel some way, as in the raven. The fore border of the sternum is generally greatly convex from side to side, and then, in those birds which have prominent pleurostea, produced in angular costal processes. This border is also thickened, and presents on each side a well-marked, smooth-faced groove, in which the expanded feet of the coracoid bones are instepped and firmly articulated. These deep grooves commonly meet in the middle; are occasionally continuous from one side to the other; sometimes each crosses to the other side a little way. The costal processes on each side also have thickened edges, with a series of articular facets for the ribs, which gives this border a fluted or serrate profile. Generally the fore half, or rather less, of the side border of the sternum is thus articular; and it is only such *costiferous* (rib-bearing) extent of sternum which corresponds to the whole body of the bone in a mammal, all the rest being " xiphoid." The singular carinate sternum of *Notornis*, and the ratite bone of *Apteryx*, are concave crosswise along the front border, and bear the coracoids far apart, at the summits of antero-lateral projections.

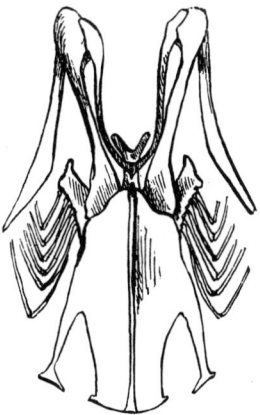

Fig. 58. — Typical passerine sternum, pectoral arches, and sternal ends of ribs; from the robin, *Turdus migratorius*, nat. size; Dr. R.W. Shufeldt, U.S.A. Sternum single-notched, with prominent costal processes and forked manubrium; five ribs reaching sternum, one rib "floating."

A sternum is generally concavo-convex in each direction, bellying downward; somewhat rectangular, it may be long and narrow, or short, broad, and squarish. It is commonly longer than broad, with convex front border, a median beak, which is often forked, prominent antero-lateral corners, pinched-in sides (bulging in tinamou) and indeterminate hind border. The keel usually drops down lowest in front, sloping or curving gently up to the general level behind, with a concave (rarely protuberant) vertical border, and pronounced apex, to which the clavicles may or may not be anchylosed, as they are in a pelican for instance. In *Opisthocomus*, the clavicles anchylose with the manubrium of the sternum. The external surface, both of body and keel, is ridged in places, indicating lines of attachment of the different pectoral muscles. In a few birds, notably swans and cranes, the keel is expanded and hollowed out to receive folds of the windpipe in its interior (see figs. 99, 100). — But the numberless modifications of the sternum in details of configuration belong to systematic ornithology, not to rudimentary anatomy.

3. THE PECTORAL ARCH.

The Pectoral Arch (Lat. *pectus*, the breast; figs. 1, 2, 56, 58, 59) is that bony structure by which the wings are borne upon the axial skeleton. It is to the fore limb what the pelvic arch is to the hind limb; but is disconnected from the back-bone and united with the breast-bone, whereas the reverse arrangement obtains in the pelvic, which is fused with the sacral region of the spine. Each pectoral arch of birds consists (chiefly) of three bones: the *scapula* and *coracoid*, forming the *shoulder-girdle* proper, or *scapular arch*; and the accessory *clavicles*, or right and left half of the *clavicular arch*. There is also at the shoulder-joint of most birds an insignificant sesamoid ossicle, called *scapula accessoria* or *os humero-scapulare* (fig. 56, *ohs*); and in many a rudiment of a bone called *procoracoid*, which occurs in reptiles, but in birds is united with the clavicle. From the ribs, the scapula; from the sternum, the coracoid; from its fellow, the clavicle, converges to meet each of the two other bones at the point of the shoulder. The lengthwise scapular arches of opposite sides are distinct from each other; the clavicular arch is crosswise, and nearly always completed on the middle line of the body; by which union of the clavicles the whole pectoral arch is coaptated. The coracoid bears the shoulder firmly away from the breast; the scapula steadies the shoulder against the ribs; the clavicles keep the shoulders apart from each other. The scapular arch is always present and complete; the clavicular is sometimes defective or wanting. There are two leading styles of scapular arch, corresponding to the ratite and carinate sternum. (1) In *Ratitæ* the axes of the coracoid and scapula are nearly coincident (for the most part in a continuous right line) and anchylosed together; the clavicles are usually wanting, or defective; and the coracoids are instepped on the sternum far apart. (2) In all *Carinatæ*, the axes of the coracoid and scapula form an acute or scarcely obtuse angle (fig. 56, *sglc*); normally these bones are not anchylosed; perfect clavicles are present, anchylosed with each other, but free from the other bones; and the coracoids are instepped close together. Decided exceptions to these conditions, as in *Notornis*, are anomalous; though incompletion of the clavicles repeatedly occurs, as noted below.

The Coracoid (Gr. κόραξ, *korax*, a crow; εἶδος, *eidos*, form: the corresponding bone of the human subject, which is the stunted "coracoid process of the scapula," being likened to a crow's beak; no applicability in the present case; figs. 56, *c*, 59, *c*) is a stout, straight, cylindric bone, expanded at each end, extending forward, outward, and upward from the fore border of the sternum to the shoulder. Its foot is flattened and splayed to fit in the articular groove of fore border of the sternum already described; it often overlaps that of its fellow on the median line; is narrower and remote from its fellow in *Ratitæ*. The head of the bone, irregularly expanded, articulates or anchyloses with the end of the scapula, and also usually with the clavicle. It bears externally a smooth demi-facet, which represents the share it takes in forming the *glenoid* (Gr. γλήνη, *glene*, a shallow pit; fig. 59, *gl*) *cavity*, which is the socket of the humerus. This articular expansion is the *glenoid process* of the coracoid: the *clavicular process* is that by which the bone unites with the clavicle. The relation between the heads of the three bones (each uniting with the other two) is such that a pulley-hole is formed, through which plays the tendon of the pectoral muscle which elevates the wing. The coracoid is a very constant and characteristic bone of birds.

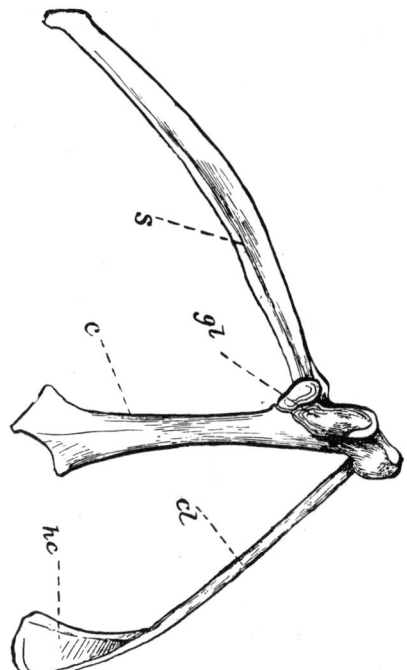

FIG. 59.— Right pectoral arch of a bird, *Pediœcetes phasianellus*, nat. size, outside view; Dr. R. W. Shufeldt, U. S. A. *s*, scapula; *c*, coracoid; *gl*, glenoid, the cavity for head of humerus; *cl*, clavicle; *hc*, hypocleidium. *In situ*, the right end of the figure should tilt up a little; see fig. 56.

The Scapula (Lat. *scapula*, the shoulder-blade; figs. 56, 59, *s*) merits in birds its name of "blade-bone," being usually a long, thin, narrow, sabre-like bone, which rests upon the ribs — usually not far from parallel with the spinal column, and near it; but in *Ratitæ* otherwise. It seldom gains much width, and is quite thin and flat in most of its length; but it has a thickened head or handle, expanding outwards into a *glenoid process* which unites with that of the coracoid to complete the glenoid cavity, and dilated inward to form an *acromial* (Gr. ἀκρώμιον, *akromion*, point of the shoulder) *process* for articulation with the clavicle (as it does in man), when that bone exists. The other end is usually sharp-pointed, but may be obtuse, or even clubbed, as in a woodpecker. The scapula is broadest and most plate-like in the penguins, in which birds all the bones of the flipper-like wing are singularly flattened. In *Apteryx* it reaches in length over only a couple of ribs; in most birds, over most of the thorax; and in some its point overreaches the pelvis.

THE ANATOMY OF BIRDS.—OSTEOLOGY.

The Clavicles, or Furculum (Lat. *clavicula*, a little key: *furculum*, a little fork; figs. 56, 59, *cl*), or the clavicular arch, are the pair of bones which when united together form the object well known as the "merry-thought" or "wish-bone," corresponding to the human "collar-bones." They lie in front of the breast, across the middle line of the body like a V or U; the upper ends uniting as a rule both with scapula and coracoid. For this purpose, in most birds, the ends are expanded more or less; such expansion is called the *epicleidium* (Gr. ἐπί, *epi*, upon; κλειδίον, *kleidion*, the collar-bone); in Passerine birds it is said to ossify separately, and is considered by Parker to represent the *procoracoid* of reptiles. At the point of union below, the bones often develop a process (well shown in the domestic fowl) called the *hypocleidium* (Gr. ὑπό, *hypo*, under; fig. 59, *hc*), supposed to represent the *interclavicle* of reptiles. The clavicles are as a rule present, perfect, anchylosed together, articulated at the shoulder; in a few birds anchylosed there; in several, there and with the keel of the sternum; in *Opisthocomus* there and with the manubrium of the sternum. In various birds, chiefly Picarian and Psittacine, they are defective, not meeting each other. They are wanting in *Struthio*, *Rhea*, *Apteryx*, and some *Psittacidæ*. Besides curving toward each other, the clavicles have usually a fore-and-aft curvature, convex forward. In general, the strength of the clavicles, the firmness of their connections, and the openness of the V or U, are indications of the volitorial or natatorial power of the wings. The end of the furculum is hollowed for a fold of the windpipe in the crested pintado (Owen).

4. THE PELVIC ARCH.

The Pelvis (Lat. *pelvis*, a basin, fig. 60), is that posterior part of the trunk which receives the uro-genital, and lower portion of the digestive, viscera. It consists of the "sacral" vertebræ on the middle dorsal line, flanked on each side by the bones of the *pelvic arch*, which supports the hind limb. In vertebrates generally the pelvic basin is completed on the ventral aspect by union (*symphysis*; Gr. σύν, *sun*, together; φύσις, growth) of the bones from opposite sides. Excepting only *Struthio*, which has a pubic symphysis; and *Rhea*, which has an ischiac symphysis just below the sacral vertebræ, the pelvis of a bird is entirely open below and behind; each pelvic arch anchylosing firmly with the sacral vertebræ to form a roof over the viscera above named. This sacro-iliac anchylosis is commonly coextensive with the confluence of the many vertebræ which make the "sacrum" of ordinary language, that is, from the first dorso-lumbar to the last uro-sacral. The whole roof-like affair looks something like a keelless sternum inverted. The pelvic arch of each side consists of three bones, *ilium*,

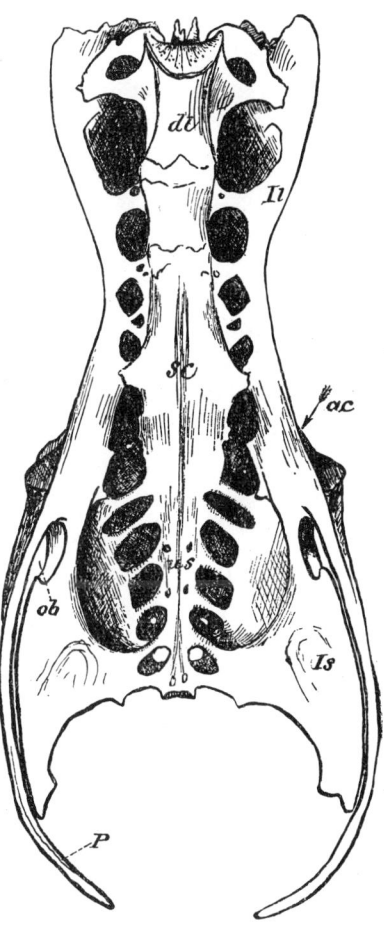

FIG. 60.—Pelvis of a heron (*Ardea herodias*), nat. size, viewed from below; from nature by Dr. R. W. Shufeldt, U.S.A. *dl*, dorso-lumbar vertebræ to and including the last one, *sc*; below *sc*, for the extent of the *large* black spaces (opposite the arrow) are the true sacral vertebræ; *us*, urosacral vertebræ (opposite the five oval black spaces; *Il*, ilium; *Is*, ischium; *P*, pubis; *ob*, obturator foramen. The arrow flies into the acetabulum.

ischium, and *pubis*, which have independent ossific centres, but become firmly consolidated together to form the haunch-bone or *os innominatum*. Each of these bones unites with the other two, somewhere near the middle of the whole affair, at a ring-like structure called the *acetabulum* (Lat., a vinegar-cruet, fig. 56, *a;* fig. 60, arrow *ac*), which all three consequently contribute to the formation of, and which is the socket for the head of the thigh-bone (*femur*, p. 125). When free ribs issue from under cover of the pelvis, they are commonly anchylosed with the ilia; and all the abortive pleurapophyses of the lumbar and uro-sacral vertebræ have likewise iliac anchylosis, as explained in treating of the sacrum (p. 146). As a whole, the pelvis varies like the sternum in relative length, breadth, and degree of convexity; and especially in the configuration of its posterior border; but few zoölogical characters are derived from this structure.

Viewed from below, the pelvis is seen to be much hollowed or excavated for the lodgment of the kidneys, and cross-cut into compartments by the sacral rafters; the series of sacral bodies forming a ridge-pole along the middle line. Above, the series of sacral spinous processes represent the ridge-pole; anteriorly, the somewhat spoon-shaped iliac bones are applied, concavity outward, to the dorso-lumbars; posteriorly, in the middle line, is a more or less flattened horizontal expansion, and laterally are the more expanded sides of the ischiac roof, finished along the eaves and behind by the slender pubic bone, which commonly projects backward, and inclines toward its fellow of the opposite side. The most prominent formation of the side wall of the pelvis is the thick-lipped smooth articular ring, the *acetabulum*, converted in the natural state into a cup by a membrane. The postero-superior segment of the rim is prominent, to form the *antitrochanter* (Gr. ἀντί, anti, against; τροχαντήρ, trochanter of the femur) against which the shoulder of the femur abuts when the head is in the ring.

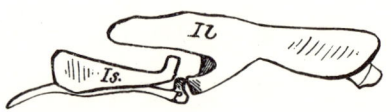

FIG. 61. — Pelvis of *young* grouse, showing three distinct bones. *Il, Is, P.* ilium, ischium, pubis. In front of former a dorsal vertebra protrudes. (Dr. R. W. Shufeldt, U. S. A.)

It is normal to recent Carinate birds to have the ischium fused with the ilium, however distinct the pubis may remain; but to Cretaceous birds (even the carinate *Ichthyornis*), and the existing *Ratitæ*, to have both ischium and pubis distinct in most of their extent.

The Ilium (Lat. *ilium*, haunch-bone; pl. *ilia;* adj. *iliac;* figs. 56, *I;* 60, 61, *Il*) is the median, most anterior and longest of the haunch-bones, and the only one which extends in advance of the acetabulum. Such anterior prolongation of this bone is the specialty of the avian pelvis: it commonly overlies one or more ribs, and is often overreached by the end of the scapula. It is longest and narrowest and flattest in some of the lower swimmers; the reverse among the highest birds. Its relations and connections have been sufficiently indicated. The bone is almost always separated from its fellow by the sacrum, though the approximation may be very close over the back of the pelvis, along the middle line.

The Ischium (Gr. ἰσχίον, ischion, the haunch-bone; pl. *ischia;* adj. *ischiadic, ischiatic,* better *ischiac;* figs. 56, 60, 61, *Is*) lies entirely post-acetabular, or behind the socket which it contributes to form, and composes most of the side-wall of the pelvis thence to the end. It is generally a thin, plate-like bone. Among Cretaceous birds and existing *Ratitæ* it only unites with the ilium at and just behind the acetabulum, whence a deep *ilio-ischiac* fissure between the two exists, as in the *young* grouse, fig. 61; but in ordinary adult birds this fissure is converted into a fenestra or window of large size, just behind the acetabulum, by union of the two bones behind it. This vacuity, whether a notch or a hole, corresponds to the "sacro-sciatic notch" of human anatomy (fig. 56, *in*). The ischia of opposite sides are distinct, except in *Rhea*.

The Pubis (Lat. *pubis*, bone of the front of the human pelvis where the hair grows at *puberty ;* pl. *pubes ;* adj. *pubic ;* figs. 56, 60, 61 *P*), beginning at its share of the acetabular ring, is a long slender bone which runs along the lower border of the ischium, sometimes for a short distance only, often for the whole length of the ischium, and usually projecting behind ; more or less perfectly parallel with, applied to, or united with, the inferior ischiac border. When separate, a long deep fissure results ; when united at the end, a long narrow foramen is formed ; when incompletely united in any part of its ischiac continuity, a fissure and a foramen, in the ostrich two foramina, result. All these conditions occur ; in any case, such ischio-pubic interval corresponds to the *obturator* foramen (fig. 56, *o;* fig. 60, *ob*) of human anatomy ; it is greatest in Cretaceous birds and existing *Ratitæ*. The free ends of the pubes may be more or less expanded. In the ostrich only there is a pubic symphysis of the ends of the bones ; in the same bird a separate ossicle, situated upon the lower border of the pubes, and called *epipubic*, is considered to represent a "marsupial" bone (Garrod). In various birds, among them our ground cuckoo, *Geococcyx californianus*, the pubis projects a little forward, under the acetabulum : this prominence is the *propubis.* Separation of the pubes is supposed to be for amplification of the pelvic strait to facilitate the passage of the large chalky eggs birds lay.

5. THE SKULL.

The Skull of a Bird is a poem in bone — its architecture is the "frozen music" of morphology ; in its mutely eloquent lines may be traced the rhythmic rhymes of the myriad amœbiform animals which constructed the noble edifice when they sang together.[1] The poësy (ποίησις, *poiesis*, a making) of the subject has been translated with conspicuous zeal and success by Mr. W. K. Parker ; its zoölogical moral has been similarly pointed by Professor Huxley ; and the young ornithologist who would not be hopelessly unfashionable must be able to whistle some bars of the cranial song — the pterygo-palatine bar at least.

The rapid progress of ossification soon obliterates most of the original landmarks of the skull, fusing the distinct territories of bone in one great indistinguishable area. Thus the brain-box of almost any mature bird is apparently a single solid bone, and most parts of the jaw-scaffolding similarly run together. Aside from the bones of the tongue, which are collectively separate from those of the skull proper ; and of the compound lower jaw, which is freely articulated with the rest of the skull ; only two or three other bones of the skull, as a rule, are permanently and perfectly free at both ends. These are the quadrate bones — the anvil-shaped pieces by which the lower jaw is slung to the skull ; the pterygoids, articulating the palate with the quadrate ; and sometimes the vomer. Traces only of the bones of the face and jaws are usually found ; but even such vestiges disappear, as a rule, from among the bones of the brain-box. It is necessary to any intelligent understanding of the construction of a bird's skull, to learn somewhat of its mode of development in the embryonic stage ; this being the only clue to the individual bones of which it is composed, and so to any correct idea of its morphology. One theory is, that the skull consists of four modified vertebræ ; and the principal bones have been named and described by some in terms indicating the elements of a theoretical vertebra. It is true that the skull is segmented, or may be segmented off, like a chain of several vertebræ ; that it continues the vertebral axis forward ; that it has a *basis cranii* like a series of vertebral centrums, above which rises a segmented neural arch enclosing the great nervous mass, and below which depends a set of bones enclosing visceral parts like a hæmal arch. The hindmost cranial segment, the occipital bone, resembles a vertebra in many physical characters, and even in mode of development. But if the serial homology of the skull with

---

[1] Bone-tissue chiefly consists of the aggregated skeletons of *Osteamœbœ* — a kind of uni-cellular protozoan animals which inhabit in myriads the bodies of nearly all the *Vertebrata*, possessing the faculty of feeding upon phosphate of lime and other earthy matters they find in the blood, and afterward excreting them in the form of multiradiate exoskeletons of their own, collectively forming the whole skeleton of their host.

the back-bone be real and true, it is so obscured by the extraordinary modifications to which the vertebral elements have been subjected that the fact of such homology cannot be demonstrated; and to interpret the skull as something super-imposed upon, and morphologically different from the spinal column, is perfectly warranted if not required by the known facts of its constructive development. This is the view taken by the rulers of to-day's science. As already said (p. 143) the relation between cranial and vertebral parts is rather the analogy of adaptive modification than a true homology of structure.

Before proceeding to describe the mature skull, it will be best to consider its mode of development. In this I shall closely follow Parker, often using the words of that master, and illustrating the early stages of the embryo with figures borrowed from the same safe source. In the fewest words possible, I wish to convey an idea of the embryonic skull up to Parker's "third stage," at which it begins to ossify. Here, however, I will first insert a figure, kindly drawn for me by Dr. R. W. Shufeldt, of the U. S. Army, which shows most of the cranial bones, and will give the student a preliminary notion of the "lay of the land." I advise him to contemplate this picture till he has learned the names printed on it by heart, and can apply them to the identification of the parts of the real skull he should have in hand at the same time. He may also meditate on fig. 63.

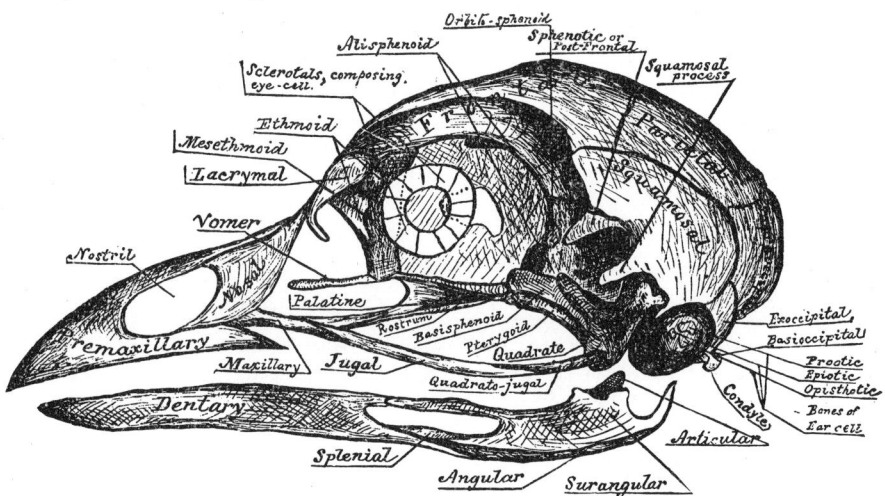

FIG. 62. — Skull of common fowl, enlarged; from nature by Dr. R. W. Shufeldt. U. S. A. The names of bones and some other parts are printed, requiring no explanation; but observe the following points: The distinction of *none* of the bones composing the brain-case (the upper back expanded part) can be found in a mature skull. The brain is contained between the *occipital, sphenoidals, squamosals, parietals* and part of *frontal;* the *ethmoidals* belong to the same group of cranial bones proper. All other bones, excepting the three *otic* ear-bones, are bones of the face and jaws. The lower jaw, of five bones, is drawn detached; it articulates by the black surface marked *articular* with the prominence just above — the *quadrate bone.* Observe that from this quadrate a series of bones — *quadrato-jugal, jugal, maxillary* — makes a slender rod running to the *premaxillary;* this is the *zygoma,* or *jugal bar.* Observe from the quadrate also another series, composed of *pterygoid* and *palatine* bones, to the premaxillary; this is the *pterygo-palatine bar;* it slides along a median fixed axis of the skull, the *rostrum,* which bears the loose *vomer* at its end. The under mandible, quadrate, pterygoid, and vomer are the only movable bones of this skull. But when the quadrate rocks back and forth, as it does by its upper joint, its lower end pulls and pushes upon the under mandible, by means of the jugal and pterygo-palatine bars, setting the whole scaffolding of the upper jaw in motion. This motion hinges upon the elasticity of the bones of the forehead, at the *thin* place just where the reference-lines from the words "lacrymal" and "mesethmoid" cross each other. The dark oval space behind the quadrate is the external orifice of the *ear;* the parts in it to which the three reference-lines go are diagrammatic, not actual representations; thus, the quadrate articulates with a large *pro-otic* as well as with the *squamosal.* The great excavation at the middle of the figure, containing the circlet of unshaded bones, is the left *orbital cavity, orbit,* or socket of the eye. The *mesethmoid* includes most of the background of this cavity, shaded diagonally. The upper one of the two processes of bone extending into it from behind is the *post-frontal* or *sphenotic process;* the under one (just over the quadrate) is the *squamosal process.* A bone not shown, the *presphenoid,* lies just in front of the oval black space over the end of *basisphenoid.* This black oval is the *optic foramen,*

through which the nerve of sight passes from the brain-cavity to the eye. The black dot a little behind the optic foramen is the orifice of exit of a part of the *trifacial* nerve. The black mark under the letters " o n " of the word "frontal" is the *olfactory foramen*, where the nerve of smell emerges from the brain-box to go to the nose. The nasal cavity is the blank space behind *nasal* and covered by that bone, and in the oval blank before it. The parts of the beak covered by horn are only *premaxillary, nasal*, and *dentary*. The *condyle* articulates with the first cervical vertebra; just above it, not shown, is the *foramen magnum*, or great hole through which the spinal medulla, or main nervous cord, passes from the skull into the spinal column. The *basioccipital* is hidden, excepting its condyle; so is much of the *basisphenoid*. The prolongation forward of the basisphenoid, marked "rostrum," and bearing the vomer at its end, is the *parasphenoid*, as far as its thickened under border is concerned. Between the fore end of the pterygoid and the basisphenoidal rostrum, is the site of the *basipterygoid process*, by which the bones concerned articulate by smooth facets; further forward, the palatines ride freely upon the parasphenoidal rostrum. In any Passerine bird, the *vomer* would be thick in front, and forked behind, riding like the palatine upon the rostrum. The palatine seems to run into the maxillary in this view; but it continues on to premaxillary. The *maxillo-palatine* is an important bone which cannot be seen in the figure because it extends horizontally into the paper from the maxillary about where the reference line "maxillary" goes to that bone. The general line from the condyle to the end of the vomer is the *cranial axis, basis cranii*, or base of the cranium. This skull is widest across the post-frontal; next most so across the bulge of the jugal bar.

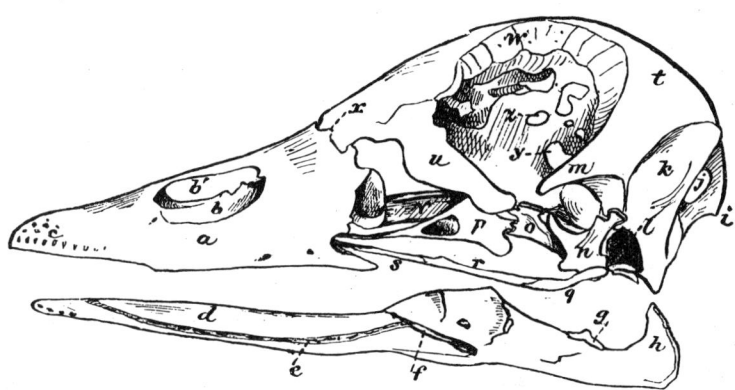

FIG. 63. — Skull of a duck (*Clangula islandica*), nat. size; Dr. R. W. Shufeldt, U. S. A. *a*, premaxillary bone; *b*, partly ossified internasal septum; *b'*, pervious part of nostril; *c*, end of premaxillary, perforated for numerous branches of second division of the fifth cranial nerve; *d*, dentary bone of under mandible; *e*, groove for nerves, etc.; *f*, a vacuity between dentary and other pieces of the mandible; *g*, articular surface; *h*, recurved "angle of the jaw;" *i*, occipital protuberance; *j*, vacuity in supraoccipital bone; *k*, muscular impression on back of skull; *l* is over the black ear-cavity; *m*, post-frontal process; *n*, quadrate bone; *o*, pterygoid; *p*, palatine; *q*, quadrato-jugal; *r*, jugal; *s*, maxillary; *t*, fronto-parietal dome of the brain-cavity; *u*, the lacrymal bone, immense in a duck, nearly completing rim of the orbit by approaching *m*; *v*, vomer; *w*, supra-orbital depression for the nasal gland (see p. 163); *x*, cranio-facial hinge; *y*, optic foramen; *z*, etc., interorbital vacuities.

Development of the Fowl's Skull (figs. 64 to 69). — In the chick's head cartilage is formed along the floor of the skull by the fifth day of incubation. This cartilaginous basilar plate is formed on each side of th *notochord*, fig 64, *c* (Gr. νῶτον, *noton*, back ; χορδή, *chorde*, a chord), a rod-like structure, the primordial axis of the body, around which, along the spinal column, the bodies of the vertebræ are formed, and which runs in the middle line of the floor of the skull as far as the *pituitary space, pts*. The basilar plate is the *parachordal* (Gr. παρά, *para*, by the side of) cartilage. In this, at the earliest stage, are already planted certain parts of the ear, the *cochlea, cl*, (Lat. *cochlea*, a snail-shell), and the horizontal one of the three *semi-circular canals, hsc*. Opposite the end of the notochord, the border of the parachordal plate is notched, 5 ; this notch afterward forms the *foramen ovale*, for the passage of parts of the *fifth* or *trifacial* nerve. Near the middle line, posteriorly, the plate is perforated for the passage of the twelfth or *hypoglossal* nerve, *q*. At each lateral corner is the separate *quadrate* cartilage, to form the quadrate bone. Anteriorly, the plate connects by a strap or bridge of cartilage, the *lingula, lg* (Lat. *lingula*, a little tongue) with the *trabeculæ, tr* (Lat. *trabecula*, a little beam), which enclose the *pituitary space, pts* (Lat. *pituita*, mucus : no applicability here). In front of this pituitary interval the trabeculæ come together to form an *inter-*

nasal plate, which is so arched over downward as to disappear from this view, as seen in fig. 65, where *fn* is the fronto-nasal process, and *n* is the future external nostril. After uniting in the inter-nasal plate, the fore ends of the trabeculæ separate and become free ; their free ends are the under extremities of this *first visceral arch* (first and only pre-oral arch).

The same chick's head, now viewed from below, fig. 65, shows the squarish aperture, *m*, of the future mouth ; the three post-oral arches, with their respective cartilaginous bars, out of which are to be formed the bones of the jaws and tongue. 1, 2, 3, are the corresponding *visceral clefts*, between the arches ; the first of these is to be modelled into the ear-passages (outer and middle ear and eustachian tube) ; the others will disappear. The *quadrate cartilage*, *q*, is the same that was seen in fig. 64; it is already nearly in position, between the hind ends of the scaffolding of the upper and under jaw. The curved *subocular* or *maxillo-palatine* bar, *mxp*, developed in the first post-oral arch, already indicates anteriorly *palatine*, *pa*, and posteriorly, *pterygoid*, *pg*, parts ; it will form the bones so named, and others of the

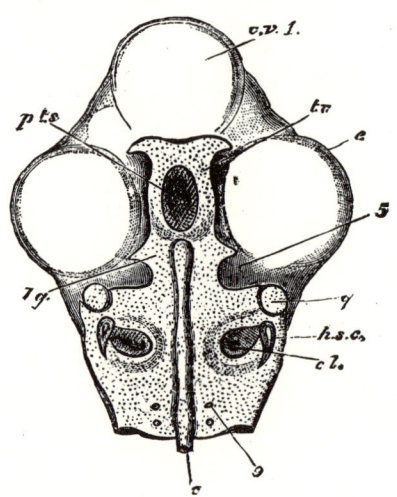 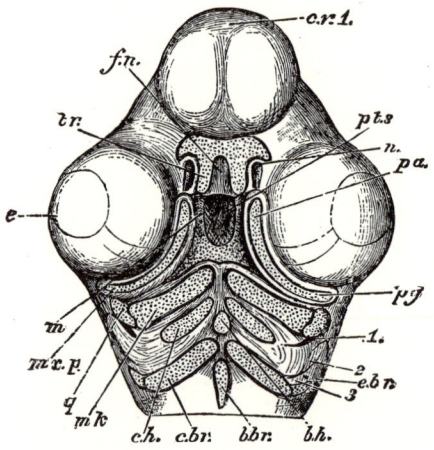

FIG. 64. — Skull of chick, fifth day of incubation, ×9 diameters. Seen from above, the membranous roof of the skull and the brain removed. *cv*1, anterior cerebral vesicle ; *e*, eye ; *c*, notochord, running through the middle of the basilar plate or parachordal cartilage, in which are already visible the rudimentary ear-parts, *cl*, the cochlea, *hsc*, the horizontal semicircular canal ; *pts*, the pituitary space, bounded by *tr*, the trabeculæ, which come together before it to form the fronto-nasal plate, *fn*, in fig. 65; *lg*, *lingula* or bridge connecting trabeculæ with parachordal cartilage ; 5, notch afterward becoming foramen ovale for passage of parts of the fifth (trifacial) nerve ; 9, foramen for hypoglossal nerve ; *q*, separate cartilage forming the future quadrate bone. (After Parker, in *Ency. Brit.*)

FIG. 65. — Same as fig. 64, but seen from below. *cv*1, anterior cerebral vesicle ; *e*, eye ; *m*, mouth ; *pts*, pituitary space; *fn*, fronto-nasal plate ; *tr*, ends of the trabeculæ, free again after their union and bent strongly from the original axis of the trabeculæ ; *n*, external nostril ; *mxp*, subocular bar of cartilage, or pterygo-palatine rod, to form *pa*, palatine, and *pg*, pterygoid bone, and other parts of the upper jaw, as the maxillary, jugal and quadrato-jugal ; *q*, quadrate cartilage, same as seen in fig. 64; *mk*, meckelian cartilage, to form lower jaw ; these parts are in the first post-oral visceral arch ; *ch*, cerato-hyal, and *bh*, basihyal, of second post-oral arch ; *cbr*, cerato-branchial, *ebr*, epi-branchial, *bbr*, basi-branchial, of third post-oral arch ; the parts of the second and third arch all going into the yoid bone. 1, 2, 3, 1st, 2d, 3d visceral clefts, whereof the 1st is to be modified into the ear-passages, and the others are to be obliterated. (After Parker.)

upper jaw. This subocular bar is an antero-superior part of the first post-oral arch, of which *q* and *mk* are a postero-inferior portion ; the cleft of the future mouth is to lie between them. The lower jaw bone, or *mandible*, is entirely developed from *mk*, its several bones developing around this rod of cartilage, the *meckelian* cartilage ; it is to become movably articulated with the bone, the *quadrate*, into which *q* will be transformed. Thus the postero-inferior part of the first post-oral arch (second of the whole series of arches) begins in two pieces, one of which is to become the *suspensorium*, or suspender of the mandible, and the other the mandible itself. The rest of the pieces belong to the *second* and *third* post-oral arches, and all together make up the very composite *hyoid* bone, or bone of the tongue (figs. 72, 73, 74). The pieces *ch* and *bh* are in the *second* arch, and form respectively the *ceratohyal* and *basihyal* bones; the pieces *cbr*, *ebr*, and *bbr* are in the third arch, and form respectively the *ceratobranchial*, *epibranchial* and *basibranchial* bones. These pieces of the third arch have already outgrown those of the second arch, and they will form the greatest part of the hyoid bone.

In the *second stage*, after the fifth day of incubation, but before any ossification has begun, a vertical section shows the appearances represented in fig. 66. The parachordal and trabecular cartilages are applied to each other unconformably, the latter rising high between second and third cerebral vesicles to form the posterior pituitary wall, *pcl*, in which the axial skeleton properly ends. There are other changes in the parachordal cartilages. The internasal plate, formed by the union of the trabeculæ in front of the pituitary space, has become a vertical median wall between the olfactory and optic chambers of the right and left sides (*pn* and *eth*, to *ps* and *alc*). This partition, besides forming finally the *interorbital septum* which divides the right and left orbits, will undergo further notable changes in direction, and will develop lateral plates and processes, which will make up the nasal labyrinth and the partition between the cavity of the nose and that of the eye, when any exists. Such lateral developments of the ethmoid plate are the *aliethmoid*, *aliseptal*, and *alinasal*. This plate extends backward in mid-line to the optic foramen, 2, ending in the *anterior clinoid* wall, *asc*, separated from the (parachordal) *posterior clinoid* wall by the original pituitary space, now the opening through which the carotid arteries, *ic*, enter the brain cavity. Besides ethmoidal parts proper, the plate develops at what will be the end of the upper beak a *prenasal cartilage*, *pn*, to become the axis of the beak. The mouth is become already better formed, the axis of its cavity pointing more forward than downward; and great changes are undergoing in parts of the ear at the back corner of the mouth. The quadrate and meckelian cartilages are assuming much of their true form. The quadrate develops an *orbital* process, which extends free into the orbit, and an *otic* process which articulates with the auditory sac and parts of the exoccipital cartilage. The relations at

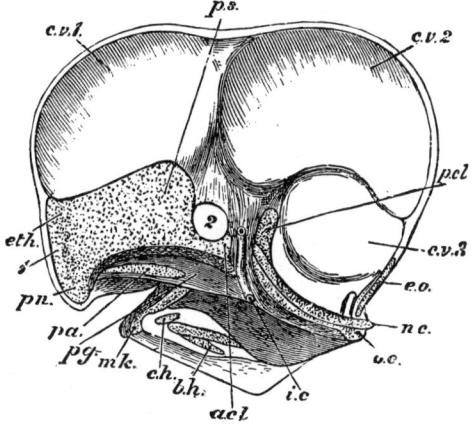

FIG. 66. — Head of a chick, second stage, after five days of incubation, section in profile; × 6 diameters. *cvl*, *cv2*, *cv3*, first, second, and third cerebral vesicles; 1, place of the first nerve, the olfactory; 2, place of second nerve, the optic; *ic*, internal carotid artery, running into skull at what was originally the pituitary space, now an opening bounded in front by the anterior, *acl*, behind by the posterior, *pcl*, clinoid walls; *nc*. notochord; *oc*, occipital condyle, thence to *pcl* being the original parachordal cartilage, here seen in profile; *eo*, exoccipital; *eth*, ethmoid, with *ps*, its presphenoid region posteriorly, and *pn*, pre-nasal part; this whole plate afterward developing into parts of the nose and the partition between the eyes; *pa*, palatine; *pg*, pterygoid region; *pa* and *pg* reference lines are in the chick's mouth; *mk* meckelian cartilage (lower jaw); *ch* and *bh*, ceratohyal and basihyal parts of the hyoid or tongue bone. (After Parker.)

this stage have not been made out in the fowl, but are figured and described from the corresponding stage of the European house martin (*Chelidon urbica*). In fig. 67, *mk* is the cut stump of the meckelian cartilage, of which *ar* is the articular part; *q* is the quadrate, of which a backward process is seen articulating with *teo*, the tympanic wing of the exoccipital. Just below and behind this otic process of the quadrate, exactly where in riper embryos is the *fenestra ovalis* in which is fitted the foot of the *stapes* or stirrup-bone of the middle ear, there appears a trowel-shaped projection of cartilage, the handle of which is continuous with the substance of the ear-capsule; the sickle-shaped piece behind which is the tympanic wing of the exoccipital (*teo*). This trowel of cartilage is the upper anterior segment of the hyoidean (second post-oral) arch, being to that arch what the pterygo-palatine bar is to the mandibular (first post-oral) arch. Several parts of this *stapedial* cartilage are recognized, as named in the fine print under the figure. If the connections of the second post-oral arch were completed, as those of the first are, the tongue bone would be slung to the skull as the lower jaw is; but they are not, the tract represented by the dot-line from the *stylo-hyal, sth*, to the *cerato-hyal, chy*, being, like *ist*, above *sth*, only soft connective tissue. This defect of connection is made up for by the great development of the hyoidean parts of the third post-oral arch, *br* 1 and *br* 2, which retain the tongue-bone in position, without however articulating it with the skull. The hand of the trowel of cartilage soon segments itself off from the ear-capsule, bringing away with it a small oval piece of the periotic wall, which piece is the true stapes, and the oval space in which it fits is the *fenestra ovalis* leading into the inmost ear (the *cochlea*). The broad part of the trowel-blade is the extra-stapedial part, on which the *membrana tympani*, or ear-drum, will be stretched. The stylo-hyal, *sth*, will join the extra-stapedial plate, and the afterward chondrified band of union will be the *infra-stapedial, ist*. (Figs. 71, *st*, and 83.)

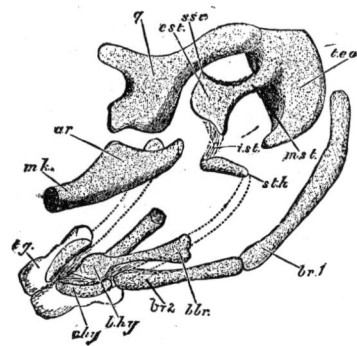

FIG. 67.—The post-oral arches of the house martin, at middle of period of incubation, lateral view, × 14 diameters. *mk*, stump of meckélian or mandibular rod, its articular part, *ar*, already shapen; *q*, quadrate bone, or suspensorium of lower jaw, with a free anterior orbital process and long posterior otic process articulating with the ear-capsule, of which *teo*, tympanic wing of occipital, is a part; *mst, est, sst, ist, sth*, parts of the suspensorium of the third post-oral arch, not completed to *chy*; *mst*, medio-stapedial, to come away from *teo*, bringing a piece with it, the true *stapes* or *columella auris*; the oval base of the stapes fitting into the future *fenestra ovalis*, or oval window looking into the *cochlea*; *sst*, supra-stapedial; *est*, extra-stapedial; *ist*, infra-stapedial, which will unite with *sth*, the stylo-hyal; *chy* and *bhy*, cerato-hyal and basi-hyal, distal parts of the same arch; *bbr, br* 1, *br* 2, basi-branchial, epi-branchial and cerato-branchial pieces of the third arch, composing the rest of the hyoid bone; *tg*, tongue. (After Parker.)

Returning now to the chick's head, which we left to examine the intricate ear-parts at the proximal end of the second post-oral arch, we see by fig. 68 how rapidly the parts are shaping themselves at the end of this second stage of development. This figure shows the cartilaginous skull, in which no trace of ossification has appeared, excepting in the under mandible. The brain and membranous parts of the cranium have been removed. The roof of the skull never becomes cartilaginous, bone there growing directly from the membrane; and the whole of the chondro-cranium, as shown in the figure, is one continuous cartilaginous structure (like the whole skull of an adult shark or skate), excepting the parts of the post-oral arches, which are separate. The auditory capsule is environed by occipital cartilage, *eo*, stretching over the back of the skull, and by wing-like growths (*alisphenoids, as*) which wall most of the brain-box in front. The high orbito-nasal septum is a continuous vertical plate of cartilage, upgrowing from the tract of the conjoined trabeculæ. Lateral developments of this ethmoidal wall, in

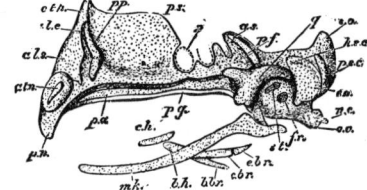

FIG. 68.—Skull of chick, second stage, in profile, brain and membranes removed to show cartilaginous formations, × 4 diameters. *eth*, ethmoid, forming median nose-parts and inter-orbital septum; developing lateral parts, as *ale*, aliethmoid, *als*, aliseptum, *aln*, alinasal, *pp*, partition between nose and eye; *pn*, prenasal cartilage; *ps*, presphenoidal part of mid-ethmoid; 2, optic foramen; *as*, alisphenoid, walling brain-box in front; *pf*, post-frontal, bounding orbit behind; *pa, pg*, palatine and pterygoid; *q*, quadrate; *so*, supra-occipital; *eo*, ex-occipital; *oc*, occipital condyle, borne upon basi-occipital, and showing *nc*, remains of notochord; these occipitals bound the foramen magnum, and *eo* expands laterally to form a tympanic wing, circumscribing the external auditory orifice behind and below; *hsc, psc*, horizontal and posterior vertical semicircular canals of ear; *fr, st*, fenestra rotunda and fenestra ovalis, leading into inner ear, latter closed by foot of the stapes; *mk, ch, bh, bbr, cbr, ebr*, parts of jaw and tongue, as named in figs. 65, 66 and 67. (After Parker.)

front, are divided into several recognizable parts, *ale*, *als*, *aln*, the latter being the external nostril; *pp* is a transverse partition between the orbital and nasal chambers. The nasal cartilages ultimately become much convoluted to form the nasal labyrinth, among the convolutions of which will be the superior and inferior turbinal cartilages, in addition to those already noted. The ethmoidal wall ends behind at *ps*, the presphenoidal region, where the brain case begins; below and behind, it is deeply notched for the *optic* foramen, 2. The pituitary space forms a circular foramen, through which the carotid arteries enter. The site of the orbit of the eye is bounded behind and below by the postfrontal process of the alisphenoid wing, *pf* of *as*. The pterygo-palatine rod is seen along the under border of the skull, *pg* and *pa*. The quadrate, *q*, has acquired nearly its shape, and the rest of the mandibular and hyoidean parts are clearly displayed, *mk*, etc. The proximal hyoidean element, *st*, is freed from the periotic cartilage, leaving the fenestra ovalis (see last paragraph). Below the general outline, *pa* to *oc*, is not shown a mat of soft tissue, in which are to be developed the *basitemporal* and *parasphenoid* bones which underfloor the whole skull, — the former making a plat between the ears, fig. 69, *bt*, the latter forming the thickened under edge of the *rostrum* of the skull *rbs*.

At the third stage, about the middle of the second week of incubation, the cartilaginous parts already described are neatly finished, and the skull is beginning to *ossify*. The occipital parts are well formed; the condyle is perfect; the foramen magnum is circumscribed by the ex- and supra-occipitals, *eo* and *so*, fig. 69. Investing bones, formed in membrane without previous cartilage, are becoming apparent. The basitemporal, *bt*, and parasphenoid, *rbs*, are engrafting upon the base of the skull. The *prenasal cartilage, pn*, now at its fullest growth, is beginning to decline; on each side of it is formed a three-forked bone, the premaxillary, *px*, having superiorly nasal, and laterally palatal and dentary processes. This bone is to grow to great size, forming most of the upper beak, and starving out the maxillary, which in mammals is the principal bone of the upper jaw. The palatal, *pa*, and pterygoid, *pg*, bones are ossified, and the quadrate, *q*, is ossifying. Between the premaxillary and the quadrate are the bones forming the *zygoma*, or jugal bar, developed in the outer part of the maxillo-palatine bar of the earlier embryo. They are the weak *maxillary*, *mx*, with its ingrowing process, the *maxillo-palatine* bone, *mxp;* next the *jugal*, *j;* then the *quadrato-jugal*, *qj;* the

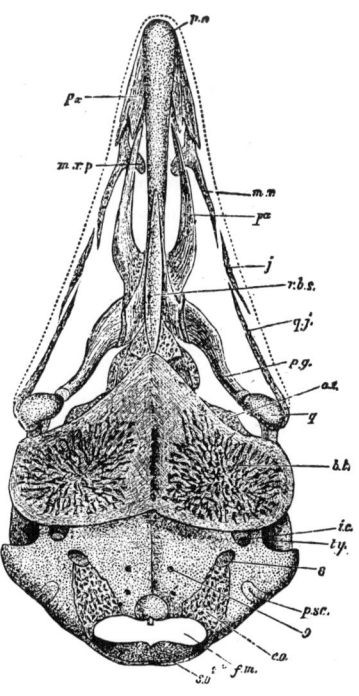

FIG. 69. — Skull of chick, third stage, viewed from *below*, × 6¾ diameters. *pn*, prenasal cartilage, running behind into the septum nasi ; on each side of it the premaxillary, *px*, of which the (inner) palatal and (outer) dentary processes are seen (the upper nasal process hidden) ; *mx*, the maxillary, developing inner process, the maxillo-palatine, *mxp ; pa*, the palatal, well-formed, articulating behind with *rbs*, the sphenoidal rostrum, its thickened under border, the parasphenoid; this will bear the vomer at its end when that bone is developed; *j*, jugal, joining *mx* and *qj*, the quadrato-jugal, joining *j* and *q*, the quadrate ; *mx* to *q*, the jugal bar or zygoma ; *pg*, the pterygoid, making with *pa* the pterygo-palatine bar, joining *q* and *px ; bt*, the basitemporal, great mat of bone from ear to ear, underflooring the skull proper, as *rbs*, a similar formation, does further forward; *ic*, outer end of carotid canal, to run between the *bt* plate and true floor of skull, and enter brain cavity at original site of pituitary fossa (figs. 64, 66, *ic*); *ty*, tympanic cavity — external opening of ear; *as*, alisphenoid, bounding much of brain-box anteriorly, and orbital cavity posteriorly; *psc*, posterior semicircular canal of ear, in opisthotic bone, which will unite with the spreading *eo*, exoccipital, which will reach the condyle shown in the middle line, above the foramen magnum, *fm*, completed above by *so*, supra-occipital; 8, foramen lacerum posterius, exit of pneumogastric, glosso-pharyngeal and spinal accessory nerve; 9, exit of hypoglossal nerve, in basi-occipital. (After Parker.)

whole forming an outer lateral rod from quadrate to premaxillary, like a duplicate of the pterygo-palatine rod from the same to the same.

Among occurrences of later stages are to be noted the development in membrane in the middle line below of the *vomer*, borne upon the end of the rostrum; the roofing in of the whole skull by the *parietal, squamosal, frontal* and *nasal* bones; the completion of the *periotic bones* as the *proötic, epiotic* and *opisthotic*, which form the *otic capsule*; the development of *lacrymal bones*, bounding the orbits of the eyes in front. Absorption of the middle wall of cartilage between the nasal and orbital cavities nicks off the nose parts from those of the orbit (fig. 70, between *ntb* and *eth*); and certain changes in the orbital septum develop the *orbito-sphenoids*. Very nearly all the bones of a bird's skull having thus been accounted for, we may next consider them in their adult condition. Reference should now be made to figs. 62, 63, 70, 71.

The Occipital Bone (fig. 62, 70, 71) forms the back part of the floor of the skull, and lower part of the back wall of the skull; neither its boundaries nor its composition is visible in adult skulls. It is formed by the *basioccipital, bo*, below in the middle line; the *supra-occipital so*, above in the middle line; the *exoccipital, eo*, on either side. These bound the *foramen magnum* (fig. 69, *fm*), where the nerve mass makes its exit from the cavity of the cranium into the tube of the spinal column. At the lower part of the foramen is the protuberant occipital condyle (figs. 68, 71, *oc*), borne chiefly upon the basioccipital, but to the formation of which the exoccipitals also contribute; the latter flare widely on each side, into the tympanic wings, which bound the external auditory meatus behind. The true basioccipital is mostly covered by the underlying secondary bone, the *basitemporal* (69, 70, *bt*), which extends from one tympanic cavity to the other, and more or less forward in the middle line to the sphenoidal rostrum. Openings to be observed in the occipital region, besides the great foramen, are those for the hypoglossal nerve, 9, near the condyle; for the parts of the vagus nerve, 8, more laterally, and the carotid canal, *ic*: also, above the foramen magnum, openings for veins, sometimes of great size, as in fig. 63, *j*.

The Parietals (figs. 62, and 70, *p*, 71). — Proceeding up over the brain-box, the next bones are a pair of parietals, between the occipital behind, the frontal before, and the squamosal beside; but their limits are rarely if ever to be seen in adult skulls. They are relatively small in birds; simply squarish plates, bounded as said, coming together in the midline.

The Frontals (fig. 62, and 70, *f*, 71), originally paired, soon fuse together, and with surrounding bones of the skull, though maintaining some distinction from those of the nose and jaw. These roof over much of the brain cavity, close in much of it in front, and form the roof and eaves of the great orbital sockets. Anteriorly in the middle of the forehead line the feet of the nasal process of the premaxillary are implanted upon the frontal, usually distinctly; more laterally, the nasal bones are articulated or anchylosed; this fronto-naso-premaxillary suture forming the fronto-facial hinge, (fig. 63, *x*) by the elasticity or articulation of which the upper jaw moves upon the skull, when acted on by the palatal and jugal bars. In the midst of the forehead the two halves of the frontal sometimes separate, as they do in the fowl, allowing a little of the mesethmoid to come to the front. In the middle line, underneath, the frontals fuse with whatever extent there may be of the mesethmoid which forms the lengthwise inter-orbital septum, and often a crosswise partition between the orbital and nasal cavities. To the antero-external corners of the frontal are articulated or anchylosed the lacrymals. The *post-frontal process*,[1] morphologically the post-frontal or sphenotic bone, bounds the rim of the orbit behind;

---

[1] There is apparently some ambiguity in the use of the term "post-frontal" process by different authors. It would appear that this process, bounding the rim of the orbit behind, may be a projection of the frontal bone, and therefore strictly a post-frontal process. Or that, as said by Owen for *Rhea*, it may be a separate bone, and there-

it is usually quite prominent. The frontal rim of the orbit in many birds shows a crescentic depression (very strong in a loon and many other water birds; fig. 63, *w*), for lodgment of the **supra-orbital gland, the secretion of which lubricates the nasal passages.** The cerebral plate of the frontal is often imperfectly ossified, showing large "windows" besides the regular openings for the exit of nerves which are always found at the back of the orbit. View from above, the frontal is vaulted and expanded behind, over the brain cavity, then pinched more or less, sometimes extremely narrow over the orbits, then usually somewhat expanded again at the fronto-facial suture. The extent of the frontal between the orbits and face, in the lacrymal region, is very great in the duck family, as seen in fig. 63.

The Squamosal (Lat. *squama*, a scale ; figs. 70, 71, *sq.*) bounds the brain-box laterally, between occipital, parietal, frontal and sphenoidal bones, its distinction from all of these being obliterated in adult life. It is situated near the lower back lateral corner of the skull, forming some part of the cranial wall just over the ear-opening, and a strong eaves for that orifice. It is firmly united also to the bones of the ear proper, and receives the larger share of the free articulation which the quadrate has with the skull. It often develops a strong forward-downward spur, the squamosal process (fig. 62), looking like a duplicate post-frontal process ; between these two is the *crotaphyte depression*, corresponding to the "temporal fossa" of man, in which lie the muscles which close the jaws. It scarcely or not enters into the orbit, the adjacent part of the orbit being alisphenoidal.

The Periotic Bones (Gr. περί, *peri*, about; οὖς, ὠτός, *ous, otos*, the ear; fig. 70) are those that form the *petrosal bone* (Lat. *petrosus*, rocky, from their hardness), or bony periotic capsule, containing the essential organ of hearing. When united with each other and with the squamosal, they form the very composite and illogical bone called "temporal" in human anatomy. There are three of these otic bones, — an anterior, the *pro-otic;* a posterior and inferior, the *opisthotic* (Gr. ὄπισθε, *opisthe*, behind) and a superior and external, the *epiotic*. They can only be studied in young skulls, upon careful dissection ; they do not appear upon the outside of the skull at all, excepting a small piece of the opisthotic, which there fuses indistinguishably with the exoccipital. But somewhat of these bones are seen on looking into the cavity of the outer ear, and if the fenestra ovalis can be recognized, it determines a part of the boundary between the proötic and opisthotic bones, while the fenestra rotunda lies wholly in the latter. The cavity of the periotic bone is hollowed for the labyrinth of the internal ear, including the cochlea, which contains the essential nervous organs of hearing, and the three semicircular canals — so much of them as does not invade surrounding bones. In the young fowl's skull viewed internally (fig. 70), Parker figures a very large proötic portion (*po*) of the periotic, perforated by the internal auditory meatus (7) for the entrance from the brain of the auditory nerve ; below and behind the proötic a small opisthotic (*op*), in relation with the exoccipital, upon the surface of which it also appears, outside (fig. 69, at *psc*), and with which it blends; a very small epiotic centre (*ep*), between the proötic and supraoccipital ; and the anterior semicircular canal (*asc*) embedded in the latter. In Dr. Shufeldt's figure the otic elements are merely noted diagrammatically. According to Huxley's generalization, the epiotic is in special relation with the posterior semicircular canal; the proötic with the anterior vertical canal, between which and the foramen ovale (5) for the lower divisions of the trifacial nerve it lies. That part on which the inner foot of the quadrate is implanted is proötic. Below the drooping eaves of the squamosal, before the flaring wing of the exoccipital, and behind the quadrate bone, is the always decided and considerable cavity of the ear, bounded pretty sharply by the squamosal and exoccipital rim,

fore properly a *post-frontal bone*. Or, again, that it may have nothing to do with the frontal bone, but belong to the alisphenoid, as a process of the latter or a separate ossification; in which case it would be properly the *sphenotic*. In no event has it anything to do with the *squamosal* process lettered as such in fig. 62.

sloping with less distinction in front toward the orbital cavity. In this auditory hollow may be seen several openings: the *meatus* or proper ear-passage, through which, in one direction, a bristle may be passed to emerge at or near the middle line of the base of the skull, about the root of the basisphenoidal rostrum. Such a passage is through the *first visceral* cleft of the early embryo, modified into *meatus auditorius* and *eustachian tube*, which latter communicates with the back part of the mouth. Besides the other ear-passages proper, may be found other openings of air-passages leading into the interior diploic tissue of bones of the skull, and especially into the lower jaw bone. The ear-parts are immensely developed in owls, in many species of which they are unsymmetrical, that is, not sized and shaped alike on right and left sides of the head.

The **Sphenoid** (Gr. σφήν, *sphen*, a wedge; εἶδος, *eidos*, form; figs. 62, 70, 71) is a compound bone, not easy to understand as it occurs in birds, as much of it is hidden from the outside, some of it is very slightly developed, and all of it is completely consolidated with surrounding bones in the adult. It is wedged into the very midst of the cranial bones proper, with its body in the middle line below, next in front of the basioccipital, and its wings spread on either side in the orbital cavity. A sphenoid consists essentially of the *basisphenoid*, or main part of the bone (fig. 62); the *alisphenoids* or "wings," on either side (figs. 70, 71, *as*); the obscure *presphenoid*, (*ps*) in the middle line in front of and above the main body; and the small *orbito-sphenoids*, which are in fact the wings of the presphenoid. The body is usually covered in by the underflooring of the basitemporal; it is a flat triangular plate, produced more or less forward in the middle line as the *basisphenoidal rostrum*, or beak of the skull. This *rostrum* is an important thing. It forms, in fact, the central axis of the base of the skull; with the mesethmoid plate the inferior border of the interorbital septum, usually

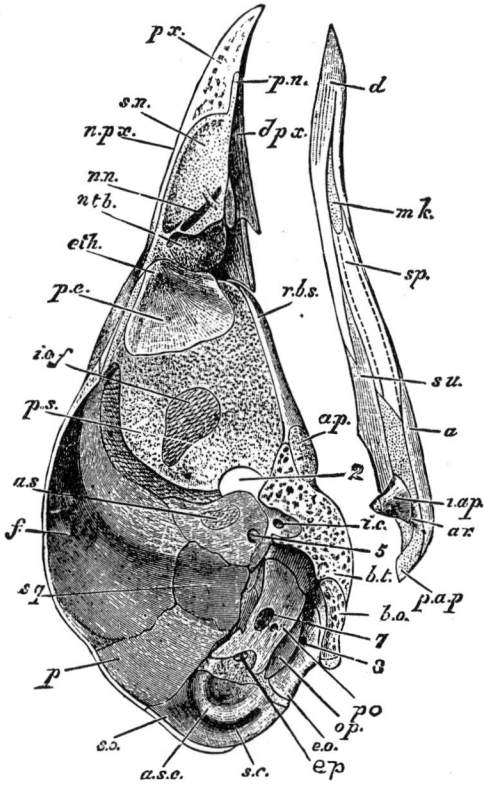

FIG. 70. — Ripe chick's skull, longitudinal section, viewed *inside*, × 3 diameters; after Parker. In the mandible are seen: *mk*, remains of meckelian rod; *d*, dentary bone; *sp*, splenial; *a*, angular; *su*, surangular; *ar*, articular; *iap*, internal articular process; *pap*, posterior articular process. In the skull: *pn*, the original prenasal cartilage, upon which is moulded the premaxillary, *px*, with its nasal process, *npx*, and dentary process, *dpx*; *sn*, septo-nasal cartilage, in which is seen *nn*, nasal nerve; *ntb*, nasal turbinal; the reference line crosses the *cranio-facial suture*, the face parts and cranial parts being nearly separated here by the nick seen in the original cartilaginous plate; *eth*, ethmoid; *pe*, perpendicular plate of ethmoid, which will spread nearly throughout the dotted cartilaginous tract in which it lies, to form nearly all the interorbital septum; transverse thickening (in some birds) below the reference line *eth* will form the pre-frontal, or orbito-nasal septum; *iof*, inter-orbital foramen; *ps*, pre-sphenoidal region, just above which is the orbito-sphenoidal region; 2, optic foramen; *as*, alisphenoid, with 5, foramen for divisions of the 5th (trifacial) nerve; *f*, frontal; *sq*, squamosal; *p*, parietal; *so*, superoccipital; *asc*, anterior semicircular canal; *sc*, a sinus (venous canal); *ep*, epiotic; *eo*, exoccipital; *op*, opisthotic; *po*, proötic, with 7, meatus auditorius internus, for entrance of 7th nerve; 8, foramen for vagus nerve; *bo*, basioccipital; *bt*, basitemporal; *ic*, canal (in original pituitary space; fig. 66 *ic*) by which carotid artery enters brain cavity; *ap*, basipterygoid process; *ap* to *rbs*, rostrum of the skull, being the parasphenoid bone underflooring the basisphenoid and future perpendicular plate of ethmoid. (The scaffolding of the upper jaw not shown, excepting *px*, &c.).

THE ANATOMY OF BIRDS. — OSTEOLOGY. 165

thickened by the underflooring of the *parasphenoid* (fig. 70, *rbs*). The rostrum often bears on each side a *basipterygoid process* (*ap*), — a smooth facet with which the pterygoid articulates. These processes may be very strong, and far back on the basisphenoid body, when the pterygoids articulate with them near their own posterior ends, as in the struthious birds and tinamous (fig. 75, *btp*) ; or they may be further along on the rostrum, and the pterygoids then articulate near or at their fore-ends. The rostrum may be produced far forward, beyond the maxillo-palatines and vomer even, as in an ostrich ; or it may bear the vomer at its end ; or may be embraced by forks of the vomer ; the palatines may glide along it, or be remote from it on either side. In any event, whatever its production, whatever part may be ethmoidal, or basisphenoidal, or parasphenoidal thickening, pterygo-faceting, etc., this "beak" of the basisphenoid is always in the axis of the base of the skull, and at the bottom of the interorbital plate ; it may be horizontal, or obliquely ascending forward ; and the variety of its relations with the pterygopalatine and vomerine mechanism furnishes important zoölogical characters, as we shall see when we come to treat of palatal structure particularly. Just at the base of the beak, where it widens into the main body of the bone, may commonly be seen, coming from between the sphenoidal body and the lip of the basitemporal underflooring, the orifices of the eustachian tubes, and often also the anterior ends of the carotid canal. If a bristle, passed into a questionable foramen here, comes out of the ear, it

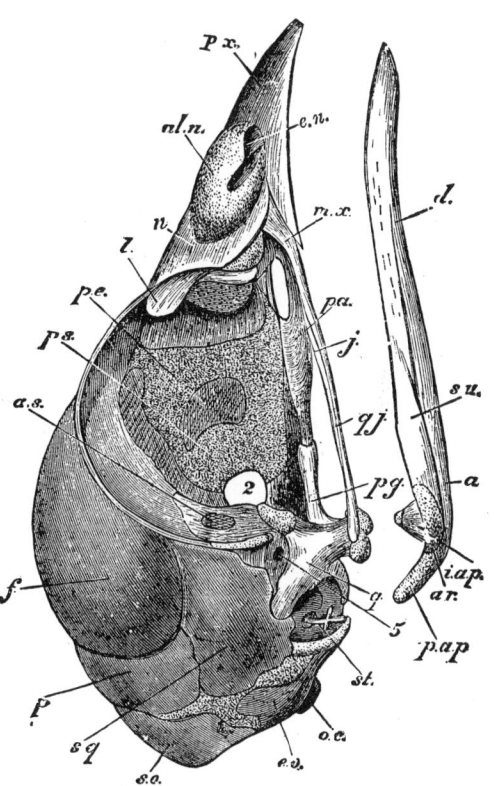

FIG. 71. — Ripe chick's skull, in profile, × 3 diameters; after Parker. *px*, premaxillary; *aln*, ali-nasal cartilage; *en*, septonasal: *n*, nasal bone ; *l*, lacrymal; *pe*, perpendicular plate of ethmoid, as in fig. 70; *ps*, presphenoidal region; *as*, alisphenoid; *f*, frontal; *p*, parietal; *sq*, squamosal; *so*, superoccipital; *eo*, exoccipital; *oc*, occipital condyle; *st*, the cross-like object, the stapes, whose foot fits fenestra ovalis, see fig. 83; *q*, quadrate; *pg*, pterygoid; *qj*, quadrato-jugal; *j*, jugal; *pa*, palatine; *mx*, maxillary. In the mandible: *d*, dentary; *su*, surangular; *a*, angular; *ar*, articular; *iap*, internal angular process; *pap*, posterior angular process. 2, optic foramen; 5, foramen ovale, for inferior divisions of the 5th nerve. (Compare fig. 70.)

has gone through the eustachian tube; if it comes out below the ear, on the floor of the skull, outside, it has run in the carotid canal. The extent of the *alisphenoids* (figs. 70, 71, *as*) cannot be determined in old skulls. They lie at the back lower border of the orbital cavity, closing in most of the brain box that is not foreclosed by the frontal bone. You will *always* find at the back of the orbit, close to the mid-line, and rather low down, the very large *optic foramina* (any figs., 2); alisphenoid should not extend in front of these orifices. A little below and behind the optic foramina, and much more laterally, not far from the quadrate itself, is a considerable foramen, quite constant, for transmission of the inferior divisions of the *fifth* (*trigeminal* or *trifacial*) nerve. This is the *foramen ovale* (any figs., 5) ; it is either *in* the alisphenoid, or between that bone and the proötic ; it must not be mistaken for one of the several smaller holes, usually seen close about the optic foramen, which transmit the nerves (oculo-motor, pathetic, and abducent) which move the muscles of the eyeball; these holes being collectively about equivalent to the *foramen lacerum anterius* of human anatomy. Parts about the optic foramen, before and above, are presphenoidal (figs. 70, 71, *ps*) and orbito-sphenoidal; but they are obscure to all but the embryologist, and practically furnish no zoölogical characters.

The Ethmoid (Gr. ἠθμός, *ethmos*, a sieve; from the way it is perforated in the human species; fig. 62) is the bone of the mid-line of the skull, in front of the sphenoidal elements and below the frontal; it is in special relation with the olfactory nervous apparatus, or sense of smell. This is not an easy bone to "get the hang of" in birds. Referring to figs. 66, 68, *eth*, the student will see in the early embryo a high thin plate of cartilage, the *mesethmoid* cartilage, which is developing lateral processes to form the convoluted walls of the nasal passages. By the uprising and forth-growing of the prenasal cartilage, the mesethmoidal plate is tilted backward, as it were, under the frontal. Next, by absorption of tissue just opposite the future cranio-facial suture, the plate is nicked apart, the portion in front of the nick elaborating the nasal chambers, which usually remain cartilaginous, and the portion behind this nick becoming the permanent plate, fig. 70, *eth*, *pe*, to which the name *mesethmoid* or mid-ethmoid is more strictly applicable. Practically, a bird's ethmoid is chiefly the inter-orbital septum, in vertical mid-line between the orbits, with such flange-like processes or lateral plates as may be developed to form an *orbito-nasal* septum separating the eye-socket from the nose-chamber. In general, the permanent ethmoidal plate becomes nearly coincident with this orbital wall, and pretty well cut off from the osseous or cartilaginous developments, when any, in the nasal cavities. It is then fairly under cover of the frontal, with which, as with the sphenoidal elements posteriorly, it becomes completely fused. When this inter-orbital septum is fully developed, it completely divides the right and left orbital cavities, and its lower horizontal border, fused with the basisphenoidal rostrum, may like the latter be thickened by bearing its share of the parasphenoidal splint. Oftener, however, this lower border slopes upward and forward, from the sphenoidal base to the roof of the skull about the site of the cranio-facial suture; and usually the septum is incomplete, having a membranous fenestra somewhere near its middle (fig. 70, *iof*). Along the upper border of the mesethmoid plate, or just in the crease between it and the overarching frontal may usually be seen a long groove, which, beginning behind at the *olfactory foramen* of the brain-box, conducts the thence-issuing olfactory nerve to the nasal chambers. Sometimes there is another such groove, from a similar foramen near by in the sphenoidal parts, which similarly traces the course of the ophthalmic (first) division of the trifacial nerve. Occasionally, as in the fowls, the two halves of the frontal bone separate a little at the extreme forehead, allowing the mesethmoid plate there to come up flush with the outer surface of the skull.

In some birds, as the low ostrich, for example, the original mesethmoidal cartilage-plate does not nick apart into orbital and nasal moieties, but ossifies as a continuous sheet of bone, dividing right and left halves of the skull far towards the point of the beak (see fig. 75, beyond *R* to *Pmx*). A nasal septum, separated from the orbital septum, may persist to ossify; forming, as in the raven, a vertical plate separate from all surroundings, and liable to be mistaken for a free *vomer* (see fig. 79, where the reference line *v* goes to it, instead of to the truncate vomer); or, as in many birds, a plate variously anchylosed with its surroundings. But these formations, as well as the various *turbinal* (Lat. *turbo*, a whorl) scrolls and whorls formed in this part of the skull, belong rather to the organ of smell than to the skull proper.

The Cranial Bones proper are all those thus far described, excepting the nasal ossifications just noted, which belong to the first pre-oral arch; and the stapedial parts of the ear, which belong to the hyoidean apparatus (second post-oral arch). Intermediate in some respects between the proper cranial bones and

The Facial Bones proper is the Vomer. — By "facial bones," as distinguished from "cranial" bones, is meant the entire bony scaffolding of the upper and lower jaws, and of the tongue, — parts developed in the pre-oral or maxillary, and first, second, and third post-oral, or mandibular, hyoidean proper, and branchial, arches.

The Vomer (Lat. *vomer*, a ploughshare; figs. 62, 63, 75 to 80, *v*) is considered, by those who hold the vertebral theory of the skull, to be the body of the foremost (fourth from behind — the basioccipital, basisphenoid, and presphenoid being the other three) cranial vertebra. So far from having any such morphological significance, it is one of the late secondary bones, developed, if at all, apart from the general make-up of the skull, as a special superaddition underlying the ethmoidal region, as the parasphenoid and basitemporal underlie the skull further back. Its character is extremely variable in the class of birds, though usually constant in the several natural divisions of the class, — a fact which confers high zoölogical value upon this anomalous bone. A vomer is a symmetrical mid-line bone of the base of the skull, found if at all at or near the end of the rostrum. It is originally double, i. e., of right and left paired halves. These halves persist distinct in the woodpeckers, and are remote from each other, one on each side of the mid-line (fig. 80). The vomer is wanting entirely in the Columbine birds, as the pigeons and some of their allies, as the sand grouse (*Pterocletes*) and bush quails (*Hemipodes*) of the old world, and in certain of the true *Gallinæ*. Its connections are various. It may be borne free upon the end of the rostrum. It may be applied like a splint by a grooved upper surface to the under side of the rostrum, and so fixed there; or, in such situation, it may glide along the rostrum according to the movements of the palatal parts with which it may connect. Thus, in the ostrich (fig. 75), it saddles the rostrum below, and is joined by the maxillo-palatines. Or, it may be united with separate ossifications, the septo-maxillaries, which in some birds bridge across the palate (fig. 80). The commonest case is its deep bifurcation behind (fig. 79), each fork uniting with the palate bone of its own side, and sometimes also with the pterygoid. Such is usually the fixture of the bone behind, and it then rides along as well as simply bestrides the rostrum. The anterior end of the vomer may be perfectly free, projecting into the floor of the nasal chambers (figs. 62, 77), or the fore end may be variously steadied or connected with maxillary processes (fig. 78). When free in front, and often when not, the vomer is a simple share-like plate, more or less expanded vertically, quite thin laterally, and "spiked," i. e., running forward to a point; under these circumstances it may or may not bifurcate behind, and be there attached to the palatines or not. But the commonest case of vomer, shown by the great Passerine group, which comprise the majority of recent birds, is different from this, the vomer being in front thickened, flattened and expanded laterally, and connected with nasal cartilages and ossifications (alinasals and turbinals). Such a vomer, deeply cleft behind to join the palatals, is endlessly diversified in the configuration of its fore end, which may be notched, lobed, clubbed, etc. The *general case* of such a vomer is indicated by the expression " vomer truncate in front," as distinguished from the simply pointed or "spiked" vomer. (For further details see description of the several patterns of palate-structure, beyond.)

The Quadrate Bone (Lat. *quadratus*, squared; figs. 62; 63, *n;* 64, 65, 68, 69, 71, *q;* 75, *Qu*), with which we may begin the jaw-bones proper, is the suspensorium of the lower jaw, — the perfectly constant and characteristic bone by means of which the mandible proper articulates with the skull. Its rudiment is seen in the earliest embryos, at the corners of the primordial parachordal cartilages. It belongs to the mandibular (first post-oral) arch, of which it is the proximal element. Its general morphology has caused much dispute. From the fact that in birds one of its functions is to support, in part, the tympanum of the ear, it has been identified with the *tympanic* bone of mammals, — that which in man forms the bony tube of the external auditory meatus. The view now generally accepted is, that the bird's quadrate represents, certainly in part, probably in whole, the little bone of the middle ear called the *malleus* in mammals. Anyhow this may be, the quadrate of a bird bears the proximal ends of *both* jaws, carrying their final (posterior) articulation up to the squamosal and petrosal bones. Thus, the foot of the quadrate forms the free hinge of the lower jaw, and also movably articulates the back end of both the zygomatic and the pterygo-palatine bars or "arcades." The head of the quadrate freely articulates with the squamosal, just in front of the tympanity cavity, which it thus bounds in front; and there is usually a shoulder which furthermore articulates with the anterior periotic bone, the proötic; Struthious birds do not have these two distinct facets. A long *pedicle* or *orbital process* extends forwards, inwards, and upwards in the orbit; this non-articular handle is for advantageous muscular traction. So circumstanced, the quadrate is a stocky bone, of a shape reminding one of an anvil; it rocks freely to and fro upon its cranial socket, pulling and pushing upon the whole maxillary and mandibular mechanism, with such effect that when the lower jaw drops, the zygomatic and palatal bars are automatically shoved forward, tending to make the upper jaw rise, and so increase the opening of the mouth. Such mobility of the upper jaw automatically with the movement of the lower is very free in parrots, whose cranio-facial connections are quite articular in character; it is well shown also in ducks; and probably nearly all birds have some such motion of the upper jaw upon the skull. In nearly all birds, the mandibular articular facet of the quadrate is divided by a lengthwise impression into inner and outer protuberances, or condyles, fitting corresponding depressions on the articular face of the lower jaw; in some birds the articular surface is single. The zygomatic articulation with the quadrate is made by the balled end of the quadrato-jugal socketed in a cup at the *outer* side of the mandibular facet (with various minor modifications in different birds). The palatal articulation is made by a little condyle of the quadrate, at the *inner* side of the main facet, socketed into the cupped end of the pterygoid (with minor modifications).

The Quadrato-jugal and Jugal Bones (Lat. *jugum*, a yoke; figs. 62, 63, *q*, *r;* 69, 71, *qj*, *j*) form most of the *outer* arcade — the *jugal* or *zygomatic* bar — leading from the quadrate bone to the beak. The quadrato-jugal is posterior, reaching a variable distance forward; at its fore end it is obliquely sutured to the jugal, a splint-rod which carries the bar forward to the maxillary bone, with which it is in like manner obliquely sutured. The whole affair is almost always a slender rod, which with its fellow of the opposite side forms the outermost lateral boundary of the skull for a great distance. It corresponds in general with the "zygomatic arch" of a mammal, which is made up of a "zygomatic process of the squamosal" and a malar or "cheek-bone." The whole zygomatic arch, including the maxillary bone itself, is developed from the outer part of the primordial pterygo-palatine bar (see fig. 65). In parrots the zygoma is movably articulated before as behind.

The Maxillary Bone (Lat. *maxilla*, upper jaw bone; figs. 62; 63, *s;* 69, 71, 75, *mx*), forming so much of the upper jaw of a mammal, is in birds greatly reduced, being starved out by the predominant premaxillaries which form most of the upper beak. The shape of this stunted bone varies too much to be concisely described. Its connections are, ordinarily, with the jugal behind by a long slender splint-like process, and with the premaxillary and usually the nasal bones in front and externally. Internally, it may or may not connect with the palatal and vomer. The zoölogical interest of this bone centres in certain inward (palate-ward) processes, often its most conspicuous parts, and apparently corresponding to the plate which in a mammal roofs the hard palate anteriorly. Though these are mere processes from the main maxillary, they are so distinct and important that they are commonly described as if they were independent bones, under the name of the *maxillo-palatines*. They are flange-like or scroll-like plates, or large spongy masses of delicate bone-tissue, — endlessly varied in configuration and context (see the various figures of base of skull, *mxp*, beyond, where the palate-patterns are described).

Certain other inward maxillary processes, which may or may not unite with the vomer, and so bridge over the palate, are called *septo*-maxillaries (fig. 80, *smx*) ; and in some woodpeckers yet other palate-processes appear (fig. 80, *pmx*).

The Pterygoid Bones (Gr. πτέρυξ, *pterux*, wing; εἶδος, *eidos*, form ; figs. 62 ; 63, *o;* 65, 66, 68, 69, 71, 80, *pg;* 75 to 79, *Pt*). Returning now to the quadrate, and going along the inner arcade, we first encounter the *pterygoid*,—a generally rod-like, but variously twisted, crooked, or expanded bone which makes the connection between the quadrate behind and the palate bone before. The pterygoid is always freely jointed at both ends; its posterior quadrate articulation has been noted above; its anterior connection is usually by little nipper-like claws by which it " catches on" to the hind end of the palatine. In the ostrich (fig. 75, *Pt*) the pterygoid expands into a scroll-like plate ; but its rod-like shape is usually preserved. Besides passing very obliquely inward as it goes forward from the wide-apart quadrates to the narrow rostrum in the axis of the skull, the pterygoid often bellies or elbows inwards in its course to join the basisphenoidal beak, and be movably articulated therewith. In the majority of birds, there is no such rostral articulation, or the pterygoid only touches the rostrum at its fore end where it joins the palatal. In many, however, special *articular facets*, called *basipterygoid processes* (fig. 70, *ap*), are developed on the rostrum for the pterygoids to abut against and glide over. In Carinate birds, excepting the tinamous (*Dromæognathæ*), these processes are forward on the beak, and the pterygoids articulate at or near their own fore ends, as well shown in the fowl or duck, figs. 77, 78, *Pt*. In Ratite birds and tinamous, the basipterygoids are very long, flaring transverse processes, far back on the rostrum, at the sphenoidal base, and the pterygoids articulate therewith at or near their own posterior ends (figs. 75, *Btp*, and 76).

The Palatal or Palatine Bones (Lat. *palatum*, roof of the mouth ; figs. 62 ; 63, *p;* 65, 66, 68, 69, 71, 77, 78, 80, *pa; 75*, 76, 79, *Pl*) are a pair, approximately parallel and near the mid-line, forming that part of the " hard palate " or roof of the mouth which is not constructed by the palatal processes of the maxillaries, or vomer. They are nearly always long thin bones, among the most conspicuous parts when the dried skull is viewed from below. Sometimes, as in the ostrich (fig. 75, *pl*), they are remote from the axis of the skull and only connected in front with the maxillaries and maxillo-palatines. In many birds they skip the maxillary parts in going forward to be fused with the premaxillaries ; in most, probably, they form anterior connections in one or another fashion with palatal parts both of maxillaries and of premaxillaries. Behind, they always correctly articulate with the pterygoid. The mid-line connections made in most Carinate birds (not in Dromæognathæ) are variously with the vomer, with the rostrum, with each other, or some or all of these relations at once. A long deeply-cleft vomer may by its posterior forks attach itself to the whole palatal mid-line, excluding the palatals from the rostrum ; less extensive attachment of the same kind may permit the palatals to touch each other and the rostrum posteriorly, while cutting them off anteriorly ; also, a non-cleft vomer may attach itself to the posterior extremity of the palatals, and bear them off the rostrum. The whole hard palate may fuse into an indistinguishable mass ; and in almost any case the relations of the palatals to each other and their connections afford some of the most valuable zoölogical characters of great groups of birds. (Details figured and described beyond.) Though very variable in configuration, as well as in connections, certain parts of a palatal may usually be recognized, and conveniently named for descriptive purposes. Anteriorly, in the great majority of birds, of whatever technical kind of palatal structure, the palatals are simply prolonged as flat strap-like or lath-like bars running past the maxillary to the premaxillary region ; and such simple band-like character may be preserved behind. Ordinarily, however, the palatals expand posteriorly, becoming more or less laminar ; and in this plate-like part three surfaces may usually be recognized. One, more or less horizontal, flaring outward, is the *external lamina*. It is well shown in a Passerine or Raptorial bird, where the *postero-external angle* (between the outer border and the posterior end) of the palatal is well-marked, or may be acutely produced; there is no such lamina in a fowl, where the palatals are for the most part slender and rod-like. An internal, more or less vertically produced, plate to make the mid-line rostral or vomerine connection is the *superior internal lamina*, or *medio-palatine process;* very strong, for example, in a fowl, where it forms all the expanded part of the bone, and ends anteriorly as a sharp *inter-palatine spur*. The medio-palatine is probably to be regarded as the main body of the bone, being the most axial part, of the most extensive and varied connections. A third lip or plate of the palatal is the *inferior internal lamina*, looking downward; it is generally very evident, but in a duck or fowl is reduced to a mere ridge, indicating where the superior internal and external laminæ meet. A duck's palatals are quite different in appearance from those of most birds, all the posterior parts just distinguished being reduced and constricted, while the fore ends, running abruptly into the hard-boned beak, are much expanded horizontally (fig. 78). The postero-external angles of the palatal (formed by the external lamina), even when much produced, may not reach as far back as opposite the pterygo-palatine articulation; or they may surpass these limits, and when they do, such backward prolongation is called *post-palatiue*, the palate being considered to end at the pterygoids. In like manner, the maxillary processes of the palatals, or the palatal strips as prolonged into the premaxillary region, are called *pre-palatines*. The inner posterior process, by which the palatine is articulated with the pterygoid, is its *pterygoid process*.

The Premaxillary Bones (figs. 62; 63, *a;* 69, 70, 71, 80, *px;* 75 to 79, *pmx*), also called **Intermaxillaries**, form most of the upper beak, attaining enormous development in birds, and reversing the usual relative size of premaxillary and maxillary. Mainly determining as they do the form of the upper mandible, their shapes are as various as the bills themselves of birds; but their generalized characters can be easily given. Each premaxillary, right and left, forms its half the bill; the two are always completely fused together in front, commonly preserving traces at least of their original distinction behind. They are commonly called one bone, *the* premaxillary. Each is a triradiate or 3-pronged bone; one upper prong, the most distinct, called the *nasal* or *frontal process*, forms with its fellow the culmen (p. 109, fig. 26, *b*) of the bill. These processes, side by side, run clear up to the *frontal* bone in birds, driving the nasal bones apart from each other. Such a *median* fronto-premaxillary suture, with lateral fronto-nasal and naso-premaxillary sutures, is highly characteristic of birds, — an arrangement probably exceptionless. Two other horizontal prongs on each side, extensively distinct from the frontal process in most birds, but less separate from each other, run horizontally along the side and roof of the mouth for a variable distance. These horizontal prongs are an *external* or *dentary* process (fig. 80, *dpx*), forming the tomium (p. 109) of the bill, and reaching back to join the dentary part of the maxillary; and an *internal* or *palatal process* (fig. 80, *ppx*), running along the commencement of the bony palate. With this latter the anterior ends of the palatal bones unite, — either on the side toward the mid-line of the beak, or between the palatal and dentary processes, as in a woodpecker (fig. 80). Great laminar expansions inward of these palatal parts of the premaxillaries roof the hard part of the mouth anteriorly, though there is usually a vacancy between the premaxillary hard palate and that formed farther back by the maxillo-palatines and palatines. The posterior extremities at least of the frontal processes of the premaxillaries are commonly distinguishable from each other, as well as from the frontal and nasal bones — in fact, these fronto-naso-premaxillary sutures are among the most persistent of all. The divergence of the frontal from the palatal and dentary processes bounds the external nostril in part, the circumscription of that orifice being completed by the prongs of the nasal bones. The superficies of the premaxillary bone, like that of the dentary piece of the lower jaw bone, is commonly sculptured with the impressions of the vessels and nerves which ramify beneath the horny integument; and in birds with very sensitive bills, as a snipe or duck, the end is perforated sieve-like with little holes, into which the skin shrinks in drying, producing the familiar "pitted" appearance (fig. 63, at *c*).

The Nasal Bones (figs. 62; 71, *n*) might have been described next after the *frontals*, as they continue forward the general roofing of the skull; but are conveniently considered in the present connection, being in birds rather "facial" than "cranial." They are of large size in birds, and pronged, — one fork, the *superior process*, being applied for a variable distance along the outer side of the frontal process of the premaxillary, the other, *inferior*, descending to or towards the dentary border of the maxillary or premaxillary, or both; the divergence of these two processes bounding the nostril behind. The base of the nasal, uppermost and posterior, anchyloses (usually) or sutures (often) or articulates (as in parrots) with the antero-external border of the frontal bone; its frequent collateral connections being with the lacrymal or ethmoid, or both of these. The nasals are very variable in shape, as well as in the extent of their connections. When expansive, they may wall in much of the nasal cavity, as well as bound the nostrils. These latter openings, as far as the bony boundaries are concerned, are usually much more extensive than they seem to be from the outside, being much contracted by membrane and integument. Ordinarily, each forms a great vacuity, which the descending prong of the nasal bone separates from a similar vacancy between itself and the lacrymal, the lacrymal in turn interposing between this and the orbital cavity. The descending process of the nasal, in fact, is a marked object at the side of the base of the upper mandible of most birds, though slight or rudimentary in the Ratitæ. A character of the nasals has been employed in classification by Mr. Garrod. A bird having the bones as above generally described, with moderate forking, so that the angle of the fork, bounding the nostrils behind, does not reach so far back as the fronto-premaxillary suture, is termed *holorhinal* (Gr. ὅλος, *holos*, whole; ῥίς, ῥινός, *rhis*, *rhinos*, nose; fig. 62). But in the *Columbidæ*, and in a great many wading and swimming birds, whose palates are cleft (*schizognathous*), the nasal bones are *schizorhinal* (σχίζω, *schizo*, I cut); that is, cleft to or beyond the ends of the premaxillaries; such fission leaving the external descending process very distinct from the other, almost like a separate bone. Pigeons, gulls, plovers, cranes, auks, and other birds are thus split-nosed. The value of the character, except as an auxiliary, is doubtful.

The Lacrymal (Lat. *lacryma*, a tear; from the relation of the human bone to the tear-duct; figs. 62; 63, *u;* 71, *l*) is one of several splint-like membrane-bones of the skull, having little intimacy of relation with the general morphology of the cranium, though quite constant in birds, and often very conspicuous. It is situated at or near the anterior outer corner of the orbit, near the nasal but behind that bone; sometimes anchylosed, sometimes very loosely attached, oftener firmly sutured with the frontal; and may also have connection with the nasal and ethmoid. It is generally a claw-like affair, depending from the front outer corner of the frontal, and consequently bounding the orbit anteriorly; it may be variously twisted, crooked, hooked, etc. It is singularly elongated and distorted in the ostrich. In the duck tribe, in which the lacrymo-frontal region of the skull is greatly elongated, the lacrymal has coextensive attachment to the frontal bone, and is broadly laminar, with a downward process; in some ducks bounding at least a fourth of the orbital brim, and almost completing the circle by extending toward the very protrusive post-frontal process, as in fig. 63, *u*. In some parrots, the rim of the orbit is completed below, and even sends a bony bar to bridge over the temporal fossa behind the post-frontal. In some birds, the lacrymal is quite free, and even in more than one free piece. The *os uncinatum*, or *os lacrymo-palatinum*, would appear to be a palatine bone distinct from the lacrymal; it has been observed in the *Musophagidæ* and many other picarian birds, in *Tachypetes* and certain *Procellariidæ*. The lacrymal bone seems to be the principal relic, in birds, of a set of splint-bones which lie about the edges of the orbits in many *Sauropsida*. Another is the post-frontal or sphenotic, usually a process of the frontal, often a separate ossification. In some birds, as various *Raptores*, there are one or more loose supra-orbital plates of bone, serving to eke out the brim of the orbits; thus forming the "orbital shields" so prominent in many hawks, and causing their eyebrows to project. Were such a a chain of splint-bones complete (lacrymal, superorbitals, post-frontal, and squamosal, to quadrate), it would form an arcade of bones *over* the orbit, like the actual zygomatic arch (maxillary, jugal, quadrato-jugal, to quadrate) which lies under the orbit; and such a double series is very perfectly illustrated in many of the *Sauropsida* below birds (Huxley).

Other special ossifications have been described in some birds, but I am obliged to pass them over. I have already far exceeded intended limits, and have yet to describe the mandibular and hyoidean arches, and the zoölogical characters of the palate as a whole.

The Mandible, or Lower Jaw Bone (figs. 62, 63, 70, 71) is a collection of bones developed in the first post-oral visceral arch. Each half of the compound bone (right and left) consists normally of *five* bones, which become immovably anchylosed, but traces of the original distinction of which commonly persist for an indefinite period, — in some birds throughout their lives. In an embryo whose skull has passed to the cartilaginous stage, a long slender rod of cartilage appears in the first post-oral visceral arch; this is *Meckel's cartilage*, or the *meckelian rod* (figs. 65, 66, 68, 70, *mk*), so named after a famous anatomist. Around this rod, which subsequently disappears, the several bones of the mandible are developed. The anterior one of these is the *dentary* (*d*), forming the scaffold of the horny part of the external under mandible. It usually unites by anchylosis, sometimes only by suture, with its fellow of the opposite side. This union in the middle line is the *symphysis* (Gr. σύν, *sun*, with; φύσις, *phusis*, growth). The line of union is externally the *gonys* (see p. 109), the length and other characters of which are determined by the mode of symphysis, as is the general shape of the tip of the lower mandible. The union generally makes an angular Λ, but may be an obtuse Π; the symphysis is very short and imperfect, as in a pelican, for instance, or the opposite, as in a woodpecker and a multitude of birds. Behind the dentary, each ramus of the jaw continues with pieces called *splenial, angular* and *surangular* (*sp, a, su*); there is often a fenestra between them, by imperfection of bony union, as shown in fig. 62, or 63, *f*, which also sufficiently indicates the relations of these parts. The articulation of the jaw with the quadrate bone is furnished by a fifth piece called *articular* (*ar*) from its function. As a whole the mandible is a pronged bone, forking with a variable degree of divergence from its obtuse or acute point, sometimes quite parallel-sided, as in a duck, oftener very open; such prongs may be straight, or variously curved or bent either in the vertical or the horizontal plane; are generally stout and stanch, sometimes so slender as to be quite flexible. The articular part, always expanded horizontally, presents a smooth irregularly *cupped* superior surface for reception of the protuberances of the foot of quadrate. In general, the concave articular surface is divided into an inner and outer cup separated by a protuberance, corresponding to similar inequalities of the opposing surface of the quadrate. Cupping of the mandibular articulation is characteristic of birds as compared with mammals, in which latter the lower jaw has always a knobbed articular surface (condyle). In many birds the angle of the jaw is prolonged back of the articulation as a *posterior articular process* (fig. 63, *h*, 70, 71, *pap*), which may be long, slender and up-curved, as is well shown in a fowl, duck, or plover. Such birds are said to have the "angle of the mandible recurved;" the opposite condition is "angle truncated" (cut off). Usually also, an *internal angular process* (figs. 70, 71, *iap*) is produced inward from the articular part of the jaw, as in the fowl, duck. Between the dentary and articular parts, the ramus of the jaw is usually vertically produced as a thin raised crest, which, when prominent, is called the *coronoid process*; it corresponds to the strong process so called in a mammal, and relates to the advantageous insertion of the temporal or masseteric muscles which effect closure of the jaw. It is scarcely evident in the fowl, fig. 62, but well marked in the duck, fig. 63, over f. At the back of the articular surface is the *pneumatic* foramen for entrance of air, when any; on the inner surface of the ramus, about the splenial bone, is the opening conveying the vessels and nerve.

The Hyoid Bone (Gr. letter v, $hu =$ hy, εἶδος, *eidos*, form; figs. 65–68, 72–74) is the skeleton of the tongue; a very composite structure, consisting of several distinct bones, developed in the second and third post-oral visceral arches (see fig. 65, where *ch* and *bh* are the original elements of the second arch, making the *basihyal* and *cerato-hyal* bones, and *bbr*, *cbr*, and *ebr* are the original elements of the third arch, making the *basibranchial*, *cerato-branchial*, and *epibranchial* bones). The whole affair is somewhat ʌ- or ⋂-shaped, lying loosely, point forward, between the forks of the lower jaw, with its long slender prongs curving up behind the hind head more or less; but not definitely connected with any other bones of the skull. The connection which exists between the hyoid and other cranial bones in a mammal is in birds broken by non-development of certain links of bone developed in the mammalian second post-oral arch, as the stylo-hyal, epihyal, etc.; though birds have a rudimentary stylo-hyal, at least in the embryo, among the several proximal parts of the second arch which form the intricate bones within the ear-passages (fig. 67). The visible parts of a bird's hyoid are usually: the body of the bone, *basihyal* (*bh*, and fig. 72, *c*), single and median, commonly quite short and stocky, sometimes long and slender. The basihyal bears in front a pair of *cerato-hyals* (*ch*; not shown in fig. 72, where they have been absorbed in *b*) usually movably articulated with the basihyal. They commonly appear as little " horns" or processes of the next piece, the *glosso-hyal* (fig. 72, *b*) or bone chiefly supporting the substance of the tongue. It may be a stout and apparently single bone, as that of the goose figured; but oftener appears as a pair of slender bones, side by side, whose backward ends are the cerato-hyals. The glossohyal may or may not bear at its fore end a cartilaginous tip, as in fig. 72, *a*. All the foregoing are hyal, i. e., belonging to the second visceral arch; the following are branchial, of the third arch: The *basi-branchial* (*bbr*, fig. 72, *d*) is a single median piece, projecting backward from the basihyal, with which it may be perfectly consolidated, as it is in the figure, or separately articulated; it may be wanting; it is usually tipped and prolonged backward with a thread of cartilage. The basibranchial is oftener called " urohyal," but had better be allowed its strict morphological name. On either side, the basihyal bears the separately articulated *cerato-branchials* (*cbr*, fig. 72, *e*), long slender bones diverging as they pass backward, and bearing upon their ends the *epi-branchials* (*ebr*, fig. 72, *f*), which finish off the hyoid bone behind, or may be in turn tipped with cartilaginous threads. The cerato- and epi-branchials together are badly called the " thyro-hyals," and in still more popular language the " greater cornua" or " horns" of the hyoid. All these bones vary in different birds in size and shape and relative development; the branchial elements are the most constant in their length and slenderness. The

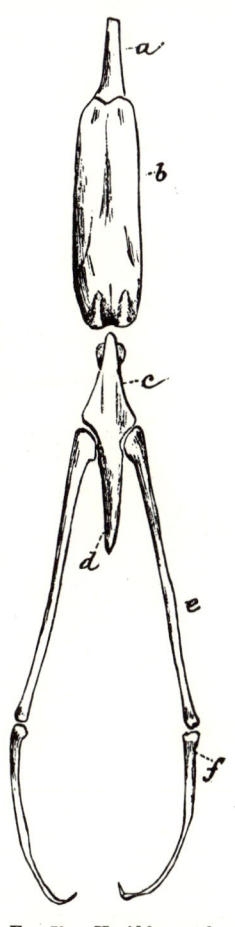

FIG. 72. — Hyoid bones of a goose, nat. size; by Dr. R. W. Shufeldt, U. S. A. *a*, cartilaginous end-piece of *b*, the great glosso-hyal, which has absorbed or replaced cerato-hyals or " lesser cornua"; *c*, basihyal, movably articulated with *b*, and combined completely with *d*, basibranchial, commonly called " urohyal;" *e*, ceratobranchial; *f*, epibranchial; *e* and *f* are together known as " thyro-hyals," or " greater cornua."

whole hyoid apparatus of the woodpeckers is specially modified; the basihyal is very long and slender, bearing stunted cerato- and glosso-hyals at its extreme end; there is no urohyal, or only a rudiment; the cerato-branchials are long, and the epibranchials so extraordinarily elongated in some species as to curl up over the back of the skull and forward along the top of the skull to a variable distance; sometimes, as in fig. 73, curling around the orbit of the eye, or, as in fig. 74, running into the nostril to the tip of the beak. In such cases they bundle together in passing forward over the skull, and go obliquely to one side. (Derivation of the terms in this paragraph: *hyal* is another form of *hyoid*; *branchial*, Lat. *branchiæ*, gills; *basi-*, Lat. *basis*, base; *cerato-*, Gr. κέρας, κέρατος, *keras, keratos*, horn; *epi-*, Gr. ἐπί, *epi*, upon; *stylo-*, Lat. *stylus*, a pen; *glosso-*, Gr. γλῶσσα, *glossa*, tongue; *uro-*, Gr. οὐρα, *oura*, tail; *thyro-*, Gr. θυρεός, *thureos*, a shield.)

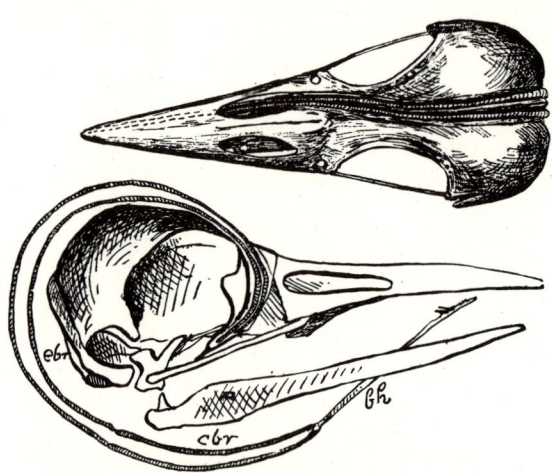

FIGS. 73, 74. — Under fig. side view of a woodpecker's (*Picus*) skull, showing the long slender basihyal (*bh*), bearing slight elements at its fore end, no uroyhal, and extraordinarily long thyrohyals (*cbr, ebr*) curving up over back of skull and curling together around orbit of the right eye. Upper fig. top view of skull of *Colaptes*, showing thyrohyals running along the skull and into right nostril to end of the bill. (Dr. R. W. Shufeldt, U. S. A.)

Other Bones of the Skull. — The articulation of the lower jaw with the quadrate may have certain *sesamoids*. Thus, there are two such *sclerosteous* or ligament-bones in the external lateral ligament of the raven's jaw-joint, and the long occipital style of the cormorant and snake-bird is of the same character, being an ossification in the nuchal ligament of the neck. The siphon-like tube which conveys air from the outer ear-passage to the hollow of the mandible may ossify, as it does in an old raven, resulting in a neat tubular "air-bone" or *atmosteon* (Gr. ἄτμος, air).

Types of Palatal Structure. — The arrangement of the bones of the palate in birds results in several types of structure, first defined by Huxley and applied to the classification of birds. These are the *dromæognathous, schizognathous, desmognathous* and *ægithognathous*; to which Parker has added the *saurognathous*. Huxley proposed to make the primary division of Carinate birds upon this score; and since the plan could not be made to work in his hands, it is certainly futile for any one else to demonstrate again the impossibility of establishing the higher groups of birds upon any one set of characters, — upon the modifications of any one structure. Nevertheless, when duly co-ordinated with other characters, palatal structure becomes of the utmost importance in defining large groups of birds. It is necessary, therefore, for the student to clearly understand this matter, which I will lay before him as nearly as possible in the words of the authors just mentioned.

Dromæognathism (Gr. δρομαῖος, *dromaios*, a runner: genus-name of the *emeu*). — All the Ratite birds, and the tinamous alone of Carinate birds, are *dromæognathous*. "The posterior ends of the palatines and the anterior ends of the pterygoids are very imperfectly, or not at all, articulated with the basisphenoidal rostrum, being usually separated from it, and supported by the broad, cleft, hinder end of the vomer. Strong basipterygoid processes, arising from the

THE ANATOMY OF BIRDS. — OSTEOLOGY.

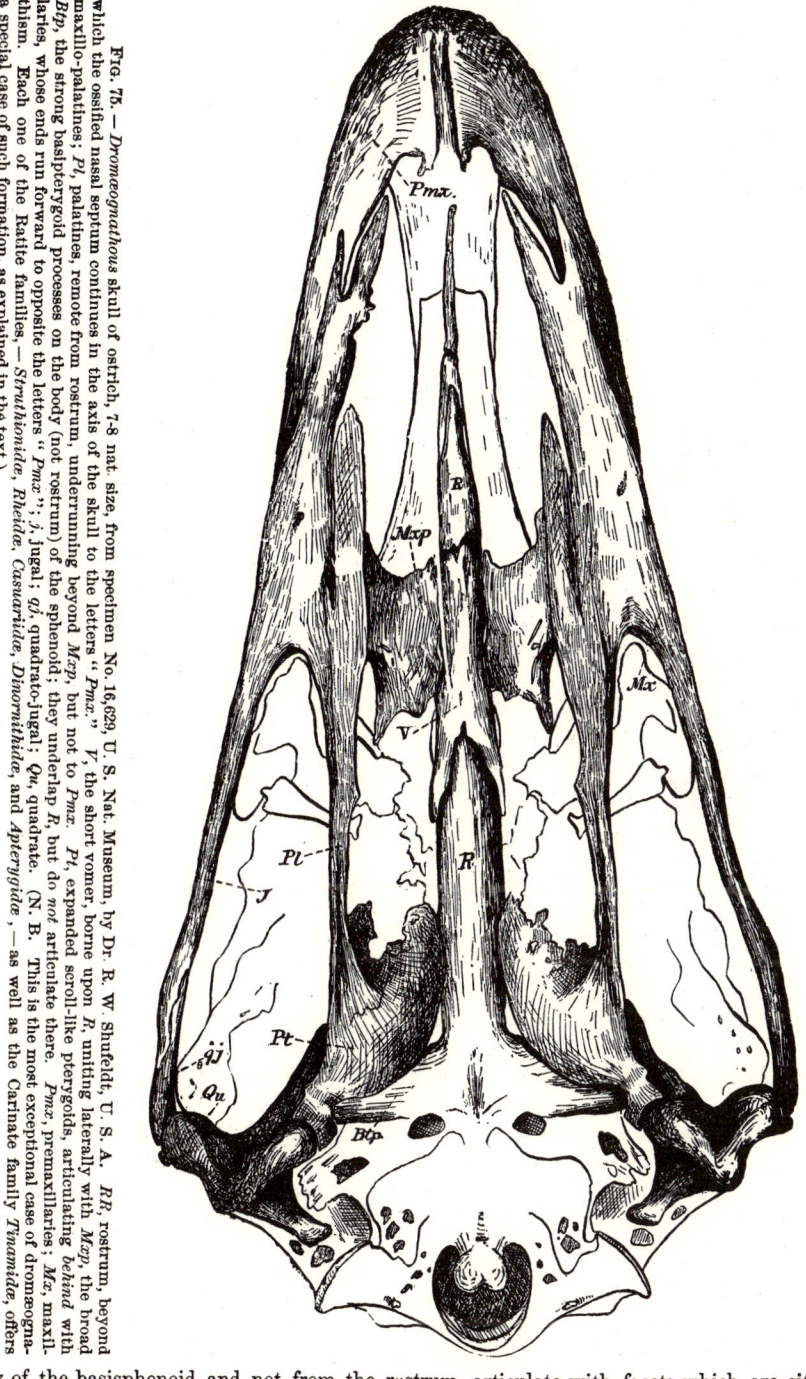

FIG. 75. — *Dromæognathous* skull of ostrich, 7-8 nat. size, from specimen No. 16,629, U. S. Nat. Museum, by Dr. R. W. Shufeldt, U. S. A. *RR*, rostrum, beyond which the ossified nasal septum continues in the axis of the skull to the letters "*Pmx*." *V*, the short vomer, borne upon *R*, uniting laterally with *Mxp*, the broad maxillo-palatines; *Pl*, palatines, remote from rostrum, underrunning beyond *Mxp*, but not to *Pmx*. *Pt*, expanded scroll-like pterygoids, articulating behind with *Btp*, the strong basipterygoid processes on the body (not rostrum) of the sphenoid; they underlap *R*, but do *not* articulate there. *Pmx*, premaxillaries; *Mx*, maxillaries, whose ends run forward to opposite the letters "*Pmx*"; *j*, jugal; *Qj*, quadrato-jugal; *Qu*, quadrate. (N. B. This is the most exceptional case of dromæognathism. Each one of the Ratite families, — *Struthionidæ, Rheidæ, Casuariidæ, Dinornithidæ*, and *Apterygidæ*, — as well as the Carinate family *Tinamidæ*, offers a special case of such formation, as explained in the text.)

body of the basisphenoid and not from the rostrum, articulate with facets which are situated nearer the posterior than the anterior ends of the inner edges of the pterygoid bones." This is the gist of *dromæognathism;* it is exhibited in several ways. (*a*) In *Struthio* alone, fig. 75, the very short vomer, borne upon the rostrum, articulates neither with palatines nor with pterygoids, but with the maxillo-palatines; and the palatines, which are remote from the rostrum, advance beyond the maxillo-palatines, as in most birds. (*b*) In *Rhea,* the vomer is as long as usual in birds, and articulates behind with the palatines and pterygoids, but does not join the maxillo-palatines in front; the short palatines unite with the inner and posterior edges of the thin fenestrated maxillo-palatines. (*c*) In *Casuarius* and *Dromæus* (cassowary and emeu), the long vomer articulates behind with the palatines and pterygoids, and unites in front with the maxillo-palatines; these are flat, imperforate, and solidly joined to the premaxillæ; the palatines are short. (*d*) The extinct *Dinornis* had flat imperforate maxillo-palatine plates uniting solidly with the premaxillæ, and probably with the vomer, as in *Dromæus*. (*e*) In *Apteryx,* the long vomer unites with palatines and pterygoids behind; short broad palatines suture obliquely with flat imperforate maxillo-palatine plates, which unite both with premaxillary and vomer. (*f*) The tinamous, *Dromæognathæ* (fig. 76) "have a completely struthious palate"; vomer very broad, uniting in front with broad maxillo-palatine plates as in *Dromæus;* behind articulating with posterior ends of palatines and anterior ends of pterygoids, both of which are thus prevented, as in all *Ratitæ,* from any extensive connection with the rostrum; basipterygoid processes springing from body of sphenoid, not from its rostrum, articulating with pterygoids very near the posterior or outer ends of the latter; head of quadrate with a single articular facet, as in *Ratitæ.*

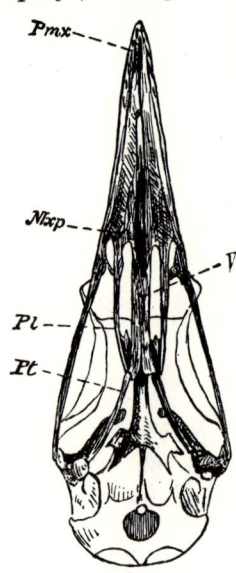

FIG. 76. — *Dromæognathous* skull of tinamou (*Tinamus robustus*); copied by Shufeldt from Huxley. Letters as before; *Mxp,* maxillo-palatine.

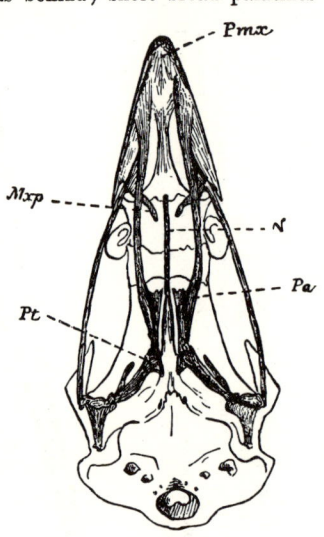

FIG. 77. — *Schizognathous* skull of common fowl, nat. size, from nature, by Dr. R. W. Shufeldt, U. S. A. Letters as before; *Pa,* palatine.

Schizognathism (Gr. σχίζω, *schizo,* I cleave) is the kind of "cleft palate" shown by the columbine and gallinaceous birds, by the waders at large, and many of the swimmers (see fig. 77). In this general case, the vomer, whether large or small, tapers to a point in front, while behind it embraces the basisphenoidal rostrum, between the palatines; these bones and the pterygoids are directly articulated with one another and with the basisphenoidal rostrum, not being borne upon the divergent posterior ends of the vomer; the maxillo-palatines, usually elongated and lamellar, pass inwards over [*under,* when the skull is viewed upside-down, as it usually is] the anterior part of the palatines, with which they unite and then bend backwards, along the inner edge of the palatines, leaving a broader or narrower fissure between themselves and the vomer, on each side, and do not unite with one another or with the vomer. It follows from this that in the dry skull of a plover, for instance, which shows the schizognathous arrangement extremely well, "the blade of a thin knife can be passed, without meeting with any bony obstacle, from the posterior nares alongside the vomer to the end of the beak." There are several groups of birds which exhibit the schizognathous plan, with ulterior modifications of palatal and other characters. (*a*) The columbine birds (*Peristeromorphæ* of Huxley's arrangement): maxillo-palatines elongate and spongy; basipterygoid processes narrow, but prominent. (*b*) The gallinaceous birds (*Alectoromorphæ*): maxillo-palatines varying greatly in size, but always lamellar; palatines long and narrow, with rounded off postero-external angles; basipterygoid processes oval, flattened, sessile upon the rostrum, articulating with the pterygoids. (*c*) The penguins (*Spheniscomorphæ*): maxillo-palatines concavo-convex and lamellar; no basipterygoid processes; pterygoids flattened. (*d*) In the gulls, petrels, loons, grebes, and auks, constituting the *Cecomorphæ* of Huxley, the maxillo-palatines are usually lamellar and concavo-convex, but may be spongy, tumid, and closely approximated to the vomer; and basipterygoid processes are absent or present. (*e*) In the cranes, rails, and their allies (*Geranomorphæ*), the maxillo-palatines are concavo-convex and lamellar, and basipterygoid processes are usually absent. (*f*). In the plover-snipe group, or limicoline *Grallæ* (*Charadriomorphæ*), the maxillo-palatines are always concavo-convex and lamellar; the basipterygoid processes narrow and prominent. Excepting perhaps group *d*, which does not hang together so well, the schizognathous groups here noted correspond very closely with recognized orders or suborders of birds; in all of them, the maxillo-palatines are perfectly distinct from one another and from the vomer, *and* the latter is slender and usually pointed. There are plenty of other birds in which the former factor in the case obtains; but in these the vomer is broad and usually truncate in front (see *Ægithognathism*, beyond).

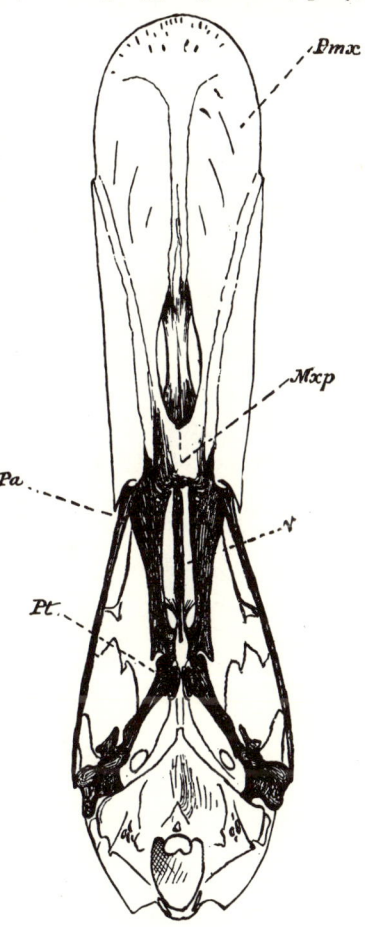

FIG. 78 — *Desmognathous* skull of mallard duck, *Anas boscas*, nat. size, from nature, by Dr. R. W. Shufeldt, U. S. A. Letters as before.

Desmognathism (Gr. δεσμός, *desmos*, a bond) is exhibited in one or another style by those swimming and wading birds which are not schizognathous, by the birds of prey, and various non-passerine perching birds. It does not fadge so well as any other one of the palatal types of structure with recognized groups of birds based on other considerations. In the "bound-palate" type, the vomer is either abortive, or so small that it disappears; when existing it is usually slender and tapers to a point in front; the maxillo-palatines are united across the median line, either directly or by means of ossifications in the nasal septum; the posterior ends of the palatines and the anterior ends of the pterygoids articulate directly with the rostrum (as in schizognathism). This type is simply and perfectly exhibited by a duck (fig. 78) in which the maxillo-palatine is a broad flat plate united with its fellow in mid-line; the oval sessile basipterygoid facets are far forward, opposite the very ends of the pterygoids. In the flamingo, ibis, spoon-bill, stork, heron, the united maxillo-palatines are tumid and spongy, filling the base of the beak; basipterygoids are wanting (rudimentary in the flamingo). In totipalmate swimmers (pelican, cormorant), desmognathism is carried to an extreme by union of the palate bones also across the mid-line; the general arrangement is as before. The birds of prey exhibit several special conditions of desmognathism. The parrots are another case; among other cranial characters of these birds is to be noted the *articulation* of the palate bones with the upper beak, like that of the zygoma. The multifarious Picarian birds, or non-passerine Insessores, are desmognathous, excepting the schizognathous trogons (*Trogonidæ*) and the "saurognathous" woodpeckers. Parker has established the following categories of desmognathism: (*a*) *Perfect direct*, the maxillo-palatines uniting below at the mid-line; either with the nasal septum free from such bony bridge, as in a duck; or anchylosed therewith, as in many birds of prey. (*b*) *Perfect indirect*, very common, as in eagles, vultures, owls; maxillo-palatines separated from each other by a chink, but anchylosed with nasal septum. (*c*) *Imperfectly direct*; maxillo-palatines sutured together, but not anchylosed. "In young falcons and hawks the palate is at first indirect, is then imperfectly direct, and at last perfectly direct." (*d*) *Imperfectly indirect*; maxillo-palatines closely articulated with, and separated by, the "median septo-maxillary;" but there is no anchylosis. (*e*) *Double*: the palatines united as well as the maxillo-palatines; as in the pelican and cormorant above noted, in certain Caprimulgine birds, horn-bills, etc. (*f*) *Compound*: when the properly *ægithognathous* skull of a passerine bird becomes also desmognathous.

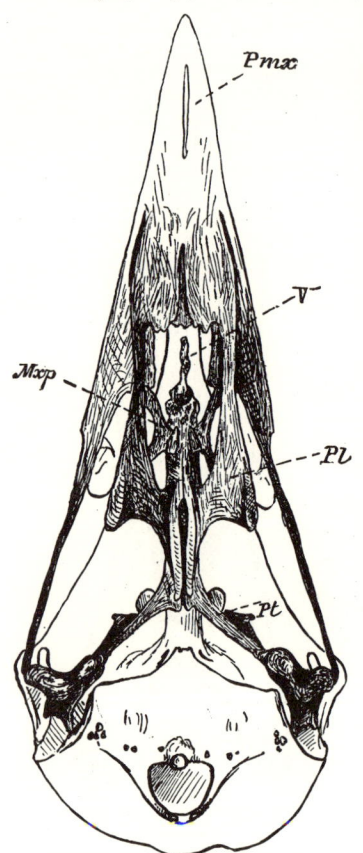

FIG. 79. — *Ægithognathous* skull of raven, *Corvus corax*, nat. size, from nature, by Dr. R. W. Shufeldt, U. S. A. Letters as before. N. B. The reference line, *V*, goes to the ossified nasal septum borne upon the end of the vomer, which latter bone begins at the thickest part of the central projection. *Mxp* underlies *V* and overlies *Pl*, but touches neither.

Ægithognathism (Gr. αἰγιθαλός, *aigithalos*, some small bird) is exhibited almost unexceptionally by the great group of Passerine birds; it is also nearly coincident with *Passeres*, though a few other birds, notably the swifts (*Cypselidæ*), also exhibit it. Huxley's term *Coracomorphæ*, nearly synonymous with *Passeres*, relates to the palatal structure exhibited by a raven (fig. 79), as typical of that of *Passeres* at large. The vomer is a broad bone, truncate in front and deeply cleft behind, embracing the sphenoidal rostrum in its forks. The palatines have produced postero-external angles. The maxillo-palatines are slender at their origin, extending inwards and backwards over the palatines and under the vomer, where they end free, being united neither with each other nor with the vomer. This disconnection of the maxillo-palatines is *quoad hoc* "schizognathous," of course; but such condition, in *association with* the peculiarities of the vomer, is ægithognathous. The nasal septum in front of the vomer is often ossified in ægithognathism, and the interval between it and the premaxillæ filled up with spongy bone; but no union takes place between this ossification and the vomer (Huxley). According to Parker, the distinguishing character of the ægithognathous type is the union of the vomer with the alinasal wall and turbinals. He distinguishes four styles: (*a*) *Incomplete;* very curiously exhibited by the low *Turnix*, which stands near the gallinaceous birds. (*b, c*) *Complete*, as represented under two varieties, one typified by the crow, an Oscine Passerine, the other by the Clamatorial Passerines *Pachyrhamphus* and *Pipra*. (*d*) *Compound*, i. e., mixed with a kind of desmognathism, as noted above. "Vomer truncated in front" is the general expression for the condition of that bone in the ægithognathous type; it is frequently massive in that direction, and of endlessly varied configuration.

Saurognathism. — (Gr. σαῦρος, *sauros*, a lizard; fig. 80). According to Huxley the woodpeckers exhibit a "degradation and simplification of the ægithognathous structure." The peculiarities of the palate of these birds (including *Picidæ*, *Picumnidæ* and *Iyngidæ*) are so decided that Parker proposes to call them *saurognathous*. The structure is very difficult to make out, and may be understood best by study of the accompanying figure, copied from Parker. The maxillo-palatines, *mxp*, are very slight, not extending inward beyond the outer margin of the palatines, and being sometimes quite rudimentary. In front of them, an additional little palatal plate of the maxillary, *pmx*, is developed. The vomers, *v*, are delicate paired rods on each *side* of the median line. The postero-external angle of the palatine is either rounded off or obtuse-angled. Where the broad main part of the palatine suddenly narrows is developed an interpalatine process, *ipa*. The ethmo-palatine plates, *epa*, or internal superior plates of the palatine, which are of variable length, are connected by the most marked *medio-palatine* ossification, *mpa*, seen in the class of birds. Bridges of bone are deposited along the inner borders of the palatines; such are the septo-maxillaries, *smx*, and other formations which, like the medio-palatine, serve to bind the palate halves together. The nasal chambers are unusually simple; there are peculiarities of the tympanic cavity and quadrate bone.

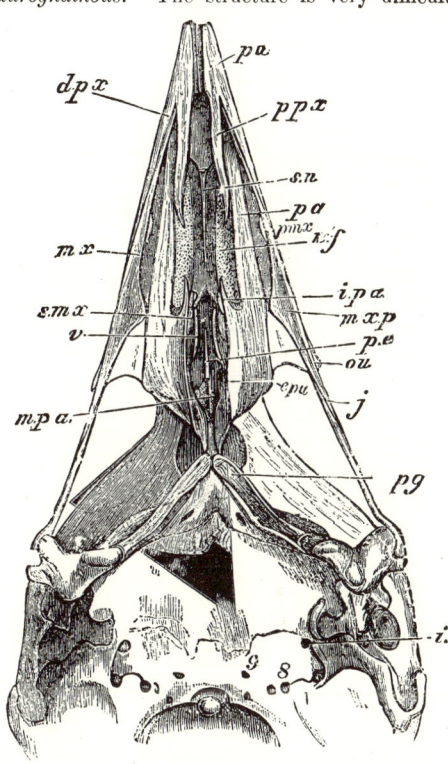

FIG. 80. — *Saurognathous* skull of nestling *Picus minor*, × 4 diameters, after Parker. *Px*, premaxillary: *dpx*, its dentary process; *ppx*, its palatal process; *sn*, septo-nasal; *pa*, palatine; *pmx*, peculiar palatal plate of maxillary of a woodpecker; *nf*, nasal turbinal; *mx*, maxillary; *ipa*, interpalatal spur of palatine bone; *mxp*, rudimentary maxillo-palatine, scarcely reaching palatine; *smx*, septo-maxillary, in several pieces; *v*, right vomer, its fellow opposite; *pe*, lower border of perpendicular plate of ethmoid, between vomers; *epa*, ethmoidal (inner) plate of palatine; *mpa*, medio-palatine; *pg*, pterygoid; *i*, foramen for internal carotid; 8, for vagus nerve; 9, for hypo-glossal nerve.

"**All these things being considered,**" says Parker, in conclusion, "it will seem contradictory now to assert the great uniformity of the skulls of Birds, and indeed of Birds themselves. Yet so it is; and the countless modifications that offer themselves for observation are gentle in the extreme. One form is often seen to pass into another by almost insensible gradations. . . . In the rest of the Birds' organization abundant evidence of the same specialization will be seen. The mind fails to desire more beauty or to contemplate more exquisite adaptations. An almost infinite variety of Vertebrate life is to be found in this class. Of its members some dig and bury their germs, which rise again in full plumage, whilst others watch and incessantly feed their tender brood in the shady covert or 'on the crags of the rock and the strong place.' In locomotion some walk, others run, or they may wade, swim, plunge, or dive, whilst most of them 'fly in the open firmament of heaven.'" (*Ency. Brit.* 9th ed. Art. Birds, p. 717.)

b. Neurology; The Nervous System; Organs of Special Senses.

The **Nervous System** of any Vertebrate determines the form of such an animal; in fact, the beautiful skeleton we have examined is simply a sketch in bone of the *cerebro-spinal nervous system*, conformably with which the whole bony framework of the body is erected. A brain and spinal chord and their lateral prolongations or nerves are the commanding superadditions, in a vertebrate, to any such nervous system as an invertebrate may or does possess. Besides the vertebrate or main nervous system, all brainy vertebrates retain a *sympathetic system* of nerves, supposed to represent a modified inheritance of the whole nervous system of Invertebrates. Thus the cerebro-spinal and sympathetic are the two distinct nervous systems of nearly all vertebrates, — of all vertebrates which have a skull and brain. The former presides over the *animal* life of the creature, — its sensations, perceptions, and voluntary actions; the latter more especially over its *vegetative* functions, as digestion, respiration, circulation, and reproduction, which are more or less involuntary. But the two are inseparably connected, anatomically and physiologically, so that no distinct line can be drawn between them. Nerve-tissue consists of an aggregation of nerve-cells and their investing substance, — the bodies of a myriad *Neuramœbæ* agglutinated by their secretions. They are of two species: *Neuramœba cinerea* and *N. candida*. The former are usually multiradiate, inosculating cells of nerve-substance, which form the "gray matter" of the brain and spinal chord and the *ganglia* (knots) of nerves; the latter are white, thready, and form the connections of the ganglionic masses and the whole substance of ordinary nerve-chords. The gray amœbas are the immediate communicants between the mind and the body of the creature; the white amœbas are the mediators between the body and outward things. The gray amœbas translate thought in terms of matter, and conversely; the white convey the translation. How this is done, no one knows, but the fact is manifest. In ordinary language, gray nerve centres receive from white tracts impressions made upon the periphery of the nervous system; and, with or without the knowledge and consent of the animal, convert these impressions into appropriately responsive actions. This is called the "reflex action" of the nervous system. Some think such reflection is the principal or only activity of the nerve-tissue, taking animals to be mere automata, the mechanism of which is only set in motion by external stimulation. Others think that animals, and even human beings, have in their *consciousness* an inner spring of action, vaguely called "spiritual," whose operations upon the matter of their bodies manifests what is called by some "mind," by others "soul." I am satisfied of the correctness, in the main, of the latter view; but, however this may be, it is quite certain that white nerve tissue is a means of carrying something to and fro, which something is called a "nerve impulse," for want of knowing what it is. White nerves have therefore an *efferent* function, when they carry impulses outward from gray centres, and an *afferent* function, when they bring impulses in to gray centres. The former is their *motor* function; the latter is their *sensory* function. In nerves at large, impulses of both kinds travel in the same tracts without interference; such mixed nerves are therefore called *sensori-motor*. Thus, each spinal nerve has a posterior sensory ganglionated root, and an anterior motor simple root, which soon blend in one chord, in which both functions coexist. Some nerves seem to be entirely motor, as those which move muscles of the face and tongue. The purest sensory nerves are those of "special sense," as the olfactory, optic, and auditory. Some nerves are so "mixed" as to combine functions of special sense, common sensation, and motion, as that called glosso-pharyngeal, which moves, feels, and tastes. The motor effluence of nerve tissue upon itself and other parts of the body is literally *animation;* the sensory influence is nominally *materialization*. The physical mechanism of these occult processes in a bird is as follows: —

THE ANATOMY OF BIRDS.—NEUROLOGY.

The Brain (Lat. *cerebrum;* Gr. ἐγκέφαλον, *egkephalon;* frontisp.) is the anterior dilatation and complication of the main nervous axis of the body, contained within the skull. It resembles a soap-bubble blown at the end of a pipe, being not less beautiful in its iris-quality, and not less lasting. It is primarily triune, or three-fold, beginning as three such bubbles, called the *anterior*, *middle*, and *posterior cerebral vesicles*, corresponding to what are afterward the forebrain, mid-brain, and hind-brain, or *prosencephalon, mesencephalon,* and *opisthencephalon.* The birth and multiplication of gray neuramœbas causes thickenings of the bladdery membranes in various places and ways; all such gray deposits are the *ganglia* of the brain, and the great peripheral ganglion is the *cortical* layer or "bark of the brain." Similar deposits of white neuramœbas connect all these ganglionic colonies, furnishing the various *commissures* of the brain. The cavity of the original bubbles, continuous with the hollow of the pipe-stem or spinal chord (which was at the outset a furrow along the back of the embryo, not a tube) becomes partially divided up into several communicating hollows; these are the *ventricles* (little bellies) of the brain. Actual prolongations of brain-tissue, or nervous threads more like the ordinary spinal nerves, pass out of the brain-box; these are *cerebral nerves*, oftener called *cranial nerves*; there are twelve pairs of them. At the pituitary space (see p. 157; the notochord ends just behind it; fig. 64) is developed a remarkable structure, the *pituitary body*: its nature is unknown. This lies under the brain; opposite it, on top of the brain, is another curiosity, the *pineal body;* it has been considered the special seat of the soul by some, though others have located that throne of animal grace in the solar plexus of the sympathetic system, which is in the belly. The pituitary and pineal are also called respectively the *hypapophysis* and *epapophysis cerebri*. They lie respectively at the bottom and top of one of the cavities of the brain, arbitrarily called the *third ventricle;* the anterior wall of this ventricle is the *lamina terminalis*, or terminal sheet of the brain, with which, morphologically speaking, the brain ends in front; though, in its actual growth, the prosencephalon crowds ahead of this formation. As the brain-cells multiply, the prosencephalon outgrows the associated parts, and becomes nearly separated into lateral halves; these are the *hemispheres of the cerebrum*, or "halves of the great brain"; they retain their ventricles, which intercommunicate through a passage-way, which also leads into the third ventricle; this is the *foramen of Munro*. Each sends out in front a hollow process; these processes are the *olfactory lobes*, or *rhinencephalon* ("nose-brain"). A great ganglionic thickening of gray matter in the interior of each hemisphere is the *corpus striatum;* these "striped bodies" are connected by the *anterior commissure* of the brain. The rest and greater part of the original anterior cerebral vesicle makes up by ganglionic thickening of its sides into what are called misleadingly the *optic thalami*, since these tracts have nothing to do with the sense of sight. The thalami and associate parts behind the lamina terminalis (third ventricle, etc.) compose what is called the *thalamencephalon*, or "bed-brain." The original middle cerebral vesicle makes up underneath into longitudinal commissural fibres, called the *crura cerebri* or "legs of the brain," connecting fore and aft parts; but especially composes the ganglionic centres called *corpora bigemina*, or "twin bodies." These are the *optic lobes*, or "eye-brain." They are connected by transverse commissure. The optic ganglia and commissure, the cerebral crura, and contained cavities, essentially compose the *mesencephalon* or "mid-brain." The original posterior cerebral vesicle (opisthencephalon) becomes separated into two parts: The fore part of it is moulded into the considerable mass of the *cerebellum* ("little brain"); which, with its connections of white substance (pons varolii, peduncles, etc.) and the hollow underneath it ("fourth ventricle") constitutes the *metencephalon* or "after-brain." The hind part of it tapers off into the spinal chord; this tapering part is the *medulla oblongata*, or "oblong marrow," also called the *myelencephalon*, or "marrow-brain." This description is pertinent to brains at large, representing the general plan of structure; any fairly developed encephalon shows the parts specified; and most complicated brain, as that of man, only shows what elaborate finishing touches may be given to the simple structure thus outlined, when cells, both white and gray, but especially the latter, are profusely furnished, to the ornamentation of the mind's estate with race-tracks great and small, and the place of fornication, — fruits of the olive, and of the arbor vitæ. The membranes, or *meninges*, which hide all this from the uninitiated, are three. The *pia mater*, or "tender mother," which immediately invests the brain, is very vascular, and furnishes the blood supply; not only by small arteries which immediately penetrate the substance of the brain, but by enfolded sheets which enter the ventricles, and are called *choroid plexus*. The *arachnoid*, or "cobweb," comes next; a serous fluid which it secretes bathes the brain, and meets concussion with its gentler fluctuation. The *dura mater*, or "stern mother," is a dense outer membrane which enwraps and holds the whole firmly. These meninges descend into the spinal column, and answer the same purpose there, maintaining the same disposition around the spinal chord.

The Bird's Brain offers the following comparative characters: It is compact, having nothing of the straggling apart of its elements seen in low vertebrates, and completely fills the cranial cavity. Its long axis is about transverse to the axis of the spinal column. The cerebral hemispheres are well developed, but do not cover the cerebellum or optic lobes; from their dome the rhinencephalon protrudes like a porte-cochère. Their surface is quite smooth (devoid of the gyri and sulci of most mammalian brains); even the sylvian fissure is barely indicated. The optic lobes are of immense size, relatively to those of most vertebrates, and relatively to the rest of the encephalon; they appear much loosened from their surroundings, at the *sides* and *lower part* of the mid-brain; they retain their ventricles, as does also the rhinencephalon. The corpora striata are very large. The *fornix* is rudimentary. The cerebellum is well developed and deeply sulcate, with transverse fissures, but is not divided into right and left lobes; a "fleecy" lobule on each side, the *flocculus*, is well defined, and received in a special recess of the inner wall of the skull. Parts of the medulla oblongata notable in mammals are obscure or obsolete. There is no *pons varolii*, or superficial transverse commissure of the cerebellum, nor any *corpus callosum*, — that great white commissure of the cerebral hemispheres, characteristic of all but the lowest mammals.

The Spinal Chord, or *medulla spinalis* ("spinal marrow") is the main nerve-axis of the body, running in the series of neural arches of the vertebræ from head to tail; it directly continues the medulla oblongata. It retains its primitively tubular character in part at least, and consists as usual of white matter enclosing gray matter. The chord is fissured into lateral columns, as these are also to some extent into anterior and posterior tracts. The latter diverge in ascending the medulla oblongata, to throw the central tube into the cavity of the fourth ventricle; and especially in the sacral region, where a sort of ventricle, known as the avian *sinus rhomboidalis*, is similarly formed. The calibre of the chord increases at the root of the neck, where large nerves are to be given off from the brachial plexus to the wings, and again in the sacral region, with the same reference to nerve supply of the legs; after which the chord continues to the end of the spinal canal as a terminal thread.

The Cranial Nerves are twelve pairs, as in mammals, the highest vertebrate number. **1,** the *olfactory* nerve of special sense (smell); origin from rhinencephalon; exit from cranial cavity by olfactory foramen, high up in orbital cavity; conducted along a groove to final escape between perpendicular and lateral plates of ethmoid into the nasal chambers; distributed to the investing mucous membrane of the septal and turbinal bones of the nose. The exit is through a sieve-like or *cribriform* plate only in *Apteryx* and *Dinornis* (Owen). **2,** the *optic*, nerve of special sense (sight); origin from optic lobe and thalamus; of great size, and forming a *chiasm* (decussation) with its fellow; exit by optic foramen, a large hole in back of orbital cavity between centres of orbito-sphenoid and alisphenoid, close to or in common with its fellow. This nerve forms the retina of the eye. **3, 4, 6**, the *oculi-motor, pathetic, abducent*, collectively the motor nerves of the eye, supplying the muscles moving the eye-ball; **3**, to all these muscles excepting superior oblique and external rectus; origin from crura cerebri, base of mesencephalon; **4**, to the superior oblique, origin behind optic lobes, upper surface of metencephalon; **6**, to external rectus (also to muscles of the third eyelid in birds); origin between met- and myel-encephalon, base of brain; **3, 4, 6**, exits from cranial into orbital cavity by several small, not constant, foramina near optic foramen; or by this foramen sometimes all the nerves which enter the orbit pass out of brain cavity through one great hole. **5**, great *trifacial* or *trigeminal*, sensori-motor; feeling skin of head, moving muscles of jaws; origin (double) from myelencephalon; leaves brain from sides of metencephalon; sensory root has gasserian ganglion; motor root simple. This nerve has three divisions, whence its name: **5a**, *ophthalmic* division, the most distinct; exit from cranial into orbital cavity by separate foramen above and to outer side of optic foramen; grooves orbital wall in passing; *ciliary* ganglion; distribution mainly to lacrymal and nasal parts; traceable to end of upper mandible; **5b**, *superior maxillary;* exit by foramen ovale, in alisphenoid or between that and proötic centre; distribution to side of upper jaw; *meckelian* ganglion; **5c**, inferior maxillary, derived chiefly from motor root; exit same as **5b**; distribution to lower jaw (muscles, substance of bone, integument); no *special sense* (gustatory) function; no *otic* ganglion. **7**, *facial* or *portio dura*, motor; origin from myelencephalon; enters periotic bone, escapes from ear behind quadrate bone, by what corresponds to stylo-mastoid foramen of mammals; communicates with **5c** by *chorda tympani* nerve, with **9, 10, 12**, and sympathetic system; distribution to skin-muscles and others of lower jaw and tongue, etc. **8**, *auditory* or *portio mollis*, nerve of special sense (hearing); origin with **7**; no exit from skull; enters meatus auditorius internus of periotic bone; forms auditory apparatus in labyrinth of ear. **9**, *glosso-pharyngeal*, mixed nerve, sensori-motor and gustatory (taste); origin myelencephalon; exit by foramen in exoccipital bone, behind basitemporal, near lower border of tympanic recess; distribution to muscles and membranes of gullet, throat, tongue, etc. **10**, *pneumogastric*, sensori-motor; origin and exit next to **9**; distribution to windpipe, lungs, gullet, stomach, heart, etc.; has recurrent syringeal to vocal organs. **11**, *spinal accessory*, sensori-motor; origin upper part of spinal chord; exit with **9, 10**; distribution to these nerves and to muscles of neck. **9, 10, 11**, are intimately connected with one another, and with other nerves, especially **10** with sympathetic. The several foramina in a bird's skull which may be seen in the place indicated at **8**, figs. 69, 70, are for the divisions of this composite *vagus* or "wandering" nerve of respiration, circulation, digestion, etc.; they represent morphologically a *foramen lacerum posterius*, between exoccipital and opisthotic centres. **12**, *hypoglossal*, motor nerve of the tongue; origin from myelencephalon; exit by anterior condyloid foramen in front of the occipital condyle. Thus the plan of the cranial nerves of birds is nearly coincident with that of mammals.

The Spinal Nerves, in pairs, correspond in a general way to the vertebræ, between which they pass out by *intervertebral foramina*, to supply the body at large. They are sensori-motor; arise from the spinal chord by anterior motor and posterior sensory (ganglionated) roots which unite before leaving the spinal canal; in the sacral region the main branches leave by separate foramina. They form *plexuses* or interlacements. The principal of these is the *brachial plexus;* constituted by several lower cervical nerves, and one or two usually counted as dorsal, which combine to form a single chord, whence the nerves of the wing are derived. Similar network of three to five true sacral nerves furnishes the nerves of the leg.

The Sympathetic System consists of a pair of nervous chords running lengthwise below the bodies of the vertebræ, one on each side in the trunk, and in corresponding relations with cranial bones. An extensive and intricate series of communications is effected with the nerves of the cerebro-spinal system, excepting the special-sense nerves of smell, sight, and hearing. The points of communication form a chain of sympathetic ganglia; from these knots, the most conspicuous features of the system, nervous chords pass to their distribution in the motory mechanism of the heart and blood-vessels and other viscera. The anterior sympathetic nerves are the *iridian*; the ganglia are the *spheno-palatine* or *meckelian*, intimately connected with cranial nerves. The system ends behind in the caudal region of the spine by a *ganglion impar*.

Sense of Smell: Olfaction. — The sense of smell is effected by terminal branches of the olfactory (1st cranial) nerve, ramifying in the mucous (pituitary or schneiderian) membrane of the nasal cavities. Owing to the comparatively small size and little complexity of the foldings and pleatings of bone or cartilage in the nasal chambers, the sensory surface being correspondingly limited, it is not probable that birds possess this sense in a high degree. Besides the cartilaginous or osseous *septum*, generally more or less complete in birds, there are lateral scrolls and whorls of bone in endless diversity in most birds, which may be ossified, or remain gristly. The general cavity is mostly bounded and enclosed by the bony beak; floored by the anterior part of the hard palate; defended on each side by the descending prong of the nasal bone; in the dry skull, it either seems continuous with the great orbital cavity on each side behind, or is separated therefrom by lateral ethmoid (pre-frontal) or lacrymal ossifications, or both. Outwardly the nasal chambers open upon the beak by the external nostrils — orifices of great zoölogical diversity, as already indicated (p. 109), bounded by prongs of the premaxillary and nasal bones. These openings are minute or quite obliterated in some *Steganopodes*, as pelicans and cormorants. The nasal cavities always communicate with the back part of the mouth, or the *posterior nares* (Lat. *naris*, a nostril); generally paired, that is, with a partition between them, sometimes united in one median aperture. The olfactory nerve, which is rather a prolongation of the rhinencephalon itself than an ordinary nerve, escaping from the brain-box by a special foramen, traversing the upper part of the interorbital septum in a groove or canal, enters the nasal cavity by a single orifice (excepting *Apteryx* and *Dinornis*), instead of the numerous apertures in a cribriform plate by which its filaments reach their destination in mammals. The true sensitive membrane in which the nervous filaments end is that investing *ethmoidal* (septal and turbinal), not maxillary parts. An associate structure of the olfactory organ is the *nasal gland*, sometimes called the *superorbital* gland, from its position in many birds. Thus it is of great size in a loon, and lodged in large deep crescentic depressions on top of the skull over the orbits (fig. 63, *w*); these crescents nearly meeting each other in the middle line. In other birds it is smaller, and within the cavity of the orbit, but never in that of the nose itself, its secretion being poured into the nasal chamber by a special duct.

Sense of Sight: Vision. — The eye is an exquisitely perfect optical instrument, like an automatic camera obscura which adjusts its own focus, photographs a picture upon its sensitized retinal plate, and telegraphs the molecular movements of the nervous sheet to the optic "twins" of the brain, where the result is "biogenized;" that is, translated from the physical terms of motion in matter to the mental terms of consciousness. But no part of the nervous tract, from the surface of the retina to the optic centre, sees or knows anything about it, being simply the apparatus through which the Bird looks, sees, and knows. In this class of Vertebrates, the optic organs, both cerebral and ocular, are of great size, power, and effect; their vision far transcends that of man, unaided by artificial instruments, in scope and delicacy. The faculty of *accommodation*, that is, of adjusting the focus of vision, is developed to a marvellous degree; rapid, almost instantaneous, changes of the visual angle being required for distinct perception of objects that must rush into the focal field with the velocity at least of the bird's flight.

Birds are therefore far-sighted or near-sighted (presbyopic or myopic) according to the degree of *tension* the nerve-tide excites in the eye by the mechanism described further on; and the transition from one to the other state is effected with great quickness and correctness. Observe an eagle soaring aloft until he seems to us but a speck in the blue expanse. He is far-sighted; and scanning the earth below, descries an object much smaller than himself, which would be invisible to us at that distance. He prepares to pounce upon his quarry; in the moment required for the deadly plunge he becomes near-sighted, seizes his victim with unerring aim, and sees well how to complete the bloody work begun. A humming-bird darts so quickly that our eyes cannot follow him, yet instantaneously settles as light as a feather upon a tiny twig. How far off it was when first perceived we do not know; but in the intervening fraction of a second the twig has rushed into the focus of distinct vision, from many yards away. A woodcock tears through the thickest cover as if it were clear space, avoiding every obstacle. The only things to the accurate perception of which birds' eyes appear not to have accommodated themselves are telegraph-wires and light-houses; thousands of birds are annually hurled against these objects to their destruction.

The *orbital cavity, orbit*, or socket of the eye, has been almost sufficiently described (p. 156; see also any figs. of skull in profile) as that great recess in the side of the skull bounded above by the roofing frontal bone, behind by this and sphenoidal elements, in front, if at all, by lateral ethmoidal elements (pre-frontal), and separated from its fellow more or less completely by the inter-orbital septum, which is chiefly the perpendicular plate of the mesethmoid, but may be also in part orbito-sphenoidal and pre-sphenoidal. The brim is completed in few birds, by union of lacrymal and post-frontal; in quite a number of birds, however, it is nearly perfected by the approximation of these same bones, as in fig. 63, *u* and *m*, and in some the rim is carried out by extra supra-orbital and infra-orbital ossification. There is no bony floor, or only such slight scaffolding as the expansion of the palatine and pterygoid may afford. The zygoma itself, in many dry skulls, seems like the threshold of the orbital chamber. The bony walls may be also defective in some places by great vacuities in the inter-orbital septum (fig. 70, *iof*, and fig. 63, *z*), and others in the cerebral wall, aside from the regular foramina which the nerves pass through. The 1st — 6th nerves (p. 182) inclusive usually enter the orbit: of their foramina, the *optic* (figs. 66, 68, 70, 71, 2, and fig. 63, *y*) is much the largest and most constant, generally blended with its fellow. Those for nerves 1 and 5 (p. 183) are next most obvious and constant; others are often, and all may be, thrown into one large opening. In such a socket as this the eye-ball rests upon a cushion of muscle, fat, gland, and connective tissue; and large as is the chamber, the ball fits and nearly fills it. A bird's eye-ball is *much* larger than the opening of the eye-lids (see p. 30, note).

As to its development: "the *Eye*" says Huxley "is formed by the coalescence of two sets of structures, one furnished by an involution of the integument, the other by an outgrowth of the brain. The opening of the tegumentary depression, which is primarily [in the very early embryo] formed on each side of the head in the ocular region becomes closed, and a shut sac is the result. The outer wall of this sac becomes the transparent *cornea* of the eye; the epidermis of its floor thickens, and is metamorphosed into the *crystalline lens;* the cavity fills with the *aqueous humor*. A vascular and muscular ingrowth taking place round the circumference of the sac, and dividing its cavity into two segments, gives rise to the *iris*. The integument around the cornea, growing out into a fold above and below, results in the formation of the eyelids, and the segregation of the integument which they enclose, as the soft and vascular *conjunctiva*. The pouch of the conjunctiva very generally communicates, by the *lacrymal duct*, with the cavity of the nose. It may be raised, on its inner side, into a broad fold, the *nictitating membrane*, moved by a proper muscle or muscles. Special glands — the *lacrymal* externally, and the *harderian* on the inner side of the eye-ball — may be developed in connection with, and pour their secretion on to, the conjunctival mucous membrane. The posterior chamber of the eye has a totally distinct origin. Very early that part of the anterior cerebral vesicle which eventually becomes the vesicle of the third ventricle, throws out a diverticulum, broad at its outer, narrow at its inner end, which applies itself to the base of the tegumentary sac. The posterior, or outer, wall of the diverticulum then becomes, as it were, thrust in, and forced towards the opposite wall by an ingrowth of the adjacent connective tissue; so that the primitive cavity of the diverticulum, which, of course, communicates freely with that of the anterior cerebral vesicle, is obliterated. The broad end of the diverticulum acquiring a spheroidal shape, while its pedicle narrows and elongates, the latter becomes the optic nerve, while the former, surrounding itself with a strong fibrous *sclerotic* coat, remains as the posterior chamber of the eye. The double envelope, resulting from the folding of the wall of the cerebral optic vesicle upon itself, gives rise to the *retina* and the *choroid* coat, the plug or ingrowth of connective tissue gelatinizes and passes into the *vitreous humor*, the cleft by which it entered becoming obliterated." (Anat. Vert., 1871, p. 79.)

Birds alone, of all animate beings, may be truly said to "fall asleep" in death. When the "silver cord" of a bird's life is loosed, the "windows of the soul" are gently closed by unseen hands, that the mysterious rites of divorce of spirit from matter may not be profaned. When man or any mammal expires, the eyes remain wide open and their stony stare is the sign of dissolution. Only birds close their eyes in dying. At the same moment, the eye sinks and seems to collapse, by the ebbing of its waters. The closure is chiefly effected by the uprising of the lower lid. These are the principal external differences between the eyes of birds and mammals. The movements of the upper lid in most birds are much more restricted than those of the lower. The few exceptions are chiefly furnished by night birds, as owls, whippoorwills, and others of their respective tribes. The lids consist externally of common skin, internally of a layer of *conjunctival* (joining) mucous membrane, with interposed connective tissue: the lower is also stiffened with a smooth plate, the *tarsal cartilage*. The upper is raised by a small muscle, called from its office *levator palpebræ superioris*, arising from the bony orbit. There is no special lowering nor lifting muscle of the under lid; the lids close together by the action of the *orbicularis oculi*, which nearly surrounds the eye, and whose chief office is to lift the lower lid; the latter has a small distinct *depressor* muscle. Birds have no true hairs, but in some kinds modified filiform feathers answer to eye-lashes. When wide open the orifice of the lids is circular, that is, without the inner and outer corners (*canthi*) of almond-eyed creatures like man. There is a *third* inner eyelid, highly developed and of beautiful mechanism : this is the *nictitating* membrane, or "winker" (*nictito*, I wink), a delicate, elastic, translucent, pearly-white fold of the conjunctiva. While the other lids move vertically and have a horizontal commissure, the winker sweeps horizontally or obliquely across the ball, from the side next the beak to the opposite. If we menace a bird's eye with the finger, it is curious to see the winker rush out of the corner to protect the ball. Owls habitually sit in the daytime with this curtain shading

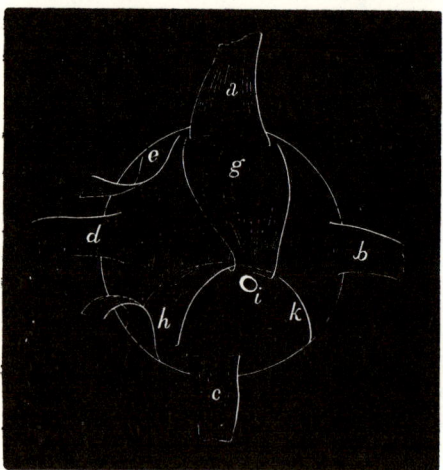

FIG. 81. — Right eye-ball, seen from behind, showing the muscles: *a*, rectus superior; *b*, rectus externus; *c*, rectus inferior; *d*, rectus internus; *e*, obliquus superior; *f*, (not lettered) obliquus inferior; *g*, quadratus; *h*, pyramidalis, with its tendon, *k*, passing through a pulley in the quadratus (as shown by the dotted line) to keep it off the optic nerve, *i*, then passing around the edge of the ball to its insertion in the nictitating membrane.

the eyes from the glare of light; and doubtless the eagle throws the same screen over its sight when soaring towards the sun. When not in action, the winker lies curled up in the corner of the eye, like those patent window shades which stay up of themselves till pulled down. The ingenious mechanism of the movement of the winker across the lid may be understood with the help of fig. 81, which represents the *back* of the eye-ball. The winker lies in front, on the left hand of the picture, and is to be pulled across the front by the slender tendon, *k*, of the *pyramidalis* muscle, *h*. As *h* contracts it pulls on *k*, and *k*, winding round to the front, pulls the winker to the right hand. But *i* is the optic nerve, entering the ball; *k* would press upon it, were it not fended off by passing, as seen by the dotted line, through a pulley in the end of the *quadratus* muscle, *g*. The harder *h* pulls, the harder does *g* also pull, their consentaneous action at once giving the proper direction to the tendon *k*, and keeping it off the nerve.

Beneath the eye-lids, upon the ball, is a delicate filmy membrane not easily recognized on ordinary inspection : this is the *conjunctiva*, so called because it joins the eye to the lids. The *ocular* layer is transparent where it passes over the cornea : it is then reflected away from the ball, to form the *palpebral* layer, — a folding between being the nictitating membrane. The conjunctiva is highly vascular, but the blood-vessels are too small to be seen unless they become congested, when the eye presents the well-known appearance called blood-shot. Though birds can hardly be said to cry, they have a well-developed apparatus for the manufacture of tears. The lacrymal are two small glands lying one in each corner of the eye, inner and outer. The former, called the *harderian* gland, is the smaller, deeply seated behind the winker, upon which it pours a glary fluid : it is an oil-can which not only supplies but applies the fluid to the winker, which needs constant lubricating to work well. The lacrymal gland proper is the outer one, which prepares the tears to moisten and cleanse the conjunctiva; after which they are drained off by the lacrymal duct into the cavity of the nose, which thus becomes a sort of cesspool to receive the refuse waters of the eye. A third gland about the orbit has been already mentioned (p. 184) as pertaining to the nose, not to the eye. Its site is shown in the crescentic super-orbital depression; fig. 63, *w*.

The motions of the eye-ball, though more restricted than in mammals, owing to the shape of the ball and its close socketing, are nevertheless subserved by the usual number of *six* muscles. Of these four are called the *recti*, or straight muscles, and two the *obliqui*, or oblique muscles; though they are all "straight" enough, the terms applying to their lines of traction. The four *recti* arise from the bony orbit, near together, about the optic foramen, and pass to be inserted in the eye-ball at as many nearly equidistant points on its circumference; the *musculus rectus superior*, fig. 81, *a*, on top; *m. r. inferior*, *c*, below, antagonizing *a*; the *m. r. externus*, *b*, and *internus*, *d*, respectively to the outer and inner (hindward and forward) sides, also antagonizing each other. The two oblique muscles arise further forward in the bony orbit, near each other, and then diverge obliquely upward, *m. o. superior*, *e*, and downward, *m. o. inferior*, *f*, to be inserted near the margin of the globe of the eye, close by the respective insertions of superior and inferior rectus. All the motions of the ball result from consentaneous or dissentaneous action along these six lines of traction; the muscles acting as ropes to pull the ball about, and to steady it in any direction of its axis. The peculiarity of mechanism in a bird is, that the superior oblique goes straight to its insertion, instead of passing through a pulley which changes its line of action in mammals. The special nerves presiding over these muscles (3, 4, 6) have been pointed out already (p. 183). In the figure, the cut orbital ends of them all are reflected away from the ball to disclose the underlying muscles of the winker: the reader must mentally bring the six loose ends together and fasten them to the bony orbit at points near about opposite *i*, as above said of their origins.

The above are the principal circumstances and accessories of the optic apparatus; we may now examine the eye itself, of which fig. 82 gives an enlarged view, in longitudinal vertical section, — the nerve, marsupium, and ciliary processes not indeed lying as shown in this section, but so introduced as to show them up intelligibly. A bird's eye-ball is not nearly so spherical or globular as a mammal's. The globe of the human eye is about a five-sixths segment of a large sphere (sclerotic) with a one-sixth segment of a smaller sphere protruding in front (corneal). The anterior part of the sclerotic of a bird is so prolonged as to be in some cases almost tubular or cylindric, and the corneal protuberance is very convex: the result may be likened to an acorn which has a short blunt kernel in a heavy shallow cup, or to a thick old-fashioned watch with a very convex crystal. This characteristic shape is fairly shown in the figure; but some birds' eyes are much more tubular in front, — owls' for example. The eye-ball being hollow and filled with fluids which press in all directions, it is hard to see at first how such a peculiar shape is maintained. But the sclerotic coat is very dense, almost gristly in some cases; and it is reinforced by a circlet of *bones*, the *sclerotals*, *h*, *h*; see also fig. 62, where the circlet is shown. These are packed alongside each other all around the circumference of one part of the sclerotic, like a set of splints. The large discoidal segment of a bird's

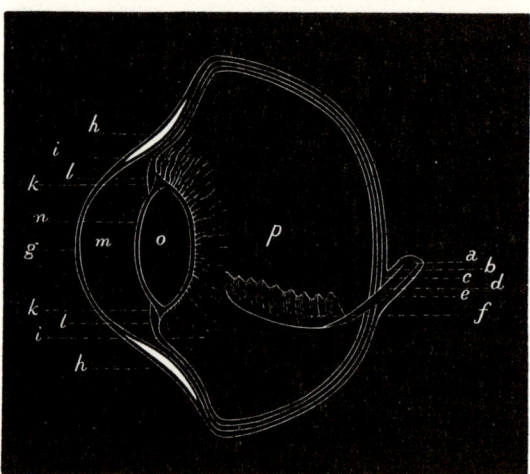

FIG. 82. — Vertical antero-posterior section of eye-ball: *a*, optic nerve; *b*, sclerotic, its outer coat; *c*, sclerotic, its middle and inner coats; *d*, choroid; *e*, hyaloid; *f*, marsupium; *g*, cornea; *h, h*, bony plates between sclerotic layers; *i, i*, corrugations of choroid, forming ciliary processes; *k, k*, canal of Petit; *l, l*, iris; *m*, anterior chamber of eye; *n*, capsule of the lens; *o*, lens; *p*, posterior chamber of eye. Neither the retina, nor the peculiar sheathing of the optic nerve, is shown. The nerve, marsupium, and ciliary processes, not falling in this section, can only be arbitrarily shown.

eye is mostly composed of the membrane called from its hardness the *sclerotic*, — thick, tough, and strong, of a glistening livid color. Three sclerotic coats or layers may be demonstrated by careful dissection; in the figure *b* is the outer, *c* the combined middle and inner ones, — much exaggerated as to their distinctness. The bony plates lie between the outer and middle coats anterior to the greatest girth of the eye-ball, extending from the rim of the disc nearly or quite to the edge of the cornea. They are a dozen to twenty in number, of oblong squarish shape, tapering toward the cornea, around which they are thus circularly disposed; they are pretty closely bound together, but the circlet as a whole enjoys some little motion back and forward with the varying convexity of the *cornea*, *g*. This last is the thin transparent membrane completing the eye-ball in front, like the crystal over the face of a watch. It is very protuberant in birds, — even a hemisphere, or almost tubular. Its structure is not peculiar in birds; but it is remarkable in this class of creatures not only for its convexity, but for the wide range of the variability in convexity which increased or diminished pressure of the contained humors may effect, and its collapse in death.

The sclerotic coat is lined with the *choroid membrane*, *d*, loosely woven of cellular tissue, replete with blood-vessels, and painted pitch-black with a heavy deposit of pigment-cells. It lines the whole globe as far forward as the edge of the sclerotal bones, where it splits in two layers. The *inner* choroid layer turns away from the wall of the eye, toward the interior, and in so reflecting becomes plaited, as a bag is puckered by pulling the strings. These pleats converge upon the rim of the delicate capsule enclosing the lens of the eye, *n*, and there adhere, forming the *ciliary processes*, *i*, *i*. The *outer* layer also starts away from the circumference of the sclerotic wall, as if to pass directly across the cavity, but ends in the *iris*..

Around the circumference of the iris, where sclerotic, corneal, and choroid coats come together, is a circular band of fibres, the *ciliary ligament;* and on the outer surface of the choroid is a similar band of circular and radiating contractile fibres, the *ciliary muscle.* These ciliary structures are supposed to be the agents of the accommodating faculty of the eye, acting upon the lens to alter its shape or its position, or both. It is a difficult matter to settle, when such delicate structures are in question.

The *iris, l, l,* or rainbow of the eye, is an exquisite structure hanging like a many-colored curtain vertically between the two compartments of the eye; a highly ornamental framework of the eye's window, being both sash and blind to the pupil. It is suspended vertically in the *aqueous* humor, just in front of the lens. Viewed in front, from the outside, the iris appears as a colored circular band around the pupil, and seems to come to the surface of the eye. But this is not so, for the conjunctiva, the cornea, and the aqueous humor of the front chamber of the eye, are between us and it. It may be likened to the dial-plate of a watch, which we look at without noticing the interposed crystal. Similarly, the *pupil* of the eye, which shows us our own reflection, diminished to the size of the "eye-baby," may be likened to the round central hole in the dial-plate through which protrudes the shaft that bears the hands of a watch. The "pupil" is the round black spot within the colored rim of the iris; but it is not a thing — it is a hole in a thing — the hole in the iris through which we may look and see the black choroid coat behind. The quivering iris is very similar in texture to the choroid, being a delicate tissue of interlacing fibres and vessels; but it is highly mobilized by circular and radiating sets of contractile fibres, by which the curtain is tightened and loosened, with corresponding change in the size of the central orifice — the pupil. Although the iridian movements are largely automatic, depending upon the stimulus of light, they are to some extent voluntary, as any one may satisfy himself who observes owls in confinement. During these expansions and contractions of the iris, the pupil in birds preserves its circularity; and even when the movement is freest and most voluntary, as in owls, the contracted pupil never appears as a vertical oval figure, or a slit, like that of cats. The round pupil of the great horned owl ranges from the diameter of a finger ring down to that of a small split-pea. The iridian colors are often striking in birds. Though black and brown are the commonest, yellow is quite frequent; red is often seen, blue and green are rarer; the eyes of cormorants are of the latter color. The iris is sometimes pure white, as it is in our common "white-eyed" greenlet, *Vireo noveboracensis.* In the Californian woodpecker, *Melanerpes formicivorus,* the eyes are indifferently (or at different ages of the bird, or seasons) brown, bluish, pink, rosy, or yellow.

The *crystalline lens, o,* is a transparent biconvex disc, like a common magnifying glass, apparently set in the iris like a mirror in its frame, but really hanging a little back of that structure. It is enclosed in a capsular membrane, *n,* of extreme delicacy and transparency, which is in turn set between two layers of the hyaloid membrane to be presently noticed. Where these layers of hyaloid separate around the rim of the capsule to form the investment, a small space is left between them; this circular tube around the lens is the *canal of Petit, k, k.* The lens is stationed in the axis of vision; some suppose it to be equally stationary in any transverse axis. It is, however, difficult to understand how an object thus suspended in fluctuating humors should be insusceptible of some motion backward or forward, as well as of alteration in its degree of convexity; both of which may be factors in the focusing process. From what has preceded, it is evident that the cavity of the eye is divided into anterior and posterior compartments, or chambers, by the reflection, from the sclerotic wall, of the choroid, hyaloid and iridian structures, which with the lens form a vertical partition. Each chamber is filled with a fluid of different density and consistence. That in the anterior or corneal chamber is thin and watery, and therefore called the *aqueous humor;* that in the sclerotic cavity is more dense and glassy, and for this reason known as the *vitreous humor.* There is much less aqueous than vitreous; but birds have comparatively more of the former than usual, owing to the relatively greater size and convexity of the cornea. The waters are enclosed in exceedingly delicate membranes; the vitreous in the *hyaloid membrane*, *e*, which, besides lining the posterior chamber and enclosing the lens as already said, sends thin partitions all through the vitreous humor to steady these glassy waters.

The *optic nerve*, *a*, of birds is peculiar. In mammals, as a rule, the nerve is a smooth cylinder, proceeding straight to the sclerotic, penetrating the coats of the eye-ball directly, near the middle point behind, and then spreading out on the inside of the ball as a large circular concave mirror. This thin, saucer-like expansion of nerve-tissue is the *retina*. In birds the optic nerve is a fluted column, which approaches the eye-ball quite obliquely, strikes it at a point eccentric from the axis of the eye, and does not at once pierce the sclerotic. Tapering to a fine point, and running still obliquely, downward and forward, in a deep groove in the sclerotic that would be a tube were it not split, and through a similar slit in the choroid, a fluting of the nerve rises to attain the cavity of the eye, and the retina spreads out from the sides and end of this fold. But the prime peculiarity of a bird's eye is the "purse" or "comb," *marsupium, pecten, f;* a very vascular structure, like the choroid, and likewise painted black; apparently "erectile," that is, capable of increasing and diminishing in size by influx and efflux of blood. It is attached behind to the nervous structure; is suspended in the vitreous humor, and runs forward obliquely a part or the whole of the way to the lens, to the envelope of which it may be attached in some cases. Its office is not fully determined. Its great resemblance to the choroid proper suggests a similar function in the absorption of light. If it be turgid and flaccid by turns it must occupy a variable space in the vitreous humor, and in the former state press the waters upon the most yielding part of their walls, — that where the lens is situated, even to the extent of altering the position of the latter; and if so, of changing the focus of the eye. It is difficult to account for the bird's eyes' powers of accommodation by the action of the ciliary muscle in only changing the *shape* of the lens, thus throwing out of account as impossible any change in the position of that refracting medium, or of the density of the refracting humors, or of the convexity of the cornea. The peculiar course of the optic nerve may be simply an anatomical convenience, or may have something to do with a bird's ability to see straight ahead though its eyes be laterally positioned. (See Am. Nat., ii, 1868, p. 578; Pr. Bost. Soc. Nat. Hist., xii, Apr. 21, 1869.)

Sense of Hearing: Audition. — This is enjoyed to a high degree by the "musical class" of the *Vertebrata*, — birds being the only animals besides man whose emotions are habitually aroused, stimulated, and to some extent controlled by the appreciation of harmonic vibrations of the atmosphere. Most birds express their sexual passions in song, sometimes of the most ravishing quality to our ears, as that of the nightingale or the bluebird, and it cannot be supposed that they themselves do not experience the effect of music in an eminent degree of pleasurable perturbations. Otherwise, they would cease to sing. The capability of musical expression resides chiefly in the more spiritualized male sex; the receptive capacity of musical affections is better developed in the female, who chiefly furnishes the plastic material which is to be moulded into the physical manifestation of the male principle. Quickness of ear is extraordinary in such birds as those of the genus *Mimus*, which correctly render any notes they may chance to hear, with greater readiness and accuracy than is usually within human possibility. It may be reasonably doubted that any others than some of the world's greatest musical composers have a higher experience in acoustic possibilities than many birds. Birds' ears have nevertheless a comparatively simple anatomical structure, on the whole much more like that of reptiles than of mammals. Such simplicity is seen in the ligulate or strap-shaped cochlea, the essential organ of hearing, figs. 84, 85, 86, 87, as compared with the helicoid curvation of the mammalian cochlea. The openness of the ear-parts which lie outside the tympanum is seen in fig. 62, at the place where the reference-lines "ear-cells" reach the skull; and especially in fig. 71, where the stapes, *st*, is seen lying in the ear-cavity, the tympanum having been removed.

There is ordinarily no external ear, in the sense of a fleshy conch or auricle, though owls at least have a considerable flap which overlies the auditory aperture. The place of an auricle is filled by a set of peculiarly modified feathers surrounding and overlying the opening, called in ornithology the ear-coverts, or *auriculars* (p. 102; fig. 25, [36]). The outer ear or *meatus auditorius externus* is a considerable shallow roundish depression in the skull, at the extreme lower lateral corner. Its ordinary boundaries are the movably articulated quadrate bone in front, the expanded rim of the squamosal above, the tympanic wing of the exoccipital behind and below; the termination of the basitemporal also usually contributing to the under boundary. (See fig. 71, at *st;* 63, under *l;* fig. 62, where reference lines " bones of ear cell " go.) On removing the quadrate from the dry skull, the general tympanic depression is seen to be more or less continuous with the alisphenoid; the boundary is best marked behind and below by the broad thin sharp-edged shell of the tympanic wing of the exoccipital. To the brim indicated is attached the *tympanum,* or drum of the ear — that membrane being, from the configuration of the parts, quite superficial, — not at the bottom of a tube-like meatus, as in man. The membrane proper is invested externally by modified common integument which readily peels off. Thus this wide shallow depression overlaid with feathers or a slight flap is all there is to represent the " outer ear-passage." The tympanic membrane sometimes develops slight ossification, which then represents the " tympanic bone," or " external auditory process " of human anatomy. Did not this membrane occlude the way, the passage through the ear to the mouth would be pervious. This passage is the modified persistence of the *first visceral cleft* or " gill-slit " of the embryo. Just within the tympanic membrane is the *cavity of the tympanum* or *middle ear,* which may be very extensively exposed by merely removing the membrane. Looking into this cavity, as may readily be done from the outside, in carefully cleaned dry skulls, many objects of interest are presented; among them, a number of foramina — openings leading in various directions. In the first place there are some (inconstant and not readily identified) holes, which are *pneumatic* openings, conveying air from the middle ear-passage to the interior of bones of the skull and lower jaw. Next is observed a large orifice in the lower anterior part of the cavity, — the mouth of the *eustachian tube.* This tube continues the ear-passage to the mouth; opening at the back of the hard palate by a median orifice in common with its fellow. In clean skulls of any size a bristle, or even a wooden tooth-pick, will pass through the eustachian tube, and appear upon the floor of the skull in mid-line or nearly there, under the basisphenoid, over the basitemporal. The foregoing passages have not conducted us to the *inner ear* or proper acoustic cavity. There will be observed, in the side-wall of the tympanic cavity, two definite openings near the eustachian orifice. One of these, anterior and superior to the other, larger usually, and oval, is the *fenestra ovalis;* it lies in the obliterated suture between the proötic and opisthotic bones; and when the membranous curtain which closes it in life is gone, you look through this " oval window " into the *vestibular cavity* of the ear proper. The lower, posterior, circular orifice is the *fenestra rotunda;* through which round window in the opisthotic bone you look into the *cochlear cavity* of the ear proper. Fenestra ovalis and f. rotunda are generally close together, — only divided by a little bridge of bone, or a mere bony bar. To the circumference of the fenestra ovalis is fitted the expanded oval foot of the trumpet-shaped *columella auris,* — the *stapes,* or " stirrup-bone," as it is called in mammals (fig. 83, *st*). This is an elegant little bone, which establishes mechanical connection between the membrane closing the fenestra ovalis and the tympanic membrane, — something on the principle of the " sounding-post " inside a violin. It is shown magnified greatly in its embryonic condition, in fig. 67, and there seems to be primitively and morphologically the proximal connection of the *hyoid bone* (by cerato-hyal elements) with the bony capsule of the ear; but no trace of this relation persists. Fig. 83 shows the mature stapes of a fowl, and indicates its several elements which have received special names. In skulls prepared with sufficient care, the stapes may be seen *in situ*, as in fig. 71, *st*, — an extremely delicate rod, stepped into the fenestra ovalis by its foot, the other end protruding freely, and bearing in many cases its hammer-like or claw-like stapedial elements. A stapes I have just picked out of an eagle's ear is a fourth of an inch long, with a stout foot, but a stem as fine as a thread of sewing silk, and at the tympanic end a still finer hair-like process half as long as the main stem, from which it stands out at a right angle. The ossification is perfect, and there appears to have been another similar process which has broken off from the cross-like figure shown in fig. 71, *st*. In a raven's skull before me the stapes has fallen into the fenestra ovalis, and lies there with its head sticking out, though perfectly loose. I cannot withdraw it intact, as the expanded foot fits the hole too closely to pass through in any position I have succeeded in placing it. It appears to be about as large as the eagle's. Close examination at a point somewhere about the fenestra ovalis, or between that and the eustachian orifice, will discover a minute foramen, corresponding to the "stylo-mastoid" foramen of mammals. It transmits cranial nerve 7 (see p. 183), or the *facial nerve*, which has burrowed through the bony acoustic capsule from the brain-cavity and entered the tympanic cavity on its way to the surface. There are sometimes *two* such minute foramina, close together, both conducting to the brain cavity (neither in common with the internal auditory meatus); as in the eagle, in which large bird a fine bristle just passes through each. Thus in the dry skull of a bird, all the hard parts of the middle ear or tympanic cavity, as well as the eustachian tube, can readily be inspected from the outside; even the limits of the opisthotic and proötic bones can be determined to some extent, and the *ossiculum auditûs* be seen *in situ*. There will also be noted, in most birds, the articular facet upon the proötic bone for the inner head of the quadrate, as well as upon the squamosal for the outer head of the quadrate; however these may shift in position, in different birds, they cannot easily be overlooked or mistaken. Details of mere size and configuration aside, the above general description will apply pretty well to any bird, and should suffice for the identification of the objects seen on looking into the ear, though the number and variety of the irregular *pneumatic* openings may be puzzling at first. To see these things clearly in a *mammal's* ear would require special preparation of the parts, as they lie inside a tympanum which is itself at the bottom of a contracted tube. In such an ear, properly laid open, would be found a chain of *three* ossicles crossing the tympanic cavity from the inner surface of the tympanic membrane to the opposite surface of the membrane closing the fenestra ovalis — the *malleus, incus*, and *stapes*, or "hammer," "anvil" and "stirrup;" and the latter would be stirrup-shaped, not trumpet-like with a cross-bar at the mouth-piece. Some mammals would also show a hyoid bone which would have what are the cerato-hyals of a bird produced up toward the ear-parts, and continued to these by a bone called *stylo-hyal*, or "styloid process of the temporal"; and any mammal's jaw would articulate directly with the squamosal, — the chain of three ossicles being entirely inside the ear. As to comparing the parts now: the mammalian stapes is the stapes or columella of a bird, — its stem and foot at least; the incus of a mammal is represented by one of the claws of the cross-bar of a bird's stapes (the *supra*-stapedial element; fig. 83, *sst*); the malleus of a mammal is the great quadrate bone of a bird; the stylo-hyal of a mammal is not fairly developed in a bird, unless contained in or represented by another claw of the stapes (an *infra*-stapedial element, *ist*); and in these facts is the reason why a bird's lower jaw is articulated indirectly to the skull by means of the quadrate, and also why a bird's hyoid bone is not articulated or in any way

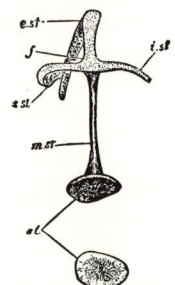

FIG. 83. — Mature stapes of fowl, about × 4; after Parker. *st*, its foot, fitting fenestra ovalis; *mst*, main shaft, or medio-stapedial element; *sst*, supra-stapedial; *est*, extra-stapedial; *ist*, infra-stapedial, its end representing a rudimentary stylo-hyal; *f*, a fenestra in the extra-stapedial. (See *st* in situ, fig. 71, and its embryonic formation, fig. 67.)

directly connected with the skull — excepting when, as in a woodpecker, elongated *branchial* elements of the hyoid bone take on such office by curling over the cranium (figs. 73, 74).

Section of the bone is required for further examination of the ear-parts. On longitudinally bisecting the skull, or otherwise gaining access to the brain-cavity, the internal surface of the *periotic* bone is brought into view (fig. 70, *po, op, ep*). It is the same bone we have seen in the tympanic cavity, now viewed upon its cerebral surface. In a skull of any size, as that of the eagle before me (from which the rest of my description will be taken), there is no difficulty in making out the parts, although the periphery of the periotic bone is completely consolidated with its surroundings. The periotic, or *petrosal* (Lat. *petrosus*, stony — from its hardness), or "petrous part of the temporal," is the bony capsule of the inner ear, enclosing the *labyrinth* or essential organ of hearing, — in fact, it is the skull of the ear, sometimes therefore called the *otocrane* — just as ethmoidal parts form the "skull of the nose," and the sclerotal bones represent a "skull of the eye." The periotic consists of the three bones already often mentioned, — the *proötic, po, epiotic, ep*, and *opisthotic, op*, or anterior, superior, and posterior otocranial bones, completely consolidated together, as well as with surrounding bones. The petrosal appears as an irregular protuberance in the inner wall of the brain-cavity, at the lower back part. It seems to be more extensive than it really is, because the great superior semicircular canal, too large to be entirely accommodated in the petrosal, has invaded the occipital bone, — the track of its bed in that bone being sculptured in bas-relief (fig. 70, *asc*). Behind this semicircular trace, the deep groove of a venous sinus is engraved in the bone, making the tract of the canal still more prominent (fig. 70, *sc*). The top of the petrosal and contiguous occipital is the floor of a recess or *fossa* in which is lodged the great optic lobe of the brain, partly divided from the general cavity for the cerebral hemisphere by a bony tentorium, like that which in mammals separates the cerebellar from the cerebral fossæ. On the vertical face of the petrosal, or on the corresponding occipital surface, is a large smooth-lipped orifice, at least $\frac{1}{10}$ of an inch in longest diameter; it leads to a tongue-like excavation of the bone, in which the *flocculus* of the cerebellum is lodged. In front, between the petrosal and alisphenoid (or in the conjoined border of one or the other of these bones) is a considerable foramen, conducting the second and third divisions of cranial nerve 5 (see p. 183; figs. 70, 71, ^{5}) into the *orbit*. Below the petrosal (in fact, between the opisthotic and the exoccipital), near the border of the foramen magnum, is a foramen (which may be subdivided into foramina), representing the *foramen lacerum posterius* of mammals, transmitting cranial nerves 9, 10, 11 (see p. 183; fig. 70, ^{8}). The general space under description is continued to the margin of the foramen magnum by the exoccipital (fig. 70, *eo*). Now on the vertical face of the petrosal itself — behind foramen for 5, above that for 9, 10, 11, in front of the large floccular orifice, will be seen a smooth-lipped depression, the *meatus auditorius internus* (fig. 70, ^{7}), at the bottom of which are at least *two separate* small foramina. A bristle passed in the upper (or anterior) one of these two holes emerges outside the skull, in the tympanic cavity, near the tympanic end of the eustachian tube; it has traversed the interior of the petrosal, in a track known as the *fallopian nerviduct;* it transmits cranial nerve 7 — the *facial*, or *portio dura*. A bristle passed into the other of the two foramina may also be made to come out in the tympanic cavity, but by a different track, for it emerges through either the fenestra ovalis or the fenestra rotunda; it has traced the course of cranial nerve 8, — the *auditory nerve* or *portio mollis*. Both bristles have entered the common internal auditory meatus, but the second one has traversed the ear-cavity proper, through the *labyrinth* of the ear, and come out at the tympanic *vestibular* orifice (fenestra ovalis), or at the tympanic *cochlear* orifice (fenestra rotunda). Either passage is easily made, without breaking down or indeed meeting with any bony obstacle, which would not be the case with a mammal. Cranial nerves 7 and 8 were formerly counted as one (seventh); hence the name *portio dura* ("hard portion") for the former, and *portio mollis* ("soft portion") for the latter. The former, as said, traverses the petrosal bone and escapes upon the face; the latter, which is the true acoustic nerve, or nerve of hearing, remains in the bone, being expended upon the labyrinthine structures within — the *vestibule, semicircular canals,* and *cochlea,* which constitute the walls of the cavities in which the essential organ of hearing is snugly encased.

If now, with a very fine saw — the saws now so much used for fancy scroll-work will answer the purpose — the whole periotic mass be cut away from the skull, and then divided in any direction, the labyrinth can be studied. It is best to make the section in some definite plane with reference to the axes of the whole skull, — the vertical longitudinal, or vertical transverse, or horizontal, — as the direction and relations of the contained structures are then more easily made out. Four or five parallel cuts will make as many thin flat slices of bone, affording eight or ten surfaces for examination; the whole course of the labyrinthine cavity can be seen in sections which, when put together in the mind's eye, or held a little apart in their proper relations and visibly threaded with bristles, afford the required picture very nicely. It is extremely difficult to chisel out the affair from the bone in which it is embedded. At first glance the slices show a bewildering maze, — a continuous net-work or lattice-work of bone, in which the unaccustomed eye will recognize nothing but confusion. All this *cancellated* structure, however, is pneumatic — the open-work tissue of the bone, containing air derived from the tympanic or eustachian cavities, and having nothing to do with the ear-passages proper. Parts of the *bony labyrinth* will soon be recognized by their firm smooth walls and definite courses, as distinguished from the irregular interstices of the pneumatic bone-tissue. The bony labyrinth consists of an irregular central cavity, the *vestibule;* of a cavity, projecting like a beak downward and backward from the vestibule, the *cochlea;* and of three horseshoe-shaped tubular cavities, above, behind, and below the vestibule, the *semicircular canals,* the ends of whose hollows all open into the vestibule. Imagine three hollow horseshoes, with their ends melted into a hollow inflation (vestibule), the opposite wall of which is a hollow projection (cochlea) — or a hollow flat-iron (vestibule) with a long nose (cochlea) and three hollow handles (the canals). Or, see figs. 84 to 87, representing the contained *membranous* labyrinth, to which the containing bony labyrinth very closely conforms, as it is simply the bony cavity whose walls encase the membranous and other soft structures. According as the sections have been made, numerous cross-cuts of the canals will be seen here and there as circular orifices; the canals themselves lying curled like worms in the petrosal and occipital substance, their ends finally converging to the vestibular cavity. As compared with those of man, the parts are of great size; in the eagle, the whole affair is as large as that part of one's thumb covered by the nail; the whole length of the superior semicircular canal is an inch or more; its calibre, I should judge, being absolutely about as great as in man. The cochlea, however, though not diminutive comparatively, is in a rudimentary condition as far as complexity of structure is concerned, in all *Sauropsida,* representing only the beginning of the cochlear structure of mammals. In the latter class, the cochlea is spirally coiled or whorled on itself like a snail-shell (whence the name — *cochlea,* a snail), making at least one turn and a half, sometimes five (two and a half in man); with a centre-post or *modiolus* around which winds a bony flange, the *lamina spiralis,* a membranous extension of which to the cochlear out-wall divides the cavity into two compartments or *scalæ* (*scala,* a flight of stairs); it is just like a spiral stairway, only an inclined plane instead of a series of steps. The membranous extension of the bony spiral lamina to the side-wall obviously throws the cavity, as just said, into *two* spirals, which only intercommunicate at the top, where the modiolus ends in a funnel-shaped expansion, the *infundibulum,* beneath the apex of the snail-shell, the *cupola.* A marble rolling down the *upper* stairway would fall into the *vestibular* cavity; this division of the cochlea is therefore the *scala vestibuli.* The marble starting from the other side of the infundibulum would roll along the under stairway, and if nothing stopped the way, would fall through the fenestra rotunda into the tympanic cavity; this is therefore the *scala tympani.* The first marble would also eventually reach the tympanum, through the vestibule, and out of the fenestra ovalis, if the foot of the

THE ANATOMY OF BIRDS.—NEUROLOGY. 195

stapes were unstepped (in life, of course, both these "windows" are closed by membranous curtains). Now in birds the cochlear cavity and its bony or cartilaginous contents are only the beginnings of such structure—a strap-shaped or tongue-like protrusion from the vestibule, as if a part of the first mammalian whorl, and very incompletely divided into scala vestibuli and scala tympani by a gristly structure (representing the modiolus and its lamina), which proceeds from the bony bar or bridge between fenestra ovalis and fenestra rotunda. (See figs. 84, 85.) This structure is the most intimate and essential part of the organ of hearing, for upon it spread the terminal filaments of the auditory nerve. A human or any well-developed mammalian cochlea is a thing of marvellous beauty, even as to its bony shell—there is nothing to compare with its exquisite symmetry; while the spiral radiation of the nervous tissue introduces yet other and more wondrous "curves of beauty."

The *vestibule* hardly requires special description; it is simply the central chamber common to the cochlear and canalicular cavities; receiving the mouth of the scala vestibuli of the cochlea; the several mouths of the separate or uniting semicircular canals; opening into tympanum by fenestra ovalis; conducting to meatus auditorius internus by the course of the auditory nerve. In the eagle, if its irregularities of contour were smoothed out, it would about hold a pea.

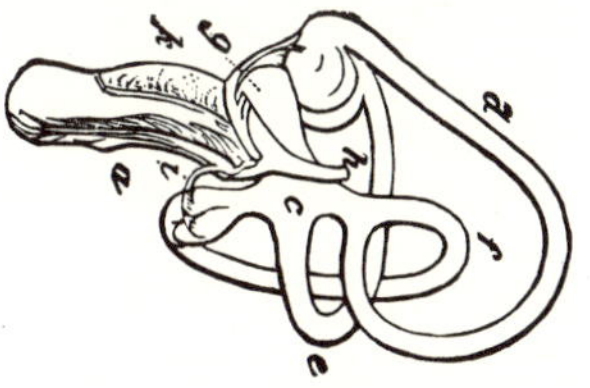

In the language of human anatomy, the three *semicircular canals* are the (*a*) anterior or superior vertical, the (*b*) posterior or inferior vertical, and the (*c*) external or horizontal; and the planes of their respective loops are approximately mutually perpendicular, in the three

FIG. 84.
FIG. 85. 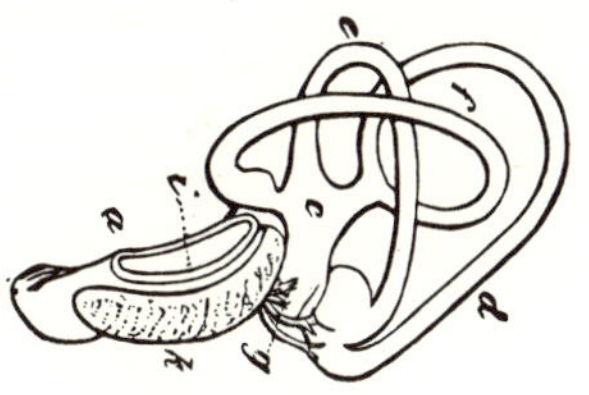
FIG. 86. 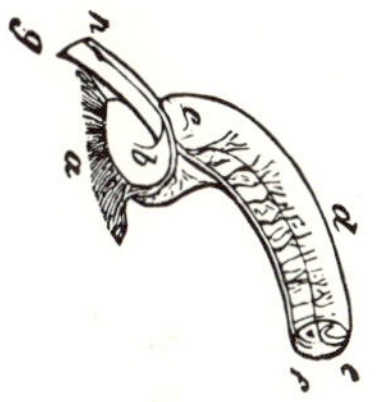
FIG. 87. 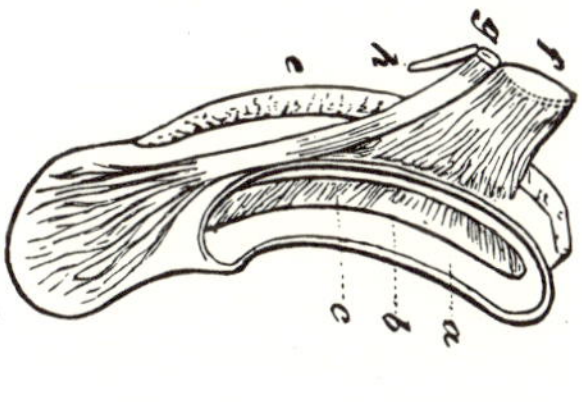
FIG. 88. 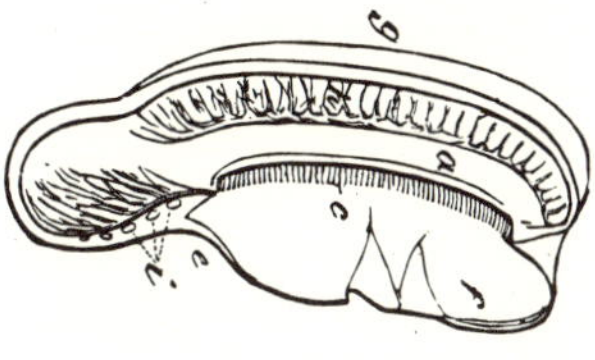

Figs. 84, 85, membranous labyrinth of *Haliaëtus albicilla*, ×2. *a*, *b*, cochlea; *b*, its saccular extremity (or lagena) *c*, vestibule; *g*, its utricle; *d*, anterior or superior vertical semicircular canal; *e*, external or horizontal semicircular canal; *f*, posterior or inferior vertical semicircular canal; *h*, membranous canal leading into aqueduct of the vestibule; *k*, vascular membrane covering the scala vestibuli; opposite this, at *i*, are seen the edges of the cartilaginous prisms in the fenestra rotunda; from the edges of these cartilages proceeds the delicate membrane closing the opening of the cochlea (not shown in the fig.). Fig. 86, part of the superior vertical semicircular canal, showing its ampulla (which is the dilatation of the base of any semicircular canal), nerve of ampulla, artery and connective tissue of the perilymph, ×3. *a*, that part of the vestibule (alveus) next to the ampulla; *b*, the dilatation of the ampulla at its vestibular opening; *c*, where it passes into the canal proper; *d*, the canal, furnished with connective tissue of the perilymph along its concave border and sides, as appears clearly at the sections *e* and *f*; *g*, nerve of the ampulla; *h*, artery of the connective tissue, running beneath it, remote from the wall of the duct. Fig. 87, cochlea, ×3. *a*, vascular membrane; *b*, external; *b*, internal, cartilaginous prism; *c*, membranous zone; *d*, saccular extremity of the cochlea, or lagena; *e*, vascular membrane; *f*, auditory nerve, its middle fascicle penetrating the internal cartilaginous prism, to reach the membranous zone by its terminal filaments; *g*, auditory nerve, its posterior fascicle, running to the most posterior part of the lagena; *h*, filament to ampulla of posterior or inferior vertical semicircular canal. Fig. 88, section of the cochlea, ×3. *a*, vestibular surface of external cartilaginous prism, extending into *d*, the lagena; *c*, section of the membranous zone; *e*, Huschke's process of the fenestra, which, with the margins of the cartilaginous prisms, affords attachment to the blind sac *f*, occluding the fenestra of the cochlea; *g*, spongy vascular membrane of the scala vestibuli; *h*, auditory lamellæ of Treviranus; *i*, canals in posterior wall of the lagena, by which the nervous filaments enter its cavity.

(From Ibsen's Anatomiske Undersøgelser over Fréts Labyrinth. 4to, Kjøbenhavn, 1881, p. 17, pl. 1, figs. 13-17.)

planes of any cubical figure. In birds these terms do not apply so well to the situation of the canals with reference to the axes of the body, nor to the direction of the loops; neither is mutual perpendicularity so nearly exhibited. The whole set is tilted over backward to some degree, so that the (*a*) "anterior" (though still superior) loops back beyond either of the others; the (*b*) "posterior" loops behind and below the (*c*) horizontal, which tilts down backward; the verticality of the planes of (*a*) and (*b*) is better kept. The canals may be better known as the (*a*) superior (vertical), and (*b*) inferior (vertical), and (*c*) internal (horizontal). Whatever its inclination backward, there is no mistaking (*a*), much the longest of the three, looping high up over the rest, exceeding the petrosal and bedded in the occipital, the upper limb and loop of the arch bas-relieved upon the inner surface of the skull (fig. 70, *asc*). It makes much more than a semicircle — rather a horse-shoe. The inferior vertical (*b*) loops lowest of all, though little if any of it reaches further backward than the great loop of (*a*); it is the second in size; in shape it is quite circular, — rather more than a half-circle. Its upper limb joins the lower limb of (*a*), as in man, and the two open by one orifice in the vestibule; but it is not simple union, for the two limbs, before forming a common tube, twine half-round each other (like two fingers of one hand crossed). The loop of (*b*) reaches very near the back of the skull (outside). The canal (*c*) is the smallest, and, as it were, set within the loop of (*b*), though its plane is nearly the opposite of the plane of (*b*); and the cavities of (*b*) and (*c*) intercommunicate at or near the point of their greatest convexity, farthest from the vestibule. This decussation of (*b*) and (*c*), like the twining inosculation of (*a*) and (*b*), is well known. It may not be so generally understood that there is (in the eagle if not in birds generally) a *third* extra-vestibular communication of the canals. My sections show this perfectly. The great loop of (*a*), sweeping past the decussating-place of (*b*) and (*c*), is thrown into a cavity common to all three. Bristles threaded either way through each of the three canals can all three be seen in contact, crossing each other through this curious extra-vestibular chamber, which may be named the *trivia*, or "three-way" place. (The arrangement I make out does not agree well with the figure of the owl's labyrinth given by Owen, *Anat. Vert.*, ii, 134. The trivia is at the place where, in fig. 84 or 85, the three *membranous* canals cross one another. It does not follow, however, that these contained membranous canals intercommunicate, and it appears from Ibsen's figures that they do not. Study of these admirable illustrations, with the explanations given under them, should make the details perfectly clear to the reader.)

All that precedes relates to the *bony* labyrinth, — the scrolled cavity of the periotic bone. The *membranous labyrinth* is a sac lying loosely in the hollow of the bone, and shaped just like it, lining the hollow of the vestibule and tubes of the semicircular canals. Withdrawn intact, it would be a perfect "cast" of the labyrinth. Originally, this sac is also continuous with one in the cavity of the cochlea, called the *membranous cochlea*, which afterward becomes shut off from the main sac. This shut-off cochlear part lies between the scala tympani below and the scala vestibuli above; its interior is the *scala media*. If demonstrable in birds, it must be quite as rudimentary as the other scalae. The membrane is not attached to the bony walls of the labyrinth, but is separated by a space containing fluid, the *perilymph*, which also occupies the scala vestibuli and scala tympani. A similar fluid, the *endolymph*, is contained in the cavity of the membranous labyrinth, and scala media of the cochlea; in it are found concretions, or *otoliths*, of the same character as the great "ear-stones" so conspicuous in many fishes. This lymph has a wonderful office — that of *equilibration*, enabling the animal to preserve its equilibrium. The labyrinth and its contained fluid may be likened to the glass tubes filled with water and a bubble of air, by a combination of which a surveyor, for example, is enabled to adjust his theodolite true to the horizontal. Somehow a bird knows how the fluid stands in the self-registering levelling-tubes, and adjusts itself accordingly. Observations made on pigeons show that "when the membranous canals are divided, very remarkable disturbances of equilibrium ensue, which vary in character according to the seat of the lesion. When the horizontal canals are divided rapid movements of the head from side to side, in a horizontal plane, take place, along with oscillation of the eyeballs, and the animal tends to spin round on a vertical axis. When the posterior or inferior vertical canals are divided, the head is moved rapidly backwards and forwards, and the animal tends to execute a backward somersault, head over heels. When the superior vertical canals are divided, the head is moved rapidly forwards and backwards, and the animal tends to execute a forward somersault, heels over head. Combined section of the various canals causes the most bizarre contortions of the head and body." (Ferrier, Funct. of the Brain, 1876, p. 57.) Injury of the canals does not cause loss of hearing, nor does loss of equilibrium follow destruction of the cochlea. Two diverse though intimately connected functions are thus presided over by the acoustic nerve, — audition and equilibration.

Senses of Taste and Touch: Gustation and Taction. — The hands of birds being hidden in the feathers which envelop the whole body — their feet and lips, and usually much if not all of the tongue, being sheathed in horn, these faculties would appear to be enjoyed in but small degree. While it is difficult to judge how much appreciation of the sapid qualities of substances birds may be capable of, we must not be hasty in supposing their sense of taste to be much abrogated. One who has had the toothache, or teeth "set on edge" by acids, or painfully affected by hot or cold drinks, may judge how sensitive to impressions an extremely dense tissue can be. Persons of defective hearing may be assisted to a kind of audition by an instrument applied to the teeth; and it is not easy to define the ways in which sensory functions may be vicariously performed or replaced. Birds are circumspect and discriminative, even dainty, in their choice of food, in which they are doubtless guided to some extent by the gustatory sensations they experience. As, however, only some human beings make these an end instead of a natural and proper means to an end, the selection of food by birds may be chiefly upon intuitions of what is wholesome. Such purely gustatory sense as they possess is presided over by the branches of the *glosso-pharyngeal* nerve which go to the back part of the tongue and mouth. Though the chorda tympani nerve exists, there is no lingual (gustatory) branch of the third division of the fifth cranial nerve. Yet the latter, which goes in mammals to the anterior part of the tongue, is less effectually gustatory than the glosso-pharyngeal; as we know by the fact that the sensation of taste is not completely experienced until the sapid substance passes to the back of the mouth. Gustation is likewise connected with olfaction; the full effect of nauseous substances for example, being not realized if the nose is held. From these alternative considerations, each one may estimate for himself how much birds know of sapidity; remembering also, how soft, thick, and fleshy are the tongue and associate parts in some birds, as parrots and ducks, in comparison with birds whose mouths are quite horny.

The beak is doubtless the principal tactile instrument; nor does its hardness in most birds preclude great sensitiveness; as witness the case of the teeth, above instanced. Sensation is here governed by the branches of the fifth nerve. In some birds, in which also the terminal filaments of this nerve are largest and most numerous, the bill acquires exquisite sensibility. Such is its state in the snipe family, in most members of which, as the woodcock, true snipe, and sandpipers, the bill is a very delicate nervous probe. The *Apteryx* also feels in the mud for its food, enjoying moreover the unusual privilege of having its nose at the end of its long exploration. Ducks dabble in the water to sift out proper food between the "strainers" with which the sides of their beaks are provided; and the ends of the maxillary and mandibular bones themselves are full of holes, indicating the abundance of the nervous supply (fig. 63).

The senses of birds and other animals are commonly reckoned as five — a number which may be defensively increased — as by a sixth, the muscular sense, which gives consciousness of strain or resistance, apart from purely tactile impressions; and perhaps a seventh, the faculty of equilibration, which has a physical mechanism of its own, at least as distinct and complete as that of hearing. The ordinary "five senses" are curiously graded. *Taction* connotes qualities of matter in bulk, as density, roughness, temperature, etc. *Gustation*, matter dissolved in water — fluidic. *Olfaction*, matter diffused in air — aeriformed. *Audition*, atmospheric air in undulation. *Vision*, an ethereal substance in undulation. All animals are probably also susceptible of *biogenation*, which is the affection resulting from the influence of biogen; a substance consisting of self-conscious force in combination with the minimum of matter required for its manifestation.[1]

c. Myology: the Muscular System.

Muscular Tissue consists of more or fewer amœbiform animals; separate colonies of which creatures, isolated in various parts of the body, compose the individual different muscles. They are enveloped in fibrous tissue, the sheets of which are called *fasciæ*, and the ends of which, usually attached to bones by direct continuity with the periosteal covering of the latter, form tendons and ligaments. The muscle-animals belong to a genus which may be termed *Myamœba*, differing from other genera of the amœbiforms which compose the body of a bird less in their physical character of being elongated and spindle-shaped, or even filiform, than in their physiological character of *contractility*. Under appropriate stimulus, as the passage of a current of electricity, or the wave of biogen-substance which constitutes a "nerve-impulse," *Myamœbæ* shorten and thicken, tending towards a state of tonic contraction which, if completed and long sustained, would cause them to become encysted as spherical bodies; but extreme contraction is never long continued. By alternate contraction and relaxation all the motions of the body in bulk are effected. The capacity of, or tendency to, contraction is called the *tonicity* of muscular fibre. The simultaneous contraction of any colony of *Myamœbæ* pulls upon the attachment of the muscle at each of its ends; in some cases approximating both ends; oftener moving the part to which one end is attached, the other being fixed. The action of a muscle is upon the simplest mechanical principles, — nothing more or less than pulling upon a part, as by a rope, the line of traction being exactly in the line of contraction of the muscle; though it is often ingeniously changed by the passage of tendons around a corner of bone, or through a loop of fibrous tissue, as if through a pulley. Such movements as those of a turtle protruding its head, or a bird thrusting its beak forward, where muscle seems to *push*, are fallacious; when analyzed, the motion is invariably resolved into simple *pulling*. The swelling up of a muscle in contracting must indeed impinge upon neighboring parts and shove them aside; but that is an extrinsic result. Muscles contract most powerfully under resistance to their turgescence: what is effected by the fasciæ which bind them down; — what the athlete seeks to increase by bandaging his swelling *biceps*. There are two species of *Myamœba*. *M. striata* is the ordinary striped fibre of voluntary motion, and also of some motion not under control of the will, as that of the heart. This species is usually of a rich red color (pale pink in many birds of the grouse family), and is the ordinary " flesh " of the body. The other species, *M. lævis*, composes the pale or colorless smooth fibre of the involuntary muscles, as those of the intestines, the gullet, etc. A species of contractile tissue commonly referred to the genus *Desmamœba* (indifferent connective-tissue cells) is very near *Myamœba lævis*; example, mammalian *dartos*. The movements of erectile organs, as the neat combs over the eyes of grouse, or the turkey's caruncles, are not in any sense *myamœbic*, but depend mechanically upon influx of blood.

The Muscular System of Aves can only be touched upon; it is impossible in my limits to even name all the muscles, much less describe them. I can only note the leading peculiarities, and present a figure in which the principal muscles are named.

[1] The reader who may be interested to inquire further in this direction is referred to a publication entitled: — Biogen: A Speculation on the Origin and Nature of Life. Abridged from a paper on the " Possibilities of Protoplasm," read before the Philosophical Society of Washington, May 6, 1882. By Dr. Elliott Coues, etc. Washington, Judd & Detweiler. 8vo, pp. 27. Second ed., Boston, Estes & Lauriat, 1884.

The subcutaneous sheet of muscle (of which the human "muscles of expression" and *platysma myoides* are segregations) is broken up in birds into a countless number of little slips which agitate the feathers collectively, and especially the great quills of the wings and tail. There are estimated to be 12,000 in a goose. The prime peculiarity of birds' musculation is the enormous development of the *pectorales*, or breast muscles, which operate the wings. The great pectoral, *p. major* or *p. primus*, arises from the sternal keel, when that special bony septum between the fellow-pectorals exists, and from more or less of the body of the sternum, passing directly to the great pectoral or outer ridge of the humerus, near the upper end of that bone. Its origin may even exceed the limits of the sternum, invading the clavicle, etc. ; it may unite with its fellow. It is the depressor of the humerus, giving the *downward* stroke of the wing. The next pectoral, *p. secundus* or *p. medius*, arises from much or most of the sternum not occupied by the first, under cover of which it lies; it passes also the humerus, but by an interesting way it has of running through a pulley at the shoulder it elevates that bone, giving the *upward* wing-stroke. A third pectoral, *p. tertius* or *p. minimus*, arising from sternum, and often contiguous parts of the coracoid bone, passes directly to the humerus, supplementing the action of the first. A fourth muscle in many birds acts upon the humerus from the sternum or coracoid, particularly the latter. These four differ greatly in their relative development. Such extent of the sternum and pectoral muscles correspondingly reduces that of the belly-walls, and the abdominal muscles are consequently scanty. Fixity of the spinal column in the dorsal region diminishes the musculation of that part, the spinal muscles being much better developed in the cervical region; where, in cases of some of the long-necked birds, there are curious contrivances for the mechanical advantage of the muscle in flexing and extending this mobile part of the body. Muscles of the hyoidean apparatus acquire a singular development in woodpeckers. The lower jaw is depressed particularly by muscle inserted into the end of the mandible; the upper is elevated by particular muscles operating the pterygoid and quadrate bones. Temporal, masseteric, and ordinary pterygoid muscles close the jaws. They are unsymmetrical in *Loxia*.

The *diaphragm*, the musculo-membranous partition which in mammals divides the thoracic from the abdominal cavity, is only represented in birds in a rudimentary condition. Macgillivray has figured that of the rook as consisting of three fleshy slips, *v*, *v*, *v*, passing from as many ribs, 4, 5, 6, to the pleural sac of the lungs, *t*, *t*, in fig. 101, p. 212. It is best developed in the *Apteryx*.

The remarkable specialization of both limbs, — the former for flight, the latter for the perfectly bipedal locomotion which only birds besides man enjoy, — results in corresponding peculiarities of the muscular mechanism. Muscles beyond the shoulder are greatly reduced in number and complexity from an ordinary quadrupedal standard; those of the legs are rather increased, and their configuration, relative size, and to some extent their relations are so much changed, that great difficulty is experienced in identifying them with the corresponding muscles of quadrupeds. The result is, great confusion in their nomenclature, which is still shifting, though much has been done of late to give it precision. Attention has recently been called by Garrod to the classificatory value of certain muscles of the limbs. The *tensor patagii*, that muscle or muscles which may have elastic tendons, and by which the folds of skin in the angles of the wing bones are regulated, may have different characters in different groups of birds. It has long been known that particular muscles of the hind limb are in direct and important relation to the prehensile power of the toes, and consequently co-ordinated with the insessorial or the reverse character of the foot. In the highest birds, *Passeres*, the foot grasps with great facility, owing to the distinctness or individuality of the *flexor longus hallucis*, or bender of the hind toe. The *ambiens* (Lat. *ambiens*, going around) is a muscle of which Garrod has even made so much as to divide all birds into two primary groups according to whether they possess it or not. The ambiens arises from the pelvis about the acetabulum, and passes along the inner side of the thigh; its tendon runs over the *convexity* of the knee to the outer side, and ends by

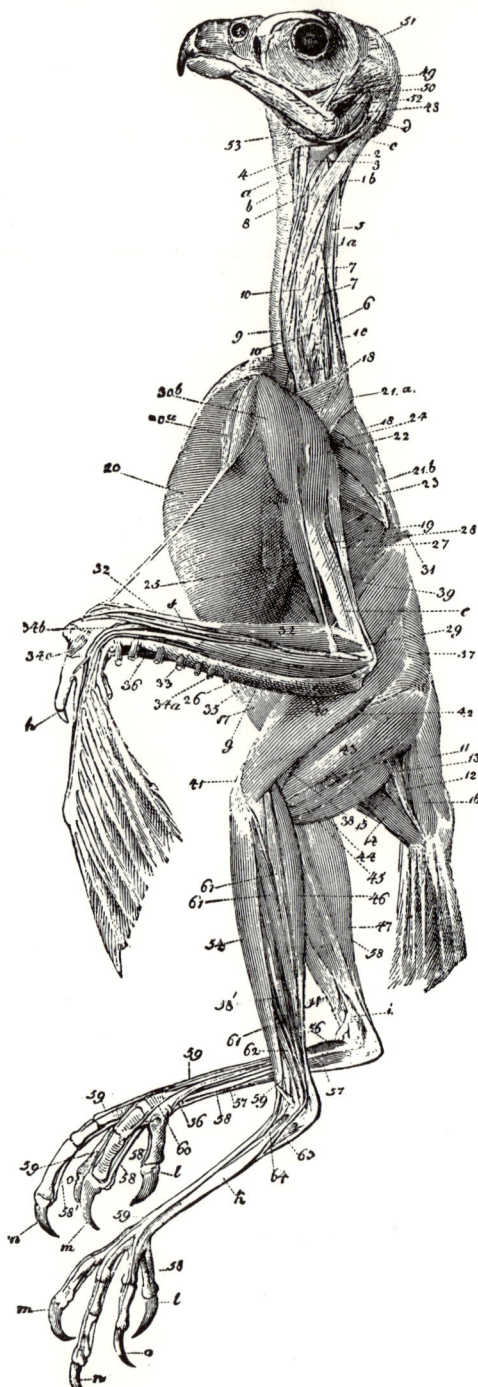

Fig. 89.—Muscles of a bird (*Accipiter nisus*), after Carus, *Tab. Anat. Comp.*, j, 1828, pl. 4. *a*, pharynx ; *b*, trachea; *c*, hyoid bone; *d*, ear ; *e*, humerus; *f*, radius; *g*, ulna; *h*, radial finger; *i*, tibia ; *k*, metatarsus *l*, hind toe ; *m*, inner toe ; *n*, middle toe ; *o*, outer toe. 1, biventer cervicis, with central tendon 1 *a*, and upper 1 *b*, and lower 1 *c*, belly. 2, complexus. 3, flexor capitis lateralis. 4, flexor longus capitis. 5, extensor magnus cervicis. 6, descendens cervicis. 7, 7, semispinales. 8, flexor superior capitis. 9, flexor inferior or longus capitis. 10, 10, intertransversales. 11, levator coccygis. 12, depressor coccygis. 13, cruro-coccygeus (ilio-coccygeus?). 14, pubo-coccygeus. 15, ischio-coccygeus. 16, lateralis quartus (quadratus coccygis, to tail-feathers). 17, obliquus externus abdominis. 18, cucullaris (trapezius). 19, serratus magnus. 20, pectoralis major. 21, *a*, *b*, latissimus dorsi. 22, deltoid. 23, suprascapular. 24, coraco-brachialis. 25, biceps brachii. 26, supinator longus. 27, anconeus longus (part of "triceps"). 28, anconeus brevis. 29, anconeus brevissimus. 30 *a*, 30 *b*, tensor patagii, carpal and radial parts. 31, tensor patagii posterior. 32, extensor metacarpi longus. 33, extensor metacarpi brevis. 34 *a*, flexor digitorum sublimis. 34 *b*, flexor digitorum profundus. 34 *c*, flexor brevis pollicis. 35, flexor metacarpi radialis. 36, flexor (meta-) carpi ulnaris. 37, glutæus max·mus. 38, adductor femoris primus. 39, sartorius. 40, latissimus femoris. 41, gracilis = *ambiens* : only its tendon in sight. 42, vastus ; 43, biceps cruris. 44, semimembranosus. 45, semitendinosus. 46, 46, 47, gastrocnemius. 48, digastricus (chief opener of the mouth). 49, temporal. 50, long ligament. 51, cutaneous muscle of scalp. 52, masseter. 53, a muscle of the hyoid bone. 54, tibialis anticus. 55, tibialis posticus. 56, extensor hallucis. 57, flexor hallucis. 58, flexor digitorum profundus or perforans, seen n various places : long and short head, and several tendons. 59, extensor longus digitorum, tendons seen in various places. 60, abductor digiti interni. 61, 61, 61, flexores dig·torum perforati. 62, peronæus. 63, abductor minimi digiti 64, abductor hallucis.

connecting with the *flexor digitorum perforatus*, — one of the muscles which bend the toes collectively. When this arrangement obtains, the result is that when a bird goes to roost, and squats on its perch, the toes automatically clasp the perch by the strain upon the ambiens that ensues as soon as the leg is bent upon the thigh, and the tarsus upon the leg, the weight of the bird thus holding it fast upon its perch. The effect is as if an elastic cord were tied to the hip joint, thence directed over the front of the knee and back of the heel and so on to the ends of the toes. Obviously, such a cord would be strained when the limb is bent, relaxed when the limb is straightened out. The reader may observe a corresponding effect of the muscular arrangement of his forearm by throwing the hand as far back as possible; the fingers tend to close by the strain on the flexors in passing over what is a convexity of the wrist when the hand is in that position. *Passeres* have no ambiens, the perfection of their feet in other respects answering all purposes. Birds having it are termed *homalogonatous* or "normally-kneed" (Gr. ὁμαλός, *homalos*, from ὁμός, *homos*, like, even, etc.; γόνυ, γόνατος, *gonu, gonatos*, knee); those wanting it are called *anomalogonatous*, "abnormally-kneed." The distinction prevails with much applicability to various large groups of birds, and does good duty in diagnosis when duly connected with other characters; but surely should not give name to primary groups founded upon it! Other muscles of the leg much used by the same sagacious and zealous anatomist are the *femoro-caudal, accessory femoro-caudal, semitendinosus*, and *accessory semitendinosus*. The whole five of these muscles "vary; any one or more than one may be absent in different birds; . . . the constancy of the peculiarities in the different individuals of each species, or the species of each genus, and very generally in the genera of each family, makes it evident to any one working at the subject that much respecting the affinities of the different families of birds is to be learnt from the study of their myology, in connection with the peculiarities of their other soft parts; and that these features will, in the long run, lead to a more correct classification than one based on the skeleton alone, becomes almost equally certain." (Garrod, *P. Z. S.*, 1873, p. 630.) I quote in justice of this author, a modern Macgillivray in sincerity and love of truth; and very generally, in constructing my characters of the higher groups of birds in the body of this work, I shall be as glad to use the myological formulæ of Garrod, as I am here to pay this slight tribute to his memory.

d. Angeiology: the Vascular or Circulatory Systems.

Blood and Lymph are the two media by the circulation of which throughout the body the various amœboid animals which compose the tissues are fed, their waste repaired, and their dead parts removed. Each species of *Amœba* has the faculty of selecting from the constituents of blood and lymph its appropriate food; and of converting such nourishment into its own proper substance. Refuse matters are either drained off by the kidneys and voided as excrement, or swept by the current of blood into the lungs and there cremated. The stream of lymph is a feeder to the blood, and when the mingled currents are no longer distinguishable has become blood. The machinery of circulation is two sets of vessels — the *hæmatic*, or vascular system proper, consisting of the heart, arteries, veins and capillaries for the blood-circulation; and the *lymphatic*, consisting of lymph-hearts and vessels, for the flow of lymph. The lymphatics, converging from all parts of the body, and especially from the intestines, end in vessels which pour the lymph into the veins of the neck. The heart is the central organ of the blood-circulation, by which that fluid is pumped into all parts of the body through the *arteries* or *efferent* vessels; straining through the network of *capillaries*, it returns to the heart through the *veins*, or *afferent* vessels. The set of efferent vessels is the *arterial system;* that of afferent vessels is the *venous system*. The blood in arteries excepting the *pulmonary* is bright red; that in veins excepting the pulmonary is dark red. The change from bright to dark occurs in the capillaries of the system at large; the change from dark to bright only in the capillaries of the lungs and air-sacs. The *systemic* blood circulation is completely separated from the *pulmonic* in all animals in which, as in birds, the right and left sides of the heart are separated from each other; such circulation is said to be *double;* that is, arterial and venous blood only mingle in the capillaries, whether of the lungs or others, and therefore at the *periphery* of the vascular system: the heart being the centre of that system. Blood, in all or some of its constituents, permeates absolutely every tissue of the body. Those tissues whose capillaries are large enough for the passage of all the constituents of blood are said to be *vascular;* those which only feed by sucking up certain constituents of the blood, and have no demonstrable capillaries, are called *non-vascular.* But nutrient fluid penetrates the densest tissue, as the dentine of teeth; no permanent tissues are really non-vascular, or they would soon die, as do feathers, which require to be renewed once a year or oftener.

Lymph and the lymphatics are noticed further on. Blood consists of water in which several ingredients are dissolved, and certain solid bodies are suspended. Its water is salted, albuminated, fibrinated, and corpusculated. The proportions, which vary in different birds and at different times in the same bird, are in round numbers: water 80, fibrine and corpuscles 15, albumen and salts $5 = 100$ parts. Withdrawn from the body and allowed to settle, blood separates into two parts, *serum* and *coagulum.* The serum is the clear yellowish salty albuminous water; the clot is the fibrine, in the meshes of which are mired the corpuscles, reddening the whole mass. The *plasma*, plasm or plastic material of the blood, is its substance dissolved in water; that is to say, *minus* the solid corpuscles. These latter interesting little bodies are a myriad of minute animals, which swim in the life-current, and are named *Hæmatamœba cruentata.* They have been supposed to be of two species; but the so-called white blood corpuscles, or *leucocytes*, indistinguishable from lymph corpuscles, are simply the formative stages of the red blood-discs. In its early colorless stage, the *Hæmatamœba* is a nucleated mass of protoplasm (*protoplasm* is the indifferent substance out of which all animal tissue is derived), of no determinate size or shape, exhibiting active amœboid movements. Later in the life of the minute creature, it passes into a sort of encysted state, in which it reddens and acquires definite dimensions and configuration. In birds, these "blood-discs" are flat, elliptical, and *nucleated*, that is, containing a kernel; they average in the long diameter $\frac{1}{2100}$, in the short $\frac{1}{3800}$, of an inch. Thus they differ decidedly from the flat, circular, non-nucleated, red blood-discs of *Mammalia*, which latter are supposed to be rather *free nuclei* than perfected *Hæmatamœbæ.* The red color of blood is entirely due to the presence of these unicellular animals. The energy of respiration, and corresponding activity of circulation in birds, make them *hæmatothermal*, or hot-blooded; the pulse is quickest, the blood hottest, and richest in organic matter, in these of all animals.

The Heart is a hollow muscular organ, at the physiological centre of the hæmatic vascular system. Its muscle presents the principal exception to the rule, that the contractility of *Myamœba striata* (see p. 198) is subject to voluntary control. It is the most industrious organ of the body, never ceasing its rhythmic *systole* and *diastole*, or contraction and dilatation, from the moment of the first pulsation in the contractile vesicle which begins it, to that when the "muffled drum" gives the last beat of the "funeral march to the grave." The arteries are the elastic thick-walled branching tubes which leave the heart on their way to the body at large; their pulsations, over which the vaso-motor nervous system presides, are isochronous with the heart-beats, and arterial blood thus flows in jets. The veins are the vessels converging from all parts; thin-walled, less elastic, with more equable current. The capillaries are the communicating vessels, of such size as just to permit the Hæmatamœbas to pass through; their network represents the terminations of arteries and the commencements of veins. The heart in adult birds is completely double; *i. e.*, the right and left sides are perfectly separated. It is also completely four-chambered; *i. e.*, there is an *auricle* and a *ventricle* on each side, which communicate; in embryonic life the two auricles communicate by the *foramen ovale*, which then closes. Arteries proceed from the strong muscular ventricles; veins are received by the weaker auricles. The course of the blood is: From the body excepting the lungs it comes, dark and heavy with products of decomposition, through the *caval* veins into the *right auricle;* from right auricle through the auriculo-ventricular opening into *right ventricle;* from right ventricle through the *pulmonary arteries* to the lungs; in the capillaries of which it is relieved of its burden. There decarbonized and oxygenized, the bright red aerated blood returns through the *pulmonary veins* to the *left auricle;* through the corresponding auriculo-ventricular opening to the *left ventricle,* which pumps it out through the *aorta* and other arteries to the capillaries, and so to the veins and heart again. Thus the pulmonary *arteries* convey black blood, the pulmonary *veins* red blood; the reverse of the usual course. Before lungs come into play, in the egg, the blood is purified in the *allantois*, an embryonic organ which then sustains a respiratory function. Besides the pulmonary there is another special circulatory arrangement, the *hepatic portal* system of veins, by which blood coming from the *chylopoetic* viscera (stomach, intestines, etc., which make chyle in the process of digestion), strains through the liver before reaching the heart. There is no *renal* portal system in birds.

The heart of birds is not peculiar in its conical shape, but is more median in position than in mammals. There being no completed diaphragm, the pericardial sac which holds it is received in a recess between lobes of the liver. The right ventricle is much thinner-walled than the left; the auricles have less of the elongation which has caused their name ("little ears" of the heart) in mammals. The *right* auriculo-ventricular valve, which prevents regurgitation of blood, instead of being thin and membranous, is a thick fleshy flap which during the ventricular systole applies itself closely to the walls of the cavity. The pulmonary artery and the aorta are each provided at their origination with the ordinary *three* crescentic or "semilunar" valves, as in mammals. The pulmonary artery arises single, forking for each lung. The pulmonary veins are *two*. The systemic veins, or *venæ cavæ*, bringing blood from the body at large, are *three* — two *pre-caval*, from head and upper extremities, one *post-caval*, from trunk and lower extremities. The *aorta*, almost immediately at the root of that great trunk, figs. 90–95, *h*, divides into three primary branches; right, *ri*, and left, *li*, *innominate* arteries, conveying blood to the neck, head and upper extremities; and main *aortic*, *a*, which curves over to the *right* (left in mammals) and supplies the rest of the body. More precise statement is, perhaps, that the aortic root, *h*, first gives off the left innominate, *li*, then at once divides into right innominate, *ri*, and main aortic trunk, *a*, (right). It represents the *fourth* primitive aortic arch of the embryo. On the whole, the avian heart is a great improvement on that of most reptiles, though nearly resembling that of *Crocodilia;* it is substantially as in any mammal, though differing in its fleshy right auriculo-ventricular valve, two instead of one pre-caval vein, right instead of left aortic arch, and mode of origin of the primary aortic branches.

The zoölogical interest of the avian blood-vessels centres in the *carotid arteries*, which, with the *vertebral* arteries, supply the neck and head. The carotids may be single or double; and other details of their disposition correspond well with certain families and orders of birds. They are the first branches of the innominates. In most birds, there is but one carotid, the left; in a few, one, formed by early union of two; in many, two, long distinct. The arrangement will be perceived by the diagrams taken from Garrod's admirable paper (*P. Z. S.*, 1873, p. 457). In nearly the words of this author: 1. In what may be termed the *typical* arrangement (though it is not the usual one), two carotids, of equal size or nearly so, run up the front of the neck, converging till they meet in the middle line, and so continue up to the head, on the front of the bodies of the cervical vertebræ, in the hypapophysial canal. Birds with this arrangement Garrod calls *aves bicarotidinæ normales* (fig. 90). 2. In most birds, the carotid branch of the right innominate being not developed, only the *left*, of larger size, traverses the hypapophysial canal; but it bifurcates before reaching the head, thus producing two carotids, distributed as if there had been two all the way up. Such birds are said to have a left carotid, and are termed *aves lævo-carotidinæ* (fig. 91). 3. In certain parrots only, with two carotids, the right is as in (1), but the left runs superficially along the neck with the jugular vein and pneumogastric nerve; such birds are *aves bicarotidinæ abnormales* (fig. 92). 4. Two carotids, arising normally, unite almost immediately, and the single trunk runs to near the head, just as if there were two as in (1); then it bifurcates, as in birds with left carotid only (2). Such birds are termed *aves conjuncto-carotidinæ*. Special cases of (4) are: in the bittern, the two roots are of nearly equal size (fig. 93); in the flamingo, the left is very small (fig. 94); in a cockatoo, the right is very small (fig. 95). Parrots display all four of the arrangements; the cases of the bittern and flamingo are unique. The question is thus for nearly all birds narrowed to whether there be two normal carotids (1), or the left only (2). Observations upon three hundred genera show two in one hundred and ninety-three, in one hundred and seven the left only; but the

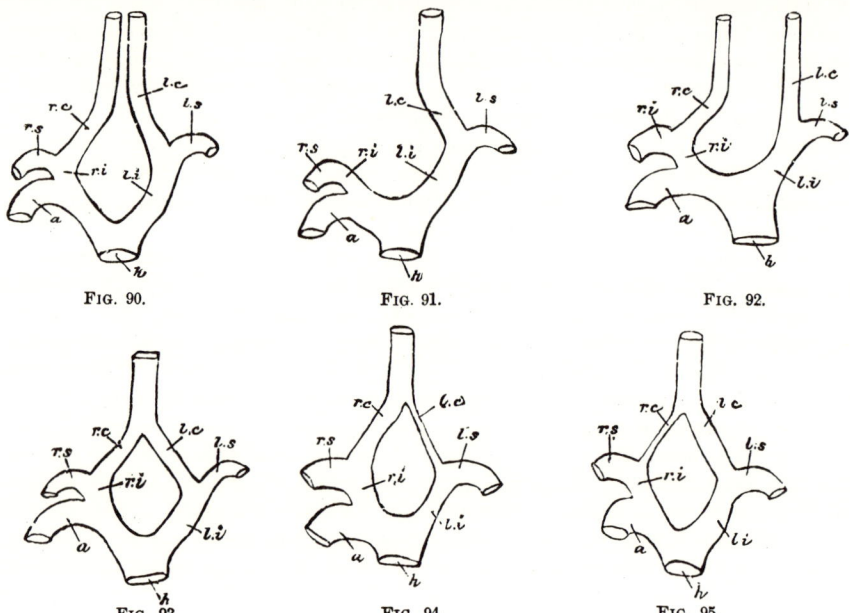

FIGS. 90-95. — Diagrams of carotid arteries of birds: *h*, root of aorta; *a*, arch of aorta, to the right side; *li*, left innominate; *ri*, right innominate; *ls*, left subclavian; *rs*, right subclavian; *lc*, left carotid; *rc*, right carotid (1) Fig 90. *Aves bicarotidinæ normales*, with two carotids, both alike. (2) Fig. 91. *Aves lævo-carotidinæ*, with left carotid only. (3) Fig. 92. *Aves bicarotidinæ abnormales*, certain parrots, with two carotids, not alike. (4, 5, 6) *Aves conjuncto-carotidinæ*, with two carotids, which speedily unite in one. (4) Fig. 93, bittern, both alike. (5) Fig. 94, flamingo, left very small. (6) Fig. 95, cockatoo, right very small. (Copied by Shufeldt from Garrod.)

numerical proportion of *Passerine* genera makes (2) the most frequent arrangement. There is but one carotid in all *Passeres* as far as known; in most *Cypselidæ*; in *Trogonidæ, Meropidæ, Upupidæ, Rhamphastidæ*, some *Psittaci*, the *Turnicidæ, Megapodidæ, Podicipedidæ, Alcidæ, Rheidæ, Apterygidæ*. Thus in *Passeres, Columbæ, Accipitres, Grallæ*, and *Anseres*, the carotid arrangement is an *ordinal* character, all but the first named of these great groups having two. The character separates most of the *families* of "Picarian" birds, and also distinguishes the families *Phœnicopteridæ, Megapodidæ, Cracidæ, Turnicidæ, Podicipedidæ*, and family groups of the *Ratitæ*, from among one another. It is apparently only a generic character in *Psittaci*, and in *Cypselidæ, Ardeidæ* and *Alcidæ*.

Reaching the skull, the carotids burrow in the bone, between the basitemporal plate and the true floor of the skull, and enter the cranial cavity by the "sella turcica" (the original pituitary space); their anastomosis furnishes a sort of "circle of Willis." (Figs. 66, 69, 70, *ic*.)

Both limbs of birds have a prime peculiarity of their arteries as compared with mammals. In the fore limb, the blood supply being chiefly absorbed by the immense pectoral muscles, vessels which in mammals are small *axillary* branches appear like the main continuation of the *subclavian* trunk, and the brachial arteries are correspondingly reduced. In the leg, the main source of supply is the great *ischiac* artery, the femoral being small. This ischiac artery corresponds to the twig which in man accompanies the great sciatic nerve (*comes nervi ischiatici*); and the rare human anomaly of a *posterior* main vessel of the thigh is therefore a reversion (atavism) to the avian rule. There is no single proper *renal* artery to the kidney.

The **Lymphatics** of birds consist chiefly of a deep set accompanying the main blood-vessels, forming various *plexus*, — nodes, "glands," or "lymph-hearts" in their course. Superficial lymphatics, so prominent in mammals, are little developed, though lymphatic glands are found in the arm-pit and groin of some birds. These are the *systemic* vessels; a special set, the *lacteals*, arise by numberless twigs in the course of the small intestine, uniting and re-uniting to form at length *two* (not one as in mammals) main tubes, which lie along either side of the spinal column. These are the *thoracic ducts;* which terminal trunks of the whole lymphatic system empty into the right and left *jugular* veins at the root of the neck. The contents of the vessels differ correspondingly. Pure lymph is a pale, limpid, albuminous fluid, containing when maturely elaborated a number of irregular amœboid bodies, indistinguishable from the white formative corpuscles of the blood (p. 202). It is strained out of the tissues at large, being that material, not yet effete, which is still fit for feeding the blood. The lacteals contain *chyle*, — the other kind of lymph, drained off by the mucous membrane of the intestine from the prepared food in that tube; an albuminous fluid, milky or cloudy from the abundance of oil-globules, which, after mingling with the systemic lymph, is poured directly into the current of the blood, in the manner above said. Since the lacteals do not appear to begin with open mouths, the chyle must soak into them through the lining membrane of the intestines; and as this consists of a layer of amœba-like animals, through whose bodies the chyle passes, it is quite true to say that the whole organism is nourished upon the excrement of amœbas.

e. Pneumatology: the Respiratory System.

The **Organs of Respiration** provide for the ventilation of the body. Since the respiratory process is also calorific, they likewise furnish a heating apparatus. They consist essentially of air-passages and air-spaces connected with lung-tissue, being therefore *pulmonary* organs. No other animals are so thoroughly permeated as birds with the atmospheric medium in which they live; in no others are the respiratory functions so energetic and effectual. The lung may be likened to a blast-furnace for the combustion of decayed animal matter; purification of the blood and warming of the body being two inseparable results obtained. Dark blood flowing to the lungs, heavy with effete carbonaceous matters, is there relieved of its burden and aërated by the action of oxygen; the products of combustion being exhaled in the form of carbonic dioxide and water. Aside from the proper lung-tissue, the capillary substance of the immense air-sacs tends to the same result. There is likewise, in birds, a lesser system of ventilation, by which air is admitted to cranial bones through the eustachian tubes; but this is unconnected with the proper respiratory office. Pulmonary tissue consists chiefly of a wonderful net (a *rete mirabile*) of capillaries, interlacing in every direction, bound together and supported by fine connective tissue, and invested with membrane so delicate that their walls seem naked, their exposure to the air being thus very thorough. Air gains such intimacy with the capillaries through the *larynx, trachea* (fig. 101, *o*), and *bronchial tubes* (*r, r*), these being the primary air-passages. But all the bronchial tubes do not subdivide into the ultimate air-cells; some large ones run through the lung, pierce its surface (as at *u, u*, fig. 101), and end in that system of enormous air-spaces for which the respiratory system of birds is so remarkably distinguished, — like a heap of soap-bubbles, blown up *en masse* from a bowl of fluid; the extra-pulmonary air-spaces being the larger superficial bubbles, the minute vesicles of lung-tissue proper being little bubbles just formed. In this way air penetrates even the hollow skeleton of most birds (p. 141).

The Lungs of Birds (fig. 101, *t, t*), notwithstanding their heated energy of respiration, are anatomically more like those of reptiles than of mammals. They are not shut by a diaphragm in a special division of the great thoracic-abdominal cavity of the body, but extend from the apex of the chest as far as the kidneys, in the pelvic region. They are not divided into lobes, as in mammals, nor do they as in that class float freely in the chest by their mooring at their roots; nor, again, are they completely invested by a serous membrane forming a closed pleural cavity. They are fixed in the dorsal region of the general cavity, covered in front with pleura, with which slips of the rudimentary diaphragm (*v, v, v*) are connected; but on the dorsal surface are accurately moulded to the intercostal spaces, showing the impressions of the ribs and vertebræ, — just as the lobulated kidneys are stamped with the sacral inequalities of surface. They are, as usual, two, right and left; their "roots" are the bronchi (*r, r*), the pulmonary arteries and veins, nerves, and connective tissue.

The Pneumatocysts. — A bird is literally inflated with these great membranous receptacles of air, and draws a remarkably "long breath," — all through the trunk of the body, in several pretty definite compartments; in many, or most, or all, of the bones; in many intermuscular spaces; in some birds also throughout the cellular tissue immediately beneath the skin. They vary so much in extent and disposition as to be not easily described except either in the most general terms already used, or with particularity of detail for different species. According to Owen, however, the usual disposition is: An *inter-clavicular* air-space, quite constant: this, with its *cervical* prolongations, furnishes the great "air-drums" of our pinnated grouse and cock-of-the-plains. *Anterior thoracic*, about the roots of the lungs. *Lateral thoracic*, prolonged to *axillary*, and to spaces and passages in the wings, including the hollow humerus. Large *hepatic* or *posterior thoracic*, about the lower part of the lung and the liver. *Abdominal*, right and left, of great size, from the lower part of the lung where the longest bronchial tubes open very freely; extending to *pelvic* and *inguinal* compartments, whence *femoral* sacs, the hollow of the femur, etc. The *subcutaneous* cells are enormously developed in the pelican and gannet; the extensive areolar tissue being thoroughly pneumatic, and furnished with an arrangement of the cutaneous muscle (*panniculus carnosus*) whereby, apparently, the air may be rapidly and forcibly expelled by compression. A similar muscle develops in some birds in connection with the interclavicular air-space. (For pneumaticity of the skeleton, see p. 141.)

The purpose of this extensive respiratory apparatus is thus dwelt upon by the great "Newton of Anatomy" just cited: "The extension from the lungs of continuous air-receptacles throughout the body is subservient to the function of respiration, not only by a change in the blood of the pulmonary circulation effected by the air of the receptacles on its repassage through the bronchial tubes; but also, and more especially, by the change which the blood undergoes in the capillaries of the systemic circulation which are in contact with the air-receptacles. The free outlet to the air by the bronchial tubes does not, therefore, afford an argument against the use of the air-cells as subsidiary respiratory organs, but rather supports that opinion, since the inlet of atmospheric oxygenated air to be diffused over the body must be equally free. A second use may be ascribed to the air-cells as aiding mechanically the action of respiration in birds. During the act of inspiration the sternum is depressed [lowered from the back-bone in horizontal position of a bird], the angle between the vertebral and sternal ribs made less acute, and the thoracic cavity proportionally enlarged; the air then rushes into the lungs and thoracic receptacles, while those of the abdomen become flaccid; when the sternum is raised or approximated towards the spine, part of the air is expelled from the lungs and thoracic cells through the trachea, and part driven into the abdominal receptacles, which are thus alternately enlarged and diminished with those of the thorax. Hence the lungs, notwithstanding their fixed condition, are subject to due compression through the medium of the contiguous air-receptacles, and are affected equally and regularly by every motion of the sternum and ribs. A third use, and perhaps the one which is most closely related to the peculiar exigencies of the bird, is that of rendering the whole body specifically lighter; this must necessarily follow from the desiccation of the marrow and other fluids in those spaces which are occupied by the air-cells, and by the rarification of the contained air from the heat of the body. . . . A fourth use of the air-receptacles relates to the mechanical assistance which they afford to the muscles of the wings. This was suggested by observing that an inflation of the air-cells in the gigantic crane (*Ciconia argala*) was followed by an extension of the wings, as the air found its way along the brachial and anti-brachial cells. In large birds, therefore, which, like the argala [or our wood ibis, *Tantalus loculator*], hover with a sailing motion for a long-continued period in the upper regions of the air, the muscular exertion of keeping the wings outstretched will be lessened by the tendency of the distended air-cells to maintain that condition. It is not meant to advance this as other than a secondary and probably partial service of the air-cells. In the same light may be regarded the use assigned to them by Hunter, of contributing to sustain the song of birds and to impart to it tone and strength. It is no argument against this function that the air-cells exist in birds which are not provided with the mechanism necessary to produce tuneful notes; since it was not pretended that this was the exclusive and only office of the air-cells." (Owen, *Anat. Vert.*, ii, 1866, p. 216.)

Though nothing like them exists in mammals, it must not be inferred that these air-pouches are unique in birds. The general pulmonary mechanism is reptile-like, and the ornithic development is simply a logical extreme of arrangements found in reptiles and lower vertebrates, — even to the swim-bladder of a fish, which is morphologically and homologically pulmonary, though fishes' *gills* are functionally, and therefore analogically, their lungs; *i. e.*, their respiratory apparatus.

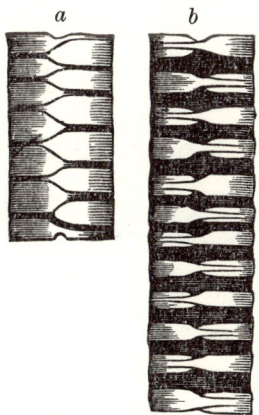

Fig. 96. — *a*, an inch of trachea, contracted to the utmost, the rings looking like alternating half-rings; *b*, the same, stretched to two inches, the rings evidently complete, with intervening membrane. (After Macgillivray.)

The Trachea (Gr. τραχεῖα, *tracheia*, rough) or "asper-artery" answers perfectly to its English name, wind-pipe. It is the tube which conveys air to and from the lungs (fig. 101, ^{1}, *o* to *q*). It commences at the root of the tongue by a chink in the floor of the mouth (fig. 101, ^{3}, *c*), runs down the neck in front between the gullet and the skin, and ends below by forking into right and left *bronchus* (fig. 101, ^{1}, *r, r*). It is composed of a series of very numerous gristly or bony rings connected together by elastic membrane. Lengthening and shortening, effected by muscles to be presently noted, is permitted by a very ingenious and interesting construction of these rings, which will be clearly understood with the help of the figures (96, *a, b*, 97 ^{1}, ^{2}) borrowed from Macgillivray's admirable account. When contracted, the rings look like an alternating series of lateral half-hoops, as in fig. 96, *a;* when stretched to the utmost, as in fig. 96, *b* they are clearly seen to be annular, or completely circular. The curious bevelling of the right and left sides of each ring alternately is shown in fig. 97, ^{1}, ^{2}; and fig. 97, ^{1}, ^{2}, represents the same two rings put together. The principle by which any two rings slip partly over each other on alternate sides is something like that upon which a cooper fastens the ends of any one barrel-hoop without any nailing or tying. The rings are in some birds perfectly cartilaginous: in most they become osseous. The trachea is moved by lateral muscles, which not only shorten the tube by approximating the rings, but also drag the whole structure backward, by their attachment to the clavicle and sternum. The strip, or two strips, of muscle lying upon each side of the trachea, is the *contractor tracheæ* (fig. 101, 1, *ss, ss*); the most anterior, when there are two, as soon as it leaves the tube to go to the clavicle, becomes the *cleido-trachealis*, or *cleido-hyoid*, fig. 101, 1, *f, f;* the other is similarly the *sterno-trachealis*. The latter may be a direct continuation of the contractor, as in fig. 101, 1, the loose strips under *q*, or apparently arise separately from the side of the lower end of the tube, as in fig. 101, 16, *e*. (Other muscles are to be described with the larynx superior and inferior.) The trachea is long in birds, proportionate to the extension of the neck; it is very flexuous, following with ease the bends of the neck in which it lies so loosely. Its cross section is oval or circular; but all that relates to the configuration and course of the pipe requires special description, — so variable is the organ in different birds. It is subject to dilatations and contractions in any part of its extent, and to deviations from its usual direct course to the lungs. Minor modifications must be passed over. The most remarkable expansions of the lower part of the tube occur in many sea-ducks and mergansers (*Fuligulinæ* and *Merginæ*), and some other birds; several lower rings of the trachea being enormously enlarged and welded together into a great bony and membranous box, of wholly irregular, unsymmetrical contour. Such a structure, represented in figs. 3 and 98, is termed a *tracheal tympanum*, or *labyrinth*.

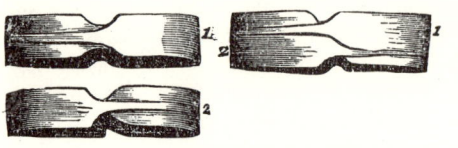

FIG. 97. — 1, 2, left hand, two tracheal rings, separate, as in fig. 96, *b;* 1, 2, right hand, the same put together, as in fig. 96, *a*. (After Macgillivray.)

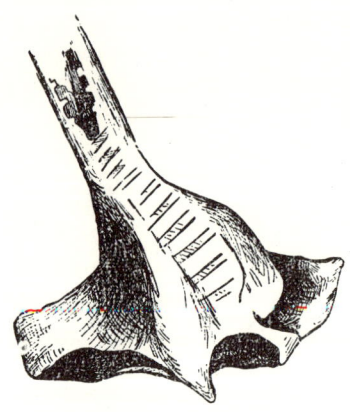

FIG. 98. — Bony labyrinth at the bottom of the trachea of the male of *Clangula islandica*, seen from behind, nat. size. Dr. R. W. Shufeldt, U. S. A.

It is not a part of the voice-organ proper, but may act as a reverberatory chamber to increase the volume of the sound, without however modulating it. Being chiefly developed in the male, it is a kind of secondary sexual organ. The vagaries of the wind-pipe are still more remarkable. Very generally, in cranes and swans, the trachea enters the keel of the sternum, which is excavated to receive it, and where it forms one or more coils before emerging to pass to the lungs. This curious winding is carried to an extreme in our *Grus americana*, the whooping crane, in which the wind-pipe is about as long as the whole bird, and about half of it — over two feet of it! — is coiled away in the breast-bone (fig. 99). The same thing occurs in *G. canadensis* to a less extent (fig. 100). In a Guinea-fowl, *Guttera cristata*, a loop of the trachea is received in a cup formed by the apex of the clavicles. In various birds, as some of the curassows (*Cracidæ*), the capercaillie (*Tetrao urogallus*), a goose, *Anseranas semipalmata*, and the female of the curious snipe, *Rhynchæa australis*, the trachea folds between the pectoral muscles and the skin.

The Larynx (the Gr. name, λάρυγξ, *larugx*) is the peculiarly modified upper end of the trachea (fig. 101, 1, and 3 to 12). In mammals it is a complicated voice-organ, containing the vocal chords and other consonantal apparatus; in birds the construction is simpler, as the larynx merely modulates the sound already produced in the lower end of the tube. It lies in

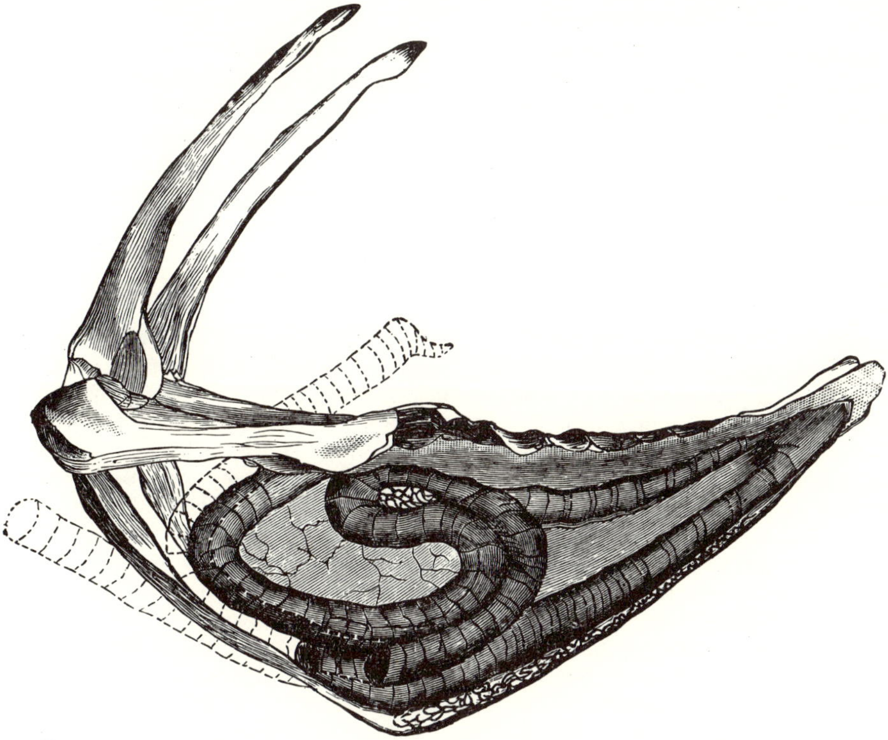

THE ANATOMY OF BIRDS.—PNEUMATOLOGY.

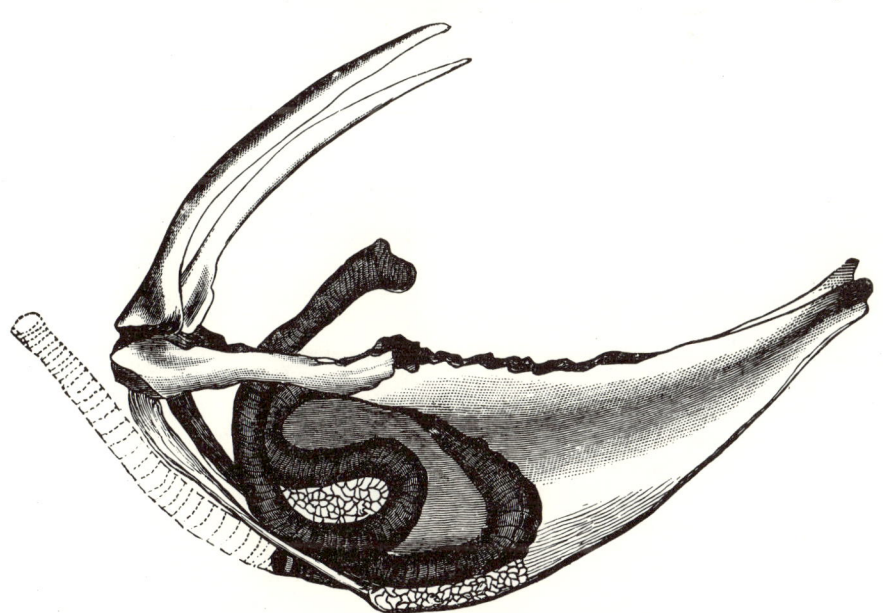

FIG. 99. — Coiling of the windpipe in the sternum of *Grus americana*; reduced. (From Amer. Nat.)

FIG. 100. — Coiling of the windpipe in the sternum of *Grus canadensis*; reduced. (From Amer. Nat.)

the floor of the mouth, at the root of the tongue, between the forks of the hyoid bone, resting upon the uro-hyal. Besides its attachments of mucous and other membrane, it is connected with the hyoid bone by a pair of *thyro-hyoid* muscles (^{8}, ^{1,1}), and usually with the rest of the trachea by prolongations of the sterno- and cleido-tracheales. It is usually a small, simple, conical "mouth-piece" of the pipe (^{4}, *a*), without the dilatation which renders the corresponding structure — the "Adam's apple," — so conspicuous in the human throat. Below, it communicates directly with the pipe: above, it opens into the mouth by the *glottidean fissure*, or *rima glottidis* (^{3}, *c*), a median lengthwise chink, which opens and shuts as its sides diverge or close together, and which is further defended in front by a folding of the mucous membrane of the mouth, constituting a rudiment of that curious trap-door arrangement which, when fully developed, is called the *epiglottis* (^{3}, *d*, *e*). Exclusive of two *broken* upper rings of the trachea (^{6}, *g*), the cartilages (or oftener bones, — for they generally ossify) of the larynx are five. One is a large single median and inferior piece, the *thyroid*, or shield-piece (^{4}, ^{6}, ^{7}, *a*), forming the most substantial part of the structure. It is somewhat triangular or oblong, running to an obtuse end in front; and with sides and posterior angles which curl upward behind. To its lateral posterior corner is attached on each side the small "horns" or *cornicula laryngis* (^{5}, ^{6}, ^{7}, *b*). There is a small median upper posterior piece, supposed to represent all there is of the *cricoid* (^{5}, ^{7}, *c*), which in man makes a ring around the larynx below the thyroid. To the cricoid, as to a base, are attached a pair of straight slender *arytenoids* (^{6}, ^{7}, *d*), projecting forward along the upper surface of the larynx: these form the *rima glottidis*, — the fissure of the glottis being between them. The arytenoids are attached in front by slender ligaments to the end of the thyroid (^{5}, the little slips between *d* and *e*), and they are supplemented by cartilaginous edges (^{6}, *f, f*); but there are no true vocal chords. Besides the extrinsic thyro-hyoid muscles, which pass from the larynx to the tongue-bone, the laryngeal parts are operated by intrinsic muscles, the sum of the motion given by which is the opening and shutting of the glottis by drawing apart or pulling together the arytenoids. Four pairs of such muscles are described for some birds. As named and figured by Macgillivray for the rook, there are: the *thyro-arytenoids*, which are the openers of the glottis (^{9}, ^{2,2}); the *oblique arytenoids* (^{10}, ^{3,3}); the *thyro-cricoids* (^{11}, ^{4,4}); and the *posterior thyro-cricoids* (^{11} and ^{12}, ^{5,5}).

The Syrinx (Gr. σύριγξ, *surigx*, a pipe) or Lower Larynx is the voice-organ of birds; in most respects a more complicated structure than the larynx proper, and one so differently constructed in different birds that it affords characters of great significance in classification. The highest group of *Passeres*, for example, is signalized by the elaboration of this musical organ, the marvellously adroit fingering of the keys of which by the little muscular performers sends through the tracheal sounding-pipe the tuneful messages of bird's highest estate. A few degraded or disgraced birds, as the ostrich and the American vultures, have no bucolic organ at all, the trachea forking as simply as possible. Others, as the common fowl, have a fair syrinx, but no muscles whatever to modulate their pastoral lays. Others have one, two, or three pairs of intrinsic muscles; to which may or may not be added a sterno-tracheal with syringeal attachment. It is not so much the bulk or mere fleshiness of the syrinx that indicates musical ability; but the distinctness of the several muscles, and the mode of their insertion, which result in endless combinations of rotating and rocking movements of the parts, whereby an infinite modulation of the musical tones becomes possible. In *Oscines*, there are normally five or six pairs of muscles, without counting the extrinsic sterno-tracheales; and the gist of the arrangement, in these melodious Passeres, is the attachment of the muscles to the *ends* of the upper bronchial half-rings, as far as the third one. As Professor Owen remarks with appreciative feeling, "the manifold ways in which the several parts of the complex vocal organ in *Cantores* may be affected, each of the principal bony half-rings, as one or the other end may be pulled, being made to perform a slight rotatory motion, are incalculable; but their effects are delightfully appreciable by the rapt listener to the singularly varied kind and quality of notes trilled forth in the stillness of gloom by the nightingale."

I should be able to make the plan of the syrinx clear to the student with the assistance of Macgillivray's beautiful figures. These are drawn from the rook, — a corvine croaker, indeed, but one whose syrinx is in good order, though he has never learned to play. As the modifications affect principally the soft parts covering and moving the music-box, one description of the latter is applicable to most birds. The last lower ring, or piece composed of several fused rings, of the trachea, at its bifurcation into bronchi, is enlarged or otherwise modified (fig. 101, ^{13}, *aba*), and crossed below from front to back by a bony bar, the *pessulus* (^{13}, at *b;* ^{15}, *a*), or bolt-bar, which, dividing it into lateral halves (as at ^{14}), forms thus two lateral openings instead of one median tube, — the beginnings of each bronchial tube. A membranous plate, strengthened by cartilage, rises vertically into the tracheal tube, forming a *septum,* or median partition, between the orifices of each bronchus. The free curved upper margin of this septum, extending of course, from front to back of the orifice, is called the *semilunar membrane;* being the edge of a partition common to both bronchi, it forms, in fact, the *inner lip* of each bronchial orifice; that is to say, the inner *rima glottidis syringis,* or lip of the syringeal mouth-piece. This membrane vibrates with the column of air, and is, in fact, one of the "vocal chords." Now the bronchial rings which succeed are not annular, circumscribing the bronchial tube, but are half-rings (^{15}, *b, b*), or arcs of circles to be completed by membrane, which forms more or less (scarcely or not half) of the circumference of the tube; this membranous part, termed the *internal tympaniform membrane* (^{15}, *c* to *c*), being on the side of the bronchus which faces its fellow, while the hard bronchial half-rings complete the rest of the cylinder. The membrane is attached to the pessulus above. This accounts for the whole bronchial tube and its vocal septum from its fellow. Now the concavity of the upper two or three bronchial half-rings, on the outer wall of the tube, but in its interior, is the place where is developed a certain fold of the mucous membrane, projecting into the tube opposite the septum, and forming the outer lip of the syringeal glottis; for this membranous fold, like the semilunar membrane, is set quivering in vocalization. The upper tracheal rings which enter into this arrangement are enlarged and otherwise modified. Thus are formed two "vocal chords," upon the vibrations of which the harmonious or discordant notes of the bird depend. The cords are struck by the hand of air indeed, but endless musical variations result from the play of the muscles in increasing or diminishing and variously combining the tension of the several parts of the instrument. In giving four pairs of intrinsic syringeal muscles (anterior external, anterior internal, intermediate, and posterior, besides the extrinsic sterno-tracheales), as figured in ^{16}, *a, b, c, d* and *e*, Macgillivray is said to have understated the full oscine number, which is five or six. In the raven, Owen describes *five,* without counting the sterno-trachealis: *broncho-trachealis anticus,* anterior external; *broncho-trachealis posticus,* posterior external; *broncho-trachealis brevis,* posterior internal; *bronchialis anticus,* anterior internal; and *bronchialis posticus.* The general arrangement, however, is fairly indicated by Macgillivray in ^{16}, where on the side of the syrinx, the muscles are seen to diverge from the tracheal lateral line to go to *ends* of the bronchial semi-rings.

The student will understand that my description is particular only as regards the oscine syrinx; that in birds at large every possible modification, almost, of lower tracheal and upper bronchial rings occurs, and with various musculation, or with none. The non-oscine rule for the muscles is, one on each side, if any; and insertion into mid-parts, not ends, of the bronchial half-rings. The latter character chiefly distinguishes the non-oscine syrinx when it has several muscles. As to situations of the syrinx, three have been recognized: the ordinary *broncho-tracheal,* in formation of which both bronchi and trachea take part; the *tracheal,* only known to occur in some American Passeres, as in *Thamnophilus* and *Opetiorhynchus,* situated wholly in the trachea, the lower part of which is extensively membranous; and the *bronchial,* wholly in the bronchi, as in *Crotophaga* and *Steatornis.*

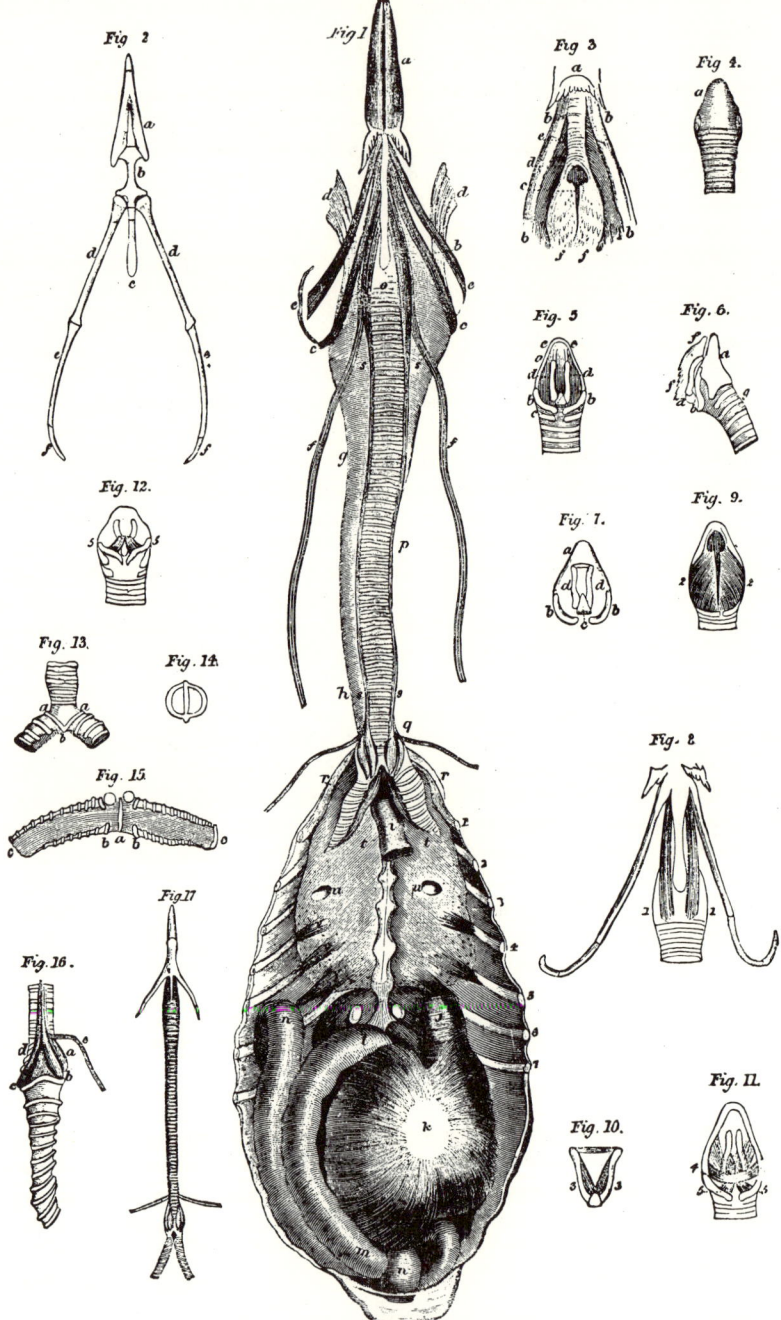

Fig. 101. — Respiratory and vocal organs of the Rook, *Corvus frugilegus*, an Oscine Passerine bird; nat. size, after Macgillivray. 1. *a*, tongue; *b*, basi-branchial, commonly called uro-hyal; *c, c*, horns of hyoid bone; *d, d*, genio-hyoid muscles; *e, e*, stylo-hyoid muscles; *f, f*, cleido-hyoid muscles; *g, h, i*, œsophagus; *j*, proventriculus; or secretory stomach; *k*, gizzard, or gigerium, the muscular stomach; *l, m, n, n*, intestine, duodenum to rectum;

o, p, trachea, or windpipe; *q*, inferior larynx, or syrinx ; *r, r*, right and left bronchus ; *ss, ss*, contractor muscles of trachea; *t, t*, lungs, with *u, u*, apertures communicating with thoracic air-cells ; *v, v, v*, three pairs of muscular slips answering to a rudimentary diaphragm; 1, 2, 3, 4, 5, 6, 7, as many ribs. — 2. Hyoid bone; *a*, glosso-hyal, tipped with cartilage, its posterior horns being cerato-hyals proper; *b*, basi-hyal; *c*, basi-branchial proper, commonly called uro-hyal ; *d, d*, cerato-branchials proper, commonly called apo-hyals ; *e, e*, epibranchials proper, commonly called cerato-hyals, tipped with cartilage, *f, f*. — 3. Glottis, or opening of trachea in the mouth; *a*, base of tongue; *b, b*, horns of hyoid bone ; *c*, rima glottidis, cleft or chink of the glottis; *d*, a triangular vacuity ; *e*, an elastic ligament ; *d* and *e* represent an epiglottis ; *f, f*, a papillose surface. — 4. Larynx viewed from before (below); *a*, thyroid bone or cartilage. — 5. Larynx viewed from behind (above); *a*, thyroid bone; *b, b*, its appendages ; *c*, cricoid; *d, d*, arytenoids; *e, e*, anterior border of thyroid, to which *d, d* are connected by two arytenoid ligaments. — 6. Larynx viewed from right side ; *a*, thyroid ; *b*, appendage ; *c*, cricoid ; *d*, arytenoid ; *f, f*, cartilage attached to arytenoid ; *g*, a tracheal ring. — 7. Larynx viewed from behind ; *a*, thyroid ; *b, b*, its appendages ; *c*, cricoid ; *d, d*, arytenoids. — 8, 9, 10, 11, 12. Muscles of the larynx ; 1, 1 (fig. 8), thyro-hyoids ; 2, 2 (fig. 9), thyro-arytenoids, or openers of the glottis; 3, 3 (fig. 10), oblique arytenoids ; 4, 4 (fig. 11), thyro-cricoids ; 5, 5 (figs. 11 and 12), posterior thyro-cricoids. — 13. Bifurcation of trachea; *aba*, last entire tracheal ring. — 14. Last entire tracheal ring, viewed from below, crossed by the pessulus. — 15. Bifurcation of trachea, and bronchi, viewed from below ; *a*, pessulus, the bolt-bar, or " bone of divarication "; *b, b*, next succeeding tracheal half-rings. — 16. *a, b, c, d*, inferior laryngeal or syringeal muscles, not well made out in this figure; see text. But the typical oscine arrangement (acromyodian) is perceived, inasmuch as anterior (*a*) and posterior (*d*) intrinsic muscular masses go to *ends* of the first tracheal half-ring, at *b* and *c ;* the extrinsic slip *e* passing to sternum ; compare fig. 1, at *q*. — 17. Trachea, etc., of the nightingale, nat. size. (Compare figs. 3, 67, 72, 73, 74.)

The Song of Birds unlocks the great secret of Genesis to those who can hear the keynote. It is the closest approach, in animate nature, to the ringing of the hydrogen bells in the physics of light. The musical instrument figured (101, ¹⁷) is the identical pipe the " great god Pan " first fashioned for a legacy to all time, as so sweetly said by Mrs. Browning : —

" He tore out a reed, the great god Pan,
 From the deep cool bed of the river.
The limpid water turbidly ran,
 And the broken lilies a-dying lay,
 And the dragon-fly had fled away,
 Ere he brought it out of the river.

" ' This is the way,' laughed the great god Pan,
 (Laughed while he sate by the river!)
The only way since gods began
 To make sweet music, they could succeed.'
 Then dropping his mouth to a hole in the reed,
 He blew in power by the river.

" Sweet, sweet, sweet, O Pan,
 Piercing sweet by the river!
Blinding sweet, O great good Pan!
 The sun on the hill forgot to die,
 And the lilies revived, and the dragon-fly
 Came back to dream on the river."

But the sad sequel, felt by Keats, when poor Psyche has seen and known, and Eros has found his wings : —

" So did he feel who pulled the boughs aside,
 That we might look into a forest wide,
 To catch a glimpse of Fauns, and Dryades
 Coming with softest rustle through the trees;
 And garlands woven of flowers wild and sweet,
 Upheld on ivory wrists, or sporting feet:
 Telling us how fair trembling *Syrinx* fled
 Arcadian Pan, with such a fearful dread.
 Poor Nymph, — poor Pan, — how he did weep to find
 Naught but a lovely sighing of the wind
 Along the reedy stream! a half heard strain
 Full of sweet desolation, balmy pain."

The blessed blue-bird, "bearing the sky upon her back," is burthened with the same " light load of song " —

Have you listened to the carol of the bluebird in the spring?
Has her gush of molten melody been not poured forth in vain?
Ah! then the pulse has quickened, and a sigh, perhaps, has risen,
From the breast the bluebird's music stirs to thoughts that lack expression —
So tender, so tumultuous are the fancies thus aroused.
The bluebird's song breathes gladness — breathes the sweet and solemn triumph
Love feels when all love's passion melts in its own fruition.
Exquisitely subtile are the chords the bluebird touches —
Chords that quiver now in ecstasy, now thrill in fond expectancy,
Now die in dreams of all that might have been.
Hers is language to interpret, and translate in accents rhythmic,
All the yearning of young love to claim his own —
Of young love that trembles on the threshold of the passions,
And shrinks before the images his ardor calls to life.
Thus to the maiden musing come thronging thoughts unbidden,
When she hears this speaking echo of the hopes that glow within;
And the tell-tale blushes redden to the rose-tint on the bosom
Of the bird that dares to breathe her secret joy.
Thus to the youth impetuous, whose life is set to music —
Let love but laugh and beckon from afar —
Fulfilment sends a greeting in the soft voluptuous languor
That steals upon the senses if the bluebird's song be heard —
This song of wondrous gladness, ever bubbling, welling, gushing,
From a fountain full of promise, inexhaustible, divine!
Sweeter far these liquid accents when the buds of hope are blighted,
And the tree of knowledge bears its bitter fruit;
When memory sits brooding on the ashes of her birthright,
And sackcloth shrouds a heart that once was young;
For a silver chord is quickened where was greedy, silent sorrow —
Responding to a sympathetic touch:
The bird sings true and tender, with a precious burden laden,
With the tidings of a love that never dies.
So in the timid spring-time, when the world wears wreaths of roses,
Ring clear the joyous melodies of hope!
So in the summer season, when the wine of pleasure reddens,
Ring passionate the triumphs of the heart!
So in the sad, still autumn, when life bends beneath its burden,
When what might have been has never come to pass,
Rings once again this music on the crushed and wounded spirit,
Bringing light where all was dark and drear before:
All is not lost if the music that the bluebird bears be heeded,
For her mission is to tell us love is God.

Though it is a fact that "the *Chenomorphæ* are not provided with intrinsic syringeal muscles," there may be much truth in treatises *de cantu Cycni morituri* which have appeared from time to time, and to the number of which I may be pardoned for adding: —

> How sadly sweet, how soft and low
>   Is the music born of pain —
> How mournful sounds the ebb and flow,
> What measured beats, what throb and throe,
>   In the wild swan's dying strain!
>
> The archer, Death, and the twanging bow,
>   And the fateful shaft on-sped,
> All state and grace and pride laid low,
> Disordered plumes and crimson flow —
>   For the white swan's heart has bled.
>
> But hear the mournful cry that rings
>   On the startled air of night!
> As a spirit form in the darkness wings
> Its way unseen, the wild swan sings
>   His psalm of life and light.

> How sadly sweet the solemn strain —
> The dirge of the dying swan!
> That wondrous music, child of pain,
> That requiem, sounding once again —
> And a bird's soul passes on.

f. Splanchnology: the Digestive System.

The Alimentary Canal, or digestive tract, is a tube which passes through the body from mouth to anus, conveying food, the nutritious qualities of which are drawn off by the lacteals *in transitu* and assimilated, the refuse being voided. This is *digestion*. The canal is really a tube within a tube, being contained in the cavity below the bodies of the vertebræ, formed by the series of *hæmal arches* (p. 141). Birds are fast livers, their digestive operations, like the processes of respiration and circulation, being very active and effectual; they require proportionally great quantities of food. The voracity of the cormorant is proverbial, but it is probably not greater than that of the ethereal nightingale. Birds as a class are omnivorous; many species are as nearly omnivorous as any animals can well be; but the majority are either vegetarian or flesh-feeding. Very many birds feed upon fruits, hard or soft; but even these, when in the nest, are nourished for the most part upon the bodies of insects; and it may be truly said, that the great majority of birds are insectivorous. Birds seem to be the great controlling agency in the economy of nature, of the increase of insect life; agriculture would be difficult if not impracticable without them, and their economic value is simply incalculable. Insectivorous birds cannot be much interfered with, without destroying one of the most important and consequential of nature's many beautiful adjustments. The bird cries perpetual "échec!" to the insect. Even those birds which are mainly flesh-eaters, as the hawks and owls, are similarly beneficial, for the creatures they chiefly prey upon are the small rodents so fateful to husbandry. The carrion-eaters contribute largely to make tropical regions habitable to man. Various tribes of birds feed almost exclusively upon fish; and these sometimes reach the dignity of diplomatic and other political interests of mankind: nations have gone to war over the dung of such birds, guano-beds being to some of the South American powers a large item of their revenue. Chili and Peru have been fighting lately, and the United States have been wrangling, over the excrements of the alimentary canal of sea-birds. This tube, in general, is shortest, simplest, and most direct in the flesh- and fish-eaters, the nature of whose food assimilates already more nearly to the substance of their bodies than does that of the vegetarians. The tube is modified in different portions of its extent, for the prehension, retention, saturation, maceration, and comminution of food, and the mixture with it of other solvent fluids than those secreted by the mucous membrane of the alimentary canal itself. Hence arise the various modifications of its length, dilatation here, contraction there; the presence in its lining membrane of numerous follicles; and the annexation of various glandular organs. Being always longer than the body, the tube is necessarily coiled away in certain places; this folding taking place chiefly in the intestinal part of the tract. Modifications of structure make recognizable parts, as the mouth, gullet, crop, stomach, gizzard, intestine, cloaca, anus. Annex organs are the salivary glands, the liver, and the pancreas, all of which pour their secretions into the canal. This tube also receives the terminations of other systems of organs: the auditory organ of special sense; the respiratory system, which is at first a mere bud or off-set from the digestive; the urinary and the generative, which, though originally distinct, primitively and permanently open into the lower bowel. The intestine is also continuous with the cavity of the umbilical vesicle of the embryo, a primitive structure which disappears as the chick matures; and with that of the allantois, another embryotic organ which begins by budding from the intestinal cavity. Its connection with the system of blood-vessels is direct through the lacteals and thoracic ducts (p. 205). Its operations are automatic and spontaneous, of the "reflex" order;

that is, excited by the presence of food, — having work to do making it work, so to speak. Its innervation is chiefly by the pneumogastric and sympathetic nerves; and digestion is the most purely vegetative function, dealing with the raw materials of nutrition and consequently of the growth and repair of the whole body. The active factors in this transaction are several species or varieties of small creatures, called *Enteramœbæ;* they are all derived by descent with modification from the hypoblastic cells of the early embryo. Those of the canal itself form all the mucous epithelium of that structure, with its various secretory crypts, follicles, and villi; similar creatures, perhaps of different genera, form the lining of the salivary, hepatic, and pancreatic glands. Blood-vessels, in intimate connection with the digestive organs, form that special venous arrangement by which the blood coming from that part of the intestinal tract where chyle is made is collected in a *portal* system and sent through the liver, — in the embryo a sort of " great dismal swamp" which interrupts the ordinary current. The tube within the tube is fixed not only at its ends, but by various membranous connections, among them the *mesenteries.* We will notice the several departments of the alimentary canal, and its annexes; reference should be made to the colored frontispiece, and to fig. 101, where most parts of the digestive system are shown.

The Mouth and Tongue. — The most anterior of the special cavities in which the tube is divided, and the " manual" organ it contains. The mouth in general corresponds to the shape of the jaws, already sufficiently noted (pp. 105, 168). The anterior part is much hardened, like the beak; in fact, this hardness of the buccal cavity, and the absence, or very slight distinction, of a "soft palate," are among the peculiarities of a bird's mouth. There is consequently little distinction, if any, between mouth proper and *fauces*, or *pharynx*, which is the posterior part, leading directly into the gullet. Besides this communication the mouth receives the terminations of four special cavities. 1. The *posterior nares*, on the roof of the mouth posteriorly, generally a median slit, leading into the nasal chambers. 2. The generally single and median and more posterior opening of the *eustachian tubes*, which lead into the tympanum, and are the remains of the first post-oral visceral cleft of the early embryo. 3. The *glottis* (fig. 101, ³, *c*), a slit at the base of the tongue, the opening of the windpipe, and so of the whole respiratory system, which is defended by a rudimentary trap-door, the *epiglottis*, if any. 4. One or several pairs of orifices, the openings of the ducts of the *salivary glands*. These structures, corresponding to the parotid, submaxillary, and sublingual glands of mammals, vary extremely in their development. In woodpeckers, for example, and some *Raptores*, elaborate special salivary glands occur, having a glomerate structure, and a special " stenonine" duct. In many other birds, similarly compound but less elaborate submaxillary glands pour their secretion into the mouth by a series of pores. In most birds, however, the salivary glands are small, simple, and less distinct from various other sets of mucous crypts which open into the mouth. In the great bustard (*Otis tarda;* fig. 102) there is a singular buccal structure; a great pouch opening beneath the tongue, susceptible of distension during those amatory antics termed the "showing-off" of the creature. It is in fact an air-sac, but not of the kind already considered (p. 200), having no connection with the respiratory system. The narial, eustachian and glottidean apertures are commonly defended by retrorse papillæ; and other such

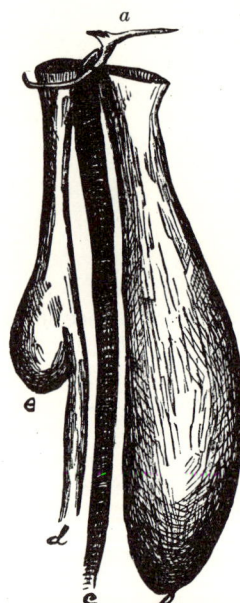

FIG. 102. — Gular pouch of bustard; copied by Shufeldt from Garrod. *a*, tongue; *b*, the pouch, opening under *a*, hanging in front of *c*, the trachea, behind which is the œsophagus, *d*, with its crop, *e*.

processes of mucous membrane, knobbed or acute, may occur elsewhere in lines and patches. The roof of the mouth is nearly all "hard palate," as already said; its soft floor is the mucous membrane and skin between the jaws, with muscular or other intervening structures. The principal flooring muscle is the *mylo-hyoid;* the *genio-hyoid* (fig. 101, [1], *d*) is another, which passes, like the first, from the mandibular to the hyoid bone; a third is the *stylo-hyoid* (*e*). The floor in some cases forms a pouch, which, as in the case of the pelican, is of great extent and susceptible of enormous dilatation (fig. 669).

The handler of the mouth, or lingual organ, is the tongue, which answers the same purpose as in other creatures: it is tactile, to some extent gustatory, sometimes prehensile, nearly always manipulatory. In some birds, as the pelican and ibis, and also the kingfisher, it is very slightly developed, — scarcely more than a pad at the bottom of the mouth, enjoying the most limited motion or other function. In some birds, as the parrot and duck tribes, and also the flamingo, the tongue is large, thick, and fleshy, quite filling the mouth. In the first-named of these, it is dexterously manipulatory; the morsel of food is managed between the tongue and upper beak; the tactile certainly and perhaps the gustatory sense is highly developed; and the fleshiness of the tongue may affect that power of articulate speech for which some parrots are justly noted. In the Lamellirostres just mentioned the tongue has lateral processes corresponding to the denticulations of the beak, and the under surface is horny at the end, like a human finger-nail. In the woodpeckers (figs. 73, 74) the tongue itself (glosso-hyal part of the hyoid) is reduced to a slight horny and spiny tip of the lingual apparatus; but other parts of that mechanism are so extraordinarily developed that the "tongue" appears as a *lumbriciform* (worm-like), spear-headed organ usually capable of great protrusion from the mouth, and therefore acting as a prehensile instrument, being bedewed for that purpose with tenacious saliva from the great salivary glands; while it is actuated in protrusion and retraction by specially developed muscles. In the snipe and many of the long slender-billed waders, the tongue is similarly slender, but not protrusible. The long narrow tongue of the toucans (*Rhamphastidæ*) is beset with slender processes, so that it seems feathery. The tongue of the humming-bird is very singular, — delicately thready, yet double-barrelled, — two tubes placed side by side, serving as siphons to extract the nectar of flowers. These and other interesting extremes aside, the ordinary style of a bird's tongue is flat, narrow, more or less sagittate or lanceolate, and tipped or sheathed in horn, commonly with lateral backward processes like the barbs of an arrow head, — the whole *glossal* structure upborne pretty distinctly upon the end of the basihyal bone. (See fig. 101, where [1], *a*, is such an ordinary tongue; and [2], *a–f*, is its whole skeleton.) Such horny tongues are commonly bifid at the extreme tip or there variously lacerate, or laciniate, or thready, — and even the fleshy tongue of some parrots, as the lories, is brushy at the end. The bony foundation of the tongue is the composite hyoid bone, already often mentioned (see p. 173); the free lingual part proper is based upon the glosso-hyal and its terminal cartilage; the roots curve more or less extensively about the base or more of the skull. The tongue is moved by some intrinsic muscles, as well as by those extrinsic ones by which it is connected to the skull, jaw, and windpipe (fig. 101, [1] and [8]).

The Œsophagus. — After comminution, if any, by the beak, and insalivation in the mouth, food passes directly through the pharynx into the *œsophagus* or gullet, — a musculo-membranous tube connecting mouth with stomach (fig. 101, [1], *g, h, i*). This is composed (besides its mucous membrane) of circularly disposed *constrictor* fibres, and longitudinal *contractor* fibres, of *Myamœba*, of the pale, smooth species (*M. lævis*). It has generally a pretty straight course, but may be diverted to one side or the other; and, in particular, is subject to various dilatations and contractions, permanent or temporary, aside from the mere distension caused by the passage of food. When the floor of the mouth is wide and loose, the gullet partakes of the same character above; the extreme case is afforded by the pelicans, especially *P. fuscus*. But the gullet of many small birds, as various genera of *Fringillidæ* and *Corvidæ*, is much more distensible than is commonly supposed, and may be found crammed with seeds which there find resting-place for some time. The fish-eating birds, as herons, cormorants, loons, and others, have also capacious gullets. The Australian bustard, *Eupodotis australis*, has an œsophagus capable of such extraordinary distension that it hangs down in front of the breast when inflated with air, as it is in the amatory display in which that species is wont to indulge. Aside from mere distensibility of transient character, the œsophagus of many birds becomes modified anatomically into a special pouch, — the crop or craw, *ingluvies*, where the food is detained to be macerated in a special secretion before passing on to the true stomach. Such definite crops occur in birds of prey, which gorge such masses of food in their irregular voracious banquets that it cannot all be received into the stomach at once; and likewise throughout the orders of Columbine and Gallinaceous birds, which habitually feed upon seeds and other fruits so hard that they are advantageously macerated as a preliminary to true digestion. The common fowl furnishes a good illustration of a large, definite, single and median crop; in pigeons it is a pair of lateral dilatations. In these latter birds, when they are rearing their young, the secretion of the ingluvies, always copious, becomes still more so, and of a milky character in consequence of the activity of the altered mucous surface; it is regurgitated into the mouths of the young, along with the macerated grains. "This phenomenon is the nearest approach in the class of Birds to the characteristic mammary function of a higher class; and the analogy of the 'pigeon's milk' to the lacteal secretion of the Mammalia has not escaped popular notice." Various other birds also feed their young by regurgitation of elaborated food; and very many similarly reject indigestible portions of their ingesta. Such vomiting is best known to be the wont of birds of prey, which habitually throw up the hair, feathers and bones of their victims, made up into the boluses called "castings"; but the practice is far from being confined to these flesh-eaters. The extreme case of emesis offered by birds is witnessed in the horn-bills (*Bucerotidæ*) which have been known to throw up the coat of their stomach without discomfort, — what a blessing it would be to some old topers if they could do the same, and grow another with equal ease! In fact, in consequence of the capacity and directness of the gullet, vomiting is very easy to birds, and with some it is a means of self-defence, — very effectual for instance in the cases of our vultures (*Cathartides*). Fish-eating birds, as herons, gulls, petrels, habitually vomit when wounded or otherwise molested.

The Proventriculus. — The tube just considered ends below in a special tract, variously dilated or not, but always peculiar in the presence of certain gastric follicles which secrete the digestive fluid proper. The "stomach" of a bird, in fact, is compound, consisting of a glandular or digestive portion, and a muscular or grinding part. The former is the *proventriculus;* whatever its size or shape, or whatever its magnitude in comparison with the grist-mill, it is recognized by the presence in its mucous surface of these gastric follicles, secreting the peptic fluid which *chymifies* the food. The follicles are perhaps always large enough for this part of the tube to be recognized by the naked eye, — the mucous membrane having here a thickened, velvety, vascular appearance. The glands are of various sizes and shapes, — usually simply tubular, sometimes clubbed or conical, or variously racemose (like a bunch of grapes). They are disposed in a zone around the tube, or in patches upon part of its surface, — in the darter (*Plotus*), very singularly in a separate lateral compartment looking like a crop. Details of the grouping of these solvent glands are interminable. Whatever its anatomical variations, and however like the end of the œsophagus it may simply appear to be, this *ventriculus glandulosus* is the bird's proper stomach (fig. 101, ^{1}, *j*).

The Gizzard. — Mixed with the salivary, ingluvial, proventricular and other secretions of the mucous surface, and already chymified, the food of birds next passes directly into the gizzard, *gigerium*, or muscular division of the stomach, sometimes called the *ventriculus bulbosus*. The two are sometimes separated by a tract, sometimes immediately consequent. In the muscular gizzard, the food-grist is ground fine. To this end, the walls of the cavity become developed into a more or less powerful muscular apparatus, and the mucous membrane changes to a tough, thick, horny, occasionally even bony, lining; this callous cuticular lining being often very loosely attached, and even deciduous in some cases. The muscular arrangement is chiefly in two great masses, called the *lateral muscles*, converging to a central tendon; between them intermediate fibres may form a more or less distinct muscular belly. In the most powerful gizzards, the muscular tissue is very dense and dark-colored; the tendons brilliantly glistening, and the contained "millstones" extremely callous. Such a gizzard is well displayed by the common fowl or the goose. The opposite extreme is afforded by the carnivorous and especially the piscivorous birds, whose soft food requires little trituration, — it is all a matter of degree. How readily this part of the canal responds to the regimen of the bird, is witnessed in our cock-of-the-plains (*Centrocercus urophasianus*), — a bird whose gizzard is so slightly muscular as to appear like a membranous bag, though its gallinaceous relatives have extremely strong grinders. Its food is chiefly the buds and leaves of the wild sage (*Artemisia*), and grasshoppers. Increased muscularity of the gizzard has even been artificially produced. Birds whose grist is heavy habitually swallow gravel, that these small stones may mechanically aid in the grinding process. The action is so energetic, that in "auscultating" a fowl when the mill is in full blast, the noise of the grinding can be distinctly heard. The pebbles, in fact, have a function which leaves "hens' teeth" not entirely mythical. The kind of motion impressed upon the opposing pads of cuticle is alternating, — a rubbing back and forth to a slight extent. Peculiar dispositions of the callous surfaces are found in some pigeons, with corresponding peculiarity of the cross-section of the gizzard. In some of the cuckoos a matting of impacted hairs of lepidopterous insects has been mistaken for a coat of the gizzard itself. In the darter, which has a pyloric division or compartment of the gizzard, this is nearly filled with a mass of matted hairs, a peculiar modification of the epithelial lining, serving to guard the pyloric orifice. Folds of the lining membrane form a pyloric valve in many birds. The *pylorus*, or the *pyloric* orifice, is that opening by which food leaves the gizzard for the intestines; the orifice of entrance from the œsophagus is the *cardiac*. The two are always near together, and sometimes adjoining. (In fig. 101, ^{1}, k is on the central tendon of the moderately muscular gizzard; the cardiac orifice is between j and k, and pylorus between l and k.)

The Intestine continues the alimentary canal to the cloaca. Any difference in the length of the whole tract, relatively to that of the bird, is chiefly produced by the foldings of the intestine, especially in the upper portion of its course. The extremes of proportionate length are perhaps not ascertained; but known to be from less than 2:1, to more than 8:1. In birds there is little or no distinction between "small" and "large" intestine, as to the calibre of the tube, nor is the latter succulated as in mammals. The former is considered to extend from the pylorus to the *cæca* (structures to be presently noticed). Above the cæca the intestine commonly receives its foldings and windings; below them it usually proceeds more directly, or quite straight, to the cloaca, forming literally a "rectum"; but in the ostrich this ultra-cæcal tract is longer than the rest, and convoluted. The cis-cæcal portion is conventionally divided into *duodenum, jejunum*, and *ileum*; there is, however, no positive anatomical distinction of these parts in any animal with which I am acquainted. In birds, a "duodenum" is perhaps as distinct as ever; it forms the most constant duplication of the intestine, the pancreas being lodged in this *duodenal fold* (fig. 101, 1, *l, m, n*). The course of the intestine is otherwise very various in different birds. The upper end, near the pylorus, receives the hepatic ducts; and food is *chylified* after impregnation with the biliary and pancreatic fluids; a process furthered by the proper secretions of the intestinal follicles. The *chyle* is drawn off by the *lacteals* already described (p. 205), and the unassimilable refuse of the food becomes excrementitious.

Cæca (Lat. *cæcus*, blind; in the nom. pl. *cæca*; sing. *cæcum*). — The "blind guts," so called because they end in *culs-de-sac*, are of two kinds. One is the *umbilical cæcum*, or *vitelline cæcum*, a rudimentary, or rather vestigial, structure, the remains of the open duct by which the cavity of the umbilical vesicle (an embryonic organ) communicated with that of the intestinal tract. It is ordinarily not to be noted at all; but it is said by Owen to have been found half an inch long in the gallinule, an inch in the bay ibis, and dilated into a sac an inch in diameter in the *Apteryx*. The structures ordinarily called *cæca*, or *cæca coli*, for they are usually paired, are pouches or diverticula which set off from the intestine proper at the junction of the ileum with colon; but there is nothing in the intestine itself to mark this point, so that when cæca are absent, as frequently happens, no distinction of ileum from colon or rectum is appreciable. No part of the intestinal tract is so variable as the cæcal; so that presence or absence of these appendages furnishes zoölogical characters now-a-days taken very commonly into account in framing genera and families. There are no cæca, as in the turkey-buzzard and some pigeons; there is a single small cæcum in herons. From a condition of extremely small size, like little buds upon the intestine, cæca are found to elongate to extraordinary dimensions; and the large specimens are frequently saccate or clubbed, with slender roots. In geese and swans the cæca are a foot long, more or less; in some grouse they are said to be a yard long. In the ostrich, the mucous membrane is thrown into a spiral fold. However developed, the physiology of these intestinal appendages is, the detention of food until all its nutritive qualities are absorbed, and increase of the absorbent surface.

The Cloa′ca (fig. 101, [1] *k*) or "sewer," very well named, is the termination of the bowel, — an oval or globular enlargement of the rectum, of sufficient capacity at least to contain the completely shelled egg. For, not as in placental mammals, the uro-genital and digestive organs are behind-hand in their evolution, and do not entirely lose connection with each other. Nor is there in birds any distinct bladder; but a cavity, originally that of the allantois of the embryo, persists in common with that of the intestines, and is the *cloaca*. Such incomplete distinction between the two as there may be, by a folding of mucous membrane or partial compartment of the whole, results in cloaca proper and *urogenital sinus*, in which latter are the papillose orifices of the *ureters*, one on each side, from the kidneys; and of the single oviduct ($♀$) or paired sperm-ducts ($♂$), from ovary or testes. The urine of birds not being liquid requires no more of a bladder than the sinus furnishes. The same cavity contains the penis of those birds, as the ostrich and drake, which are provided with an organ of copulation. A peculiar anal gland, the *bursa fabricii* (see frontisp.), also opens into the cloaca. Refuse of digestion, the renal excretion, the spermatic secretion, and the product of conception, are discharged by a single anal orifice, the two former *en masse*.

Being intimately related to dietetic regimen, and so to the habits of birds, the alimentary canal varies greatly, — even more than my slight sketch shows, — and consequently affords good zoölogical characters in the details of its construction. But of all the anatomical systems, this is the one most variable as a matter of *physiological adaptation* (see p. 67). Its characters, even when they seem weighty, are therefore peculiarly liable to be fallacious as indices of natural affinities, and must be applied with discreet caution to morphological classification. Such are commonly only of generic significance. Thus in pigeons the cæca and even the gall-bladder may be present or absent in neighboring genera.

Alimentary Annexes. — Some of these, as the salivary glands, have been noticed already. The two most important bodies connected with the digestive tract, and properly considered adjuncts, are the pancreas and the liver. The former is that kind of lobulated salivary gland which in mammals is called the "sweetbread." It lies in the duodenal loop, along which its loosely aggregated lobes extend. Its ducts, formed by the successive union of smaller efferent tubes, are two or three in number; they pierce the intestine a little below its commencement at the pylorus, and pour into the canal the pancreatic juice, which has the property of emulsionizing fat. The *liver* is a well-known glandular organ of very special structure and function, secreting the fluid called *bile*, also received into the intestine. It is of moderate size in birds, and deeply divided into two principal (right and left) lobes: in some birds there is also a smaller lobe; and one of the large lobes may also be divided. The lobes dispart above to receive between them the apex of the heart; they are held in place by pleuro-peritoneal folds contributing to form the thoracic-abdominal air-cells. The viscus receives venous blood from the extensive portal system of birds; two hepatic veins then conduct it to the post-caval. The emunctory ducts, carrying off the bile, are two or three in number. One at least goes directly to the intestine, and another to the gall-bladder, when that cyst exists; in which case there is a separate cystic duct from the bladder to the intestine, no *ductus communis choledochus*, or duct common to the hepatic substance and its cyst, being formed in birds. Two hepatic ducts may coexist with a cystic duct, making three to the intestine, all separate; two is the rule when there is no gall-bladder. These emunctories commonly enter the intestine some distance apart, and after the pancreatic ducts. The gall-bladder is generally present, frequently absent; it may occur or not in closely related genera of birds.

g. Oölogy: the Uro-Genital Organs.

The Urinary and Generative Organs may be conveniently considered together, not only on account of their close anatomical relations, but because their physiological functions, totally diverse in adult life, are primitively related in the most intimate manner. For it is a singular fact that the mean office of straining urine out of the system is at first sustained by a structure (wolffian body), in closest connection with which, in the female, actually as a part of which, in the male, are later developed those organs (ovary and testis) whose exalted office is creative; for these permanent genital glands procreate the microscopic creatures called *Dynamamœbæ*, the marriage of which results in the reproduction of a complex organism like the male or female parent. (See figs. 103, 104, and following.)

The Wolffian Bodies, or *primordial kidneys*, are a pair of tubular structures which appear very early in the progress of development of the embryo, beneath the spinal column, in front of the fore end of the future kidneys; with each of them is developed a duct, the *wolffian duct*, which carries their excretion into the cavity of the allantois (the future cloaca). Upon the appearance of the true kidneys, the transitory wolffian bodies and ducts lose their urinary function; they ultimately disappear from the female, for the most part, leaving only a trace of their former existence in certain vestigial structures (*parovaria*, etc.); in the male, likewise, they atrophy, but not to the same extent; for a portion of the bodies persists as an accessory (epididymal) portion of the testicle, and their ducts persist as the sperm-ducts, or *vasa deferentia*. Meanwhile, in closest connection with the wolffian bodies, appears a pair of organs, the *genital glands*, for a while exactly alike. If the new creature is to become *female*, the *genital gland* develops to a certain complexity of tissue and becomes the *ovary*; while a certain duct, the *müllerian duct*, developed coincidently to connect such ovary with the cloaca, becomes the *oviduct*. In birds usually only one ovary and oviduct (the left) becomes functional. If the new creature is to become *male*, the same *genital gland* develops to a higher degree of complexity, acquires a tubular structure, and becomes the *testicle;* it connects with remains of the wolffian body, and the wolffian duct becomes the permanent sperm-duct, conveying the product of the male function to the cloaca, just as the oviduct conveys the product of the female function to the same sewerage. Thus the testicle of the male and the ovary of the female are homologous, in fact primitively identical organs, upon which sexual difference is impressed by the greater complexity of structure acquired if the sex is to be male; a female being, anatomically and physiologically, simply an imperfect male, arrested at one stage of her physical progress to male perfection of structure; and the whole nature of the female bears out the same relation of inferiority. But the oviduct of the female, and the sperm-duct of the male, though physiologically identical, having the same function of conveying the products of generation from the genital gland to the light of day, are not anatomically the same; for in the case of the female, whose wolffian duct has disappeared, the müllerian is the oviduct; in the case of the male, in which no müllerian duct appears, the wolffian is the sperm-duct. The two are analogous, not homologous (a good illustration — see p. 68). But it must be further observed that while the sperm-duct conveys only the masculine essence from centre to periphery, the oviduct conveys the feminine material from centre to periphery, and *also* the male essence in the opposite direction; for, upon coitus, which is direct in all birds, the spermatozoa, deposited in the cloaca of the female, find their way up through her oviduct to the ovary, there to accomplish impregnation of the ovarian ova, the fecund product then passing down by the same avenue. All that relates to the mysteries of generation, — both the structure and function of the reproductive organs, and the maturation of the product of conception, is properly *Oölogy* (Gr. ὠόν, *oon*, an egg); though the term is vulgarly used to signify merely a description of the chalky substance in which the egg of a bird is finally invested. The anatomy of the egg is *Embryology*. An egg, or *ovum*, is simply the product of conception up to the time that product acquires an independent existence; while still connected with the female tissue of the ovary, and before or after it amalgamates with the male element, it is an *ovarian ovum;* more or less incompletely matured, it is an *embryo* or *fœtus*, — the former term being commonly applied to the unhatched young of birds. The only difference between the "egg" of a "viviparous" mammal and that of an "oviparous" bird, is in the albuminous and cretaceous envelopes of the latter, and its speedy expulsion from the body of the female to be hatched outside, without anatomical connection with the mother after the hard shell is formed; whereas, in most mammals, the ovum is retained in a dilated part of the müllerian duct (uterus or womb) until it "hatches"; but mammal and bird alike "lay eggs," the essential germinative part of which is identical. Appreciation of these facts, and a proper idea of the relations of the mature sexual organs to the wolffian bodies is necessary to any understanding of the parts and processes concerned in reproduction.[1] We have here to consider the permanent as distinguished from the transitory kidneys, and may then recur to the subject of generation.

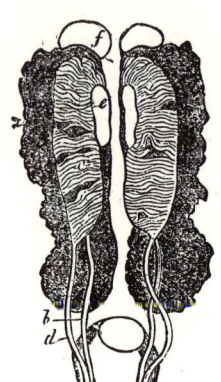

FIG. 103. — Uro-genital organs of male embryo bird; from Owen, after Müller. *a*, kidneys: *b*, ureters; *c*, wolffian bodies; *d*, their ducts, to be sperm-ducts; *e*, genital glands, to become testicles; *f*, adrenals.

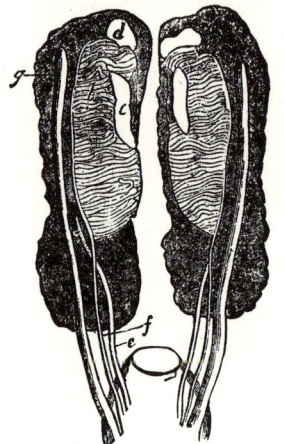

FIG. 104. — Uro-genital organs of female embryo bird; from Owen, after Müller. *a*, kidneys; *b*, wolffian bodies; *c*, genital gland, to become ovary; *d*, adrenals; *e*, ureters; *f*, wolffian ducts, to disappear; *g*, müllerian ducts, to become oviducts.

---

[1] The matter may be further illustrated by the two figures borrowed from Owen (after Müller). In both figs., the large dark masses, *a*, are the permanent kidneys, whose ducts, *b* in fig. 103, *e* in fig. 104, are the ureters, emptying into the cloaca. In fig. 103, male, *c* is the wolffian body, whose duct, *d*, persists as the sperm-duct, conveying

THE ANATOMY OF BIRDS.—OÖLOGY.

The Kidneys (Lat. *renes*, Engl. *reins*, adj. *renal;* figs. 103, 104, *a;* 105, *x*) differ much from those of mammals in physical characters, though identical in function, — that of straining off from the blood certain deleterious substances in the form of urea; whence they are sometimes called *emulgent* organs. Their office of purification is analogous to that of the lungs, which decarbonize the blood, and to some extent vicarious, as is that of excretory organs in general. As the lungs are closely bound down to the thoracic region of the trunk, so are the kidneys impacted in the pelvic region, being moulded to the sacral inequalities of surface (p. 147). They are paired, but sometimes connected across the median line by renal tissue; they have no special renal artery, but derive their blood from various sources; and blood from them takes part in the hepatic portal system, no reniportal being accomplished. They have little or nothing of the particular mammalian configuration which has made "kidney-shaped" a common descriptive term; being elongated, somewhat parallel-sided and rectangular, flattened bodies, lobated into a few large compartments, and lobulated into many lesser divisions; their figure depends much upon that of the pelvis. They are very dark-colored, rather soft, easily lacerable, and appear to the naked eye to be of a granular substance, without distinction of "cortical" and "medullary" portions. Nor is there any "pelvis" of the kidneys in which the uriniferous tubules empty together by numerous ducts as into a common basin. Each *ureter* (figs. 103, *b;* 104, *e;* 105, *y*), or excretory duct, is formed by reiterated reunion of the *tubuli uriniferi*, after the manner of a pancreatic duct; each ureter passes down behind the rectum and opens into the lower back part of the cloaca, — much like a mammalian ureter into the base of the bladder. The original cavity of the allantois remains to furnish no more of a *urinary bladder* than some special dilatation of the cloaca represents; but this rudimentary bladder, as distinguished from the uro-genital sinus in which the ureters terminate alongside the sperm-ducts, is well marked in some birds; being in the ostrich, for example, a considerable enlargement of the cloaca between the termination of the rectum proper and the uro-genital compartment of the sewer. The renal excretion is not watery as in mammals, but semi-solid, and voided with the fæces, of which it forms part.

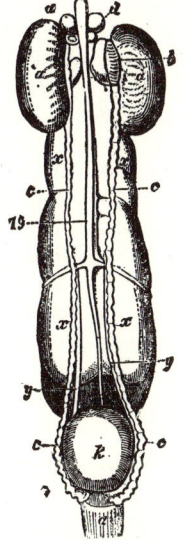

FIG. 105.— Uro-genital organs of the domestic cock; after Owen. *a*, testis; *b*, epididymis; *c*, sperm-duct or vas deferens; *d*, adrenal; *k*, cloaca; *x*, kidney; *y*, ureter.

The kidneys are capped by a pair of small yellowish bodies, the *supra-renal capsules* or **adrenals** (figs. 103, *f;* 104, 105, *d*), the nature of which is undetermined. They are chiefly interesting to the practical ornithologist in their liability to be mistaken for testes in examining specimens for sex (see p. 46).

Male Organs of Generation.— The *testis* (Lat. *testis*, pl. *testes*, a witness; fig. 105, *a*) or *testicle* has been already sufficiently noticed as to its general appearance and position (p. 46). As said above, it is the essential male organ, consisting of the primitive indifferent genital gland (fig. 103, *e*) in its highest state of development as a tubular secretory organ, connected with the remains of the wolffian body as a part of its efferent structure (*epididymis;* fig. 105, *b*) and with the original wolffian duct as its *vas deferens* (figs. 103, *d;* 105, *c*), or efferent duct, by which the semen is conveyed to the cloaca. The original glands normally remain paired, and both are usually functionally developed to corresponding size, shape, and activity; they remain in their embryonic situation in front of the upper part of the kidneys; and such difference

---

semen from *e*, the testis. In fig 104, *b* is the wolffian body, whose duct, *f*, disappears; and *g* is the müllerian duct, becoming the oviduct, to convey the egg from *c*, the ovary. Thus *e*, fig. 103, and *c*, fig. 104, are the homologous genital glands, becoming either testis or ovary: but the sperm-duct, *d*, fig. 103, is not the oviduct, *g*, fig. 104.

of appearance as they present under different circumstances is mainly seasonal. For birds, as a rule, procreate only at particular times of the year, rarely having more than one or two broods of young: the functional activity and quiescence of the testes correspond, as the enormous swelling of the gland during the breeding season is one of the peculiarities of the bird's organ. This may be related to the absence, in birds, of specially formed *vesiculæ seminales*, or seminal reservoirs; though certain contortions and dilatations of the sperm-ducts which are to be observed may imperfectly answer to detain the secretion until circumstances render it available. The passage of the sperm-duct is along the face of the kidneys, generally in company with the ureters; the opening is by a papilla upon the surface of the uro-genital sinus. These papillose terminations of the sperm-ducts are erectile to a degree, and answer the purpose of paired penes in those birds which are not provided with better-formed copulatory parts. *In coitu*, the cloacal chambers containing the orifices of the genital ducts are opened, and the more or less protruded papillæ come in contact or close juxtaposition. In cases in which a penis or two penes are developed, the urethral passage is a groove, never a tube, though cavernous and even muscular tissue may be developed; and in any case of such an intromittent apparatus, it has cloacal invagination when not operative (see p. 891). These organs, in all their variety, are of the sauropsidan, not mammalian, type; though in some respects the structure approaches that seen in the non-placental mammals. No prostate or cowperian glands exist in birds.

The sole office of the testis, or *oöphoron masculinum*, is the secretion of *semen*, associate structures being simply accessory, for the conveyance of that vital substance and its transference to the opposite sex. The seminal *fluid* itself is merely the vehicle of transport of the *spermatozoa*, in which their activity may be freely exercised in their intuitive struggles to gain access to their mates in the ovary. It is literally a "sea of life" in which the minute creatures swim in shoals to their destiny, — and their fate in any case is death. If they successfully buffet the waves of fate they find a watery grave in the ovum at last; if that haven be not reached they simply perish in mid-ocean. The spermatozoa, or seminal animalcules, or male *Dynamamœbæ* (figs. 106, 107), are the exact counterparts of ovarian ova, in so far as they are single-celled animals of a very low grade of organization; but their activity and intelligence is marvellous, and still more so is the mysterious attribute with which they are endowed of assimilating their protoplasmic substance with that of the ovum; with the result that the thus fecundated ovum is capable of procreating itself by fission for a period until a mass of similar creatures is engendered; from which mass is then speedily evolved the complex body of the Bird. The corresponding female *Dynamamœbæ* (ovarian ova) are simple spherical animalcules, physically indistinguishable from an ordinary encysted *Amœba*; but the spermatozoa are remarkably distinguished in appearance, furnishing probably the best marked case of sexual characters to be found among the *Protozoa*, to which class of animals they belong. The spermatozoa resemble flagellate infusoria or ciliated endothelium cells, though they each have but a single whip. They are of extremely minute size, much smaller than their females, and filamentous; more or less thickened and sometimes wavy at their nucleated heads, whence protrudes an excessively delicate thready tail, endowed with great vibratory energy. They may be likened to diminutive attenuated tadpoles, which swim by lashing the tail in the seminal fluid. Under the microscope shoals of these curious creatures may be seen swimming in the sea, nosing about in search of the ovum, butting their heads in wrong places, backing out and trying again in another direction; with such success that out of myriads a score or so may gain their end. It

FIG. 106. — Spermatozoa of domestic cock, greatly magnified; from Owen, after Wagner and Leuckart.

FIG. 107. — Spermatozoa of sparrow, greatly magnified; from Owen, after Wagner and Leuckart.

will be seen that they have a long journey to accomplish; for, liberated in the cloaca of the female, they have to swim through the whole length of the oviduct to the ovary. Besides such physical difference between the male and female *Dynamamœbæ* as I have indicated, they differ in their place and mode of birth; and in this difference lies the very gist of sex. The original indifferent genital gland above described, arrested, as said, at a certain stage of development and therefore female — the ovary — produces its eggs from its surface-cells, which subside into the ovarian tissue, and are quietly packed away there as ovarian ova, ready to ripen and awaken to impregnation in due course. The same gland, further developed into a testis, gives active birth to the spermatozoa in the tubules of its complicated interior tissue. In the former case, the superficial cells slowly ovulate; in the latter, the cells lining the interior speedily spermate; in a word, the testis is as literally *viviparous* as is the ovary *oviparous*, — and these conditions are certainly no insignificant indices of relative development in the scale of being. The spermatozoa appear in some animals to be set free in myriads from the walls of the seminal tubules whence they directly issue; in birds, they are described as appearing coiled or otherwise packed in delicate sperm-cells, which speedily rupture and discharge the creatures in the current of the seminal fluid, where they take up the course and display the energetic actions above noted. Either case has its parallel among ordinary Protozoans; the former corresponding to the process of budding or gemmation, the latter to that of interior fission and discharge of numerous progeny by rupture of the envelope. The final conjugation of spermatic filaments with ovarian ova is simple fusion, such as any ordinary sexless amœboid animal may practise to blend its protoplasmic substance with that of another. But there is this difference, that in the case of *Dynamamœba* it is a true sexual congress, usually *polyandrous*, and still more of a one-sided affair in that the female *Dynamamœba* is at the time in a more or less quiescent, *encysted* state.

Female Organs of Generation. — The connection between the male and female organs of generation is naturally so close that in what has preceded it has been scarcely possible to speak of the former without reference to the female counterparts. I have thus far endeavored to state clearly the nature of the originally sexless genital gland; the difference in the same gland when afterward sexed male or female; and the character of the spermatic offspring of the male gland. In reading that lesson the novitiate in such Eleusinian mysteries must not mistake the language I have used to describe the male *Dynamamœba*, or spermatozoön, as applicable to anything in the development of the female *Dynamamœba*, or ovum, into the chick; for all said thus far only relates to the bringing of the spermatozoön into contact with the ovum, preliminary to the initial step of the ovum in its course of development. It is this female *Dynamamœba* — this primitive ovarian ovum, the germ of the chick, which corresponds to and is the counterpart of the male *Dynamamœba*, on meeting and mingling with which fecundation is accomplished; the impregnated ovum being then empowered to take up its marvellous march. Conjugation of the opposite *Dynamamœbæ* occurs either in the ovary or upper part of the oviduct, — most probably the former. One or several spermatozoa — usually more than one — accomplishing their journey up the oviduct, and finding their affinity, insinuate themselves into the substance of the ovum, and die there, dissolved in amorous pain; that is to say, they melt into the substance of the ovum. The now fertile result, consisting of the mingled protoplasm of the opposite amœbas, is to all appearance precisely the same as the original infecund ovum — yet there is all the difference in the world, as the result shows.

The general character of the ovary of a bird has been already indicated (p. 46). The principal superficial difference in appearance when the ovary is in functional activity, from the corresponding organ of a mammal, is that the ova develop to such a size, in ripening in the ovary before leaving it for the oviduct, that the organ looks like a bunch of grapes, — very large and conspicuous. The oviduct is the musculo-membranous tube (modified müllerian duct) which conveys the ripened ovum, and in its passage provides it with a quantity of white albumen, and finally a chalk shell. A bird's oviduct is the strict morphological homologue (p. 68) of a mammal's fallopian tube, uterus and vagina, — more accurately, of one fallopian tube, one half of a uterus, and one half of a vagina; for the uterus and vagina of a mammal result from the union of both müllerian ducts; whereas in a bird only one — the left usually — is normally developed. Functionally, the oviduct is also analogous (p. 68) to the mammalian uterus, inasmuch as it transmits the product of conception, and detains it for a while, in the initial stage of its germination, as we shall see in the sequel; though all but the very first steps in the development of the chick are taken during incubation, the egg having so hastily left its uterine matrix. These structures — ovary and oviduct, fig. 108, — are most conveniently described as we trace the course of the ovum from its origination to its maturity. This record differs considerably from the corresponding course of events in a mammal, inasmuch as the ovum of a bird, though primitively identical with that of any other animal, acquires special albuminous and cretaceous envelopes which the mammalian ovum, developed in the body of the parent, does not require. The process is termed *ovulation*. Ovulation, which is the formation of an egg in the bird, must not be confounded with *germination*, which is the formation of a bird in the egg. The former can be accomplished by the virgin bird, which may lay eggs scarcely differing in appearance from those which have been fecundated, but germination in which is of course impossible. The course of ovulation, and afterward of germination, is now to be traced.

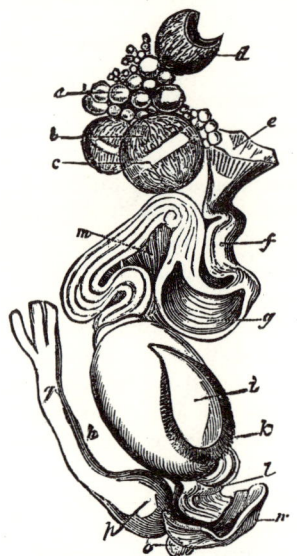

FIG. 108. — Female organs of domestic fowl, in activity; from Owen, after Carus. *a, b, c, d*, mass of ovarian ova, in all stages of development; *b*, a ripe one; *c*, its stigma, where the ovisac or calyx ruptures; *d*, a ruptured empty calyx, to be absorbed; *e*, infundibulum, or funnel-shaped orifice of the oviduct; *f*, next portion of oviduct; *g*, follicular part of oviduct; *m*, mesometry, membrane steadying the oviduct; the reference-line, *m*, crosses the constricted part or isthmus of the oviduct; these parts secrete the white of the egg; *k*, shell-forming or uterine part of oviduct, in which is a completed egg, *i*; *l*, lowest or vaginal part of oviduct, opening into uro-genital sinus of the cloaca, *n*; *o*, anus.

Ovulation. — The *ovum* begins as a microscopic point in the ovary, the *stroma* or tissue of which is packed with these incipient eggs. It is primitively just like any other female *Dynamamœba*, from that of a sponge up to that of a woman, — a naked simple cell, capable of exhibiting active amœboid movements. It consists of a finely granular protoplasm, the *vitellus*, or *yelk*, enclosed in a delicate structureless cell-wall, the *vitelline membrane*, called the *zona pellucida* from its appearance under the microscope. Imbedded in the vitellus is a nucleus, or kernel, the *germinal vesicle;* in this is a nucleolus, or inner kernel, the *germinal spot*. The ovum occupies a tiny space in the ovary, the cellular walls of which constitute an *ovisac*, or *graafian follicle*. Now if such an ovum as this were mammalian, it would, without material change, burst the ovisac, be received into the fallopian tube and conveyed to the uterus; where, supposing it already fertilized, the whole of its contents would develop into the body of the embryo. It would therefore be *holoblastic* (Gr. ὅλος, *holos*, the whole; βλαστικός, *blastikos*, germinative). It is different with a bird or other "oviparous" animal, the egg of which has to hatch outside the body; for provision must be made for the nourishment of the developing chick, thus separated from the tissues of its mother. Such provision is made by the accumulation about the ovum of a great quantity of granular protoplasmic substance, which forms nearly all the large yellow ball called in ordinary language "the yelk" of an egg. None of this adventitious substance goes to form the embryo; it is what the embryo feeds on during its formation. A bird's egg is therefore *meroblastic* (Gr. μέρος, *meros*, a part, and βλαστικός), and we must carefully discriminate between the great mass of yellow *food-yelk*, as it may be called, and a small quantity of "white yelk," the true *germ-yelk*, which alone is transformed into the body of the chick. The latter forms the *cicatricle*, vulgarly called the "tread"; that small disc, visible in most birds' eggs to the naked eye, which appears upon the surface of the great yellow ball, floating in a pale thin yelk which penetrates the denser and yellower food-yelk by a cord of its own substance leading to a central cavity, the false yelk-cavity, around which the food-yelk is deposited in a series of concentric layers like a set of onion-skins. The whole mass is surrounded by a delicate structureless yelk-skin, called the *vitelline membrane* (whether this be the original vitelline membrane of the *Dynamamæba* or not; i. e., whether the food-yelk has accumulated inside or outside the original *zona pellucida*). All this enormous accumulation, effecting what is called a *metovum* or after-egg, to distinguish it from the *protovum*, or primitive state of the egg, goes on in the ovary, and in the ovisac of each ovum; with the ripening of the ovum, the ovisacs become distended to a corresponding size, and the whole ovary acquires the familiar bunch-of-grapes appearance. With such maturation of the fruit, the connection with the rest of the ovary lengthens into a stalk, or *pedicel*, by which the ripe ovum hangs to its

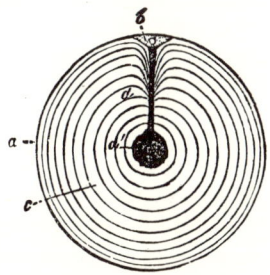

FIG. 109. — Meroblastic ovum (yelk) of domestic fowl, nat. size, in section; after Haeckel. *a*, the thin yelk-skin, enclosing the yellow food-yelk, which is deposited in concentric layers, *c*, *d*; *b*, the cicatricle or tread with its nucleus, whence passes a cord of white yelk (here represented in black) to the central cavity, *d'*.

stock, like any fruit upon its stem, ready to burst its skin and fall into the open mouth of the oviduct. Such rupture of the graafian follicle (ovisac), in its now distended state known as the *capsule* or *calyx*, occurs along a line where the numerous blood-vessels which ramify upon its surface appear to be wanting, called the *stigma*: this is rent; the ovum slips out of its calyx, like the substance of a grape pinched out of its skin, and falls into the oviduct. After this discharge, the empty calyx collapses, shrivels, and ultimately disappears by absorption. (See expl. of fig. 108).

The ovum thus acquires the full size of its yelk in the ovary, — becoming, as in the case of the hen, a yellow sphere an inch in diameter.[1] Notwithstanding its enormous distension with food-yelk, it is still morphologically a simple cell, affording the maximum dimension of any known protozoan or single-celled animal. Entering the oviduct, the germ-yelk part of the whole mass is fertilized by spermatozoa, unless this process has before occurred in the ovary, and in its passage through that tube the yelk-ball becomes invested successively with the mass of transparent albumen known as the "white" of the egg, and finally by the chalk shell — both secreted by the mucous membrane lining the oviduct.

During its functional activity, the left oviduct (there being usually only this one) becomes highly developed, both as to its muscular walls, which by their contractility embrace the ovum closely and squeeze it along, and as to its mucous secretory surface. It is supported by peritoneal folds forming a *mesometry*, like the mesentery of the intestines; its whole structure and office are quite like those of a length of intestine. The upper end of the singularly serpentine oviduct is dilated into an *infundibulum*, or funnel-like mouth, corresponding to the fimbriated extremity of the mammalian fallopian tube, and constituting a *morsus diaboli*, or "devil's grip,"

---

[1] How great this is can only be appreciated by comparison. The human egg, on escaping from the graafian follicle, is said to be from $\frac{1}{240}$ to $\frac{1}{120}$ of an inch in diameter. Taking it at $\frac{1}{200}$, there would be 40,000 in a square inch, and in a cubic inch 8,000,000. The largest bird's egg known, that of the *Æpyornis*, is said to have a content of about a gross of hen's eggs — 144. Supposing the yelk of the *Æpyornis* egg to bear the usual proportion to the other contents of the shell, and allowing for the difference in bulk between a sphere and a cube of equal diameters, there would still be somewhere about a billion human eggs in one *Æpyornis* egg-yelk, — roundly, a mass of them equal to that of the germs of more than one-half of the present population of the globe.

which gets hold of the ovum to drag it down to the common lot of mortals from its high ovarian birth. The infundibulum receives from the mesentery a delicate tunic of unstriped muscular fibres, which are so disposed as to dilate that orifice for the reception of the ovum; and during the venereal orgasm the mouth of the tube is supposed to seize upon the ripest egg. The actual anatomy of the arrangement, and the whole operation, is strangely suggestive of one of the oldest myths respecting the serpent which bore the egg of the world in its jaws. The mucous lining of the oviduct consists of a layer of ciliated epithelium; the membrane has a different character in successive portions of its extent. Above, when the tube is not distended with its burthen, the lining is thrown into lengthwise folds, which lower down become spirally disposed, and then longitudinal again before they cease. This rugous portion of the tube is beset with mucous follicles, which secrete "the white." The oviduct, after contracting at a point called the *isthmus*, enlarges to a calibre sufficient to accommodate the egg in its shell; for this is the shell-forming part, homologous with the mammalian uterus (a sinister semi-uterus at least), lined with large villi, and beset with the follicles whose secretions calcify the egg-shell, and decorate it with pigment. The rest of the tube is vaginal, being merely the passage-way by which the perfected ovum is discharged into the cloaca, to be expelled *per anum*. The muscular walls of the oviduct consist of both circular and longitudinal unstriped fibres, like those of intestine, — the latter especially in upper portions and at the infundibulum, the former more conspicuously below, where they form a sort of *os tincæ* at the bottom of the calcific portion, and a kind of *sphincter vaginæ* at the end of the tube. A recognizable *clitoris* is developed in many birds.

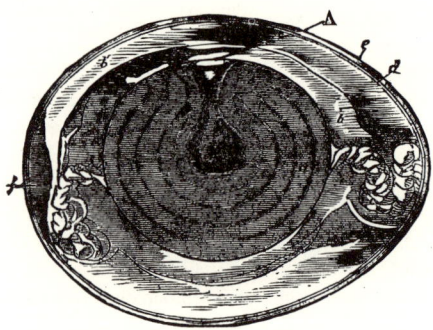

FIG. 110. — Hen's egg, nat. size, in section; from Owen, after A. Thompson. *A*, cicatricle or "tread," with its nucleus, of white germ-yelk, floating on surface of pale thin nutritive yelk, leading to central yelk-cavity, *x*; *a*, the yellow yelk-ball, deposited in the successive layers, forming a set of *halones*, and enveloped in the chalaziferous membrane which is spun out at opposite poles into the twisted strings, chalazæ, *c, c*; *b, b'*, successive investments of softer white albumen; *d*, membrana putaminis, the "soft shell" or egg-pod, between layers of which at the great end of the egg is the air space, *f*; *e*, the shell.

The deposition of the white and of the shell remains to be noticed. The first deposit upon the yelk-ball consists of a layer of dense and somewhat tenacious albumen, called the *chalaziferous membrane* (Gr. χάλαζα, *chalaza*, a tubercle, and Lat. *fero*, I bear). As the egg is urged along by the peristaltic action of the tube, it acquires a rotation about the axis of the tube; the successive layers of soft albumen it receives are deposited somewhat spirally; and the chalaziferous membrane is drawn out into threads at opposite poles of the egg. These threads, which become twisted in opposite directions during the rotation of the egg, are called *chalazæ*; they are the "strings," rather unpleasantly evident in a soft boiled egg, but serve the important office of mooring and steadying the yelk in the sea of white by adhesions eventually contracted with the membrane which immediately lines the shell. They are also intrusted with the duty of ballasting, or keeping the yelk right side up. For there is a "right side" to the yelk-ball, being that on which floats the cicatricle, or "tread." This side is also the lightest, the white yelk being less dense than the yellow; and the chalazæ are attached a little below the central axis. The result is, that if a fresh egg be slowly rotated on its long axis, the tread will rise by turning of the yelk-ball in the opposite direction, till, held by the twisting of the chalazæ, it can go no farther; when, the rotation being continued, the tread is carried under and up again on the other side, resuming its superior position as before. After all the spiral layers of soft white are laid on, a final covering of dense albumen is deposited at the isthmic part of the oviduct. This forms a tough tunic called the *membrana putaminis* (Lat.

putamen, a peel, rind), or "egg-pod"; it is the final envelope of such a "soft-shelled egg" as a hen drops when deprived of the lime required to enable her to secrete a hard shell. In the uterine dilatation of the oviduct a thick white fluid charged with earthy matter is exuded; this condenses upon the egg-pod and forms the shell. The composition of this earth is chiefly carbonate of lime (common chalk), with some carbonate of magnesia, and phosphates of both of these bases — thus like that of bone as to ingredients, but in very different proportions. The shell does not simply overlie the pod in a distinct sheet, but is intimately coherent, the microscopic crystals or other particles of the earthy matter being deposited in the matted fibrous texture of the pod. The connection is most intimate in fresh eggs; after a while, layers of the pod separate at the butt of the egg, forming the large air-space which every one has noticed in that situation. The shell being very porous, readily admits air. The air space enlarges during incubation, and the pod becomes more and more distinct from the shell, which latter also increases in porosity and fragility towards "full term." The rough or smooth appearance of an egg-shell, the pores which may be visible to the naked eye, and other physical characters, are due to the impression made upon it by the lining membrane of the "uterus." The superficial deposit of chalk is so heavy, in some cases, as those of cormorants, etc., that it may be scraped off without interfering with the texturally firm shell-substance underlying. All the coloration of egg-shells, which frequently makes them pretty objects, is simply the deposit of pigment granules in or upon the shell. Such deposit may be perfectly uniform, as it is in the bluish-green egg of a robin, for instance, but it is oftener spotty — either upon a white or a whole-colored ground. The browns and neutral tints are the usual colors, particularly a bright reddish-brown; the same, lying in instead of upon the shell, gives the grays, "lilacs," and "lavenders" so well known. In ptarmigan, the pigment is so heavily deposited that the egg comes out pasty on the surface; a sign of "fresh paint!" one must not disregard if he would not spoil the decoration.

Oviposition. — The energy and rapidity with which the processes involved in the manufacture of so complex a product as a bird's egg is now seen to be are extraordinary. A domestic fowl may lay an egg every day for an indefinite period. It is difficult to say how quickly an egg may ripen in the ovary; for, during the activity of that organ, several or many are to be found in all stages of immaturity, and the date of the initial impulse cannot well be determined. As there is probably but one egg at a time in the oviduct, the whole process of finishing off the yelk-ball with its chalaziform, soft albuminous, putaminous, and calcareous envelopes may go on in twenty-four hours, most of which time is consumed in the shell-formation. The number of eggs matured by the human female is or should be thirteen annually; this is no large number for many of the gallinaceous and anatine birds to deposit in about as many days. But a probable average number is five or six. Defeat of the procreative instinct from any accident is commonly a stimulation to renewed endeavors to reproduce; and very many birds rear two or three broods annually, though one clutch of eggs is the rule. Many, such as auks, petrels, and penguins, lay a single egg. Two eggs is the rule in humming-birds and pigeons. Three is normal to gulls and terns, though these often have but two. Four is the rule among the small waders of the limicoline groups. Some of the small *Oscines* lay over the average, having eight or ten; among these, the European sparrow, *Passer domesticus*, is probably the most prolific. The parasitic cuckoos are said to lay the relatively smallest eggs; that of the *Apertyx* is said to be the largest, weighing one fourth as much as the bird. The usual *shape* of an egg has given us the common names *oval*, *ovate*, and *ovoidal*, for the well-known figure. Some, as those of owls, woodpeckers, kingfishers, and others, more or less nearly approach a spherical shape. Eggs of grebes, herons, Totipalmate birds and various others are rather elliptical, or equal-ended, and narrow in proportion to their length. Eggs of the limicoline group are generally pyriform, — very broad at one end and narrow at the other. But the eggs of all birds vary more in size and shape than some of the devotees of theoretical oölogy admit in their practice. The variation so well known in any breed of domestic fowl is scarcely above a normal rate. The short diameter, corresponding to the calibre of the oviduct, is less variable than the long axis; for when the quantity of food-yelk and white, upon which the difference in bulk depends, varies with the vigor of the individual, the scantiness or redundancy is expressed by the shortening or lengthening of the whole mass. The egg traverses the passage small end foremost, like a round wedge, with obvious reference to ease of parturition by more gradual dilatation of the outlet.

Germination. — Leaving now all the accessory parts of an egg, let us confine attention to the *germ-yelk*, or "tread," which is alone concerned in the germinative process. Recurring to the female *Dynamamœba*, consisting of granular protoplasm (vitellus) included in its cell-wall (vitelline membrane) and including its nucleus and nucleolus (germinal vesicle and germinal spot), we will trace it up to the time it begins to take shape as an embryo chick. At first, as I have observed before, it is like any other amœba; the first step of development is probably a retrograde one; for if there ensues, when the spermatozoa melt into the ovum, the result affirmed for mammalian ova, the original germinal vesicle and germinal spot *disappear*, and the whole content of the ovum proper is simply a homogeneous mass of granular protoplasm. In this retrograde step, the organism, at the lowest possible round of the ladder of evolution, is called a *monerula*. The germinal vesicle and spot, however, are speedily reconstructed, and the ovum looks precisely as it did before. But observe that the actual difference is enormous; for it now consists of the blended substance of the original ovum and of the spermatozoa; and in this duplex or bisexed state, before any further step is taken, the creature is called a *cytula*, — the parent cell of the entire future organism. In the former state it could reproduce nothing, not even itself; for it is the strange physiological law of a *Dynamamœba* that it cannot reproduce like an ordinary cell, but must evolve an entire organism, like both of those two whose vital forces it concentrates, summarizes, and embodies, — or nothing.

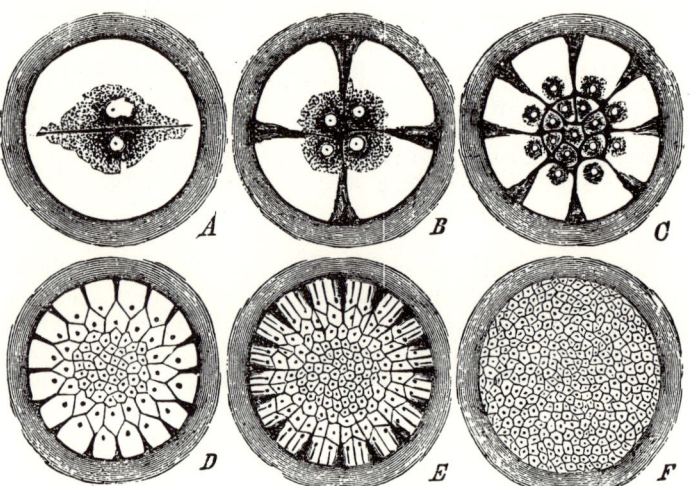

FIG. 111. — Segmentation of the vitellus by discoidal cleavage, diagrammatic, × about 10 times, after Haeckel. Only the "tread," cicatricle, or germ-yelk (figs. 109, *b*, 110, *A*) is represented, as no other part of the whole yelk-ball undergoes the process. *A*, separation into 2; *B*, into 4; *C*, into 16, by 8 radial and 1 concentric furrow; *D*, into many parts, by 16 radial and about 4 concentric furrows; *E*, 64 radial and about 6 concentric furrows; *F*, the whole tread broken up into a mulberry-mass (*morula*) of cells.

The first change in the parent-cell is that by which it becomes broken up into a mass of cells, each of which is just like itself. This process is called *segmentation of the vitellus;* each one of the numerous resulting cells is called a *cleavage-cell*. The nucleus of the parent-cell divides into two; each attracts its half of the yelk; the halves furrow apart and there are now *two* cleavage-cells in place of the one parent-cell. A furrow at right angles to the first, and redivision of the nuclei, results in *four* cleavage-cells. Radiating furrows intermediate to the first two bisect the four cells, and would render *eight* cells, were not these simultaneously doubled by a circular furrow which cleaves each, with the result of sixteen cleavage-cells. So the subdivision goes on until the parent-cell becomes a mass of cells. This particular kind of cleavage, by radiating and concentric furrowing, is called *discoidal*, and the resulting heap of little cells assumes the figure of a thin, flat, circular disc. Segmentation of the vitellus, in whatever manner it may go on, results in a mulberry-like mass of cleavage-cells; and the original cytula has become what is called a *morula*. This process and result are clearly shown in fig. 111, *A–F*.

The morula or mulberry-massed germ of which the "tread" of a bird's egg at this moment consists increases by multiplication of cells, and the disc is lifted a little away from the mass of yellow food-yelk upon which it rests, like a watch-crystal from the face of a watch. This disposition of the greatly multiplied cells in a *layer* and their coherence forms of course a *membrane*, — the *blastodermic membrane*, or *blastoderm*, fig. 112, *B, b*. The cavity between the blastoderm and the mass of food-yelk is called the *cleavage cavity, s*. At the stage when the blastodermic membrane and cleavage-cavity are formed, the germ is called a *blastula*, or *germ-vesicle*,[1] and the process by which the morula becomes a blastula is called *blastulation*. Next, from the thickened rim, *w*, of the watch-crystal-like blastula a layer of large *entoderm* cells, fig. 112, *C, i*, separates, and grows toward the centre: when it gets there, of course the original cleavage-cavity, *s*, is shut off from the surface of the food-yelk; a second crystal having grown under the first one. The second adheres to the first, obliterating the original cleavage-cavity; the germ is now obviously *two-layered;* the rising of the inner layer to meet the outer results in a cavity between itself and the food-yelk, *D, d*. This cavity exactly resembles the original cleavage-cavity, but it is a very different thing, being the primitive *intestinal cavity*.

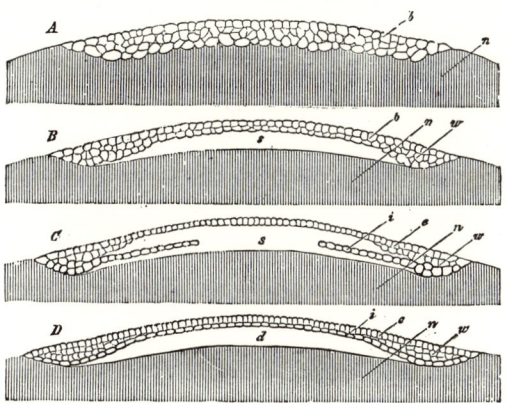

FIG. 112. — Further development of hen's egg; after Haeckel: *A*, the mulberry mass of cleavage cells, *b*, same as seen on top in fig. 111, *F*, here viewed in profile in section, resting upon *n*, the simply-shaded part of the figure, to represent conventionally the mass of food-yelk. *A*, morula stage (as before); *B*, blastula stage, the mass of cells, *b*, forming the blastoderm, uplifted from the food-yelk, leaving the cleavage-cavity, *s; w*, the thickened rim of the germ-disc; *C*, the blastula in process of inversion, by which a layer of entoderm-cells, *i*, growing from periphery to centre, will apply itself to the layer of exoderm-cells, *e*, obliterating the cleavage-cavity, *s*; *D*, the disc-gastrula completed, by union of entoderm, *i*, with exoderm, *e*, leaving the primitive intestinal cavity, *d*, which is quite similar in appearance to the cleavage cavity, *s*, but morphologically quite different.

The blastula, or germ-vesicle, has become converted into a *gastrula*, by the invaginating process just described, known as *gastrulation*. The gastrula of a bird has the circular discoidal form which causes it to be termed a *discogastrula*. This process of forming a single blastodermic layer, with a cleavage-cavity (blastula, or true germ-vesicle), then two blastodermic layers, with obliteration of the cleavage-cavity and substitution of a primitive intestinal cavity (gastrula), is common to all animals which consist of more than single cells, under various modifications and disguises; the process described is that occurring in meroblastic eggs which have a discoidal cleavage and form a discogastrula.[2]

[1] Not to be confounded with the original "germinal vesicle" of the parent-cell, which long since disappeared.
[2] The so-called "germ-vesicle" of the holoblastic mammalian egg is subsequent to gastrulation, not prior, and is therefore not a blastula proper.

What we have got now is a tread or germ consisting of a circular concavo-convex disc of two layers of blastoderm, resting by its rim upon the great yellow ball of food-yelk, from which it is separated by a cavity, as a watch-crystal from its face. All these changes, up to completion of gastrulation, may go on *before the egg is laid*, the tread of a perfectly fresh egg being already a multicellular discogastrula. Since the earlier stages of the embryo (cytula, morula, blastula, and gastrula) are actually accomplished while the egg is still in the body of the parent, the analogy of the oviduct to uterus, etc., as well as its strict homology to the parts of a müllerian duct so named, is not so fanciful as some appear to think. The outer of the two blastodermic layers is the *ectoderm* or *epiblast*, *C* or *D, e;* the inner is the *endoderm* or *hypoblast, i.* By multiplication of cells between the two arises the *mesoblast*. The mesoblastic layer of cells subsequently splits into two, of which the outer is the *somatopleura*, or body layer, the inner the *splanchnopleura* or visceral layer. The two-layered germ has then become four-layered. Up to the time of formation of four layers, the cells are all alike, or only differ slightly in size, color, or consistency. Now, however, ensues that marvellous process by which the indifferent cells of the blastodermic layers are to become *differentiated in form and specialized in function*,— a sort of division-of-labor system in the infant colony of cells, by which some are to learn to move, others to digest, others to procreate, others to think and feel, with corresponding modifications of form by which are generated the *Osteamœbæ, Myamœbæ, Neuramœbæ,* — the bone-cells, muscle-cells, nerve-cells, and all others of the complex organism which is in a few days to come into being from such simple beginnings. This of course opens up the whole field of embryology, which we cannot here enter upon. I will only add, that from the epiblast is derived the integument, and its inversions, as those of the eye and ear, and the brain and spinal chord. From the hypoblast is derived the lining of the alimentary canal and of its annexes and offsets, as liver, lungs, etc. The rest of the embryo comes from the mesoblast, and most of it from the somatopleural layer. The fissure between the two layers of the mesoblast becomes the great pleuro-peritoneal cavity.

In explaining the early embryo, I have closely followed the great German morphologist, Haeckel; and the illustrations are from the same high source.

Incubation. — To induce the wonderful metamorphoses just hinted at, it is only necessary to keep a bird's egg at a pretty even temperature of about 100° F. Nearly all birds secure this result by the process of *incubation*. In many cases the sun's rays relieve the parent of some part of the duty. In a few, the heat evolved from vegetable ferment or decomposition is utilized for the same purpose. This seems to be the case to some extent with grebes; but these incubate. "The exception to the rule of incubation is given by the Megapodial birds of the Australasian Islands. A huge mound of decaying vegetable matter is raised; the eggs are deposited vertically in a circle at a certain depth, near the summit, and the chick is developed with the aid of the heat of fermentation. The large size of the egg relates to affording a supply of material sufficing for an unusually advanced state of development of the chick at exclusion; whereby it has strength to force its way to the surface of the hatching-mound, with wings and feathers sufficiently developed to enable it to take a short flight to the nearest branch of an overshadowing tree" (Owen). The period of incubation has been ascertained with precision for few birds; it is known to range from ten days (perhaps less), as in case of the wren, to fifty or sixty for the ostrich. The female is usually the sitter. Frequently both sexes incubate in turn; such unnatural care for the young by the male is termed *double monogamy*. In most or all *Ratitæ*, in the family *Phalaropodidæ*, and some other Limicoline genera, the male incubates. Most birds attend to their own eggs; many cuckoos (*Cuculidæ*) and the species of *Molothrus*, are parasitical, laying in the nests of other birds, which are thus forced to become foster-parents of alien offspring, generally to the destruction of their own. This seems to result from some peculiarity of the egg-laying process, which does not permit several eggs to be incubated and hatched simultaneously. It is not so unusual among American cuckoos as generally supposed. The degree of development to which birds attain in the egg has been already discussed (p. 88). They break the shell by pecking at it, and struggling; for the former operation the bill is often tempered at the tip by a hard knob which is afterward absorbed. The necessity of providing a receptacle for eggs, in which they may be incubated, results in *nidification* or nest-building; and the extraordinary taste and ability many birds display in this matter, as well as the wide range of their habitudes, furnishes one of the most delightful departments of ornithology, called *caliology* (Gr. καλιά, *kalia*, a bird's nest; see p. 54, note). Many birds burrow in the ground; others in trees; the most beautiful and elaborate nests are furnished by various members of the *Oscines*, the weaver-birds of Africa (*Ploceidæ*) probably taking the lead. The male sometimes constructs his own "nest" apart from that in which the female incubates. " Certain conirostral *Cantores* still practise in the undisturbed wilds of Australia the formation of marriage-bowers distinct from the later-formed nesting-place. The satin bower-bird (*Ptilonorhynchus holosericeus*), and the pink-necked bower-bird (*Chlamydodera maculata*), are remarkable for their construction on the ground of avenues, over-arched by long twigs or grass-stems, the entry and exit of which are adorned by pearly shells, bright-colored feathers, bleached bones, and other decorative materials, which are brought in profusion by the male, and variously arranged to attract, as it would seem, the female by the show of a handsome establishment" (Owen). The extraordinary nests of the *Crotophaga*, used in common by a colony of the birds, are noted at p. 471. " Edible birds'-nests," constructed by swifts of the genus *Collocalia*, consist chiefly of inspissated saliva. Perhaps the most remarkable of all the receptacles of eggs is that which the penguin makes of its own body, the egg being carried in a sort of pouch formed by the integument of the belly, something like that of a marsupial mammal.

§ 5. DIRECTIONS FOR USING THE ARTIFICIAL KEYS.

These "Keys" differ from natural analyses in being wholly arbitrary and artificial. They are an attempt to take the student by a "short cut" to the name and position in the ornithological system of any specimen of a North American bird he may have in hand and desire to *identify*. The plan has been much used in Botany, though seldom if ever employed for a whole Fauna, before the original edition of this work. It will serve a good purpose, rightly used; but it must be remembered there *is no* "royal road to learning"; nobody can be smuggled into sound erudition, either. Nor must too much be expected of me here; I can take the student nowhere until he has learned the difference between the head and the tail of a bird, at any rate. That is what the preceding pages undertake to teach; but, until such technicalities have been mastered, progress in ornithology is out of the question.

The original " Key to the Genera" proved scarcely so satisfactory as I hoped it would be. It undertook too much, to conduct the student at once down to the intricacies of the very many modern genera, not all of which can by any possibility be characterized intelligibly in a line of type. I have probably simplified and expedited matters by preparing on the same plan Keys to the Orders and Sub-orders, and to the Families. Then in the body of the work, under each head, further analyses are given when such seems to be required, — of families under their orders or sub-orders, of genera under their families, and of species under their genera. These ulterior analyses are for the most part rather natural than artificial, though I never hesitate to seize upon *any* character that may furnish the desired clue to identification.

The artificial Keys immediately following will take the student to the *families*, with reference to the page of the work where such groups come; on turning to which, further analyses will be found, generally down to species and even varieties. They are to be used as follows (*after the preceding lessons have been learned*) : —

We have in hand a bird we do not know, and the name of which we wish to ascertain. Suppose it to be that common species which builds the nest of mud upon the bough of the apple-tree and lays greenish-blue eggs. To what *family* does it belong?

The Key opens with an arbitrary division of our birds according to the number and position of their toes. Our specimen, we see, has four toes, three in front, one behind. It therefore comes under IV. Going to IV., we read : —

> Hind toe — inserted above the level of the rest, etc.
> — not inserted above the level of the rest. . . . (Go to B.)

Our specimen has the hind toe not inserted above the level of the rest. Going to B, we find five alternatives. Our bird presents no one of the special characters of the first four alternatives, and this determined takes us to *g*. There we find : —

> (*g*) Primaries — 10 ; the 1st (never spurious), etc.
> — 10 ; the 1st (spurious or), etc. . . . (Go to *i*)
> — 9 ; the 1st (never spurious), etc.

In this case the bird has obviously a spurious first primary, not nearly two-thirds as long as the longest. Going to *i* ; —

> (*i*) Tarsus — "booted" ; wings — shorter than, etc.
> — longer than tail ; tail — double rounded.
> — not double rounded . . . . . TURDIDÆ, p. 240.

Thus (provided we have taken the trouble to inform ourselves what "spurious first primary" and "booted tarsus" mean), the key conducts to a family, by presenting in succession certain *alternatives*, on meeting with each of which, we have only to determine which one of the two or more sets of characters agrees with those afforded by our specimen. There will not, it is believed, be any trouble in determining whether a given character *is so*, or *is not so*, since only the most tangible, definite, and obvious features have been selected in framing the key. After each determination, either the name of a family is encountered, or else a reference-letter leads on to some new alternative, until by a gradual process of elimination the proper family is reached. After a few trials, with specimens representing different groups, the process will be shortened, for the main divisions will have been learned ; still the student must be careful how he strikes in anywhere except at the beginning, for a false start will soon set him hopelessly adrift. The key has been tested so thoroughly that there is little danger of his running off the track except through carelessness, or misconception of technical terms ; but there is no excuse for the former, and the latter may be obviated by the Glossary at the end of the book, and especially the foregoing General Ornithology, § 3, which should be consulted when any doubt arises. Time spent upon the preliminary lessons will be time saved in the end.

At page 240, as indicated, the family *Turdidæ* is fully characterized, and its sub-families and genera are analysed. The bird in hand should answer all the characters of the family and those of one of the sub-families, *Turdinæ*, and one of the genera, *Turdus*. The analysis of the species of *Turdus* should show the specimen to be *Turdus migratorius*, the Robin. Under the head of that species, No. 1 of the List, will be found a fair description and various other particulars.

If there be any difficulty in going at once to the family, the student may try the key to the orders and sub-orders, and get on the track in that way.

Directions for measurement have already been given (p. 24). In comparing measurements made with those given in the Synopsis, absolute agreement must not be expected ; individual specimens vary too much for this. It will generally be satisfactory, if the discrepancy is not beyond certain bounds. A variation of, say, five per cent. may be safely allowed on birds not larger than a robin: from this size up to that of a crow or hawk, ten per cent.; for larger birds even more. Some birds vary up to twenty or twenty-five per cent., in their total length at least. So if I say of a sparrow for instance, "length six inches," and the specimen is found to be anywhere between five and three-fourths and six and one-fourth, it will be quite near enough. But the relative proportions of the different parts of a bird are much more constant, and here less discrepancy is allowable. Thus "tarsus longer than the middle toe," or the reverse, is often a matter of much less than a quarter of an inch; and as it is upon just such nice points as this that a great many of the generic analyses rest, the necessity of the utmost accuracy in measuring, for the use of the keys, becomes obvious. When I find it necessary to use the qualification "about" (as, "bill *about* = tarsus") I probably never mean to indicate a difference of more than five per cent. of the length of the part in question.

It may be well to call attention to the fact, that most persons unaccustomed to handling birds are liable to be deceived in attempting to *estimate* a given dimension; they generally make it out *less* than measurement shows it to be. This seems to be an optical effect connected with the solidarity of the object, as is well illustrated in drawing plates of birds, which, when made exactly of life-size, always look larger than the original, on account of the flatness of the paper. The ruler or tape-line, therefore, should always be used, and particularly in those cases where analyses in the key rest upon dimensions. It is hardly necessary to add, that in taking, approximately, the total length from a prepared specimen, regard should be had for the "make-up" of the skin. A little practice will enable one to determine pretty accurately how much a skin is stretched or shrunken, and to make the due allowance in either case.

The measurements used in this work are all in English inches and decimals.

There are probably no signs or abbreviations not self-explanatory or not already explained in "Field Ornithology."

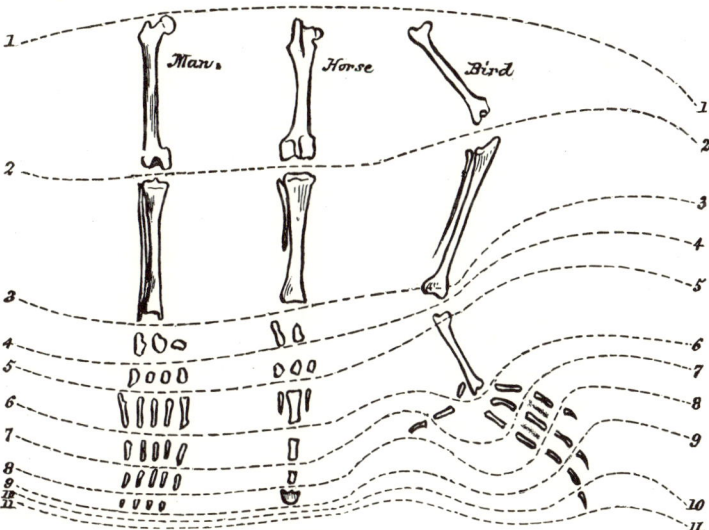

FIG. 112 *bis*. — Diagram of corresponding segments of hind limbs of man, horse, and bird. The lines 1-11 are *isotomes*, cutting the limbs into morphologically equal parts, or *isomeres*.

ARTIFICIAL KEY TO THE ORDERS AND SUBORDERS.

		Page
I. Toes 3; 2 in front, 1 behind . . . . *Pici of* PICARIÆ		537
II. Toes 3; 3 in front. Toes — cleft or semipalmate . . . . LIMICOLÆ		762
— palmate. Nostrils — tubular . . . . TUBINARES		1021
— not tubular . . . . PYGOPODES		1046
III. Toes 4; 2 in front, 2 behind. Bill — cered and hooked . . . . PSITTACI		611
— neither cered nor hooked. Tail feathers — 8 or 10 *Coccyges of* PICARIÆ		537
— 12. Bill — chisel-like *Pici of* PICARIÆ		537
— dentate *Trogones of* PICARIÆ		537
IV. Toes 4; 3 in front, 1 behind.		
Toes — syndactyle . . . . *Halcyones of* PICARIÆ		537
— totipalmate (all four full-webbed) . . . . STEGANOPODES		951
— palmate. Bill — curved up . . . . LIMICOLÆ		762
— not curved up — lamellate . . . . LAMELLIROSTRES		887
— not lamellate. Hallux — lobate . . . . . . PYGOPODES		1046
— not lobate . . . . LONGIPENNES		973
— lobate. Tail — rudimentary . . . . PYGOPODES		1046
— perfect. — A horny frontal shield . . . . PALUDICOLÆ		844
— No frontal shield . . . . LIMICOLÆ		762
— semipalmate; joined by evident movable basal web (go to **A**).		
— cleft to the base or there immovably coherent (go to **B**).		
A. Hind toe — elevated. Tibiæ — feathered below. Nostrils — perforate . . . . *Cathartides of* RAPTORES		617
— imperforate. Gape — reaching below eye *Coraciæ of* PICARIÆ		537
— not reaching below eye GALLINÆ		719
— naked below. Nostrils — perforate . . . . PALUDICOLÆ		844
— imperforate. Tarsi — scutellate in front LIMICOLÆ		762
— reticulate. Head — bald HERODIONES		863
— feathered LIMICOLÆ		762
— not elevated. Tibiæ — naked below . . . . HERODIONES		863
— feathered below. Bill — cered and hooked . . . . . . RAPTORES		617
— not cered. Nasal — membrane soft COLUMBÆ		705
— scale hard GALLINÆ		719
B. Hind toe — elevated. Gape — reaching below eye . . . . *Cypseli of* PICARIÆ		537
— not below eye. 1st primary — emarginate or about = 2d . . . LIMICOLÆ		762
— not emarginate and shorter than 2d PALUDICOLÆ		844
— not elevated. Nostrils — opening beneath soft swollen membrane . . . . COLUMBÆ		705
— otherwise. Bill — cered and hooked . . . . RAPTORES		617
— otherwise. Secondaries — only six *Trochili of* PICARIÆ		537
— more than six (go to a).		
a. Primaries — 10; 1st more than ⅔ as long as the longest . . . . *Clamatores of* ⎱ PASSERES		244
— 10; 1st not ⅔ as long as the longest ⎱ . . . . *Oscines of* ⎰		
— 9 only . . . . ⎰		

ARTIFICIAL KEY TO THE FAMILIES.

                                                                          Page

TOES 3, — 2 IN FRONT, 1 BEHIND . . . . PICIDÆ 576
TOES 3, — 3 IN FRONT. (Go to **II**.)
TOES 4, — 2 IN FRONT, 2 BEHIND. (Go to **III**.)
TOES 4, — 3 IN FRONT, 1 BEHIND. (Go to **IV**.)

II. [TOES 3, — 3 IN FRONT.]

Toes — completely webbed. Nostrils — tubular (Albatrosses) . . . . DIOMEDEIDÆ 1022
                      — not tubular (Auks, etc.) . . . . ALCIDÆ 1059
— incompletely or not webbed. Legs — about as long as wings. Bill subulate (Stilt) . RECURVIROSTRIDÆ 789
                      — much shorter than wings (go to **a**).
  (**a**) Tarsus — scutellate in front, about as long as bill (Sanderling) . . . . SCOLOPACIDÆ 798
              — reticulate in front — shorter than red chisel-like bill (Oyster-catcher) . . . HÆMATOPODIDÆ 787
                      — longer than bill (Plovers) . . . . CHARADRIIDÆ 767

III. [TOES 4, — 2 IN FRONT, 2 BEHIND.]

Bill — cered and strongly hooked. Tarsus granulated (Parrot) . . . . ARIDÆ 616
— not cered; inner hind toe — 3-jointed; plumage iridescent (Trogon) . . . . TROGONIDÆ 575
                — 2-jointed; tail of — 8 or 10 soft feathers (Cuckoos, etc.) . . . . CUCULIDÆ 602
                      — 12 (apparently only 10) rigid acuminate feathers
                              (Woodpeckers) . . . . PICIDÆ 576

IV. [TOES 4, — 3 IN FRONT, 1 BEHIND.]

HIND TOE — INSERTED ABOVE THE LEVEL OF THE REST (AND ALWAYS SHORTER THAN THE SHORTEST FRONT TOE). (Go to **A**.)
      — NOT INSERTED ABOVE THE LEVEL OF THE REST (AND GENERALLY BUT NOT ALWAYS NOT SHORTER THAN
        THE SHORTEST FRONT TOE). (Go to **B**.)

A. [*The hind toe elevated.*]

Feet — TOTIPALMATE (*all 4 toes webbed; hind toe semi-lateral and barely elevated*). (Go to **A.**)
     — PALMATE (3 *front toes full-webbed, hind toe well up, or else connected by slight webbing to base only of inner
        toe*). (Go to **B.**)
     — LOBATE (3 *front toes partly webbed or not, and conspicuously bordered with plain or scalloped membranes;
        hind toe free, and simple or lobed*). (Go to **C.**)
     — SEMIPALMATE (2, or 3, *front toes webbed at base only by small yet evident membrane; hind toe well up,
        simple*). (Go to **D.**)
     — SIMPLE (*front toes with no evident membranes; hind toe simple*). (Go to **E.**)
(**A.**) Tarsus — feathered, partly; tail deeply forked; bill epignathous (Frigate-bird) . . . . . FREGATIDÆ 969
        — naked; bill —> tail, hooked at tip, furnished with enormous pouch (Pelicans) . . PELECANIDÆ 956
                — < tail; throat — feathered; middle tail feathers filamentous (Tropic-birds)
                                                      PHAËTHONTIDÆ 971
                — naked; tail — pointed, soft; tomia subserrate (Gannets) SULIDÆ 953
                      — rounded, stiff; bill — paragnathous (Anhinga)
                                                      ANHINGIDÆ 968
                      — epignathous (Cormorants)
                                                PHALACROCORACIDÆ 959
(**B.**) Bill — curved up, extremely slender and acute (Avocet) . . . . RECURVIROSTRIDÆ 789
        — bent abruptly down, very stout, lamellate (Flamingo) . . . . PHŒNICOPTERIDÆ 888
        — lamellate; mostly membranous, with nail at end (Swans, Geese, Ducks, etc.) . . . . ANATIDÆ 890
        — not lamellate; nostrils — tubular; hind toe very small (Petrels) . . . . PROCELLARIIDÆ 1026
                      — not tubular; hind toe — free, not lobed; bill — cered (Jaegers)
                                                  STERCORARIIDÆ 975
                                        — not cered (Gulls, Terns,
                                                   etc.) LARIDÆ 982
                                    — not free, lobed (Loons) . . . . . GAVIIDÆ 1047
(**C.**) Tail — rudimentary; lores naked (Grebes) . . . . PODICIPEDIDÆ 1051
        — perfect; forehead — covered with a horny shield (Coots) . . . . RALLIDÆ 850
                    — feathered (Phalaropes) . . . . PHALAROPODIDÆ 793

		Page
(**D**.) Mid-claw — pectinate ; 4th toe 4-jointed ; plumage lax (Goatsuckers)	CAPRIMULGIDÆ	561
— not pectinate ; hind toe — versatile ; plumage compact (Swifts)	MICROPODIDÆ	555
— not versatile ; head — naked (go to **b**).		
— feathered (go to **c**).		
(**b**.) Nostrils — imperforate ; naked leg and foot shorter than tail (Turkey)	MELEAGRIDIDÆ	726
— perforate ; naked leg and foot — shorter than tail (Turkey-buzzards) . .	CATHARTIDÆ	700
— longer than tail (Cranes)	GRUIDÆ	847
(**c**.) Nostrils — feathered, or scaled, in deep fossa of stout hard bill ; shank — more or less feathered		
(Grouse)	TETRAONIDÆ	730
— entirely bare and scaly		
(Partridges and Quail)	PERDICIDÆ	749
— not feathered nor scaled, in groove of softish bill ; tarsus — reticulate (Plover)		
	CHARADRIIDÆ	767
— scutellate in front (Snipe, etc.)		
	SCOLOPACIDÆ	798
(**E**.) Wing — spurred	JACANIDÆ	765
— not spurred ; forehead — covered with a horny shield (Gallinules)	RALLIDÆ	850
— feathered ; length — 2 feet or more	ARAMIDÆ	849
— under 2 feet ; 1st primary — attenuate (Woodcock)		
	SCOLOPACIDÆ	798
— not attenuate — much		
shorter than 2d (Rails)		
	RALLIDÆ	850
— about equal to 2d (Snipe,		
etc.)	SCOLOPACIDÆ	798
or APHRIZIDÆ	783

B. [*The hind toe not elevated.*]

TOES SYNDACTYLOUS ; tibiæ naked below ; bill straight, acute (Kingfishers)	ALCEDINIDÆ	571
TIBIÆ NAKED BELOW. (Go to **d**.)		
NOSTRILS OPENING BENEATH SOFT SWOLLEN MEMBRANE. (Go to **e**.)		
BILL HOOKED AND FURNISHED WITH A CERE. (Go to **f**.)		
BIRDS WITHOUT THE ABOVE CHARACTERS. (Go to **g**.)		
(**d**.) Middle claw — pectinate (Herons)	ARDEIDÆ	871
— simple ; tarsus — scutellate in front (Ibises)	IBIDIDÆ	864
— reticulate ; bill — flat, spoon-shaped (Spoonbill) . .	PLATALEIDÆ	868
— not flat, stout, tapering (Wood Ibis)	CICONIIDÆ	869
(**e**.) Bird over 18 inches long, greenish (Texan Guan)	CRACIDÆ	721
Birds under 18 inches long (Pigeons)	COLUMBIDÆ	709
(**f**.) Eyes — lateral, not surrounded by a disc ; nostrils *in* the cere (Hawks, Eagles, etc.) . .	FALCONIDÆ	649
or PANDIONIDÆ	698
— anterior ; face more or less disc-like ; nostrils at edge of cere (Owls) ; middle claw — simple		
	STRIGIDÆ	623
— jagged		
	ALUCONIDÆ	621
(**g**.) PRIMARIES — 10 ; the 1st (never spurious) *always more than* ⅔ *as long as longest* (go to **h**).		
— 10 ; the 1st (spurious or) *at most not* ⅔ *as long as longest* (go to **i**).		
— 9 ; the 1st (*never spurious*) of variable length (go to **k**).		
(**h**.) Tail — 12-feathered ; tarsus — exaspidean (Flycatchers)	TYRANNIDÆ	510
— pycnaspidean (Cotingas)	COTINGIDÆ	534
— 10-feathered ; secondaries — only 6 ; bill subulate (Humming-birds)	TROCHILIDÆ	543
— more than 6 ; bill small, very short (Swifts) . .	MICROPODIDÆ	555
(**i**.) Tarsus — "booted" ; wings — shorter than tail, both much rounded ; plumage very lax	CHAMÆIDÆ	266
— longer than tail ; wing — over 3 inches ; rictus — bristled (Thrushes, etc.)		
	TURDIDÆ	247
— unbristled (Dippers)		
	CINCLIDÆ	260
— not over 3 inches (Kinglets, etc.)	SYLVIIDÆ	261
— scutellate ; nostrils — concealed ; bill — strongly epignathous, toothed and notched (Shrikes)		
	LANIIDÆ	369
— paragnathous ; — over 7 inches long (Crows and		
Jays) CORVIDÆ	484
— not 7 inches ; bill — nearly = head		
(Nuthatches) SITTIDÆ	276
— scarcely or not		
½ = head (Tits) PARIDÆ	267

ARTIFICIAL KEY TO THE FAMILIES. 239

	Page
— exposed; length — over 9 inches; color brown or blue . . CORVIDÆ	484
— 8¼ inches; glossy green and blue, speckled; bill yellow . . . . STURNIDÆ	502
— 7-8 inches; crested; ♂ glossy black AMPELIDÆ	357
— 4¾-6¼ inches; bill distinctly hooked; tail soft, without black . . . . VIREONIDÆ	361
— 4½-5½ inches; bill slender, curved, tail stiff, acute CERTHIIDÆ	278
— under 6 inches; colors bluish, black and white (Gnatcatchers) . . . . SYLVIIDÆ	261
— Birds without these characters (Wrens, Thrashers, etc.) . . . . TROGLODYTIDÆ	280

(**k.**) Tarsus — scutelliplantar; hind claw straight (Horned Larks) . . . . ALAUDIDÆ 503
— laminiplantar; bill — metagnathous, both mandibles falcate, their points crossed
    FRINGILLIDÆ 373
— paragnathous, tomia of up. mand. toothed or lobed near middle
    (Tanagers) TANAGRIDÆ 347
— epignathous, notched and hooked at tip. Length 5¼-6¼ VIREONIDÆ 361
— various. Quills — tipped with red horny appendages; head
    crested (1st quill minute) . . AMPELIDÆ 357
— not appendaged; bill — fissirostral (go to **l**).
    — dentirostral or tenuirostral (go to **m**).
    — conirostral (go to **n**).

(**l.**) Bill triangular-depressed, about as wide at base as long, gape twice as long as culmen, reaching about
opposite eyes, tarsus not longer than outer toe and claw (Swallows) . . . . . . HIRUNDINIDÆ 350
(**m.**) Longest secondary nearly reaching end of primaries in closed wing; hind claw (usually) little curved,
nearly twice as long as middle claw (Wagtails and Pipits) . . . . MOTACILLIDÆ 300
Longest secondary not nearly reaching end of primaries in closed wing; hind claw well curved, not
nearly twice as long as middle claw (Warblers, etc.) . . . CŒREBIDÆ 346, or MNIOTILTIDÆ 304
(**n**) Bill usually thick, stout, and with evident angulation of the commissure . . . . ICTERIDÆ 463
    or FRINGILLIDÆ 373

Note. — These two families *cannot* be concisely distinguished. ICTERIDÆ contains the Blackbirds, Orioles, Meadow Starlings, Bobolinks, and Cowbirds. FRINGILLIDÆ, our largest family, includes all kinds of Grosbeaks, Buntings, Linnets, Finches, and Sparrows.

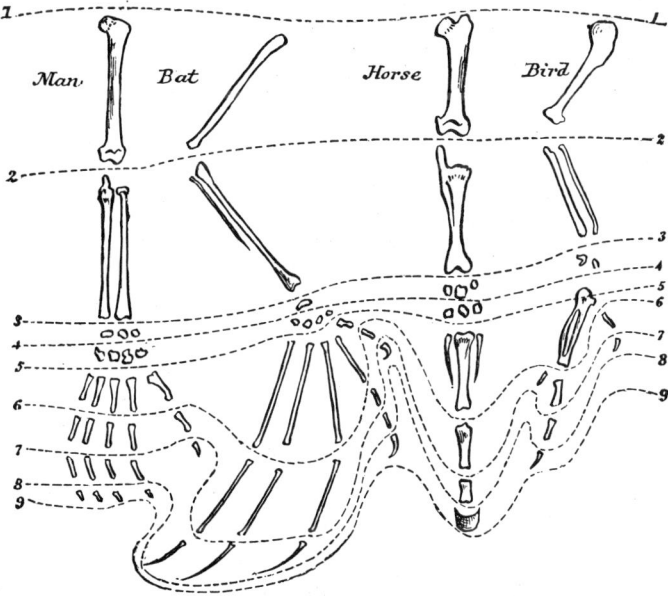

FIG. 112 *ter.* — Diagram of fore limbs of man, bat, horse, and bird. The lines 1-9 are *isotomes*, cutting the limbs into morphologically equal parts, or *isomeres*.

TABULAR VIEW OF THE GROUPS HIGHER THAN GENERA

ADOPTED IN THIS WORK FOR THE

CLASSIFICATION OF NORTH AMERICAN BIRDS.

Subclass CARINATÆ: Carinate Birds.

Orders (14)	Suborders (26)	Families (71)	Subfamilies (71)
I. PASSERES	1. OSCINES	1. *Turdidæ*	1. Turdinæ.
			2. Myiadestinæ.
		2. *Cinclidæ*	
		3. *Sylviidæ*	3. Sylviinæ.
			4. Regulinæ.
			5. Polioptilinæ.
		4. *Chamæidæ* (?)	
		5. *Paridæ*	6. Parinæ.
		6. *Sittidæ*	
		7. *Certhiidæ*	7. Certhiinæ.
		8. *Troglodytidæ*	8. Miminæ.
			9. Troglodytinæ.
		9. *Motacillidæ*	
		10. *Mniotiltidæ*	
		11. *Cœrebidæ*	
		12. *Tanagridæ*	
		13. *Hirundinidæ*	
		14. *Ampelidæ* (?)	10. Ampelinæ.
			11. Ptilogonatinæ.
		15. *Vireonidæ*	
		16. *Laniidæ*	12. Laniinæ.
		17. *Fringillidæ*	
		18. *Icteridæ*	13. Agelæinæ.
			14. Sturnellinæ.
			15. Icterinæ.
			16. Quiscalinæ.
		19. *Corvidæ*	17. Corvinæ.
			18. Garrulinæ.
		20. *Sturnidæ*	19. Sturninæ.
		21. *Alaudidæ*	
	2. CLAMATORES	22. *Tyrannidæ*	20. Tyranninæ.
		23. *Cotingidæ*	21. Tityrinæ.
II. PICARIÆ (?)	3. TROCHILI	24. *Trochilidæ*	
	4. CYPSELI	25. *Micropodidæ*	22. Micropodinæ.
			23. Chæturinæ.
	5. CORACIÆ	26. *Caprimulgidæ*	24. Caprimulginæ.
	6. HALCYONES	27. *Alcedinidæ*	25. Alcedininæ.
	7. TROGONES	28. *Trogonidæ*	
	8. PICI	29. *Picidæ*	26. Picinæ.
	9. COCCYGES	30. *Cuculidæ*	27. Crotophaginæ.
			28. Neomorphinæ.
			29. Cuculinæ.
III. PSITTACI	10. EUPSITTACI	31. *Aridæ*	30. Conurinæ.
IV. RAPTORES	11. STRIGES	32. *Aluconidæ*	
		33. *Strigidæ*	
	12. ACCIPITRES	34. *Falconidæ*	31. Circinæ.
			32. Milvinæ.
			33. Accipitrinæ.

Orders (14)	Suborders (26)	Families (71)	Subfamilies (71)
			34. Falconinæ.
			35. Polyborinæ.
			36. Buteoninæ
		35. *Pandionidæ*	
	13. Cathartides	36. *Cathartidæ*	37. Sarcorhamphinæ.
			38. Cathartinæ.
V. Columbæ	14. Peristeræ	37. *Columbidæ*	39. Columbinæ.
			40. Zenaidinæ.
			41. Starnœnadinæ.
VI. Gallinæ	15. Peristeropodes	38. *Cracidæ*	42. Penelopinæ.
	16. Alectoropodes	39. *Phasianidæ*	43. Phasianinæ.
		40. *Meleagrididæ*	
		41. *Tetraonidæ*	
		42. *Perdicidæ*	44. Perdicinæ.
			45. Odontophorinæ.
VII. Limicolæ		43. *Jacanidæ*	
		44. *Charadriidæ*	46. Charadriinæ.
		45. *Aphrizidæ*	47. Aphrizinæ.
			48. Arenariinæ.
		46. *Hæmatopodidæ*	
		47. *Recurvirostridæ*	
		48. *Phalaropodidæ*	
		49. *Scolopacidæ*	
VIII. Paludicolæ	17. Grues	50. *Gruidæ*	
		51 *Aramidæ*	
	18. Ralli	52. *Rallidæ*	49. Rallinæ.
			50. Gallinulinæ.
			51. Fulicinæ.
IX. Herodiones	19. Ibides	53. *Ibididæ*	
		54. *Plataleidæ*	
	20. Ciconiæ	55. *Ciconiidæ*	52. Tantalinæ.
			53. Ciconiinæ.
	21. Herodii	56. *Ardeidæ*	54. Ardeinæ.
			55. Botaurinæ.
X. Lamellirostres	22. Odontoglossæ	57. *Phœnicopteridæ*	
	23 Anseres	58. *Anatidæ*	56. Cygninæ.
			57. Anserinæ.
			58. Anatinæ.
			59. Fuligulinæ.
			60. Merginæ.
XI. Steganopodes		59. *Sulidæ*	
		60. *Pelicanidæ*	
		61. *Phalacrocoracidæ*	
		62. *Anhingidæ*	
		63. *Fregatidæ*	
		64. *Phaëthontidæ*	
XII. Longipennes		65. *Stercorariidæ*	
		66. *Laridæ*	61. Larinæ.
			62. Sterninæ.
			63. Rhynchopinæ.
XIII. Tubinares		67. *Diomedeidæ*	
		68. *Procellariidæ*	64. Fulmarinæ.
			65. Puffininæ.
			66. Procellariinæ.
			67. Oceanitinæ.
XIV. Pygopodes	24. Gaviæ	69. *Gaviidæ*	
	25. Podicipedes	70. *Podicipedidæ*	
	26. Alcæ	71. *Alcidæ*	68. Fraterculinæ.
			69. Phaleridinæ.
			70. Allinæ.
			71. Alcinæ.
14 Orders.	26 Suborders.	71 Families.	71 Subfamilies.

PART III.

SYSTEMATIC SYNOPSIS

OF

NORTH AMERICAN BIRDS.

CLASS **AVES**: BIRDS.

THIS CLASS OF ANIMALS, while sharply distinguished from Mammals, is so closely related to Reptiles, that the presence of feathers in the former, and their absence from the latter, is the most obvious if not the only positive character by which the two classes are separable.

Though the species of birds are numerous (some 11,500 are known), the structural diversity of the Class is comparatively so slight, that the characters upon which the primary divisions are based seem insignificant in view of those upon which the major groups of Mammals or Reptiles may be founded. With strict regard for equivalency of taxonomic groups, based on morphological considerations, the conventional "class" of Birds is scarcely or not of higher value than an order of Reptiles, with which Birds are associated under the name SAUROPSIDA. But it is not proven that a given structural character may not have classificatory value in one case, different from that which may properly be attributed to it in another; so that, though the most diverse birds may be more alike than are extremes among Lizards for example, we may still continue to speak of a *class Aves*, to be primarily divided into sub-classes or orders.

All known Birds, living and extinct, are divisible into the following primary groups, which may be termed sub-classes :

    I. SAURURÆ. — Birds with teeth. Vertebræ biconcave (amphicœlous). Sternum keeled. Wings small, with separate metacarpals. Tail longer than body, its vertebræ not pygostyled, its feathers arranged in distichous series. (One species, *Archæopteryx lithographica*, from the Jurassic of Europe. Fig. 14.)

    II. ODONTOTORMÆ. — Birds with teeth, implanted in sockets. Vertebræ biconcave. Wings large, with anchylosed metacarpals. Sternum keeled. Tail short. (Typified by the genus *Ichthyornis*, from the Cretaceous of North America. Fig. 16.)

III. ODONTOLCÆ. — Birds with teeth, implanted in grooves. Vertebræ saddle-shaped (heterocœlous). Wings rudimentary, wanting metacarpals. Sternum without keel. Tail short. (Typified by the genus *Hesperornis*, from the Cretaceous of North America. Fig. 15.)

IV. RATITÆ. — Birds without teeth. Vertebræ (some) saddle-shaped. Wings rudimentary, or at most unfit for flight, with anchylosed metacarpals. Sternum without keel (as in *Odontolcæ*, fig. 15). Tail short. (Embracing the extinct Moas, and the living Ostriches, Cassowaries, Emeus, and Kiwis.)

V. CARINATÆ. — Birds without teeth. Vertebræ (some) saddle-shaped. Wings developed, with rare exceptions fit for flight, with anchylosed metacarpals. Sternum keeled. Tail short (as to its vertebræ, which are usually pygostyled). (Embracing all living birds excepting the *Ratitæ*.)

AVES CARINATÆ: ORDINARY BIRDS.

The essential characters of this group, which includes all living birds excepting Ostriches and their allies (*Ratite* or *Struthious* birds), are absence of teeth, saddle-shaped faces of the best-developed vertebræ, and keeled breast-bone (fig. 56), in combination with perfection of wing-structure in adaptation to aerial (or aquatic) flight. The metacarpals and three metatarsals are anchylosed (figs. 27, 34); the scapula and coracoid meet at less than a right angle (very rarely more), and the furculum is usually perfect (fig. 59). (In the flightless parrot of New Zealand (*Stringops habroptilus*), the sternal keel is rudimentary.) The caudal vertebræ are few, and the last few (pygostyle, fig. 56) are peculiarly modified to support the tail-feathers in fan-like array. There is normally extensive post-acetabular anchylosis of the pelvic bones, which are normally separate there in the other groups (compare figs. 56 and 15).

The division of Carinate birds has always exercised the judgment and ingenuity of ornithologists; no system that has been proposed has been universally adopted. The orders of *Carinatæ*, therefore, are still provisional. But a great assemblage of birds have been ascertained to agree (with few exceptions) in possessing a certain combination of characters, upon which may be based the

Order PASSERES: Insessores, or Perchers Proper.

The feet are perfectly adapted for grasping by length and low insertion of the hind toe, great power of apposing which to the front toes, and great mobility of which, are secured by separation of its principal muscle (flexor longus hallucis) from that which bends the other toes collectively (flexor profundus digitorum).[1] The hind toe is always present, perfectly incumbent, and never turned forward or even sideways; its claw is as long as, or longer than, the claw of the middle toe. The feet are never zygodactyl, or syndactyl, or semipalmate, or palmate; the front toes are usually immovably joined to each other at base, for a part, or the whole, of the basal joints. No one of the front toes is ever versatile. The joints of the toes are always 2, 3, 4, 5, counting from 1st (hind one) to 4th (outer front one). The toes are always 4 in number (excepting *Cholornis* with 4th toe abortive). (Figs. 36, 37, 42, 43.) Various as are the shapes of the wings, these members agree in having the great row of coverts not more than half as long as the secondaries; the developed primaries either 9 or 10 in number, and the secondaries more than 6. (Fig. 30.) The tail, extremely variable in shape, has

---

[1] The notable exception to this statement is the Broadbill family, *Eurylæmidæ*, which have a plantar vinculum; for which reason some authors make them a prime division of *Passeres* under the name of *Desmodactyli*, all other *Passeres* being then called *Eleutherodactyli*.

12 rectrices (with certain anomalous exceptions: none in *Pnoëpyga*; 10 in *Xenicus*, *Acanthisitta*, *Phrenotrix*, *Edolius*; 16 in *Menura*). The bill is too variable in form to furnish characters of groups higher than families; but its covering is always hard and horny, in part or wholly — never extensively membranous, as in many wading and swimming birds, or softly tumid, as in Pigeons, or cered, as in Parrots and birds of prey. The nostrils do not openly communicate with each other. The oil-gland (elæodochon, p. 89) is nude, and of a characteristic shape. Besides these external characters, which the student may readily examine without dissection, there are some more important anatomical ones. The sternum (with few exceptions) is cast in a particular mould, having a forked manubrium (except *Eurylæmidæ*), prominent costal processes, and each side of the posterior border single-notched (neither entire, nor deeply nor doubly notched, nor fenestrate; fig. 58). The bony palate has a peculiar structure, called *ægithognathous* (fig. 79), but in some cases a sort of desmognathism occurs; there are no basipterygoids; the nasal bones are holorhinal. The atlas is perforated by the odontoid process of the axis. Beddard has called attention to a disposition of the abdominal septa which may be a passerine character: the oblique septa being either free from the sternum, or sharing their attachment thereto with the falciform ligament. There is but one carotid artery, the left (fig. 91). Cœca coli are present, though small. The plumage is aftershafted, as a rule (except *Eurylæmidæ*). There is a peculiarity in the method of insertion of the tensor patagii brevis; "the tendon of the muscle does not end upon the tendon of the extensor, as it does in the picarian bird, but, though attached to it firmly, retains its independence, and runs back to be attached near it to the extensor condyle of the radius" (*Beddard*); there is no biceps slip, nor any expansor of the secondaries. Besides possessing the separate flexor of the hind toe already mentioned, *Passeres* are anomalogonatous (p. 201) — that is, the ambiens is absent; so is the accessory femorocaudal; the femorocaudal and semitendinosus are present, as is usually also the accessory semitendinosus. The formula is therefore A X Y (rarely A X).

No North American Passerine bird shows any of the exceptions noted in the foregoing paragraph; all are normally passerine.

Physiologically, the nature of *Passeres* is altricial and psilopædic (p. 91); that is, the young are hatched weak and naked, and require to be fed for some time in the nest by the parents. They represent the highest grade of physiological development, as well as the most perfect physical organization of the class of birds. Their nervous irritability is great, coördinate with rapidity of respiration and circulation; they consume the most oxygen, and live the fastest, of all birds. They habitually reside above the earth, in the air that surrounds it, among the plants that with them adorn it; not on the ground, nor on "the waters under the earth."

Pas'seres were named by Cuvier in 1798 as an order of birds; the name is simply the plural of Lat. *passer*, a sparrow. But the group as established by him included many forms which were first properly excluded by the celebrated Nitzsch, who in 1829 limited the group as now accepted. Besides being one of the best defined, it is by far the largest group of its grade in ornithology. For example, of the 888 birds enumerated as North American in my last Check List, no fewer than 394 are *Passeres*; as are more than half of all known birds, or about 6,000 out of some 11,500 species.

Passeres are primarily divisible into two groups, commonly called suborders, mainly according to the structure of the vocal organ — the lower larynx, or *syrinx*. In one of these groups, the musical apparatus is highly developed, with several distinct pairs of intrinsic muscles, inserted into the ends of the upper three half-rings of the bronchial tubes. In the other, the voice-organ is less complex, with less specialized muscles inserted into the middle portions of the upper bronchial half-rings. The former arrangement is termed *acromyodian*, the latter *mesomyodian;* the two are also contrasted as *polymyodian* and *oligomyodian*, with reference to number of syringeal muscles. Birds which exhibit this difference of structure are respectively called *Passeres acromyodi* and *Passeres mesomyodi*, or *Oscines* and *Clamatores*.[1] (See p. 212, fig. 101.)

Associated with the acromyodian or oscine type of syrinx is a peculiar condition of the tarsal envelop. In nearly all *Oscines*, the tarsus is covered on each side with a horny plate, nearly or quite undivided, meeting its fellow in a sharp ridge behind. This condition of the tarsus is called *bilaminate*, and birds showing it are *laminiplantar* (figs. 37, 42, 43). In some cases fusion of the tarsal envelop proceeds so far that the front of the tarsus likewise presents a nearly or quite undivided surface, the whole tarsus being then encased in a "boot," as it is called. A "booted" tarsus may be said to be *trilaminate* (fig. 36). The principal exception to association of a bilaminate or trilaminate tarsus with an acromyodian syrinx is afforded by *Alaudidæ*, which have the tarsus scutellate and blunt behind; and, with very few exceptions, no bird which is not acromyodian has a bilaminate tarsus. A third important feature characterizes *Oscines*, as a rule. This is reduction in length of the 1st primary, which never equals the longest primary in length, is rarely over ⅔ as long as the longest, is so short as to be called spurious, or is quite rudimentary and apparently wanting, leaving apparently only 9 primaries (fig. 30).

Associated with the mesomyodian or clamatorial type of syrinx is seen (with few exceptions) an opposite condition of the tarsus, the sides and back of which, as well as the front, are covered with variously arranged scutella, so that there is no sharp undivided ridge behind. In such cases there are also 10 fully developed primaries, the 1st of which, if not equalling or being itself the longest, is at least ⅔ as long. (See p. 510, fig. 343.)

These combinations of characters may be contrasted for the purpose of dividing the great group *Passeres* into two sections, conventionally denominated suborders.

1. Suborder ACROMYODI, POLYMYODI, OR OSCINES : Singing Birds.

Fig. 113. — Thrushes: European Redwing (*Turdus iliacus*) and Fieldfare (*T. pilaris*). From Dixon.

Syrinx with 4 or 5 distinct pairs of intrinsic muscles, inserted at ends of 3 upper bronchial half-rings, and thus constituting a highly complex and effective musical apparatus. Each side of tarsus covered with a horny plate meeting its fellow in a sharp ridge behind; front of tarsus also sometimes laminate. Primaries apparently 10, the 1st short or spurious; or apparently only 9.

Here belong all the North American families of *Passeres*, except *Tyrannidæ*, or Flycatchers, and *Cotingidæ*, which are clamatorial (mesomyodian). The only North American exceptions to the diagnosis given are afforded by *Alaudidæ*, or Larks, and certain *Troglodytinæ*, which, with an oscine syrinx and wing-structure, do not have a bilaminate tarsus.[2] Of our nearly 550 Passerine species and subspecies, no fewer than 500 are Oscine. The name is the Lat. *os'cen*, n. pl. *os'cines*, divining-birds — those whose notes were regarded as augural.

It is a question, which one of the numerous Oscine families should be placed at the head

---

[1] I do not wish to modify this statement, made in former editions of the Key, notwithstanding what is said of *Eurylæmidæ* in the note on p. 244.

[2] The most abnormal *Oscines* are the Australian Scrub-birds and Lyre-birds, *Atrichiidæ* and *Menuridæ*. In these the syringeal muscles are reduced, the furculum is rudimentary, there are more than the typical number of rectrices, etc. — so that these families are sometimes made a prime division, *Pseudoscines* or *Abnormales*, contrasted with *Normales*.

of the series. Largely, perhaps, through the influence of those ornithologists who hold that fusion of the tarsal envelop into one continuous plate indicates the acme of bird-structure, the place of honor has of late been usually assigned to the Thrushes. It seems to me most probable that this character, though unquestionably of high import, should be taken as of less value than reduction of number of primaries from 10 to 9; and I am inclined to believe that eventually some Oscine family with only 9 primaries — as the Finches or Tanagers — will take the leading position. Some contend for the headship of the Crows. Here, however, I follow usage in the sequence of North American families, as follows: — *Turdidæ, Cinclidæ, Sylviidæ, Chamæidæ, Paridæ, Sittidæ, Certhiidæ, Troglodytidæ, Motacillidæ, Mniotiltidæ, Cœrebidæ, Tanagridæ, Hirundinidæ, Ampelidæ, Vireonidæ, Laniidæ, Fringillidæ, Icteridæ, Corvidæ, Sturnidæ, Alaudidæ.*

Family TURDIDÆ: Thrushes, etc.

The essential character of this great group of *Oscines* is booted tarsi and 10 primaries, the 1st spurious. But *Turdidæ* do not show this combination exclusively as birds of some other families also possess it. Though it be as natural as any other Oscine family of equal extent and variety, and equally close relationships with other groups, it is insusceptible of perfect definition in concise terms. The North American representatives, however, may readily be circumscribed in a manner enabling the student to assure himself of the family to which they belong, though no line whatever can be drawn between *Turdidæ* and *Sylviidæ*. The vast assemblage of Old World Warblers are in fact thoroughly Thrush-like.

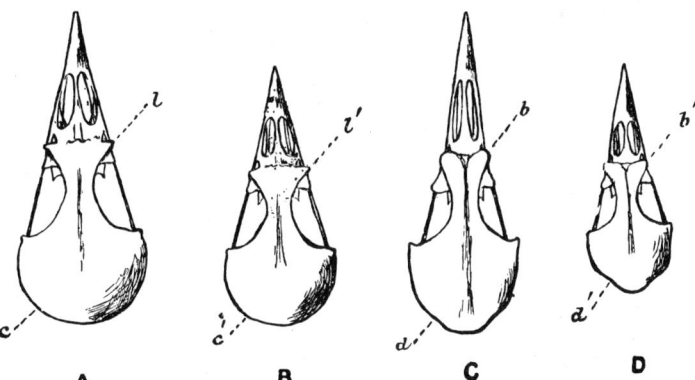

FIG. 114. — Skulls of *Turdidæ, Miminæ,* etc., nat. size; after Shufeldt. A, *Oroscoptes montanus;* B, *Sialia mexicana;* C, *Cinclus mexicanus;* D, *Siurus nævius.* Observe likeness between A and B, at points marked c, c', l, l'; and between C and D, at points marked b, b', d, d'.

Wing of 10 primaries, of which the 1st is spurious or quite short. Wing more or less elongate and pointed, longer than tail. Inner secondaries never long and flowing as in *Motacillidæ*. Bill never stout and conical, nor with angulated commissure, nor flattened with gape reaching under eyes; usually slender, straight or little curved, more or less compressed, subulate and acute, usually notched at end of upper mandible (but the nick frequently obsolete), and thus of a character which is called *grypaniform*. Nostril oval or roundish rarely linear, exposed in conspicuous nasal fossæ; nearly or quite reached or overreached by frontal feathers, but never concealed by a dense ruff as in *Paridæ* and *Sittidæ*. Rictus bristled or with bristle-tipped feathers. Tarsus normally booted, the anterior scutella, excepting a few below, being fused in a continuous plate. On the sides and behind, tarsus strictly laminiplantar (compare *Alaudidæ* and some *Troglodytinæ*). Tarsus usually also long and slender; never decidedly shorter than middle toe and claw, often decidedly longer. Anterior toes deeply cleft; inner to its very base, outer adherent to middle for only length of its basal joint (compare *Troglodytinæ*).

Hind claw never lengthened and straightened as usual in *Motacillidæ*. Tail-feathers 12; tail normally much shorter than wings, sometimes about equal; **never** cuneate, or deeply forked.

Any North American bird, except the Dipper, showing **booted** tarsi, 10 primaries, the 1st spurious, — and wing over 3 inches long, — is one of the *Turdidæ*.

OBS. — In determining character of tarsus, whether booted or **scutellate, it is** necessary to examine adult birds; for the fusion of the anterior scutella is progressive, and only accomplished **perfectly at** maturity. And in general, in using artificial keys to genera and species, the student must agree with **the author in** understanding that specimens fairly illustrating normal adult characters are in hand.

Subfamily TURDINÆ: Typical Thrushes.

Tarsus, in the adult, "booted" or enveloped in a continuous plate, formed by fusion of all tarsal scutella except 2 or 3 just above base of toes (fig. 36). Toes deeply cleft — inner to the very base; outer coherent with middle only for the length of its basal joint. Wings more or less pointed, longer than tail; 1st primary spurious; 2d longer than 6th. Bill moderate, shorter than head, straight, more or less subulate, little depressed at base, with bristly rictus. Nostrils oval, nearly or quite reached by the frontal feathers. (Fig. 116.) Tail-feathers widening somewhat toward ends; tail as a whole somewhat fan-shaped, neither decidedly forked nor much graduated. Upward of 250 species are now usually assigned to *Turdinæ*. They are nearly cosmopolitan, and have a great development in the warmer parts of America, where they are mainly represented by types closely allied to *Turdus* proper; more aberrant forms, constituting very distinct genera, occur in the Old World. We have 6 genera in North America, and a species of *Catharus* occurs very near if not actually over our Mexican border. Some of the leading genera which are not represented in North America are *Oreocincla*, which presents the passerine abnormality of 14 tail-feathers; *Geocichla*, an extensive group of Ground Thrushes, with several subdivisions; *Mimocichla* and *Catharus*, of Neotropical America; *Erithacus*, typified by the Robin-redbreast of Europe, *E. rubecula*, familiar in the traditions of the nursery; *Aëdon*, which contains the famous Nightingale, *A. luscinia;* *Ruticilla*, based on the European Redstart, *E. Phœnicura;* and *monticola*, including the Rock Chats, such as *M. saxatilis*. The *Turdinæ* are diffused over all woodland parts of our country; all are strictly migratory, insectivorous birds, though feeding also upon berries and other soft fruits. Though not truly gregarious, some, as the Robin for instance, often collect in troops at favorite feeding places, or migrate in companies. They build rather rude nests, often plastered with mud, never pensile, but saddled on a bough, fixed on a fork, or set on the ground; and lay 4-6 green or blue eggs, sometimes plain, sometimes spotted. All are vocal; some, like the Wood Thrush, are exquisitely melodious.

FIG. 115. — A typical Thrush, the European Blackbird (*Turdus merula*). From Dixon.

These birds may be taken in illustration of a character which runs through other groups of *Turdidæ* besides *Turdinæ* proper. The young, in their first feathering, which is worn but a short time, are curiously speckled and streaked, in a manner quite different from the adults. This feature is well shown by a young Robin, or Bluebird; it disappears entirely, from the upper parts at least, but continues to characterize the under parts of our Wood Thrush and its allies of the genus *Hylocichla*. Our genera of *Turdinæ* (including those formerly referred to *Saxicolinæ*) may be recognized by the following artificial

TURDIDÆ — TURDINÆ: THRUSHES.

Analysis of Genera.

Tail-feathers not particolored; no blue anywhere.
  Neither spotted nor banded below, but throat streaked . . . . *Merula*
  Banded crosswise below; slate-colored above . . . . *Hesperocichla*
  Spotted or streaked below on a white or whitish ground.
    European . . . . *Turdus*
    Native American . . . . *Hylocichla*
Tail-feathers black and white; no blue anywhere . . . . *Saxicola*
Tail-feathers brown and chestnut; throat blue and chestnut . . . . *Cyanecula*
Tail-feathers blue, like back . . . . *Sialia*

MER'ULA. (Lat. *merula*, a kind of Thrush, the European Blackbird, *Turdus merula* LINN., type of the genus: LEACH, Syst. Cat. Brit. Birds, 1816, p. 20.) EUROPEAN BLACKBIRDS. AMERICAN ROBINS. Characters of *Turdus* proper, but tail relatively longer, being thrice as long as tarsus. Tarsus a little longer than middle toe and claw. Bill notched near end, little widened at base. Of large, stout form. Sexes similar; beneath mostly unicolor, with streaked throat, but breast not spotted or collared. We have two good species, and a subspecies of one of these. (Given as a subgenus of *Turdus* in former eds. of the Key, p. 243.)

Analysis of Species and Subspecies.

Upper parts slate-colored; breast chestnut.
  Outer tail-feather with white tipping. (Eastern.) . . . . *migratoria*
  Outer tail-feather without decided white tipping. (Western.) . . . . *M. propinqua*
Upper parts grayish-ash; breast yellowish-buff. (Cape St. Lucas.) . . . . *confinis*

M. migrato'ria. (Lat. *migratoria*, migratory; *migrator*, a wanderer. Figs. 36, 58, 116.) ROBIN. Adult ♂, in summer: Upper parts slate-color, with a shade of olive. Head black; eyelids and spot before eye white; throat streaked with white. Quills of the wings dusky, edged with hoary-ash, and with color of back. Tail blackish; outer feather usually tipped with white. Under parts to vent, including under wing-coverts, chestnut. Under tail-coverts and tibiæ white, showing more or less plumbeous. Bill yellow, often with a dusky tip; mouth yellow; eyes dark brown; feet blackish; soles yellowish. Length about 10.00; extent 16.00; wing 5.00–5.50; tail 4.00–4.50; bill 0.80; tarsus, or middle toe and claw, 1.25. Adult ♀, in summer: Similar, but colors duller; upper

FIG. 116. — Robin, nat. size. (Ad. nat. del. E. C.)

parts rather olivaceous-gray; chestnut of under parts paler, the feathers skirted with gray or white; head and tail less blackish; throat with more white. Bill much clouded with dusky. ♂ ♀, in winter, and young: Similar to adult ♀, but receding somewhat farther from ♂ in summer by duller colors, paleness and restriction of the chestnut, its extensive skirting with white, lack of distinction of color of head from that of back, tendency of white spot before eye to run into a superciliary streak, and dark color of most of bill. Very young birds have the back speckled, each feather being whitish centrally, with a dusky tip; cinnamon of under parts spotted with blackish; greater coverts tipped with white or rufous, frequently persistent, as are also some similar markings on lesser coverts. N. Am. at large, W. to the Rocky Mts., to the Pacific in Alaska, and to eastern Mexico; an abundant and familiar bird, migratory, breeding from middle portions of the U. S. northward to the Arctic Ocean, and wintering from Canada and the northern States irregularly to the middle districts, abundantly in the Southern States; casual in the Bermudas; accidental in Europe. Nest in trees usually, saddled on a horizontal bough, composed largely of mud; eggs 3–6, about 1.18 × 0.80, uniform greenish-blue, normally unspotted.

M. m. propin'qua. (Lat. *propinqua*, neighboring; as related to the last.) WESTERN ROBIN. Quite like *M. migratoria;* averaging slightly larger; wing up to 5.60; tail up to 4.70, not so blackish as that of *M. migratoria*, the outer feather without white, or merely a narrow edging. A scarcely distinguished race, of the Rocky Mt. region and westward in the U. S. to Lower California and Mexico.

M. confi'nis. (Lat. *confinis*, allied or related; as to *M. migratoria*.) ST. LUCAS ROBIN. Adult ♂ ♀. Upper parts, including sides of head and neck, uniform grayish-ash, with slight olive shade, scarcely darker on head; chin and throat white, streaked with ashy-brown; breast, sides, and lining of wings pale yellowish-buff, belly white, flanks ashy. A distinct white superciliary stripe; lower eyelid white. Feathers of jugulum and sides with ashy tips; greater wing-coverts tipped with whitish; bill yellowish, upper mandible and tip of lower tinged with dusky; feet pale brown. Wing 5.10, tail 4.10; tarsus 1.20; middle toe and claw 1.07. Lower California; has occurred at Hayward, Cal., exceptionally. General appearance of a young dull-colored common Robin, but quite distinct.

FIG. 117. — Varied Thrush.

HESPEROCICH'LA. (Gr. ἕσπερος, *hesperos*, Lat. *vesperus*, of the evening or time of sunset, hence western; κίχλα, *kichla*, a kind of thrush: BAIRD, Rev. Am. B., i, 1864, p. 12.) VESPER THRUSHES. In general, similar to *Merula* and *Turdus* proper. Tarsus no longer than middle toe and claw. Bill unnotched; nostrils partly overhung by feathers which fill the nasal fossæ. Sexes subsimilar; ♂ with a black pectoral collar. One strongly marked species, by one author referred to the extensive Old World genus *Geocichla*. (Given as a subgenus of *Turdus* in former eds. of the Key.)

TURDIDÆ — TURDINÆ: THRUSHES. 251

H. næ'via. (Lat. *nævia*, spotted, varied; *nævus*, a birth-mark. Figs. 117, 118.) VARIED THRUSH. OREGON ROBIN. Adult ♂, in summer: Entire upper parts dark slate-color, varying in shade from blackish to plumbeous slate, in less perfect specimens with a slight olive tinge; wings and tail blackish, with more or less of plumbeous or olive shade, according to age of the quills; greater and lesser wing-coverts, tipped with orange-brown, forming two cross-bars, and quills edged in two or three places with the same; quills also white at base on inner webs, this marking not visible from the outside; one or several lateral tail-feathers tipped with white. A broad black collar across breast, mounting on side of neck and head. Stripe behind eye, lower eyelid, and under parts orange-brown, gradually giving way to white on lower belly; vent and crissum mixed white, orange-brown, and plumbeous. Bill black; feet and claws dull yellowish. Length 9.50–10.00; extent about 16.00; wing 5.00; tail 3.75; bill 0.80; tarsus, or middle toe and claw, 1.25. Adult ♀, in summer: Upper parts olivaceous-plumbeous (almost exactly the shade of the common Robin in winter); wings and tail scarcely darker; pectoral collar narrow, like the back in color; other under parts like those of the ♂, but duller, paler, and rather rusty than orange-brown, with more white on lower belly. Markings of head, tail, and wings exactly as in the ♂. Young: Like adult ♀, in many respects; duller; no white on belly and crissum. Upper parts in many cases with a decided umber-brown wash; feathers of breast and throat with blackish edgings; lesser wing-coverts with angular rusty spots, but no fully speckled stage, like that of the very young Robin, has been observed, though August specimens have been examined. In young ♂, black pectoral bar at first indicated by interrupted blackish crescents on individual feathers. Young ♀ ♀ sometimes show scarcely a trace of collar. At all ages, markings of head and wings are much the same. Pacific coast region, Alaska to Mexico, abundant, migratory; accidental in Mass., N. J., and Long Island. Nest in bushes, of twigs, grasses, mosses, and lichens; eggs 1.12 × 0.80, light greenish-blue, speckled with dark brown. Breeds S. to Humboldt Co., Cal.

FIG. 118. — Varied Thrush, nat. size. (Ad. nat. del. E. C.)

TUR'DUS. (Lat. *turdus*, a thrush.) TRUE THRUSHES. A large genus, even when taken in its most restricted sense, including many species, occurring in most parts of the world, of medium and small size. Tail rather short, not thrice as long as tarsus, which is decidedly longer than middle toe and claw. Bill notched near end, more or less widened and depressed at the bases. Sexes indistinguishable, or at least quite similar, extensively streaked or spotted on the under parts. The type of *Turdus* LINN., Syst. Nat., i, 1758, p. 168, is now taken to be *T. viscivorus*, the Missel Thrush of Europe, with which such species as *T. musicus*, the Mavis or Song Thrush, and *T. iliacus*, the Redwing, are strictly congeneric.

T. ili'acus. (Lat. *iliacus*, relating to the flanks, which are reddish. Fig. 113.) RED-WINGED THRUSH. WIND THRUSH. REDWING. WINNARD. Upper parts hair-brown with an olive shade, darker on head, paler on rump. Wing-quills deep brown; coverts and inner secondaries tipped with whitish. Tail dark brown, the outer feather usually white-tipped. Lore blackish; eyelids and superciliary stripe whitish; auriculars streaked with light and dark brown. Throat yellowish-white, streaked with brownish-black; breast and belly grayish-white; lower tail-coverts whitish, streaked with brown. Sides and under wing-coverts light reddish. Bill brownish-black; basal half of lower mandible orange-yellow; iris brown; feet flesh-colored. Sexes alike. Length 8.50; extent 14.00; wing 4.50; tail 3.25; bill 0.75; tarsus, or middle toe and claw, 1.15. European; only N. American as occurring accidentally in Greenland.

HYLOCICH'LA. (Gr. ὕλη, *hule*, woods, forest; κίχλα, *kichla*, a thrush.) AMERICAN WOOD THRUSHES. Chiefly distinguished from *Turdus* proper by length and slenderness of tarsus, which is more than ¼ as long as wing, and longer than middle toe and claw. Bill comparatively weak, small, depressed and broad at base. Species of small size, and not robust form; sexes similar; adults not spotted or streaked above, but spotted or streaked below on a white or whitish ground; young with whitish or buff markings on upper parts. A beautiful genus of woodland vocalists, included under *Turdus* in former eds. of the Key and A. O. U. Lists. *Hylocichla* BD., Rev. Am. B., i, 1864, p. 12, type *mustelinus*, the well-known Wood Thrush, besides which the genus contains the Veery, the Hermit, the Olive-back, the Gray-cheek, and their several subspecies, as follows:

FIG. 119. — Wood Thrush.

Analysis of Species and Subspecies.

Upper parts not of one color from head to tail. Eggs not spotted.
    Upper parts tawny, shading to olive on rump. Wood Thrush. (Eastern.) . . . . *mustelina*
    Upper parts olive, shading to rufous on rump. Hermit Thrushes.
        Of medium size. (Eastern.) . . . . *aonalaschkæ pallasi*
        Of largest size. Rocky Mts. . . . . *aonalaschkæ auduboni*
        Of smallest size. Pacific coast . . . . *aonalaschkæ*
Upper parts of one color throughout.
    Eggs not spotted.

Upper parts tawny; spots below few, pale, chiefly on jugulum; no buff eye-ring. Tawny Thrush, or Veery.
(Eastern.) . . . . *fuscescens*
Upper parts russet-olive; spots below as before; no buff eye-ring. Willow Thrush. (Western.) . *fuscescens salicicola*
Eggs spotted.
Upper parts russet-olive; spots below numerous, invading white breast; a buff eye-ring. Russet-backed Thrush.
Pacific coast, northerly . . . . *ustulata*
Pacific coast, southerly . . . . *ustulata œdica*
Upper parts dark pure olive; spots below as before.
A buff eye-ring. Olive-backed Thrush. (Eastern.) . . . . *ustulata swainsoni*
No buff eye-ring. (Eastern.)
Of general distribution. Alice's Thrush . . . . *aliciæ*
Of local distribution. Bicknell's Thrush . . . . *aliciæ bicknelli*

H. musteli'na. (Lat. *mustelinus*, weasel-like; *i. e.* tawny in color: *mustela*, a weasel. Figs. 119, 120.) WOOD THRUSH. WOOD ROBIN. BELLBIRD. GERALDINE. Adult ♂ ♀: Upper parts, including surface of closed wings, tawny-brown, purest and deepest on head, shading insensibly into olivaceous on rump and tail. Below, pure white, faintly tinged on breast with buff, and everywhere, except on throat, middle of belly, and crissum, marked with numerous large, well-defined, rounded or subtriangular blackish spots. Inner webs and ends of quills fuscous, with white or buffy edging toward base; under wing-coverts mostly white. Auriculars sharply streaked with dusky and white. Bill blackish-brown, with flesh-colored or yellowish base; feet like this part of the bill. Length 7.50-8.00; extent about 13.00; wing 4.00-4.25; tail 3.00-3.25; bill 0.75; tarsus 1.25; middle toe and claw less. Young: Speckled or streaked above with pale yellowish or whitish, especially noticeable as triangular spots on wing-coverts.

FIG. 120. — Wood Thrush (*T. mustelinus*), nat. size. (Ad. nat. del. E. C.)

But these speedily disappear, when a plumage scarcely different from that of the adult is assumed. The most strongly marked species of the genus; in no other are the spots below so large, sharp, numerous, and generally dispersed. In the Hermit, our only other Thrush showing both tawny and olive on upper parts, the position of the two colors is reversed, tawny occupying the rump, olive the head. Eastern U. S., N. to Massachusetts, Michigan and Southern Canada, W. to the Plains, S. in winter to Guatemala; Cuba; a famous vocalist, common in low damp woods and thickets; migratory; breeds throughout its U. S. range. Nest in bushes and low trees, of leaves, grasses, etc., and mud; eggs usually 3-4, plain greenish-blue like the Robin's, but smaller: 1.08 × 0 70.

H. fusces'cens. (Lat. *fuscescens*, less than *fuscus*, dark.) WILSON'S THRUSH. TAWNY THRUSH. VEERY. PINE SPIRIT. Adult ♂ ♀: Upper parts reddish-brown, with slight olive shade; no contrast of color between back and tail; quills and tail-feathers darker and purer brown, former with white or buff spaces at concealed bases of inner webs (as usual in this genus). *No* light ring around eye; auriculars only obsoletely streaky. Below, white; sides shaded with hoary-gray or pale grayish-olive; jugulum buff-colored, contrasting with white of breast, and marked with a few small brown arrow-heads; chin and middle line of throat, however, nearly white and immaculate. A *few* obsolete grayish-olive spots in white of breast; but otherwise markings confined to the buff area. Bill dark above, mostly pale below, like feet. ♂: Length 7.25-7.50; extent about 12.00; wing 4.00-4.25; tail 3.00-3.25; bill 0.60; tarsus 1.20. ♀ smaller; average of both sexes: length 7.35; extent 11.75; wing 3.90; tail 2.85; tarsus 1.12. Chiefly Eastern U. S., but N. to Canada; common, migratory, nesting in northerly parts of its range. Wintering mostly extralimital, but sparingly in Florida. Nest on ground or near it, of leaves, grasses, etc., but no mud; eggs 4-5, greenish-blue like the Wood Thrush's, normally unspotted, 0.90 × 0.60. A delightful songster, like others of the genus, found in thick woods and swamps; of shy and retiring habits.

H. f. salici'cola. (Lat. *salix*, a willow; *colere*, to inhabit or cultivate.) WILLOW TAWNY THRUSH. Like *fuscescens*, but averaging larger; upper parts less decidedly tawny; jugulum less distinctly buff. Wing 3.80–4.25, av. 4.02; tail 2.95–3.40, av. 3.20; bill 0.55–0.60; tarsus, av. 1.17; middle toe without claw, av. 0.69. Rocky Mt. region, U. S., N. to British Columbia, S. to Brazil in winter, occasionally E. to Illinois and South Carolina. This subspecies is clearly referable to *fuscescens*; but it bears an extraordinary resemblance to *ustulata*, in the russet-olive color of the upper parts, and only slightly buff tinge of the jugulum. It is distinguished from *ustulata* by lack of the buff orbital ring so characteristic of *ustulata* and *swainsoni*, and other characters by which *fuscescens* differs, notably the few if any spots on the white breast back of the buff area, and pale hoary gray instead of sordid olive-gray shading of the sides. The nest and eggs are like those of *fuscescens*, not like those of *ustulata* or *swainsoni*.

H. aonalasch'kæ. (Of Aonalaschka, Oonalashka, Oonalaska, Ounalashka, Unalachka, Unalashka, Unalaska, etc., one of the largest islands of the Aleutian chain in Alaska. Unfortunately, this barbarous name, of unsettled orthography, was given to the Western form of Hermit Thrush by GMELIN in 1788, before the common Eastern form had been described; for it thus takes precedence as the specific term. In the 2d–4th eds. of the Key I softened the outlandish word into the Latin-looking form of *unalascæ*; but by our rigid rules it must be restored to its original terrors.) DWARF HERMIT THRUSH. WESTERN HERMIT THRUSH. In color absolutely like the common Eastern Hermit below described; in size slightly less on an average; length scarcely 7.00; wing 3.30; tail 2.50; tarsus 1.15. Pacific coast region of N. A., Alaska to Lower California and Western Mexico, breeding from the Sierra Nevadas northward, and in migration found in the Great Basin. Nest and eggs not distinguishable with certainty from those of the Eastern Hermit.

H. a. aud'uboni. (To J. J. Audubon.) AUDUBON'S HERMIT THRUSH. In color absolutely like the common Eastern Hermit; in size larger on an average; length about 7.75; wing 4.20; tail 3.30; tarsus 1.20. Inhabits Rocky Mt. region of the U. S., westward in the Great Basin to Southern California, S. in winter through Mexico to Guatemala. A better marked variety than the last; besides the larger size, on an average, the general tone is rather duller or grayer, and the rufous of the tail is not so bright. Nest and eggs as in the common Hermit.

NOTE. *T. sequoiensis* BELDING, Proc. Cala. Acad., ii, June, 1889, p. 18, breeding at Big Trees, Calaveras Co., Cal., is deemed inadmissible, as noted in the Key, 4th ed., p. 897. It resembles other Western Hermits in the rufous tail, unspotted eggs, etc.; the ascribed dimensions are intermediate between those of the two preceding forms.

H. a. pal'lasi. (To Peter S. Pallas, the celebrated Russian traveller and naturalist (CABANIS, 1845). *T. unalascæ nanus* of 2d–4th eds. of the Key, based on *T. nanus* AUD. There is much to be said in favor of this name, but I waive my contention in deference to the A. O. U. committee. For synonymy of all our *Hylocichlæ*, see COUES, B. Col. Vall., i, 1878, pp. 22–28.) EASTERN HERMIT THRUSH. SWAMP ANGEL. ♂ ♀, in summer: Upper parts olivaceous, with a brownish cast, and therefore not so pure as in *swainsoni*; this color changing on rump and upper tail-coverts into rufous of tail, in decided contrast with back. Under parts white, shaded with grayish-olive on sides; breast, jugulum, and sides of neck more or less strongly tinged with yellowish, and marked with numerous large, angular, dusky spots, which extend back of the yellowish-tinted parts. Throat immaculate. A yellowish orbital ring. Bill brownish-black; most of under mandible livid whitish; mouth yellow; eyes brown; legs pale brownish. ♂: Length 7.00–7.25; extent 11.00–12.00; wing 3.50–3.75; tail 2.75–3.00. ♀ smaller: Length 6.75–7.00; extent 10.75–11.25; wing 3.25–3.50. Averages of both sexes are: Length 7.00; extent 11.25; wing 3.50; tail 2.75; tarsus 1.15. The dimensions thus overlap those of both *aonalaschkæ* and *auduboni*, and no positive discrimination is possible; the differences, when any, being of averages, not of extremes either way. ♂ ♀, in winter: The olivaceous of upper parts assumes a more rufous cast, much like that of *ustulata*, and the yellowish wash of under parts and sides of head and neck is more strongly pronounced. But the most rufous specimens are readily distinguished from *fuscescens* by the strong contrast between the color of the tail and other upper parts. Very young: Most of the upper parts marked with pale yellowish longitudinal streaks, with clubbed extremities, and dusky specks at the end; feathers of belly and flanks often skirted with dusky in addition to the numerous blackish spots of other under parts. Chiefly the Eastern Province of North America; abundant; migratory, and found in all woodland, but breeds only northerly, from Massachusetts and corresponding latitudes and northern Alleghanies; winters in the Southern States. Nest and eggs not distinguishable from those of the Veery.

H. ustula'ta. (Lat. *ustulata*, scorched, singed; referring to the warm russet coloration.) OREGON OLIVE-BACKED THRUSH. RUSSET-BACKED THRUSH. Quite like the Eastern Oliveback (*swainsoni*) in uniformity of color of whole upper parts, presence of buff orbital ring, and general character of the shading and spotting of under parts; but olive of upper parts not pure, having a decided rufous tinge, resulting in a russet-olive of exactly the shade of that of the upper parts of the Western subspecies of *fuscescens* (*salicicola*); from which distinguished by the buff orbital ring, and very different size, and marking of under parts (compare *salicicola*); there being, as in *swainsoni* proper, much olive-gray spotting of the white breast back of the buff area, and much shading of the same olive-gray on the sides. Size of *swainsoni*. Nest in bushes, and eggs spotted, as in the latter. Pacific coast region of the U. S. and British Columbia, from Alaska S. in winter to Guatemala; abundant.

H. u. œdica. (Gr. ᾠδικός, *oidikos*, fond of singing, musical, vocal.) TUNEFUL OLIVE-BACK. Described as like *ustulata*, with upper parts and flanks paler. Ascribed to California and S. Oregon. A very slight local race, included under *ustulata* in former eds. of Key. OBERHOLSER, Auk, Jan. 1899, p. 23; A. O. U. Suppl. List, *ibid.*, p. 127.

H. u. swain'soni. (To Wm. Swainson, an English naturalist.) SWAINSON'S THRUSH. OLIVE-BACKED THRUSH. EASTERN OLIVE-BACK. Adult ♂ ♀: Above, clear olivaceous, of exactly the same shade over all upper parts; below, white, strongly shaded with olive-gray on sides and flanks; throat, breast, and sides of neck and head strongly tinged with yellowish, the fore parts, excepting throat, marked with numerous large dusky spots, which extend backward on breast and belly, there rather paler, and more like the olivaceous of upper parts. Edges of eyelids yellowish, forming a strong buff orbital ring; lores the same. Mouth yellow; bill blackish; basal half of lower mandible pale; iris dark brown; feet pale ashy-brown. Length of ♂ 7.00–7.50; extent 12.00–12.50; wing 3.75–4.00; tail 2.75–3.00; bill 0.50; tarsus 1.10. ♀ averaging smaller: Length 6.75; extent 11.50–12.00, etc. Eastern N. Am., W. to Colorado in migrations; winters in Cuba, C. and S. Am.; breeds in Canadian fauna, S. in the Alleghanies to West Virginia. Nest in bushes and low trees, thus in situation like that of the Wood Thrush, but no mud in its composition; eggs unlike those of *mustelinus*, *fuscescens*, and *aonalaschkæ*, in being freely speckled with different shades of brown on a greenish-blue ground; size 0.90 × 0.66; number 3–4.

NOTE.— *H. u. almæ* OBERH., Auk, Oct. 1898, p. 304, from the R. Mts., Utah, and E. Nevada, is recognizably different from *swainsoni*. North to Yukon Basin, south in winter to Mexico.

H. ali'ciæ. (To Miss Alice Kennicott, sister of Robert Kennicott.) GRAY-CHEEKED THRUSH. ALICE'S THRUSH. Similar to *swainsoni* in uniformity and purity of the olive of upper parts, which is as dark and pure (no tendency to the rufous of *ustulata*); but sides of head lacking the yellowish or buffy suffusion seen in *swainsoni*, being thus like the back, or merely grayer; *no* buff ring around eye; breast slightly if at all tinged with yellowish. Rather larger than *swainsoni*, about equalling *mustelina*: Length 7.50–8.00; extent 12.50–13.50; wing 4.00–4.25; tail 3.00–3.25; bill over 0.50; average dimensions about the maxima of *swainsoni*. Distribution and nesting the same, but breeding range more northerly, being beyond the U. S. to the Arctic coast; occurs in Alaska, and even in Siberia; S. in winter to Central America. *T. aliciæ* BD., 1858; *T. swainsoni aliciæ* COUES, Key, orig. ed., 1872, p. 73; *T. ustulatus aliciæ* of the Key, 2d-4th eds., 1884-90, p. 248, the specific distinctness there indicated now confirmed.

H. a. bick'nelli. (To E. P. Bicknell of New York.) BICKNELL'S THRUSH. A local race, described as smaller on an average, with the bill usually slenderer; colors exactly those of *aliciæ* proper. Breeding in the Catskill Mts. of New York, the White Mts. of New Hampshire, and in Nova Scotia, migrating S. in winter to parts unknown, because nobody can recognize as different from *aliciæ* specimens found away from the ascribed breeding range. *Hylocichla a. bicknelli* RIDGW., Proc. U. S. Nat. Mus., iv, Apr. 1882, p. 377; *Turdus a. bicknelli* COUES, Key, 2d-4th eds., 1884-90, p. 248.

SAXIC'OLA. (Lat. *saxum*, a rock; *colo*, I inhabit. Fig. 121.) STONE-CHATS. Bill shorter than head, slender, straight, depressed at base, compressed at end, notched. Wings long, pointed; tip formed by 2d-4th quills; 1st spurious, scarcely or not $\frac{1}{4}$ as long as 2d. Tail much shorter than wing, square. Tarsi booted, but with 4 scutella below in front, long and slender, much exceeding middle toe and claw; lateral toes of about equal lengths, very short, the tips of their claws not reaching base of middle claw; claws little curved; feet thus adapted to terrestrial habits. A large and widely distributed Old World genus, of some 30 species, inhabiting Europe, Asia, and especially Africa. With some authors, it gives

FIG. 121.—Generic details of *Saxicola*.

name to a subfamily *Saxicolinæ* of *Turdidæ*, or even of a family *Saxicolidæ*. I have presented such a group in earlier eds. of the Key, after a fashion then prevalent, but with the remark that "it has never been defined with precision, being known conventionally by the birds ornithologists put in it." (Key, 2d ed., 1884, p. 256); and I am now glad to abandon it altogether, with the sanction of the A. O. U.

S. œnan'the. (Gr. οἰνάνθη, *oinanthe*, name of a bird, from οἴνη, *oine*, the grape, and ἄνθος, *anthos*, a flower.) STONE-CHAT. WHEAT-EAR. Adult ♂ : Ashy-gray; forehead, superciliary line and under parts white, latter often brownish-tinted; upper tail-coverts white; wings and tail black, latter with most of the feathers white for half or more of their length; line from nostril to eye, and broad band on side of head, black; bill and feet black. ♀ more brownish-gray, the black cheek-stripe replaced by brown. Young without the stripe: above, olive-brown; superciliary line, edges of wings and tail, and all under parts cinnamon-brown; tail black and white as in the adult. Length of ♂ 6.75; extent 12.50; wing 3.75; tail 2.50; tarsus 1.00; middle toe and claw 0.75. ♀ smaller: length 6.50; extent 11.50, etc. Europe, Asia, and N. Africa; Atlantic coast, from Europe *via* Greenland; also N. Pacific and Arctic coast, from Asia. Common in Greenland, and probably also breeds in Labrador; straggles S. to Nova Scotia and other parts of Canada, New England, New York, even to the Bermudas, and New Orleans, La.; also Colorado (at Boulder, May 14, 1880). Nest in holes in the ground or rocks, crevices of stone walls, etc. Eggs 4-7, 0.87 × 0.60, greenish-blue, without spots.

SIA'LIA. (Gr. σιαλίς, *sialis*, a kind of bird.) BLUEBIRDS. Primaries 10; 1st spurious and very short. Wings pointed; tip formed by 2d, 3d, and 4th quills. Tail much shorter than wings, emarginate. Bill $\frac{1}{2}$ as long as head or less, straight, stout, wider than deep at base, compressed beyond nostrils, notched near tip; culmen at first straight, then gently convex to end; gonys slightly convex and ascending; commissure slightly curved throughout. Nostrils overhung and nearly concealed by projecting bristly feathers; lores and chin likewise bristly.

Gape ample; rictus cleft to below eyes, furnished with a moderately developed set of bristles reaching about opposite nostrils. Feet short, rather stout, adapted exclusively for perching (in *Saxicola* the structure of the feet indicates terrestrial habits). Tarsus not longer than middle toe; lateral toes of unequal lengths; claws all strongly curved. Blue is the principal color of this beautiful genus, which contains 3 species and several subspecies. They are strictly arboricole; frequent the skirts of woods, coppices, waysides, and weedy fields; nest in holes, and lay whole-colored eggs; readily become semi-domesticated; feed upon insects and berries; and have a melodious warbling song. Polygamy is sometimes practised by them, contrary to the rule among *Oscines*. Bluebirds are peculiar to America, and appear to have no exact representatives in the other hemisphere.

Analysis of Species.

♂ Rich sky-blue, uniform on back; throat and breast chestnut; belly white . . . . *sialis*
♂ Rich sky-blue, including throat; middle of back and breast chestnut; belly whitish . . . . . . *mexicana*
♂ Light blue, paler below, fading to white on belly; no chestnut . . . . *arctica*

S. si'alis. (Gr. σιαλίς, *sialis*, a kind of bird. Fig. 122.) EASTERN BLUEBIRD. WILSON'S BLUEBIRD. BLUE ROBIN. ♂, in full plumage: Rich azure-blue; ends of wing-quills blackish; throat, breast, and sides of body chestnut; belly and crissum white or bluish-white. The blue sometimes extends around the head on sides and fore part of chin, so that the chestnut is cut off from bill. Length 6.50–7.00; extent 12.00–13.00; wing 3.75–4.00; tail 2.75–3.00; bill 0.45; tarsus 0.70. ♂, in winter, or when not full-plumaged: Blue of upper parts interrupted by reddish-brown edging of the feathers, or obscured by a general brownish wash. White of belly more extended; tone of other under parts paler. In many Eastern specimens, the reddish-brown skirting of the feathers blends into a dorsal patch; when this is accompanied by more than ordinary extension of blue on throat they closely resemble *S. mexicana*. ♀, in full plumage: Blue mixed and obscured with dull reddish-brown; becoming bright and pure on rump, tail, and wings. Under parts paler and more rusty-brown, with more abdominal white than in ♂. Little smaller than ♂. Young, newly fledged: Brown, becoming blue on wings and tail; back sharply marked with whitish shaft-lines. Nearly all the under parts closely and uniformly freckled with white and brownish. A white ring round eye; inner secondaries edged with brown. From this stage, in which the sexes are indistinguishable, to the perfectly adult condition, the bird changes by insensible degrees. Eastern U. S., and Canada; abundant and familiar, almost domestic; W. often to the Rocky Mts. Migratory, but breeds throughout its range, wintering in the Middle States and beyond, whence it comes as one of the early harbingers of spring, or during mild winter weather, bringing its bit of blue sky with cheery, voluble song. Nest in natural or artificial hollows of trees, posts, or bird-boxes, loosely constructed of the most miscellaneous materials; eggs 4–6, pale bluish, occasionally whitish, unmarked, 0.80 × 0.60; two or three broods in one season.

FIG. 122. — Bluebird, nat. size. (Ad nat. del. E. C-)

S. s. azu'rea. (New Lat. adj. *azureus*, azure, sky-blue; Middle Lat. noun *azura*, *azurum*, *lazur*, *lazurius*, *lazulus*, a blue stone, the *lapis lazuli*, Gr. λαζούριον, *lazourion*, from Arabic *lazward*; Persian *lazhward*; said to be named from the mines of Lajwurd.) AZURE BLUEBIRD. Similar to *S. sialis*; the blue of a greenish shade; breast paler chestnut; crissum buffy; tail about 3.00. S. Arizona and southward. A slight variety, first described by BAIRD, Rev. Am. Birds, 1864, p. 62; taken into Key, 3d ed., 1887, p. 866; A. O. U. List, No. 766 *a* (wrongly accredited to SWAINSON, and the date of BAIRD's Review misprinted 1884).

S. mexica'na occidenta'lis. (Lat. *mexicana*, of Mexico; *occidentalis*, of the occident or setting sun, *i. e.*, Western.) TOWNSEND'S WESTERN BLUEBIRD. MEXICAN BLUEBIRD. ♂, adult : Rich azure-blue, including head and neck all around; a patch of purplish-chestnut on upper back, more or less completely divided into a pair of patches; breast and sides rich chestnut; belly and vent dull blue or bluish-gray. Bill and feet black. Size of the last species. ♀, and young: Changes of plumage coincident with those of the Eastern bluebird. Immature birds may usually be recognized by some difference in color between middle of back and other upper parts, and between color of throat and of breast ; but birds in the streaky stage could not be determined if the locality were unknown. In typical adult ♂, the dorsal patch is restricted, or broken into two scapular patches with continuous blue between; the chestnut of breast sometimes divides, permitting connection of the blue of throat and belly (see *anabelæ* below). Specimens with little trace of the dorsal patch are scarcely distinguished from those of *S. sialis*, in which there is much blue on the throat, the grayish-blue of the belly, instead of white, being a principal character. Pacific Coast region of the U. S. and British Columbia. E. occasionally in migrations to Utah, Nevada, New Mexico, etc. Abundant; habits, nest, and eggs identical with those of *S. sialis*. *S. occidentalis* TOWNS., Journ. Acad. Nat. Sci. Philada., vii, 1837, p. 188. *S. mexicana occidentalis* RIDGW., Auk, Apr. 1894, pp. 151, 154; A. O. U. List, 2d ed., 1895, p. 322, No. 767 ; as subspecifically distinguished from the typical Mexican form with which it had before been considered identical, and from the following :

S. m. baird'i. (To S. F. Baird.) CHESTNUT-BACKED BLUEBIRD. BAIRD'S BLUEBIRD. In typical adult ♂ the patch of chestnut on the back forming a single solid area, well defined against blue surroundings. Rocky Mountain region of the U. S., S. into Mexico. Auk, Apr. 1894, p. 151, p. 157 ; A. O. U. List, 2d ed., 1895, p. 323, No. 767 *a*.

S. m. anab'elæ. (To Mrs. Anabel Anthony, wife of W. A. Anthony.) ANABEL'S BLUEBIRD. SAN PEDRO BLUEBIRD. Chestnut of breast divided by blue of throat, and thus restricted to lateral pectoral patches ; that of scapulars almost entirely absent; size at a maximum. San Pedro Martir Mts. of Lower California. ANTHONY, Proc. Cala. Acad. Sci., 2d ser., ii., Oct. 1889, p. 79 (see Key, 4th ed., 1890, p. 897); A. O. U. List, 2d ed., 1895, p. 323, No. 767 *b*. (This and the last subspecies recognized by name in no former ed. of the Key, being both included under *Mexicana*, with express statement, however, of their respective peculiarities.)

S. arc'tica. (Lat. *arctica*, arctic ; Gr. ἄρκτος, *arctos*, a bear; *i. e.*, near the constellation Ursa Major.) ARCTIC BLUEBIRD. ROCKY MOUNTAIN BLUEBIRD. ♂, in perfect plumage : Above azure-blue, lighter than in the two foregoing, and with a faint greenish hue ; below, paler and more decidedly greenish-blue, fading insensibly into white on the belly and under tail-coverts. Ends of wing-quills dusky ; bill and feet black. Larger ; length 7.00 or more ; extent 13.00 or more ; wing 4.50 ; tail 3.00. ♀ : Nearly uniform rufous-gray, lighter and more decidedly rufous below, brightening into blue on rump, tail, and wings, fading into white on belly and crissum ; a whitish eye-ring. Young : Changes parallel with those of the other species. Birds in the streaky stage may be known by superior size, and greenish shade on the wings and tail. N. America from the W. portions of the Great Plains and E. spurs and foothills of the Rocky Mts. to the Pacific, chiefly in high open regions, abundant ; resident southerly, migratory N. to Great Slave Lake, S. into Mexico. Habits those of the others ; nesting the same, but eggs larger, about 0.92 × 0.70.

CYANEC'ULA. (A diminutive form of Gr. κυάνεος, *kuaneos*, Lat. *cyaneus*, blue ; as we should say, "bluet.") BLUETHROATS. Bill much shorter than head, slender, compressed throughout, acute at tip, with obsolete notch (as in *Saxicola*, but slenderer). Feet, as in *Saxicola*, long and slender ; tarsus much longer than middle toe and claw ; lateral toes of unequal lengths; outer longer, but tip of its claw falling short of base of middle claw; claws little curved, the hinder fully as long as its digit. Wings long (less so than in *Saxicola*), pointed by

3d, 4th, and 5th quills; 2d about equal to 6th; 1st spurious, about ⅛ as long as the longest. Tail of moderate length, slightly rounded; particolored with chestnut; throat and breast with azure-blue and chestnut. An Old World genus, one species of which occurs casually in Alaska.

C. sue'cica. (Lat. *Suecica*, Swedish.) BLUE-THROATED REDSTART. RED-SPOTTED BLUE-THROAT. ♂, adult : Entire upper parts dark brown with a shade of olive (about the color of a Titlark, *Anthus pensilvanicus*); feathers of crown with darker centres; rump and upper tail-coverts rather lighter, mixed with bright chestnut-red. Wings like back, with slightly paler edgings of the feathers. Middle tail-feathers like back, or rather darker, the rest blackish, with the basal half or more of their length bright chestnut-red or orange-brown. Lores dusky; a whitish superciliary line. Chin, throat, and forebreast rich ultramarine blue, enclosing a bright chestnut throat-patch; the blue bordered behind by black, this again by chestnut mixed with white. Rest of under parts white, washed on sides, lining of wings and under tail-coverts with pale fulvous. Bill and feet black. ♀ and young similar, the throat-markings imperfect. Length 5.75–6.00; wing 3.10; tail 2.25–2.50; bill 0.50; tarsus 1.00; middle toe and claw 0.75. A beautiful and interesting bird, widely distributed in northerly parts of the Old World, casually found at St. Michael's in Alaska.

Subfamily MYIADESTINÆ: Fly-Catching Thrushes; Solitaires.

Bill very short, much depressed and widened at base; rami of under mandible deeply cleft, with short gonys (only ⅛ as long as culmen), tarsus *booted*, and toes deeply cleft, as in other *Turdidæ*. Feet weak; lateral toes unequal; tip of inner claw falling short of base of the middle one. Wing of 10 primaries; 1st spurious, 2d about = 6th; tip formed by 3d–5th. Tail long, about equalling wing, *double-rounded*, being forked centrally, graduated externally, all the feathers tapering. Head subcrested; plumage sombre, variegated on wings; sexes alike; young *spotted*. Containing a dozen or more species, chiefly of the genus *Myiadestes ;* others of *Cichlopsis* and *Platycichla;* all except one are birds of C. and S. Am. and the W. Indies. Though

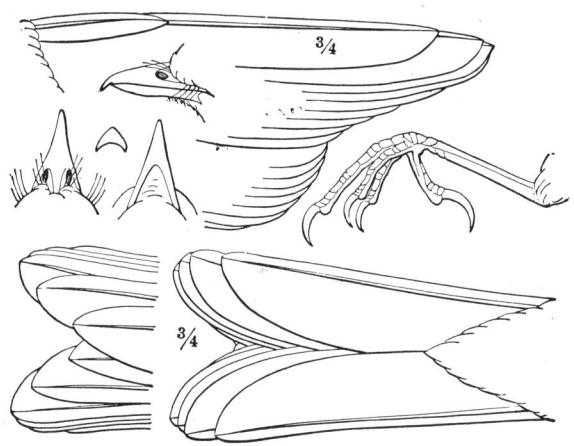

FIG. 123. — Generic details of *Myiadestes* (*M. townsendi;* bill and foot nat. size, wing and tail ¾). (From Baird.)

our species was formerly called "Ptilogonys," it has nothing to do with that genus; and though it has usually been placed near *Phainopepla*, and referred with the latter to *Ampelidæ*, it is no member of that family. As stated in the Key, 2d ed., 1884, p. 325, "the *Myiadestinæ* are near the true Thrushes," to which they have been since referred by common consent of American writers, and with all the authority of the A. O. U. I avail myself of the first opportunity to make the required transposition to the *Turdidæ* of the matter on pp. 328 and 329 of 2d–4th eds. of Key.

MYIADES'TES. (Gr. μυῖα, *muia*, a fly, and ἐδεστής, *edestes*, an eater.) FLY-CATCHING THRUSHES. SOLITAIRES. Characters those of the subfamily as above given.

M. town'sendi. (To J. K. Townsend.) TOWNSEND'S FLY-CATCHING THRUSH, or

SOLITAIRE. ♂ ♀ : General color dull brownish-ash, paler below, bleaching on throat, lower belly, and crissum. Wings blackish; inner secondaries edged and tipped with white, nearly all the quills extensively tawny or fulvous at the base, and several intermediate ones again edged externally toward their ends with the same color. In the closed wing, the basal tawny shows upon the outside as an oblique spot in the recess between the greater coverts and the bastard quills, separated by an oblique bar of blackish from the second tawny patch on the outer webs of the quills near their ends. Tail like wings (the middle pair of feathers more nearly like the back); outer feather edged and broadly tipped, next one more narrowly tipped, with white. A white ring around eye. Bill and feet black. Eyes brown. Length about 8.00; wing and tail about equal, 4.00-4.50; the latter forked centrally, graduated laterally; bill 0.50; tarsus 0.75; middle toe and claw rather more. Young: speckled at first, like a very young Thrush; each feather with a triangular or rounded spot of dull ochraceous or tawny, edged with blackish. Western U. S., from eastern foot-hills of the Rocky Mts. to the Pacific; N. to British Columbia and upper Yukon, S. to Sonora and L. Cala.; breeds from New Mexico, Arizona, and Southern California northward. A bird not less strange and unlike anything seen in the East than *Phainopepla;* inhabiting woodland and shrubbery, feeding on insects and berries, and capable of musical expression in an exalted degree. Nest on ground or in rubbish near it, loosely made of grasses; eggs 3 to 6, bluish-white, freckled with reddish-brown, 0.95 × 0.67.

Family CINCLIDÆ : Dippers.

Primaries 10; 1st spurious, and, like the others, falcate; 2d entering into point of wing; wing short, stiff, rounded, concavo-convex. Tail still shorter than wing, soft, square, of 12 broad, rounded feathers, almost hidden by the coverts, which reach nearly or quite to the end, the under ones especially long and full. Tarsus booted, about as long as middle toe and claw. Lateral toes equal in length. Claws all strongly curved. Bill shorter than head, slender and compressed throughout, higher than broad at nostrils, about straight, but seeming to be slightly recurved, owing to a sort of upward tilting of the superior mandible; culmen at first slightly concave, then convex; commissure slightly sinuous, to correspond with culmen, notched near end; gonys convex. Nostrils linear, opening beneath a large scale partly covered with feathers. No rictal vibrissæ, nor any trace of bristles or bristle-tipped feathers about nostrils. Plumage soft, lustreless, remarkably full and compact, water-proof. Body stout, thick-set. Habits aquatic. A small but remarkable group, in which characters shared by the *Turdidæ* and *Sylviidæ* are modified in adaptation to the singular aquatic life the species lead. There is only one genus, with about 12 species, inhabiting clear mountain streams of most parts of the world, chiefly the Northern Hemisphere; easily *flying under water,* and spending much of their time in that element, where their food, of various aquatic animal substances, is gleaned. (Subfamily *Cinclinæ* of former eds. of the Key, now raised to family rank.)

CIN'CLUS. (Gr. κίγκλος, *kigklos,* Lat. *cinclus,* a kind of bird. Fig. 114.) DIPPERS. Characters those of the family, as above given.

C. mexica'nus. (Lat. *mexicanus,* Mexican. AMERICAN DIPPER, or WATER OUZEL. ♂ ♀, adult, in summer: Slaty-plumbeous, paler below, inclining on the head to sooty-brown. Quills and tail-feathers fuscous. Eyelids usually white. Bill black; feet yellowish. Length 6.00-7.00; extent 10.00-11.00; wing 3.50-4.00; tail about 2.25; bill 0.60; tarsus 1.12; middle toe and claw rather less. Individuals vary much in size. ♂ ♀, in winter, and most immature specimens, are still paler below, all the feathers of the under parts being skirted with whitish; quills of the wing also tipped with white; bill yellowish at base. Young: Below, whitish, more or less so according to age, frequently tinged with pale cinnamon-brown; whole under parts sometimes overlaid with whitish ends of the feathers, shaded with rufous posteriorly; throat usually nearly white; bill mostly yellow; white tipping of wing-feathers at a maximum; in some cases the tail-feathers similarly marked. Mountains of Western N. A., from Alaska to Guatemala; E. in the U. S. to the eastern bases and spurs of the Rocky Mts., as in the Black Hills of S. Dakota; a sprightly and engaging resident of clear mountain streams, usually observed flitting among the rocks; has a fine song. Nest a pretty ball of green moss lined with grasses, with a hole at the side, hidden in the rift of a rock, or other nook close to the water: eggs about 5, 1.04 × 0.70, pure white, unmarked.

Family SYLVIIDÆ : Old World Warblers, Kinglets, etc.

A large family of chiefly Old World birds, mainly represented in America by the genera *Regulus* and *Polioptila*. They belong to the Turdoid series, and the line between *Turdidæ* and *Sylviidæ* is not a hard and fast one. The fact that young *Sylviidæ* are not spotted like young Thrushes is probably the best character that can be ascribed; this seems to be correlated with the double annual moult which normal *Sylviidæ* undergo, in spring and fall, as contrasted with the single moult of true *Turdidæ*. The tendency of *Sylviidæ* is toward booted tarsi, as in *Turdidæ*, and fusion of scutella is usually extensive, as in *Sylviinæ* and *Regulinæ;* but in some groups, as *Polioptilinæ*, the scutellation is plain. There is no difficulty, however, in recognizing any North American bird of the family as here given, by the very diminutive size (length under 6.00, usually 5.00 or less); 10 primaries, the 1st spurious; slender bill, more or less notched or even hooked at tip; and greenish or bluish coloration. Our 3 genera fall in as many subfamilies, recognition of which is convenient, but a mere conventionality. (The *Sylviidæ* are brought under *Turdidæ* in 2d–4th eds. of Key, but we have the authority of the A. O. U. for separating them as a family, and thus reverting to the arrangement given in the orig. ed., 1872.)

Analysis of Subfamilies.

Tarsus more or less booted.
    Colors greenish; no crest. (Old World. N. Am. only in Alaska.) . . . . *Sylviinæ*
    Colors greenish, with a red or flaming crest. No black on wings or tail . . . . *Regulinæ*
Tarsus distinctly scutellate.
    Colors bluish and white, with much black on wings and tail . . . . *Polioptilinæ*

Subfamily SYLVIINÆ: Old World Warblers.

Characters sufficiently indicated for present purposes in the above analysis, as the subfamily cuts no figure in America, though it is a large and important group in Europe, Asia, and Africa, with numerous (over 100) species of many modern genera, among which are *Sylvia* proper, *Phyllopseustes* (or *Phylloscopus*), *Hypolais*, *Acrocephalus*, *Locustella*, *Lusciniola*, and *Cettia* — this last exhibiting the passerine anomaly of only 10 rectrices, and perhaps standing as type of a different subfamily *Cettiinæ*. We have here to do only with the genus

PHYLLOPSEUS'TES. (Gr. φύλλον, *phullon*, a leaf; ψσεύστης, *pseustes*, a liar, cheat; application not obvious. Meyer, 1815. *Phylloscopus* of most authors, as of 2d–4th eds. of Key; Phyllopneuste of the orig. ed.) OLD WORLD WOOD-WARBLERS. WILLOW WARBLERS. Bill shorter than head, slender, straight, depressed at base, compressed and notched at tip; nostrils exposed, though reached by the frontal feathers. Tarsus longer than middle toe and claw, booted or indistinctly scutellate; wings longer than tail; pointed by 3d and 4th quills; 5th much shorter, 6th shorter still, 2d between 5th and 6th; 1st spurious, very short, exposed less than 0.50. Tail about even. Size diminutive and coloration simple. Includes numerous (about 25) Old World species, one of them occurring in Alaska.

P. borea'lis. (Lat. *borealis*, northern; *boreas*, the north-wind.) ARCTIC WILLOW WARBLER. KENNICOTT'S WARBLER. ♂ ♀, adult: Above, olive-green, clear, continuous, and nearly uniform, but rather brighter on rump; quills and tail-feathers fuscous, edged externally with yellowish-green; a long yellowish superciliary stripe; under parts yellowish-white; lining of wings and flanks yellow; wings crossed with two yellowish bars, that across ends of greater coverts conspicuous, the other indistinct; bill dark brown, pale below; feet and eyes brown. Length 4.75; extent 6.00; wing 2.25-2.50; tail 1.75-2.00; tarsus 0.70; middle toe and claw 0.55. Europe; Asia; casually N. Am. in Alaska.

Subfamily REGULINÆ: Kinglets.

Characters sufficiently indicated in the following diagnosis of our only genus.

REG'ULUS. (Lat. *regulus*, diminutive of *rex*, a king.) KINGLETS. Tarsus booted, very slender, longer than middle toe and claw. Lateral toes nearly equal to each other. 1st primary spurious, its exposed portion less than half as long as 2d. Wings pointed, longer than tail, which is emarginate, with acuminate feathers. Bill shorter than head, straight, slender, typically Sylviine, not hooked at end, well bristled at rictus, with nostrils overshadowed by tiny feathers. Coloration olivaceous, paler or whitish below, with red, black, or yellow, or all three of these colors, on head of adult. About 10 species, of Europe, Asia, and America; elegant and dainty little creatures, among the very smallest of our birds excepting Hummers. They inhabit woodland, are very agile and sprightly, insectivorous, migratory, and highly musical.

Analysis of Species and Subspecies.

Head with a scarlet patch, but no black or yellow. A tuft of bristly feathers over nostrils. (*Subgenus* PHYLLOBASILEUS: Ruby-crowns.)
  The ordinary bird of N. Am. at large . . . . *calendula*
  A dark insular species of Guadalupe Island, L. Cala . . . . *obscurus*
Head with black and orange or yellow. A single tiny feather over each nostril. (REGULUS *proper:* Gold-Crests.)
  The ordinary form of N. Am. at large . . . . *satrapa*
  A brighter form of the Pacific coast region . . . . *S. olivaceus*

(*Subgenus* PHYLLOBASILEUS.)

R. (P.) calen'dula. (New Lat. *calendula*, dimin. of Ital. *calandra*, Fr. *calandre*, Eng. *calender*, a kind of Lark, *Melanocorypha calandra;* so called from Lat. *caliendrum*, a headdress of false hair, chignon, wig. In botany, *calendula*, a word of identical form but different derivation, is the name of the genus of marigolds.) RUBY-CROWNED KINGLET. ♂, adult: Bill and feet black. Upper parts greenish-olive, becoming more yellowish on the rump; wings and tail dusky, strongly edged with yellowish; whole under parts dull yellowish-white, or yellowish- or greenish-gray (very variable in tone); wings crossed with two whitish bars, and inner secondaries edged with the same. Edges of eyelids, lores, and extreme forehead hoary whitish. A rich scarlet patch, partially concealed, on the crown. This beautiful ornament is apparently not gained until the 2d year, in some cases, as it is absent from some adult ♂ ♂ in the spring, or of a yellow instead of flaming color; but as a rule it is present in young ♂ ♂ the first autumn. It is never present in the ♀, as a normal character, though possibly to be found in some individuals of that sex which have taken on the ♂ dress in consequence of age or sterility. Length 4.10-4.50; extent 6.66-7.33; wing 2.00-2.33; tail 1.75; bill 0.25; tarsus 0.75. Young of the year: Quite like the adult of each sex; *i. e.*, ♀ wanting the scarlet patch, usually present in the ♂, sometimes wanting, or merely yellowish. In a newly fledged specimen wings and tail as strongly edged with yellowish as in adult; but general plumage of upper parts rather olive-gray than olive-green, and under parts sordid whitish; bill light colored at base, and toes appear yellowish. N. Am. at large, breeding far N. and in mountains of the west to S. California and Arizona, wintering in the Southern States and beyond to Guatemala. An exquisite little creature, famous for vocal power, abundant in wooded regions. Nest a large mass of matted hair, feathers, moss, straws, etc., placed at or near the end of a bough, usually of a coniferous tree; eggs numerous, 0.54 × 0.42, creamy white, sparsely speckled with brown, chiefly about the larger end.

R. c. grinnel'li. (To Joseph Grinnell.) SITKAN KINGLET. Sooty-olive above, blackening along sides of the vermilion patch; throat and breast dusky gray; belly yellowish-white. Bill acute, with wide base. Sitka, Alaska; a dark coast form. W. PALMER, Auk, Oct. 1897, p. 399.

R. obscu'rus. (Lat. *obscurus*, obscure, dark.) DUSKY KINGLET. Resembling the common Ruby-crown, but with darker and more plumbeous shade of upper parts, and some slight differences in proportions. A dark insular form from Guadalupe Island, Lower California. *R. c. obscurus* RIDGW., Bull. U. S. Geol. Surv. Terr., Apr. 1876, p. 184; COUES, Key, 2d ed., 1884, p. 260; *R. obscurus* RIDGW., Bull. Nutt. Club, July, 1877, p. 59; A. O. U. Lists, 1886-95, No. 750.

(*Subgenus* REGULUS.)

FIG. 124. — Golden-crested Kinglet. (After Audubon.)

R. satra'pa. (Gr. σατράπης, Lat. *satrapes*, a ruler; alluding to the bird's golden crown. Figs. 124, 125.) GOLDEN-CRESTED KINGLET. ♂, adult: Upper parts olive-green, more or less

FIG. 125. — Golden-crested Kinglet.

bright, sometimes rather olive-ashy, always brightest on rump; under parts dull ashy- or yellowish-white. Wings and tail dusky, strongly edged with yellowish, the inner wing-quills with whitish. On the secondaries, this yellowish edging stops abruptly in advance of ends of coverts, leaving a pure blackish interval in advance of white tips of greater coverts; this, and similar tips of median coverts, form two white bars across wings; inner webs of quills and tail-feathers edged with white. Superciliary line and extreme forehead hoary-whitish. Crown black, enclosing a large flame-colored space, bordered with pure yellow. The black reaches across forehead; but behind, the yellow and flame-color reach the general olive of upper parts. Or, top of head may be described as a central bed of flame-color, bounded in front and on sides with clear yellow, this similarly bounded by black, this again in same manner by hoary-whitish. Smaller than *R. calendula;* overlying nasal plumes larger. Length 4.00; extent 6.50–7.00; wing 2.00–2.12; tail 1.67. ♀, adult, and young: Similar to adult ♂, but central field of crown entirely yellow, enclosed in black (no flame-color). N. Am. at large; another exquisite, abundant in woodland and shrubbery, breeding in various mountains of U. S. and from northern parts northward, wintering in most of the U. S., and also extending S. to Central America. Nest pensile or not, of moss, hair, feathers, etc., about 4.50 in diameter, on high or low bough of a tree, preferably evergreen; eggs 6–10, 0.50 × 0.40, white, fully speckled.

R. s. oliva'ceus? (Lat. *olivaceus,* olivaceous; *oliva,* an olive.) WESTERN GOLDEN-CRESTED KINGLET. Said to be of livelier coloration than the last. Pacific coast region of California and northward.

OBS. — *R. cuvieri,* AUD., Orn. Biogr., i, 1832, p. 288, pl. 55, and B. Am., ii, 1841, p. 163, pl. 131; NUTT., Man., i, 1832, p. 416; Schuylkill River, Pa., June 8, 1812, said to have two black stripes on each side of head, continues unknown; A. O. U. Hypothetical List No. 26. — *R. tricolor,* NUTT., Man., i, 1832, p. 420, is *R. satrapa;* so is his *R. cristatus,* which latter is the name of the European Gold-crest, not found in N. Am.

Subfamily POLIOPTILINÆ: Gnat-catchers.

A small group of one genus and about a dozen species, South American; peculiar to America. *Polioptila* has been associated with *Paridæ,* but differs decidedly and is apparently Sylviine. Some authors believe it to be Muscicapine, and would place it near the Old World genus *Stenostira* (see SHARPE, Cat. B. Brit. Mus., x, 1885, p. 440). Should this view be correct, it would add to North America the family *Muscicapidæ,* which has been supposed to be absent from the New World. Characters those of the single genus.

POLIOP'TILA. (Gr. πολιός, *polios,* hoary; πτίλον, *ptilon,* a feather; the primaries being edged with whitish.) GNAT-CATCHERS. Tarsi scutellate. Toes very short, the lateral only about half as long as tarsus; outer a little longer than inner. 1st primary spurious, about ½ as long as 2d. Wings rounded, not longer than the

FIG. 126. — Blue-gray Gnat-catcher, nat. size. (Ad nat. del. E. C.)

graduated tail, whose feathers widen toward their rounded ends. Bill shorter than head, straight, broad and depressed at base, rapidly narrowing to the very slender, distinctly notched and hooked end — thus Muscicapine in character. Rictus with well-developed bristles. Nostrils entirely exposed. Coloration without bright tints; bluish-ash, paler or white below; tail black and white. Delicate little woodland birds, peculiar to America, not over 5.00 long; migratory, insectivorous, very active and sprightly, with sharp squeaking notes.

Analysis of Species (adult males).
Forehead and line over eye black; outer tail-feather white . . . . *cœrulea*
Whole crown black; outer web of outer tail-feather white . . . . *plumbea*
Whole crown black; outer web of outer tail-feather only edged with white . . . . *californica*

P. cœru'lea. (Lat. *cœrulea*, cerulean, blue. Figs. 126, 127, *b*.) BLUE-GRAY GNAT-CATCHER. ♂, adult: Grayish-blue, bluer on crown, hoary on rump; forehead black, continuous with a black superciliary line. Edges of eyelids white, and above these a slight whitish stripe bordering the black exteriorly. Below white, with a faint plumbeous shade on breast. Wings dark brown; outer webs, especially of inner quills, edged with hoary, and inner webs of most bordered with white. Tail jet-black; outer feather entirely or mostly white, next one about half white, 3d one tipped with white. Bill and feet black. Length 4.50–5.00; extent 6.25–7.00; wing 2.00–2.20; tail about the same. ♀: Like ♂, but duller and more grayish above; head like back, without any black. Bill usually in part light-colored. Eastern U. S., N. to New York, Great Lakes, and S. New England, casual to Minn. and Me., W. to Col.; breeds in most of range, and winters on the S. border and southward to the Bahamas, Cuba, and Guatemala; abundant in woodland. Nest a model of bird-architecture, compact-walled and contracted at the brim, elegantly stuccoed with lichens, fixed to slender twigs at a varying height from 10 to 50 or 60 feet; eggs 4–5, about 0.60 × 0.45, whitish, fully speckled with reddish and umber-brown and lilac.

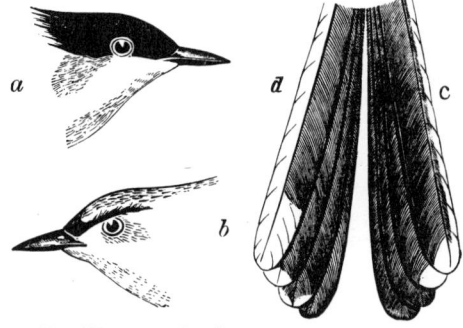

FIG. 127. — *a*, head of *Polioptila californica; b*, of *P. cœrulea; c*, tail of *P. californica; d*, of *P. plumbea;* all nat. size.

P. c. obscu'ra. (Lat. *obscurus*, obscure, dark.) OBSCURE GNAT-CATCHER. WESTERN BLUE-GRAY GNAT-CATCHER. Resembling the last: coloration less clear. S. W. U. S., in Arizona, Southern and Lower California, and Western Mexico. RIDGW., Pr. U. S. Nat. Mus., v, Mar. 1883, p. 535; COUES, Key, 4th ed., 1890, p. 897; A. O. U. List, 2d ed., 1895, p. 315, No. 751 *a*. The distinction of this form is so slight that it was ignored in the first three eds. of the Key, 1872–87, and in the A. O. U. List, 1st ed., 1886.

P. plum'bea. (Lat. *plumbeus*, plumbeous, lead-colored. Fig. 127, *d*.) BLACK-CAPPED GNAT-CATCHER (*adult*). PLUMBEOUS GNAT-CATCHER (*young*). ♂, adult: Upper parts like those of *P. cœrulea*, but duller and more grayish; whole top of head black, involving lores and auriculars; under parts white, with an ashy shade on the sides. Outer tail-feather with whole outer web and tip white (like the *second* feather of *P. cœrulea*); next two feathers tipped with white. Size of *P. cœrulea;* tarsi rather longer — about 0.70. Immature ♂: acquiring the black cap by degrees, beginning with a small black stripe on each side, over a white superciliary line, and gradually spreading. ♀: Like ♂; upper parts still duller, frequently with a decided brownish shade; no black on head; distinguished from ♀ *cœrulea* by less white on tail. Nest high in a tree, saddled on a limb, small, neat, compact, with contracted brim, composed of various downy substances and cobwebs; size outside about 2.50 in diam. × 1.75 deep, with a cavity of 1.75 × 1.25; eggs about 4, 0.58 × 0.45, bluish-white, speckled with reddish brown, umber, and lilac; laid in March and April. Valley of the Gila and Colorado, in Arizona and Southern and Lower California; also, valley of the Upper Rio

Grande, in New Mexico and Western Texas. As stated in the Key, 2d, 3d, and 4th eds., p. 262, *P. melanura* LAWR., Ann. Lyc. N. Y., vi, 1856, p. 168, and of other authors referring to the bird of the above given habitat, is the adult of *P. plumbea* BAIRD, Proc. Philada. Acad., 1854, p. 118; as is also *Culicivora atricapilla* of LAWR., Ann. Lyc. N. Y., v, 1851, p. 124, and of CASS., Ill., 1854, pl. 27 (not of SWAINSON).

P. califor'nica. (Lat., *californian*.) BLACK-TAILED GNAT-CATCHER. CALIFORNIA BLACK-CAPPED GNAT-CATCHER. ♂, adult: As compared with *P. plumbea*, upper parts decidedly plumbeous instead of bluish; throat, breast, and sides dull ashy instead of ashy-white; lower belly and crissum fulvous or even pale chestnut; light edging of tail-feathers confined to outer pair, with sometimes slight tipping of next pair (as in my Fig. 127, *c*.); lining of wings pearly-ash, not white; secondaries and tertials edged with light brown. No pure white anywhere; general aspect of under parts nearly as dark as those of a Cat-bird. Whole crown glossy black. Length 4.50; extent 6.10; wing 1.80–1.90; tail 1.90–2.20; tarsus 0.73; bill 0.50. ♀: Similar, but no black on crown; belly and crissum pale chestnut; outer webs of 2d pair of rectrices edged with white. Changes of plumage of young ♂ in reaching maturity like those of *P. plumbea*. This is mainly a Pacific coast form; *P. melanura* of authors referring to that region, but not of LAWR., 1856. It extends from Southern California into Lower California, where its range reaches that of *P. plumbea*. BREWSTER, Bull. Nutt. Orn. Club, Apr. 1881, p. 103; COUES, Key, 2d–3d eds., 1884–87, p. 262; RIDGW., Man., 1887, p. 570; A. O. U. Lists, 1886–95, No. 753.

Family CHAMÆIDÆ: Wren-tits.

Framed for a single species, much like a Titmouse in general appearance, but with tarsus not evidently scutellate in front; rounded wings much shorter than graduated tail; lores bristly; plumage extraordinarily soft and lax. With the general habits of Wrens, with which the species was formerly associated. The position and valuation of the group are still uncertain, probably to be determined upon anatomical characters. I have little doubt that *Chamæa* will yet be found referable to some other recognized family of birds, and suspect that it might be assigned to the Old World *Timeliidæ*, with at least as much propriety as some other American groups which have been relegated to that ill-assorted assemblage. In the A. O. U. Lists, 1886–95, *Chamæa* is referred to *Paridæ* as type of a subfamily *Chamæinæ*, which curiously combines such dissimilar forms as *Chamæa*, *Psaltriparus*, and *Auriparus* — "inadvertently," as one member of the Committee has remarked (Man. N. A. Birds, 1887, p. 558). When doctors disagree like this, it is useless to exchange one dubiosity for another, and safest to continue the treatment the unfortunate patient has survived for some years. I therefore retain the family *Chamæidæ* of the 2d–4th eds. of the Key, 1884–90.

CHAMÆ'A. (Gr. χαμαί, *chamai*, on the ground.) WREN-TITS. Form and general aspect combining features of wrens and titmice. Plumage extraordinarily lax, soft, and full. Coloration simple. Tarsal scutella obsolete, or faintly indicated, at least outside. Toes coherent at base for about $\frac{1}{2}$ the length of proximal joint of middle one. Soles widened and padded, much as in *Paridæ*. Primaries 10; 6th longest, 3d equal to longest secondaries, 1st about $\frac{2}{3}$ as long as longest; wing thus extremely rounded, and much shorter than tail (about $\frac{3}{5}$ as long). Tail very long, constituting more than $\frac{1}{2}$ the entire length of the bird, extremely graduated, with soft, narrow feathers, widening somewhat toward tips, rounded at end; lateral pair not $\frac{2}{3}$ as long as the middle. Bill much shorter than head, very deep at base, straight, stout, compressed-conical, not notched, with ridged and very convex culmen, but nearly straight commissure and gonys; nostrils naked, scaled, linear, gape strongly bristled. Frontal feathers reaching nasal fossæ, but no ruff concealing nostrils as in *Paridæ*.

C. fascia'ta. (Lat. *fasciata*, striped; *fascis*, a bundle of faggots.) GAMBEL'S WREN-TIT. Adult ♂ ♀: Dark brown with an olive shade; top of head clearer and somewhat streaky; wings and tail purer brown, obscurely fasciated with numerous cross-bars; below, dull cinnamon-brown, paler on belly, shaded with olive-brown on sides and crissum; throat and breast obscurely streaked with dusky; bill and feet brown; iris white. Length about 6.00; wing 2.25-2.50; tail 3.25-3.50, much graduated, lateral feathers 1.00 or more shorter than middle ones; bill 0.40; tarsus 0.90-1.00; middle toe and claw 0.75. First primary nearly 1.00 shorter than longest one. California coast region, N. to Humboldt Bay at least. A remarkable bird, resembling no other, common in shrubbery; nest in bushes, of twigs, grasses, and feathers, neither roofed over nor purse-like; eggs 3-5, 0.70 × 0.52, plain greenish-blue.

C. f. hen'shawi. (To H. W. Henshaw.) HENSHAW'S WREN-TIT. Much lighter and duller colored; above, grayish-ash, with slight olive shade (about the color of a *Lophophanes*); below, scarcely rufescent upon a soiled whitish ground, shaded on sides with color of back; bill and feet smaller. Interior of California, including W. slopes of the Sierra Nevadas, from the valley of the Sacramento River S. to Lower California.

Family PARIDÆ: Titmice, or Chickadees.

Ours are all small (under 7.00 long) birds, having 10 primaries, 1st much shorter than 2d; wings barely or not longer than tail; tail-feathers not stiff nor acuminate; tarsus scutellate, longer than middle toe; anterior toes much soldered at base; nostrils concealed by dense tufts; bill compressed, stout, straight, unnotched, and much shorter than head — characters that readily marked them off from all their allies, as Wrens, Creepers, etc. They are hard to distinguish, technically, from Jays; but all our Jays are much over 7.00 long.

Titmice are distributed over North America, but the crested species are rather southern, and all but one of them western. Most of them are hardy birds, enduring the rigors of winter without inconvenience, and consequently none are properly migratory. They are musical, after a fashion of their own, chirping a quaint ditty; are active, restless, and very heedless of man's presence; and eat everything. Some of the western species build astonishingly large pensile nests, like a bottle or purse with a hole in one side, as represented in Fig. 134; others live in knot-holes, and similar snuggeries that they usually dig out for themselves. They are very prolific, laying numerous eggs, and raising more than one brood a season; the young closely resemble the parents, and there are no obvious seasonal or sexual changes of plumage. All but one of our species are plainly clad; still they have a pleasing look, with their trim form and the tasteful colors of the head.

Subfamily PARINÆ: True Titmice.

Exclusive of certain aberrant forms, usually allowed to constitute a separate subfamily, and sometimes altogether removed from *Paridæ*. Titmice compose a natural and pretty well defined group, to which the foregoing diagnosis and remarks are particularly applicable, and agree in the following characters: Bill very short and stout, straight, compressed-conoid in shape, not notched nor with decurved tip, its under as well as upper outline convex. Rictus without true bristles, but base of bill covered with antrorse tufts of bristly feathers, entirely concealing nostrils. Feet stout; tarsi distinctly scutellate, longer than middle toe; toes rather short, the anterior soldered together at base for most of the length of basal joint of middle one. Hind toe with an enlarged pad beneath, forming, with consolidated bases of anterior toes, a broad firm sole. Primaries 10; 1st very short or spurious, scarcely or not ½ as long as 2d; wing as a whole rounded, scarcely or not longer than tail, which latter is rounded or graduated, composed of 12 narrow soft feathers, with rounded or somewhat truncated tips. Plumage long, soft, and loose, without bright colors or well-marked changes according to sex, age, or season (excepting *Auriparus*). There 'may be about 75 good species of *Parinæ* as thus restricted, most of them falling in the genus *Parus*, or its immediate neighborhood. With few exceptions they are birds of the Northern Hemisphere, abounding in Europe, Asia, and North America. The larger proportion of the genera and species inhabit the Old World. All those of the New World occur within our limits.

Analysis of Genera.

Crested.
    Wings and tail rounded, of about equal lengths. No red or yellow . . . . *Lophophanes*
Not crested.
    Wings and tail rounded, of about equal lengths. No red or yellow . . . . *Parus*
    Wings rounded, shorter than the graduated tail. No red or yellow . . . . *Psaltriparus*
    Wings pointed, longer than the even tail. Head yellow; bend of wing red . . . . *Auriparus*

LOPHOPH'ANES. (Gr. λόφος, *lophos*, a crest; φαίνω, *phaino*, I appear.) CRESTED TITMICE. Head crested. . Wings and tail rounded, of about equal lengths. Bill conoid-compressed, with upper and under outlines both convex. No yellow on head or red on wing. Plumage lax, much the same in both sexes at all ages and seasons. Average size of the species at a maximum for *Parinæ*. Nests excavated in trees; eggs spotted (except in *L. wollweberi*).

    OBS. This genus is reduced to a subgenus of *Parus* by the A. O. U. But it is quite a good genus, as genera go nowadays, and I need not disturb the position it has held in the Key since 1872 — in fact, among most American writers since 1858. The user of the Key has only to read " *P.*" for *L.* in the following paragraphs if he wishes to be perfectly orthodox on the subject of *Parus* (*Lophophanes*).

Analysis of Species.

Frontlet black; sides washed with rusty. Eastern . . . . *bicolor*
Crest like rest of upper parts; no rusty on sides. Southwestern . . . . *inornatus*
Crest entirely black; rusty on sides. Texan . . . . *atricristatus*
Head with several black stripes; no rusty on sides. Southwestern . . . . *wollweberi*

L. bi'color. (Lat. *bis*, twice; *color*, color. Fig. 128.) TUFTED TITMOUSE. PETO. ♂ ♀, adult: Entire upper parts ashy; back usually with a slight olivaceous shade; wings and tail rather purer and darker plumbeous, the latter sometimes showing obsolete transverse bars. Sides of head and entire under parts dull whitish, washed with chestnut-brown on sides. A black frontlet at base of crest. Bill plumbeous-blackish; feet plumbeous. Length 6.00–6.50; extent 9.75–10.75; wing and tail 3.00–3.25; bill 0.40; tarsus 0.80; middle toe and claw 0.75. ♀ smaller than ♂. Young: Crest less developed; little if any trace of black frontlet; sides scarcely washed with rusty. Eastern U. S., rather southerly; scarcely N. to New England; resident, abundant in woodland and shrubbery. It is a hardy, sprightly bird, fond of reiterating its loud ringing " peto, peto." Nest in holes; eggs 6 or 8, 0.75 × 0.56, white, dotted with reddish-brown and lilac.

FIG. 128. — Tufted Titmouse, nat. size. (Ad. nat. del. E. C.)

L. b. texen'sis. (Lat., of Texas.) TEXAN TUFTED TITMOUSE. Paler than the last, with chestnut instead of black frontlet at base of crest; this chestnut corresponding in tint to that which suffuses sides of body. Tarsus 0.85; bill 0.45. Southeastern Texas. *Parus bicolor texensis* SENN., Auk, Jan. 1887, p. 29; RIDGW., Man., 1887, p. 561; A. O. U. List, 2d ed., 1895, p. 306, No. 731 *a*. *Lophophanes bicolor texensis* COUES, Key, 3d ed., 1887, p. 866; 4th ed., 1890, p. 897.

L. inorna'tus. (Lat. *in*, as signifying negation, and *ornatus*, adorned; *orno*, I ornament.) PLAIN TITMOUSE. TOPPY. ♂ ♀, adult: Entire upper parts dull leaden-gray, with a slight olive shade; wings and tail rather purer and darker. Below, dull ashy-whitish, without any rusty wash on sides. No black on head; extreme forehead and sides of head obscurely speckled with whitish. No decided markings anywhere. In size rather less than *L. bicolor;* length usually under 6.00; wing and tail under 3.00. Young quite like adults, which closely resemble the young of *L. bicolor;* but in the latter there are traces at least of the reddish of the sides or black of the frontlet, or both; the general coloration is purer, with more distinction between upper and under parts, and the size is rather greater. The speckled appearance of the sides of head and lores of *L. inornatus* is peculiar. Abundant, resident. The typical form is from the coast region of California and Oregon; a rather larger, stouter-billed form, lighter leaden-gray with scarcely any olive shade, from the Great Basin, is

L. i. gris'eus. (Lat. *griseus*, grisly. GRAY TITMOUSE. Said to differ from ordinary *inornatus* in rather larger size and decidedly grayer color. Wing 2.90; tail 2.55. Middle Province of the U. S.; Colorado, Utah, Nevada, New Mexico, Arizona, and California E. of the Sierra Nevadas. *L. i. griseus* RIDGW., Pr. U. S. Nat. Mus., v, Sept. 1882, p. 344; COUES, Key, 2d ed., 1884, p. 264, in text; 3d and 4th eds., 1887 and 1890, p. 866; *P. i. griseus* A. O. U. Lists, No. 733 *a*.

L. i. cinera'ceus. (Lat. *cineraceus*, somewhat cinereous or ashy in color.) ASHY TITMOUSE. Another local race, described as grayer above and paler below than *L. i. griseus*, with smaller bill, black in color. Lower California. *L. i. cineraceus* RIDGW., Pr. U. S. Nat. Mus., vi, Oct. 1883, p. 154; COUES, Key, 3d and 4th eds., p. 866; *P. i. cineraceus* A. O. U. List, No. 733 *b*.

L. atricrista'tus. (Lat. *ater*, black, *cristatus*, crested; *crista*, a crest.) BLACK-CRESTED TITMOUSE. ♂ ♀, adult: Plumbeous, with a shade of olive; wings and tail rather darker and purer, edged with color of back, or a more hoary shade of the same. Beneath, dull ashy-whitish, especially on breast; abdomen whiter; sides chestnut-brown as in *L. bicolor*. Extreme forehead and lores whitish; entire crest glossy black. Bill blackish-plumbeous; feet plumbeous. Small: Length about 5.00; wing and tail 2.75. Valley of the Lower Rio Grande, S. E. Texas, and N. E. Mexico. Nest in natural cavities of trees, usually including cast snake-skins among its materials; eggs 0.75 × 0.58, white, spotted with reddish-brown in fine dots over the general surface, boldly blotched at large end, but not distinguishable from those of *L. bicolor*.

L. a. castan'eifrons. (Lat. *castaneus*, of chestnut color; *frons*, forehead.) CHESTNUT-FRONTED TITMOUSE. Resembling the last: upper parts plumbeous, faintly tinged with olive; under parts pale ashy, washed with chestnut on sides, with faint trace of the same on breast and crissum. Crest thin, 1.00 long, dark brown and ashy instead of black, and with a chestnut frontlet; lores white; bill black; feet dark plumbeous. Size of *L. bicolor*, the bill even larger. Wing 3.12; tail 2.95; tarsus 0.77; bill 0.42. Bee County, Texas. *P. a. castaneifrons* SENN., Auk, Jan. 1887, p. 28; RIDGW., Man., 1887, p. 561. *L. a. castaneifrons* COUES, Key, 3d and 4th eds., 1887 and 1890, p. 866; not admitted in the A. O. U. List; a dubious form, whose characters suggest hybridism between *L. bicolor* and *L. atricristatus*.

L. wollweb'eri. (To one Wollweber. Fig. 129.) BRIDLED TITMOUSE. ♂ ♀, adult: Upper parts olivaceous-ash; wings and tail darker, edged with color of back, or even a brighter tint, sometimes nearly as yellowish as in *Regulus*. Under parts sordid ashy-white. Crest black, with a central field like the back. Whole throat black, as in species of *Parus*. A black line runs behind

FIG. 129. — Bridled Titmouse, nat. size. (Mex. B. Survey.)

eye and curves down over auriculars, distinguished from black of crest and throat by white of side of head and white superciliary stripe; a half-collar of black on nape, descending on sides of neck, there separated from the black crescent of auriculars by a white crescent, which latter is continuous with the white of superciliary line; considerable whitish speckling in black of forehead and lores. Bill blackish-plumbeous; feet plumbeous. Smallest: Length 5.00 or less; wing 2.70; tail 2.40-2.65; bill 0.33; tarsus 0.60-0.70. Young: Chin narrowly or imperfectly black, and some of the above described head-markings obscure or incomplete. The singularly variegated markings of the head of this species at once distinguish it. W. Texas, southern New Mexico, and Arizona, S. in Mexico to Orizaba. Abundant, going in troops, in woods and shrubbery. Eggs 5-7, 0.65 × 0.50, white, unmarked.

PA'RUS. (Lat. *parus*, a titmouse.) TYPICAL TITMICE. CHICKADEES. Head not crested. Wings and tail rounded, of approximately equal lengths. Bill typically parine (see foregoing characters). No bright colors (in any North American species). Head in most species with black. Plumage lax and dull, without decided changes with age, sex, or season. Size medium in the family. Nest excavated. Eggs spotted.

Analysis of Species and Subspecies.

Species definitely black-capped and black-throated.
  A white superciliary stripe. Western . . . . *gambeli* (formerly called *montanus*)
  No white superciliary stripe. Eastern and Western.
    Tail not shorter than wing; feathers of both with much hoary-whitish edging.
      Larger; tail at maximum length; coloration most hoary. Rocky Mts. . . . . . . *a. septentrionalis*
      Smaller; tail moderate; coloration less hoary. Eastern . . . . *atricapillus*
      Size of *atricapillus*; coloration darker. Pacific . . . . *a. occidentalis*
    Tail shorter than wings; whitish edgings of wings and tail obsolete.
      Rather smaller than *atricapillus*. South Atlantic States (and Texas) . . *carolinensis* (and *c. agilis*)
      Rather smaller than *atricapillus*; coloration very dark. Mexican border . . . . . . *meridionalis*
Species brown-capped, or with crown quite like back, and blackish throat.
  Cap hair-brown; back little different.
    White confined to side of head. Eastern and Arctic . . . . *hudsonicus*, etc.
    White spreading over sides of neck. Arctic, Alaska, and Siberia . . . . *cinctus alascensis*
  Cap dark wood-brown; back chestnut.
    Back and sides rich chestnut alike. Pacific, northerly . . . . *rufescens*
    Back chestnut, but sides only washed with rusty. Pacific, southerly . . . . *r. neglectus*

FIG. 130.—Black-capped Chickadee, reduced. (Ad. nat. del. E. C.)

P. atricapil'lus. (Lat. *ater*, black; *capillus*, hair. Fig. 130.) BLACK-CAPPED TITMOUSE. COMMON CHICKADEE. ♀, adult: Crown and nape, with chin and throat, black, separated by white sides of the head. Upper parts brownish-ash, with slight olive tinge, and a rusty wash on rump. Under parts more or less purely white or whitish, shaded on sides with a brownish or rusty wash. Wings and tail like upper parts, the feathers moderately edged with hoary-white. Average dimensions: Length 5.25; extent 8.00; wing and tail, each, 2.50; tarsus 0.70. Extremes: Length 4.75-5.50; extent 7.50-8.50; wing and tail 2.35-2.65; tarsus 0.65-0.75. Eastern N. Am., from the Middle States northward, very abundant, well-known by its familiar habits and peculiar notes. Nest in holes of trees, stumps, or fences, natural or excavated by the bird, made of grasses, mosses, hair, fur, feathers, etc.; eggs 6-8, 0.58 × 0.47, white, fully sprinkled with reddish-brown dots and spots.

P. a. septentriona'lis. (Lat. *septentrionalis*, northern; *septentriones*, the constellation of seven stars, the dipper.) LONG-TAILED CHICKADEE. Similar to *P. atricapillus*; averaging larger, and especially longer-tailed; tail rather exceeding wing in length. Coloration clear and pure;

wings and tail very strongly edged, especially on secondaries and outer tail-feathers, with hoary-white, which usually passes around their tips. Cap pure black and very extensive on nape; black of throat reaching breast; sides of head and neck snowy-white. Bill and feet dark plumbeous. Average dimensions about the maxima of *P. atricapillus*: Length 5.25–5.50; extent 8.50; wing 2.50–2.75; tail 2.60–2.80, sometimes 3.00. This style reaches its extreme development in the region of the Upper Missouri and Rocky Mts., there apparently to the exclusion of *P. atricapillus* proper.

P. a. occidenta'lis. (Lat. *occidentalis*, western; *occido*, I fall; *i. e.*, where the sun sets.) WESTERN CHICKADEE. OREGON CHICKADEE. Similar to *P. atricapillus*; of same average size; presenting the opposite extreme from *P. septentrionalis* in minimum edging of wing- and tail-feathers with hoary, heavy brownish wash of sides, and general dark sordid coloration. Pacific coast region, California to Alaska.

P. carolinen'sis. (Lat. of Carolina.) CAROLINA CHICKADEE. Averaging smaller than *P. atricapillus*, with relatively as well as absolutely shorter tail, which is rather shorter than wings; wings and tail very little edged with whitish. Average dimensions about the minima of *P. atricapillus*. Length about 4.50; wing 2.50; tail 2.25. S. Atlantic and Gulf States; N. to New Jersey, Illinois, and Missouri. Nesting like *P. atricapillus*; eggs similar, rather smaller.

P. c. a'gilis. (Lat. *agilis*, agile, active.) PLUMBEOUS CHICKADEE. Differs from *P. carolinensis* proper by more plumbeous shade of upper parts, whiter under parts, which lack any decided buffy wash, and somewhat longer tail in comparison with other dimensions; wing and tail of about the same length — 2.40. Eastern and Central Texas. SENNETT, Auk, Jan. 1888, p. 46; COUES, Key, 4th ed., 1890, p. 898; A. O. U. List, 2d ed., 1895, p. 308, No. 736 a.

P. meridiona'lis. (Lat. *meridionalis*, southern.) MEXICAN CHICKADEE. Differs decidedly from *P. atricapillus* in having the under parts merely a paler shade of the ashy of the upper, instead of white, without any brownish wash on sides; wing-coverts and tail lacking any hoary edging, though the wing-quills have a slight grayish-white edging. Thus quite like *P. gambeli* in color, but no white superciliary stripe. Length 4.80–5.20; extent 8.00–8.70; wing 2.67–2.90; tail 2.40–2.67. Mexico, from Orizaba to Arizona.

FIG. 131. — Mountain Chickadee, nat. size. (Ad. nat. del. E. C.)

P. gam'beli. (To Wm. Gambel, its original describer as *P. montanus*. Figs. 131, 132.) MOUNTAIN CHICKADEE. GAMBEL'S CHICKADEE. Upper parts ashy-gray, with scarcely a shade, and only on rump, of the ochraceous seen in most other species; under parts similarly grayish-white, without rusty tinge; middle of belly nearly white, the rest more heavily shaded. Wings and tail with comparatively little whitish edging — tail with no more than that of *P. carolinensis*. Sides of head and neck white; top of head, and throat, black. A conspicuous white superciliary stripe in the black cap, usually meeting its fellow across forehead. Length about 5.00; extent 8.30; wing 2.50–2.75; tail rather less; bill 0.38, slender; tarsus 0.66. U. S., from Eastern foot-hills of the Rocky Mts. to the Pacific,

FIG. 132. — Mountain Chickadee.

British Columbia to Lower California, chiefly in alpine regions. Eggs 0.62 × 0.57, spotted as usual in this genus, or not. *P. montanus* GAMBEL, of all former eds. of the Key, as of most writers; *P. gambeli* RIDGW., Man., 1887, p. 562; A. O. U. Lists, No. 738.

P. rufes'cens. (Lat. *rufescens*, rufous, reddish.) CHESTNUT-BACKED TITMOUSE. ♂ ♀, adult: Crown and nape dark wood-brown, becoming sooty along sides, separated from the sooty-black of throat by a large white area extending back on sides of neck. Entire back and sides of body rich dark chestnut, contrasting strongly with the brown of head. Breast and central line of under parts, with lining of wings, whitish. Wing- and tail-coverts more or less washed with rusty-brown. Quills and tail-feathers scarcely or slightly edged with whitish. Bill black; feet dark; iris brown. Young with throat brown, like crown, instead of sooty. Length 4.75; extent, 7.50; wing 2.30; tail about 2.00. A strongly marked species, with chestnut back and sides contrasting with dark brown cap and sooty throat. Pacific coast region of Oregon, Washington, British Columbia, and Southern Alaska, very abundant in coniferous woods and shrubbery; resident. Nest in hollow of a tree, 10-40 feet up, of moss, hair, feathers, etc.; eggs 6-8, 0.61 × 0.42, minutely speckled with reddish, rarely immaculate; May, June.

P. r. neglec'tus? (Lat. *neglectus*, neglected, *i. e.*, not chosen; *nec*, not, and *lego*, I gather, choose.) Quite similar: crown, throat, and back the same, but sides not extensively chestnut, being simply washed with rusty-brown. Coast region of California.

P. hudson'icus. (Lat. *hudsonicus*, of Hudson's Bay; after Henry Hudson, the navigator. Fig. 133.) HUDSONIAN TITMOUSE. ♂ ♀, adult: Crown, nape, and upper parts generally clear hair-brown, or ashy-brown with a slight olive shade; coloration quite the same on back and crown, and not separated by any whitish nuchal interval. Throat quite black, in restricted area, not extending backward on sides of neck; separated from the brown crown by silky white on side of head, this white not reaching back of auriculars to sides of nape. Sides, flanks, and under tail-coverts washed with dull chestnut or rusty-brown; other under parts whitish. Quills and tail-feathers lead-color, as in other Titmice, scarcely or slightly edged

FIG. 133. — Hudsonian Titmouse.

with whitish. Little or no concealed white on rump. Bill black; feet dark. Size of *P. atricapillus*, or rather less. Wing 2.50-2.60; tail rather less; tarsus 0.60. N. New England and Great Lake region of the U. S., and British America generally; common in coniferous woods.

P. h. ston'eyi. (To Lieut. Geo. M. Stoney, U. S. N.) STONEY'S TITMOUSE. KOWAK CHICKADEE. Like *P. hudsonicus*; grayer above; sides of neck purer ashy-gray; sides paler rusty, and throat clear slaty-black instead of sooty-black. Size of *P. hudsonicus*; wing 2.55-2.75, averaging 2.62; tail 2.62; tarsus 0.62-0.70. Kowak River, N. W. Alaska. *P. stoneyi*, RIDGW., Man., 1887, p. 591; *P. hudsonicus stoneyi*, A. O. U. List, 1st Suppl., 1889, p. 17; Key, 4th ed., 1890, p. 897; A. O. U. List, 2d ed., 1895, p. 309, No. 740 *a*.

P. h. columbia'nus. (Lat., Columbian.) RHOADS' TITMOUSE. COLUMBIAN CHICKADEE. Like *P. hudsonicus*, but presenting the opposite variation from that of *P. h. stoneyi* in general darker coloration: throat jet black; lores and frontal area sooty-black; crown and neck slaty, with little or no brownish tinge. Size slightly larger than that of the typical form; wing 2.70; tail 2.64; tarsus 0.67. Type locality, Field, British Columbia; range extending in the Rocky Mts. from Liard River S. to Montana. RHOADS, Auk, Jan. 1893, p. 23; A. O. U. List, 2d ed., 1895, p. 310, No. 740 b.

P. h. evu′ra. (Gr. εὖ, well; οὐρα, tail.) WELL-TAILED TITMOUSE. ALASKAN CHICKADEE. Like *P. hudsonicus* in color; larger, with especially longer tail; tail nearly 3.00. The variation in this case corresponds to that of *P. septentrionalis* as compared with *P. atricapillus*. COUES, Key, 2d ed., 1884, p. 267; not recognized in A. O. U. Lists.

P. cinc′tus alascen′sis. (Lat. *cinctus*, girdled, from *cingo*, I bind about; *alascensis*, of Alaska.) SIBERIAN TITMOUSE. In general, similar to *P. hudsonicus*, but quite distinct. ♂ ♀, adult: Throat sooty-blackish; crown and nape dark hair-brown, bordered laterally with dusky, appreciably different in tone from the brighter brownish of back, from which also separated to some extent by whitish of cervix. Sides of head and neck pure white, in a large area widening behind, this white of opposite sides nearly meeting across cervix. Back ashy overlaid with flaxen-brown; rump light brown with much concealed white. Under parts whitish centrally from the black throat, but heavily washed on sides, flanks, and crissum, sometimes quite across belly, with light brownish. Wings and tail slate-color, as usual in the genus, with much whitish edging, especially on secondaries. Bill plumbeous-blackish; feet plumbeous. Length 5.30–5.60; wing 2.60–2.80; tail rather more; tarsus 0.65. Eggs 0.65 × 0.50. A large stylish Chickadee, lately ascertained to inhabit Arctic America, especially Alaska. It is very near the E. Siberian form of the Lapp titmouse (true *P. cinctus* of BODDAERT, *P. sibiricus* GM., or *P. lapponicus* LUNDAHL), from which *P. obtectus* CAB., J. f. O., 1851, p. 237, is by some considered specifically distinct. Compare *Pœcila submicrorhynchus* of BREHM, Naumannia, 1856, p. 369. Our Alaskan bird is *P. cinctus* of former eds. of the Key; *P. c. obtectus* RIDGW., Pr. U. S. Nat. Mus., viii, 1885, p. 354; Man., 1887, p. 564; A. O. U. Lists, 1886 and 1895, No. 739; name now changed, after *Pœcila cincta alascensis* PRAZAK, Orn. Jahrb., Mar.–Apr. 1895, p. 92, to *Parus cinctus alascensis*, A. O. U. Suppl. List, Jan. 1897, p. 132, No. 739.

PSALTRIP′ARUS. (Gr. ψάλτρια, Lat. *psaltria*, a lutist; *parus*, a tit.) BUSH-TITS. Dwarfs among pygmies! 3.75–4.25 long; wing 2.00 or less, tail 2.00 or more. Ashy or olive-gray, paler or whitish below; neither crown nor throat black; no bright colors. Head not crested; wings rounded, shorter than the long narrow graduated tail. Nest large, woven, pensile, with lateral entrance (fig. 134). Eggs 6–9, white, unmarked. The 4 species are Western; they are notable for their diminutive size, scarcely equalling a *Polioptila* in bulk.

Analysis of Species and Subspecies.

Crown brown, unlike back; no black on side of head. Pacific coast region . . . . *minimus*
    N. California and northward . . . . *minimus* proper
    N. and S. California . . . . *m. californicus*
    Lower California . . . . *m. grindæ*
Crown like back; no black on side of head. S. Rocky Mt. region . . . . *plumbeus*
Crown ash, unlike back; a black stripe on side of head.
    ♂ with the black stripe narrow and occipital only . . . . *santaritæ*
    ♂ with the black stripe broad and long . . . . *lloydi*

P. min′imus. (Lat. *minimus*, least, smallest.) LEAST BUSH-TIT. ♂ ♀, adult: Dull lead-color, frequently with a brownish or olivaceous shade; top of head abruptly darker — clove-brown or hair-brown. Below sordid whitish, or brownish-white. Wings and tail dusky, with slight hoary edgings. Bill and feet black. Length 4.00 or less; wing scarcely or not 2.00; tail 2.00 or more; bill 0.25; tarsus 0.60. Young birds do not differ materially. There is considerable variation in the precise shade of the body, but the brown cap always differs in color from the rest of the upper parts. Eggs 0.55 × 0.40. The typical dark Northern form inhabits the coast region of N. California, Oregon, and Washington, shading insensibly into the following:

P. m. califor'nicus? (Lat., Californian.) CALIFORNIAN BUSH-TIT. Lighter colored than the last, on an average. This form inhabits the greater part of the coast region of California, and is intermediate between the last and the next variety. Pr. Biol. Soc. Wash., ii, Apr. 1884, p. 89; A. O. U. Lists, 1886 and 1895, No. 743 a; not admitted in any former ed. of the Key.

FIG. 134.—Least Bush-tit and nest, about ⅔ nat. size. (Ad. nat. del. H. W. Elliott.)

P. m. grin'dæ. (To Don Francisco C. Grinda.) GRINDA'S BUSH-TIT. Adult: Cap pale brown, lightening on sides of head into white on chin and throat; other under parts exactly as in *P. minimus.* Upper parts light plumbeous-gray, well contrasted with brown of nape. Bill and feet black. Wing 2.00; tail 2.30, graduated 0.50; bill 0.20. A further local variation, combining to some extent the characters of *minimus* and *plumbeus.* Lower California. *P. grindæ,* BELDING, Pr. U. S. Nat. Mus., vi, Oct. 1883, p. 155; *P. m. grindæ,* RIDGW., *ibid.,* viii, 1885, p. 354; COUES, Key, 3d ed., 1887, p. 867; A. O. U. List, 2d ed., 1895, No. 743 b.

P. plum'beus. (Lat. *plumbeus,* lead-colored.) PLUMBEOUS BUSH-TIT. ♂♀, adult: Clear plumbeous, with little or no olive or brownish shade; top of head not different from back, its sides pale brownish. Under parts as in *P. minimus,* but clearer. Tail longer than wings. Eyes yellow or dark brown. Length about 4.25; wing 1.88–2.1 tail 2.25–2.50; bill 0.25; tarsus 0.60. Closely related to *P. minimus,* but readily distinguishable. Total length greater, owing to elongation of tail, which sometimes exceeds wings by 0.50. General coloration clearer and purer; crown not different in color from back, but cheeks brownish in obvious contrast. Rocky Mt. region, from Wyoming and Oregon southward; common in Arizona.

P. santari'tæ. (Lat., of the Santa Rita (mountains). *Santa Rita* is a Spanish phrase, meaning "holy creek," *rita* being a diminutive form of *rio,* river.) SANTA RITA BUSH-TIT. ♂, similar to the last; sides of head paler, and marked with a lateral occipital blackish line over the auriculars, as in the ♀ of *P. lloydi*. Said to be smaller than *P. plumbeus*, but the ascribed dimensions do not bear out the statement. Santa Rita Mts. of Southern Arizona. RIDGW., Pr. U. S. Nat. Mus., x, Sept. 1888, p. 697; COUES, Key, 4th ed., 1890, p. 898; A. O. U. List, 2d ed., 1895, p. 312, No. 744.1.

P. lloyd'i. (To Wm. Lloyd.) LLOYD'S BLACK-EARED BUSH-TIT. ♂, adult: Sides of head broadly black with greenish lustre, the bands meeting narrowly across chin, and nearly meeting on nape. Crown and nape clear ash. Back hair-brown. Wings and tail fuscous, with narrow pale ashy edgings of the feathers; outer webs and tips of outer tail-feathers, and inner webs of many wing-feathers, whitish. Below, white, pure on throat and sides of neck, thence passing through lavender-gray to rusty-brownish on flanks and crissum. Bill and feet black; iris brown. ♀, adult: Black of head reduced to a streak along each side of the occiput, leaving sides of head light brown. Young ♂ quite similar to the adult, having glossy black on head before it is fully feathered; but the black marks do not at first meet on chin. Length about 4.00; wing 1.90; tail 2.25; bill 0.25, compressed, with very convex culmen and nearly straight under outline; tarsus 0.60; middle toe and claw 0.45. Northern Mexico, from Sonora and Chihuahua into Arizona, New Mexico, and W. Texas. *P. melanotis* of earlier eds. of the Key, ascertained to be different from true *melanotis* of Mexico; *P. lloydi* SENNETT, Auk, Jan. 1888, p. 43; Key, 4th ed., 1890, p. 898; A. O. U. List, 2d ed., 1895, p. 312, No. 745. Nest pensile, pear-shaped, with large end downward, about 6 inches long, made of mosses, lichens, and plant-stems, lined with feathers; eggs white, unmarked, 0.58 × 0.42. The bird has been found breeding in pineries, at an altitude of over 6,000 feet; a nest was affixed to twigs of a cedar tree, 7 feet from the ground.

AURIP'ARUS. (Lat. *auri*, of gold, and *parus*, a tit; from the yellow head.) GOLD-TITS. Head not crested. Wings pointed; 2d quill little shorter than 3d; 1st spurious. Tail little rounded, decidedly shorter than wings. Bill not typically Parine — extremely acute, with straight or slightly concave under outline, and barely convex culmen, thus resembling that of a *Helminthophila*; longer and slenderer than usual in *Parinæ*; nostrils scarcely concealed by the imperfect ruff. Tarsus relatively shorter than in preceding genera. Bright colors on head (yellow) and wing (red). Plumage comparatively compact; sexes alike, but young very different from adults. Size very small. General form Sylvicoline. Nest globular, woven. Eggs spotted. One species.

A. flav'iceps. (Lat. *flaviceps*, yellow-head.) GOLD-TIT. VERDIN. YELLOW-HEADED TITMOUSE. ♂♀, adult: Upper parts ashy; under parts whitish; wings and tail dusky, with hoary edging. Whole head rich yellow. Lesser wing-coverts chestnut-red. Bill dark plumbeous; feet plumbeous. Length 4.00-4.25; wing 2.00; tail 2.25. Young without red on wing or yellow on head; thus obscure objects, known, however, by their generic characters. Adults vary in having the yellow heightened to orange, or dull and greenish; the red sometimes hæmatitic; and shade of the ashy clear and pure, or dull and brownish. Valleys of the Rio Grande and Colorado, N. to S. Nevada and S. W. Utah, S. extensively in Mexico; resident in most of its range; abundant in chaparral, building in bushes a great globular or purse-like nest of twigs, lined with down and feathers; eggs 4-6, 0.60 × 0.45, pale bluish speckled with brown.

A. f. lamproceph'alus. (Gr. λαμπρός, *lampros*, bright; κεφαλή, *kephale*, head.) BRIGHT-HEADED TITMOUSE. Like *flaviceps;* head clearer yellow; wings shorter; tail much shorter. Lower California. Included under *flaviceps* in former eds. of the Key. *A. f. ornatus* BRYANT, Zoe, i, 1890, p. 149 (nec *Conirostrum ornatum* LAWR.). *A. f. lamprocephalus* OBERHOLSER, Auk, Oct. 1897, p. 391; A. O. U. Suppl. List, Auk, Jan. 1899, p. 126.

Family SITTIDÆ: Nuthatches.

Bill subcylindrical, tapering, compressed, slender, acute, nearly or about as long as head; culmen and commissure about straight; gonys long, convex, ascending (giving a sort of recurved look to a really straight bill). Nostrils rounded, concealed by bristly tufts. Wings long, pointed, with 10 primaries, 1st very short or spurious; tail much shorter than wings, broad, soft, nearly even; tarsus shorter than middle toe and claw, scutellate in front; toes all long, with large, much curved, compressed claws; 1st toe and claw about equal to 3d; 2d and 4th toes very unequal in length. Plumage compact; body flattened; tongue horny, acute, barbed. Nuthatches are amongst the most nimble and adroit of creepers; they scramble about and hang in every conceivable attitude, head downward as often as otherwise. This is done, too, without any help from the tail — the whole tarsus being often applied to the support; and there is in their movements something so suggestive of mice that they are sometimes called *Treemice* — a term which contrasts very well with *Titmice*, of the neighboring family *Paridæ*. They are chiefly insectivorous, but feed also on hard fruits; and gained their English name from their habit of sticking nuts and seeds in cracks in bark, and hammering away with the bill till they break the shell. They are very active and restless little birds, quite sociable, often going in troops which keep up a continual noise; lay 4–6 white, spotted eggs, in hollows of trees. The family is a small one, of less than 30 species, among them a single remarkable Madagascar form (*Hypositta*), and a genus peculiar to Australia (*Sittella*); but is chiefly represented by the genus *Sitta*, with some 15 species of Europe, Asia, and North America. The genera *Xenicus* and *Acanthisitta* of New Zealand, long supposed to be *Sittidæ*, are now known to belong elsewhere. The A. O. U. reduces *Sittidæ* to a subfamily of *Paridæ* — in my judgment a needless if not unwise step I am not prepared to take. The change is of no practical consequence.

FIG. 135. — European Nuthatch, *Sitta cæsia* (resembling *S. pusilla*), nearly nat. size. (From Brehm.)

SIT'TA. (Lat. *sitta*, Gr. σίττα, name of a bird. Fig. 135.) TYPICAL NUTHATCHES. TREE-MICE. Characters practically those given under head of the family.

Analysis of Species and Subspecies.

White below; crissum washed with rusty-brown; cap glossy black, without stripes.
    Bill stouter, 0.18–0.20 deep at base. Inner secondaries boldly variegated with black. Eastern.
        Eastern U.S. and British Provinces . . . . *carolinensis*
        Florida to South Carolina coastwise . . . . *c. atkinsi*
    Bill slenderer, 0.12–0.16 deep at base. Inner secondaries scarcely variegated with blackish. Western.
        . . . . *c. aculeata*
Rusty-brown below; cap glossy black with white stripes, or color of back . . . . *canadensis*
Rusty-brown or brownish-white below; cap brown, unlike back, without stripes.
    Crown clear hair-brown; a white spot on nape; middle tail-feathers plain. Southeastern . . . *pusilla*
    Crown dull-brownish, with darker border; middle tail-feathers with black.
        Southwestern; little or no white on nape . . . . *pygmæa*
        Lower California; more white on nape . . . . *p. leuconucha*

SITTIDÆ: NUTHATCHES.

S. carolinen'sis. (Lat., of Carolina. Figs. 136, 137.) CAROLINA NUTHATCH. WHITE-BELLIED NUTHATCH. TREEMOUSE. DEVIL DOWNHEAD. QUANK. ♂, adult: Upper parts, central tail-feathers, and much edging of wings, clear ashy-blue; whole crown, nape, and back of neck glossy black. Under parts, including sides of neck and head to above eyes, dull white, more or less marked on flanks and crissum with rusty-brown. Wings and their coverts blackish, much edged as already said, and with an oblique bar of white on outer webs of primaries toward their ends; concealed bases of primaries white; under wing-coverts mostly blackish; bold bluish and black variegation of inner secondaries. Tail, excepting the two middle feathers, black, each feather marked with white in increasing amount; outer web of lateral feather mostly white. Bill blackish-plumbeous, pale at base below. Feet dark brown. Iris brown. Length 5.50–6.00; extent 10.50–11.00; wing 3.50; tail 1.75; bill about 0.66 long, 0.18–0.20 deep at base. ♀: Similar; black of head imperfect, mixed or overlaid with color of back, or altogether restricted to nape. Eastern U. S. (except S. Atlantic coast region) and British Provinces, resident, abundant in woodland, where its curious *quank, quank, quank* may often be heard as the nimble bird hops up and down the tree-trunks. Nest in holes, often excavated by the birds with infinite labor, lined with fur, feathers, grasses, etc.; eggs 5–8, 0.80 × 0.60, white, profusely speckled with reddish and lilac.

FIG. 136. — Carolina Nuthatch, nat. size. (Ad. nat. del. E. C.)

S. c. at'kinsi. (To John W. Atkins, of Key West, Fla.) FLORIDA WHITE-BREASTED NUTHATCH. Said to be smaller than the last, to the extent of 0.15–0.20 in average length of wing; bill said to be longer, 0.69–0.78; wings and tail said to have less white. ♀ with crown pronounced black, not easily distinguished from the ♂. Florida, and coastwise to S. Carolina. SCOTT, Auk, Apr. 1890, p. 118; A. O. U. List, 2d ed. 1895, No. 727 *b* (de minimis curat lex!).

FIG. 137. — White-breasted Nuthatch.

S. c. aculea'ta. (Lat. *aculeata*, sharpened; referring to the slender bill.) SLENDER-BILLED NUTHATCH. Like *carolinensis;* bill slenderer, 0.12–0.16 at base. Inner secondaries scarcely or not variegated with blackish, and general tone of coloration duller. Woodland of Middle and Western provinces of the U. S., and S. into Mexico; common, replacing *carolinensis*. The two forms are separated for the most part by the treeless plains where neither occurs, and are well marked. Specimens which I lately shot in the Black Hills of S. Dakota, were however somewhat equivocal. Nesting as in other species; eggs 5–9, ordinarily 7–8, March and later.

S. canaden'sis. (Lat., of Canada; an Iroquois word. Fig. 138.) RED-BELLIED NUTHATCH. CANADA NUTHATCH. ♂, adult: Upper parts leaden-blue (brighter than in *carolinensis*); central tail-feathers the same; wings fuscous, with slight ashy edgings and concealed white bases of primaries. Entire under parts rusty-brown, very variable in shade, from rich fulvous to brownish-white, usually palest on throat, deepest on sides and crissum; tail-feathers, except middle pair, black, the lateral marked with white. Whole top and side of head and neck glossy black, that of the side appearing as a broad bar through eye from bill to side of neck, cut off from that of crown by a long white superciliary stripe, which meets its fellow across forehead. Bill dark plumbeous, paler below; feet plumbeous-brown. Length 4.50–4.75; extent 8.00–8.50; wing 2.60; tail 1.50; bill 0.50. ♀: Crown like back; lateral stripe on head merely blackish. The under parts average paler than those of the ♂, but there is no constancy about this. Young birds resemble the ♀. Temperate N. Am., range on the whole more northerly than that of *carolinensis*, breeding from the northern tier of States northward, and further south only in mountainous regions; winters S. through the S. States; common in woodland; habits like those of the Carolina Nuthatch; eggs similar, smaller, 0.65×0.54 down to 0.60×0.45.

Fig. 138. — Canada Nuthatch, nat. size. (Ad. nat. del. E. C.)

S. pusil′la. (Lat. *pusilla*, puerile, petty. Fig. 139.) BROWN-HEADED NUTHATCH. ♂ ♀, adult: No black cap or white stripe on head. Upper parts dull ashy-blue; under parts sordid or muddy whitish. Cap clear hair-brown. A decided spot of white on middle of nape, in the brown cap, which on sides of head includes eyes, and is bordered with dusky. Middle tail-feathers like back, without black, and with little or no white. Length scarcely 4.00; extent about 8.00; wing 2.50; tail 1.25; tarsus 0.60; bill about 0.50. S. Atlantic and Gulf States, N. to Virginia, Ohio, and Missouri. Habits of the other species: eggs 0.60×0.50, very heavily speckled with dark reddish-brown.

Fig. 139. — Brown-headed Nuthatch, nat. size. (Ad. nat. del. E. C.)

S. pygmæ′a. (Gr. πυγμή, *pugme*, the fist; Lat. *pygmæus*, a pygmy, fistling, or tom-thumb.) PYGMY NUTHATCH. ♂ ♀, adult: Upper parts ashy-blue; wings with slight if any markings (as in *canadensis*), though some outer primaries may be narrowly edged with white. Whole crown, nape, and sides of head to below eyes, olive-brown, the lateral borders of this patch blackish; an obsolete whitish patch on nape. Central tail-feathers like back, but with a long white spot, and their outer webs black at base; other tail-feathers blackish, with white marks, often also tipped with color of back. Entire under parts ranging from muddy-white to smoky-brown or rich rusty, nearly or quite as intense as in *canadensis;* flanks and crissum shaded with a dull wash of color of back. Bill and feet dark plumbeous, the former paler at base below. Iris black. Size of the last. Young: Differs much as ♀ *canadensis* does from ♂, having top of head like back. U. S. from Rocky Mts. to the Pacific, abundant, chiefly in pine woods; N. to British Columbia, S. into Mexico. Eggs 6–7, 0.62×0.50; white, profusely speckled with reddish.

S. p. leuconu′cha. (Gr. λευκός, *leucos*, white, and Lat. *nucha*, nape.) WHITE-NAPED NUTHATCH. Like the last; nuchal spot more conspicuous; under parts whiter; head grayer; bill larger. San Pedro Mts., Lower California. ANTHONY, Pr. Cala. Acad., 2d ser., ii, Oct. 1889, p. 77; COUES, Key, 3d ed., 1890, p. 898; A. O. U. List, 2d ed., 1895, p. 305, No. 730 *a*.

Family CERTHIIDÆ: Creepers.

A very small, well-marked group, of about 12 species, and 4 or 5 genera, which fall in 2 sections, commonly called subfamilies; one of these, *Tichodrominæ*, is represented by the well-known European Wall Creeper, *Tichodroma muraria*, and several (chiefly Australian) species of the genus *Climacteris;* while the genus *Certhia*, with 5 or 6 species or subspecies, and certain allied genera (all but one Old World) constitute the

Subfamily CERTHIINÆ: Typical Creepers.

Our species may be known on sight, among North American *Oscines*, by its *rigid, acuminate* tail-feathers, like a Woodpecker's. Besides: bill about equal in length to head, extremely slender, sharp, and decurved; nostrils exposed; no rictal bristles; tarsus scutellate, shorter than 3d toe and claw, which is connate for the whole of the 1st joint with both 2d and 4th toe; lateral toes of unequal lengths; 1st toe shorter than its claw; claws all much curved and very sharp; wing 10-primaried, 1st primary very short, not ½ the 2d, which is less than 3d; point of wing formed by 3d, 4th, and 5th; tail rounded, equal to or longer than wing, of 12 stout, elastic, curved, acuminate feathers. Restless, active, little forest birds that make a living by picking bugs out of cracks in bark. In scrambling about they use the tail as Woodpeckers do, and never hang head downward like Nuthatches. Lay numerous white, speckled eggs; are not regularly migratory; have slight seasonal or sexual changes of plumage; are chiefly insectivorous, and not noted for musical ability.

FIG. 140. — Common Brown Creeper, *Certhia familiaris*, nearly nat. size. (From Brehm.)

CER′THIA. (Lat. *certhius*, a creeper. Fig. 141.) Characters as above. The stock-form of this genus varies according to locality. European varieties sometimes recognized are *C. costæ* and *C. britannica.* The N. Am. bird, when separated from the European, has been called *C. rufa* (BARTRAM, 1791), *fusca* (BARTON, 1799), and *americana* (BP. 1838), for Eastern specimens; *C. montana* for those from the Rocky Mt. region; *C. occidentalis* for those from the Pacific coast region; and *C. mexicana* (or *alticola*) and *C. albescens* for the Mexican forms. The differences between any of these forms are slight; but if they are to be recognized by name, all the American ones must be specifically separated from those of Europe; for we adopt the fact of intergradation, not any degree of difference, as our touchstone of subspecificality, and it is a physical impossibility for any of our creepers to intergrade now with any European ones. Therefore our birds should stand as *C. americana, C. a. montana, C. a. occidentalis,* and *C. a. albescens.* But I forbear to make the change, in deference to the A. O. U. committee over which I had the honor of presiding in our attempts to confer immutability upon nomenclatural permutability.

FIG. 141. — Head, foot, and tail-feather of *Certhia*, nat. size. (Ad. nat. del. E. C.)

C. familia'ris americana. (Lat. *familiaris*, from *familia*, family; domestic, home-like: Fig. 140.) BROWN CREEPER. ♂ ♀: Upper parts dark brown, changing to rusty-brown on the rump, everywhere streaked with ashy-white. An obscure whitish superciliary stripe. Under parts dull whitish, sometimes tinged with rusty on flanks and crissum. Wing-coverts and quills tipped with white; inner secondaries with white shaft-lines, which, with the tips, contrast with the blackish of their outer webs. Wings also twice crossed with white or tawny-white; anterior bar broad and occupying both webs of feathers, other only on outer webs near their ends. Tail grayish-brown, darker along shaft and at end of feathers, sometimes showing obsolete transverse bars. Bill blackish above, mostly flesh-colored or yellowish below; feet brown; iris dark brown. Length of ♂ 5.25–5.75; extent 7.50–8.00; wing 2.50, more or less; tail usually a little longer than the wing, sometimes not so, 2.50 to nearly 3.00; tarsus about 0.60; bill 0.65–0.75; ♀ averaging smaller than ♂. Eggs 5–9, 0.60 × 0.45; white speckled with reddish-brown, especially about the large end. Eastern N. Am., in woodland; migratory to some extent, as it breeds chiefly from northerly or mountainous parts of the U. S. northward, and winters chiefly further S.; abundant, generally seen winding spirally up the trunks and larger branches of trees. *C. fusca* BARTON, Frag. N. H. Penna., 1799, p. 11, nec GM., 1788. *C. familiaris fusca* COUES, B. N. W., 1874, p. 230; A. O. U. Suppl. List, Auk, Jan. 1899, p. 126. *C. familiaris* of 2d–4th eds. of Key, 1884–90, p. 273. *C. familiaris americana*, A. O. U. Lists, 1886–95, No. 726.

C. f. monta'na. (Lat. *montanus*, of mountains.) ROCKY MOUNTAIN CREEPER. Grayer above than the last, with more distinctly contrasted tawny rump, and longer bill, wings, and tail. Rocky Mt. region of the U. S., including Alaska. Not recognized in any former ed. of the Key. RIDGW., Proc. U. S. Nat. Mus., v, July, 1882, p. 114; Man., 1887, p. 558; not recognized in A. O. U. List, 1886; but *ibid.*, 2d ed., 1895, No. 726 b.

C. f. occidenta'lis. (Lat. *occidentalis*, of the occident or setting sun, western.) CALIFORNIAN CREEPER. The darker form, from the Pacific coast region, from southern California to Alaska. Not recognized in any former ed. of the Key, nor in A. O. U. List, 1886; RIDGW., Proc. U. S. Nat. Mus., v, 1882, p. 114; Man., 1887, p. 558; A. O. U. List, 2d ed., 1895, No. 726 c.

C. f. albes'cens. (Lat. *albescens*, somewhat white, whitish.) MEXICAN CREEPER. Differs in lacking light tips of the primary coverts, and general richer coloration, the brown more rusty; rump bright chestnut; under parts grayish. Mexico, to S. W. border of the U. S., in the mountains of Arizona. This is *C. f. mexicana* of previous eds. of the Key; but the name *mexicana* cannot stand in this genus, as there is a prior *Certhia mexicana* (GM.). See MILLER, Auk, Apr. 1895, p. 186, where the Mexican creeper is named *C. f. alticola*, the same being No. 726 a of the A. O. U. List, 2d ed., 1895; and OBERHOLSER, Auk, Oct. 1896, p. 315, where the same is divided into two races, the northern one, which occurs over our border, being regarded as the *C. m. albescens* of BERLEPSCH, Auk, Oct. 1888, p. 450, renamed as *C. f. albescens*. Suppl. List, Auk, Jan. 1897, p. 132.

Family TROGLODYTIDÆ: Wrens; Thrashers, etc.

Embracing a number of forms assembled in considerable variety, and difficult to define with precision. Closely related to the last three families; known from these by non-acuminate tail-feathers and exposed nostrils. Distinguished from typical Turdine and Sylviine birds by the not strictly spurious character of the 1st primary, short as it may be and generally is; as a rule, by the shortness of the rounded wings in comparison with the length of the usually rounded or graduated tail; and especially, by the distinctly scutellate instead of booted tarsi. (Compare diagnoses already given of *Turdidæ, Sylviidæ, Cinclidæ;* and observe that the dubious family *Chamæidæ* is wren-like in most respects.) In former editions of the Key, the *Miminæ* or so-called Mocking "Thrushes" were brought under *Turdidæ*, as a subfamily of the latter, with the express statement, however, that they were "an aberrant group, related to the *Troglodytidæ*" (2d ed., 1884, p. 242), "departing from the prime characteristic of the family in having the tarsi scutellate in front" (*ibid.*, p. 248). I now avail myself of the first opportunity, incident to the resetting of the type for the present edition, to remove the *Miminæ* from *Turdidæ* to *Troglodytidæ* — the position assigned them in the A. O. U. Lists, 1886-95. This is a happy atavism — a reversion to the stand taken by BAIRD in 1858, when he combined the Mockers with the Wrens under the family name of "*Liotrichidæ*," after the example set by CABANIS in the Museum Heineanum of 1850. In so far as American forms are concerned, the *Troglodytidæ* here given are precisely Baird's *Liotrichidæ* under another name: and they correspond exactly to what are recognized by SHARPE (Cat. B. Brit. Mus., vi and vii, 1881 and 1883) as the subfamilies *Troglodytinæ* and *Miminæ* of the so-called "*Timeliidæ* or Babbling Thrushes" — that vast assemblage of some 1,100 species of chiefly Old World birds which makes a sort of ornithological waste-basket for want of any satisfactory classification. To discipline that unruly mob is not our present purpose; we have only to recognize by name a family group to contain our own Mockers and Wrens; and as *Troglodytes* VIEILL., 1807, antedates both *Leiothrix* SWAINS., 1831, and *Timalia* HORSFORD, 1820 (or *Timelia* SUND., 1872), we use *Troglodytidæ* instead of "*Liotrichidæ*" or "*Timeliidæ*" without prejudice to any question of the relationships of American Wrens and Mockers to the various Old World birds concerned in the case, and with the assurance that in any event *Troglodytidæ* is a straitly orthodox name for the family with whose members we have here to do.

In 1858 BAIRD divided his *Liotrichidæ* = *Troglodytidæ* into four subfamilies — *Miminæ*, *Campylorhynchinæ*, *Troglodytinæ*, and *Chamæinæ*. There is much to be said in favor of this arrangement, especially regarding the position thus assigned to the refractory genus *Chamæa*. In 1884 I had no difficulty in distinguishing *Campylorhynchinæ* from *Troglodytinæ* (see Key, 2d ed., p. 274), upon consideration of the North American genera alone. But other American forms obliterate the dividing line between them, so that they must be combined in one, to be called *Troglodytinæ*. Upon this understanding, our *Troglodytidæ* now consist of two subfamilies, which may be easily recognized, as follows:

Analysis of Subfamilies.

Size large, and general aspect thrush-like. Length 8.00 or more, wing 3.50 or more. Rictal bristles evident. Tarsal scutellation moderate, in some cases obsolete. Inner toe free to its base from middle toe. Represented by Mockingbirds, Catbirds, Thrashers . . . . *Miminæ*

Size small, and general aspect wren-like. Length 8.00 or less, wing 3.50 or less (usually *much* less). Rictal bristles not evident. Tarsal scutellation moderate, in some cases excessive. Inner toe extensively coherent with middle toe. Represented by all species of Wrens . . . . *Troglodytinæ*

Subfamily MIMINÆ: Mockingbirds; Thrashers.

Birds of maximum size among *Troglodytidæ*, simulating *Turdidæ* in some respects; distinguished from *Troglodytinæ* by greater size, rictal bristles, different nostrils, and more deeply cleft toes. Tarsi scutellate in front (the scutella sometimes fusing, however, as in the Catbird). Wings short and rounded, about equal to tail only in *Oroscoptes*; 1st primary short, but not spurious; 2d primary shorter than 6th. Tail large and rounded or much graduated, usually decidedly longer than wings. Tarsus about equal to middle toe and claw; feet stout, in adaptation to somewhat terrestrial life. Bill various in form, usually longer or at least more curved than in Thrushes; in *Harporhynchus* attaining extraordinary length and curvature. As a group the *Miminæ* are rather southern, hardly passing beyond the U. S.; few species reaching even the Middle States, and the maximum development being in Central and South America. They are peculiar to America, where they are represented by *Oroscoptes*, *Mimus*, *Galeoscoptes*, *Harporhynchus*, and 5 or 6 related genera, with upward of 40 recorded species. About one-half of these fall in *Mimus* alone; nearly all the species of *Harporhynchus* occur in the U. S. In their general habits they resemble Wrens, habitually residing in shrubbery near the ground, relying for concealment as much upon the nature of their resorts as upon their own activity and vigilance. They are all melodious, and some, like the immortal mockingbird, are as famous for their powers of mimicry as for the brilliant execution of their proper songs. In compensation for this great gift of music, perhaps that they may not grow too proud, they are plainly clad, grays and browns being the prevailing colors. The nest is generally built with little art, in a bush, and the eggs, 2–6 in number, are blue or green, plain or speckled.

Analysis of Genera.

Smallest: bill shortest; wing about equal to tail. Adults speckled below . . . . *Oroscoptes*
Medium: bill moderate; wing a little shorter than tail. Adults plain below.
  Ashy above, white below, with much white on wings and tail . . . . *Mimus*
  Blackish-ash above, no white anywhere; crown black . . . . *Galeoscoptes*
Largest: bill immoderate; wing much shorter than tail. Plain or spotted below . . . . . . *Harporhynchus*

OROSCOP'TES. (Gr. ὄρος, ογος, a mountain; σκώπτης, *scoptes*, a mimic.) MOUNTAIN MOCKERS. Wing and tail of about equal length; former more pointed than in other genera of *Miminæ*; 1st quill not half as long as 2d, which is between 6th and 7th; 3d, 4th, and 5th about equal to one another, and forming the point of the wing. Tail nearly even, its feathers but slightly graduated. Tarsus longer than middle toe and claw, anteriorly distinctly scutellate. Bill much shorter than head, not curved, with obsolete notch near end. Rictal bristles well developed, the longest reaching beyond nostrils. *O. montanus* is the only known species.

O. monta'nus. (Lat. *montanus*, of a mountain. Fig. 142.) MOUNTAIN MOCKINGBIRD. SAGE THRASHER. ♂ ♀, in summer: Above, grayish- or brownish-ash, the feathers with obsoletely darker centres. Below, whitish, more or less tinged with pale buffy-brown, everywhere marked with triangular dusky spots, largest and most crowded across breast, small and sparse, sometimes wanting, on throat, lower belly, and crissum. Wings fuscous, with much whitish edging on all the quills, and two white bands formed by tips of greater and median coverts. Tail like wings; outer feather edged and broadly tipped, and all the rest, excepting usually the middle pair, tipped with white in decreasing amount. Bill and feet black or blackish, the former often with pale base. Length about 8.00; wing and tail, each, about 4.00; tarsus 1.12; bill 0.75. Young: Dull brownish above, conspicuously streaked with dusky; the markings below streaky and diffuse. Plains to the Pacific, U. S.; also Texas and Lower California; an interesting species, resembling an undersized young Mockingbird, abundant in the sage-brush of the W. Nest on ground or in low bushes; eggs usually 4, 1.00 × 0.72, light greenish-blue, heavily marked with brown and neutral tint.

FIG. 142. — Sage Thrasher.

MI'MUS. (Lat. *mimus*, a mimic.) MOCKINGBIRDS. Bill much shorter than head, scarcely curved as a whole, but with gently-curved commissure, notched near the end. Rictal vibrissæ well developed. Tail rather longer than wing, rounded, the lateral feathers being considerably

graduated; wing rounded. Tarsus longer than middle toe and claw. Above ashy-brown, below white; lateral tail-feathers and bases of primaries white. (Tarsal scutella always distinct.)
M. polyglot'tus. (Lat. *polyglottus*, many-tongued; Gr. πολύς, *polus*, many; γλῶττα, *glotta*, tongue. Fig. 143.) MOCKINGBIRD. ♂, adult: Upper parts ashy-gray; lower parts soiled white. Wings blackish-brown; primaries, except the 1st, marked with a large white space at base, usually restricted on outer quills to half or less of these feathers, but occupying nearly all of inner quills. The shorter white spaces show as a conspicuous spot when the wing is closed, the longer inner ones being hidden by the secondaries. Wing-coverts also tipped and sometimes edged with white; and there may be much edging or tipping, or both, of the quills themselves. Outer tail-feather white; next two white, except on outer web; next usually white toward end; the rest sometimes tipped with white. Bill and feet black, the former often pale at base below; soles dull yellowish.

FIG. 143.—Mockingbird, about ⅔ nat. size. (After Wilson.)

Length about 10.00 (9.50-11.00); extent about 14.00 (13.00-15.00); wing 4.00-4.50; tail 4.50-5.00; bill 0.75; tarsus 1.25. ♀, adult: Similar, but colors less clear and pure; above rather brownish- than grayish-ash, below sometimes quite brownish-white, at least on breast. Tail and wings with less white than as above described. But the gradation in these features is by imperceptible degrees, so that there is no infallible color-mark of sex. In general, the clearer and purer are the colors, and the more white there is on the wings and tail, the more likely is the bird to be a ♂ and prove a good singer. ♀ also smaller than ♂ on an average, generally under and rarely over 10.00; extent usually less than 14.00; wing little if any over 4.00; tail about 4.50. Young: Above decidedly brown, and below speckled with dusky. U. S. from Atlantic to Pacific, southerly; rarely N. to New England (Maine, Am. Nat., v, 1871, p. 121, Auk, 1897, p. 224), and not common N. of 38°, though known to reach 42°; thronging the groves of the S. Atlantic and Gulf States. Nest in bushes and low trees, bulky and inartistic, of twigs, grasses, leaves, etc.; eggs 4-6, averaging 1.00 × 0.75, bluish-green, heavily speckled and freckled with several brownish shades. Two or three broods are generally reared each season, which in the South extends from March to August. When taken from the nest, the "prince of musicians" becomes a contented captive, and has been known to live many years in confinement. Naturally an accomplished songster, he proves an apt scholar, susceptible of improvement by education to an astonishing degree; but there is a great difference with individual birds in this respect.

GALEOSCOP'TES. (Gr. γαλέη or γαλῆ, *galee* or *gale*, an animal of the weasel or marten kind known to the ancients, commonly later translated "cat," and σκώπτης, *skoptes*, a mocker.) CATBIRDS. Characters of *Mimus* proper, of which given as a subgenus in former eds. of the Key, and best distinguished by color: Blackish-ash, scarcely paler below, no white anywhere, crown black, crissum reddish. (Tarsal scutella sometimes obsolete.) (*Lucar* BARTRAM, Trav., 1791, p. 291 *bis*: see COUES, Pr. Phila. Acad., 1875, p. 349, and Auk, Jan. 1897, p. 97.)

G. carolinen'sis. (Of Carolina: *Carolus*, Charles IX., of France. Figs. 37, 144, 145.)
CATBIRD. ♂ ♀: Slate-gray, paler and more grayish-plumbeous below; crown of head, tail, bill, and feet black. Quills of wing blackish, edged with the body-color. Under tail-coverts rich dark chestnut or mahogany-color. Length 8.50–9.00; extent 11.00 or more; wing 3.50–3.75; tail 4.00; bill 0.66; tarsus 1.00–1.10. Young: More sooty above, with little or no distinction of a black cap, and comparatively paler below, where the color has a soiled brownish cast. Crissum dull rufous. (Specimens in which the black cap does not come snug to the bill, leaving the forehead gray, are *grisifrons* of MAYN., B. E. N. Am., pt. 40, p. 710: see Auk, Jan. 1897, p. 133.) U. S. and adjoining British Provinces, chiefly Eastern. W. to the Rocky Mts., even to Washington; migratory, but resident in the Southern States, and breeds throughout its range; nest of sticks, leaves, bark, etc., in bushes; eggs 3–6, oftenest 4–5, 0.95 × 0.70, deep greenish blue, not spotted; they resemble Robins' eggs, but are smaller and more deeply tinted. An abundant and familiar inhabitant of our groves and briery tracts, remarkable for its harsh cry, like the mewing of a cat (whence its name), but also possessed, like all its tribe, of eminent vocal ability.

FIG. 144. — Catbird.

HARPORHYN'CHUS. (Gr. ἅρπη, *harpe*, a sickle; ῥύγχος, *hrygchos*, beak; *i. e.* bow-billed.) THRASHERS. Bill of indeterminate size and shape; in one extreme straight and shorter than head; in the other exceeding the head in length and bent like a bow (see figs. 146, 152.) Feet large and strong, indicating terrestrial habits; tarsus strongly scutellate anteriorly, equalling or slightly exceeding in length the middle toe with its claw. Wings and tail rounded; latter decidedly longer than former. Rictus with well developed bristles. Viewing only extreme shapes of bill, as in *H. rufus* and *H. crissalis*, it would not seem consistent with the minute subdivisions which now obtain in ornithology to place all the species in one genus; but the gradation of form is so gentle that it seems impossible to dismember the group without violence, though two subgenera may be conveniently recognized. Most of our species represent the subgenus *Methriopterus*, which contains the common Thrasher, *Harporhynchus* proper being restricted to the three species which have the most arcuate bills. Arcu-

FIG. 145 — Catbird, nat. size. (Ad. nat. del. E C.)

ation of the bill proceeds *pari passu* with its elongation, the shortest bills being the straightest, and conversely; very young birds of the most bow-billed species are straight-billed. There is also a curious correlation of color with shape of bill; the short-billed species being the most richly colored and heavily spotted, while the bow-billed ones are very plain, sometimes with no spots whatever on the under parts. Our 11 forms of the genus are with one exception Southwestern, focusing in Arizona.

TROGLODYTIDÆ—MIMINÆ: THRASHERS.

Analysis of Species and Subspecies.

Bill not longer than head (0.87–1.12), little or not curved. Breast spotted. (*Subgenus* METHRIOPTERUS.)
   Bill 1.00, quite straight. Above rich rusty-red; below whitish, heavily spotted and streaked with dark brown. Eastern . . . . *rufus*
   Bill 1.12, slightly curved. Above dark reddish-brown, below whitish, heavily spotted and streaked with blackish. Texas . . . . *longirostris sennetti*
   Bill 1.12, curved. Above ashy-gray, below whitish, breast with round spots of the color of the back. Mexican border and Arizona . . . . *curvirostris* and *c. palmeri*
   Bill 0.87, scarcely curved. Above grayish-brown, below brownish-white, breast alone with arrow-heads of the color of the back. Arizona . . . . *bendirei*
   Bill 1.12, curved. Above ashy-gray, below whitish, with profuse distinct blackish-brown spots. Lower California . . . . *cinereus* and *c. mearnsi*
Bill longer than head (1.50), arcuate. Breast not spotted. (HARPORHYNCHUS *proper*.)
   Dark oily olive-brown, below paler, belly and crissum rufescent. Coast of California . . . . . *redivivus*
   Pale ash, paler still below, lower belly and crissum brownish-yellow. Arizona and Lower California. . . . . *lecontei* and *l. arenicola*
   Brownish-ash, paler below, crissum chestnut in marked contrast. Arizona, New Mexico, and California . . . . *crissalis*

(*Subgenus* METHRIOPTERUS.)

H. ru′fus. (Lat. *rufus*, rufous, reddish. Figs. 146, 147.) THRASHER. GROUND THRUSH. BROWN THRUSH. RED THRUSH. FERRUGINOUS MOCKINGBIRD. SANDY MOCKINGBIRD. FRENCH MOCKINGBIRD. MAVIS. ♂ ♀, adult: Upper parts uniform rich rust-red, with a bronzy lustre. Concealed portions of quills fuscous; greater and median wing-coverts blackish near ends, then conspicuously tipped with white; bastard quills like the coverts. Tail like back, the lateral feathers with paler ends. Under parts white, more or less strongly tinged, especially on breast, flanks, and crissum, with tawny or pale cinnamon-brown; breast and sides marked with a profusion of well-defined spots of dark brown, oval in front, becoming more linear posteriorly. Throat immaculate, bordered with a necklace of spots; middle of belly and under tail-coverts likewise unspotted. Bill quite straight, black, with yellow base of the lower mandible; feet pale; iris yellow or orange.

FIG. 146. — Thrasher, nat. size. (Ad. nat. del. E. C.)

Young sufficiently similar to be unmistakable. Length about 11.00; extent 12.50–14.00; wing 3.75–4.25; tail 5.00 or more; bill 1.00; tarsus 1.25. Eastern U. S. chiefly, but N. to adjoining British Provinces and W. to the Rocky Mts.; migratory, but breeds throughout its range, and winters in the Southern States. A delightful songster, abundant in thickets and shrubbery. Nest in bushes (sometimes on ground), bulky and rude, of sticks, leaves, bark, roots, etc.; eggs 3–5, sometimes 6, 1.05 × 0.80, whitish or greenish, profusely speckled with brown.

H. longiros′tris sen′netti. (Lat. *longus*, long, and *rostris*, from *rostrum*, beak; *i. e.*, long-billed. To George B. Sennett.) TEXAS THRASHER. SENNETT'S THRASHER. Similar to *H. rufus;* upper parts dark reddish-brown, instead of rich foxy-red; under parts white, with little if any tawny tinge, the spots large, very numerous, and blackish instead of brown; ends of rectrices scarcely or not lighter than the rest of these feathers; bill almost entirely dark-colored. Besides these points of coloration, there is a decided difference in shape of bill. In

H. rufus, the bill is quite straight, and only about 1.00; the gonys is straight, and makes an angle with the slightly concave lower outline of the mandibular rami. In *H. longirostris sennetti*, the bill is over 1.00, and somewhat curved; the outline of the gonys is a little concave,

FIG. 147. — Brown Thrasher.

making with the ramus one continuous curve from base to tip of bill. Size of *H. rufus.* Eggs 1.05 × 0.75. Lower Rio Grande valley, from Corpus Christi and Laredo, Tex., southward in Tamaulipas and Nuevo Leon. *H. longirostris* (LAFR.) of BD., B. N. A. 1858, p. 352; *H. rufus longirostris,* COUES, Key, 1872, p. 72, 1884–87, p. 251; *H. longirostris sennetti* RIDGW., Proc. U. S. Nat. Mus., x, Aug. 1888, p. 506; COUES, Key, 4th ed., 1890, p. 897.

H. curviros'tris. (Lat. *curvus,* curved, and *rostris,* bow-billed.) CURVE-BILLED THRASHER. ♂ ♀ : Above, uniform ashy-gray (exactly the color of a Mockingbird); wings and tail darker and purer brown. Below, dull whitish, tinged with ochraceous, especially on flanks and crissum, marked with rounded spots of the color of the back, most numerous and blended on the breast. Throat quite white, immaculate, without maxillary stripes; lower belly and crissum mostly free from spots. No decided markings on side of head. Ends of greater and median wing-coverts white, forming two decided cross-bars; tail-feathers distinctly tipped with white. Bill black, curved, stout; feet dark brown. Length of ♂ about 11.00; wing 4.25–4.50; tail 4.50–5.00; bill 1.12; tarsus 1.25; middle toe and claw 1.33. ♀ averaging rather smaller. Mexico, reaching the U. S. border of Texas and New Mexico.

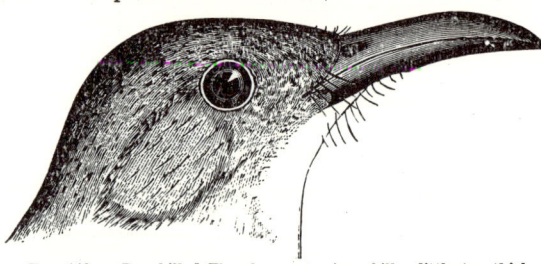

FIG. 148. — Bow-billed Thrasher, nat. size; bill a little too thick. (Ad. nat. del. E. C.)

H. c. pal'meri. (To Edw. Palmer. Fig. 148.) BOW BILLED THRASHER. Above, grayish-brown, nearly uniform; wing-coverts and quills with slight whitish edging; edge of wing itself white; tail-feathers with slight whitish tips. Below, a paler shade of the color of the upper parts; throat quite whitish; crissum slightly rufescent; breast and belly with obscure dark gray spots on the grayish-white ground; no obvious maxillary streaks, but vague speckling on cheeks. Bill black; feet blackish-brown. Length 10.75; bill 1.12; wing 4.25; tail 5.00; tarsus 1.25; middle toe and claw 1.30. ♀ smaller; wing 3.75; tail 4.50; tarsus 1.20; middle toe and claw 1.12; bill barely 1.00. Although the differences from the typical form are not easy to express, they are readily appreciable on comparison of specimens. The upper parts are quite similar; but the under parts, instead of being whitish, with decided spotting of the color of the back, are grayish, tinged with rusty, especially behind, and the spotting is nebulous. The white on the ends of the wing-coverts and tail-feathers is reduced to a minimum or entirely suppressed. The bill is slenderer and apparently more curved. Arizona and Sonora, common, in desert regions. Nest in cactus, mesquite and other bushes; eggs usually 3, 1.10 × 0.80, pale greenish-blue profusely dotted with reddish-brown.

H. bendi'rei. (To Capt. Chas. Bendire, U. S. A. Fig. 149.) ARIZONA THRASHER. BENDIRE'S THRASHER. ♂ ♀: Bill shorter than head, comparatively stout at base, very acute at tip; culmen quite convex; gonys just appreciably concave. Tarsus a little longer than middle toe and claw. 3d and 4th primaries about equal and longest, 5th and 6th successively slightly shorter, 2d equal to 7th, 1st equal to penultimate secondary in the closed wing. Entire upper parts, including upper surfaces of wings and tail, uniform dull pale grayish-brown, with narrow, faintly-rusty edges of wing-coverts and inner quills, and equally obscure whitish tipping of tail-feathers. No maxillary nor auricular streaks; no markings about head except slight speckling on cheeks.

FIG. 149. — Arizona Thrasher, nat. size. (Ad. nat. del. E. C.)

Under parts brownish-white, palest (nearly white) on belly and throat, more decidedly rusty-brownish on sides, flanks, and crissum, the breast alone marked with numerous small arrow-head spots of the color of back. Bill light-colored at base below. ♂: Length about 9.25; wing 4.00; tail 4.25; bill 0.87; along gape 1.12; tarsus 1.25; middle toe and claw 1.12. ♀ rather smaller; wing, 3.75, etc. Young birds are quite rusty or tawny on the wings and rump, and at all ages the species is a plain dull one. Arizona and Sonora, less common than *palmeri*, with which it is associated; has been found also W. to Agua Caliente, Cal., and N. to Colorado Springs, Col.; also breeds commonly about Rouse Junction, Col., in May and June (see Osprey, Sept. 1897, p. 7). Nest in bushes; eggs 2-3, rarely 4, about 1.00 × 0.73, elliptical rather than oval, whitish, spotted and blotched with reddish-brown.

FIG. 150. — St. Lucas Thrasher, nat. size. (Ad. nat. del. E. C.)

H. ciner'eus. (Lat. *cinereus*, ashy; *cinis*, *cineris*, ashes. Fig. 150.) ST. LUCAS THRASHER. ♂ ♀, adult: Upper parts uniform ashy-brown; wings and tail similar, but rather purer and darker brown; wings crossed with two white bars formed by the tips of the coverts; tail tipped with white. Below, dull white, often tinged with rusty, especially behind, and thickly marked with small, sharp, triangular spots of dark brown or blackish. These spots are all perfectly distinct, covering the lower parts excepting the throat, lower belly, and crissum; becoming smaller anteriorly, they run up each side of the throat in a maxillary series bounding the immaculate area. Sides of head finely speckled, and auriculars streaked; bill black, lightening at base below, little longer than that of *H. rufus*, though decidedly curved. Length of ♂ about 10.00; wing 4.00; tail 4.50; bill 1.12; tarsus 1.25; middle toe and claw 1.25. ♀ averaging rather smaller. Young: Upper parts strongly tinged with rusty-brown, this color also edging the wings and tipping the tail. The resemblance of this species to the Mountain Mockingbird (*Oroscoptes montanus*) is striking. It is distinguished from any others of the U. S. by the sharpness of the spotting underneath, which equals that of *H. rufus* itself, the small and strictly triangular character of the spots, together with the grayish-brown of the upper parts, and inferior dimensions. Lower California, from Cape St. Lucas N. to about lat. 30°, common. Nest a slight shallow structure of twigs in cactus and other bushes; eggs 1.12 × 0.77, greenish-white, profusely speckled.

H. c. mearns'i. (To Dr. E. A. Mearns, U. S. A.) MEARNS' THRASHER. Like *H. cinereus*; differing in much darker upper parts, becoming bister-brown on the rump and upper tail-coverts, rustier flanks and crissum, larger and blacker spots on under parts, and less curved bill. San Quintin, L. Cala. ANTHONY, Auk, Jan. 1895, p. 53; A. O. U. List, 2d ed., 1895, No. 709 *a*. (Included under *cinereus* proper in all earlier eds. of the Key, 1872-90.)

(*Subgenus* HARPORHYNCHUS.)

H. redivi'vus. (Lat. *redivivus*, revived; the long-lost species having been rediscovered and so named. Fig. 151.) CALIFORNIA THRASHER. ♂ : No spots anywhere; wings and tail without decided barring or tipping. Bill as long as head or longer, bow-shaped, black. Wings very much shorter than tail. Above, dark oily olive-brown; wings and tail similar, but rather purer brown. Below, a paler shade of color of upper parts; belly and crissum strongly rusty-brown; throat definitely whitish in marked contrast, and not bordered by decided maxillary streaks. Cheeks and auriculars blackish-brown,

FIG. 151. — California Thrasher, nat. size. (Ad. nat. del. E. C.)

with sharp whitish shaft-streaks. Length 11.50; wing 4.00 or rather less; tail 5.00 or more; bill (chord of culmen) nearly or quite 1.50; tarsus 1.35; middle toe and claw about the same. ♀ similar, rather smaller. Coast region of California from the valleys of the Sacramento and Russian rivers southward. Abundant in dense chaparral; nest a rude platform of twigs, roots, grasses, leaves, etc., in bushes; eggs 2-4, 1.15 × 0.85, bluish-green, with olive and russet-brown spots.

H. r. pasadenen'sis. (Lat., of Pasadena, a place in California.) PASADENA THRASHER. A very slightly differentiated race, continuing the distribution of the species southward to about lat. 30° in Lower California. General coloration "ashier or less distinctly brown" than in *redivivus* proper; throat nearly pure white. No appreciable difference in dimensions. Included under *redivivus* in former eds. of Key. GRINNELL, Auk, July, 1898, p. 237; A. O. U. Suppl. List, Auk, Jan. 1899, p. 123.

H. lecon'tei. (To Dr. John L. Le Conte, the entomologist.) YUMA THRASHER. LE CONTE'S THRASHER. Size and proportions nearly same as in *redivivus*; differs very notably in the pallor of all the coloration. Excepting the slight maxillary streaks, there are no decided markings anywhere; and the change from the pale ash of the general under parts to the brownish-yellow of the lower belly and crissum is very gradual. Valley of the Gila and Lower Colorado in contiguous portions of W. Arizona, N.W. Sonora, N.E. Lower California; S. California in the Mojave River desert, Death Valley, San Joaquin Valley, etc.; N. to S. Nevada, and the extreme S.W. corner of Utah. Specimens were for many years very rare, and the species was long regarded as a bleached desert race of *redivivus*, as in all former eds. of the Key. The type specimen from Fort Yuma, and another which I took near Fort Mojave in 1865 were long the only ones known. But we now have plenty of them, and their specific character is confirmed. Young birds have the bill very short and quite straight, the elongation and arcuation being gradually acquired as they come to maturity. Nest in bushes, bulky, loose, deep; eggs 2-4, 1.15 × 0.77, pale greenish, rather sparsely dotted with reddish-brown. For various observations on the life-history of this interesting bird, see Auk, 1884, pp. 253-258; 1885, p. 197 and pp. 229-231; 1886, pp. 299-307; 1895, pp. 54-60.

H. 1. arenic′ola. (Lat. *arena*, sand, sandy place; *colere*, to inhabit, or *incola*, an inhabitant.) DESERT THRASHER. Like the last; darker above and on tail, grayer on breast; tail perhaps shorter. Locally developed in the sand dunes of Rosalia and Playa Maria Bays, Lower California, in common with Mearns' Thrasher. ANTHONY, Auk, Apr. 1897, p. 167; A. O. U. Suppl. List, Auk, Jan. 1899, p. 124.

H. crissa′lis. (Lat. *crissalis*, relating to the *crissum*, or under tail-coverts. Fig. 152.) CRISSAL THRASHER. ♂ : Brownish-ash, with a faint olive shade, the wings and tail purer and darker fuscous, without white edging or tipping. Below, a paler shade of color of upper parts. Throat and side of lower jaw white, with sharp black maxillary streaks. Cheeks and auriculars speckled with whitish. Under tail-coverts rich chestnut, in marked contrast with surrounding parts. Bill black, at the maximum of length, slenderness, and curvature; feet blackish.

FIG. 152. — Crissal Thrasher, nat. size. (Ad. nat. del. E. C.)

Length about 12.00; wing 4.00-4.25; tail 5.50-6.00; its lateral feathers 1.50 shorter than the central ones; bill 1.50; tarsus 1.33; middle toe and claw 1.25. This fine species is distinguished by the strongly chestnut under tail-coverts, the contrast being as great as that seen in the Catbird. The sharp black maxillary streaks are also a strong character. The bill is extremely slender, the tail at a maximum of length, and the feet are notably smaller than those of *H. redivivus*. Western Texas, New Mexico, Arizona, some parts of Nevada and Utah, and California in the Colorado Valley, common in chaparral; nest in bushes near the ground, of twigs lined with vegetable fibres; eggs usually 3, 1.10 × 0.75, emerald green, unspotted.

Subfamily TROGLODYTINÆ: Wrens.

For characters in comparison with *Miminæ*, see the analysis on p. 281. The *Troglodytinæ* are small birds, only exceptionally over 6 inches long, and nearly all may be recognized on sight by any one familiar with our common House Wren. In comparison with any member of the *Sylviidæ*, observe that in *Regulus* the tarsus is booted; that in *Polioptila* the colors are bluish, black, and white. In comparison with *Paridæ* or *Sittidæ*, observe that Wrens have a different character of the nostrils and nasal plumules; with reference to *Certhiidæ*, that the tail is not rigid and acuminate; while as regards any small 9-primaried birds like the Warblers

(*Mniotiltidæ*), the wrens have 10 primaries. Furthermore: "the inner toe is united by half its basal joint to the middle toe, sometimes by the whole of this joint; and the second joint of the outer toe enters wholly or partially into this union, instead of the basal only." Nostrils narrowly or broadly oval, exposed, overhung by a scale; bill moderately or very slender, straight or slightly decurved, from half as long to about as long as head, unnotched in all our genera; no evident rictal bristles; wings short, more or less rounded, with 10 primaries, the 1st short, but not strictly spurious; tail of variable length, much or little rounded, of broad or narrow feathers, often held over the back. Tarsus scutellate, sometimes behind as well as in front.

Excluding some Old World forms of doubtful affinity, and excepting some species of *Anorthura* proper, the *Troglodytinæ* are confined to America. About 100 species and sub-species are recognized, usually referred to about 16 genera, most of which belong to tropical America, where the group reaches its maximum development, — over 20 species of *Heleodytes* being described, for instance. Of North American genera, *Heleodytes*, *Catherpes*, and *Salpinctes* are confined to the West, and represent a section distinguished by breadth of tail-feathers, which widen toward the end. Species of all our other genera are common and familiar Eastern birds, much alike in disposition, manners, and habits; the House Wren typifies these. They are sprightly, fearless, and impudent little creatures, apt to show bad temper when they fancy themselves aggrieved by cats or people, or anything else that is big and unpleasant to them; they quarrel a good deal, and are particularly spiteful towards martins and swallows, whose homes they often invade and occupy. Their song is bright and hearty, and they are fond of their own music; when disturbed at it they make a great ado with noisy scolding. Part of them live in reedy swamps and marshes, where they hang astonishingly big globular nests, with a little hole in one side, on tufts of rushes, and lay 6 or 8 dark-colored eggs; the others nest anywhere, in shrubbery, knotholes, hollow stumps, and other odd nooks. Nearly all are migratory; one is stationary; one comes to us in the fall from the north, the rest in spring from the south. Insectivorous, and very prolific, laying several sets of eggs each season. Plainly colored, the browns being the usual colors; no red, blue, yellow, or green in any of our species.

Analysis of Genera.

Fan-tailed Wrens. Feet not strictly laminiplantar; lateral plates divided, or not perfectly fused in one.
    Tail broad, fan-shaped, the individual feathers widening toward the end.
        Very large; length about 8 inches. Tarsus decidedly scutellate behind. Lateral toes of equal lengths.
            Above streaked with white, below spotted with black . . . . *Heleodytes*
        Smaller, about 6.00 long. Tarsus scutellate behind. Lateral toes of unequal lengths . . . . . . *Salpinctes*
        Smaller about 5.50 long. Tarsus scarcely scutellate behind. Lateral toes of unequal lengths . . . *Catherpes*
Thin-tailed Wrens. Feet strictly laminiplantar, as usual in *Oscines*. Tail thin, with narrow parallel-edged feathers.
    Wings and tail more or less completely barred crosswise.
        Large. Upper parts uniform in color, without streaks or bars; rump with concealed white spots. Belly unmarked; a conspicuous superciliary stripe.
            Tail shorter or not longer than wing, all the feathers brown, distinctly barred *Thryothorus* (*T. ludovicianus*)
            Tail decidedly longer than wing, blackish, not fully barred on all the feathers . *Thryomanes* (*T. bewicki*)
        Small. Upper parts not uniform; back more or less distinctly barred crosswise; wings, tail, and flanks fully barred.
            Tail about equal to wing; outstretched feet reaching scarcely or not beyond its end *Troglodytes* (*T. aëdon*)
            Tail decidedly shorter than wing; outstretched feet reaching far beyond its end . *Anorthura* (*A. hyemalis*)
        Small. Upper parts not uniform; back streaked lengthwise; flanks scarcely or not barred.
            Bill scarcely or not $\frac{1}{2}$ as long as head; crown streaked, like whole back . . . . *Cistothorus* (*C. stellaris*)
            Bill about $\frac{2}{3}$ as long as head; crown plain; streaks of back confined to interscapular region
                                                                                              Telmatodytes (*C. palustris*)

HELEOD'YTES. (Gr. ἕλος, gen. ἕλεος, *helos*, *heleos*, a marsh, meadow, or lowland; δύτης, *dutes*, a diver, used as in *Troglodytes*, etc., simply as one who enters in upon, or inhabits:

CAB., Mus. Hein., i, 1850, p. 80. *Campylorhynchus* (SPIX, 1824) of all previous eds. of the Key — a name which proves unavailable by our rules. See Auk, Jan. 1893, p. 86.) CACTUS WRENS. Of largest size in subfamily; length about 8.00. Tarsus *scutellate behind*. Lateral toes of equal lengths. Wings and tail of about equal lengths. Tail broad, with wide feathers. Tarsus a little longer than middle toe and claw. Upper parts with sharp white streaks on a brown ground; under parts boldly spotted with black on a white ground; tail-feathers barred with black and white. A neotropical genus of numerous species, one of which overreaches our Mexican border.

H. brunneicapil'lus. (Lat. *brunneus*, brown; *capillus*, hair.) BROWN-HEADED CACTUS WREN. ♂, adult: Back grayish-brown, marked with black and white, each feather having a central white field several times indented with black. Whole crown of head and nape rich dark wood-brown, immaculate. A long white superciliary stripe from nostril to nape. Beneath, nearly pure white anteriorly, gradually shading behind into decided cinnamon-brown; throat and fore part of breast marked with large, crowded, rounded black spots; rest of under parts with small, sparse, oval or linear black spots, again enlarging on crissum. Wings darker and more fuscous-brown than back; all the quills with a series of numerous white or whitish indentations along edges of both webs. Central tail-feathers like wings, with numerous more or less incomplete blackish bars; other tail-feathers blackish, the outer with several broad white bars on both webs, the rest with usually only a single complete white bar near end. Bill dark plumbeous, paler below; iris orange. Length near 8.00; wing 3.50; tail rather longer; bill 0.80; tarsus 1.00; middle toe and claw 0.90. ♀, adult: Quite like ♂, but spots on throat and breast rather smaller, therefore less crowded, and less strongly contrasting with the sparse speckling of the rest of under parts. Young: Similar to adult on upper parts, but throat whitish with little speckling; scarcely any spots on the rest of under parts, which are, however, as decidedly cinnamon as those of the adults. Southwestern U. S., Texas, New Mexico, Arizona, southern Utah and Nevada, portions of California, and S. into Mexico; common in cactus and chaparral, building a large purse-shaped nest in bushes; eggs about 6, 1.00 × 0.68, white, but so uniformly and minutely dotted with reddish-brown as to produce a nearly flat salmon-color. (If not *Picolaptes brunneicapillus* LAFR., this will stand as *H. b. couesi*, after SHARPE, Cat. Br. Mus., vi, 1882, p. 196.)

H. b. bryant'i. (To W. E. Bryant.) BRYANT'S CACTUS WREN. Intermediate in all respects between *brunneicapillus* and *affinis*; thus connecting the two, and making it necessary to reduce *affinis* to the grade of a subspecies; tail fully barred, and under parts pale, but heavily spotted. Lower California, N. into S. California; a form best developed about San Telmo, 50 miles N. of San Quentin, L. Cala. ANTHONY, Auk, July, 1894, p. 212; July, 1895, p. 280; A. O. U. List, 2d ed., 1895, No. 713 a. The describer discusses the question whether, after all, *this* be not the form upon which Lafresnaye based his *brunneicapillus* from California; if it be, *bryanti* becomes a strict synonym.

H. b. affi'nis. (Lat. *affinis*, affined, allied; *ad*, and *finis*.) ST. LUCAS CACTUS WREN. Similar to the last. Cap reddish-brown, lighter instead of darker than back. Markings of back very conspicuous, in strong streaks of black and white, these two colors bordering each other with little or no indentation. Under parts nearly white, the black spots, though conspicuous, not enlarged and crowded on breast, but more regularly distributed. All the lateral tail-feathers, instead of only the outer ones, crossed on both webs with numerous complete white bars. The variations with sex and age correspond with those of *H. brunneicapillus*. Lower California, Cape St. Lucas and northward. Nest and eggs as before. (According to SHARPE, *l. c.*, p. 197, this is **P. brunneicapillus** LAFR.) *Campylorhynchus affinis* of former eds. of Key; A. O. U. List, 1st ed., 1886, No. 714. *Heleodytes affinis* A. O. U. List, Sixth Suppl., Auk, Jan. 1894, p. 48. *Heleodytes brunneicapillus affinis* ANTHONY, Auk, July, 1895, p. 280; A. O. U. List, 2d ed., 1895, No. 713 b.

SALPINC′TES. (Gr. σαλπιγκτής, *salpigktes*, a trumpeter.) ROCK WRENS. Bill about as long as head, slender, compressed, straight at base, then slightly decurved, acute at tip, faintly notched. Nostrils conspicuous, scaled, in a large fossa. Wing longer than tail; exposed portion of 1st primary about half as long as 2d, which is decidedly shorter than 3d. Tail rounded, of 12 broad plane feathers, with rounded or subtruncate ends. Feet small and weak; tarsus longer than middle toe, *scutellate posteriorly*. Hind toe and claw shorter than middle one; lateral toes of unequal lengths, outer longest, both very short; tips of their claws falling short of base of middle claw. Two species.

FIG. 153. — Rock Wren, nat. size. (Ad. nat. del. E. C.)

S. obsole′tus. (Lat. *obsoletus*, unaccustomed; *ob*, and *soleo*, I am wont; hence obsolete, effaced, the coloration being dull and diffuse. Figs. 153, 154.) ROCK WREN. ♂ ♀, adult: Upper parts pale brownish-gray, minutely dotted with blackish and whitish points together, and usually showing obsolete wavy bars of dusky. Rump cinnamon-brown; a whitish superciliary line. Beneath, soiled white, shading behind into pale cinnamon; throat and breast obsoletely streaked, and under tail-coverts barred, with dusky. Quills of wings rather darker than back, with similar markings on outer webs. Middle tail-feathers like back, with many dark bars of equal width with the lighter ones; lateral tail-feathers similarly marked on outer webs, plain on inner webs, with a broad subterminal black bar on both, and cinnamon-brown tips, the latter usually marbled with dusky; outer feathers with several blackish and cinnamon bars on both webs. Bill and feet dark horn color, the former paler at base below. Length 5.50–6.00; wing 2.60–2.80; tail 2.20–2.40; bill 0.66–0.75; tarsus 0.75–0.80. Most of the markings blended and diffuse. Shade of upper parts variable, from dull grayish to a more plumbeous shade, often with a faint pinkish tinge. Specimens in worn and faded plumage may fail to show the peculiar dotting with black and whitish; but in these the cross-wise dusky undula-

FIG. 154. — Rock Wren.

tion, as well as the streaks on the breast, are commonly more distinct than in fresher-feathered examples. The rufous tinge of the under parts is very variable in shade; that of the rump, however, being always well marked. Western U. S., and adjoining British provinces, W. to the Pacific, E. to Iowa; S. on the Mexican table-lands to Central America; breeds throughout its range; migratory in the U. S., except along the southern border; common, haunting rocky places, where it is conspicuous by its restlessness and loud notes; nest of any rubbish in a rocky nook; eggs 5–8, of crystalline whiteness, sparsely sprinkled with reddish-brown dots, 0.75 × 0.62.

S. guadalupen′sis. (Lat., inhabiting the island off the coast of L. California called in Spanish Guadalupe, and not known by the French name of Guadeloupe.) GUADALUPE ROCK WREN. Resembling the last; darker colored, with more distinct speckling; wings and tail somewhat shorter; bill and tarsi rather longer. Wing 2.50–2.70; tail 2.00–2.30; tarsus 0.80–0.90.

Guadalupe Island, Lower California. *S. obsoletus guadeloupensis* (by error for *Guadalupensis*) RIDGW., Bull. U. S. Geol. Surv., ii, Apr. 1876, p. 185; *S. obsoletus guadalupensis* COUES, Key, 2d–4th eds., 1884–90, p. 867; *S. guadalupensis*, RIDGW., Bull. Nutt. Club, ii, July, 1877, p. 60; *S. guadeloupensis* [*sic*], RIDGW., Man., 1887, p. 548; A. O. U. Lists, 1886 and 1895, No. 716.

CATHER'PES. (Gr. καθερπής, *katherpes*, a creeper; κατά, *kata*, down, ἕρπω, *herpo*, I creep.) CAÑON WRENS. Bill singularly attenuate, about as long as head, nearly straight in all its outlines, with such direction of its axis that the bill as a whole appears continuous with the line of the forehead. Tarsus not longer than middle toe and claw, with tendency to subdivision of the lateral tarsal plate. Lateral toes of unequal lengths, the outer longest. Wings and tail as in *Salpinctes;* and system of coloration much the same. One known species, of which 3 subspecies occur in the U. S.

C. mexica'nus al'bifrons. (Lat. *mexicanus*, Mexican. Lat. *albifrons*, white-fronted; *albus*, white; *frons*, front, forehead.) TEXAN CAÑON WREN. Similar to the form next described; much darker colored both above and below, with sharper contrast of the white throat; white speckling mostly restricted to back and wings; black tail-bars broader and more regular; light markings of wings mere indentations instead of complete bars. Bill straight, more abruptly decurved at extreme tip. Feet stouter, dark brown. Length about 6.00; wing 2.80; tail 2.40; bill nearly 1.00 long, only about 0.12 deep at base. Specimens vary much in sharpness and extensiveness of speckling of upper parts. In best-marked cases, the spots quite white, almost lengthened into streaks, each one completely set in black; in other examples, small, sparse, and restricted, these specimens also showing wavy transverse bars of blackish. Lower Rio Grande of Texas, and southward in Tamaulipas and Nuevo Leon. *C. mexicanus* of all former eds. of Key, and of A. O. U. Lists, 1886–95. *Certhia albifrons* GIRAUD, Sixt. Sp. Tex. B., 1841, pl. XVIII. *Catherpes mexicanus albifrons* NELSON, Auk, Apr. 1898, p. 160; A. O. U. Suppl. List, Auk, Jan. 1899, p. 124.

C. m. consper'sus. (Lat. *conspersus*, speckled.) SPECKLED CAÑON WREN. ♂ ♀, adult: Upper parts brown, paler and grayer anteriorly, behind shading insensibly into rich rufous, everywhere dotted with small dusky and whitish spots. Tail clear cinnamon-brown, crossed with numerous very narrow and mostly zigzag black bars. Wing-quills dark-brown; outer webs of primaries and both webs of inner secondaries barred with color of back. Chin, throat, and fore breast, with lower half of side of head and neck, pure white, shading behind through ochraceous-brown into rich deep ferruginous, and posteriorly obsoletely waved with dusky and whitish. Bill slate-colored, paler and more livid below; feet black; iris brown. Length about 5.50; extent 7.50; wing 2.30; tail 2.12; tarsus 0.65; bill 0.80. California, New Mexico, Arizona, and portions of Texas, Colorado, Utah, Nevada, Wyoming, Idaho, and Oregon; also S. in Mexico to Aguas Calientes; resident in most of its range, and common in suitable localities. A remarkable bird, famous for its ringing notes, inhabiting cañons and other rocky places. Nesting and eggs like those of the Rock Wren; eggs 5 or more, 0.75 × 0.55, crystal white, fairly sprinkled and blotched with reddish-brown.

C. m. punctula'tus. (Lat. *punctulatus*, dotted.) DOTTED CAÑON WREN. Smaller than either of the foregoing: Length about 5.00; wing 2.10; tail 1.90; bill 0.75. Coloration intermediate; upper parts most like those of *conspersus*, and wings as completely barred; but under parts posteriorly dusky ferruginous (dark mahogany color), and tail-bars broad, firm, and regular, as in *mexicanus* proper. Coast region of California and Oregon; resident in most of its range. The type specimen, the only one I have seen, for some years in my cabinet and now No. 82,715, Mus. S. I., seems to be recognizably distinct; but all the forms of the genus intergrade. RIDGW., Pr. Nat. Mus., v, Sept. 1882, p. 343; disallowed by A. O. U. Committee, 1886; Key, 2d ed., 1884, p. 276; see also 4th ed., 1890, p. 896; A. O. U. List, 2d ed., 1895, p. 297, No. 717 *b*.

THRYOTHO'RUS. (Gr. θρύον, *thruon*, a reed, and θοῦρος, *thouros*, leaping. This is the spelling given and etymology indicated by VIEILLOT, Anal., 1816, p. 70, but on p. 45 he first spells the word *thriothorus*.) REED WRENS. CAROLINA WRENS. Of largest size in this group; length up to 6.00. Tail decidedly shorter than wings. Back uniform in color, without streaks or bars; wings and tail more or less barred crosswise; belly unmarked; a long superciliary stripe; rump with concealed white spots. Eggs colored.

Analysis of Species and Subspecies.

Reddish-brown above, rusty whitish below; tail regularly barred throughout. Wing 2.40 . . . . . *ludovicianus*
Similar; more heavily colored; rusty-brown below. Wing 2.75. Florida . . . . . . *ludovicianus miamensis*
Grayish-brown, more extensively barred on flanks, barring of tail irregular; small; wing 2.25. S. E. Texas
ludovicianus lomitensis
Darker brown, most extensively barred on flanks, barring of tail irregular; small; wing 2.25. N. E. Mexico
ludovicianus berlandieri

T. ludovicia'nus. (Lat. *Ludovicianus*, Louisiana; of Ludovicus, Louis XIV., of France. Fig. 155.) GREAT CAROLINA WREN. Upper parts uniform reddish-brown, brightest on rump, where are concealed whitish spots; a long whitish superciliary line, usually bordered with dusky streaks; upper surfaces of wings and tail like back, barred with dusky; outer edges of primaries and lateral tail-feathers showing whitish spots. Below, rusty or muddy whitish, clearest anteriorly, deepening behind, the under tail-coverts reddish-brown barred with blackish. Wing-coverts usually with dusky and whitish tips. Feet livid flesh-colored. Length 6.00; extent nearly 7.50; wing 2.40; tail 2.25; bill 0.65; tarsus 0.75. Eastern U. S., southerly; N. regularly to the Middle States, rarely to Massachusetts and Ontario; Michigan; Nebraska; resident in most of its range. A common and well-known inhabitant of shrubbery, with a loud ringing song; shy and secretive. Nest in any nook about out-buildings, trees or stumps, or in shrubbery, when in the latter usually roofed over, of the most miscellaneous materials; eggs 4–7, 0.72 × 0.60, white, profusely speckled and blotched with shades of reddish, brown, and purplish.

FIG. 155. — Great Carolina Wren, reduced. (From Nuttall, after Audubon.)

T. l. miamen'sis. (Of the Miami River, Florida.) FLORIDA WREN. Similar: larger, stouter, and more deeply-colored, especially below, where nearly uniform rusty-brown. Wing 2.75; tail 2.60; bill 0.90; tarsus 0.95. Florida; a local race.

T. l. lomiten'sis. (Of Lomita ranch, Hidalgo Co., Tex., where the types were taken.) LOMITA WREN. Similar to *ludovicianus* proper; rather smaller; length about 5.25; wing 2.25; tail 2.05: shade of the upper parts rather grayish-brown than reddish-brown; barring of the tail broken and irregular, giving a mottled appearance; in this respect, as well as in a tendency to barring of the flanks, approaching *berlandieri*. S. E. Texas, on the Rio Grande. SENNETT, Auk, Jan. 1890, p. 58 (*T. l. lomitæ* COUES, Key, 4th ed., 1890, p. 898, by slip of the pen for *lomitensis*), A. O. U. List, 2d ed., 1895, p. 298, No. 718 *b*.

T. l. berlan'dieri. (To Dr. Louis Berlandier.) BERLANDIER'S WREN. Similar: smaller; length 5.25; wing 2.25; tail 2.12. Coloration darker than in typical *ludovicianus*, especially below; flanks as well as crissum barred with dusky; tail-bars broken up into irregular nebulation. Valley of the Rio Grande; a local race of N. E. Mexico, which is admitted in neither of the A. O. U. Lists, 1886 and 1895, perhaps as being extralimital, or in some uncertainty regarding its subspecific status. But it seems to be as well marked as the others, and for the present I let it stand, as in all the previous eds. of the Key.

THRYO'MANES. (Gr. θρύον, *thruon*, a reed; μάνης or μανῆς, *manes*, a kind of cup.) BEWICK'S WRENS. Similar to *Thryothorus*, but tail not decidedly shorter than wings — usually decidedly longer — blackish, not fully barred. Coloration not reddish-brown above. (Included under *Thryothorus* in former eds. of the Key, and in A. O. U. Lists till 1899.)

Analysis of Species and Subspecies.

Tail decidedly longer than wings.
    Eastern U. S. southerly . . . . *bewicki*
    Western U. S. except Pacific coast region. Grayer above, whiter below . . . . *b. leucogaster*
    Pacific coast region. Like the last, with less evident contrast between middle tail-feathers and back *b. spilurus*
    San Clemente Island. Superciliary stripe more conspicuous . . . . *leucophrys*
Tail about as long as wings, both under 2.00.
    Guadalupe Island . . . . *brevicauda*

T. be'wicki. (To Thomas Bewick.) BEWICK'S WREN. Above, dark grayish-brown; below, ashy-white, with a brownish wash on flanks. Rump with concealed whitish spots. A long whitish superciliary stripe from nostrils to nape. Under tail-coverts dark-barred; two middle tail-feathers like back, with numerous fine black bars; others black with whitish markings on the outer webs and tips. Length about 5.50; extent 6.75; wing 2.00–2.12; tail 2.35; bill 0.50; tarsus 0.75. Eastern U. S., southerly, N. to the Middle States and Minnesota, W. to the edge of the Great Plains; resident in most of its range. Not very common in the Atlantic States, but so abundant as to replace the House Wren in some parts of the interior. Nest in holes in trees, stumps, fences, etc.; eggs 5–9, 0.65 × 0.50, white, finely dotted and spotted, resembling those of *Catherpes* or *Salpinctes*.

T. b. leucogas'ter. (Gr. λευκός, *leukos*, white; γαστήρ, *gaster*, belly.) BAIRD'S WREN. WHITE-BELLIED WREN. Above, uniform clear ashy-brown; below, clear ashy-white; pure white on middle under parts. A long, strong, white superciliary stripe; auriculars speckled with white. Concealed white spots on rump. Quills of wings fuscous, the inner ones very obsoletely waved with color of back. Two middle tail-feathers closely barred with pure dark ash and black well contrasted with the ashy-brown of the back; others black, with irregular white or ashy-white tips; outer web of exterior one barred with white. Length 5.50–5.75; extent 6.75; wing 2.00–2.33; tail 2.25–2.50; bill 0.50; tarsus 0.75. A well-marked geographical race, inhabiting the Great Plains and Great Basin, from Kansas and Colorado to Utah, Texas, New Mexico, Arizona, California, E. of the Sierra Nevadas, and S. into Mexico. *Thryothorus (Thryomanes) bewickii*, var. *leucogaster* BD., Rev. A. B., 1864, p. 127 (not *Troglodytes leucogastra* of GOULD, P. Z. S., 1836, p. 89, as Baird supposed it was, for Gould's bird is *uropsila leucogaster* of SCL. and SALV., *Cyphorhinus pusillus* SCL., *Heterorhina pusilla* BD.); *Thryothorus bairdi* SALV. and GODM., Biol. Cent.-Am., i, Apr. 1880, p. 95; *T. bewickii bairdi* RIDGW., Pr. U. S. Nat. Mus, viii, 1885, p. 354; A. O. U. Lists, 1886 and 1895, No. 719 *b;* but no rule of nomenclature requires us to change the subspecific name *leucogaster* BD., 1864. The fact that *Troglodytes leucogastra* GOULD, 1836, is an entirely different bird, belonging to another genus, does not outlaw *Thryothorus bewicki leucogaster*, or in any way affect nomenclature in the genus *Thryothorus*. See COUES, Auk, Oct. 1896, p. 345. The point is conceded in A. O. U. Suppl. List, Auk, Jan. 1897, p. 131, where the Key name is restored, as above.

T. b. spilu'rus. (Gr. σπῖλος, *spilos*, spotted; οὐρα, *oura*, tail.) VIGORS' WREN. SPECKLED-TAILED WREN. Similar to *T. bewicki* in color; upper parts more uniform dull bistre rather than umber brown, with little contrast in shade between back and middle tail-feathers; bill averaging slightly longer. This is "Bewick's" Wren of the Pacific Coast region, from British Columbia to Southern and Lower California, and Western Mexico.

T. leu'cophrys. (Gr. λευκός, *leucos*, white; ὀφρύς, *ophrus*, eyebrow.) SAN CLEMENTE WREN. Resembling *T. b. spilurus;* upper parts with a decided grayish wash; superciliary stripe white, very conspicuous; under tail-coverts less heavily barred; bill longest. San Clemente Island, 75 miles off coast of California, common in cactus and other bushes. ANTHONY, Auk, Jan. 1895, p. 52; A. O. U. List, 2d ed., 1895, p. 299, No. 719.1.

T. cerroen'sis. (Lat., of Cerros Isl., Spanish *cerro*, a mountain or large hill.) CERROS ISLAND WREN. Like the last; darker above; less gray on the flanks; bill shorter. Cerros Island, Lower California. ANTHONY, Auk, Apr. 1897, p. 166.

T. brevicau'dus. (Lat. *brevis*, short; *cauda*, tail.) GUADALUPE WREN. Resembling *T. bewicki leucogaster*, but distinct. Above grayish-brown, grayest on tail, brownest on rump; few if any concealed white spots on the rump; wing-feathers obsoletely and tail-feathers distinctly cross-barred with dusky; the 3 outermost of the latter pale dull gray at ends, with one or two broad dusky bars. A strong white superciliary stripe, below which a grayish-brown loral and auricular stripe. Below, white, shaded into ashy on belly and sides; crissum with broad black bars. Wing 1.85–1.90; tail 1.80; bill 0.45–0.50; tarsus 0.70–0.75. Guadalupe Island, Lower California. *Thryomanes brevicauda* RIDGW., Bull. U. S. Geol. Surv., ii, Apr. 1876, p. 186; *Thryothorus brevicaudus* COUES, Key, 2d–4th eds., 1884–90, p. 868; *Thryothorus (Thryomanes) brevicaudus* RIDGW., Man., 1887, p. 551. *Thryothorus brevicauda* SHARPE, Cat. B. Brit. Mus., vi, 1881, p. 227. *Thryothorus (Thryomanes) brevicauda*, A. O. U. Lists, 1st and 2d eds., 1886–95, No. 720.

TROGLOD'YTES. (Gr. τρωγλοδύτης, *troglodutes*, a cave-dweller.) HOUSE WRENS. Of small size; no decided superciliary line. Upper parts not uniform in color; back more or less distinctly barred crosswise; wings, tail, and flanks fully barred crosswise; tail about equal to wing in length, the outstretched feet scarcely or not reaching beyond its end. Eggs colored.

Analysis of Species and Subspecies.

Umber-brown on back, little barred there. Eastern U. S. . . . . *aëdon*
Grayish-brown on back, more barred there. Western U. S. . . . . *aëdon aztecus*
Brown on back, most barred. Pacific coast . . . . *aëdon parkmani*

T. aëdon. (Gr. ἀηδών, *áedon*, the songstress, applied by Hesiod to the Nightingale; in Homer as a proper name, Ἀηδών, daughter of Pandareus, changed into a Nightingale.) EASTERN HOUSE WREN. Brown, brighter behind; below rusty-brown, or grayish-brown, or even grayish-white; everywhere waved with darker shade, very plainly on wings, tail, flanks, and under tail-coverts, breast apt to be darker than either throat or belly; bill shorter than head, about 0.50; wings and tail nearly equal, about 2.00, but ranging from 1.90 to 2.10; total length 4.50–5.25, averaging about 4.90; extent about 6.75. Exposed portion of 1st primary about ½ as long as longest primary. Eastern U. S., N. to Canada, W. to Dakota; very abundant anywhere in shrubbery, gardens, and about dwellings, where its active, sprightly, and fearless demeanor, together with its hearty trilling song, bring it into friendly notoriety. Nest of any trash in a hole of a building, fence, tree, or stump; eggs 6–9, 0.65 × 0.55, profusely and uniformly studded with minute points of brown, often rendering an almost uniform color; two or three broods each season. Resident in the South, migratory elsewhere. (*T. domesticus* of 2d–4th eds. of the Key, 1884–90, after BARTRAM.)

T. a. az'tecus. (Lat., Aztec, as this form was originally described from Mexican specimens.) WESTERN HOUSE WREN. Brown above, little brighter on rump, nearly everywhere waved with dusky, strongest on wings and tail, but usually appreciable on the whole back. Below brownish-white, nearly white on belly, obscurely variegated with darker markings, which on flanks and crissum become stronger bars, alternating with brown and whitish ones. Bill blackish above, pale below; feet brown. Length 5.00–5.25; extent 6.75; wing and tail about 2.10. Exposed portion of 1st primary about ½ as long as 2d primary. Western U. S., chiefly from the Plains and Rocky Mt. region, E. to Illinois, N. to Manitoba, S. into Mexico, abundant, there replacing *T. aëdon*, to which it is so similar; but on an average paler and grayer, with rather longer wings and tail. Parkman's Wren was originally described by Audubon from the Columbia River, and the name therefore belongs to the next variety, as stated in the Key, 4th ed., 1890, p. 898: see ALLEN, Auk, Jan. 1888, p. 164. The present form is *T. aëdon aztecus* BAIRD, Rev. A. B., 1864, p. 139; *T. aëdon parkmanni* COUES, Key, 1st ed., 1872, p. 87; *T. domesticus parkmani*, Key, 2d–4th eds., 1884–90, p. 278; *T. aëdon aztecus*, A. O. U. Lists, 1886–95, No. 721 *b*.

T. a. park′mani. (To Dr. George Parkman, of Boston, murdered Nov. 23, 1849, by Prof. John W. Webster, Professor of Chemistry in Medical College of Harvard, Cambridge, Mass.) PARKMAN'S WREN. PACIFIC HOUSE WREN. More heavily colored than typical *aëdon*, with tendency to more extensive barring. Pacific coast region, British Columbia, Washington, Oregon, and Northern California, grading directly into the last form. *T. parkmanii*, AUD., Orn. Biogr., v, 1839, p. 310; *T. aëdon parkmanni*, in part, COUES, Key, 1872, and later eds.; *T. aëdon parkmanii*, A. O. U. List, 2d ed., 1895, p. 300, No. 721 *a*.

ANORTHU′RA. (Gr. ἀν, *an*, signifying negation; ὀρθός, *orthos*, straight; οὖρα, *oura*, tail.) WINTER WRENS. Like *Troglodytes* proper, but tail decidedly shorter than wings, the outstretched feet reaching far beyond its end. Eggs colored.

Analysis of Species and Subspecies.

Length about 4.00; bill about 0.40.
    Eastern . . . . *hiemalis*
    Western . . . . *h. pacificus*
Length about 4.50; bill 0.50–0.60.
    Alaskan . . . . *alascensis*

A. hiema′lis. (Lat. *hiemalis*, wintry; *hiems*, winter. Fig. 156.) WINTER WREN. Above brown, darker before, brighter behind, most of back, together with tail and inner wing-quills, banded with dusky; markings obsolete on back, where usually accompanied by whitish specks, strongest on wings and tail. Outer webs of several primaries regularly barred with brownish-white, in marked contrast with other bars of the wings. An inconspicuous whitish superciliary line. Below brownish, paler or whitish anteriorly; belly, flanks, and crissum heavily waved with dusky and whitish bars. Bill slender, straight, decidedly shorter than head. Tail much shorter than wings. Length 3.90–4.10; extent 6.00–6.50; wing 1.75; tail 1.25; bill 0.40; tarsus, middle toe,

FIG. 156. — Winter Wren, little reduced. (Baird's figure of *A. alascensis*.)

and claw together, about 1.12. Eastern N. Am., common, migratory, breeding from N. New England and corresponding latitudes northward, in the Alleghanies S. to N. Carolina, wintering in the U. S. from about its southern limit of breeding southward; the strict representative of the European Wren, *A. troglodytes*. Nest of twigs, moss, lichens, hair, feathers, etc., usually in a stump or log close to the ground; eggs 5–8, 0.65 × 0.50, pure white, minutely dotted with reddish-brown and purplish, but not nearly so heavily marked as those of House Wrens and long-billed Marsh Wrens, sometimes very sparingly sprinkled. A sly, secretive little bird, less often seen than other Wrens no less common; voice strong and highly musical. *Anorthura troglodytes hiemalis* of 2d-4th eds. of Key. *Anorthura hyemalis* COUES and PRENTISS, 1862. *A. hiemalis* A. O. U. Suppl. List, Auk, Jan. 1899, p. 125.

A. h. pacif'icus. (Lat. *pacificus*, pacific, peace-making; *pax*, peace, and *facio*, I make, do ; alluding to "the stilly sea.") WESTERN WINTER WREN. Like the last; darker, in lack of whitish specks of upper parts, and of whitish bars on outer webs of primaries; but very slightly distinguished. Pacific Coast region, from Southern Alaska (Sitka) to southern California; E. to Idaho. *Anorthura troglodytes pacificus* of 2d-4th eds. of Key. *A. hiemalis pacifica* A. O. U. Suppl. List, Jan. 1899, p. 125.

A. alascen'sis. (Of Alaska.) ALASKAN WINTER WREN. Like the common species in form and coloration; larger, size of a House Wren; wing 2.00-2.20; tail 1.50; tarsus 0.75; tarsus, middle toe, and claw together 1.40; bill 0.65. Culmen, gape, and gonys almost perfectly straight, latter slightly ascending. Aleutian and Pribylov Islands, Alaska. Well distinguished from the common form, and nearer the Japanese *A. fumigatus*. *Anorthura troglodytes alascensis* of 2d-4th eds. of Key.

TELMATO'DYTES. (Gr. τέλμα, *telma*, a swamp; δύτης, *dutes*, an inhabitant.) MARSH WRENS. Small. Upper parts not uniform; back streaked lengthwise with white and black; flanks scarcely or not barred; crown plain; bill $\frac{1}{2}$-$\frac{2}{3}$ as long as head. Eggs dark chocolate-brown. Nest globular, bulky, with a hole in the side, affixed to reeds in swamps or marshes.

Analysis of Species and Subspecies.

Basal third of lower mandible flesh color.
    Middle tail feathers and upper tail coverts indistinctly barred, if at all. Eastern . . . . *palustris*
    Middle tail feathers and upper tail coverts more distinctly barred. Western . . . . . *palustris paludicola*
Lower mandible almost entirely horn color.
    General coloration paler and more uniform than in *palustris*, but breast clouded. S. Carolina and Georgia
        palustris griseus
    General coloration darker and less uniform than in *palustris;* black of upper parts extensive, brown of under parts with an olive shade, breast clouded, barring of tail and its coverts well marked. W. coast of Florida
        marianæ

T. palus'tris. (Lat. *palustris*, marshy; *palus*, a marsh. Figs. 157, 158.) LONG-BILLED MARSH WREN. Above clear brown, unbarred; middle of back with a large black patch sharply streaked with white (these white stripes sometimes deficient). Crown of head usually darker than back, often quite blackish and continuous with black interscapular patch. A dull white superciliary line. Wings fuscous; inner secondaries blackish on outer webs, often barred or indented with light brown. Tail evenly barred with fuscous and color of back. Under parts white, usually quite pure on belly and middle line of breast and throat, but much shaded with brown on sides, flanks, and crissum. Bill blackish above, pale below; feet brown. Length about 5.00; extent 6.50; wing 1.75-2.00; tail about the same; bill 0.50 or more; tarsus 0.66-0.75. Eastern U. S. and British Provinces, even casually to Greenland. Breeds throughout its usual range, and winters chiefly in the Southern States, sometimes N. to New England; an abundant bird, colonizing reedy swamps and marshes in large numbers, its great globular nests of plaited rushes, with a hole in the side, being affixed to the swaying herbage; eggs 5-10, 0.58 × 0.45, very dark-colored, being so thickly dotted with chocolate-brown as to appear almost uniformly of this color. *Telmatodytes palustris* of 2d-4th eds. of the Key.

FIG. 157. — Long-billed Marsh Wren, nat. size. (Ad. nat. del. E. C.)

T. p. gris'eus. (Lat. *griseus*, gray.) WORTHINGTON'S MARSH WREN. A local race of *T. palustris;* paler and grayer, yet with dark under mandible, clouded breast, and some other features of *T. marianæ*; markings of wings and tail less pronounced than in typical *palustris*. Coast of South Carolina and Georgia. BREWSTER, Auk, July, 1893, p. 216; A. O. U. List, 2d ed., 1895, p. 302, No. 725 *b*.

T. marian'æ. (To Mrs. Marian J. Scott, wife of W. E. D. Scott.) MARIAN'S MARSH WREN. Differing from *T. palustris* in general darker coloration; black of back and crown extensive; the brown parts of an olivaceous rather than rufous shade; breast clouded; upper and under tail coverts and flanks decidedly barred; lower mandible dark. West coast of Florida, Tarpon Springs to Cedar Keys; apparently resident. Intermediates between this supposed species and *T. palustris*, through *T. p. griseus*, may be expected to occur, but none such are as yet forthcoming. SCOTT, Auk, Apr. 1888, p. 188; A. O. U. List, 2d ed. 1895, p. 303, No. 725. 1. *Telmatodytes marianæ* COUES, Key, 1890, 4th ed., p. 898; *Cistothorus palustris marianæ* BREWSTER, Auk, July, 1893, p. 219.

T. p. paludi'cola. (Lat. *paludicola*, a marsh-inhabiter; *palus*, a marsh, and *colo*, I cultivate.) TULE MARSH WREN. Bill averaging shorter; tail and its coverts more distinctly barred. Western United States and British Provinces from the Rocky Mountains to the Pacific, and S. in Mexico; breeds throughout its U. S. range, and winters from Oregon southward. It abounds in many localities in the tule (*Scirpus validus*) marshes, whence its Spanish vernacular name is derived. See Key, 4th ed., 1890, p. 898. Disallowed in A. O. U. List of 1886; admitted in 2d ed., 1895, p. 302, No. 725 *a*. This has lately been split into *paludicola* proper of the Pacific Coast, and *C. p. plesius* OBERHOLSER, Auk, Apr. 1897, p. 188, supposed to be paler, etc., and to inhabit the rest of the region just said.

FIG. 158.—Long-billed Marsh Wren. (From The Osprey.)

CISTOTHO'RUS. (Gr. κίστος, *kistos*, a shrub · θοῦρος, *thouros*, leaping.) MARSH WRENS. Like *Telmatodytes*; whole back and crown streaked with white. Bill scarcely or not one-half as long as head. Eggs white.

C. stella'ris. (Lat. *stellaris*, starry; *i. e.*, speckled. Fig. 159.) SHORT-BILLED MARSH WREN. Upper parts brown; crown and most of back blackish, streaked with white. Below, whitish, shaded with clear brown across breast, along sides, and especially on flanks and crissum, the latter more or less indistinctly barred with dusky (often inappreciable). A whitish line over eye. Wings and tail marked as in the last species; upper tail-coverts decidedly

Fig. 159. — Short-billed Marsh Wren, nat. size. (Ad. nat. del. E. C.)

barred. Bill blackish above, whitish below, extremely small, scarcely ½ as long as head; feet brown. Length 4.50; extent 5.75–6.00; wing and tail each about 1.75; bill 0.35–0.40; tarsus, middle toe, and claw together, about 1.12. The streaking of head and that of back are usually separated by a plain nuchal interval; but these are often run together, the whole bird above being streaked with whitish and blackish upon a brown ground. The wings, tail, and entire under parts are much like those of *palustris*, from which the species is distinguished by the markings of the upper parts and extremely short bill. Eastern U. S. and adjoining British Provinces; N. to New Hampshire, Michigan, Ontario, and Manitoba, W. to the Great Plains. Migratory; winters in the Southern States. Frequents marshy places like *palustris*, but is not common. Nesting similar, but eggs pure white, 0.65 × 0.45; nest typically a ball of green grass hung in meadow grass.

Family MOTACILLIDÆ: Wagtails and Pipits.

Bill shorter than head, very slender, straight, acute, notched at tip. Nostrils not concealed by feathers, which however reach into nasal fossæ. Rictus not notably bristled. Primaries 9; 1st about as long as 2d; first 3, 4, or 5, forming point of the wing; inner secondaries enlarged, the longest one nearly or quite equalling primaries in the closed wing. Tail lengthened, averaging about equal to wing. Feet long and slender; tarsus scutellate, usually longer than middle toe and claw; inner toe cleft to the very base, but basal joint of outer toe soldered with middle one; hind toe bearing a long and little curved claw (except in *Motacilla* proper). A well-defined group of about 60, chiefly Old World, species, which may be termed terrestrial Sylvias, all living mostly on the ground, where they run with facility, like Larks, never hopping like most *Oscines*. They are usually gregarious; are insectivorous and migratory. They have gained their name from the characteristic habit of moving the tail with a peculiar see-saw motion, as if they were using it to balance themselves upon unsteady footing. They may be distinguished from all the foregoing birds by having only 9 primaries; from all the following *Oscines* except *Alaudidæ*, by having long flowing inner secondaries; and from *Alaudidæ*, with which they agree in this respect, as well as in usually having a lengthened, straightish hind claw, by having the tarsal envelop as in *Oscines* generally, slender bill, exposed nostrils and double moult. Two subfamilies have been generally recognized, but the distinctions are scarcely more than generic. They hold pretty well for the few forms found in America, but break down when the Old World genera are considered. I therefore banish them from the Key, as the A. O. U. does from its List, following Sharpe, Cat. B. Brit. Mus., x, 1885, p. 456.

Analysis of Genera.

Wagtails. Point of wing formed by first 3 primaries. Tail longer or not obviously shorter than wings, with narrow tapering feathers. Hind claw variable in length and curvature. Coloration black and white, or yellow and greenish.
  Tail decidedly longer than wings, doubly emarginate. Hind claw of ordinary length and curvature.
    Colors black, ashy, and white, in masses . . . . *Motacilla*
  Tail, if anything, shorter than wings, nearly even. Hind claw lengthened and straightened. Colors yellow and green, in masses . . . . *Budytes*
Pipits. Point of wing formed by first 4 or 5 primaries. Tail decidedly shorter than wings, its feathers not tapering. Hind claw lengthened and straightened. Coloration brownish; under parts streaked, upper usually also variegated
 . . . . *Anthus*

MOTACIL'LA. (Lat. *mota-cilla*, wag-tail; name of some small bird.) Water Wagtails. Tail much longer than wings, of 12 narrow, weak, tapering or almost linear feathers. First 3 primaries about equal and longest; longest secondary (when full grown) about reaching their ends when the wing is closed; these flowing secondaries narrow and tapering. Tarsus long and slender; lateral toes of about equal lengths; hind claw not particularly lengthened or straightened; with its digit much shorter than tarsus. Form remarkably lithe and slender; coloration black, ashy, and white, in large masses.

M. al'ba. (Lat. *alba*, white.) WHITE WAGTAIL. ♂, in summer: Head black, with a broad mask of white across forehead and along sides; black extending on fore breast; wings blackish, with much white edging and tipping of quills and greater coverts; tail black, the two lateral feathers on each side mostly white; back and sides ashy; lower parts mostly white; bill and feet black. In winter the black more restricted, that on the fore breast forming a crescent. ♀ similar; black still more restricted, in part replaced by gray. Young gray above, grayish-white below, with a gray or blackish crescent on the fore neck. Length about 7.00; wing 3.25; tail 3.75; tarsus 0.90; hind toe and claw 0.60; bill 0.50. A species of wide distribution in Europe, Africa, and Asia, occasional in Greenland. Nest on the ground; eggs 3–5, 0.80 × 0.60, white, fully speckled with brown.

M. ocula'ris. (Lat. *ocularis*, ocular.) SIBERIAN WAGTAIL. SWINHOE'S WAGTAIL. Closely resembling *M. alba*. Larger; length 7.00–7.50; wing 3.50–3.60; tail 3.50–4.00. A black eye-stripe in the white mask; wing-coverts mostly white, forming a large wing-patch; upper parts mostly gray. Young with the transocular fascia indicated by a dusky line. N. E. Siberia and southward; accidental in Lower California; probably also occurring in Alaska. This fine species agrees with *M. lugens* in the head-markings, but in the latter the back is black. SWINH., Ibis, 1860, p. 55; see RIDGW., Proc. U. S. Nat. Mus., iv, 1882, p. 414; NELSON, Cruise of the Corwin, 1883, p. 62, plate 2; COUES, Key, 2d ed., 1884, p. 284; SHARPE, Cat. B. Brit. Mus., xv, 1885, p. 471, pl. 4, figs. 5, 6, showing the difference

FIG. 160. — Siberian Yellow Wagtail.

between *ocularis* and *lugens;* which latter may also be looked for in Alaska.

BU'DYTES. (Gr. βουδύτης, *boudutes*, some small bird.) FIELD WAGTAILS. Characters of *Motacilla;* tail shorter, not exceeding the wing in length; hind claw lengthened and straightish; hind toe and claw nearly as long as the tarsus. Coloration chiefly yellow and greenish.

FIG. 161. — Yellow Wagtail, nearly nat. size. (After Baird.)

B. fla'vus leucostria'tus. (Lat. *flavus*, yellow. Gr. λευκός, *leucos*, white; and Lat. *striatus*, striped, striated. Figs. 160, 161.) SIBERIAN YELLOW WAGTAIL. HOMEYER'S QUAKETAIL. Adult: Above yellowish-green; below, yellow, shaded with greenish on sides, with dusky on breast, and bleaching on chin. Top of head bluish-gray; a long white superciliary stripe; a dusky area from corner of mouth through eye to ear-coverts. Quills of wing dusky; lesser coverts edged with color of back; median and greater coverts showing whitish wing-bars; inner secondaries edged with the same. Tail dusky; middle feathers edged with color of back; outer two on each side mostly white. Bill and feet black. Length about 6.50; wing 3.00; tail about 2.75; bill 0.50; tarsus 0.90; hind toe and claw 0.65. Nest on the ground; eggs usually 4–6, 0.75 × 0.55, whitish, thickly speckled with brown. *B. flavus* is a protean species of Europe, Africa, and Asia, occurring abundantly in Alaska, in a form with whole side of head, below the white stripe, slaty-blackish, and some dusky markings on breast; this is the Asiatic subspecies given in 2d–4th eds. of the Key as "flavus?", and considered the same by SHARPE, of extensive dispersion in Siberia, Kamtschatka, and southward. *B. leucostriatus* HOM., J. F. O., 1878, p. 128; *B. f. leucostriatus* ST., Orn. Expl. Kamtsch., 1885, p. 280; RIDGW., Man., 1887, p. 535; A. O. U. Lists, No. 696.

AN'THUS. (Gr. ἄνθος, *anthos*, Lat. *anthus*, a kind of bird.) PIPITS. Bill shorter than head, about as wide as high at base, compressed in most of its extent, acute at tip, where distinctly notched; culmen slightly concave between base and terminal convexity; rictus slightly bristled. Wings longer than tail, usually tipped by first 4 primaries, 5th abruptly shorter. Tarsus not shorter or rather longer than hind toe and claw; inner lateral toe rather longer than outer, or the two about equal; hind claw always lengthened and straightened (as in the figure beyond given of *Anthus pensilvanicus*). Coloration "niggled"—that is to say, broken up in streaks and spots. The species of *Anthus* make up about half the family; there are several genera. In typical *Anthus* the wing is longer than the tail, and its point is formed by the outer 4 primaries, the 5th being abruptly shorter; the hind claw is nearly straight, and nearly or quite equals its digit in length. *Neocorys* only differs in having the feet larger and tail shorter. In certain S. Am. forms (*Pediocorys* and *Notiocorys*) the wing is more rounded, and 4 or even 5 primaries enter into tip of wing; in several European subgenera only 3 primaries are abruptly longer than succeeding ones. *Anthus pensilvanicus* is strictly congeneric with the European *A. spinoletta*, type of the genus. About 50 species (among them six or eight Central and South American ones) have been ascribed to *Anthus ;* the true number is less than 40. They are terrestrial and more or less gregarious birds, migratory and insectivorous; nest on the ground, a large compact structure of grasses, mosses, hairs, feathers, etc. Eggs so heavily specked and clouded with brown as to present a nearly flat dark tone.

Analysis of Subgenera and Species.

Tarsus not shorter (rather longer) than hind toe and claw. Tail moderately shorter than wing, the outstretched feet not reaching beyond its end (ANTHUS *proper*).
 Markings of upper parts distinct —
  Except on rump and upper tail-coverts. Europe ; Greenland . . . . *pratensis*
  Including rump and upper tail-coverts. Asia ; Alaska ? L. Cala. . . . . *cervinus*
 Markings of upper parts obscure. North America . . . . *pensilvanicus*
Tarsus shorter than hind toe and claw. Tail only about two-thirds as long as wing, the outstretched feet reaching beyond its end (*Subgenus* NEOCORYS).
 Markings of upper parts distinct . . . . *spraguei*

A. praten'sis. (Lat. *pratensis*, relating to *pratum*, a meadow.) MEADOW PIPIT. Adult: Upper parts pale greenish-brown, distinctly marked with blackish-brown centres of the feathers; wing-quills and coverts clove-brown, edged with greenish-gray. Tail-feathers dark brown, edged with the greenish shade of the back; outer one obliquely white for nearly half its length, and others with white at end. Cheeks olivaceous, speckled with dusky. Under parts brownish-white with a tinge of green, marked on breast and sides with brownish-black streaks running forward as a maxillary chain; chin, belly, and under tail-coverts unmarked. Bill dusky above and at end, the rest livid flesh-color; feet obscure flesh-color; iris blackish. Length about 6.00; extent 9.50; wing 3.00; tail 2.50; bill 0.50; tarsus 0.75. Eggs 0.78 × 0.58. Europe ; Africa ; North American as occurring in Greenland, and also, it is said, in Alaska. I have seen Alaskan Pipits, certainly not *pensylvanicus*, but too young and in too bad condition to furnish decisive characters.

A. cervi'nus. (Lat. *cervinus*, fawn-colored.) RED-THROATED PIPIT. Adult: Above, light grayish-brown, fully streaked with dusky, the streaks broadest and darkest on the back. Wings and tail dusky, the feathers edged with pale brown, the long inner secondaries with buff; ends of middle and greater wing-coverts whitish; outer tail-feathers with much white on both webs, and next feather with a white spot at end of inner web. A pale and more or less buffy superciliary and malar stripe. Below, whitish, more or less suffused with fawn-color on chin and throat, the throat, breast, and sides broadly streaked or longitudinally spotted with brownish-black, aggregated into a stripe on each side of throat; chin, belly, and vent immaculate. Bill black, with yellowish base of lower mandible; feet dark brown. Wing 3.36; tail 2.50; bill 0.45; tarsus 0.85. A species of extensive distribution in northerly parts of the Old World, probably occurring in Alaska, and accidental in Lower California: see Pr. U. S. Nat. Mus., vi, Oct. 1883, p. 156. COUES, Key, 3d and 4th eds., 1887-90, p. 868; RIDGW., Man., 1887, p. 537; A. O. U. Lists, 1st and 2d eds., 1886-95, No. [699].

A. pensilva'nicus. (Properly spelled *pennsylvanicus*, conformably with the name of the State; originally called "Penn's Wood" after Wm. Penn, its founder; Lat. *silvanus* or *sylvanus*, pertaining to *silva* or *sylva*, a forest, woods. Fig. 162.) PENNSYLVANIAN PIPIT. AMERICAN TITLARK. BROWN LARK. Adult ♂ ♀: Upper parts dark brown with an olive shade, most of the feathers with dusky centres, giving an obscure streaky or nebulous appearance; eyelids, superciliary line, and all under parts brownish-white, or pale buffy or ochrey brown, very variable in shade from muddy white to rich buff, the breast and sides of the body and neck thickly streaked with dusky; wings and tail blackish, the inner secondaries pale-edged, and 1-3 outer tail-feathers white wholly or in part. Bill blackish, pale at base below; feet brown. Length 6.25-6.75, sometimes 7.00; extent 10.25-11.00; wing 3.25-3.50; tail 2.75-3.00; bill 0.50; tarsus 0.90. Young hardly differ appreciably from adults. N. Am., everywhere; an abundant and well-known bird of fields and plains; migratory; in the U. S. seen chiefly in flocks in fall, winter, and early spring; breeds in high latitudes, and in the Rocky Mts. above timber line as far south as Colorado; accidental in Europe; lays 4-6 very dark-colored eggs, 0.80 × 0.60, in a mossy or grassy nest on the ground; voice querulous, gait tremulous, flight vacillating. (*A. ludovicianus* of all former eds. of the Key, as of most writers, after *Alauda ludoviciana* GM., 1788; but the name *A. pensilvanicus* (LATHAM, Syn. Suppl. i, 1787, p. 287) has priority.

FIG. 162.—Titlark, nat. size. (Ad. nat. del. E. C.)

Subgenus NEOCORYS.

(This section has been given full generic rank in all former eds. of the Key: for characters see foregoing analysis of *Anthus*.)

A. (N.) spra'guei. (To Isaac Sprague, of Mass.) SKY PIPIT. SPRAGUE'S PIPIT. MISSOURI TITLARK. Adult ♂ ♀: Above, variegated with numerous streaks of dark brown and gray, in largest pattern on back, smallest on nape, the gray constituting the edging of the feathers. Below, dull whitish, more or less brownish-shaded across breast and along sides; breast sharply streaked, sides less distinctly so, with dusky; a more or less evident series of maxillary spots. Quills dark grayish-brown; inner ones, and wing-coverts, edged with grayish-white, corresponding to pattern of back. Middle tail-feathers like back; next ones blackish-brown; two outer pair wholly or mostly pure white; 3d pair from the outside usually touched with white near the end. With reduction of the gray edgings of the feathers of the upper parts by wearing away in summer, the bird becomes darker above, with narrower and sharper variegation, and the pectoral streaks are fainter. Bill blackish above; below, like the feet, pale flesh-color; iris black. After the fall moult the colors again become pure; the streaking of the upper parts is strong and sharp, and the under parts acquire a ruddy-brown shade. Young: Edgings of feathers of upper parts buffy, giving a rich complexion to the plumage; feathers of back with pure white edging, forming conspicuous semicircular markings; greater wing-coverts and long inner secondaries broadly tipped with white; primaries broadly edged and tipped with white or buff. Ear-coverts buffy-brown, forming a more conspicuous patch than in the adult. Under parts strongly tinged, except on throat and middle of belly, with buffy-brown, the pectoral and lateral streaks large and diffused. Sexes indistinguishable; ♀ rather smaller than ♂. Length 6.25–6.75, rarely 7.00; extent 10.00–11.00, generally about 10.50, rarely 11.50; wing 3.00–3.30; tail 2.25–2.40; bill 0.50; tarsus 0.80–0.90; middle toe and claw 0.90; hind toe and claw nearly 1.00, the claw alone about 0.50. Central portions of the U. S., and adjoining British Provinces. from E. edge of the high Central Plains to the Rocky Mts., from the valleys of the Red River of the North and of the Saskatchewan to Texas and the table lands of Mexico; accidental in South Carolina; breeding in profusion in Dakota and Montana; nest on the ground, of fine dried grasses, sometimes arched over; eggs 4–5, 0.90 × 0.60, grayish-white minutely flecked with dark tints, giving a purplish-brown cast. General habits and manners of Titlarks; but the soaring flight of the Sky Pipit when singing, and the song itself, possess all the qualities which have made the European Skylark famous, and are no less worthy of celebration in poetry: see Birds of the N. W., 1874, p. 42.

Family MNIOTILTIDÆ: American Warblers.

(Commonly called SYLVICOLIDÆ.*)*

Primaries 9; rectrices 12; tarsi scutellate; inner secondaries not enlarged, nor hind toe lengthened and straightened, as in the preceding family; bill without a lobe or tooth near middle of commissure, as in *Piranga;* not strongly toothed and hooked at end, as in *Lanius* and *Vireo* (which may have 10 primaries), nor greatly flattened with gape reaching to eyes, as in *Hirundinidæ*, nor strictly conical with angulated commissure, as in *Fringillidæ*. The family presents such a number of minor modifications of form, that it seems impossible to characterize it, except negatively; in fact, it has never been satisfactorily defined. But doubtless the student will be able to assure himself that his specimen is sylvicoline by its not showing the peculiarities

FIG. 163. — Black-throated Green Warbler, nat. size. (Ad. nat. del. E. C.)

of our other nine-primaried *Oscines*. All the Warblers are *small* birds; excepting *Icteria*, and perhaps a species of *Siurus*, not one is over 6.00 long, and they hardly average over 5.00. With few exceptions they are beautifully clothed in variegated colors; but the sexes are generally unlike, and the changes of plumage, with age and season of the year, are usually strongly marked, so that different specimens of the same species may bear to each other but little resemblance; this of course requires careful discrimination. The usual shape of the bill may be called conoid-elongate (something like a slender minié bullet in miniature), but the variations in precise shape are endless. The rictus is usually bristled; the bristles sometimes have an extraordinary development, and are sometimes wanting. The wings are longer than the tail, except in *Geothlypis*, *Icteria*, and a few exotic genera; neither wing nor tail ever presents striking forms; the head is never crested. The feet have no special peculiarities, though they show some slight modifications corresponding to somewhat terrestrial, or more strictly arboricole, habits. The nidification is endlessly varied, more or less artistic or artless nests being built in trees, bushes, holes, or on the ground. Musical proficiency might be expected from the agreeably suggestive name of the family, but as a rule the " Warbler's " singing is rather " quaint and curious " than very skilfully modulated or highly melodious — to which statement, however, there is signal exception to be taken, as in the case of the *Siuri*. Some Warblers have the habits of Titmice or Wrens; others of Creepers or Nuthatches; the *Siuri* closely resemble Titlarks in some respects, and have even been placed in *Motacillidæ*; while *Setophaginæ* simulate *Tyrannidæ* (of a different suborder) so perfectly that they used to be classed with these clamatorial Flycatchers. Warblers grade so perfectly toward Tanagers that they have all been made a subfamily of *Tanagridæ* (where possibly they belong). The affinity of some of them with *Cœrebidæ*, or Honey-creepers of the tropics, is so close that the dividing line has not been drawn. The position of *Icteria* and its two associate exotic genera, *Granatellus* and *Teretistris*, is open to question; perhaps they come nearer *Vireonidæ*. It is probable that final critical study will result in a remapping of the whole group; meanwhile, the very diversity of its forms enables us to discriminate the genera with ease. We have usually followed BAIRD in recognizing for our genera the three subfamilies " *Sylvicolinæ*," *Icteriinæ*, and *Setophaginæ*, which have been formally presented in previous eds. of the Key, and such subdivision has the merit of practical convenience. But the basis of this grouping is not scientifically strong, and I am quite ready to follow the example of the A. O. U. in ignoring subfamilies altogether in treating our North American forms.

It is unfortunate that the long-current name of this family, SYLVICOLIDÆ, which has been used in all former eds. of the Key, can no longer be used consistently with our rules; and I wish we could employ the term *Dendrœcidæ*, derived from the name of our most characteristic genus. But *Sylvicola* in ornithology is inadmissible, having been given to a genus of Mollusks long before it was applied to our Warblers; *Sylvicolidæ* must therefore be discarded in favor of *Mniotiltidæ*, because *Mniotilta* is the earliest name of exclusive pertinence to any genus of this family.

This is the second largest family of North American birds, *Fringillidæ* alone surpassing it in number of species. If not exactly " representative," in a technical sense, of the Old World *Sylviidæ*, it may be considered to replace that family in America, having much the same *rôle* in bird-economy; both families abound in species and individuals; they are small, migratory, insectivorous, and everywhere take prominent part in the make-up of the bird-fauna. There are nearly or about 140 good species of *Mniotiltidæ*, distributed over the whole of North and Middle America, and much of South America. The centre of abundance of *Setophaginæ*, or Flycatching Warblers, is in the warmer parts of America; comparatively few species reach the United States, and only two or three are extensively dispersed in this country. On the other hand, the *Mniotiltinæ* are more particularly birds of North America; very few of the species are confined to Middle or South America; and *Dendrœca*, the leading type of this group, is the largest, most beautiful, and most attractive genus of North American birds, pre-eminently characteristic of this country. The Warblers have we always with us, all in their own good time; they come out of the South, pass on, return, and are away again, their appearance and withdrawal scarcely less than a mystery; many stay with us all summer long, and some brave our winters. Some of these slight creatures, guided by unerring instinct, travel true to the meridian in hours of darkness, slipping past "like a thief in the night," stopping at daybreak from their lofty flights to rest and recruit for the next stage of the journey. Others pass more leisurely from tree to tree, in a ceaseless tide of migration, gleaning as they go; the hardier males, in full song and plumage, lead the way for the weaker females and yearlings. With tireless industry do Warblers befriend the human race; their unconscious zeal plays due part in the nice adjustment of Nature's forces, helping to bring about that balance of vegetable and insect life without which agriculture would be in vain. They visit the orchard when the apple and pear, peach, plum, and cherry are in bloom, seeming to revel carelessly amid the sweet-scented and delicately-tinted blossoms, but never faltering in their good work. They peer into crevices of bark, scrutinize each leaf, and explore the very heart of buds, to detect, drag forth, and destroy those tiny creatures, singly insignificant, collectively a scourge, which prey upon the hopes of the fruit-grower, and which, if undisturbed, would bring his care to nought. Some Warblers flit incessantly in the terminal foliage of the tallest trees; others hug close to the scored trunks and gnarled boughs of the forest kings; some peep from the thicket, the coppice, the impenetrable mantle of shrubbery that decks tiny watercourses, playing at hide-and-seek with all comers; others more humble still descend to the ground, where they glide with pretty mincing steps and affected turning of the head this way and that, their delicate flesh-tinted feet just stirring the layer of withered leaves with which a past season carpeted the ground. We may seek Warblers everywhere in their season; we shall find them a continual surprise; all mood and circumstance is theirs.

Artificial Key to the Genera and Subgenera of Mniotiltidæ.

Length 7.00 inches or more; bill very stout . . . . *Icteria*
Length 5.50 inches or more and tail-feathers plain; bill ordinary . . . . *Siurus*
Length under 5.50 or tail-feathers not plain.
  Wing shorter than tail or equal, and head ashy . . . . *Geothlypis*
  Wing longer than tail or equal, and head not ashy.
    Tarsus shorter than middle toe and claw; plumage black and white in streaks . . . . *Mniotilta*
    Tarsus not shorter than middle toe and claw.
      Rictal bristles evidently reaching far beyond nostrils.
        Tail black and orange, or black and white, or dark and yellow . . . . *Setophaga*
        Tail ashy edged with white, and head with red . . . . *Cardellina*
        Tail greenish, unmarked, or with white blotches . . . . *Wilsonia*
        Tail dusky and reddish, body carmine, ears silvery . . . . *Ergaticus*
        Tail otherwise, head striped with black and yellow . . . . *Basileuterus*
      Rictal bristles evidently not reaching far beyond nostrils, or not evident at all.
        Tail-feathers all unmarked.
          Bill at least 0.50 inch long, very acute.
            4 black stripes on head . . . . *Helmitherus*
            no black stripes on head . . . . *Helinaia*
          Bill not 0.50 inch long
            Wing over 2.50 inches; bill not acute; bright yellow below, or head ashy . . . . *Oporornis*
            Wing not over 2.50 inches; bill very acute; no bristles . . . . *Helminthophila*
        Tail-feathers blotched with white, or yellow on inner webs.
          Rictal bristles not evident.
            Bill not 0.50 inch long; whole fore parts not yellow . . . . *Helminthophila*
            Bill at least 0.50 inch long; whole fore parts yellow . . . . *Protonotaria*
          Rictal bristles very evident.
            Back blue with gold spot, throat and legs yellow . . . . *Compsothlypis*
            Head orange-brown with black bar through eye . . . . *Peucedramus*
            Coloration otherwise . . . . *Dendrœca*

Diagnostics or Characteristics of certain Genera and Subgenera of Mniotiltidæ.

Mniotilta, Compsothlypis, and *Peucedramus* are *creeping* Warblers, with certain slight modifications of the feet, enabling them to scramble about trees much like Creepers or Nuthatches.

Geothlypis and *Oporornis* are *ground* Warblers, with the feet modified in adaptation to terrestrial life. *Siurus* is similar in this respect; the species *walk* on the ground, and act in some respects like Motacillines.

Protonotaria, Helinaia, Helmitherus, and *Helminthophila* are "*worm-eating*" Warblers (the old genus Vermivora), with slight rictal bristles or none.

Setophaga, Cardellina, Wilsonia, Ergaticus, and *Basileuterus* are *fly-catching Warblers,* with strongly bristled bill and muscicapine habits, in some respects like species of *Tyrannidæ.*
Icteria is isolated by its peculiarities of form and habits, and great size for this family.
Dendrœca comprehends the wood *Warblers par excellence,*—the largest genus, with over twenty species.
BILL:—Peculiarly stout, high, and compressed in *Icteria;*—flattish, and strongly bristled in *Setophaga* and *Wilsonia;*—parine in *Ergaticus* and *Cardellina;*—large, with straightish outlines, scarcely or not bristled, and very acute in *Protonotaria, Helinaia,* and *Helmitherus;*—small, unbristled, and very acute in *Helminthophila.*
FEET:—Tarsus longest, slenderest, and usually pale-tinted in *ground Warblers;*—shortest in *creeping Warblers,* with relatively longest toes.
WINGS:—Shorter than tail in *Icteria* and species of *Geothlypis;*—about equal to tail in species of *Geothlypis, Siurus, Setophaga,* and *Cardellina;*—usually decidedly longer than tail.
TAIL:—The feathers (some or all) *blotched with white* in the following: *Mniotilta, Compsothlypis, Protonotaria,* species of *Helminthophila,* all *Dendrœcæ* (excepting *D. œstiva* and its allies), *Peucedramus,* one *Wilsonia,* one *Setophaga.* The feathers plain olivaceous, or otherwise like back, unmarked, in species of *Helminthophila,* in *Helmitherus, Oporornis, Geothlypis, Siurus, Icteria,* species of *Wilsonia, Cardellina;* yellow and dark in one *Setophaga* and one *Dendrœca;* dusky and reddish in *Ergaticus.*

MNIOTIL'TA. (Gr. μνίον, *mnion,* moss, and τίλλο, *tillo,* I pluck, or τιλτός, *tiltos,* plucked; conjectural application to the nest-building.) CREEPING WARBLERS. Coloration entirely black-and-white; tail-feathers white-blotched. Tarsus not longer than middle toe and claw; hind toe long, with large claw. Wings long, pointed, 1st primary about as long as 2d; tail nearly even, much shorter than wing. Bill nearly as long as head, slender, much compressed, with concave lateral outlines, and curved culmen and gonys, slightly notched and bristled. Only one good species.

M. var'ia. (Lat. *varia,* variegated. Fig. 164.) VARIED CREEPING WARBLER. WHITE-POLL WARBLER. BLACK-AND-WHITE CREEPER. ♂, adult: Black; edges of feathers of upper parts, coronal, superciliary, and maxillary stripes, tips of greater and median wing-coverts, outer edges of inner secondaries and inner edges of quills and tail-feathers, and spots on inner webs of lateral tail-feathers, white; under parts white, with black streaks on throat, sides, and crissum; bill and feet black. ♀ similar: less black in proportion to the white, being mostly white below. Young of both sexes resemble the ♀; at a very early age the white parts are tinged with tawny, and the black is not pure—rather gray; but the streakiness of the bird at all ages is unmistakable. Length 5.00–5.25; extent 8.25–8.75; wing 2.35–2.75; tail 2.25; bill nearly 0.50. Eastern N. Am.; N. to the

FIG. 164.—Black-and-white Creeper, nat. size. (Ad. nat. del. E. C.)

Fur Countries; W. to the Plains; accidental in California (Pasadena, Auk, 1896, p. 260); migratory; breeds throughout most of its range; winters from the southern border southward to the West Indies, Central America, and northern South America. A common bird of woodland, thicket, and swamp, generally seen scrambling actively about the trunks and larger branches of the trees, rather like a Nuthatch than like a Creeper, the tail not being used as a prop. Nest on the ground or in a stump, of bark-strips, mosses, grasses, leaves, hair, etc.; eggs 4–5, 0.70 × 0.52, white, profusely marked with reddish and other dots. (*M. v. borealis,* queried in former eds. of the Key, as based on northerly birds said to be smaller-billed, may now be disregarded.)

COMPSOTH'LYPIS. (Gr. κομψός, *kompsos,* dressy, exquisite, ornate, as these birds certainly are; θλυπίς or θραυπίς, *thlupis* or *thraupis,* some bird so called; θλαυπις is also alleged as a personal proper name.) PARULA WARBLERS. Coloration highly variegated; tail-feathers white-blotched; back bluish, with yellowish spot; throat yellow, with dark spot; feet pale. Size under 5.00. Bill short, stoutish; notch obsolete; bristles slight, though evident. Two distinct species in N. Am., and others in warmer parts of America. (*Parula* of previous eds. of the Key, and of most writers since 1858; rejected by our rules on account of the earlier *Parulus* SPIX, 1824; for synonymy see COUES, Birds Col. Vall. i, 1878, p. 206.)

C. america'na. (Lat., of America.) BLUE YELLOW-BACKED WARBLER. PARULA WARBLER. ♂, in spring: Upper parts clear ashy-blue; middle of back with a patch of greenish-yellow or brownish-golden. Lores dusky. A white spot on each eyelid. Wings blackish, crossed on ends of greater and middle coverts with two broad white bars; primaries narrowly, secondaries more broadly, edged externally with the color of the back, internally with white. Tail like wings, with much edging of outer webs like the back, the middle feathers mostly bluish; at least two outer feathers on each side with large, white, squarish patches on inner web near the end, usually 3d feather blotched with white, and a white touch on 4th or even 5th feather. Chin and throat yellow, rather narrowly confined, this yellow spreading over whole breast, but much of breast spotted or tinged with orange-brown, and jugulum showing even a blackish collar; coloration of this part very variable; sometimes reddish-brown markings along sides, much as in the Chestnut-sided Warbler. Rest of under parts white. Bill above black; below whitish or flesh-colored, drying yellowish. Legs pale. Length 4.50-4.75; extent 7.00-7.50; wing 2.10-2.30; tail 1.75. ♀, in spring: Like ♂; upper parts less brightly bluish, or with slight greenish gloss; back-patch not so well defined; less white on tail; white wing-bands narrower; dark or reddish tinting of fore breast less decided or scarcely indicated; the yellow more restricted. Young: Bluish of upper parts glossed over with greenish, sometimes to such extent as to obscure the dorsal patch, which is then not very different from the rest of the upper parts. White tail-spots smaller, generally confined to two outer feathers on each side. White wing-bands narrower. Edging of tail and wings tinged with greenish, like back. Eyelids not spotted with white. Yellow of fore under parts pale, with little or no indication of dusky across jugulum. White of under parts tinged with yellowish posteriorly, and frequently showing brownish touches along sides. Eastern U. S.; W. sometimes to the Rocky Mts.; migratory; breeds in the greater part of its U. S. range, but chiefly southerly; winters from Florida southward. An elegant, diminutive species, abundant in high open woods, where it is generally observed fluttering among the smallest twigs and terminal foliage. Nest in trees, an elaborate woven structure of mosses and lichens often placed in a bunch of Spanish moss (*Tillandsia usneoides*); eggs 4-5, 0.62 × 0.48, white with the usual sprinkling of reddish and other dots.

FIG. 165.—Northern Parula. (L. A. Fuertes.)

C. a. us'neæ. (Lat., of *usnea*, a kind of lichen hanging like moss from trees, etc. *Usnea barbata* is one of the species. Fig. 165.) NORTHERN PARULA WARBLER. Like the last, slightly larger on an average and with shorter bill. ♂, adult: More black on lores; less yellow on under parts; the collar black or blackish, and fore breast much spotted with rich dark chestnut. Eastern U. S. and British Provinces; breeding range more northerly than that of typical *americana*; nest usually almost invariably placed in the hanging-moss whence the name is derived, mostly 2-8 feet from the ground; eggs 4-5, May (best account of nesting in Auk, July, 1897, pp. 289-294). Included with the foregoing in all former eds. of the Key. BREWST., Auk, Jan. 1896, p. 44; A. O. U. List, Eighth Suppl., Auk, Jan. 1897, p. 123.

C. nigrilo'ra. (Lat. *niger*, black; *lorum*, a bridle; applied to the space between eye and bill of a bird.) SENNETT'S WARBLER. Adult ♂: Upper parts of the same ashy-blue color as in *C. americana*, with a dorsal patch of greenish-yellow exactly as in that species. Wings also as in *americana*, dusky, with grayish-blue outer and whitish inner, edgings, and crossed by two conspicuous white bars on tips of greater and middle coverts. Tail as in *americana*, but the white spots smaller and almost restricted to two outer feathers on each side. Eyelids black, *without* white marks. Lores broadly and intensely black, this color extending as a narrow frontal line to meet its fellow across base of culmen, and also reaching back to invade auriculars, on which it shades through dusky to the general bluish. Under parts yellow as far as middle of belly, a little farther on flanks, also spreading on sides of the jaw to involve part of mandibular and malar region; on fore breast deepening into rich orange, but showing nothing of the orange-chestnut and blackish of *C. americana.* Lower belly, flanks and crissum white. Bill black above, yellow below. Legs light horn-color. Length about 4.50; wing 2.00–2.20; tail 1.80–1.90; bill from nostrils 0.38–0.40; tarsus 0.62–0.65; middle toe alone 0.40. Texas, in the valley of the Lower Rio Grande, and southward in Mexico. Another little exquisite, which I added to our fauna in 1878.

PROTONOTA′RIA. (Low Lat. *protonotarius*, first notary, or scribe; why?) GOLDEN SWAMP WARBLERS. Bill of great size, nearly as long as head, compressed, conic, acute, with slightly notched tip and scarcely bristled rictus. Wings pointed, unmarked, much longer than the short, nearly even, tail. Tarsus about equal to middle toe and claw. One species.

P. cit′rea. (Lat. *citrea*, pertaining to the citron; *i. e.*, yellow.) PROTHONOTARY WARBLER. Adult ♂ ♀ : Golden-yellow, paler on belly, changing to olivaceous on back, thence to bluish-ashy on rump, wings, and tail; most of the tail-feathers largely white on inner webs; no other special markings; bill entirely black, very large, at least 0.50 long. Length about 5.50; extent 9.25; wing 2.75–3.00; tail 2.25; tarsus 0.75. Sexes similar. In highest feather the yellow of the head sometimes becomes orange-red. Eastern U. S., southerly; N. to Virginia, southern Michigan, and S. E. Minnesota, casually to Maine and Ontario, W. to Kansas, Nebraska, Indian Territory, and Texas; winters extra-limital. A beautiful species, of striking form and colors, and sedate manners, inhabiting swamps and thickets; nest in holes or other sheltered cavities in trees, stumps, and logs, of the most miscellaneous materials; eggs usually 4–5, but varying 3–7, 0.68 × 0.54, creamy white, profusely speckled with brown and gray.

HELMITHE′RUS. (Gr. ἕλμις, gen. ἕλμινθος, *helmis, helminthos*, a bug; θηρᾶν, to hunt; θήρ, an animal; *i. e.,* ἑλμινθοθήρας, *helminthotheras*, a bug-hunter; like *vermivora*, worm-eating.) WORM-EATING SWAMP WARBLERS. Bill large, conic-acute, especially high and stout at base, nearly as long as head, unnotched and scarcely or not bristled. Wings rather pointed, much longer than the little rounded tail. Tarsus about equal to middle toe and claw. Sexes similar; tail-feathers unmarked; legs pale. Two very distinct species were formerly included in this genus, as in all former eds. of the Key. It is now restricted to one of them. The name of the genus is *Helmitheros* RAF., Journ. Phys., 1819, p. 417, cited as *Helmitherus* by Baird, 1858, by me in orig. ed. of the Key, 1872, and so given in the A. O. U. Lists, 1886 and 1895; given as *Helmintherus* by me in 2d-4th eds. of the Key, 1884–1890, and in the Century Dictionary; given as *Helminthotherus* in the British Museum Catalogue, 1885; probably the most classic form we could use would be *Helminthotheras*.

H. vermi′vorus. (Lat. *vermivorus*, worm-eating; *vermis*, a worm; *voro*, I devour. Fig. 166.) WORM-EATING WARBLER. Adult ♂ ♀ : Olive, below buffy, paler or whitish on belly; head buff, with four black stripes, two along sides of crown from bill to nape, one along each side of head through eye; wings and tail olivaceous, unmarked; iris dark brown; upper mandible brown, lower mandible and feet pale; bill acute, unbristled, unnotched, at least 0.50. Length 5.50; extent 8.75; wing 2.75–3.00; tail 2.00–2.25. The distinctive head-stripes appear before the bird is fully fledged, when the upper parts are brownish, and the wing-coverts have buff tips.

FIG. 166. — Worm-eating Warbler, nat. size. (Ad. nat. del. E. C.)

Eastern U. S., rather southerly; but N. regularly to Middle States and Connecticut; west to Kansas, Missouri, and Indian

Territory; breeds throughout its U. S. range; winters from Florida southward; common in woods, shrubbery, and swamps. A bird of rather slow and sedate movements; nest on the ground, of leaves, grasses, rootlets; eggs 4–5, crystal-white, minutely dotted with reddish-brown, 0.70 × 0.50.

HELINAI'A. (Gr. ἔλος, *helos*, a marsh, and ναίω, *naio*, I dwell, abide: AUD., Syn., 1839, p. 66, where the faulty word is coined; emended to *Helonæa* by AGASSIZ, and so given by me in the Century Dict. The orig. form of the word is preserved in the A. O. U. Lists. The genus was intended by Audubon to include all the so-called "worm-eating" Warblers; but by successive restrictions it has been confined to its type species, which has usually been included under *Helmintherus*, as in all former eds. of the Key.) CANEBRAKE WARBLERS. Characters in general of *Helmitherus*. Bill larger and differently shaped, nearly as long as the tarsus, deep at base, acute at tip, with straight, sharp culminal ridge rising high on the forehead, something like a meadow-lark's. Point of wing formed by 2d and 3d quills; 1st shorter than 2d. Feet stout, with tarsus slightly longer than middle toe and claw. Coloration plain and simple. Habits terrestrial and aquatic. One species, affording a curious analogy to the "acrocephaline" type of Reed Warblers of the Old World family *Sylviidæ*.

H. swain'soni. (To Wm. Swainson.) SWAINSON'S WARBLER. Somewhat similar to the last; no long black head stripes; no strong markings anywhere. Adult ♂ ♀ : Upper parts olive, nearly uniform, but brownish on exposed surfaces of wings and tail, and quite reddish-brown on crown and nape. A long light superciliary stripe; below this a dusky loral and transocular line; sides of head below this speckled with brownish on a whitish ground; sometimes also a short median yellowish stripe on forehead. Lower parts whitish, of a creamy or pale yellowish tinge, shaded on sides with brownish-olive, and quite across breast with some nebulous markings. Specimens vary much in precise tone of coloration, some being more olivaceous, others more brownish, independently of sex and season. Bill brown above, pale below; feet flesh-color; iris brown. Young in the fall are browner than adults above, more yellowish below and on eyebrows; they show blackish lores; the first plumage is mostly dull rufous-brown becoming whitish on belly; wings and tail as in adults. A rather large Warbler; length up to 6.00 or more; extent about 9.00; wing 2.65–2.95; tail 1.85–2.15; tarsus 0.65–0.75; middle toe nearly as much; culmen 0.65–0.75. This interesting bird, long very rare in collections and supposed to be confined to the S. Atlantic States, is now well known by many specimens to extend N. to the Great Dismal Swamp in Virginia, to Indiana, Missouri, and E. Texas; in winter, S. in Mexico to Vera Cruz; Cuba; Jamaica. It is a beautiful songster, of sedate movements and retiring disposition, breeding in canebrakes, where the nest is affixed to canes over the water, like a Marsh Wren's; it is a bulky structure of twigs, leaves, mosses, rootlets, hairs, etc., sometimes 5 or 6 inches in diameter; eggs 2 or 3, 0.75 × 0.58, whitish, plain or variously marked with pale spots which may be scattered over the whole surface or wreathed about one end, laid late in June. For history of the species since the Audubonian period, see GUNDL., Journ. f. Orn., 1872, p. 412 (Cuba); MAYNARD, B. Fla., 1873, p. 47; N. C. BROWN, Bull. Nutt. Orn. Club, 1878, p. 172; RIDGW., *ibid.*, p. 163, and 1881, p. 54; A. NEWTON, P. Z. S., 1879, p. 552 (Jamaica); HOXIE, Orn. and Oöl., 1884, p. 138; COUES, Forest and Stream, Nov. 6, 1884, p. 285; and especially BREWST., Auk, Jan. 1885, p. 65; For. and Str., July 9, 1885, p. 468; Auk, Oct. 1885, p. 346.

HELMINTHO'PHILA. (Gr. ἕλμις, ἕλμινθος, *helmis, helminthos*, a bug; φιλέω, *phileo*, I love.) WORM-EATING WARBLERS. Bill slender and exceedingly acute, unnotched, unbristled (fig. 167). Wings pointed, longer than nearly even tail — in one species nearly half as long again. Tarsus longer than middle toe and claw. Tail-feathers in some species white-blotched, in others plain — the former being otherwise of bright and varied colors, the latter more simply clad. Nest on the ground or quite near it (excepting in the case of *H. luciæ*); eggs white, spotted. To the 8 established U. S. species of the genus have been added 3 others;

but one of them is almost certainly a hybrid between *H. pinus* and *Oporornis formosa*, while the other two are probably hybrids between *H. pinus* and *H. chrysoptera*. There have also been added a variety of *H. rubricapilla*, and two varieties of *H. celata*. These are enumerated beyond, but only the 8 established species are considered in the analysis of the genus. Even with this reduction, *Helminthophila* is still the second largest genus of the subfamily. It is peculiarly North American, all the known species occurring in this country, some of them not being known to occur elsewhere. The genus may be divided according to coloration into two groups, which correspond in a general way with geographical distribution. Three species (*HH. pinus, chrysoptera,* and *bachmani*), exclusively Eastern, are of variegated colors, the tail-feathers white-blotched as in *Dendrœca*. In the other five the coloration is simpler; the tail-feathers are not, or not conspicuously, blotched with white, and all but one of these species have a crown-patch; one of them is Eastern, two are Western, and two of general dispersion. The natural analysis of the species, and a shorter Key to them, are subjoined; these tables should suffice to identify adult males, but females and young, particularly of *celata, rubricapilla,* and *virginiæ*, require detailed descriptions for their recognition. (In *H. peregrina*, with tail normally plain, the outer feather is sometimes distinctly white-blotched.) *Helminthophaga* Cab., of 1st ed. of Key. *Helminthophila* Ridgw., of all later eds.

Fig. 167. — *H. chrysoptera*, nat. size. (Ad. nat. del. E. C.)

Natural Analysis of Species.

I. Tail-feathers conspicuously white-blotched. Wings with white or yellow on coverts. Head or breast with black.
  Exclusively Eastern.
  1. Bluish-ash, below white; crown and wing-bars yellow; throat and stripe on side of head black
    . . . . *chrysoptera*
  2. Olive-green; wings and tail bluish-ash, former with white or yellow bars; crown and under parts yellow; lores black . . . . *pinus*
  3. Olive-green, below yellow; throat, breast, and crown-patch black; forehead yellow . . . . . *bachmani*
II. Tail-feathers inconspicuously or not blotched with white. No decided wing-markings. No black anywhere.
  a. Crown without colored patch. Wings about half as long again as tail.
  4. Tail with obscure whitish spot on outer feather; under parts white or whitish; upper parts olive-green, brighter behind, quite ashy in front. Chiefly Eastern . . . . *peregrina*
  b. Crown with colored patch. Wings shorter.
  5. Crown-patch orange-brown; tail unmarked; upper parts olive-green, under parts greenish-yellow, both nearly uniform. Western and incompletely Eastern . . . . *celata*
  6. Crown-patch chestnut; tail unmarked; upper parts olive-green, growing ashy on head; under parts uniformly yellow. Eastern and incompletely Western . . . . *rubricapilla*
  7. Crown-patch chestnut; tail unmarked; above olivaceous-ash, below whitish; rump and under tail-coverts bright yellow; breast yellowish. Western . . . . *virginiæ*
  8. Crown-patch and upper tail-coverts chestnut; outer tail-feather with dull white patch; above pale cinereous, below white. Southwestern . . . . *luciæ*

Pass-key to the Species.

Tail-feathers white-blotched — bluish, crown yellow, throat black . . . . *chrysoptera*
  — greenish, crown and all under parts yellow . . . . *pinus*
  — greenish, crown (partly) and throat black . . . . *bachmani*
  — upper tail-coverts chestnut, crown-patch chestnut . . . . *luciæ*
Tail-feathers all unmarked — upper tail-coverts — yellow; crown-patch chestnut . . . . *virginiæ*
  — not yellow; crown-patch chestnut . . . . *rubricapilla*
    orange-brown . . . . *celata*
    wanting . . . . *peregrina*

H. pi'nus. (Lat. *pinus*, a pine-tree. Fig. 168.) BLUE-WINGED YELLOW WARBLER. Adult ♂: Fore part of crown and entire under parts rich yellow; upper parts yellow-olive, becoming slaty-blue on wings and tail (system of coloration thus like that of *Protonotaria*). Wings with two white or yellowish bars; tail with several large white blotches; under tail-coverts white; eyelids bright yellow; small stripe through eye black; bill blue-black. Female and young not very dissimilar; duller and more olivaceous. Length about 4.75; extent 7.50; wing 2.40-2.50; tail 2.00-2.10; tarsus 0.65; bill 0.45. Eastern U. S., N. to Massachusetts and Minnesota, W. to Kansas, Indian Territory, and Texas; common, migratory, breeding in most of range, wintering extralimital in Mexico and Central America. Nest on the ground; eggs 4-5, 0.67 × 0.48, white, sprinkled with reddish-brown and blackish dots chiefly abounding near the large end, laid late in May and early in June.

FIG. 168. — Blue-winged Yellow Warbler. (L. A. Fuertes.)

H. lawren'cei? (To Geo. N. Lawrence, of N. Y.) LAWRENCE'S WARBLER. Like *H. pinus;* but a large black patch on throat and breast, and broad black eye-stripe, reaching over auriculars, as in *H. chrysoptera;* thus *pinus* × *chrysoptera*, and doubtless a hybrid between the two. About a dozen specimens known, New Jersey, Connecticut, etc. A. O. U. Hypothetical List, 1896, No. 20.

H. leucobronchia'lis? (Gr. λευκός, *leucos*, white, βρόγχος, *brogchos*, becoming *bronchus,* throat.) WHITE-THROATED WARBLER. Like *H. chrysoptera;* but a black bar through eye as in *pinus*, and lacking the black breast-patch of *chrysoptera*, the entire under parts being white; thus *chrysoptera* × *pinus*, and doubtless a hybrid between the two, though up to date numerous specimens have been described, from New England, New York, New Jersey, Pennsylvania, Virginia, Michigan, etc. Figured in colors on pl. 1 of the Nuttall Club Bulletin, 1876. A. O. U. Hypothetical List, 1896, No. 21.

H. cincinnatien'sis? (Of Cincinnati, Ohio, where discovered.) CINCINNATI WARBLER. Like *H. pinus* in color; bill with evident rictal bristles; no white wing-bars or tail-blotches; no ashy-blue on wings or tail; concealed black on crown and sides of head like the incompleted black mask of *Oporornis formosa*, with which the bird otherwise closely agrees in color; thus curiously being *H. pinus* × *O. formosa*. Length 4.75; wing 2.50; tail 1.85; bill 0.44. One specimen known, Ohio. A. O. U. Hypothetical List, 1896, No. 22.

H. chrysop'tera. (Gr. χρυσός, *chrusos*, golden, and πτερόν, *pteron*, wing.) BLUE GOLDEN-WINGED WARBLER. ♂, adult: Upper parts slaty-blue, or fine bluish-gray; crown, and large wing-patch formed by confluent wing-bars, rich yellow; a broad stripe on side of head and patch on chin, throat and fore-breast, black, the eye-stripe bordered above and below with white; under parts generally, excepting the black breastplate, white, often tinted with yellowish, and shaded on the sides with ashy. Exposed surfaces of wings and tail like upper parts; great white blotches on three lateral tail-feathers; bill black; feet dark. ♀ and immature specimens have the back more or less glossed with yellowish-olive; yellow of crown obscured with greenish; black eye-stripe and breastplate veiled with gray tips of the feathers, or not at all evident. Size of *H. pinus*. A beautiful species, common in Eastern United States; N. to Southern New England, Ontario, Minnesota, etc., migratory, breeding from our middle districts northward, and in mountains S. to the Carolinas, retiring in the fall entirely to winter in Cuba, E. Mexico, Central America, and the U. S. of Colombia. Nest on the ground, like that of *H. pinus;* eggs similar, 0.65 × 0.50, white, dotted with browns in fine pattern, mostly about the larger end.

H. bach'mani. (To Rev. John Bachman, of S. C. Fig. 169.) BACHMAN'S WARBLER. Adult ♂: Upper parts yellowish-olive, including sides of head and neck, tinged with ashy on hind head; forehead and under parts bright yellow; a black band on vertex separating yellow front from ashy occiput; throat and fore breast black, this breastplate isolated in yellow surroundings. Wings dusky, glossed with color of back on all the exposed surface, the quills edged with ashy, and some of the lesser coverts yellow. Inner webs of three outer tail-feathers white-blotched. Small; length 4.50; wing 2.35; tail 2.00; bill at maximum of acuteness, and curvature. ♀ resembles ♂, but lacks the black crown, and the breastplate is dusky veiled with olive. S. Atlantic and Gulf States; N. to Virginia and S. Indiana, W. to Louisiana and Arkansas; Cuba in winter. This was long considered an extremely rare species, few specimens having been known until recently. See Auk, Jan. 1887, p. 35; Apr. 1887, p. 165; and for its rediscovery in abundance, with best biography, Auk, Apr. 1891, pp. 149-157.

Fig. 169. — Bachman's Warbler.

H. lu′ciæ. (To Miss Lucy Baird, daughter of Prof. S. F. Baird.) LUCY'S WARBLER. Adult ♂ ♀: Clear ashy-gray. Beneath white, with a faint tinge of buff on breast. A rich chestnut patch on crown, and upper tail-coverts of the same color. A white eye-ring. Quills and tail-feathers edged with the color of the back or whitish. Lateral tail-feather with an obscure whitish patch. Lining of wing white. Feet dull leaden-olive. Iris dark brown or black. Length 4.33-4.66; extent 7.00-7.50; wing 2.25-2.50; tail 1.75-2.00; tarsus 0.66; bill 0.25-0.33. Young: Lack chestnut on crown, though that of rump is present; throat and breast milk-white, without the ochrey tinge of the adults; wing-coverts edged with pale rufous. The chestnut upper tail-coverts, and absence of any trace of olivaceous or yellowish coloration, distinguish this interesting species, the general superficial aspect of which is quite like that of a *Polioptila*. Valley of the Colorado and Gila; common in Arizona, where I found it breeding at Fort Whipple in 1866; N. to Utah, S. into Sonora. The exceptional nidification of this species of the genus (Am. Nat., vi, 1872, p. 493) has been confirmed: nest in crevice behind bark of a tree or bush, or other odd nook, even some other bird's nest, of straws, leaves, hair, and feathers, such as a Wren might select; eggs 3-7, 0.58 × 0.45; not peculiar, being white dotted with reddish, chiefly wreathed about the large end, laid in May.

H. virgin′iæ. (To Mrs. Virginia Anderson, wife of the discoverer.) VIRGINIA'S WARBLER. ♂, in summer: Ashy-plumbeous, alike on back, and top and sides of head. Below dull whitish, the sides shaded with ashy. Lining and edge of wings white. Upper and under tail-coverts, and isolated spot on breast, yellow, in strong contrast with all surroundings. A white ring round eye. Wings and tail without yellowish edgings. Crown with a chestnut patch, as in *H. rubricapilla*. Length 4.75; extent 7.50; wing 2.25-2.50; tail 2.25. ♀, in summer: The yellow duller and slightly tinged with greenish; that of breast, and the chestnut of crown, more restricted. Autumnal specimens resemble the ♀; but in both sexes the plumbeous of the upper parts has a slight olive shade, and in birds of the year the crown-patch may be wanting. Southern Rocky Mt. region; N. to Colorado, Nevada, Utah, and Wyoming; S. into Mexico. Nest on ground, like others of this genus, at roots of a bush or tuft of grass, loosely made of hay, rootlets, and other fibres; eggs 4, 0.60 × 0.48, indistinguishable from those of allied species; laid in May and June. In Arizona and New Mexico the breeding range is at 5,000 feet or more.

H. rubricapil′la. (Lat. *ruber*, red; *capillus*, hair. Fig. 170.) NASHVILLE WARBLER. ♂, in summer: Upper parts olive-green or yellowish-olive, clearer and brighter on rump and upper tail-coverts. Top and sides of head and neck ashy, with a veiled chestnut patch on crown, and a white ring round eye. No superciliary stripe. Lores pale. Wings and tail fuscous, edged with color of back. Entire under parts yellow, including under wing-coverts and edge of wing; sides shaded with olive. Length 4.50–4.75; extent 7.50; wing 2.33–2.50; tail 1.75–2.00. ♀, in summer: Similar; head less purely ashy; crown-patch smaller and more hidden, if not wanting; yellow of under parts paler, whitening on belly. Autumnal specimens, of both sexes, though quite as yellow below as in summer, have the ash of the head glossed over with olivaceous, and in birds of the year the crown-patch may be entirely wanting. This species is distinguished by the rich clear yellow of the under parts at all seasons. In *H. celata*, which is next most yellow below, the color has a greenish cast; the head is little, if any, different from the rest of the upper parts, and the crown-patch is orange-brown. Eastern N. Am., W. to the Plains, N. far into the fur countries, S. in winter to Mexico and Central America. A common bird, migratory in most of U. S., breeding in latitude of S. New England (further S. in alpine regions) and thence northward. Nest on the ground, like the others, and eggs not peculiar. (*Sylvia ruficapilla* WILS., 1810, nec LATH., 1790. *H. ruficapilla* of 2d–4th eds. of Key, as of most late authors. *Sylvia rubricapilla* WILS., 1812. *H. rubricapilla* FAXON, Auk, July, 1896, p. 264; A. O. U. Suppl. List, Auk, Jan. 1897, p. 130, No. 645.)

FIG. 170. — Nashville Warbler.
(L. A. Fuertes.)

H. r. guttura'lis. (Lat. relating to *guttur*, the throat.) CALAVERAS WARBLER. Quite like the last; said to be more brightly colored; rump and upper tail-coverts more yellowish; lower parts more richly yellow; slightly larger; average size as alleged equal to the largest *rubricapilla*: ♂ wing 2.40–2.55; tail 1.90–2.00. Rocky Mts. to the Pacific, N. to Alaska (Kadiak), S. to L. Cala. and W. Mexico. I have heretofore declined to recognize this slight race, and do so now with reluctance. *Helminthophaga ruficapilla gutturalis* BD., BREW. and RIDGW., Hist. N. A. B., i, 1874, p. 191; *Helminthophila r. g.* RIDGW., Pr. U. S. Nat. Mus., viii, 1885, p. 354; A. O. U. Lists, 1886 and 1895, No. 645 *a*; *H. rubricapilla gutturalis* FAXON, Auk, July, 1896, p. 264; A. O. U. Suppl. List, Auk, Jan. 1897, p. 131, No. 645 *a*.

H. cela'ta. (Lat. *celata*, concealed, as is the orange on the crown.) ORANGE-CROWNED WARBLER. ♂♀, in summer: Upper parts olive, duller and washed with grayish toward and on head, brighter and more yellowish on rump and upper tail-coverts. Beneath greenish-white, palest on belly and throat, more olive-shaded on sides; the color not pure, but rather streaky, and having in places a grayish cast. Wings and tail edged with color of back; lining of wings like belly; inner edges of tail-feathers whitish. Orbital ring and lores yellowish. An orange-brown patch on crown, partially concealed, smaller and more hidden in ♀ than in ♂. Length 4.80–5.20; extent 7.40–7.75; wing 2.30–2.50; tail 2.00 or rather more. Resembling the last, and often difficult to distinguish in immature plumage; but a general *oliveness* and *yellowness*, compared with the ashy of some parts of *rubricapilla*, and different color of crown-patch in the two species, will usually be diagnostic. The sexes of this species scarcely differ, and young or autumnal birds are very similar to adults, except frequent or usual absence of the orange-brown crown-spot in birds of the year. The species is well distinguished from all its allies by color of crown-patch. N. Am. at large, but especially Western and Middle regions; rare or occasional in the Eastern Province; N. to Mackenzie River region and the Yukon in British America and Alaska; migratory mainly in the interior; winters in S. Atlantic and Gulf States and E. Mexico; breeds in Arctic and subarctic regions; in alpine localities S. to New Mexico; nest and eggs not peculiar.

H. c. lutes'cens. (Lat. *lutescens*, growing yellowish.) PACIFIC ORANGE-CROWNED WARBLER. Differs in being much more richly colored. It may be described simply as olive-green above, and greenish-yellow, shaded with olive on sides, below, without the qualifying terms required for precision in the case of typical *celata*. Pacific Coast region, Alaska to Lower California and W. Mexico: E. in migration to the Rocky Mts.; breeding range from S. Cala. to S. Alaska. Nest normally on ground, sometimes 3–6 feet up in a shrub or vine, built of leaves, grass, and hair; eggs laid in May and June, not peculiar.

H. c. sor'dida. (Lat., *sordid*, soiled, stained.) DUSKY ORANGE-CROWNED WARBLER. Differs in being more darkly colored; "there is an appearance of grayness about the upper plumage, owing to a leaden tinge on ends of feathers. Throat and under parts slightly streaked." San Clemente, Santa Rosa, and Santa Cruz Islands, off California. Neither this nor the last-named variety amounts to much, and both have been recognized mainly upon geographical considerations. TOWNS., Pr. U. S. Nat. Mus., xiii, 1890, p. 139; A. O. U. List, 2d ed., 1895, No. 646 *b*.

H. peregri'na. (Lat. *peregrina*, wandering, alien, foreign; *i. e.*, migratory. Fig. 171.) TENNESSEE WARBLER. Adult ♂ : Upper parts yellowish-olive, brightest posteriorly; on fore parts and head changing to pure ash, without any greenish tint whatever. No crown-patch of any different color. Lores, eye-ring, or frequently a decided superciliary stripe, whitish. Entire under parts dull white, scarcely or not tinged with yellowish. Wings and tail dusky, strongly edged with color of back; outer tail-feathers frequently with an obscure whitish spot. Bill and feet dark. Length 4.50–4.75, rarely 5.00; extent 7.50–8.00; wing about 2.75, thus long for the size of the bird, and especially in comparison with the short tail, pointed, with little difference in length between the first 3 or 4 quills; tail only 2.00 *or less*, thus remarkably short — the comparative length of wings and tail, with other characters, probably always distinguishes the species from the foregoing. Adult ♀: Quite like ♂, but ashy of head less pure

FIG. 171. — Tennessee Warbler. (L. A. Fuertes.)

and clear, and under parts more or less tinged with greenish-yellow. Young: Entire upper parts strongly and uniformly yellowish-olive, like rump of adult ♂, or even brighter, this color also tingeing eye-ring and superciliary stripe. Under parts as in adult ♀, or more decidedly greenish-yellow, leaving only belly and crissum whitish. In this condition specimens more closely resemble some other species than when adult; but the short tail, long wings, and no crown-patch should be distinctive. Chiefly Eastern N. Am., but W. to the Rocky Mts.; common, especially in the Mississippi Valley, but less so in the Atlantic States; migratory; breeds in N. New England and northern tier of States, and thence to high latitudes in British America; winters S. through E. Mexico to Central Amer. and the U. S. of Colombia. Nest and eggs as in other species of the genus.

DENDRŒ'CA. (Gr. δένδρον, *dendron*, a tree, and οἰκέω, *oikeo*, I inhabit.) WOOD WARBLERS. Bill variable in shape, usually conico-attenuate, more or less depressed at base, compressed from the middle, notched near tip, not showing the extreme acuteness of that of *Helmitherus*, *Helinaia*, *Helminthophila*, and *Protonotaria* (except in the subgenus *Perissoglossa*). Rictus with obvious bristles, which are not evident in the true "worm-eating" warblers. Tarsus longer than middle toe and claw (it is shorter, or not longer, in *Mniotilta*). Hind toe little if any longer than its claw (decidedly longer in *Mniotilta* and *Compsothlypis*). Wings much longer than tail, pointed, 1st and 2d primaries longest. Tail moderate, with rather broad feathers, nearly even, but varying to slightly rounded, or with slight central emargination. Pattern of coloration indeterminate. Tail always with white blotches (except in *æstiva* and its immediate allies, where the inner webs are yellow), never plain olivaceous. Crown never with lateral black stripes, nor under parts uniformly streaked with blackish on a pale ground, nor back with a yellow patch, nor whole head yellow. Length usually 5.00–6.00; rarely under and perhaps never over these dimensions. Nest in bushes or trees, with rare exceptions. Eggs white, spotted. It is not easy to frame a definition of this genus covering all its modifications, yet introducing no term inapplicable to any species; but the foregoing expressions considered collectively, however arbitrary or trivial some of them may seem to be, will serve to distinguish any *Dendrœca* from its allies of other genera; and, if so, the diagnosis is exclusively pertinent to the group as conventionally accepted. The coloration of the rectrices is a good clue to this genus; for all the species (excepting *D. æstiva* and its conspecies) have the tail-feathers blotched with white — a feature only shown, among North American allies, in *Mniotilta*, *Compsothlypis*, *Protonotaria*, and some species of *Helminthophila* and *Sylvania*. There is as much uniformity in the nest and eggs of *Dendrœca* as in those of *Helminthophila*. Whereas all these nest on the ground, as far as known all the *Dendrœcæ* nest in trees and bushes, with the single exception of *D. palmarum*. Excepting *D. castanea*, the eggs are essentially similar; all being white, variously speckled, dotted, or blotched with shades of reddish and darker brown, and lilac or purplish shell-spots. About 40 species are current, but not all of them are well established; notable extralimital species are: *pityophila* (Cuba), *adelaidæ* (Porto Rico), *pharetra* (Jamaica), *eoa* (Jamaica), *aureola* (Galapagoes), *capitalis* (Barbadoes), and *petechia* (West Indies) with its several tropical forms, all like our *æstiva*. Of the 26 species which have been ascribed to North America, "montana" and "carbonata" remain unknown: leaving 24 species to be treated, nearly as in the orig. ed. of the Key, there having been but two North American accessions (*olivacea* and *bryanti*) to the genus since 1872, though four varieties (respectively of *æstiva*, of *dominica*, of *palmarum*, and of *cœrulescens*) have meanwhile been described. *D. tigrina* was made type of a genus *Perissoglossa* by Baird in 1865, and I made *olivacea* type of a genus *Peucedramus* in 1876; but both of these are now reduced to subgenera of *Dendrœca*, as follows:

Analysis of Subgenera of Dendrœca.

Bill very acute, with appreciably decurved tip (much as in some species of *Helminthophaga*; tongue peculiarly fringed . . . . *Perissoglossa* (*tigrina*)
Bill very long, attenuate, culmen rather concave than convex in part, and under outline about straight. Wing half as long again as tail . . . . *Peucedramus* (*olivacea*)
Bill otherwise . . . . *Dendrœca* (proper)

The following artificial analysis will facilitate the determination of our 24 established species; I believe it to be an infallible key to the perfect male plumages, and that it will probably hold good for spring specimens of both sexes of many species; but it will fail for nearly all autumnal and most female specimens of (b). It is difficult if not impossible to meet the varied requirements of these by rigid analysis; and recourse must be had to the detailed descriptions of the species arranged in what seems to be their natural sequence. The supplementary table of certain diagnostic marks may prove of much assistance, though it is not a complete analysis.

Analysis of perfect Spring Males.

Tail-feathers edged with yellow; head — yellow . . . . *æstiva*, *æ. sonorana*, *æ. rubiginosa*
— chestnut . . . . *bryanti castaneiceps*
Tail-feathers blotched with white; a white spot at the base of primaries
head — black and blue . . . . *cœrulescens* and *c. cairnsi*
— orange-brown with black stripe . . . . *olivacea*
— no white spot at base of primaries. (a)
(a) Wing-bars not white. Below, white; sides chestnut-streaked, crown yellow . . . . *pennsylvanica*
— yellow; sides reddish-streaked, crown reddish . . . *palmarum* and *p. hypochrysea*
— black-streaked; above, ashy . . . . *kirtlandi*
— olive, reddish-streaked . . . . *discolor*
(a) Wing-bars white (sometimes fused into one large white patch). (b)
(b) Crown blue, like back; below white, sides and breast streaked . . . . *rara*
— chestnut, like throat; below, and sides of neck, buffy-tinged . . . . *castanea*
— clear ash; rump and under parts yellow, breast and sides black-streaked . . . . *maculosa*
— blackish, with median line orange-brown, like auriculars; rump yellow . . . . *tigrina*

— perfectly black ; throat black ; a small yellow loral spot . . . . *nigrescens*
— not black ; no yellow ; feet flesh color . . . . *striata*
— with yellow spot ; throat flame-color ; rump not yellow . . . . *blackburniæ*
— white ; rump and sides of breast yellow . . . . *coronata*
— yellow ; rump and sides of breast yellow . . . . *auduboni*
(b) Crown otherwise ; throat black ; back ashy, streaked, rump ash, crown yellow . . . . *occidentalis*
— black, like rump and crown . . . . *chrysoparia*
— olive ; crown like back . . . . *virens*
— not like back . . . . *townsendi*
— yellow ; back olive ; no black or ashy on head . . . . *vigorsi*
— ashy-blue ; cheeks the same ; eyelids yellow . . . . *graciæ*
— black ; eyelids white . . . . . *dominica* and *d. albilora*

Diagnostic marks of certain Warblers in any plumage.

Bill as above said for *Perissoglossa ;* rump generally yellow . . . . *tigrina*
Bill as above said for *Peucedramus ;* head orange-brown or yellowish . . . . *olivacea*
Wing-bars and belly yellow . . . . *discolor*
Wings and tail dusky, edged with yellow . . . . *æstiva* or *bryanti*
Wing-bars yellow, and belly pure white . . . . *pennsylvanica*
A yellow spot in front of eye and nowhere else . . . . *nigrescens*
A white spot at base of primaries (almost never wanting) . . . . *cærulescens*
Throat definitely yellow, belly white, back with no greenish . . . . *dominica, d. albilora* or *graciæ*
Rump, sides of breast, crown and throat more or less yellow . . . . *auduboni*
Rump, sides of breast, and crown more or less yellow ; throat white . . . . *coronata*
Wing-bars white, tail-spots oblique, at end of two outer feathers only . . . . *vigorsi*
Tail-spots at middle of nearly all the feathers, rump and belly yellow . . . . *maculosa*
Wing-bars brownish, tail-spots square, at end of two outer feathers only . . . . *palmarum* and *p. hypochrysea*
Wing-bars not very conspicuous, whole under parts yellow, back with no greenish . . . . *kirtlandi*
Tail-spots at end of nearly all the feathers, and no definite yellow anywhere . . . . *rara*
Throat, breast, and sides black or with black traces, sides of head with diffuse yellow, outer tail-feather white-edged
 externally . . . . *virens, chrysoparia, townsendi,* or *occidentalis*
Throat yellow or orange, crown with at least a trace of a central yellow or orange spot, and outer tail-feather white-
 edged externally . . . . *blackburniæ*
Bill ordinary ; and with none of the foregoing special marks . . . . *striata* or *castanea*

(*Subgenus* PERISSOGLOSSA.)

D. (P.) tigri′na. (Lat. *tigrina*, striped like a tiger, *tigris*. Fig. 172.) CAPE MAY WAR-
BLER. Adult ♂, in spring: Back yellowish-olive, spotted with black ; crown in high plumage
perfectly black, usually interrupted with olive. Rump, sides
of neck nearly meeting across nape, sides of head and entire
under parts, bright yellow ; *ear-patch orange-brown ;* a black
transocular stripe, cutting off a yellow superciliary stripe ;
lower throat and whole breast and sides thickly streaked
with black ; yellow of throat sometimes tinged with orange-
brown ; that of belly and under tail-coverts pale or whitish.
Wing-bars fused in a large white patch, formed by middle
coverts and outer webs of most of the greater coverts.
Quills and tail-feathers blackish, edged on outer webs with
olive ; tail-spots on three outer feathers near their ends,
oblique, large on outer feather, diminishing on the next
successively ; bill and feet blackish. The yellow patch on

FIG. 172. – Cape May Warbler. (L. A. Fuertes.)

the rump is conspicuous, and in high plumage that on the side of the neck is immaculate
and very bright. ♀, in spring : Similar ; lacking the distinctive head markings ; under parts
paler and less streaked, tail-spots small or obscure ; less white on wing. Young : An insig-
nificant-looking bird, resembling an overgrown Ruby-crowned Kinglet, without its crest ;
obscure greenish-olive above ; *rump yellowish ;* under parts yellowish-white ; breast and sides
with the streaks obscure or obsolete ; little or no white on wings, which are edged with
yellowish. Length 5.00–5.50 ; wing 2.75–2.85 ; tail 2.15–2.25. Eastern N. Am. to Hudson's

Bay, only known W. to the edge of the Great Plains; breeds from northern New England northward, and winters in the West Indies; resident, however, in Jamaica. The Cape May is an exquisite, resembling the Magnolia in its yellow rump and yellow black-striped under parts, but easily recognized at maturity by the orange-brown ear-coverts; possessing also the charm of rarity in most parts. The curved and very acute bill, and some anatomical peculiarities of the tongue will assist the student in recognizing the obscure ♀ and young. Nest in low trees or bushes, preferably evergreens, neatly cupped, built of small twigs, grasses, cobwebs, etc. Eggs 3–4, 0.70 × 0.50, white or whitish, marked chiefly about the larger end with the usual reddish-brown and darker spots or dots, with others of blackish and neutral tint.

(Subgenus PEUCEDRAMUS.)

D. (P.) oliva'ceus. (Lat. *olivaceus*, olivaceous in color; *oliva*, an olive. Fig. 173.) OLIVE WARBLER. Tongue much as in *Dendrœca*, but larger, with revolute edges, cleft tip, and laciniate for some distance from the end. Wings elongated, half as long again as tail (in *Dendrœca* less than half as long again), reaching, when folded, nearly to end of tail. Tail emarginate. Tarsus longer than middle toe and claw. Hallux little if any longer than its claw. Bill little shorter than tarsus (averaging little over half the tarsus in *Dendrœca*), attenuate, notably depressed, yet very little widened at base. Culmen rather concave than convex in most of its length, the under outline almost perfectly straight from extreme base to tip. Nasal fossæ very large, with a highly developed nasal scale. Rictal vibrissæ few and short. Plumage without streaks. Adult ♂: Upper parts ashy, more or less olivaceous, changing to greenish on nape. Head and neck all around orange-brown or intense saffron-yellow, with a broad black bar on side of head through eye. Wings blackish; inner webs of all the quills edged with white; outer webs of most primaries edged with whitish, and outer webs of secondaries with greenish; most of the primaries also marked with white on outer webs at base, forming a conspicuous spot (only seen elsewhere in *D. cærulescens*); middle and greater wing-coverts with white bars. Tail like wings, with greenish edging of most of the feathers, the two outer ones on each side mostly or wholly white. Belly and sides whitish, tinged with olive or brownish. Basal half of under mandible light brown. Length 4.75–5.25; extent 8.25–9.00; wing 2.75–3.10; tail 1.95–2.20; bill 0.55; tarsus 0.75. Adult ♀ and young ♂: The saffron color much clearer yellowish, and shaded with olive-green on crown; the black bar replaced by whitish, excepting a dusky patch on auriculars. A remarkable Mexican Warbler, also ascertained to inhabit S. Arizona and New Mexico, in mountainous localities; probably also Texas. It has much the habits of the Pine-creeping Warbler (*D. vigorsi*); nest very pretty, somewhat like the Blue-Gray Gnatcatcher's, high up in a coniferous tree, saddled on a limb or fixed in a forked twig, composed of moss, lichens, fir blossoms, and cobwebs, lined with fine rootlets; eggs peculiar, olive-gray, very thickly speckled with black; set of 3–4, May, June.

FIG. 173. — Olive Warbler.

(*Subgenus* DENDRŒCA.)

D. æsti'va. (Lat. *æstiva*, summery; *æstas*, summer.) SUMMER WARBLER. SUMMER YELLOW-BIRD. YELLOW-POLL WARBLER. BLUE-EYED YELLOW WARBLER. GOLDEN WARBLER. "WILD CANARY." Adult ♂ : Golden yellow ; back with a greenish tinge resulting in rich yellow-olive ; rump more yellowish ; middle of back sometimes obsoletely streaked with darker. Crown like under parts, in high plumage often tinged with orange-brown. Breast and sides, and sometimes most of the under parts, streaked with orange-brown. Quills and tail-feathers dusky, edged on both webs with yellow, occupying most of the inner webs of the tail-feathers. Bill plumbeous. Feet pale brown. Length 4.75–5.00 ; extent 7.50–7.75 ; wing 2.50 ; tail 2.00. Adult ♀ : Yellow-olive of upper parts extending on crown ; streaks below obsolete or entirely wanting. General coloration paler. Young : Like ♀, but still duller. Upper parts, including crown, pale olive, with an ochrey instead of clear yellow shade ; below ochrey-white or dull pale yellowish. Edgings of wings and tail dull yellowish. N. Am., everywhere in woodland, gardens, orchards, parks, and even city streets, a beautiful, abundant, and familiar little bird. Nests throughout its range, in fruit or shade trees, shrubbery and brushwood, building a neat, compact, and durable nest of soft vegetable and animal substances felted together ; eggs commonly 4–5, 0.64–0.69 × 0.48–0.53, grayish- or greenish-white, variously dotted and blotched with reddish-brown and lilac shades. The color of this precious gem makes a pretty spot as it flits through the verdure of the grove or plays amidst the rose-tinted blossoms of the fruit-orchard ; and its sprightly song is one of the most familiar sounds of bird-life during the season when the year renews its youth.

D. æ. sonora'na. (Lat. *sonoran*.) SONORA SUMMER WARBLER. Adult ♂ : Like the last ; upper parts, especially the rump, wings, and tail, more uniformly yellow, the rump usually pure yellow and the back and wings scarcely tinged with greenish ; light yellowish edgings of wing-quills and coverts broader ; crown with a brownish-orange tinge, and feathers of the interscapulars with shaft-stripes of purplish-chestnut, usually conspicuous ; under parts faintly and sparsely streaked. ♀ : Much paler and grayer than that of *æstiva* proper : yellowish-gray above, in contrast with more decided yellowness of wing-coverts and tail-feathers, the latter only narrowly edged with yellow ; under parts very pale straw yellow, whitening on the throat. Wing of ♂ 2.55 ; tail 1.80 ; tarsus 0.70 ; bill 0.50. Sonora, through S. Arizona and S. New Mexico to W. Texas. This would appear to be a recognizable form, especially in the streaking of the interscapulars, though this feature is not always exhibited. BREWSTER, Auk, Apr. 1888, p. 137 ; COUES, Key, 4th ed., 1890, p. 898 ; A. O. U. List, 2d ed., 1895, p. 274, No. 652 *a*. In some respects it resembles *D. æ. morcomi* COALE, Bull. Ridgw. Club, Chicago, No. 2, Apr. 1887, p. 82 ; RIDGW., Man., 1887, p. 494 ; but this Western form of the Yellow Warbler has been disallowed by the A. O. U. Committee.

D. æ. rubiginosa. (Lat. *rubiginosus*, reddish, as the streaks on the under parts of the ♂ are.) Like the last, but upper parts nearly uniform, as the olivaceous of the back extends over the crown and rump ; streaks of breast and edgings of wings and tail rather narrow. Only recognized from Alaska and British Columbia ; type specimen from Kadiak Isl. *Motacilla rubiginosa* PALL., Zoog. R.-A., i, 1811, p. 496 (Kadiak). *D. æ. rubiginosa* OBERHOLSER, Auk, Jan. 1897, p. 76 ; A. O. U. Suppl. List, *ibid.*, p. 123, No. 652 *b*.

D. bry'anti castaneiceps. (To Dr. Henry Bryant. Lat. *castaneiceps*, chestnut-headed.) CHESTNUT-HEADED GOLDEN WARBLER. MANGROVE WARBLER. Belonging to the "golden warbler" group of the genus, and resembling *D. æstiva* in general characters. Dusky predominating over yellow on tail-feathers ; tarsus about 0.72. Adult ♂: Whole head chestnut, well defined all around against the yellow ; edging of wing-coverts slight ; rufous streaks of breast and sides few and narrow. This resembles the continental *D. vieilloti*, as described by Cassin in 1860, but would appear to be well distinguished by the rufous hood which envelops the head. The form of *bryanti* here given is the Mexican race, lately ascertained to occur at La Paz, Lower California; it is *D. vieilloti bryanti* of the 2d and later eds. of the

Fig. 174. — Black-throated Green Warbler. (From the Osprey.)

Key; as above in A. O. U. Lists, both eds., No. 653. The ♀ is said to be indistinguishable from that of others of the Golden Warbler group. The extra-limital forms all differ from N. Am. *æstiva* in having longer tarsi and less yellow on the tail-feathers. (Not in the Check List, 1882. See Am. Nat., vii, Oct. 1873, p. 606; Hist. N. A. Birds, i, 1874, p. 217; Pr. U. S. Nat. Mus., iv, 1882, p. 414, and viii, 1885, p. 350.

D. vir′ens. (Lat. *virens*, growing green. Figs. 174, 175.) BLACK-THROATED GREEN WARBLER. Adult ♂, in spring: Back and crown clear yellow-olive; forehead, superciliary line, and whole sides of head rich yellow (in very high plumage, middle of back with dusky marks, and dusky or dark olive lines through eyes and auriculars, and even bordering crown); *chin, throat, and breast jet black*, prolonged behind as streaks on sides; other under parts white, usually yellow-tinged; wings and tail dusky, former with two white bars and much whitish edging, latter with outer feathers nearly all white; bill and feet blackish. ♂ in fall, and ♀ in spring: Similar, but the black restricted, interrupted, or veiled with yellow; *young* similar to ♀, but the black still more restricted or wanting altogether, except a few streaks along sides (? *Sylvia montana* WILS.). Small: Length 4.80–5.10; extent 7.60–8.00; wing 2.30–2.55; tail 2.00. Eastern U. S. and British Provinces, N. to Hudson's Bay and casually even to Greenland, W. only to the edge of the Plains; migratory, abundant; breeds from higher portions of the Middle States, in mountains even S. to the Carolinas, and plentifully from New England northward; winters extralimital in the W. I. and S. to Panama; has occurred accidentally in Europe. This jaunty bird is one of the commonest warblers of summer in New England, breeding mostly in the pineries, in June. Nest in fork of a bough, usually at some elevation, but very variable in this respect, of the most miscellaneous materials; eggs 4–5, 0.67 × 0.54 to 0.58 × 0.48, white, with the usual sprinkling or wreathing of brown and purplish markings. The nuptial song is very peculiar.

Fig. 175. — Black-throated Green Warbler. (L. A. Fuertes.)

D. town'sendi. (To J. K. Townsend.) TOWNSEND'S WARBLER. Adult ♂ : Entire upper parts yellowish-olive, rather darker than in *virens*, everywhere streaked with black, especially on crown, where black usually predominates; no hidden yellow on crown. Side of head bright yellow, enclosing a large black patch on loral, orbital, and auricular regions, in which the yellow eyelids appear. Chin, throat, and jugulum black; breast, and sides part way, yellow; sides of breast and of body streaked with black. Under wing-coverts, belly, flanks, and crissum white, the two latter slightly shaded and streaked with dusky. Wings crossed with two white bands, that of the median coverts broadest. Wings and tail fuscous, the former with pale edgings, the latter having two or three outer feathers largely blotched with white. Bill and feet blackish horn-color. Length about 5.00; extent 7.50–8.00; wing 2.25–2.50; tail 2.00. ♀ : Like ♂, but black of throat veiled with yellow, and that on top and sides of head mixed with or replaced by olive. Young : Shade of upper parts slightly brownish, and the black streaks slight, obsolete, or wanting. The dark patch on side of head olivaceous, like back. No continuous black on throat. Autumnal adults show various gradations between characters of old and young. Very closely related to *D. virens*, of which it is the western representative. Adult males readily distinguished by darker greenish upper parts, conspicuously streaked, especially on head, with black which in summer is uniform; black cheeks and auriculars; and yellow bordering black of throat laterally and spreading on breast behind. Young birds not so easily discriminated; but there are usually traces at least of the black streaks on the upper parts; there is no concealed yellow on crown; the yellow of under parts, quite as bright as in the adult, extends far along the breast, behind that part where it veils the black. Rocky Mts. to Pacific, Alaska to Guatemala; breeds in coniferous woods in U. S. and British Columbia range, from S. Cala. to upper Yukon valley, E. to Idaho ; in migrations E. to Col. and W. Texas; straggler taken in Penn. Eggs indistinguishable from those of *virens*.

D. occidenta'lis. (Lat. *occidentalis*, western; where the sun sets.) WESTERN WARBLER. HERMIT WARBLER. Adult ♂ : Above, ashy-gray, tinged with olive, especially on rump, and closely streaked with black; top and sides of head rich yellow, the former spotted with black. Below, white; central line of chin, throat, and jugulum black, ending on breast with a sharp convex outline, contrasted with the adjoining white. Wings and tail as in *virens*. Bill black. Length 4.75–5.00; extent 7.75; wing 2.50–2.75; tail 2.12–2.25; tarsus 0.66–0.75; bill 0.40. Adult ♀: Like ♂, but darker gray above, with yellow of head less extended, and throat whitish, spotted with dusky. Young: Upper parts olivaceous-ash, and yellow of top of head overlaid with olive. Sides of head pretty clear yellow, fading gradually into white of throat. No black on throat. White of under parts faintly brownish-tinged, and sides with obsolete streaks. In a September specimen I took in Arizona the dusky-olive extends over all upper parts, tinging the ashy of lower back, and reaching on crown nearly to bill, where it gradually lightens by admixture of yellow; sides of head clear yellow, soiled with some olivaceous; chin and throat the same, fading on breast into the dull white of the other under parts; sides with obsolete streaks, and a slight grayish-olive wash. There is no black whatever about head or throat, and the blackish streaks of back are obsolete. The wings are twice-barred with conspicuous white tips of greater and median coverts. Rocky Mts. to the Pacific, U. S. and British Columbia; S. in winter to L. California and Mexico to Guatemala; another of the several western warblers of the *D. virens* group. Nest high in conifers, of pine-needles, bark-shreds, rootlets, and sometimes hairs, 4.00 in outside diameter, by 2.00 inside, with a cavity about 1.00 deep; eggs 0.70 × 0.52, creamy white, marked with brown and neutral tints, as usual; laid in June.

D. chrysopari'a. (Gr. χρυσός, *chrusos*, golden, and παρειά, *pareia*, cheek.) GOLDEN-CHEEKED WARBLER. Prevailing color of upper parts black, usually mixed with olive-green ; sides of head yellow, with narrow black stripe through eye; below, with wings and tail, as in *virens;* size of that species, and changes of plumage generally parallel; very closely related. ♂, in full dress: Above, jet-black from bill to tail, anteriorly narrowing to a point on the forehead, with scarcely a trace of olivaceous toward and on rump. Entire side of head and neck golden-yellow, reaching bill, elsewhere enclosed in black, and enclosing a long black stripe through eye to side of nape, nearly cutting off a superciliary stripe from the general yellow area, which, however, is continuous on lore and side of nape. Chin, throat, and breast jet-black, this color extending backward along sides as heavy streaking; narrowing anteriorly where sharply defined against the yellow; other under parts, including lining of wings, white, squarely defined against black of breast (whole under parts thus as in *virens*). Wings blackish, with two broad white cross-bars, and whitish edging of quills, especially the inner secondaries. Tail blackish; outermost feather white with only a black shaft-line clubbed at end; next three pairs with decreasing white. Adult ♀: Above olive-green indistinctly streaked; throat yellowish more or less mixed with black. Texas and southward. Nest in upright fork, preferably of a cedar, large for the bird, compactly felted of bark strips, fine grasses, rootlets, and slender vegetable fibres and cobwebs, lined copiously with hair and feathers; eggs 0.75 × 0.55, white, dotted with reddish-brown and lavender, and blotched with darker brown, laid in May.

D. nigres'cens. (Lat. *nigrescens*, growing black. Fig. 176.) BLACK-THROATED GRAY WARBLER. Adult ♂: Above, bluish-ash, the interscapular region, and usually also upper-tail coverts, streaked with black. Below, from breast, pure white, the sides streaked with black. Entire head, with chin and throat, black; a sharply defined yellow spot before eye, a broad white stripe behind eye, and a long white maxillary stripe widening behind from corner of bill to side of neck. Wings fuscous, with much whitish edging, crossed with two broad white bars on ends of greater and median coverts. Tail like wings, the three lateral feathers mostly white, except on outer webs, the fourth with a white blotch. Bill and feet black. Size of *D. townsendi*. ♀ like ♂, but black of crown mixed with the ashy of back, and that of throat veiled with white tips of the feathers. Young: Like ♀, but crown almost entirely like back, and black of throat still more hidden. Back not streaked. Less white on tail. Bill not entirely black. Rocky Mts. to the Pacific, U. S. and British Columbia, southward in winter in Mexico, common in woodland. Quite unlike any other species; one of the five *Dendrœcæ* which are normally confined to the West. Nest, usually low, in bushes and shrubbery, small, 2.00 × 1.50, resembling that of the Summer Warbler, but lined with grasses and hairs; eggs from dull white to greenish-buff, heavily marked, 0.63 × 0.50; May, June. The breeding **range is coincident with the distribution of the bird in the U. S., it being common in summer in the mountains of** Southern Arizona up to 9,000 feet.

FIG. 176. — Black-throated Gray Warbler; nat. size. (Ad, nat. del. E. C.)

D. cœrules'cens. (Lat. *cœrulescens*, growing blue; *cœruleus*, blue. Fig. 177.) BLACK-THROATED BLUE WARBLER. Adult ♂, in spring: Above, uniform slaty-blue, the perfect continuity of which is only interrupted in very high plumages, by a few black dorsal streaks; below, pure white; sides of head to above eyes, chin, throat, and whole sides of body continuously jet-black; *wing-bars wanting* (the coverts being black, edged with blue), but *a large white spot at base of primaries:* quill-feathers blackish, outwardly edged with bluish, the inner ones mostly white on inner webs; tail with ordinary white blotches, the central feathers edged with bluish; bill black, feet dark. Young ♂: Similar, but the blue glossed with olivaceous, and the black interrupted and restricted. ♀ *entirely different:* Dull olive-greenish, with faint bluish shade, below pale soiled yellowish; but recognizable by the *white spot at base of pri-*

FIG. 177. — Black-throated Blue Warbler. (L. A. Fuertes.)

maries, which, though it may be reduced to a mere speck, is nearly always evident, at least on pushing aside the primary coverts; no other wing-markings; tail-blotches small or obscure; feet rather pale. Size of *virens*. Eastern N. Am., abundant, in woodland, its range somewhat coincident with that of *virens;* breeding range N. from northern New England and the northern border of the U. S. at large, S. in the Appalachian chain to the Carolinas and even Georgia (var. *cairnsi*); in winter found in the West Indies and S. to Guatemala; west in migrations to Rocky Mts. It is rather a bird of brake and burn than of high woods, at least in summer; and nests in bushes, close to the ground. Nest of bark-strips, mosses, lichens, rootlets, cobwebs, etc., built by the ♀, rather compact, about 1.50 deep and 2.00 across (inside measurement), affixed usually to upright supports; eggs 3–4, 0.67 × 0.48, white with buffy or even greenish tinge, well spotted and blotched with reddish-brown. This is a beautiful bird, the ♂ with black, white and blue in masses, thus resembling no other, and the olive-colored ♀ as different as possible from her mate.

D. c. cairn'si. (To John S. Cairns of Weaverville, N. C.) CAIRNS' BLACK-THROATED BLUE WARBLER. A local race of the last, with nearly black back, and rather smaller, breeding in the mountains of western N. Carolina in May and June, building in shrubs and weeds from six inches to three feet from the ground, and laying 3–4 eggs. Arrival a week or ten days earlier than that of the stock form, which latter is migrating in the same region while *cairnsi* is nesting. COUES, World's Congress on Ornith., Nov., 1896, p. 138; Auk, Jan., 1897, p. 96; A. O. U. Suppl. List, *ibid.*, p. 123, No. 654 *a*.

D. ra'ra. (Lat. *rarus*, rare. Fig. 178.) CERULEAN WARBLER. AZURE WARBLER. Adult ♂ : Entire upper parts *sky-blue*, the middle of the back streaked with black; the crown usually richer and also with dark markings. Below, *pure white*, streaked across the breast and along the sides with *dusky-blue* — the breast-streaks inclining to form a short bar, sometimes interrupted in the middle. Auriculars dusky; edges of eyelids and superciliary line white. Wings blackish, much edged externally with the color of the back; inner webs of all quills, outer webs of inner secondaries, and two broad bars across tips of greater and median coverts, white. Tail black, with much exterior edging of the color of the back, all the feathers, except middle pair, with small, white, subterminal spots on inner webs. Length 4.00–4.50; wing 2.66; tail 2.00 or less. Adult ♀: Quite different. Upper parts dull greenish, with

FIG. 178. — Cerulean Warbler. (L. A. Fuertes.)

more or less grayish-blue shade, the greenish brightest and purest on crown. Eyelids, line over eye, and entire under parts, whitish, more or less strongly overcast with dull greenish-yellow. Wings and tail dusky, with exterior edgings of the color of the back; the bars, spots, and interior edgings white, as in ♂. The ♀ is curiously similar to the same sex of *D. cœrulescens* (but in the latter the tail-spots are different; there are no white wing-bars, and instead there is a small whitish spot at base of outer primaries). The autumnal plumage of adults is said to differ in no wise from that of the spring. Young males are much like adult females, but less uniformly greenish-blue above and purer white below, with evident blackish stripes on interscapulars and sides of head. The young ♀ resembles the adult of that sex, but is still greener above, with little or no blue, and quite buffy-yellowish below. When in full dress this is a perfect little beauty, there being something peculiarly tasteful and artistic in the simple contrast of snowy-white with delicate azure-blue, without any " warm " color. Eastern U. S. to the bordering British Provinces, rarely N. to New England, and apparently not common anywhere E. of the Alleghanies; W. ordinarily to the Plains, sometimes to the Rocky Mts. in the latitude of Colorado; in winter S. through Mexico, Central America, and much of

324   SYSTEMATIC SYNOPSIS. — PASSERES — OSCINES.

S. America; rare in Cuba. Breeding range chiefly Mississippi Valley at large from middle districts northward. Nest small and neat, well cupped, placed in fork of a bough 20–50 feet from the ground, preferably in woods of deciduous trees, and composed of the usual materials; eggs 4, 0.66 × 0.47, creamy-white or with a faint greenish tinge, heavily blotched with reddish-brown, especially about the larger end. (*Sylvia cærulea* WILS., 1811, nec LATH., 1790; *Dendroica* or *Dendrœca cærulea* of authors, as of former eds. of Key. *Sylvia rara* WILS., 1811; *Dendroica rara* RIDGW., Auk, Jan. 1897, p. 97; A. O. U. Suppl. List, Auk, Jan. 1897, p. 131, No. 658.)

D. corona'ta. (Lat. *coronata*, crowned; *corona*, a crown. Fig. 179.) YELLOW-RUMPED WARBLER. YELLOW-CROWNED WARBLER. MYRTLE BIRD. Adult ♂, in spring: Slaty-blue, streaked with black; below, white, breast and sides mostly black, belly, and especially throat, pure white, immaculate; *rump, central crown-patch, and sides of breast, sharply yellow*, there being thus *four* definite yellow places; sides of head black; eyelids and superciliary line white; ordinary white wing-bars and tail-blotches; bill and feet black. ♂ in winter, and ♀ in summer, similar, but slate-color less pure, or quite brownish; *young* birds quite *brown* above, with a few obscure streaks in the whitish of under parts. It is impossible to specify the endless intermediate styles; but I never saw a specimen without the yellow rump, and at least a trace of the other yellow marks;

FIG. 179. — Yellow-rumped Warbler, nat. size. (Ad. nat. del. E. C.)

these points therefore are diagnostic. (The only other obscure brownish Warblers with yellow rump are *maculosa* and *tigrina*, when young. Resembles *auduboni*, excepting in the following points: Throat white. Breast black, mixed with white. Sides of head definitely pure black; edges of eyelids, and long narrow superciliary line, white. Wings crossed with two broad white bars, which do not fuse into one white patch, owing to narrowness or deficiency of white edging along outer webs of greater coverts.) One of the large species. Length 5.30–5.75; extent 8.80–9.40; wing 2.75–3.00; tail about 2.50. N. Am., but chiefly eastern; Alaska; Washington; California; Arizona; U. S. rarely in summer except along the northern borders, but during the migrations and in winter the most abundant of all Warblers; winters as far N. as New England, and thence through the U. S. to the West Indies, Mexico, and Central America; resident in Jamaica; seen everywhere, but is particularly numerous in shrubbery, along hedge-rows, in flocks, with troops of Sparrows, Titmice, etc. Breeds from our northern borders northward; nest generally low in evergreens; eggs 4, about 0.75 × 0.55, white with a creamy or slight buff tinge, and with the usual markings of browns, blackish and neutral tints. Moult double, there being a vernal as well as an autumnal change, the former usually effected during the spring migrations.

D. c. hoo'veri. (To Theodore J. Hoover.) HOOVER'S YELLOW-RUMP. Like the last, wing and tail longer; ♂ wing 3.00 or more; ♀ wing 2.87. Western N. Am. McGREGOR, Bull. Cooper Club, i, No. 2, Mar. 15, 1899, p. 32.

D. aud'uboni. (To J. J. Audubon.) AUDUBON'S WARBLER. WESTERN YELLOW-RUMP. Adult ♂, in summer: Upper parts clear bluish-ash, streaked with black. A central longitudinal spot on crown, the rump, *throat*, and a patch on each side of breast, rich *yellow*. Sides of head little darker than upper parts; eyelids narrowly white, but no decided superciliary white stripe; ash of upper parts extending far around sides of neck. Jugulum and breast in high plumage pure black, though usually mixed with some grayish skirting of the feathers, or invaded by white from behind, or even touched with yellow here and there. Belly and under tail-coverts white, the sides streaked with black. Wings blackish, with gray or white edging, especially on inner quills; median wing-coverts tipped, greater ones edged and tipped, with white, forming a great white blotch. Tail like wings; outer webs narrowly edged with gray or white; inner webs of all the lateral feathers with large white blotches. Bill and feet black. One of the largest species. Length 5.50–5.75; extent 8.75–9.33; wing 2.75–3.00; tail 2.25. ♀, in summer: Generally similar to ♂. Upper parts duller and browner slate-color, with less heavy dorsal streaks; crown-spot and other yellow parts paler; breast not continuously black, but variegated with black, white, and color of back. Sides only obsoletely streaked. Eyelids scarcely white, and cheeks hardly different from back. White of wing-coverts mostly restricted to two bars; white tail-spots smaller. ♂ ♀, in autumn and winter, and young: Upper parts quite brown, with obscure black marking. Yellow crown-spot concealed or wanting; yellow of throat, rump, and sides of breast paler and restricted. Under parts whitish, shaded on sides, and usually across breast, with a dilute tint of color of back, the breast and sides obsoletely streaked with darker. White of wing-coverts obscured with brownish. N. Am., from easternmost woodland of the Rocky Mts. to the Pacific; N. to British Columbia and probably to Alaska; S. in winter to Central America; accidental in Pennsylvania and Massachusetts; migratory, breeding northward and in Alpine regions; extremely abundant; nesting in no wise peculiar; nest usually high in coniferous trees, made of bark-strips, pine-needles, rootlets, mosses, hairs, and even feathers; eggs about 4, white with a greenish tinge, rather sparsely marked with the usual colors.

D. black′burniæ. (To Mrs. Blackburn, an English lady. Fig. 180.) BLACKBURNIAN WARBLER. HEMLOCK WARBLER (young). TORCH-BIRD. FIREBRAND. PROMETHEUS. Adult ♂, in spring: Entire upper parts, including wings and tail, *black,* the back varied with whitish; wings with a large *white* speculum on coverts and much white edging of coverts; lateral tail-feathers largely *white,* only a shaft-line, with clubbed extremity, being left blackish on the outer two or three pairs. Spot on fore part of crown, eyelids, line over eye spreading into a large spot behind the auriculars, with chin, throat, and fore breast, *intense orange or flame-color* — there is nothing to compare with the exquisite hue of this Promethean torch. Side of head black in an irregular patch, usually confluent with the black streaks on side of breast, isolating the orange of the side of the head from that of throat, and circumscribing the orange patch below eye. Under parts from the breast white, more or less tinged with orange or yellow, and whole sides streaked with black. Bill and feet dark. Length about 5.50; extent 8.50; wing 2.75; tail 2.00. Adult ♀, in spring: Similar to ♂ in pattern and distribution of the colors; upper parts brownish-olive, streaked with black; the fiery orange of ♂ not so intense, or merely yellow, that on crown obscure or obsolete. White speculum of wing resolved into two white bars. Sides of head like back, instead of black as in ♂, and the lateral streaks duller and more blended. ♂ and ♀, adult, in autumn, are sufficiently similar to the respective sexes in spring, but the coloration is toned down, the fiery colors of the ♂ being less intense, and the black of the back being much mixed with olivaceous, bringing about a close resemblance to the spring ♀; while the ♀ is duller still, and more impurely colored. Young: Early autumnal birds of the year are very obscure, showing no sign of the rich coloration of the adults, and are *Sylvia parus,* the Hemlock Warbler, of old authors. Above, like adult ♀, but still browner, with more obsolete dusky streaking. Usually an indication of the crown-spot in a lightening of the part. Sides of head like crown, cutting off a superciliary stripe and the eyelids, which are ochrey-white. Whole under parts white, tinged, especially on throat and breast, with yellowish, the sides with obsolete streaking. Indication of the peculiar pattern of the adults, though without their actual coloration, together with extent of white on the tail-feathers, will usually suffice for determination of the species, before any orange appears on the throat, after

FIG. 180. — Blackburnian Warbler. (L. A. Fuertes.)

which there can be no difficulty. Eastern N. Am.; W. regularly to the Plains, casually to Utah. Abundant in mixed woodland; breeds in northerly parts of its U. S. range and northward, also much further S. in the Alleghanies; winters extralimital in the Bahamas, Mexico, and Central and S. Am. One of the later migrants in spring. Nests in bushes and trees, preferably evergreens, building a rather large and flattish nest, about 5.00 broad outside and 2.00 deep, with a cavity of only about 2.00 × 1.00; eggs not peculiar, 0.70 × 0.50, greenish or bluish-white, with the usual shades of brown and neutral tint in dots and spots, chiefly at or near the larger end.

D. stria′ta. (Lat. *striata*, striped. Fig. 181.) BLACK-POLL WARBLER. Adult ♂ : Back, rump, and upper tail-coverts *grayish-olive*, heavily streaked with black; whole crown pure *glossy black*. Below, *pure white;* a double series of black streaks starts from the extreme chin, and diverges to pass one on each side to the tail, the streaks being confluent anteriorly, discrete posteriorly. Side of head above the chain of streaks pure white, including lower eyelid. Wings dusky; primaries with much greenish edging; inner secondaries with whitish edging, greater and median coverts tipped with white, forming two cross-bars. Tail like wings, with rather small white spots at ends of inner webs of two or three outer feathers. Upper mandible brownish-black; lower mandible and feet flesh-colored or yellowish. Length 5.25–5.75; extent 8.75–9.30; wing 2.70–2.90; tail 2.25. Adult ♀ : Entire upper parts, including crown, greenish-olive, with dusky streaks; below, white, much tinged with greenish-yellow, especially anteriorly, the streaks dusky and not so sharp as those of ♂, but still very evident. Bars and edgings of wings greenish-white. Tail as in ♂. Rather smaller than ♂ on an average. Young : Similar to adult ♀, but brighter and more greenish-olive above, the streakings few and chiefly confined to middle of back; below, more or less completely tinged with greenish-yellow, the streakings obsolete, or entirely wanting. Under tail-coverts usually pure white. These autumnal birds bear an extraordinary resemblance to those of *D. castanea* (though the adults are so very different), the upper parts being, in fact, the same in both. But young *castanea* generally shows traces of chestnut, or at least a buffy shade, quite different from the clear greenish-olive of *striata*, this tint being strongest on flanks and under tail-coverts, where *striata* is the most purely white. Moreover, *castanea* shows no streaks below, traces at least of which are usually observable in *striata*. N. Am., excepting Western and most of Middle Province; N. to the Arctic Ocean, Greenland, Alaska; W. to Montana and Colorado. Winters extralimital, in S. America. Breeds from northern New England and Michigan and mountainous parts of New York. Migrates late in spring, bringing up the rear-guard of the Warbler hosts; when the Black-polls appear in force the collecting season is about over! Nests low in spruce-trees and other evergreens and sometimes on the ground, in high latitudes lined with feathers; eggs 5, 0.72 × 0.50, not peculiar, being white with a creamy or buff tinge, very variably dotted, spotted, or blotched with different shades of brown, gray, and blackish.

FIG. 181.—Black-poll Warbler, nat. size. (Ad. nat. del. E. C.)

FIG. 182.—Bay-breasted Warbler. (L. A. Fuertes.)

D. casta′nea. (Lat. *castanea*, a chestnut, in allusion to the color. Fig. 182.) BAY-BREASTED WARBLER. Adult ♂, in spring : Back thickly streaked with black and grayish-olive; *forehead and sides of head black, enclosing a large deep chestnut patch;* a duller chestnut (exactly like a Bluebird's breast) occupies the whole chin and throat and thence extends, more or less interrupted, along entire sides of body; rest of under parts ochrey or buffy whitish; a similar buffy area behind ears; wing-bars and tail-spots ordinary; bill and feet blackish. ♀, in spring: More olivaceous than ♂, with the markings less pronounced; but always shows evident *chestnut* coloration: and probably traces of it persist in all *adult* birds in the fall. The young, however, so closely resemble young *striata*, that it is sometimes impossible to distinguish them with certainty. The upper parts, in fact, are of precisely the same greenish-olive, with black streaks; but there is *generally* a difference below — *castanea* being there tinged with buffy or ochrey, instead of the clearer pale yellowish of *striata;* this shade is particularly observable on belly, flanks, and under tail-coverts, where *striata* is whitest; and moreover, *castanea* is usually not streaked on the sides at all. Mature spring birds vary interminably in the extent and intensity of the chestnut. Size of *striata*. Eastern N. Am., N. to Hudson's Bay, W. to the edge of the Plains. Winters extralimital in Mexico, Centr. Am., and the U. S. of Colombia. Migratory in most of the U. S. Breeds from northern New England, Michigan, etc. northward. Nests moderately high, in conifers, usually 15-25 feet up, building a large nest of twigs, tree-moss, rootlets, fur, etc.; eggs 3-6, 0.70 × 0.52, *bluish-green*, profusely spotted with browns and neutral tints.

D. pennsylva′nica. (Of "Penn's woods;" *sylva*, a forest; *sylvanus*, sylvan. Figs. 183, 184.) CHESTNUT-SIDED WARBLER. BLOODY-SIDED WARBLER. Adult ♂, in spring: Back streaked with black and pale yellow (sometimes ashy or whitish); *whole crown pure yellow*, immediately bordered with white, then enclosed with black; sides of head and neck and whole under parts *pure white*, former with an irregular black crescent before eye, one horn extending backward over eye to border the yellow crown and be dissipated on sides of nape, the other reaching downward and backward to connect with a chain of pure *chestnut* streaks that run the whole length of the body, the under eyelid and auriculars being left white; wing-bands generally fused into one large patch, and, like the edging of the inner secondaries, much tinged with yellow; tail-spots white, as usual; bill blackish; feet brown. ♀, in spring: Quite similar; colors less pure; black loral crescent obscure or wanting; chestnut streaks thinner. Young: Above, including crown, clear yellowish-green, perfectly uniform, or back with slight dusky touches; no distinct head-markings; below, *entirely white* from bill to tail, unmarked, or else showing a trace of chestnut streaks on sides; *wing-bands* clear *yellowish* as in adult — this is a diagnostic feature, taken in connection with the continuously white under parts; bill light-colored below. Small: Length 4.80–5.10; extent 7.75–8.10; wing 2.30–2.50; tail 2.00. Eastern U. S. and adjoining British Provinces; W. to the edge of the Plains; winters extralimital; breeds abundantly in Middle and Northern States, S. to Illinois, and still further in the Appalachian ranges; nests in forks of low saplings, shrubs, and bushes; eggs 4-5, 0.68 × 0.50, with the usual markings. A pretty species chained with chestnut on snowy ground.

FIG. 183. — Chestnut-sided Warbler, nat. size. (Ad. nat. del. E. C.)

FIG. 184. — Chestnut-sided Warbler. (L. A. Fuertes.)

D. maculo′sa. (Lat. *maculosa*, full of spots; *macula*, a spot. Figs. 185, 187.) BLACK-AND-YELLOW WARBLER. BLUE-HEADED YELLOW-RUMP WARBLER. SPOTTED WARBLER. MAGNOLIA WARBLER. Adult ♂ ♀, in spring: Back black, usually quite pure and uninterrupted in ♂, more or less mixed with olive in ♀; *rump yellow;* upper tail-coverts black, often skirted with olive or ashy. Whole crown of head *clear ash;* sides of head black, including a very narrow frontlet; eyelids and a stripe behind eye, between the ash and black, white. Entire under parts rich yellow, excepting the white crissum, heavily streaked with black across breast and along sides, the streaks on the breast so thick as to form a nearly continuous black border to the immaculate yellow throat. Wings fuscous, with white lining, white edging of inner webs of all the quills, of outer webs of the inner secondaries, and with a large white patch formed by tips of median coverts and tips and outer edges of greater coverts. Tail blackish, with *square* white spots on middle of inner webs of *all* the feathers excepting middle pair. Bill blackish; feet dark. Length 4.75-5.00; extent 7.00-7.50; wing 2.25-2.50; tail 2.00-2.25. Young: Upper parts ashy-olive, grayer on head; rump as yellow as in the adult; no decided head-markings; a whitish ring around eye. Below, yellow, generally pure and continuous, sometimes partially replaced by gray; black streaks wanting, or few and confined to the sides. Wings with two bars; *tail-spots* as in the adult. While the sexes of this dainty little species are quite similar, the young require looking after; observe yellow rump (usually as conspicuous as in the species so named), small square tail-spots on middle of feathers, and extensively or completely yellow under parts. Eastern N. Am., N. to Hudson's Bay and Great Slave Lake, W. to the Rocky Mts. of Colorado, casually to British Columbia; abundant, chiefly migratory in the U. S., but breeds from our N. border northward, and S. in mountains to Pennsylvania at least, probably still further; winters wholly extralimital, in the Bahamas, Cuba, Mexico, and Central America. Builds a small neat nest in conifers at a very variable height from the ground; eggs 4-5, 0.64 × 0.48, not peculiar, and with considerable range of variation in the markings.

FIG. 185. — Black-and-yellow Warbler. (L. A. Fuertes.)

D. dis′color. (Lat. *discolor*, parti-colored; opposed to *concolor*, whole-colored. Fig. 186.) PRAIRIE WARBLER. Adult ♂ ♀: Yellow-olive; back with a patch of *brick-red spots;* forehead, superciliary line, two wing-bars, and entire under parts, rich yellow; a V-shaped black mark on side of head, its upper arm running through eye, its lower arm connecting with a series of black streaks along sides of neck and body; white tail-blotches very large, occupying most of inner web of outer feathers. The sexes are almost exactly alike, and the young only differ in not being so bright and in having the dorsal patch and head-markings obscure. Small: Length 4.75; extent 7.00-7.40; wing 2.15-2.25; tail 2.00. Eastern U. S. to Massachusetts and Michigan, W. to Kansas; an abundant bird of the Middle and Southern States, in sparse low woodland, cedar thickets, and old fields grown up to scrub-pines;

FIG. 186. — Prairie Warbler. (L. A. Fuertes.)

remarkable for its quaint and curious song; an expert flycatcher, constantly darting into the air in pursuit of winged insects, like the Redstart and the species of *Sylvania*. Breeds throughout its U. S. range; winters in Florida and the West Indies. Nest in a bush or sapling near the ground; a small, neat, compact structure; eggs 3-4, not peculiar. On the nesting of the Prairie Warbler in the vicinity of Washington, D. C., see the account by my son, Mr. E. B. COUES, Auk, Oct., 1888, pp. 405-408.

D. gra′ciæ. (To Miss Grace D. Coues, the author's sister.) GRACE'S WARBLER. Adult ♂: Entire upper parts ashy-gray, with a slaty-blue tinge; middle of back streaked with black; upper tail-coverts less conspicuously so marked; crown with crowded black arrow-heads, especially anteriorly and laterally, the tendency of these markings being to form a line along

Fig. 187. — Magnolia Warblers. (From The Osprey.)

side of crown, meeting its fellow on forehead. A broad superciliary line of yellow, confluent with its fellow on the extreme front, changing to white behind eye. Lores blackish; sides of head otherwise like back, enclosing a crescentic yellow spot below eye; edges of eyelids yellow. Chin, throat, and fore breast bright yellow, bordered with blackish streaks; yellow of throat separate from that under eye or on lores. Under parts from breast white; sides shaded with color of back, and streaked with black in continuation of the chain of shorter streaks along side of neck. Wings dusky, with very narrow whitish edging, and crossed with two white bars along ends of greater and median coverts. Tail like wings; lateral feather mostly white, excepting outer web; next two or three with white blotches, decreasing in size. Eyes, bill, and feet black; soles dirty yellowish. Length 4.90-5.25; extent about 8.00; wing 2.60; tail 2.25; bill under 0.50. ♂, in autumn: Color of upper parts obscured with a shade of brownish-olive; dorsal streaks obscure; head-markings as in summer, and yellow parts quite as bright. Adult ♀: Quite similar to ♂, in fact scarcely distinguishable in autumn, though the yellow is not quite so strong. Young: Slate-gray of upper parts much shaded with brownish-olive; black streaks wanting on back, those on crown obsolete; yellow much as in the adult but paler, and not bordered along sides of neck with black streaks; black lores poorly defined; wing-bars grayish or obsolete. The white of the under parts has an ochrey tinge, and the lateral streaks are not so heavy in color nor so well defined. Southern Rocky Mt. region of the U. S. and southward; a beautiful species, related to *dominica* and *adelaidæ*; abundant in pine woods of Arizona and New Mexico. Nest high in a coniferous tree, usually in a bunch of needles, of the usual materials; eggs 3-4, not peculiar, white dotted with reddish; May, June.

D. domin'ica. (Lat. *dominicus*, of St. Domingo. Fig. 188.) YELLOW-THROATED WARBLER. Much like the last species, with which its changes of plumage correspond; back without black streaks; no yellow in the black under eye. A white patch separating black of cheeks from bluish-ash of neck; a long superciliary stripe, usually yellow from bill to eye, thence white to nape. Forehead and sides of crown usually quite black, chin and throat rich yellow, bordered on each side by black. Rest of under parts white, the sides boldly streaked with black. Bill black, extremely compressed, almost a little decurved, very long (at least 0.50). Length 5.00 or more; extent 8.00; wing 2.70; tail 2.25. A large handsome species, with its bright yellow throat. S. Atlantic and Gulf States, common; N. sometimes to Middle States, casually to New England. Breeds in its U. S. range at large; winters in Florida and the West Indies. Nest in trees, usually pines, at varying height, often hidden in bunches of Spanish moss (*Tillandsia usneoides*), composed of the usual materials; eggs 0.70 × 0.50, white with a greenish or grayish tinge, and marked with the usual shades of brown and neutral tint, especially about the larger end.

FIG. 188. — Yellow-throated Warbler. (L. A. Fuertes.)

D. d. albilo'ra. (Lat. *albus*, white; *lorum*, the lore.) WHITE-BROWED WARBLER. SYCAMORE WARBLER. Precisely like the last; but superciliary stripe usually white, and yellow of chin cut off from bill by white; bill smaller on an average (0.45 instead of 0.50 along culmen). This slight variety (considering how variable *dominica* is in amount of yellow in the superciliary line) is the common form of the Mississippi and Ohio valley, N. regularly to Ohio, Indiana, Illinois, Michigan, etc., W. to Kansas and Texas, S. in winter to Mexico and Central America.

D. kirt'landi. (To Dr. Jared P. Kirtland, of Ohio. Fig. 189.) KIRTLAND'S WARBLER. Adult ♂: Upper parts slaty-blue; crown and back streaked with black; lores and frontlet black; eyelids mostly white. Under parts clear yellow, whitening on crissum, the breast with small spots and sides with short streaks of black; greater and middle wing-coverts, quills, and tail-feathers edged with white; two outer tail-feathers white-blotched on inner web. Length 5.50; wing 2.80; tail 2.70. Adult ♀: Upper parts dull bluish-gray, obscured with brownish on hind neck and back, marked with heavy blackish streaks on whole back; crown and upper tail-coverts with fine black shaft-lines. Sides of head and neck like upper parts, with darkened lores and whitish eye-ring. Wing-quills dusky, with slight whitish edging of both webs; coverts like back, but with large blackish cen-

FIG. 189. — Kirtland's Warbler.

tral field, and whitish edging and tipping, forming two inconspicuous wing-bars. Tail-feathers like wing-quills, only the outermost one having a small white blotch. Entire under parts dull yellow, brighter on breast, paler on throat and belly, washed with brownish on sides, with a slight necklace of brownish dots across fore breast (as in *Wilsonia canadensis*); these spots stronger on sides of breast, whence lengthening into streaks on sides and flanks; a few small sharp scratches of the same nearly across lower breast. Under tail-coverts white, unmarked. Bill and feet black. Length about 5.60; wing 2.60; tail 2.30; bill 0.40; tarsus 0.80. Eastern U. S. and Bahama Isls., the rarest of all the Warblers; only 20 U. S. specimens have thus far been taken, in Minn., Wisc., Mich., Mo., Ind., Ill., Ohio, Va., S. C., and 55 in the Bahamas. The relationships appear to be with *dominica*, *graciæ*, and *adelaidæ*. Nest and eggs still unknown in 1899: see especially Auk, Oct., 1898, pp. 289-293, pl. iv, and Jan., 1899, p. 81.

D. palma'rum. (Lat. *palmarum*, of the palms; gen. pl. of *palma*, a palm.) YELLOW RED-POLL WARBLER. PALM WARBLER. Adult ♂, in spring: Brownish-olive; rump and upper tail-coverts brighter yellowish-olive; back obsoletely streaked with dusky; *crown chestnut*; superciliary line and most under parts rich yellow, breast and sides with reddish-brown streaks, somewhat as in the Summer Warbler; a dusky loral line running through eye; *no white wing-bars*, the wing-coverts and inner quills being edged with yellowish-brown; tail spots *at very end* of inner webs of two outer pairs of tail-feathers only, and *cut squarely off* — a peculiarity distinguishing the species in any plumage. ♀ not particularly different from ♂. Young: An obscure object, brownish above like a young Yellow-rump, but upper tail-coverts yellowish-olive, and under tail-coverts apt to show quite bright yellow in contrast with the dingy yellowish-white or brownish-white of other under parts; pectoral and lateral streaks obscure; crown generally showing chestnut traces; but in any plumage, known by absence of white wing-bars and peculiarity of tail-spots. Length 5.00–5.25; extent about 8.00; wing 2.50; tail 2.25; tarsus 0.75. The Palm Warbler (including its alleged var. *hypochrysea*) is abundant in eastern North America, especially in the interior; N. to Labrador, Hudson's Bay, Fort Resolution, etc.; breeds only beyond the U. S., excepting (*hypochrysea*) in Maine. Nest *on the ground*; peculiar in this respect in the genus, as far as known (excepting some instances of groundnesting of *D. striata*); eggs not peculiar. When the bird is migrating it is usually found in fields, along hedgerows and roadsides, with Yellow-rumps and Sparrows; the most terrestrial species of the genus, often recalling a Titlark; migrates early in spring, and remains in fall latest of any, except the Yellow-rump, being observed at both these seasons in New England, with snow, in April and November; winters abundantly from the Carolinas to Texas, and in the West Indies.

D. p. hypochry'sea? (Gr. ὑπό, *hupo*, under; χρύσεος, *chruseos*, golden.) YELLOW-BELLIED RED-POLL WARBLER. Said to differ in being more brightly and continuously yellow on the under parts, with the streaks confined mostly to the sides, broadly tear-shaped instead of linear, reddish instead of dusky; lower eyelid yellow, not whitish; back brighter olive. "Atlantic States, N. to Hudson's Bay. Breeds from eastern Maine, New Brunswick, and Nova Scotia northward; winters in the South Atlantic and Gulf States," along with true *palmarum*. According to this, *hypochrysea* should be the common bird of the Atlantic States, and what is above described as true *palmarum* should be the bird of the interior. But I have little faith in the validity of the physical characters assigned, and none in the geographical distinctions sought to be established.

D. vigors'i. (To N. A. Vigors, the English quinarian naturalist.) PINE WARBLER. PINE-CREEPING WARBLER. PINE CREEPER. "VIGORS' VIREO." Adult ♂: Uniform yellowish-olive above, yellow below, paler or white on belly and under tail-coverts, shaded and sometimes obsoletely streaked with darker on the sides; superciliary line yellow; wing-bars *white;* tail-blotches *confined to two outer pairs of feathers, large, oblique.* ♀ and young: Similar, duller; sometimes merely olive-gray above and sordid whitish below, or even brownish-gray above and brownish-white below, thus making very dingy, non-committal objects. The variations in precise shade are interminable; but the species may always be known by lack of any special sharp markings whatever, except the superciliary line; and by combination of white wing-bars with large oblique tail-spots confined to two outer pairs of feathers. One of the largest species, as well as most simply colored; length 5.50–5.75; extent 8.50–9.00; wing 2.75–3.00; tail 2.40; tarsus 0.70; bill 0.45. Eastern U. S. to the Plains; N. only to Manitoba, Ontario, and New Brunswick. Breeds throughout its range, and abounds in winter in the Southern States; is nearly resident, being sometimes seen in the Middle States in midwinter, and in New England early and late, with snow. Nests high in pine-trees; eggs 0.68 × 0.52, not peculiar in ground-color or markings. *D. pinus* of authors, as of all previous eds. of the Key, after *Sylvia pinus* WILS., 1811, antedated by *S. pinus* LATH., 1790, which latter is now *Helminthophila pinus:* for full synonymy see COUES, B. Col. Vall., i, 1878, p. 251. *Sylvia vigorsii* AUD., Orn. Biog., i, p. 153, 1832, named *Vireo vigorsi* on pl. 30, and in NUTT., Man., i, 1832, p. 318; *Dendroica vigorsii* ST., Auk, Oct., 1885, p 343; A. O. U. Lists, 1886–95, No. 671.

*** Thus passing in review the 24 "solid" species of *Dendrœca*, with four varieties lately introduced, I may allude to two species described by early authors, but never identified: 1. *Sylvia montana* WILS. This I have given (in the orig. ed., p. 105) some reasons for supposing to be a young *D. virens.* 2. *Sylvia carbonata* AUD. A strongly-marked bird, the like of which has never been seen since; conjectured to be a hybrid of *D. tigrina* and *D. striata.* Perhaps it is an offspring of the imagination, stimulated by the artistic sense of its originator, as possibly *Regulus cuvieri* and certainly *Sylvia rathbonia* are also.

SIU'RUS. (Gr. σείω, *seio*, I wave or brandish; οὐρά, *oura*, tail.) WAGTAIL WARBLERS. In general form scarcely distinguishable from *Dendrœca;* larger in size, different in pattern of coloration, in habits, gait, and nidification. Bill ordinary. Rictal bristles short but evident. Wings pointed, much longer than tail. Tarsus longer than middle toe and claw. Tail nearly even, with rather acute feathers, and long, copious under coverts. Neither wings nor tail particolored. Above olivaceous, with or without head-markings, otherwise uniform; below white, buffy, or yellowish, profusely streaked. Legs slender, usually pale-colored. Habits terrestrial to some extent; nest on ground; eggs white, spotted. Vocal powers pre-eminent. Gait ambulatorial, not saltatorial, and some other traits decidedly Motacilline. (A. O. U. spells *Seiurus.*)

Analysis of Species.

Crown orange-brown, with two black stripes; no superciliary line . . . . *auricapillus*
Crown like back; a long superciliary line.
    Below, yellowish, heavily streaked; smaller; bill not over 0.50 . . . . *nœvius*
    Below, whitish, lightly streaked; larger; bill over 0.50 . . . . *motacilla*

S. auricapil′lus. (Lat. *aurum*, gold; *capillus*, hair. Fig. 190.) GOLDEN-CROWNED WAGTAIL WARBLER. GOLDEN-CROWNED ACCENTOR. GOLDEN-CROWNED THRUSH. OVENBIRD. Adult ♂ ♀: Entire upper parts, including wings and tail, uniform bright olive-green, without markings. Top of head with black lateral stripes, bounding a golden-brown or dull orange space. A white ring round eye; no white superciliary stripe. Under parts white, thickly spotted with dusky on breast, the spots lengthening into streaks on sides; a narrow black maxillary line; under wing-coverts tinged with yellow. Legs flesh-colored. Length 5.75–6.50, usually 6.00–6.25; extent 8.75-10.40, usually 9.50–10.00; wing 2.90–3.25; tail about 2.50. Varies much in size, but is remarkably constant in coloration; sexes indistinguishable, and young

FIG. 190. — Ovenbird, nat. size. (Ad. nat. del. E. C.)

scarcely to be told from the adults. Fall specimens ordinarily quite as bright-colored as those of spring; and the orange-brown crown-spot, though it may be less bright, is acquired by the young with their first full feathering. There are at first no crown-stripes; lower parts buffy, indistinctly streaked; upper parts fulvous-brown; wings and tail as in the adult. N. Am., W. to Colorado, Montana and Alaska; breeds throughout its N. Am. range; winters from our S. border southward. A pretty and engaging species, called Oven-bird from the way it has of roofing over its nest, abundant in woodland, migratory. In May the woods resound with its loud *crescendo* chant, so incessant and obtrusive that the bird was long in acquiring the reputation of musical ability to which its luxurious nuptial song entitles it not less than the Louisiana Water Thrush itself. The bird spends much of its time on the ground, trailing prettily among fallen leaves with mincing steps. Nest on the ground, of leaves, grasses, etc.; eggs 4–6, white or slightly creamy, profusely speckled with reddish-brown and lilac, 0.85 × 0.65. (Name misspelled *Seiurus aurocapillus* in A. O. U. Lists, preserving the original cacography of SWAINS., 1827.)

S. næ′vius. (Lat. *nævius*, spotted; *nævus*, a mole, birth-mark.) SMALL-BILLED WAGTAIL WARBLER. AQUATIC ACCENTOR. NEW YORK WATER THRUSH. BESSY KICK-UP. RIVER PINK. Adult ♂ ♀: Uniform dark olive-brown; wings and tail similar, unmarked; below, pale sulphury-yellow everywhere, except perhaps on middle of belly, thickly speckled or streaked with dark olive-brown, the markings smallest on throat, largest on sides. A long dull whitish superciliary line. Bill and feet dark. Length 5.50–6.00; extent 8.50–9.50; wing 2.75–3.00; tail 2.25; bill not over 0.50 along the culmen. The sexes do not differ appreciably. The shade of the upper parts varies from a decidedly olivaceous-brown to a purer, darker bistre-brown, and that of the under parts from sulphur-yellow to nearly white; but it is never of the buffy-white of *S. motacilla*. The streaking varies in amount and intensity, but has a sharp distinct character in comparison with *S. motacilla*, and is rarely if ever absent from the throat. No bill over 0.50, and this member lacks the peculiar shape, as well as size, characteristic of *S. motacilla*. The very young bird sooty-blackish, each feather of upper parts with terminal bar of ochraceous; wing-coverts tipped with the same, forming two bars; streaks below as in the adult, but broader, and not so sharply defined. Eastern N. Am. to the arctic regions, the typical form migratory especially along the Atlantic slope, but also in the Mississippi Valley at large; breeds mainly from our N. borders northward, and winters from the S.

border S. to the West Indies, and Central and S. America; a common inhabitant of thickets, swamps, and morasses, less frequently of mixed woodland. Nest usually under a stump or log, in wet places or near water, not roofed over, but simply built of mosses, leaves, and grasses, lined with rootlets; eggs 4-6, brilliant white, profusely speckled with reddish-brown surface-markings and neutral-tint shell-spots, 0.80 × 0.60. *S. noveboracensis* A. O. U.

S. n. nota'bilis? (Lat. *notabilis*, noteworthy.) WYOMING WATER THRUSH. GRINNELL'S WATER THRUSH. Identical in coloration with the last, but larger; wing 3.25; tail 2.50; bill from nostril 0.50; its depth at base 0.25; tarsus 0.83; middle toe without claw 0.56. A slight variation upon the last, originally described from Wyoming, later extended to include the small-billed Water Thrushes of Western N. Am., chiefly in the interior, E. to Illinois and Indiana (sometimes to the Atlantic coast!), with latitudinal extension from Arctic to South America. I continue to query the bird, as in former eds. of the Key. A. O. U. Lists, No. 675 *a*.

S. motacil'la. (Lat. *motacilla*, a wag-tail. See p. 300.) LARGE-BILLED WAGTAIL WARBLER. LOUISIANA WATER THRUSH. Very similar to *S. nævius*; larger; length 6.00-6.25; extent 10.00-10.75; wing 3.00-3.25; bill especially longer and stouter, over 0.50; tarsus nearly 1.00. Under parts white, only faintly tinged, and chiefly on flanks and crissum, with buff (not sulphury-yellow); the streaks sparse, pale, and not very sharp; throat, as well as belly and crissum, unmarked; legs pale. I have yet to see a specimen I cannot distinguish on sight; size of bill is by no means the only character, though it is a principal one. Eastern U. S., rather southern, and not very common; N. to Massachusetts and southern Ontario, Mich., and Minn.; W. to Kansas, Indian Territory, and Texas; more abundant in the Mississippi Valley; breeds in its U. S. range at large; winters extralimital in the West Indies, some parts of Mexico, and thence to Panama. Habits, nest, and eggs like those of *S. nævius*. A sweet and skilful songster.

GEO'THLYPIS. (Gr. γῆ or γέα, *ge* or *gea*, the earth, and θλυπις or θραυπις, *thlupis* or *thraupis*, name of some bird.) GROUND WARBLERS. Bill of ordinary Sylvicoline characters; rictal bristles short and few, but evident. Wings variable; pointed, and much longer than the tail in the subgenus *Oporornis*, with 1st quill nearly or quite the longest; short and much rounded, scarcely or not longer than the tail in *Geothlypis* proper; colored like the back, and without markings, in both subgenera. Legs stout; tarsi longer or not shorter than middle toe and claw. Of medium and rather small size for this family. Coloration plain olivaceous above, with more or less extensive yellow below and veiled with ash or blackish on the head (as in *Oporornis* and some species of *Geothlypis*) or there masked with black, ash, and white or hoary, as in ♂ of the *G. trichas* group; sexes alike in the former case, unlike the latter. Tail about even, or a little rounded, without white spots. Legs pale-colored. Habits somewhat terrestrial. Nest on the ground or near it. This genus affords numerous species more or less resembling the common Maryland Yellow-throat, chiefly of the warmer parts of America — seven of N. Am. Most of them are well distinguished from other Warblers by the extreme shortness of the wings, which are scarcely or not longer than the tail, and all of them by the size of the pale-colored legs, which indicates their somewhat terrestrial habits; in the two species of *Oporornis* the outstretched feet reach nearly or quite to the end of the tail; and they reach about as far in *G. philadelphia* and *G. macgillivrayi*, though the tail is relatively longer in the *G. trichas* group. Our species are familiar inhabitants of shrubbery, ordinarily keeping near the ground, where the nest is usually placed. (Genera *Oporornis* and *Geothlypis* of all previous eds. of the Key, the former being now reduced to a subgenus of the latter. I am glad to follow the A. O. U. example in this case, as two of our species (*philadelphia* and *tolmiei* both) connect the two species of *Oporornis* so closely with the *trichas* group, that they have been even placed in the former subgenus by one high authority.)

MNIOTILTIDÆ: AMERICAN WARBLERS.

Analysis of Species.

Wing much longer than tail, pointed, 1st quill longest or nearly so (*Subgenus* OPORORNIS).
    Head with black; line over eye and all under parts rich yellow in ♂ ♀ . . . . *formosa*
    Head without black or yellow; crown, throat, and breast ashy in ♂; a white eye-ring . . . . . . *agilis*
Wing not longer than tail, rounded, 1st quill not nearly longest (*Subgenus* GEOTHLYPIS).
    Sexes nearly alike: head and throat ashy, deepening on breast.
        No white eyelids; breast of adult ♂ quite blackish . . . . *philadelphia*
        White eyelids; breast of adult ♂ scarcely different from throat . . . . *tolmiei*
    Sexes quite unlike. ♂ with black and ash or yellow on head; ♀ with head plain.
        Black mask involving front and sides of head.
            Mask bordered with hoary ash; throat and breast only yellow . . . . *trichas*
            Mask bordered with yellow; under parts all yellow . . . . *beldingi*
        Black on sides of head only; top of head ash; eyelids white (*Subgenus* CHAMÆTHLYPIS) *poliocephala ralphi*

(*Subgenus* OPORORNIS.)

G. (O.) formo'sa. (Lat. *formosa*, shapely, comely; hence, beautiful in any way. Fig. 191.) KENTUCKY WARBLER. Adult ♂ ♀: Clear olive-green; entire under parts pure bright yellow, olive-shaded along sides; crown black, the feathers more or less skirted with ashy, separated by a rich yellow superciliary line (which curls around eye behind) from a broad black bar running from bill below eye and thence down side of neck; wings and tail unmarked, glossed with olive; feet flesh-color. Length 5.50–5.75; extent about 9.25; wing 2.75–3.00; tail 2.00–2.25; tarsus 0.85. In the fall, the black of head and neck is much overlaid by ashy or grayish tips of the feathers; the yellow of under parts is paler, and more shaded with olivaceous along sides. Young birds lack the black and yellow of head; the under parts are much duller, and the upper parts have a brownish cast; at a very early age the wing-coverts are tipped with buff. Eastern U. S., N. to the Connecticut Valley, Michigan, etc., and rarely to Quebec; W. to the Plains; not abundant.

FIG. 191. — Kentucky Warbler, nat. size. (Ad. nat. del. E. C.)

Not abundant at large, but very common in certain sections, as in Illinois, Kansas, and other portions of the Mississippi Valley. Breeds throughout its U. S. range; winters extralimital, in some of the West Indies, parts of Mexico, and S. to Panama. A beautiful object, gleaming like gold in the tangle and débris of thick dark woods and swamps. Nest on the ground, or in rubbish near it, of leaves, grasses, weed-stems and rootlets, large and shallow; eggs 4–5, 0.70 × 0.56, crystal-white, sprinkled with spots and dots of reddish, brownish, and neutral tint.

G. (O.) a'gilis. (Lat. *agilis*, agile, active. Fig. 192.) CONNECTICUT WARBLER. Adult ♂:

FIG. 192. — Connecticut Warbler.

Olive-green, becoming ashy on head; below, from the breast, yellow, olive-shaded on sides; chin, throat, and breast dark ash; a white ring around eye; wings and tail unmarked, glossed with olive; under mandible and feet pale. Length about 5.50; extent 8.50–9.00; wing 2.75–3.00; tail 2.00; tarsus 0.80. In spring males the ash of head and throat is quite pure and very dark, almost black on breast; then the resemblance to *G. philadelphia* is close; but in the latter the wings are little if any longer than the tail. The ♀ is not always distinguishable from the ♂; but the top of the head is less purely ash, being tinged with olivaceous, and the sides of the head, the chin, and throat, are light gray or even whitish. In most specimens of both sexes in the fall the upper parts from bill to tail are nearly uniform olive, and the ash of the throat is pale. Young of the year resemble the adult ♀, but are more dingy brownish; and the species lacks any very strong or decided markings, except the ♂ in full plumage. Eastern U. S. and adjoining British Provinces; known to breed in Ontario and Manitoba, and to reach northern S. Am. in winter; not commonly observed in spring; abounding in fall in some localities (whence the name of the subgenus *Oporornis* from Gr. ὀπώρα, *opora*, autumn, and ὄρνις, *ornis*, a bird); a shy, fugitive inhabitant of brushwood and thickets. Nest on ground, as usual in this genus; eggs 0.75 × 0.52; white, dotted and spotted with reddish and darker brown and with neutral tints.

(*Subgenus* GEOTHLYPIS.)

G. philadel′phia. (To the city of brotherly love; Gr. φιλέω, *phileo*, I love; ἀδελφός, *adelphos*, brother. Fig. 193.) MOURNING WARBLER. CRAPE WARBLER. Adult ♂ ♀, in spring: Bright olive, below clear yellow; on the head the olive passes insensibly into ash; in high plumage of ♂ the throat and breast black; but generally ash, showing black traces, the feathers being black veiled with ash, producing a peculiar appearance suggestive of the bird's wearing crape; wings and tail unmarked, glossed with olive; under mandible and feet flesh-color; *no white about eyes* in adult ♂. Young, and generally fall specimens: Ash of fore parts veiled with olive; sides and across breast quite olivaceous, leaving only central line of under parts yellow; blackish-ash of jugulum

FIG. 193. — Mourning Warbler. (L. A. Fuertes.)

veiled by bright yellow tips of the feathers; eyelids brownish-yellow. Young birds have little or no ash on head, and no black on throat, thus resembling *agilis* ♀ and young. The Mourning Warbler is very closely related indeed to the Connecticut Warbler; taking sex for sex and season for season, the changes of plumage are quite correspondent; but the two species are of course distinguishable by their subgeneric characters: observe shortness and rounding of wing in *philadelphia*, as compared with its length and pointedness in *agilis*, in either case as relative to length of tail. Length 5.25–5.50; extent 7.50–8.00; wing 2.25–2.50; tail 2.00–2.25; tarsus 0.80. Eastern N. Am., W. to Kansas and Dakota, rare in most localities in the Atlantic States, but abundant in the Mississippi Valley; migratory; no record of wintering in the U. S.; breeds chiefly in the northernmost tier of States and along the British border, but farther S. in mountainous portions of New England, New York, and Pennsylvania; S. in winter to Cent. and S. Am.; accidental in Greenland. Nidification like that of *G. trichas;* eggs not distinguishable.

G. tol′miei. (To Dr. Wm. Fraser Tolmie, surgeon and chief factor H. B. Co., whom J. K. Townsend met on the Columbia in 1834.) TOLMIE'S WARBLER. MACGILLIVRAY'S WARBLER. Adult ♂ ♀: Upper parts, including exposed surfaces of wings and tail, clear olive-green; below, bright yellow, shaded with olive on the sides. Head and neck all around, throat, and fore breast, clear ashy; *eyelids white;* loral region dusky or quite black, the throat with blackish centres to the feathers, veiled by their gray skirting. Upper mandible blackish;

under mandible and feet flesh-colored or pale yellowish. Size of *G. philadelphia* exactly. Seasonal and sexual differences those of *G. philadelphia*, of which it is the Western representative, differing in having white eyelids and black lores, and in never showing a decided black patch on the breast, which is conspicuous in the highly plumaged ♂ of the other form; but thus closely resembling ♀ *philadelphia*, which normally shows a whitish eye-ring, and has not the breast black. Middle and Western Provinces of the U. S., E. to the limit of trees on the plains, N. to British Columbia; abundant, migratory; breeds throughout its U. S. range; winters beyond, in L. Cala., Mexico, and Cent. Am. to the U. S. of Colombia. Nest and eggs as in others of the genus. *G. macgillivrayi* of most authors, as of all former eds. of the Key; but *Sylvia tolmiei* J. K. Towns., Journ. Phila. Acad., viii, pt. 1, read Apr. 1839 (vol. pub. 1840), pp. 149 and 159, and Narr., Apr. 1839, p. 343, has a few months' priority over *Sylvia macgillivrayi* Aud., Orn. Biog., v, June, 1839, p. 75, folio pl. 399: see Stone, Auk, Jan. 1899, p. 81; A. O. U. Suppl. List, *ibid.*, p. 122.

G. trich'as. (Gr. τριχάς, name of some bird in Aristotle. Fig. 194.) Yellow-throated Ground Warbler. Maryland Yellow-throat. Black-masked Warbler. ♂, in summer: Upper parts rich olive, inclining to grayish on head, brightest on rump. Wings and tail brown, edged with color of back. Chin, throat, and breast, with under wing- and tail-coverts, rich yellow. Middle under parts dull whitish, shaded on sides. A broad black mask on front and sides of head, bordered behind by hoary-ash. Bill black; feet flesh-colored. Length 4.75–5.00; extent 6.50–6.90; wing 1.90–2.10; tail hardly more; tarsus 0.75. ♀, in summer: Rather smaller; yellow of under parts paler and more restricted; no black or ashy markings on head, but crown usually with some concealed reddish-brown. Otherwise top and sides of head like back, with some obscure whitishness about lores and orbits. Young: Simi-

Fig. 194. — Maryland Yellow-throat, ♂, nat. size. (Ad. nat. del. E. C.)

lar to adult ♀, but the olive of upper parts with much of a brownish tinge, the yellow parts and, in fact, most of the under parts, quite buffy. The adults, in fall and winter, are similar to each other, except in the purer and stronger yellow of the ♂, as at that season the peculiar black and ashy markings of the head are wanting. Both sexes then resemble the autumnal plumage of the young in the browner shade of the olive and buffiness of the under parts. Eastern U. S. and British Provinces, N. to Labrador, W. in the Mississippi Valley; breeds throughout most of this range; winters from the S. Atlantic and Gulf States southward to the West Indies, Eastern Mexico, and Central America, but is occasionally found at this season N. to Massachusetts. An abundant and familiar inhabitant of shrubbery and underbrush, the sameness of which is enlivened by its sprightly presence and hearty song throughout the summer months. Nest on the ground or near it, usually carefully concealed, of large size and built of any rubbish; eggs 4–6, usually 0.60–0.70 long by 0.50–0.55, white, rather sparingly sprinkled, and mostly at the large end, with several shades of brown : but the markings, like the size and shape of the eggs, are very variable.

G. t. occidenta'lis. (Lat., of the occident or setting sun, *i. e.*, western.) Western Yellow-throat. Like the last; appearing somewhat larger, owing to longer tail; upper parts lighter, the olivaceous having a more yellowish hue, and the hoary ash of cap paler and more extended; under parts rich yellow, extending over the belly and sometimes farther; the shading of the flanks ochraceous rather than olivaceous. Wing and tail each about 2.30. Western N. Am., Mississippi Valley to the Pacific, British Columbia in summer to Central Am. in winter. This is a fairly well-marked form, which should have appeared in all the previous eds. of the Key. More than 30 years ago I named it *G. hypochryseus* in a monograph of the genus which was never published, but subsequently overlooked it. The extensive bright yellow of the under parts is a good feature, and the tail averages 0.25 longer than that of average *trichas*. Brews-

TER, Bull. Nutt. Orn. Club, July, 1883, p. 159; RIDGW., Man., 1887, p. 523; A. O. U. Lists, 1886 and 1895, No. 681 *a*.

G. t. igno′ta. (Lat. *ignotus*, unknown or ignored, as this form was for many years; but the name ceased to be applicable as soon as it was used.) FLORIDA YELLOW-THROAT. Like the last; with somewhat longer bill, tarsus, and tail, as usual in Florida birds; yellow of under parts bright and extensive; olivaceous of upper parts browner in shade; flanks deeply shaded; facial mask broader. Florida and Georgia. CHAPMAN, Auk, Jan. 1890, p. 11; COUES, Key, 4th ed., 1890, p. 898; A. O. U. List, 2d ed., 1895, p. 284, No. 681 *b*. This is *G. t. roscoe* of HASBROUCK, Auk, Apr. 1889, p. 167; but *Sylvia roscoe* of Audubon, Orn. Biog., i, 1831, p. 124, pl. 29, was based on an immature autumnal ♂ taken in Mississippi in September, of such equivocal character that it has thus far proved unidentifiable; the name cannot therefore be used for the resident Florida bird. S. Atlantic and Gulf coast, S. E. Va. to E. Texas.

G. bel′dingi. (To L. Belding.) BELDING'S YELLOW-THROAT. Adult ♂ : Above nearly uniform olive-green, a little browner anteriorly; below, rich yellow, paler on the vent, tinged with brown on the flanks and sides. Black mask exactly as in *G. trichas*, but bordered behind for its whole extent with rich yellow; there being thus no hoary ash on the head. Wing 2.60; tail 2.70, graduated 0.50; bill 0.50 or more; tarsus 0.90. Adult ♀ similar to ♂ in the body colors, but lacking the distinctive head markings, as usual in the *trichas* group; more brownish on the head, duller yellow below, and whitish or grayish on the belly and flanks; size less. Lower California, N. to San Ignacio, about lat. 27°. Quite distinct from any other species in this list; near the Mexican *G. melanops*. RIDGW., Pr. U. S. Nat. Mus., v, 1882, p. 344; COUES, Key, 3d ed., 1887, p. 870; A. O. U. Lists, No. 682.

(*Subgenus* CHAMÆTHLYPIS.)

G. poliocept′ala ralph′i. (Gr. πολιός, *polios*, hoary; κεφαλή, *kephale*, head. To Dr. Wm. L. Ralph.) HOARY-HEADED YELLOW-THROAT. RIO GRANDE YELLOW-THROAT. RALPH'S TRICHAS. Quite different again from any of the foregoing, and representing a section of the genus which has been called *Chamæthlypis*. Bill very stout, with strongly curved culmen hardly twice as long as the bill is deep at base. Adult ♂ : Olive-green above, becoming gray on the crown, the loral and more or less of the circumocular region black, the eyelids white; yellow below, including edge and more or less of lining of wings, paling to buffy whitish on the belly and flanks. ♀ similar, having the distinctive head-markings. Length about 5.50; wing 2.30; tail 2.60; tarsus 0.87; bill 0.47, its depth at base 0.20 or rather more. Brownsville, Texas, in Lower Rio Grande Valley. Very close indeed to *G. poliocephala* proper of western Mexico, and also near *G. palpebralis* of E. Mexico, with which it agrees closely in size and proportions, but is not entirely yellow below; said to differ from *poliocephala* proper only in rather larger size, especially of the bill, grayer upper parts, and paler yellow below. Not in any previous ed. of the Key; *G. p. palpebralis*, ALLEN, Auk, July, 1891, p. 316; *G. p. ralphi* RIDGW., Pr. U. S. Nat. Mus., 1894, p. 692; A. O. U. List, 2d ed., 1895, No. 682. 1.

ICTE′RIA. (Gr. ἴκτερος, *ikteros*, the jaundice; hence, yellowness; from the bird's golden breast.) CHATS. Bill stout, high at base (higher than broad at nostrils), thence compressed; unnotched, unbristled, with much curved culmen and commissure. Frontal feathers reaching nostrils, which are subcircular and scaled. Wings much rounded, shorter or not longer than graduated tail. Tarsus partly booted, longer than middle toe; feet stout. Inner toe cleft to the degree usually seen in this family. Of largest size for this family. Form stout. Coloration simple, chiefly olive, yellow, and white. Sexes alike. Nest in bushes. Eggs white, spotted. Probably only one species.

I. vi′rens. (Lat. *virens*, being green. Figs. 195, 196.) YELLOW-BREASTED CHAT. POLYGLOT. CLOWN. CHARLATAN. MOUNTEBANK. Adult ♂ ♀: Bright olive-green, below golden-yellow, belly abruptly white; lore black, isolating the white under eyelid from a white superciliary line above and a short white maxillary line below; wings and tail unmarked, glossed with olive; bill blue-black; feet plumbeous. Length about 7.50; extent about 10.00;

FIG. 195. — Yellow-breasted Chat.

wing about 3.00; tail about 3.25. Little difference with age, sex, or season in the plumage of this rich bird; very young have the fore under parts gray or white slashed with yellow, no black on lore, and lower mandible pale; white of belly and crissum tinged with buff. Eastern U. S., N. to Massachusetts, S. Minnesota, and S. Ontario; breeds throughout its range; winters through E. Mexico to Central America. An exclusive inhabitant of low tangled undergrowth, and oftener heard than seen, except during the mating season, when it performs the extravagant aërial evolutions for which, as well as for the variety and volubility of its song, it is noted. Nest in crotch of a bush near the ground; eggs 3–5, very variable in size and markings, 0.90 to 1.00 × 0.70 to 0.80, white, dotted, spotted or blotched with reddish-browns and the usual lilac shell-markings.

FIG. 196. — Yellow-breasted Chat, nat. size. (Ad. nat. del. E. C.)

I. v. longicau′da. (Lat. *longus*, long; *cauda*, tail.) LONG-TAILED CHAT. Adult ♂ ♀: Entire upper parts, including exposed surfaces of the wings and tail, grayish-olive. Quills of wings and tail fuscous. Fore half of body below, including lining of wings, rich yellow; hinder half white, shaded with gray on sides. Loral region black; a sharp maxillary line, another from nostril over eye, and under eyelid, white. Bill blackish-plumbeous; feet plumbeous. Size of the last; tail averaging longer. Middle and Western Provinces of the U. S. This form, in its typical manifestation, differs from *virens* in the shade of the upper parts — quite grayish instead of pure olive-green; in the dullest-colored birds there is scarcely a tinge of olive in the gray, though the yellow of the breast is as rich as that of *virens*.

WILSO'NIA. (To Alexander Wilson, "father of American Ornithology.") FLY-CATCHING WARBLERS. Bill Muscicapine, though with lateral outlines a little concave, broad and depressed at base, with many obvious rictal bristles reaching decidedly beyond nostrils; culmen and commissure nearly straight. Wings pointed, as in most *Mniotiltidæ*, longer than tail; 1st quill longer than 5th, 3d equalling or exceeding 4th. Tail narrow, even or little rounded. Middle toe without claw about ⅔ as long as tarsus. Tail unmarked, or with white blotches as in *Dendrœca*. No red or flame-color; always yellow below. Comprehends three species, well distinguished among *Mniotiltidæ* by development of rictal bristles and depressed shape of bill, though these Muscicapine characters are not pushed to the extreme seen in *Setophaga*. Nest on the ground, as in the genera *Geothlypis*, *Helminthophila*, etc. (except in case of *mitrata*); eggs white, marked after the usual fashion of Warblers'. (Genus *Myiodioctes* AUD., 1839, of most writers, as of all previous eds. of the Key; *Myioctonus* CAB., 1850; *Wilsonia* BP., 1838; not *Sylvania* NUTTALL, 1832, which is a mere synonym of *Setophaga*, including the Redstart, Blue-gray Gnatcatcher, and several species of the present genus, but untenable for any of these, though misused for them by various authors, and so by the A. O. U., 1886-95: see COUES, Auk, Apr. 1897, pp. 223, 224, where the error is exposed, and *Wilsonia* shown to be the proper name: see also COUES, Bull. Nutt. Club, 1880, p. 95. *Wilsonia* was adopted by the A. O. U. in its Ninth Suppl. List, Auk, Jan. 1899, p. 123.)

Analysis of Species.

Olive and yellow; tail-feathers white-blotched . . . . *mitrata*
Olive and yellow; tail-feathers plain . . . . *pusilla*
Ashy-blue and yellow; tail-feathers plain . . . . *canadensis*

NOTE. — The "small-headed flycatcher, *Muscicapa minuta*" of WILS., NUTT., AUD., etc. (nec GM., 1788), conjectured to belong to this genus, continues to be unknown. Its whole record is a tissue of surmises: for the synonymy, see COUES, Birds Col. Valley, i, 1878, p. 326, and add: *Sylvania microcephala* RIDGW., Pr. U. S. Nat. Mus., viii, 1885, p. 354; Man., 1887, p. 527; A. O. U. Hypothetical List, No. 25. There certainly was such a bird, for Wilson figured it, and he never drew upon his imagination; but we do not recognize his plate, nor that of Audubon. The mysterious bird has been claimed for New Jersey, Pennsylvania, Massachusetts, Wisconsin, and Kentucky. I have long believed it to be the Pine-creeping Warbler, *Dendrœca vigorsi*: see Key, orig. ed., 1872, p. 109.

W. mitra'ta. (Lat. *mitrata*, wearing a mitre, or other head-dress. Fig. 197.) HOODED FLY-CATCHING WARBLER. SELBY'S SYLVAN FLYCATCHER. Adult ♂: Clear yellow-olive above; below, rich yellow, shaded with olive along sides; whole head and neck pure black, enclosing a broad golden mask across forehead and through eyes; wings unmarked, glossed with olive; tail with large white blotches on 2 or 3 outer pairs of feathers, as in *Dendrœca*; bill black; feet flesh-colored. Length 5.00-5.50; extent about 8.50; wing 2.50-2.75; tail about 2.25. Adult ♀ and young ♂: The black restricted or interrupted, if not wholly wanting, as it is in the earlier stages, when the parts concerned are simply colored to correspond with the upper and under surfaces of the bird.

FIG. 197. — Hooded Warbler, nat. size. (Ad. nat. del. E. C.)

Hood said to be not perfected till the 3d year, and to be finally acquired, in fulness of its extent if not in purity of the black, by the ♀. Eastern N. Am., strictly, W. only to the edge of the Plains, N. regularly to the Connecticut Valley, some portions of New York, southern Ontario, and southern Michigan; migratory; breeds at large in its U. S. range; winters extralimital in some of the West Indies, eastern Mexico, and Central America. A lovely bird, reminding one of the Kentucky Warbler, common in the South in such brakes and bottoms as the Kentucky haunts, rarer northward. Nest in bushes; eggs 3-4, about 0.70 × 0.50, as usual white, sprinkled with reddish-brown, neutral gray, and sometimes darker spots and dots, chiefly about the larger ends.

W. pusil'la. (Lat. *pusilla*, puerile, petty, small. Fig. 198.) GREEN BLACK-CAPPED FLY-CATCHING WARBLER. WILSON'S SYLVAN FLYCATCHER. WILSON'S WARBLER. Adult ♂ ♀: Upper parts, including exposed edgings of wings and tail, bright yellowish-olive; under parts, including front and sides of head and superciliary line, rich yellow, shaded with olive on sides. A squarish, glossy blue-black patch on crown. Wings and tail plain fuscous, with greenish edgings, unmarked with other color. Upper mandible dark; under mandible and feet light. Length 4.75; extent 6.75–7.00; wing 2.00–2.25; tail 2.00. Young: Lacking the black cap; as sometimes also the ♀. There is very little variation in this species, according to age or season, though the adult summer birds are the more richly colored. Eastern N. Am., in wooded regions; common, migratory. Breeds from the northernmost States northward to the limit of trees; occasional west in migration in the Rocky Mountains; winters extralimital. Nest on the ground; eggs 4–5, 0.60 × 0.50, white, speckled and blotched with dark reddish-brown and lilac.

FIG. 198. — Black-capped Warbler, nat. size. (Ad. nat. del. E. C.)

W. p. pileola'ta. (Lat. *pileolata*, wearing the *pileum*, a kind of cap.) WESTERN BLACK-CAPPED FLY-CATCHING WARBLER. PILEOLATED SYLVAN FLYCATCHER. Specimens from the southern Rocky Mts. and Pacific coast region are frequently of a brighter yellow, almost orange, on head and fore parts below. Breeds from Rocky Mts. to Pacific and N. to Alaska.

W. canaden'sis. (Lat., of Canada. Fig. 199.) CANADIAN FLY-CATCHING WARBLER. BONAPARTE'S SYLVAN FLYCATCHER. Adult ♂, in spring: Bluish-ash; crown speckled with lanceolate black marks, crowded and generally continuous on forehead; latter divided lengthwise by a slight yellow line; short superciliary line and edges of eyelids yellow; lores black, continuous with black under eye, and this passing as a chain of black streaks down side of neck and prettily encircling throat like a necklace of jet; excepting these streaks and the white under tail-coverts, entire under parts clear yellow; wings and tail unmarked; feet flesh-color. ♂ in autumn with the yellow very rich, even tipping feathers of the black necklace. Length 5.25–5.50; extent 7.75–8.25; wing 2.50; tail 2.25. In the ♀ and young the black of crown, cheeks, and necklace is obscure or much restricted, and in the young the back may be glossed with olive; but they cannot be mistaken for any other species. Eastern N. Am.; an abundant and beautiful woodland species; migratory; breeding in the Alleghanies from as far S. at least as the mountains of western N. Carolina, where I have found fledglings, and at lower elevations from the Middle States occasionally, from New England regularly, northward to the limit of trees; in winter S. to Central and S Am. Nest on the ground or very close to it in the grass or weeds of wet woods; eggs 3–5 0.75 × 0.55, white, dotted and blotched with reddish-brown and other shades, as usual.

FIG. 199. — Canadian Fly-catching Warbler. (L. A. Fuertes.)

NOTE on *Wilsonia microcephala*. The small-headed Flycatcher, *Muscicapa minuta* WILS., 1812, supposed to belong to this genus, continues unknown. It was renamed *Sylvania microcephala* by Ridgway in 1885, and so stands with a query, in A. O. U. Hypothetical List, 1886 to date, p. 333.

SETO'PHAGA. (Gr. σής, σητός, *ses, setos*, an insect; φάγω, *phago*, I eat.) REDSTARTS. Bill thoroughly Muscicapine in depression and breadth at base, where wider than high, straightness of superior and lateral outlines, and development of rictal bristles, which reach far beyond nostrils. Wings pointed, not shorter than tail; 2d, 3d, and 4th quills nearly equal and longest; 1st intermediate between 4th and 5th. Tail rather long and fan-shaped, with broad flat feathers, widening at ends. Feet slender, with long tarsi indistinctly scutellate externally, and short toes, the middle one without its claw about half as long as tarsus. Coloration indeterminate. Habits arboricole and Muscicapine. The genus has been made to cover considerable variety in form among the numerous species of Fly-catching Warblers of subtropical and tropical America, where it is best represented. The diagnosis, drawn up from *S. ruticilla*, may require some little modification in order to its applicability even to *S. picta*. All the extralimital species differ in the shorter and more rounded wing and other characters. *S. ruticilla* is the only species in which the sexes are decidedly dissimilar in color; even in *S. picta*, the nearest ally, they are substantially alike; and in all the rest, in which the coloration is very various, there is no obvious difference between the sexes. Species of *Setophaga* (including *Myioborus* and *Euthlypis*), to the number of 15 or more, are recognized by late authors. *S. ruticilla* is the only one generally distributed in N. Am.

Analysis of Species.

♂ Black, white, and orange; ♀ brown, white, and yellow . . . . *ruticilla*
♂ . ♀ Black, white, and carmine-red; no chestnut . . . . *picta*
♂ ♀ Black, white, slate-gray, and vermilion red; cap chestnut . . . . *miniata*

S. ruticil′la. (Lat. *ruticilla*, red-tail; *rutilus*, reddish; "redstart" is corrupted from *rothstert*, red-tail. Figs. 200, 201.) **American Redstart. Little Oriole. Fire-tail. "Live Coals."** Adult ♂: Lustrous blue-black; belly, flanks, and crissum white. Sides of body and lining of wings rich flame-color, which often tinges the breast quite across. Basal portions of all wing-quills, excepting innermost secondaries, the same rich reddish-orange, brightest on outer webs, where it forms a conspicuous exposed spot; paler and more extensive on inner webs. All lateral tail-feathers similarly colored for half or more of their length, orange meeting black abruptly with transverse outline. Bill and feet black. Length 5.00–5.50; extent 7.50–8.00; wing 2.25–2.50; tail the same; bill 0.33; tarsus 0.66. Adult ♀: Black of ♂ replaced on upper parts with olive, growing more ashy on head, on wings with fuscous, and below with white. Sides rich yellow where ♂ is orange, this color often tingeing the breast across. Orange mark-

Fig 200. — American Redstart.

ings of wings and tail of ♂ replaced by clear yellow. Lores dusky; eyelids and slight stripe from nostrils to eye whitish. Rather smaller than ♂, about equal to the lesser several dimensions given. ♂, young: Like the ♀, but upper parts more brownish, tail quite black, and yellow of sides brighter. Males changing in spring to their final plumage are irregularly patched with black in the general olivaceous and white. The spring migration includes males in this condition, and others irregularly patched with black, as well as those in perfect dress; whence it is evident that the Redstart does not acquire his full-dress suit until in his third year (see Birds Col. Vall., p. 340). Temperate N. Am., but chiefly Eastern; W. to the Great Basin regularly, casually to Upper and Lower California. Breeds in much of its U. S., and all of its British American range, abundantly from the Northern States northward; winters in the West Indies, Mexico, and Cent. and S. Am. Nest a neat, compact structure in fork of a shrub or sapling at little elevation; eggs 3-5, averaging 0.65 × 0.50, not distinguishable from other Warbler eggs. During the nuptial ecstasies the lovely Redstart shines among the birds that throng the woodland, where his transparent beauty flashes like a lambent tongue of flame at play amidst the tender pale green foliage of the trees.

S. pic'ta. (Lat. *picta*, painted. Fig. 202.) PAINTED REDSTART. Adult ♂ ♀: Lustrous black; middle of breast and belly carmine-red; eyelids, a large patch on wings formed by greater and middle coverts, broad edging of inner secondaries, edging of inner webs of primaries toward base, lining of wings, nearly all the outer tail-feather, and a diminishing space on next two or three, together with crissum, white. Bill and feet black. Length 5.00-5.50; wing 2.75; tail 2.50; tarsus 0.66; bill 0.33-0.40. ♀ not particularly different from ♂, though rather less richly colored. In poor plumages, the black is not so lustrous; red of belly less extensive and of a more bricky-red tone; white of wings and tail more restricted. Very young: Dull black, or only slightly lustrous; white nearly as in the adult; spot on lower eyelid, patch on wing, outer edge of first primary only, outer edges of secondaries, inside of wings, axillars, crissum, tibiæ, outer tail-feather except at base,

Fig. 201. — American Redstart. (From The Osprey.)

and a diminishing space on the 2d and 3d, white. Mountains of Mexico, N. to Arizona, New Mexico, and doubtless also Texas; common in summer in Santa Rita, Santa Catalina, and

Huachuca Mts. of Arizona. Nest on ground, usually under a projecting stone, or in a bank near water; large, flat, shallow, of bark, weed-fibre, grasses, and a few hairs. Eggs 3–4, 0.65 × 0.50, pure white, speckled and wreathed with pale reddish-brown; Apr.–June.

Fig. 202. — Painted Redstart. (Ad. nat. del. H. W. Elliott.)

S. minia'ta. (Lat. *miniata*, miniated, rubricated, marked with red.) RED-BELLIED REDSTART. Adult ♂ ♀ : Dark bluish-ash or slate-gray above, and on the sides below. A square patch of chestnut on crown. Forehead and sides of head, with whole fore-neck and sides of jugulum black; breast and belly vermilion red; lining of wings and tips of under tail-coverts white. Wing-feathers dusky; tail-feathers black with the lateral one white, and more restricted white areas on the next two. Very young: Sooty blackish, little darker on the head, the dark parts of the adults much overlaid with brown; most of the under parts chocolate-brown, lighter on the belly, where the feathers have whitish bases; wing-coverts tipped with rusty brown; under tail-coverts pale fulvous. Length 5.10; wing 2.50; tail 3.00; tarsus 0.75. Highlands of Mexico to Texas. An extralimital species, admitted to the 3d ed. of the Key, 1887, p. 870, on the authority of GIRAUD; not in either of the COUES Check Lists; A. O. U. Lists, 1886 and 1895, No. [689].

CARDELLI'NA. (Apparently derived from Lat. *carduelis*, a kind of Finch; *carduus*, a thistle.) ROSE FLY-CATCHING WARBLERS. Bill Parine in shape, much shorter than head, high at base, culmen convex throughout; commissure curved. Rictal bristles stiff, but hardly reaching half-way from nostrils to tip of bill, which shows scarcely a trace of notch. Wings long and pointed; 2d, 3d, and 4th quills nearly equal and longest; 1st a little longer than 5th. Tail shorter than wings, nearly even. Feet small; tarsal scutella indistinct externally; tarsus longer than middle toe and claw.

C. ru'brifrons. (Lat. *ruber*, red; *frons*, front, forehead. Fig. 203.) RED-FRONTED FLY-CATCHING WARBLER. Adult ♂ ♀ : Upper parts ash; wings and tail rather darker, edged with ashy-white; a broader and whiter bar across ends of median coverts. Below, from breast, white, more or less shaded with ashy on sides, and tinged with rosy. Rump and a nuchal patch white, or rosy-white. Whole head, throat, sides of neck, and fore breast, bright red, with a broad black cap extending down on sides of head, involving eyes and ears, ending in a point below auriculars. The border of this cap is squarely transverse against the red of the forehead from eye to eye; behind it, the red reaches up sides of neck, but not across back of neck, the white nuchal area there meeting the ashy of back. Bill and feet dark. In the highest summer plumage, the red is rich carmine, the cap glossy-black; the under parts are much tinged with rosy; the rump is snowy-white. Less richly-feathered

Fig. 203. — Red-fronted Fly-catching Warbler.

specimens have the head plain red, the cap sooty-black. There is much difference in the character of the white on nape. Length 5.00; wing 2.66; tail 2.50; tarsus 0.66; bill 0.33, quite different in shape from that of *Setophaga*. Young, newly fledged : Ash of upper parts much shaded with brown, as is white of under parts. Rump snowy-white, as in the adult, but the nuchal patch obscure or inappreciable. Wings and tail as in the adult, but with browner edgings. Black cap restricted to top of head, and of a dull sooty cast. Red parts of the adult, including those parts of side of head which are occupied in the adult with the extension of the black cap, dull grayish-brown, tinged or irregularly slashed with red, especially on forehead and throat. Bill light brown; feet pale. Arizona, New Mexico, and doubtless Texas; S. to Guatemala; common in pineries of southern Arizona, especially during migration, and also breeding there in mountains up to 7,000 feet, May and June. Nest on ground, under a tuft of grass, of hay and leaves; eggs 4, pure white, fully speckled and blotched.

ERGA'TICUS. (Gr. ἐργατικός, *ergatikos*, able or willing to work, industrious, diligent, active.) CARMINE FLY-CATCHING WARBLERS. Bill Parine in appearance, as in *Cardellina*, and other characters much as in that genus, of which the present has often been considered a subgenus. Tail about equal to the wings, both lengthened; 1st quill about equal to the 6th. Rictus well bristled, as in other genera of this group. Plumage nearly unicolor, rich red, with white auriculars; sexes alike. One species.

E. ru'ber. (*Lat.*, red.) CARMINE FLY-CATCHING WARBLER. Adult ♂ ♀ : Rich carmine red, obscured on the back; ear-coverts silvery-white; wing- and tail-feathers dusky, edged externally with reddish; middle wing-coverts mostly pink or rosy white. Young simply rusty brown, paler and more rufous below; but known by the silver ears, which show plainly with the first feathering. Length 4.75; wing 2.40; tail 2.50, graduated 0.20; tarsus 0.75. A very beautiful extralimital species, inhabiting the highlands of Mexico, like *Setophaga miniata*, and believed to extend N. to Texas. Not in either of the COUES Check Lists; admitted to the 3d ed. of the Key, 1887, p. 870; A. O. U. Lists, 1886 and 1895, No. [691], on the authority of GIRAUD.

BASILEU'TERUS. (Apparently the comparative degree of βασιλευτός, *basileutos*, kingly, regal, regnant, from βασιλεύς, *basileus*, a king, monarch.) CROWNED FLY-CATCHING WARBLERS. Bill Muscicapine, more or less widened at base, as in *Setophaga* and *Wilsonia*, but deep, with gently curved culmen; rictal bristles variable, but obvious. Wings rather short, more or less rounded; 1st quill equal to 7th, or still shorter. Tail about equal to or longer than wings with narrow feathers unmarked in color. Coloration olivaceous and yellow, the head (in our species) marked with black stripes bordering a yellow, rufous, or orange-brown field, strikingly after the pattern of *Siurus auricapillus*. Sexes alike. A large genus of tropical and subtropical American species, two of which reach the Mexican border of Texas. These represent respectively *Basileuterus* proper of CABANIS, 1848, and the subgenus *Idiotes* BAIRD, 1865.

Analysis of Species.

Top of head with black stripes bordering a yellow or orange-brown median one ; no yellow superciliary line or any chestnut on side of head . . . . *culicivorus*
Top of head with black stripes bordering a rufous median one ; front black, a bright yellow superciliary line, and sides of head chestnut . . . . *belli*

B. culici'vorus. (Lat., gnat-eating; *culex*, a gnat, midge, mosquito; *vorare*, to devour.) BRASHER'S FLY-CATCHING WARBLER. Adult ♂ ♀ : Above, grayish-olive, or olivaceous-ash, of variable shade with age or season. Below, yellow, shaded with olive on sides. Crown yellow, varying to orange-brown, rufous, or somewhat greenish-yellow, bordered on each side with a stripe of black; some dusky loral or ocular markings, not well defined, but no bright yellow or chestnut on sides of head. No special markings of wings or tail. Length 5.00 or less; wing 2.40 or less; tail 2.00-2.20, graduated 0.15; bill 0.50; tarsus 0.75. Central

America and Mexico to Texas. An extralimital species admitted to our fauna in the 3d ed. of the Key, 1887, p. 871, on the authority of GIRAUD, 1841, who called it *muscicapa brasieri* (for *brasheri*); A. O. U. Lists, 1886-95, No. [692].

B. bell'i. (To J. G. Bell of New York.) BELL'S FLY-CATCHING WARBLER. Somewhat similar to the foregoing, but readily distinguished, and belonging to the subgenus *Idiotes*. Adult ♂ ♀: Above, plain greenish-olive, or olive-green; below, yellow, including the edge of the wing, shaded with olivaceous on sides and lining of wings. Sides of head rich chestnut, blackening on the lores; a long bright yellow superciliary stripe, extending on the side of nape; frontal bar and lateral stripe on crown black, enclosing a chestnut or rufous area. Bill black; feet pale. Length 5.10; wing 2.40; tail rather more, graduated 0.33; bill 0.50; tarsus 0.80. Central America and Mexico to Texas. Another extralimital species, admitted to our fauna in the 3d ed. of the Key, 1887, p. 871, on the authority of GIRAUD, 1841; A. O. U. Lists, 1886-95, No. [693].

Family CŒREBIDÆ: Honey Creepers.

Primaries 9, and other external characters very nearly as in the last family; but bill generally slenderer and sharper, often a little decurved. The line between the two families has never been drawn with precision, and has become more difficult of expression since some of the *Mniotiltidæ* have proven possessed of a peculiarity of the *Cœrebidæ*: deeply bifid, penicillate tongue. As commonly understood, it is a small group containing perhaps 70 species of pretty little birds, of about a dozen genera, which are arranged by Sclater (1886) in 4 subfamilies — *Diglossinæ*, *Dacnidinæ*, *Cœrebinæ*, and *Glossiptilinæ*. All are confined to tropical and subtropical America, being especially numerous in the West Indies. Our species is merely a stray visitor to Florida.

FIG. 204. — Honey Creeper (*Cœreba flaveola*; not distinguishable in a cut from *C. bahamensis*), ⅔ nat. size. (From Brehm.)

CŒ'REBA. (Brazilian name of some guitguit or small creeping bird, perhaps of this family. Fig. 204.) HONEY CREEPERS. Bill little shorter than head, stout at base, but rapidly tapering to the extremely acute tip; whole bill much curved, culmen very convex, outline of under mandible concave from base to tip. Rictus unbristled. Wings long, exceeding the short rounded tail. Tarsus longer than middle toe without claw. Contains about 19 species or varieties, mostly West Indian. (*Certhiola* of previous eds. of the Key, as of authors generally; but VIEILLOT, Ois. Am. Sept. i., 1807, p. 70, based his genus *Cœreba* upon *Certhia flaveola* LINN. and consequently *Certhiola* of SUNDEVALL, 1835, becomes a synonym.)

C. bahamen'sis. (Lat., of the Bahamas.) BAHAMAN HONEY CREEPER. Dark brown above; long superciliary line and under parts dull white; breast, edge of wing, and rump, bright yellow; wings dusky, with a white spot at base of primaries, and whitish edging of quills; tail dusky, tipped with white; bill and feet black; eyes blue. Length 4.50; wing 2.33; tail 1.75. Florida; Bahamas; closely related to the stock species, *C. flaveola*.

Family TANAGRIDÆ: Tanagers.

An extensive, brilliant family, confined to America, abounding in species between the tropics. Its position is a point at issue with ornithologists; it may naturally follow *Cœrebidæ* and *Mniotiltidæ*, though certainly no families should stand between it and *Fringillidæ*. In fact, certain tropical forms might be assigned to either indifferently. The best definition of Tanagers is that given by the distinguished ornithologist who called them "dentirostral finches;" but this generalization, like other happy epigrams, is insusceptible of application in detail, and Tanagers remain to be precisely characterized. As a consequence; the number of species can hardly be approximately estimated; but upward of 300 are usually enumerated.

Fig. 205.—Dentirostral bill of a Tanager (*Piranga hepatica*), nat. size.

The principal North American genus, *Piranga*, may be recognized among all the birds of our country by the combination of 9 primaries and scutellate tarsi with a turgid bill, notched at tip and toothed or lobed near middle of the maxillary tomia (fig. 205); though this last character is sometimes so obscure that it might be looked at without being seen. It is better marked in the Scarlet and Hepatic Tanagers than in the Summer Tanagers. The species of *Piranga* are birds of brilliant colors, with great seasonal and sexual differences of plumage. They are frugivorous and insectivorous, and consequently migratory in the United States. They inhabit woodland, lay 3–5 dark-colored, speckled eggs, about 0.95×0.65, nest in trees, and are no great songsters. In distribution they are rather southerly, scarcely passing northward beyond the U. S.

EUPHO'NIA. (Gr. εὔφωνος, *euphonos*, euphonious, sweet-voiced, musical; one of the species is *E. musica*, the Organist Tanager of the West Indies.) MUSICIAN TANAGERS. A large genus of tropical and subtropical species, one of which is supposed to occur in Texas; but no specimens are known to have been taken over our border since GIRAUD'S time. The following species may be recognized by its small size and peculiar coloration.

E. elegantis'sima. (Lat., superlative degree of *elegans*, choice, select, elegant.) BLUE-HEADED TANAGER. Adult ♂ : Above, black, with a purplish gloss ; crown and nape blue; frontlet chestnut, bordered behind by a black line. Below, deep brownish-orange, the throat black. Lining of wings and inner edges of wing-feathers white. Bill black; feet light brown. Length 4.50; wing 2.50; tail 1.50. ♀ : Upper parts olive-green with blue cap and chestnut frontlet; below, olive-yellow, brightest in middle of belly. Eggs creamy white, sparsely marked, and chiefly at the larger end, with different shades of brown. Mexico to Texas. This beautiful little Tanager was duly noted in the Key, 1872, and 1884, but first formally introduced in the 3d ed., 1887, p. 871; it is No. [606] of the A. O. U. Lists, 1886 and 1895.

PIRAN'GA. (Barbarous name of some South American bird.) SUMMER TANAGERS. Bill stout, turgid, conoidal, usually notched at tip, with one or more denticulations of cutting edge of upper mandible near middle of commissure. Rictal bristles well developed. Nostrils basal, the frontal antiæ reaching them. Wings lengthened and pointed; first 4 feathers subequal and longest. Tail moderate in length, shorter than wings, emarginate. Tarsus not shorter than middle toe; lateral toes about equal, outer coherent with middle by nearly all of the length of its basal joint. Sexes more or less unlike in color; red usually prevailing in the ♂. Habits migratory, insectivorous, arboreal; voice not musical. Eggs spotted. Four species of this beautiful genus inhabit the U. S., three of them representing, according to pattern of coloration, as many of the sections into which it is divisible. Numerous others are found in the warmer parts of America. The name of this genus has been commonly spelled *Pyranga*, after VIEILLOT, Analyse, 1816, p. 32; but as Vieillot used *Piranga* in the first instance, Ois. Am. Sept., i, 1807, p. iv, this form is to be preferred as a choice of evils in the barbarous name.

Analysis of Species.

♂ Crimson or scarlet, with black wings and tail: ♀ clear olive and yellow. No wing bars . . . . *erythromelas*
♂ Vermilion or rose-red, including wings and tail: ♀ brownish-olive and buffy-yellow. Bill light.
    Smaller: length about 7.50; wing 3.75 . . . . *rubra*
    Larger: length about 8.00; wing 4.25 . . . . *cooperi*
♂ Dusky-red above, including wings and tail. ♀ ashy-olive and yellow. Bill dark . . . . *hepatica*
♂ Yellow, with scarlet head and black back, wings and tail; two wing-bars. ♀ clear olive and yellow. *ludoviciana*
    no wing-bars, but lesser and middle coverts yellow. (Extralimital.) . . . . *rubriceps*

P. erythro'melas. (Gr. ἐρυθρός, *eruthros*, red, and μέλας, *melas*, black.) SCARLET TANAGER. BLACK-WINGED REDBIRD. Adult ♂ in summer: Crimson or scarlet; wings and tail black; bill and feet dark horn-color. Adult ♀: Above, clear olive-green; below, clear greenish-yellow; wings and tail dusky, glossed with color of back. Winter ♂ similar to ♀, but wings and tail black. Young ♂: Similar to ♀; later, when changing, patched with red, green, and black. Adult males often show abnormal coloring, the body being yellow, orange, or flame-color; or red patches appearing on wing-coverts. Length 6.75-7.90; extent 11.00-12.00; wing 3.50-3.90; tail about 3.00. Eastern U. S. and adjoining British Provinces; W. to Kansas, Indian Ter., and Texas; not common N. of Massachusetts; breeds nearly through U. S. range; winters in W. Indies, E. Mexico, Cent. Am., and northern S. Am. This brilliant creature nests in woods, groves, and orchards, upon the horizontal bough of a tree, building a rather loose and shallow fabric of twigs, fibres, rootlets, etc. Eggs 3-5, 0.95 × 0.65, dull greenish-blue, fully spotted with brown and lilac. This is *P. rubra* of authors generally, and of all former eds. of the Key; but, unfortunately, according to our rules of nomenclature, the name *rubra* must be transferred to the Summer Tanager, and the Scarlet Tanager become known as *P. erythromelas* VIEILLOT, 1819; A. O. U. Lists, No. 608.

P. rub'ra. (Lat. *rubra*, red.) ROSE TANAGER. SUMMER REDBIRD. Adult ♂: Rich rose-red or vermilion, including wings and tail; the former dusky on unexposed portions of the feathers; bill pale; feet darker. Adult ♀: Dull brownish-olive above, below dull brownish-yellow; no wing-bars. Young ♂: Like ♀. ♂ changing plumage shows red, greenish and yellowish in irregular patches, but no black. The ♀ distinguished from ♀ *erythromelas* by the dull brownish, ochrey, or buffy shades of the olive and yellowish, the greenish and yellowish of ♀ *erythromelas* being much clearer and paler; by paler bill and feet, and also by lack of any evident tooth of upper mandible, as this formation is obsolete in the present species. The tint of mature males varies greatly; from rosy to bricky red. Size of *erythromelas*, or rather larger. Eastern U. S., strictly, and rather southerly; N. rarely to Connecticut, only casually farther, as in Massachusetts, Nova Scotia, Ontario, etc.; W. to Kansas, Indian Territory, and Texas. Migratory, abundant; breeds throughout its range; winters extralimital in Cuba, Mexico, Cent. Am., and as far S. as Peru. Nesting and eggs like those of *erythromelas*. It is unluckily the fact that LINNÆUS first named the Summer Tanager *Fringilla rubra* in the 10th ed. of the Systema Naturæ, i, 1758, p. 181, and *Muscicapa rubra* in the 12th ed., 1766, p. 326; for by our rules we must accept the specific name *rubra*, and that of course debars us from using it for the Scarlet Tanager, which LINNÆUS named *Tanagra rubra* in 1766, Syst. Nat., i, 12th ed., p. 314. This necessary change caused some confusion at first, but we have already become used to it, and it is not likely to make any trouble in future. See my "Birds of the Colorado Valley," i, 1878, p. 351, where I made the point 20 years ago, stating that "the name *rubra* should stand in place of *æstiva* for the summer redbird," though I was not enough of a stickler for strictness to make in former editions of the Key a change which I am perfectly willing to follow, now that it has been made and generally adopted by other writers.

P. r. coo'peri. (To Dr. J. G. Cooper, of California.) COOPER'S TANAGER. WESTERN SUMMER REDBIRD. Characters of *P. rubra;* back rather darker than head; larger; length about 8.00; extent about 13.00; wing 4.25; tail 3.60; bill 0.75; tarsus 0.80. Southern Rocky Mt. region; Texas to Lower Colorado Valley, Cal., and southward; originally based as a full species, *Pyranga cooperi,* upon ♂ ♀ specimens which I shot at Los Pinos, N. M., on the Rio Grande, in June, 1864; *P. æstiva cooperi* of all previous eds. of the Key; *Piranga rubra cooperi* RIDGW., Pr. U. S. Nat. Mus., viii, 1885, p. 354; A. O. U. Lists, No. 610 *a*.

P. hepa'tica. (Lat. *hepar, hepatis,* the liver.) HEPATIC TANAGER. Adult ♂: Upper parts brownish-ashy, intimately mixed with dull red; top of head, upper tail-coverts, and edgings of wings and tail, brighter brownish-red. Inner webs and ends of wing-quills dusky; tail-feathers throughout decidedly tinged with red. Sides of head like back; edges of eyelids red. Below, bright red; sides and flanks shaded with color of back, many feathers often also with ashy skirting. Bill and feet blackish-plumbeous, the cutting edge of the upper mandible furnished with a tooth more prominent than in most species (fig. 205). Length about 8.00; wing 4.00; tail 3.33; bill 0.66; tarsus 0.80. Adult ♀: Bill and feet as in ♂. Upper parts greenish-olive, with an ashy-gray tinge; crown and rump clearer and more yellowish-olive. Sides of head like back. Beneath, yellow, clear and nearly pure medially, shaded on sides with color of back, sometimes brightening almost into orange on throat. Quills and tail fuscous, with olivaceous-yellow edgings, former darker than latter. Young ♂: Like ♀; in males changing, the characters of the two sexes confused. Very young: There is an earlier *streaky* stage, before the assumption of a plumage like that of ♀. Upper parts grayish-brown with an olive tinge; lower parts grayish-white with a yellowish shade; both everywhere streaked with dusky. Wings and tail like those of adult ♀, but former with ochraceous bands across ends of greater and middle coverts. Southern Rocky Mt. region and southward to Guatemala. *Pyranga hepatica* SWAINSON, Philos. Mag., i, 1827, p. 438, and of former eds. of the Key.

P. ludovicia'na. (Lat., of Louisiana, formerly of great extent in the West; name now inapplicable. Fig. 206.) CRIMSON-HEADED TANAGER. Adult ♂: Middle of back, wings, and tail black; wings crossed by two yellow or yellowish-white **bars** on ends of greater and middle coverts; inner secondaries marked with white or yellowish. Head all around scarlet or even crimson, the color extending diluted on breast. Other parts bright yellow, generally purest on rump. Iris brown; bill horn-color; legs livid bluish. Length about 7.00; wing 3.50-4.00; tail 2.75-3.25; bill 0.60; tarsus 0.75. Adult ♀: Above, olive, darker and somewhat ashy-shaded on middle of back, clearer and brighter on rump and crown. Below, greenish-yellow, shaded with olive on sides. Wings and tail fuscous, with edgings of color of upper parts; greater and median coverts tipped with white or yellowish; inner secondaries edged with

FIG. 206. — Crimson-headed Tanager.

the same. Averaging rather less than ♂. The bird lacks the **buffy shades** characteristic of ♀ *rubra*, besides being decidedly smaller. The general coloration, **in its clear** olive and yellow, is exactly that of ♀ *erythromelas*; from which distinguished by white or yellow markings on wings. The ♂ at first resembles the ♀, and in progress toward maturity every gradation between the two is presented. The distinctive dark dorsal area, and traces of the red of the head soon appear. In a usual condition of incomplete dress, the black of the back is mixed with gray or olive, the yellow of the back of the neck is obscured, that of the under parts is shaded with olive, and the head is only partly red. Western U. S., from the Great Plains and eastern foothills of the Rocky Mts. to the Pacific; British Columbia, S. in winter to Guatemala; accidental eastward, as in New York and New England. Breeds in all its N. Am. range and winters extralimital. Habits, nests, and eggs like those of our other Tanagers; eggs 0.95 × 0.65. This beautiful bird was discovered by LEWIS and CLARK, in their Camp Chopunnish, on the Kooskooskee River, in Idaho, June 6, 1806: see my ed. of their Travels, 1893, p. 1035; but it was first named and described by WILSON, Am. Orn., iii, 1811, p. 27, pl. 20, fig. 1.

P. rub'riceps. (Lat. *rubriceps*, red-headed.) GRAY'S TANAGER. Adult ♂: Whole head and more or less of the neck and breast bright red; rest of under parts rich yellow; back olive-green, changing to yellowish on rump and upper tail-coverts; tail blackish with olivaceous edgings of the feathers; wings the same, excepting the lesser and middle coverts, which are yellow. About the size of the last, and somewhat resembling it, but quite distinct; wing 3.75; tail 3.40. The sexual differences, and changes of plumage of young males, are probably coincident with those of *P. ludoviciana*. U. S. of Colombia; accidental in Dos Pueblos, Santa Barbara Co., California. *Pyranga rubriceps* G. R. GRAY, Gen. of Birds, ii, 1844, pl. 89; *Piranga rubriceps* BRYANT, Auk, Jan. 1887, p. 78; RIDGW., Man., 1887, p. 589; COUES, Key, 4th ed., 1890, p. 899; A. O. U. List, 2d ed., 1895, p. 255, No [607. 1].

Family HIRUNDINIDÆ: Swallows.

Swallows are fissirostral Oscine Passeres with nine primaries. Bill short, broad, flat, somewhat triangular, deeply cleft; gape wide, about twice as long as culmen; mouth thus opening to about beneath eyes. This is the strongest character of the family in comparison with its Oscine allies, and one perfectly distinctive, though some genera of Hirundines, especially *Progne*, approach *Ampelidæ* in form of bill. The bill narrows rapidly to the compressed acute tip. Nasal fossæ short and wide; nostrils directed laterally or upward, sometimes circular and completely exposed, sometimes scaled over. Culmen convex, scarcely a third as long as head; tip of upper mandible overhanging, usually nicked. Rictus smooth (or with a few inconspicuous bristles?). Wings extremely long and strong, the pinion bearing only 9 primaries, 1st of which equals or exceeds 2d in length, rest so rapidly graduated that 9th is scarcely or not half as long as 1st; secondaries and their coverts also very short; all these quill-feathers broad and stout. An acute, thin-bladed and somewhat falcate wing, of surpassing volatorial power, results from these modifications. Tail of 12 rectrices, perhaps abnormally only 10, usually forked, or at least emarginate, often deeply forficate, the outermost feathers being in this latter case narrowly linear for a considerable distance. Feet short, small, and weak, ill-adapted to secure foot-hold, and very badly formed for walking. Swallows scarcely use their feet for locomotion, relying mainly upon their prowess of pinion. Tarsal envelop thoroughly Oscine in structure, being scutellate in front and laminate behind; sometimes partially, or almost entirely, feathered; tarsi commonly shorter than lateral toes. The digits possess the normal number of phalanges; basal phalanx of middle digit commonly coherent with one or both lateral toes; hallux ordinary, not reversible. Digits commonly naked and scutellate, rarely feathered to the claws. Claws comparatively strong, compressed, well-curved, and acute, apt for clinging. Plumage soft, smooth, and blended, most frequently glossy or even iridescent, but sometimes lustreless. Head short, broad, and depressed; neck short. Mouth capacious, its greatest width equalling that of head.

This is a perfectly natural group, well distinguished by the foregoing characters. The Swallows alone represent, among Oscines, the fissirostral type of structure; they have a close superficial resemblance to Swifts and Goatsuckers, of another order, but the relation is one of analogy, not of affinity, though all these birds were formerly classed together in the highly unnatural " order " *Fissirostres*. (See beyond, under *Micropodidæ* and *Caprimulgidæ*.)

A hundred species of Swallows are pretty well ascertained to be genuine. They are distributed all over the world; the most generalized types, like *Hirundo* itself, are more or less cosmopolitan, but each of the great divisions of the globe has its peculiar subgenera or particular sets of species. Thus, all the American groups except *Hirundo* and *Clivicola* are peculiar to this continent.

Swallows are insectivorous, and therefore migratory in cold and temperate latitudes; unsurpassed in powers of flight, they are enabled to pass with ease and swiftness from one country to another, as the state of the weather may require. With us a few warm days in February and March often allure them northward, only to be driven back again by the cold, giving rise to the well-known adage: " One Swallow does not make a summer." No birds are better known to all classes than these, and none so welcome to man's abode — cherished witnesses of peace and plenty in the homestead, dashing ornaments of the busy thoroughfare.

The habits of Swallows best illustrate the modifying influences of civilization on indigenous birds. Formerly, they all bred on cliffs, in banks, in hollows of trees, and similar places, and many do so still. But most of our species have forsaken these primitive haunts to avail themselves of the convenient artificial nesting-places that man, intentionally or otherwise, provides. Some are just now in a transition state; thus the Purple Martin, in settled parts of the country, chooses the boxes everywhere provided for its accommodation, while in the West it retains its old custom of breeding in hollow trees. The nesting of our Swallows now presents the following categories of method: —

1. Holes in the ground, dug by the birds, slightly furnished with soft material: *Clivicola riparia, Stelgidopteryx serripennis*.

2. Holes in trees or rocks not made by the birds, fairly furnished with soft material: *Progne subis, Tachycineta bicolor, T. thalassina*.

3. Holes, or their equivalents, not made by the birds, but secured through human agency, and more or less fully furnished with soft material, according to the shallowness or depth of the retreat. (*Formerly, no species; now, all the species excepting Clivicola riparia.*)

4. Holes constructed by the birds, of mud, plastered to surfaces, whether artificial or natural, and loosely furnished with soft material. This is seen in perfection in the nesting of *Petrochelidon lunifrons*, and is imperfectly illustrated by the nidification of *Hirundo erythrogastra*.

5. Eggs pure white, unmarked: *Tachycineta bicolor, T. thalassina, Clivicola riparia, Stelgidopteryx serripennis, Progne subis*.

6. Eggs thickly speckled: *Hirundo erythrogastra, Petrochelidon lunifrons*.

Aside from three extralimital species (*Progne cryptoleuca, Petrochelidon fulva,* and *Callichelidon cyaneoviridis*), lately ascertained to occur as stragglers in Florida, the seven established North American species, referable to six genera, may readily be determined by the following

Analysis of Genera and Species.

Tail deeply forficate, with linear lateral feathers; lustrous steel-blue above, rufous below . *Hirundo erythrogaster*
Tail simply emarginate; lustrous green; beneath white . . . . *Tachycineta bicolor*
Tail simply emarginate; opaque velvety green; beneath white . . . . *Tachycineta thalassina*
Tail nearly even; lustrous steel-blue; rump rufous . . . . *Petrochelidon lunifrons*
Tarsus with tuft of feathers below; lustreless gray; below white . . . . *Clivicola riparia*
Outer edge of first primary serrate; lustreless brownish; paler below . . . . *Stelgidopteryx serripennis*
Bill very stout, curved; male entirely lustrous blue-black . . . . *Progne subis*

HIRUN'DO. (Lat. *hirundo*, a swallow. Fig. 207.) BARN SWALLOWS. Tail deeply forficate, nearly or about as long as wings; lateral feather linear-attenuate, about twice as long as middle feather. Tarsi shorter than middle toe and claw, above feathered for a little distance; basal joint of middle toe partly adherent to both lateral toes. Bill of moderate size for this family, of usual shape, with straight commissure; nostrils lateral, overarched by a membranous scale. Upper parts glossy, dark-colored; a dark pectoral collar; forehead and under parts rufous; tail spotted with white. Eggs colored. Sexes similar. This is the genuine genus *Hirundo* LINN., 1758, type *H. rustica*, the common Swallow of Europe, as restricted by SCHAEFFER, Elem. Orn., 1774: see COUES, Auk, July, 1898, p. 271; SHARPE, Monograph of *Hirundinidæ*, p. xxxv. *Hirundo* of former editions of the Key, and of A. O. U. Suppl. List, Auk, Jan. 1899, p. 122. But *Chelidon* (after FORSTER, Synop. Cat. Brit. Birds, 1817, p. 55). STEJNEGER, Proceedings U. S. National Museum, v, June 5, 1882, p. 31; A. O. U. Lists, 1st and 2d editions, 1886-95.

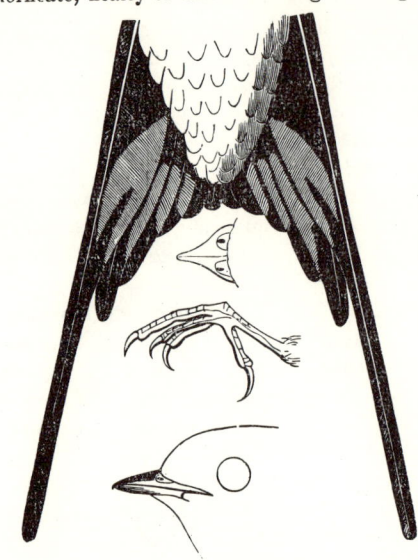

FIG. 207. — Generic details of *Hirundo* (*H. erythrogastra*, nat. size). (Ad. nat. del. E. C.)

H. erythrogas'tra. (Gr. ἐρυθρός, *eruthros*, ruddy, and γαστήρ, *gaster*, belly. Fig. 208.) AMERICAN BARN SWALLOW. Adult ♂: Deep lustrous steel-blue; forehead and entire under parts rufous, generally deepest on forehead and throat; an imperfect **steel-blue** collar. **Wings and tail blackish, with steel-blue or somewhat greenish gloss; lateral pair of tail-feathers much lengthened and filiform at the end, all but central pair with a white spot.** Length 6.00-7.00, very variable, according to development of tail; extent 12.50-13.50; wing 4.50-5.00; tail 3.00-5.00, the fork 2.00-3.00 deep. Adult ♀: Quite like ♂; colors rather less intense and lustrous; average size smaller. Young: Lacking in great measure elongation and attenuation of lateral tail-feathers, the fork being an inch or less in depth. Similar to the adults, but much duller, and with rather a greenish than steel-blue lustre — at an early age quite brown, with scarcely any lustre;

FIG. 208. — Barn Swallow.

rump and upper tail-coverts skirted with rusty; frontlet obscurely marked or reduced to a mere tawny line; under parts, especially behind the dark collar, very pale, even brownish-white. N. Am. at large; abundant; breeds throughout its range; migrates through the West Indies, and winters in Cent. and S. Am. *Hirundo erythrogaster* BODDAERT, 1783; *H. hor-*

reorum BARTON, 1799; BAIRD, 1858; COUES, 1872, in 1st ed. of the Key; *H. erythrogastra horreorum* of other eds. of the Key; *Chelidon erythrogaster*, A. O. U. List, 1886, p. 292, wrong for genus and wrong for gender; *C. erythrogastra*, A. O. U. List, 2d ed., 1895, p. 258, No. 613. This means that attempts to distinguish the North American bird from that of South America have finally failed.

TACHYCINE'TA. (Gr. ταχυκίνητος, *tachukinetos*, moving rapidly.) IRIS SWALLOWS. VIOLET-VELVET SWALLOWS. Similar to the last, but lacking elongation and attenuation of lateral tail-feathers, which also lack white spots. Tail simply emarginate. Under parts snowy white. Sexes alike. Eggs 3–6, colorless. *Iridoprocne* and *Tachycineta* of 2d, 3d, and 4th eds. of the Key, p. 322.

Analysis of Subgenera and Species.

Iridoprocne. Plumage of upper parts lustrous and unicolor . . . . *bicolor*
Tachycineta proper. Plumage of upper parts lustreless and versicolor . . . . *thalassina*

T. (I.) bi'color. (Lat. *bicolor*, two-colored. Fig. 209.) WHITE-BELLIED SWALLOW. Adult ♂ : Entire upper parts glossy dark green ; wings and tail blackish, lustrous ; lores black. Entire under parts pure white. Bill black ; feet dark. Length about 6.00 ; extent 13.00 ; wing 4.50–5.00 ; tail 2.50. ♀ : Similar, the colors rather less intense and lustrous. Young : Birds of the year slowly acquire a plumage differing only in less lustre and intensity from that of adults; but, on leaving the nest, they are dark mouse-gray or slate-color above, including wings and tail; interscapulars and inner quills tipped with rusty; white below, slightly shaded with ashy; thus curiously similar to *Clivicola riparia*. Feet yellow. The first plumage is worn longer than usual, the autumnal dress being slowly gained — one or two of the metallic-tinted feathers at a time. The quills of the wing are moulted by the young as well as by the adult, and in both, in autumn, the inner secondaries are white-tipped. Temperate N. Am. Breeds indifferently in most parts of its range, and winters abundantly on the southern border, sometimes even from South Carolina, to the West Indies and Cent. Am.

FIG. 209. — White-bellied Swallow, nat. size. (Ad. nat. del. E. C.)

T. thalas'sina. (Gr. θαλάσσινος, *thalassinos*, sea-green.) VIOLET-GREEN SWALLOW. Adult ♂ : Entire under parts, including sides of head to just above eyes, and an enlarged fluffy tuft on flanks tending to join its fellow over rump, pure silky white. Upper parts rich velvety-green, mixed with a little violet-purple ; crown of head similar, but rather greenish-brown, with purplish tinge. Cervical region, in some cases a well-defined though narrow cervical collar, and upper tail-coverts, violet-purple. These rich colors opaque, without gloss or sheen ; wings and tail blackish, with violet and purplish gloss. Bill black ; feet brownish-black, small ; iris brown ; mouth pale yellow. Length 4.50–5.50 ; extent 11.50–12.50 ; wing 4.50 ; tail 2.00, lightly forked ; bill 0.25 ; tarsus 0.40. The ♀, and immature birds in general, differ simply in less purity and intensity of colors of upper parts. In highest plumaged specimens, the back is nearly pure green, the cervical collar distinct, and the several contrasts of crown, collar, back, and upper tail-coverts are strong ; in general, the back has a brownish-purple shade, more like that of crown. Very young birds are like *T. bicolor*, though smaller, being dark mouse-gray above and white below. But traces at least of the special tints speedily appear. Young or autumnal birds usually have the inner secondaries white-tipped, as in *T. bicolor*. Middle and Western Provinces, U. S., and adjoining portions of British America; E. to the

Upper Missouri, N. to the Yukon, S. in winter to Costa Rica. Breeds throughout its range, and winters extralimital. A lovely species.

CALLICHELI'DON. (Gr. καλλι-, *kalli*-, usual combining form of καλός, *kalos*, beautiful; χελιδών, *chelidon*, a swallow.) BLUE-GREEN SWALLOWS. Resembling *Tachycineta*, especially that section of the genus in which the upper parts are not iridescent, though versicolorous; tarsus rather longer, exceeding middle toe without claw; points of folded wings reaching about to end of tail, which is forked for about 1.00. One species, a straggler from the Bahamas.

C. cyaneovir'idis. (Lat. *cyaneus*, blue; *viridis*, green.) BAHAMAN SWALLOW. Adult ♂: Upper parts beautiful soft velvety green with golden gleam but without sheen, gradually changing to bluish-green or violet on wings and tail. Entire under parts pure white, this color extending upon sides of head to include auriculars; feathers of chin and throat snowy to the very base. Adult ♀: Similar, but the white somewhat soiled with grayish on sides of head, body, and lining of the wings. Length 5.75; wing 4.50; tail 3.00, forked nearly or quite 1.00; tarsus 0.50. Bahamas, casually on the Dry Tortugas and at Tarpon Springs, Florida. A lovely species, first described as *Hirundo cyaneoviridis* by Dr. H. BRYANT, Pr. Bost. Soc. N. H., vii, 1859, p. 111, from the Bahamas, type No. 11946, Mus. Smiths. Inst.; *Hirundo (Callichelidon) cyaneoviridis* BAIRD, Rev. Am. B., i, 1865, p. 303; *Callichelidon cyaneoviridis* BRYANT MS., *ibid.*; not in previous eds. of the Key; not in A. O. U. List, 1886; first added to our fauna by W. E. D. SCOTT; see Auk, July, 1890, p. 265, and Oct. 1890, p. 303, specimen taken Apr. 9, 1890; BREWST., Auk, Apr. 1897, p. 221, Tarpon Springs, Sept. 3, 1890; A. O. U. List, 2d ed., 1895, p. 259, No. [615. 1].

PETROCHELI'DON. (Gr. πέτρα, *petra*, a rock; χελιδών, *chelidon*, a swallow. Fig. 210.) CLIFF SWALLOWS. Bill stout and deep (for this family); nostrils superior, opening without nasal scale. Tail unusually short, the tips of the folded wings reaching beyond it, about even, or only slightly emarginate, with the feathers broad to their ends. Feet much as in *Hirundo*; tarsi feathered above; toes extensively adherent at base. A bristly appearance of front and chin, different from what is seen in other groups. The tuft of crissal feathers is full, reaching nearly to end of tail. The species agree well in a special pattern of coloration, being steel-blue above, with rufous rump and nuchal band, and usually a frontlet of different color from the rest of the upper parts; under parts not continuously white as in *Tachycineta* and

FIG. 210. — Cliff Swallow.

Callichelidon. Nidification peculiar; eggs colored. Sexes alike.

P. lu'nifrons. (Lat. *luna*, the moon, or a crescent; *frons*, forehead. Fig. 211.) CLIFF SWALLOW. EAVES SWALLOW. CRESCENT SWALLOW. MUD SWALLOW. Adult ♂ ♀: Back and top of head, with spot on throat, deep lustrous steel-blue, that of crown and back separated by a grayish nuchal collar. Frontlet white or brownish-white. Shorter upper tail-coverts rufous. Chin, throat, and sides of head intense rufous, sometimes purplish-chestnut, prolonged around side of nape. Under parts dull grayish-brown, with usually a rufous tinge (rusty-gray), and dusky shaft-lines, whitening on belly; under tail-coverts gray, whitish-edged and tinged with rufous. Wings and tail blackish, with slight gloss. Bill black; feet brown. Length 5.00-5.50; extent 12.00 or more; wing 4.25-4.50; tail 2.25, nearly square. Sexes not distinguishable; both vary much in tone of coloration, especially of the rufous parts. Forehead sometimes white, sometimes quite brown. In young birds, the frontlet may be altogether wanting; upper parts lustreless dark brown, most of the feathers being skirted with whitish; rufous of throat and rump a mere tinge; spot on throat wanting, and the parts often speckled with white. N. Am. at large, abundantly but irregularly distributed, breeding in colonies wherever suitable sites may be found for its curious retort-shaped or bottle-nosed nests of mud. It has been traced N. to the limit of trees in Brit. Am., S. to Cent. and S. Am. According to SHARPE, Cat. B. Brit. Mus., x, 1885, p. 193, the proper name of our Cliff Swallow is *P. pyrrhonota* (VIEILL., 1817), with a possibility of *P. americana* (GM., 1788); both these names being based on the South American species.

FIG. 211. — Cliff Swallow, nat. size. (Ad. nat. del. E. C.)

P. ful′va. (Lat., fulvous.) CUBAN CLIFF SWALLOW. In general, like the last; differing as follows: no blue-black spot on throat; frontlet rich chestnut; chin, throat, and sides of head pale rufous, like the flanks; rump darker rufous. Smaller; length under 5.00; wing 4.50-4.75, thus relatively longer; tail 2.00 or rather less. Greater Antilles and some parts of Cent. Am. Accidental on the Dry Tortugas, Florida. This is the true *Hirundo fulva* VIEILL., Ois. Am. Sept. i. 1807, p. 62, pl. 30, though the name used to be misapplied to our Cliff Swallow; *Petrochelidon fulva* CAB., Mus. Hein. i. 1850, p. 47; A. O. U. List, 2d ed., 1895, No. [612. 1]; not in any previous ed. of the Key; see W. E. D. SCOTT, Auk, July, 1890, p. 264, specimen taken Mar. 22, 1890.

CLIVIC′OLA. (Lat. *clivus*, a slope, acclivity or declivity, such as the birds breed in; and *colere*, to inhabit, *(in)cola*, an inhabitant.) BANK SWALLOWS. Tarsus with a tuft of feathers at base below, near insertion of hind toe. Edge of wing not rough. Claws little curved, the lateral reaching beyond base of middle one. Bill very small, nostrils opening laterally, overhung by a membrane. Tail much shorter than wings, emarginate. Coloration dull and simple — lustreless brown above and across breast, white below. Eggs uncolored, laid in holes in the ground excavated by the bird. Sexes alike. (*Cotile* of all previous eds. of the Key, and of most authors; but *Riparia* and *Clivicola* of I. R. FORSTER, Syn. Cat. Brit. B., 1817, pp. 17 and 55, antedate *Cotile* of BOIE, Isis, 1822, p. 550; see COUES, Auk, July, 1898, p. 271, and OBERH., Auk, July, 1899, p. 281.)

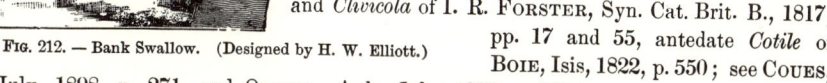

FIG. 212. — Bank Swallow. (Designed by H. W. Elliott.)

C. ripa′ria. (Lat. *riparia*, riparian; *ripa*, bank of a stream. Fig. 212.) BANK SWALLOW. Adult ♂ ♀: Lustreless mouse-brown; wings and tail fuscous. Below, white, with broad pectoral band of color of back. A dusky ante-orbital spot. Length about 5.00; extent 10.50; wing 4.00; tail 2.00. Sexes similar. Young differ chiefly in whitish edgings of the feathers, especially of wings and tail. Even in the adult, the upper parts are apt to be not quite uniform, there being paler gray edgings of most feathers. The dark pectoral band sometimes extends backward along middle of under parts (not shown in fig. 212). Autumnal specimens have the secondaries white-tipped. Very young birds have rather rusty than whitish skirting of the dark feathers, and white throat speckled with the same. Almost cosmopolitan: Europe, Asia, Africa, America; abundant in N. Am., breeding in immense troops in holes in the ground, wherever suitable sites offer, as natural embankments, railroad cuttings, gravel-pits, etc.; N. to the limit of trees, S. into S. Am.

STELGIDOP'TERYX. (Gr. στελγίς, *stelgis*, a scraper; πτέρυξ, *pterux*, wing.) ROUGH-WINGED SWALLOWS. General aspect of *Clivicola*; form and coloration much the same. Outer web of 1st primary converted into a series of stiff, recurved hooks. (Other Swallows, as *Psalidoprocne* CAB., have this peculiar wing structure, but are otherwise different.) The design of the structure is not clear, but we may readily suppose that the hooks assist the birds in crawling into their holes, and in clinging to vertical or hanging surfaces. Tarsus slightly feathered above, but lacking the curious tuft seen at base of hind toe in *Clivicola*. Lateral claws curved, and not reaching beyond base of middle. Basal joint of middle toe extensively adherent to the outer, much less so to the inner. Bill small, with oval, superior nostrils margined by membrane behind, but not overhung. Tail short and slightly emarginate. Eggs uncolored, in holes dug by the birds, or elsewhere. Sexes alike.

S. serripen'nis. (Lat. *serra*, a saw; *penna*, a feather.) ROUGH-WINGED SWALLOW. Adult ♂♀: Lustreless mouse-brown or brownish-gray, paler below, gradually whitening posteriorly. Wings and tail darker than upper parts. Rather larger than the last species. No dark pectoral band contrasting with white. No tuft of feathers at base of hind toe. Young: At a very early age, the feathers of the back, rump, and wings are suffused or edged with rich rusty-brown, while the under parts are more or less tinged with a paler shade of the same. The hooklets of the wings are only fully developed in adult birds, and are not appreciable in young ones. U. S. and adjoining British Provinces; rare in New England States; breeds throughout its N. Am. range, and in Mexico; extends in winter to Cent. Am.

PROG'NE. (Gr. Πρόκνη, *Procne*, a mythological character.) Of large size and robust form for this family. Bill long and stout, with much-curved commissure and deflected tip; culmen convex, its tomial edge concavo-convex like ↘. Nostrils circular, opening upward, without nasal scale. Feet large, with strong, much-curved claws; tarsus shorter than middle toe and claw; lateral toes about equalling each other in length; basal joint of middle toe freer from lateral toes than usual. Tail forked. Sexes dissimilar. Eggs colorless.

P. su'bis. (Lat. *subis*, name of an unknown bird. Fig. 213.) PURPLE MARTIN. Adult ♂: Intense lustrous steel-blue. Wings and tail blackish, with bluish lustre. Bill black; feet blackish. Length 7.50–8.50; extent 15.50; wing 5.50–6.00; tail 3.00–3.50, forked; bill 0.50, very stout, broad at base, somewhat decurved at end; nostrils circular, exposed, opening upward. ♀: Dark grayish-brown, glossed on back and head with steel-blue. Wings and tail fuscous, paler on inner webs, with narrow gray edgings. Beneath, whitish, shaded with dark gray in most parts, the feathers very generally with dusky shaft-line. Young birds of both sexes resemble adult ♀, though the young males are rather darker. The steel-blue appears at first in patches. Eggs 3–5, 0.95 to 1.00 × 0.70 to 0.75. U. S. and British Provinces, abundant and generally distributed; breeds throughout its range, usually in the East in boxes provided for its accommodation, in the West in holes in trees; winters extralimital, in S. Am.

P. s. hespe'ria. (Gr. ἑσπέρια, *hesperia*, feminine form of ἑσπέριος, *hesperios*, western; ἑσπέρα, *hespera*, the evening, hence western, equivalent to Lat. *vespera*, vesper; Ἕσπερος, *Hesperos*, in Lat. and English, *Hesperus*, the evening star, *i. e.*, Lucifer or Venus, when setting in the evening; Ἑσπερίδες, *Hesperides*, the nymphs who guarded the golden apples in the garden of the same name, supposed to be in Africa somewhere in the vicinity of Mt. Atlas.) WESTERN MARTIN. VESPER MARTIN. Closely resembling the last, the ♂ not satisfactorily distinguishable. ♀ differing in having the belly, vent, and crissum white, nearly or quite immaculate; flanks, breast, throat, forehead, and nuchal collar grayish-white; feathers of the back and rump with pale edgings; bend of wing and under wing-coverts spotted with white. California and Arizona, from lat. 40° S. to Nicaragua. BREWSTER, Auk, Apr. 1889, p. 92; COUES, Key, 4th ed., 1890, p. 899; A. O. U. List, 2d ed., 1895, p. 257, No. 611 *a*.

P. cryptoleu′ca. (Gr. κρυπτός, *kruptos*, hidden, concealed, occult, secret; λευκός, *leukos*, white.) CUBAN MARTIN. In general, resembling *P. subis;* smaller on an average, hardly reaching 8.00; wing about 5.50; tail about 3.00, with narrower feathers and relatively deeper forking than in *P. subis*.

FIG. 213. — Purple Martin.

Adult ♂: feathers of belly with concealed white spots or bars. Adult ♀ and young ♂: Belly and crissum quite white, in contrast with the grayish-brown of other under parts. Cuba, Florida. BAIRD, Rev. Am. Birds, i, 1865, p. 277; not recognized in the Key, 1st, 2d, and 3d eds., nor in A. O. U. List, 1st ed.; *P. subis cryptoleuca*, Key, 4th ed., 1890, p. 899; *P. cryptoleuca*, A. O. U. List, 2d ed., 1895, p. 257, No. [611. 1].

Family AMPELIDÆ: Chatterers.

This appears to be an arbitrary and unnatural association of a few genera that agree in some particulars, but are widely different in others. The composition and position of the group differ with almost every writer; some place it in *Clamatores*, next to *Tyrannidæ*. I think that the family should be dismembered; and doubtless the two subfamilies here presented may be properly dissociated. They are discriminated, so far as our forms are concerned, by the characters given under the heads of the only two genera with which we have here to do.

Subfamily AMPELINÆ: Waxwings.

Of this subfamily, as here restricted, there is only one genus with three species — one of Europe, Asia, and America, one of Asia and Japan, one peculiar to America.

AM'PELIS. (Gr. ἀμπελίς, Lat. *ampelis*, name of a bird.) WAXWINGS. Bill, short, broad, flat, rather obtuse, plainly notched near tip of each mandible, with wide and deeply-cleft gape; convex culmen and gonys less than half as long as the nearly straight commissure; width of rictus more than two-thirds the length of gape. Nasal fossæ broad, but filled with short, erect or antrorse, and close-set, velvety feathers; nostrils narrowly elliptical, overarched by a (feathered) scale. Rictal vibrissæ few and short. Wings long and pointed, much longer than tail, their point formed by 3d primary, closely supported by 2d and 4th, 5th abruptly shorter, the rest rapidly graduated. Primaries 10, but 1st spurious, so very short as readily to escape

FIG. 214. — Bohemian Waxwings, ½ nat. size. (From Brehm.)

observation, and sometimes displaced to the outer side of the 2d, — a condition like that seen among Vireos. Inner quills, as a rule, and sometimes the tail-feathers, tipped with curious red horny appendages, like sealing-wax. Tail short, narrow, even, two-thirds or less of the length of wing. Feet rather weak; tarsus shorter than middle toe and claw, distinctly scutellate with five or six divisions anteriorly and somewhat receding from strict Oscine character by subdivision of the lateral plates. Lateral toes of nearly equal lengths; ends of their claws scarcely reaching base of middle claw; hallux about as long as inner lateral toe. Basal phalanx of middle toe coherent with outer toe for about two-thirds its length, with inner toe for about half its length. Body stout. Head conspicuously crested. Plumage peculiarly soft, smooth, and silky. Tail tipped with yellow (or red, in the Japanese *A. phœnicoptera*). Sexes alike; young different. Eggs spotted. Nest on trees.

A. gar′rulus. (Lat. *garrulus*, a jay-bird : from its loquacity. Fig. 214.) BOHEMIAN WAXWING. BLACK-THROATED WAXWING. LAPLAND WAXWING. SILK-TAIL. Adult ♂ ♀ : General color brownish-ash, shading insensibly from the clear ash of the tail and its upper coverts and rump into a reddish-tinged ash anteriorly, this peculiar tint heightening on head, especially on forehead and sides of head, into orange-brown. A narrow frontal line, and broader bar through eye, with chin and throat, sooty black, not or not sharply bordered with white. No yellowish on belly. Under tail-coverts orange-brown or chestnut. Tail ash, deepening to blackish-ash toward end, broadly tipped with rich yellow. Wings ashy-blackish ; primaries tipped (chiefly on outer webs) with sharp spaces of yellow, or white, or both ; secondaries with white spaces at ends of outer webs, the shafts usually ending with enlarged, horny, red appendages. Primary coverts tipped with white. Bill blackish-plumbeous, often paler at base below; feet black. Length 7.00–8.00 ; wing 4.50 ; tail 2.50. The sexes of this beautiful bird are alike, and the principal variations, aside from mere shade of the body-color, consist in the markings of the wings. In the finest specimens, the ends of the primary quills are rich yellow, like the tips of the tail-feathers, forming broad firm spaces, in a continuous line when the wing is closed, with narrower offsets going around the ends of the quills. In less perfect specimens, these markings are simply white, are less firm, and do not appear on all the quills. The secondaries may or may not show the red "sealing-wax" tips, but in adult birds at least probably always show white markings at the ends, and the same is the case with the primary coverts. These wing-markings, with chestnut crissum, and absence of yellowish on belly, will always distinguish the species from *A. cedrorum*, independently of its much superior size. Young : There is an early streaked stage, like that of *A. cedrorum*. Northern hemisphere, northerly, wandering S. in vast troops at irregular periods. In America, S. regularly in winter to the northern tier of States; in the Rocky Mts. much farther; casually to about 35°. Rare on the Pacific coast except in Alaska. Breeds in high latitudes, but down to the U. S. border in the Rocky Mts. Nesting substantially the same as that of *A. cedrorum*, and eggs only different in their greater size — about 1.00 × 0.67.

FIG. 215. — Cedar Waxwing.

A. cedro′rum. (Lat. *cedrus*, gen. pl. *cedrorum*, the cedar. Figs. 215, 216.) CEDAR WAXWING. CAROLINA WAXWING. CANADA WAXWING. CEDAR-BIRD. CHERRY-BIRD. THE POLITE BIRD. RÉCOLLET. Adult ♂ ♀ : General color shading from clear pure ash on upper tail-coverts and rump through olivaceous-cinnamon into a richer and somewhat purplish-cinnamon on fore parts and head. On under parts, the color shades through yellowish on belly into white on under tail-coverts. There is no demarcation of color whatever, and the tints are scarcely susceptible of adequate description. Frontlet, lores, and stripe through eye, velvety-black; chin the same, soon shading into color of breast. A sharp white line on side of under jaw; a narrower one bordering the black frontlet and lores; lower eyelid white. Quills of wings slate-gray, blackening at ends, paler along edges of inner webs; without white or yellow markings, as a rule; inner quills tipped with red horny appendages. Tail-feathers like primaries, but tipped with yellow, and sometimes also showing red horny appendages. Bill plumbeous-black, sometimes paler at base below; feet black. Length 6.50–7.25; extent 11.50–12.00; wing 3.50–3.75; tail 2.25. Young: Brownish-gray, with a slight olive shade; paler below, whitening or becoming slightly yellowish on belly; everywhere streaked with dingy whitish; the markings most evident on breast and sides. Wings and tail as in adults, but usually lacking red appendages. The velvety-black and white on head imperfectly defined. Bill pale at base below; feet plumbeous. Specimens apparently mature and full-feathered frequently lack the sealing-wax tips. These are normally confined to the secondaries, but occasionally appear on one or several primaries, and some or all rectrices (as in fig. 214); a case is recorded in which an under tail-covert was similarly embellished. Both sexes possess these ornaments, but as a rule they are best developed in the ♂. The normal period of their appearance is not known — it is probably not constant; birds in the earliest known plumage may possess one or more. They are possibly deciduous, independently of moult of the feather. Their use is unknown. N. Am. at large to lat. 54° N. at least; breeds indifferently throughout its N. Am. range, and migrates or rather wanders according to food-supply; winters in most of the U. S.; goes in flocks nearly the whole year, and is especially fond of resorting to cedar thickets to feed upon the berries; breeds late (June, July), in orchards and groves; nest in trees or bushes, in crotch of a bough or saddled on a limb; eggs 3–6, livid or pale bluish, sharply and usually thickly marked with blackish surface spots and others paler in the shell; narrow and elongate, about 0.82×0.60.

FIG. 216.—Cedar-bird, nat. size. (Ad. nat. del. E. C.)

Subfamily PTILOGONATINÆ: Fly-snappers.

Bill much as in the last subfamily, but slenderer for its length: nasal scale naked; a few short bristles about base of bill. Tarsus scutellate anteriorly, sometimes also on sides; about as long as middle toe and claw; hind toe remarkably short. Wing not longer than tail, much rounded, of 10 primaries; first short, less than half as long as 2d, which is only about as long as 8th; point of wing formed by 4th, 5th, and 6th or 3d quills. Tail long, nearly even, with broad plane feathers (*Phaïnopepla*); or much graduated, with tapering central feathers (*Ptilogonys*). Head conspicuously crested; sexes (in our genus) dissimilar; young not streaked or spotted. The two leading genera of the subfamily are *Phaïnopepla* and *Ptilogonys*, the latter with two strongly marked species of Mexico and Central America — *P. cinereus* and *P. (sphenotelus) caudatus*.

PHAÏNOPEP'LA. (Gr. φαείνος, *phaeinos*, shining; πέπλος, *peplos*, a robe.) SHINING FLY-SNAPPERS. Bill somewhat as in *Ampelis*, but slenderer for its length; nostrils naked, scaled; antiæ bristly, reaching nostrils; a few short rictal bristles. Tarsus scutellate anteriorly, slightly subdivided on sides below. Hind toe very short; middle toe and claw about as long as tarsus; lateral toes a little unequal, outer the longer, reaching a little beyond base of middle claw, its basal joint adhering to middle; inner lateral toe nearly free to the base; claws all much curved. Wing not longer than tail, rounded, of 10 primaries, 1st developed, though only about half as long as 2d, which about equals length of secondaries: point of wing formed by 4th, 5th, and 6th quills. Tail long and fan-shaped, not emarginate, of broad plane feathers widening to their obtuse ends. Head with a long, thin, occipital crest. Sexes dissimilar : ♂ glossy-black, with large white wing-patch; ♀ dull-colored; young not spotted or streaked. Fine songsters. Nidification arboreal; eggs colored.

P. ni′tens. (Lat. *nitens*, shining.) SHINING FLY-SNAPPER. Adult ♂: Entirely rich lustrous black, with steel-blue or greenish reflections. Primaries with a large white space on inner webs. Bill and feet black. Length about 7.50; extent 11.50; wing 3.50–3.70; tail 3.50–4.12; bill 0.40–0.50; tarsus 0.60–0.66; middle toe and claw 0.66–0.75. Adult ♀: Crested, like ♂. Entirely brownish-gray, paler beneath; wings and tail blackish; white on inner webs of primaries much reduced or extinguished, and in its stead much whitish edging of quills and coverts, tail-feathers, and crissum. Young ♂: Like ♀; during progress to maturity every gradation between characters of the two sexes is observed. Sometimes nearly all the feathers are skirted with white. Middle and Western Provinces, U. S., from Utah, Nevada, and Colorado southward in Lower California and Mexico. A bird of remarkable characters and appearance, restless and vigilant; feeds on berries and insects; sings beautifully. Nest a slight shallow structure, about 4.00 in diameter by 2.50 high, with a cavity about 2.00 deep, saddled on a bough, loosely fabricated of twigs, plant-fibres, and down; eggs 2–5 (rarely single), averaging 0.93 × 0.65, greenish-white, distinctly and profusely speckled with blackish or dark brown.

Family VIREONIDÆ: Vireos, or Greenlets.

Small dentirostral *Oscines*, related to Shrikes, with hooked bill, 10 primaries, and extensively coherent toes. Bill shorter than head, stout, compressed, distinctly notched and hooked at tip; rictus with conspicuous bristles; nostrils exposed, overhung with a scale, but reached by small bristly erect frontal feathers. Toes soldered at base for the whole length of basal joint of middle one, which is united with basal joint of inner and two basal joints of outer, all these coherent phalanges very short. (Lateral toes unequal in the genus *Vireo*.) Tarsus equal to or longer than middle toe and claw, scutellate in front, laterally undivided, except at extreme base. Wings moderate, of 10 primaries, of which 1st is short (one-half to one-fourth the 2d), or spurious, or *apparently* wanting (being rudimentary and displaced). Size small, under 7.00; coloration simple, mostly and oftenest greenish; young not spotted or streaked.

FIG. 217. — Warbling Vireo, reduced. (From Tenney.)

This family was formerly united with *Laniidæ*, chiefly on account of resemblance in shape of bill of certain species to that of Shrikes; but the likeness is never perfect, and there are other more important characters, especially in the structure of the feet, by which the two groups may be discriminated. *Vireonidæ* are peculiar to America; they are a small family of five or six genera and nearly 70 recorded species, of which about five-sixths appear to be genuine. The typical and principal genus, *Vireo*, containing nearly 30 species, is especially characteristic of North America, though several species occur in the West Indies and Central America; one genus and species, *Laletes osburni*, is exclusively West Indian; the rest — *Cyclorhis*, *Hylophilus*, *Vireolanius*, and *Neochloe* — are, with one exception, South and Central American. In further illustration of the group, I offer some remarks under head of the only genus with which we have to do in the present connection.

VIR'EO. (Lat. *vireo*, I am green or flourishing.) GREENLETS. VIREOS. Bill like that of a Shrike in miniature, moderately or very stout, shorter than head, compressed at least toward end, distinctly hooked and notched at tip, sometimes with trace of a tooth behind notch of upper mandible, and usually a nick in under mandible also. Rictal bristles conspicuous, and others present among the frontal and mental feathers. Nasal fossæ nearly filled with short erect feathers. Toes extensively coherent at base, as explained under head of the family; lateral toes of unequal lengths; claws stout, narrowly compressed, much curved and acute. Wing at least as long as tail, more or less rounded; sometimes much longer and quite pointed; of 10 primaries, 1st usually evident, though short or spurious, but sometimes (in the section *Vireosylvia* and in *Vireo flavifrons*) rudimentary and more or less completely concealed (exceptionally obvious even in these species). Tail short, even, of narrow feathers. Size small; length usually 5.00–6.00. Coloration simple; above olivaceous or grayish, crown like back, or ashy (in one case brown, in another black); under parts white, or white and yellow, or partly olivaceous. Sexes quite indistinguishable; young similar, not spotted or streaked. Migratory in N. Am. Insectivorous, arboricole. Nest pendulous; eggs white, spotted (except in *V. atricapillus*).

The numerous species of this genus have been divided into several groups, but no violence will be done by considering them all as *Vireo* — in fact it is difficult to do otherwise. For even the seemingly substantial division into two genera, according as there is an evident spurious 1st primary or apparently none, separates species, like *gilvus* and *philadelphicus*, slightly otherwise specifically distinguishable; while another division into two genera, according to shape of wings and length of spurious 1st primary or its absence, is subject to some uncertainty of determination, and unites species, like *olivaceus* and *flavifrons*, most dissimilar in other respects. Probably the best way is to recognize three subgenera — *Vireosylvia* for *barbatulus*, *olivaceus*, *flaviviridis*, *gilvus*, and *philadelphicus*; *Lanivireo* for *flavifrons* and *solitarius*; and *Vireo* for all the rest. The fact is, that almost every single species of *Vireo* has its own peculiar form, in shape of bill, proportions of primaries, etc., and these details cannot well be considered as of more than specific value. These slight differences are perfectly tangible and surprisingly constant, rendering the determination of the *species* comparatively easy, though these birds bear to each other a close general resemblance in size and color, and some of the *subspecies* are not easily discriminated. They are all more or less *olivaceous* above, sometimes inclining to gray or plumbeous, with crown either like back, or else ashy — in one species, however, brown, and in another black; and white or whitish below, usually more or less tinged with yellow. The coloration is very constant, the sexes being indistinguishable, and young differing little, if at all, from adults. All are small birds, — about 5 or 6 inches long. As a group the student will probably have no difficulty in recognizing them by the foregoing diagnosis, as the character of the feet seems to be peculiar, among North American birds, and is at any rate diagnostic when taken in connection with the character of the bill — all those *Oscines*, as Wrens, Creepers, or Titmice, that show much cohesion of the toes, having an entirely different bill. Some of the weaker-billed species might be carelessly mistaken for Warblers; but there is no excuse for this, nor for confounding them with any of the little Clamatorial Flycatchers. Vireos were long supposed to possess either 9 or 10 primaries. But that the important character of number of primaries — one marking whole families, as we have seen — should here subside to specific value only, seemed suspicious; and the fact is that all the species really have 10, only that, in some instances, the 1st is rudimentary and displaced, lying concealed outside the base of the 2d. The North American species are distributed over the temperate portions of this continent, and several of them are abundant birds of the Atlantic States, inhabiting woodland and shrubbery. They are exclusively insectivorous, and are therefore necessarily migratory in our latitudes. They build a neat pensile nest in the fork of a branchlet, and commonly lay four or five white, speckled eggs. All are alike in this respect, the nest and eggs of none of the species (except-

VIREONIDÆ: VIREOS, OR GREENLETS.

ing *atricapillus*) being distinguishable with certainty, though differing in size with that of the parent, and somewhat in position, according as the parents are birds of woodland or shrubbery; it would be useless, therefore, to give particular descriptions for each species. Next after Warblers, Greenlets are the most delightful of our forest birds, though their charms address the ear and not the eye. Clad in simple tints that harmonize with the verdure, these gentle songsters warble their lays unseen, while the foliage itself seems stirred to music. In the quaint and curious ditty of the White-eye — in the earnest, voluble strains of the Red-eye — in the tender secret that the Warbling Vireo confides in whispers to the passing breeze — he is insensible who does not hear the echo of thoughts he never clothes in words.

Analysis of Species and principal Subspecies.

Primaries apparently 9 (the 1st rudimentary and displaced). (a)
Primaries evidently 10 (the 1st short or spurious). (b)

(a) Throat yellow . . . . *flavifrons*
— white; crown ashy, not black-edged, hardly contrasting with back . . . . *philadelphicus*
— black-edged; back olive; with maxillary streaks . . . . . *calidris barbatulus*
— no maxillary streaks; crissum merely yellowish
olivaceus
— bright yellow
flaviviridis
(b) Crown black. Eggs white . . . . *atricapillus*
— not black; the 1st quill at least ½ as long as 2d, and wing 2.50 long . . . . *vicinior*
— not ½ as long as 2d, or wing not 2.50 long (c)
(c) Wing-bands wanting: coloration similar to *philadelphicus* . . . . *gilvus*
— present; length over 5.00; back olive, contrasting with ashy-blue crown . . . . *solitarius*
— plumbeous, crown scarcely different . . . . *plumbeus*
— 5.00 or less; wing = tail, both about 2.25; 1st quill = ½ the 2d . . . . *pusillus*
— > tail; crown ashy, chin and superciliary line white . . . *belli*
— olive, chin white, superc. line yellowish . . *novebor.*
— and under parts yellowish . . . *huttoni*

(*Subgenus* VIREOSYLVIA *Bonaparte.*)

V. cal'idris barba'tulus. (Gr. καλίδρις or σκαλίδρις, *kalidris* or *skalidris*, the name in Aristotle of some small spotted water-bird known to the Greeks, of no applicability to the present species; Lat. *barbatulus*, having a little beard. Fig. 218.) BLACK-WHISKERED GREENLET. WHIP-TOM-KELLY. Similar to *olivaceus*; distinguished by a narrow dusky maxillary line, or line of spots, on each side of the chin; bill longer, 0.75–0.80; proportion of quills slightly different. (See the figs.) Cuba, Bahamas, and casually in Florida. *V. altiloquus* is the West Indian stock-form, to which I have hitherto referred our Whip-tom-kelly; but it now appears to be itself but a form of the South American *V. calidris*; hence the change of name from former eds. of the Key.

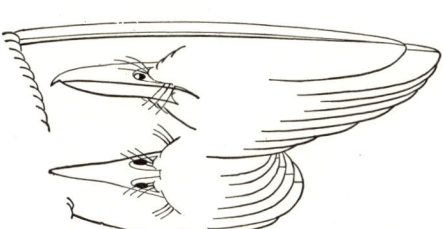

FIG. 218. — *V. c. barbatulus*, nat. size. (From Baird.)

V. oliva'ceus. (Lat. *olivaceus*, olive-colored. Fig. 219.) RED-EYED GREENLET. THE PREACHER. Above, olive-green; crown ash, edged on each side with a blackish line, below this a white superciliary line, below this again a dusky stripe through eye; under parts white, faintly shaded with greenish-yellow along sides, and tinged with the same on under wing- and tail-coverts; wings and tail dusky, the feathers edged with olive outside, with whitish inside; bill dusky above, pale below; feet leaden-blue; eyes red: no dusky maxillary streaks; no apparent spurious quill. Little different with age, sex, or season; young and fall birds the brightest colored, especially on sides, crissum, and lining of wings. Large; length 5.75–6.25; extent 9.75–10.75; wing 3.00–3.33; tail 2.33–2.50; bill about 0.66; tarsus 0.75. Eastern N. Am.; N. to Hudson's Bay and even Greenland; W. sometimes to Utah, Washington, and British Columbia; breeds throughout its N. Am. range, and winters from the Gulf States southward to northern S. Am. In most places the most abundant species of the genus, in woodland; a voluble, tireless songster. Eggs 0.80 × 0.55; nest often in a sapling.

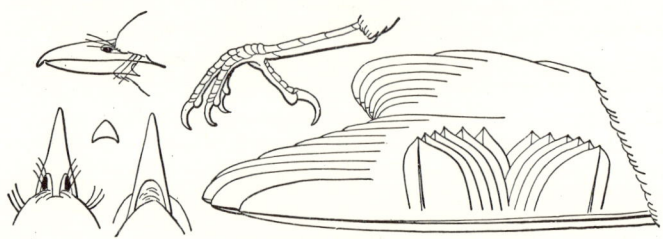

FIG. 219. — *V. olivaceus*, nat. size. (From Baird.)

V. flavivi'ridis. (Lat. *flavus*, yellow; *viridis*, green. Fig. 220.) YELLOW-GREEN GREENLET. Resembling the last; more yellowish below; under wing- and tail-coverts decidedly yellow; sides of body decidedly greenish-yellow; length 6.25–6.75. Lower Rio Grande valley of Texas and southward to Ecuador, Peru, and upper Amazon region; accidental in California and Quebec.

V. philadel'phicus. (Gr. φιλέω, *phileo*, I love; ἀδελφός, brother. Fig. 221.) BROTHERLY-LOVE GREENLET. Above, dull olive-green, brightening on rump, fading insensibly into ashy on crown, which is not bordered with blackish; a dull white superciliary line; below, very pale sulphur-yellow, whitening on throat and belly, slightly olive-shaded on sides; sometimes a slight creamy or buffy shade throughout the under parts; no obvious wing-bars; no apparent spurious quill. Length 4.80–5.10; extent 8.00–8.50; wing 2.66; tail 2.15; bill hardly or about 0.50; tarsus 0.66. Eastern N. Am., strictly; N. to Hudson's Bay; breeds from the northern tier of states northward; S. in winter to Cent. Am. A small, plainly-colored species, distinguishable from *gilvus* by apparent absence of a spurious quill; not very common in the Atlantic States, more so in the Mississippi Valley. (Best account of this species by DWIGHT, Auk, July, 1897, pp. 259–272, pl. 2.)

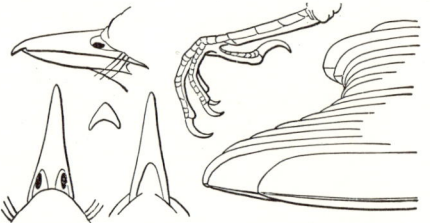

FIG. 220. — *V. flaviviridis*, nat. size. (From Baird.)

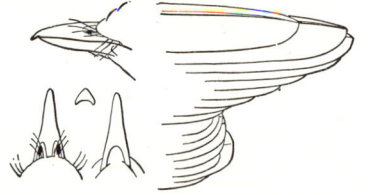

FIG. 221. — *V. philadelphicus*, nat. size. (From Baird.)

FIG. 222. — *V. gilvus*, nat. size. (From Baird.)

V. gil'vus. (Lat. *gilvus*, yellowish. Figs. 217, 222.) WARBLING GREENLET. Colors much as in the last species, but below with very little yellowish; spurious quill present and evident, $\frac{1}{4}$ to $\frac{1}{3}$ as long as 2d primary. Length 5.50–6.00; extent 8.50–9.25; wing 2.80; tail 2.25; bill 0.40; tarsus 0.65. Eastern N. Am. to the high central plains, breeding throughout its range; wintering extralimital; an abundant little bird and an exquisite songster. Its voice is not strong, and many birds excel it in brilliancy of execution; but not one of them all can rival the tenderness and softness of the liquid strains of this modest vocalist. Not born to "waste its sweetness on the desert air," the Warbling Vireo forsakes the depths of the woodland for the park and orchard and shady street, where it glides through the foliage of the tallest trees, the unseen messenger of rest and peace to the busy, dusty haunts of men.

V. g. swain'soni? (To Wm. Swainson. Fig. 223.) WESTERN WARBLING GREENLET. "Similar to *V. gilvus*, but smaller; colors paler; bill more depressed; upper mandible almost black; 2d quill much shorter than 6th." Rocky Mts. to the Pacific, U. S. This Western form has long been described as distinct, but the characters assigned will not be found constant or always appreciable. It is simply a dull-colored race, like many other birds of this region. It is ignored in both A. O. U. Lists; retained in RIDGW., Man., 1887, p. 472; and now left here with this explanation.

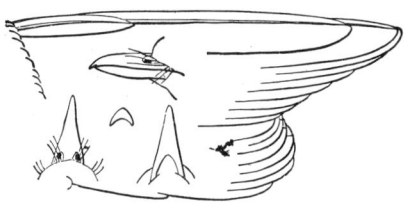

FIG. 223. — *V. g. swainsoni*, nat. size. (From Baird.)

FIG. 224. — *V. flavifrons*, nat. size. (From Baird.)

(*Subgenus* LANIVIREO *Baird.*)

V. fla'vifrons. (Lat. *flavus*, yellow; *frons*, front. Fig. 224.) YELLOW-THROATED GREENLET. Above, rich olive-green, crown the same or even brighter, rump insensibly shading into bluish-ash; below, bright yellow, belly and crissum abruptly white, sides anteriorly shaded with olive, posteriorly with plumbeous; extreme forehead, superciliary line and ring round eye, yellow; lores dusky; wings dusky, with inner secondaries broadly white-edged, and two broad white bars across tips of greater and median coverts; tail dusky, nearly all the feathers completely encircled with white edging; bill and feet dark leaden-blue; no apparent spurious quill. Length 5.75–6.00; extent 10.00; wing 3.00; tail only about 2.25. A large, stout, highly-colored species, curiously resembling *Icteria virens*, common in Eastern U. S. and adjoining British Provinces; W. to edge of the plains; winters in Florida and southward to Colombia; breeds in all its N. Am. range. Its proper name may be *V. ochroleucus*.

V. solita'rius. (Lat. *solitarius*, solitary; *solus*, alone. Fig. 225.) BLUE-HEADED GREENLET. SOLITARY GREENLET. Above, olive-green; crown and sides of head bluish-ash in marked contrast, with a broad white line from nostrils to and around (not beyond) eye, and a dusky loral line; below, pure white, flanks washed with olivaceous, and axillars and crissum pale yellow; wings and tail dusky, most of the feathers edged with white or whitish, and two conspicuous bars of the same across tips of middle and greater coverts; bill and feet blackish-plumbeous; iris brown. Length

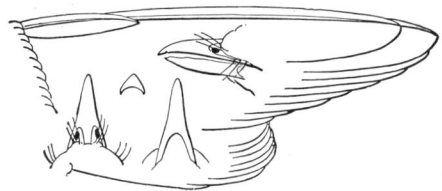

FIG. 225. — *V. solitarius*, nat. size. (From Baird.)

5.25–5.75; extent 8.50; wing 2.75–3.00; tail 2.25–2.33; bill about 0.40, stout, nearly 0.20 deep at base; spurious quill 0.50–0.66 long, about ¼ as long as 2d primary. Young and fall specimens more brightly colored. A stoutly-built species, known at a glance by the bluish cap. Eastern U. S. and Canada, N. to Hudson's Bay and Great Slave Lake; S. in winter to Guatemala; breeds from southern New England and the northern tier of States northward, and thus, like *philadelphicus*, is chiefly found as a migrant in the U. S. It is not rare, but not so common as *olivaceus*, *flavifrons*, or *noveboracensis*; inhabits woodland; a delicious songster.

V. s. alti′cola. (Lat. *altus*, high; *colere*, to inhabit, or (*in*)*cola*, an inhabitant.) MOUNTAIN SOLITARY GREENLET. Like *solitarius* proper, but larger, with stouter bill, and darker coloration; upper parts nearly uniform dark plumbeous, only tinged with olive on back, instead of being quite olivaceous contrasting with bluish ash of head. Wing 3.00–3.30; tail 2.25. Mountains of North Carolina, S. in winter to Florida. BREWSTER, Auk, Jan. 1886, p. 111; COUES, Key, 3d ed., 1887, p. 872; A. O. U. Lists, 1st and 2d eds., 1886 and 1895, No. 629 *c*.

V. s. cas′sini. (To John Cassin.) CASSIN'S GREENLET.) Like *solitarius* proper; duller and more brownish-olivaceous; under parts tinged with buff or ochrey where *solitarius* is pure white; loral line and eye-ring impurely whitish. Western U. S., especially the Pacific coast region, in which it breeds from British Columbia southward to Lower California; Arizona, and probably other portions of the Great Basin, where it is associated with *V. s. plumbeus*, but is not to be confounded with the latter.

V. s. lucasa′nus. (Lat., of or pertaining to any one named Luke or Lucas; in this case referring to Cape St. Lucas.) Like the last; rather smaller; the bill longer and stouter, the sides and flanks much yellower. Young in autumn resembling that of *solitarius* proper. Wing 2.70; tail 2.00-2.15. Lower California, apparently a very slightly marked form. BREWSTER, Auk, Apr. 1891, p. 147; A. O. U. List, 2d ed., 1895, p. 265, No. 629 *d*.

V. s. plum′beus. (Lat. *plumbeus*, lead-colored. Fig. 226.) PLUMBEOUS GREENLET. Leaden-gray, rather brighter and more ashy on crown, but without marked contrast, faintly glossed with olive on rump; a conspicuous white line from nostril to and around eye, and below this a dusky loral stripe; below, pure white, sides of neck and breast shaded with color of back; flanks, axillars, and crissum with a mere trace of olivaceous, or none; wings and tail dusky, with conspicuous pure white edgings and cross-bars. Size of *solitarius* or larger. Length 5.75–6.10;

FIG. 226. — *V. s. plumbeus*, nat. size. (From Baird.)

extent 9.75–10.25; wing 2.90–3.10; tail 2.50; bill 0.50; tarsus 0.66; middle toe the same; spurious quill exposed about 0.75, ⅓ as long as the 2d quill. Central Plains to the Great Basin, U. S., and especially southern Rocky Mts., where it is abundant; N. to Wyoming, S. in winter to Oaxaca, Mexico; accidental in New York. A large stout species, a near ally of *solitarius*, but nearly all the olivaceous of that species replaced by plumbeous, and the yellowish by white, so that it is a very different-looking bird. Fall specimens, however, are more olivaceous, and the bird evidently grades closely up to *solitarius*.

(*Subgenus* VIREO *proper*.)

V. vici′nior. (Lat. *vicinus*, neighboring.) GRAY GREENLET. With the general appearance of a small faded specimen of *plumbeus*: leaden-gray, faintly olivaceous on rump, below white, with hardly a trace of yellowish on sides; wings and tail hardly edged with white; no markings about head except a whitish eye-ring. Length 5.75; extent 8.66; wing and tail each 2.50; tarsus nearly 0.75; middle toe and claw hardly over 0.50; tip of inner claw falling short of base of middle claw; tail decidedly rounded; first primary exposed 0.75, ½ as long as 2d primary, which latter is not longer than 8th. These peculiar proportions of the original type-specimen are constant, and the species is distinct from any other. It is our plainest-colored species, resembling *plumbeus*, but more closely allied to the smaller rounder-winged species like *noveboracensis* and especially *pusillus* ; toes almost abnormally short, and tail as long as wing. Southwestern U. S., in western Texas, New Mexico, Arizona, Nevada, southern and Lower California; northwestern Mexico. My type-specimen from Fort Whipple, Arizona, long remained unique, but others have since been found, extending the known geographical range considerably. Nest in bushes; eggs 0.72 × 0.52.

V. noveboracen'sis. (Lat. *novus*, new ; *Eboracum*, York. Fig. 227.) WHITE-EYED GREENLET. THE POLITICIAN. Above, bright olive-green, including crown ; a slight ashy gloss on cervix, and rump showing yellowish when the feathers are disturbed; below, white; sides of breast and belly, with axillars and crissum, bright yellow; a bright yellow line from nostrils to and around eye;

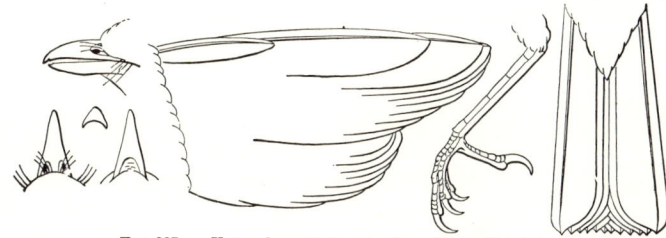

FIG. 227. — *V. noveboracensis*, nat. size. (From Baird.)

lores dusky; two broad yellowish wing-bars; inner secondaries widely edged with the same ; bill and feet blackish-plumbeous; eyes white. About 5.00; extent 8.00; wing 2.33-2.50; tail 2.05; spurious quill exposed 0.75, ½ as long as 2d, which about equals 8th ; tarsus about 0.75; middle toe and claw 0.50; bill nearly 0.50. A small, compact, brightly-colored species, abundant in shrubbery and tangle of the Eastern U. S.; W. rarely to the Rocky Mts. ; rather southerly, N. to Massachusetts and Minnesota; winters from Florida southward to Cent. Am.; resident in the Bermudas; noted for its sprightly manners and emphatic voice.

V. n. may'nardi. (To C. J. Maynard.) KEY WEST GREENLET. Coloration much as in the last, but grayer above and paler below; size and proportions as in *V. crassirostris* (an extralimital species), the bill being notably large and stout. Wing 2.20-2.50; tail 1.90-2.05; bill 0.55-0.65, its depth at nostrils 0.18-0.20. Southern Florida. BREWSTER, Auk, Apr. 1887, p. 148; COUES, Key, 3d ed., 1887, p. 872; 4th ed., 1890, p. 899; A. O. U. List, 2d ed., 1895, p. 266, No. 631 *a*.

V. hut'toni. (To Wm. Hutton, of California. Fig. 228.) HUTTON'S GREENLET. Similar to *noveboracensis*, but differing much as *flaviviridis* does from *olivaceus*, in having the under parts almost entirely yellowish. First quill rather less than half 2d, which about equals 10th ; 3d a little longer than 7th ; 4th and 5th nearly equal and longest. Tail slightly rounded, shorter than wings. Bill very small. Above olive-green; brightest behind, especially on rump and edging of tail; duller and more ashy toward and on top and sides of head and neck. Wings with two bands on coverts, and outer edges of innermost secondaries rather broadly olivaceous-white ; other quills edged externally with olive-green, paler toward outer primary, internally with whitish. Lateral tail-feathers edged externally with yellowish-white. Feathers of rump with much concealed yellowish-gray. Under parts pale olivaceous-yellowish, purest behind, lightest on throat and abdomen; breast more olivaceous, soiled with a slight buffy tinge ; sides still deeper olive-green. Axillars and crissum yellowish; inside of wings whitish. Loral region and narrow space around eye dull yellowish, in faint contrast to olive of head. Bill horn-color above, paler below; legs dusky. Length 4.70; wing 2.40; tail 2.05. Coast region of Southern and Central California, resident. (Description from Baird.)

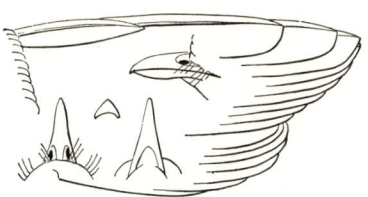

FIG. 228. — *V. huttoni*, nat. size. (From Baird.)

V. h. obscu'rus. (Lat., obscure.) OBSCURE GREENLET. ANTHONY'S VIREO. The dark form of *V. huttoni* from the Pacific coast region, breeding in British Columbia, Washington, and Oregon, and migrating S. in winter. ANTHONY, Zoe, i, Dec. 1890, p. 306; A. O. U. List, 2d ed., 1895, p. 266, No. 632 *c*; not in any previous ed. of the Key.

V. h. ste'vensi. (To F. Stephens.) STEPHENS' GREENLET. Like *V. huttoni*. Bill stout; wings 0.30–0.40 longer than tail. Above, grayish-ash; crown, vertex, and sides of head and neck nearly pure ash; back faintly tinged with olive; rump and edging on tail-feathers dull olive-green. Wings with two nearly confluent bands on coverts, and outer edges of inner secondaries broadly white; outer quills edged more narrowly with the same color. Beneath brownish or smoky-white, with a mere wash of yellowish on sides and crissum. Upper eyelid dusky-brown; remainder of orbital region, with lores, ashy-white in decided contrast with the nearly clear cinereous of the head generally. Lining of wings white. Length 5.20; extent 8.50; wing 2.55–2.90; tail 2.25; tarsus 0.73; culmen 0.50. Lower California, Arizona, and New Mexico, especially in mountain ranges. Related to *huttoni*, which has bill less stout, wing 2.40 *or less*, and is olive-green above and olivaceous-yellow below, without clear white anywhere. The differences are nearly parallel with those between *belli* and *pusillus*, — *stevensi* being grayish-ash above with no decided olive-green excepting on rump and tail, brownish-white below, untinged with yellowish excepting on sides and crissum, the wing-bands pure white and nearly confluent. (Not in Check List, 1880. Description from BREWSTER, Bull. Nutt. Club, vii, July, 1882, p. 142.)

V. bel'li. (To J. G. Bell, of New York. Fig. 229.) BELL'S GREENLET. Olive-green, brighter on rump, ashier on head, but without decided contrasts; head-markings almost exactly as in *gilvus*; below, sulphury-yellowish, only whitish on chin and middle of belly; inner quills edged with whitish; two whitish wing-bands, but one more conspicuous than the other. Hardly or not 5.00 long; wing scarcely over 2.00; tail under 2.00; spurious quill about $\frac{2}{5}$ the 2d, which equals or exceeds the 7th. A

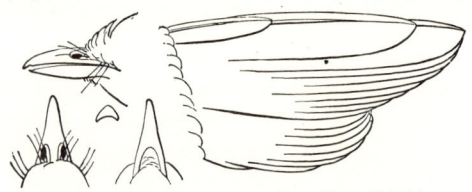

FIG. 229. — *V. belli*, nat. size. (From Baird.)

pretty little species, like a miniature of *gilvus*, but readily distinguished from that species by its small size, presence of decided wing-bars, more yellowish under-parts, and different wing-formula. Middle region of the U. S., W. to the Rocky Mts., E. to the valley of the Ohio, N. to the valley of the Red River in Minnesota and Dakota; an abundant species, inhabiting copses and shrubbery in open country, with much the same sprightly ways and loud song as those of *noveboracensis*. Nest in bushes; eggs 0.67 × 0.48.

V. b pusil'lus. (Lat. *pusillus*, puerile, petty. Fig. 230.) LEAST GREENLET. Olivaceous-gray, below white, merely tinged with yellowish on sides; head-markings obscure; wing-bands and edgings, though evident, narrow and whitish; no decided olive or yellow anywhere. Size of *belli*; wing and tail of equal lengths, little over 2.00; bill 0.33; tarsus 0.66; middle toe and claw 0.50; spurious quill about $\frac{1}{2}$ as long as 2d, which is intermediate between 7th and 8th. A small, obscure-looking bird, re-

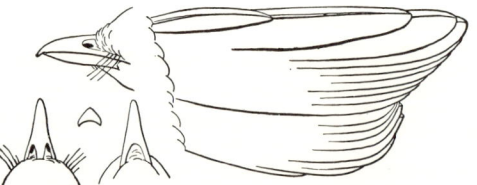

FIG. 230. — *V. pusillus*, nat. size. (From Baird.)

sembling *belli*, but much grayer, tail relatively longer, spurious quill longer, and 2d primary shorter. Questionably right to reduce this to a subspecies of *belli*, for the difference is obvious at a glance, and more decided than that separating most of the subspecies. It has held specific rank in all previous editions of the Key. Arizona and Southern and Lower California;

western Mexico; common. Eggs undistinguishable from those of *belli*, and nesting the same.

V. atricapil'lus. (Lat. *ater*, black; *capillus*, hair.) BLACK-CAPPED GREENLET. ♂ : Top and side of head black, excepting a white eye-ring and white loral stripe. Upper parts olivaceous; lower parts white, tinged with pale greenish on sides and flanks. Wings and tail blackish, edged with olivaceous, the former with two dingy whitish bars across ends of greater and median coverts; lining of wings yellowish. Bill black; feet dark; iris red. Length 4.75; extent 7.25; wing 2.25; tail nearly 2.00; bill 0.50; tarsus 0.75; middle toe and claw 0.50; 1st primary exposed 0.66. ♀ : Black of head replaced by dark slate color; upper parts duller olive, lower somewhat buffy. The black cap of the ♂ renders the species conspicuous among all its congeners. Kansas to Texas and Mexico. Nest in small trees or bushes, near the ground, pensile from a forked twig as usual in the genus, but eggs white, unmarked; 0.65–0.75 × 0.50–0.55; usually 4 in number.

Family LANIIDÆ : Shrikes.

Essentially characterized by the combination of comparatively weak, strictly passerine feet with a notched, toothed, and hooked bill, the size, shape, and strength of which recalls that of a bird of prey (fig. 231). The family comprises about 200 recorded species, referable to numerous genera and divisible into three groups, not very well defined, however, of which the following typical subfamily is the only one occurring in America : —

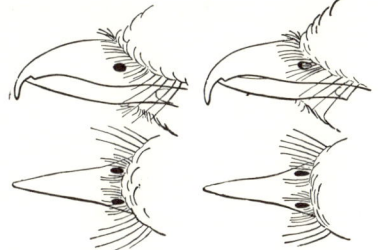

FIG. 231. — Shrikes' Bills, nat. size. (From Baird.)

Subfamily LANIINÆ : True Shrikes.

In this group the wing has 10 primaries and the tail 12 rectrices; both are much rounded and of nearly equal lengths. The rictus is furnished with strong bristles. The circular nostrils are more or less perfectly covered and concealed by dense tufts of antrorse bristly feathers. The tarsi are scutellate in front and outside — in the latter respect deviating from a usual Oscine character. Our Shrikes will thus be easily distinguished; additional features are given under head of the genus *Lanius*, the only representative of this group in America.

FIG. 232. — Butcher-bird, reduced. (From Tenney, after Wilson.)

Shrikes are bold, spirited birds, quarrelsome among themselves, and tyrannical toward weaker species; in fact, their nature seems as highly rapacious as that of the true birds of prey. They are carnivorous, feeding on insects and such small birds and quadrupeds as they can capture and overpower; many instances have been noted of their dashing attacks upon cage-birds, and their reckless pursuit of other species under circumstances that cost them their own lives. But the most remarkable fact in the natural history of Shrikes is their habit of impaling their prey on thorns or sharp twigs. They build a rather rude and bulky nest of twigs, and lay 4–6 speckled eggs. They are not strictly migratory, although our northernmost species usually retires southward in the fall. The sexes are alike, and the young differ but little. There are only two well-determined American species.

LA′NIUS. (Lat. *lanius*, a butcher. Fig. 232.) GRAY SHRIKES. Wing of 10 primaries, and tail of 12 rectrices, both rounded in shape, and of nearly equal lengths. Point of wing formed by 3d, 4th, and 5th quills; 2d not longer than 6th, 1st about half as long as 3d. Tarsus equalling or slightly exceeding in length the middle toe and claw, strongly scutellate in front, with outer lateral plate usually more or less subdivided, as is unusual among *Oscines*. Lateral toes of about equal lengths, their claws reaching base of middle claw; inner toe cleft nearly to base, outer more extensively coherent with basal joint of middle toe. Feet large and strong, but without specially " raptorial " development either of the digits or of their claws. Bill large and powerful, compressed, deep, completely notched and toothed, and strongly hooked, presenting the full accomplishment of a raptorial character. Rictus ample and deeply cleft, strongly bristled; gonys short, only about half the length of lower mandible. Nostrils circular or nearly so, placed well forward in the nasal fossæ, more or less perfectly overhung and concealed by tufts of antrorse bristly feathers. Body stout; neck short; head relatively large. Coloration simple, the black, white, and bluish or grayish tints being unrelieved by red or other bright color. In amount of dusky vermiculation of under parts the species are graded from *borealis* (most) to *excubitorides* (least or none), and each one is graded from young to old. In all, the general resemblance to a Mocking-bird is striking.

Analysis of Species and Subspecies.

Large : length 9.00 or over. Black head-stripe broken on under eyelid and across forehead. Always waved below with dusky . . . . *borealis*
Small : length under 9.00. Black head-stripe unbroken across forehead : no white on under eyelid. Adults unwaved below.
    Lighter : much white on rump and scapulars ; long white patch on primaries . . . . . . *l. excubitorides*
    Darker : little white on rump and scapulars ; short white patch on primaries . . . . . . *ludovicianus*
    Darkest : Pacific coast form . . . . *l. gambeli*

L. borea′lis. (Lat. *borealis*, northern. Figs. 233, 234.) GREAT NORTHERN SHRIKE. BUTCHER-BIRD. NINE-KILLER. SHAMBLE-STICKER. Adult ♂ ♀ : Above, clear bluish-ash, blanching on rump and scapulars; below, white, always vermiculated transversely with fine wavy blackish lines; a broad black bar along side of head, *not* meeting its fellow across forehead, interrupted by a white crescent on under eyelid, and bordered above by hoary white that also occupies extreme forehead; wings and tail black, former with a large white spot near base of primaries and white tips of most of the quills, latter with outer web of outer feather edged, and all the feathers excepting the middle pair broadly tipped, with white, and with concealed white bases; bill and feet bluish-black; eyes blackish. Length 9.00-10.00; extent 13.50-14.50; wing 5.00-5.50; tail rather more; bill 0.75; tarsus 0.90; middle toe and claw

FIG. 233. — Northern Shrike.

0.75. Young: Colors much less pure and clear. Above, grayish-brown, scarcely or not whitening on scapulars, tail-coverts, and forehead. The younger the browner, sometimes almost with a rusty tinge; grayer according to age. Below brownish-white (the younger the browner), the wavy dark markings stronger than in the adult. The bar along the head poorly defined, merely dusky, or quite obsolete. Wings and tail brownish-black, with less white than in the adult. Bill plumbeous-brown, flesh-colored at base below. At a very early age, the upper parts are probably vermiculated somewhat like the lower, as in the same stage of *L. ludovicianus;* but this state I have not observed. In old age, the dusky vermiculation of the

FIG. 234. — Butcher-bird (*L. borealis*), nat. size. (Ad. nat. del. E. C.)

under parts is much diminished, but I have never seen it absent altogether. This feature, coupled with the particular character of the head-markings, the large size, and comparatively short tarsi, will always distinguish the species from *L. ludovicianus* or *excubitorides.* N. Am., northerly; in winter, usually extends S. to about 35°. The castle of this "feudal baron and brigand bold" is built in a bush or low tree with a basement of sticks, upon which is matted and felted a thick warm superstructure of bark-strips, grasses, and soft vegetable substances: eggs 4–6, about 1.10 × 0.80, rather elliptical in shape, so profusely speckled, scratched, and marbled with reddish, brownish, and purplish shades that the greenish-gray ground color is scarcely perceptible.

L. ludovicia'nus. (Lat. *ludovicianus*, of Louisiana. Fig. 235.) LOGGERHEAD SHRIKE. Adult ♂ ♀: Above, slate-colored, slightly whitish on upper tail-coverts and ends of scapulars; below, white, sometimes a little ashy-shaded, but no wavy black lines, or only a few slight ones; white on wings and tail less extensive than in *borealis* or *excubitorides;* black bridle meeting its fellow across forehead, not interrupted by white on lower eyelid, scarcely or not bordered above by hoary white. Smaller: length 8.00–8.50; wing and tail each 4.00 or little more; tarsus at least 1.00, thus relatively longer than in *borealis;* bill about 0.50. Young: differing from the adult much as young *borealis* does, and decidedly waved below, as in that species: but the size and other characters are distinctive. Eastern and Southern U. S., resident, common; in its typical manifestation it is characteristic of the S. Atlantic States, and is known to occur in the Bahamas; but specimens more like *ludovicianus* than *excubitorides* occur N. to New England and W. to the Great Plains.

L. l. excubitori'des. (Lat. *excubitor*, a sentinel; Gr. εἶδος, *eidos*, resemblance; *i. e.*, like the European *L. excubitor.*) WHITE-RUMPED SHRIKE. COMMON AMERICAN SHRIKE. Adult ♂ ♀: Leaden-gray or light slate-color, whitening on scapulars and upper tail-coverts. Beneath white, slightly shaded with French gray on sides, but without dusky vermiculation. A narrow stripe across forehead, continuous with a broad bar along side of head, embracing eye, black, slightly, if at all, bordered with whitish. Lower eyelid not white. Wings and tail black, with white markings, much as in the last species. Bill and feet plumbeous-black. Length under 9.00; extent 12.00–13.00; wing and tail, each, about 4.00; bill 0.66; tarsus 1.00 or more. Young: Vermiculated below with dusky, upon a brownish ground, about to the same extent as is seen in very old examples of *L. borealis.* General tone of upper parts less pure than in the adult; scapulars and tail-coverts not purely white; black bar of head less firm, but as far as it goes maintaining the characters of the species. At a very early age, the upper parts, including the whitish of scapulars and tail-coverts, are finely vermiculated with dusky waves. The ends of the quills, wing-coverts, and tail-feathers often have rusty or rufous markings. Extreme examples of *excubitorides* look very different from *ludovicianus* proper, but the two are observed to melt into each other when many specimens are compared, so that no specific character can be assigned. Middle and Western N. Am. and Mexico; N. to the

FIG. 235. — Loggerhead Shrike.

Saskatchewan, E. to Ohio, New York, Canada, and even New England. It is an arbitrary distinction which assigns this variety a range restricted to Western N. Am. The nest and eggs are indistinguishable from those of *ludovicianus* proper; both resemble those of *L. borealis*, but the eggs average smaller — about 0.97 × 0.72.

L. l. gam'beli. (To Wm. Gambel.) CALIFORNIA SHRIKE. The Pacific coast form, doubtfully attempted to be distinguished by its sordid coloration. RIDGW., Man. 1887, p. 467; COUES, Key, 4th ed. 1890, p. 899; A. O. U. List, 2d ed. 1895, p. 262, No. 622 b.

L. l. antho'nyi. (To A. W. Anthony.) ISLAND SHRIKE. An insular form, said to be darker than *gambeli*, even darker than *ludovicianus* proper, and quite small. Santa Barbara Islands. *L. l. gambeli* ANTHONY, Pub. No. 1, Pasadena Acad. Sci. Aug. 1897, p. 19. *L. l. anthonyi* MEARNS, Auk, July, 1898, p. 261; A. O. U. Suppl. List, Auk, Jan. 1899, p. 122.

NOTE. — Yet another despairing aspirant for recognition is *L. l. migrans* of W. PALMER, Auk, July, 1898, pp. 244-258. This consists of 104 specimens intermediate between *ludovicianus* proper and *excubitorides*, or not exactly referable to either. When the A. O. U. Committee of 1899 reached this case, it was "deferred for lack of material"! I should say that it simply proves the position I have taken in the Key since 1872, when I first reduced *excubitorides* to a subspecies of *ludovicianus*; for these 104 intermediates attest that intergradation between the two forms which is the test of any subspecies. Mr. Palmer's 15-page painstaking discussion of 104 specimens adds nothing to what he might have learned in a few lines, if he had looked at the Key, 2d-4th eds., 1884-90, p. 338.

Family FRINGILLIDÆ: Finches, etc.

Conirostral Oscines with 9 primaries. — The largest North American family, comprising about one-seventh of all our birds, and the most extensive group of its grade in ornithology. As ordinarily constituted, it represents, in round numbers, 600 current species and 100 genera, of nearly all parts of the world, except Australia, but more particularly of the northern hemisphere and throughout America, where the group attains its maximum development. Any one United States locality of average attractiveness to birds has a bird-fauna of over 200 species; and if it be away from the sea-coast, and consequently uninhabited by marine birds, about one-fourth of its species are *Mniotiltidæ* and *Fringillidæ* together — the latter somewhat in excess of the former. It is not easy, therefore, to give undue prominence to these two families.

The *Fringillidæ* are more particularly what used to be called "Conirostral" birds, in distinction from "Fissirostres," as the Swallows, Swifts, and Goatsuckers; "Tenuirostres," as Humming-birds and Creepers; and "Dentirostres," as Warblers, Vireos, and most of the preceding families. The bill approaches nearest the ideal cone, combining strength to crush seeds with delicacy of touch to secure minute objects. The cone is sometimes nearly expressed, but is more frequently turgid or conoidal, convex in most directions or, again, so contracted that some of its outlines are concave. The nostrils are always situated *high up* — nearer culmen than cutting edge of bill; they are usually exposed, but in many, chiefly boreal, genera, the base of the bill is furnished with a ruff or two tufts of antrorse feathers more or less completely covering the openings. The cutting edges of the bill may be slightly notched, but are usually plain. There are usually a few inconspicuous bristles about the rictus, sometimes wanting, sometimes highly developed, as in our Grosbeaks. The wings are endlessly varied in shape, but agree in possessing only 9 developed primaries; the tail is equally variable in form, but always has 12 rectrices. The feet show a strictly Oscine or laminiplantar podotheca, scutellate in front, covered on each side with an undivided plate, producing a sharp ridge behind. None of these members offer extreme phases of development in any of our species.

But the most tangible characteristic of the family is *angulation of the commissure*. The commissure runs in a straight line, or with a slight curve, to or near to the base of the bill, and is then more or less abruptly bent down at a varying angle — the cutting edge of the upper mandible forming a re-entrance, that of the lower mandible a corresponding salience. In familiar terms, we might say that the corners of the mouth are drawn down — that the Finches, though very merry little birds, are literally "down in the mouth." In most cases this feature is unmistakable, and in the Grosbeaks, for example, it is very strongly marked indeed; but in some of the smaller-billed forms, and especially those with slender bill, it is hardly perceptible. On the whole, however, it is a good character, and at any rate it is the most reliable external feature that can be found. It separates our fringilline birds pretty trenchantly from other 9-primaried Oscines except *Icteridæ*, and most of these may be distinguished by the characters given beyond.

Taking their characters all together, *Fringillidæ* may be defined as 9-primaried, conirostral, laminiplantar, oscine Passeres with axis of bill at an angle with that of skull, and nostrils nearer culmen than cutting edge of bill.

When we come, however, to consider this great group of conirostral Oscines in its entirety, as compared with bordering families like the Old World *Ploceidæ*, or the *Icteridæ*, and especially the *Tanagridæ*, of the New, the difficulty, if not the impossibility, of framing a perfect diagnosis becomes apparent, and I am not aware that any attempts at rigid definition have proven successful. Ornithologists are nearly agreed what birds to call fringilline, without being so well prepared to say what "fringilline" means. The subdivisions of the family, as might be expected, are still conventional, and varying with every leading writer. Our species might be thrown into several groups, but the distinctions would be more or less arbitrary and not readily perceived. It is therefore best to waive the question, and simply collocate the genera in orderly sequence.

Fringillidæ are popularly known by several different names. Here belong all the Sparrows, with the allied birds called Finches, Buntings, Linnets, Grosbeaks, and Crossbills. The species and subspecies ascertained to occur within our limits are mostly well determined, although the number of genera to which they are customarily referred is, I think, altogether too large. Three of them, *Carduelis carduelis*, *Passer domesticus*, and *P. montanus*, are imported and naturalized. Species occur throughout our country, in every situation, and many of them are among our most abundant and familiar birds. They are all granivorous — seed-eaters, but many feed extensively on buds, fruits, and other soft vegetable substances, as well as on insects. They are not so perfectly migratory as the exclusively insectivorous birds, the nature of whose food requires prompt removal at the approach of cold weather; but, with some exceptions, they withdraw in the fall from their breeding places to spend the winter farther south, and to return in the spring. With a few signal exceptions, they are not truly gregarious birds, though they often associate in large companies, assembled in community of interest. The modes of nesting are too various to be here summarized. Nearly all the Finches sing, with varying ability and effect; some of them are among our most delightful vocalists. As a rule, they are plainly clad — even meanly, in comparison with some of our sylvan beauties; but among them are birds of elegant and striking colors. Among the highly-colored ones, the sexes are more or less unlike, and other changes, with age and season, are strongly marked; the reverse is the case with the rest.

The unpractised student will have more trouble with this family than elsewhere in identifying his specimens. In the first place, the genera and species are very numerous, and so variously interrelated that no satisfactory subfamilies have been established; they are therefore not parcelled out in sets. Secondly, all the genera cannot be discriminated in a line of type. To meet the difficulty, I have caused the family to be profusely illustrated with cuts of more than average excellence, and have attempted a tabular analysis of the genera, which, though necessarily defective, will doubtless help to some extent. Speaking roundly, there are three lots of genera: (*a*) *Loxiine*, mostly boreal birds, sexed unlike, ♂ often red, ♀ dull, no blue, colors massed or streaky, bill usually ruffed at base, wings pointed, tail forked, feet weak; (*b*) *Spizelline*, everywhere, mostly small streaked and spotted species, sexed alike, may be yellowed but are never red or blue, wings, tail, and feet various; (*c*) *Spizine*, mostly southerly, sexed unlike, ♂ often red or blue, bill unruffed, wings, tail, and feet various; — but nothing will serve to distinguish these groups unexceptionably, and I therefore refrain from presenting them formally as subfamilies. The British Museum Catalogue of 1888 arranges the *Fringillidæ* of the world in 3 subfamilies (*Coccothraustinæ*, *Fringillinæ*, and *Emberizinæ*, or Grosbeaks, Finches proper, and Buntings), with 99 genera and about 575 species or subspecies.

Analysis (partial) of Fringilline Genera.

Bill *metagnathous*, both mandibles falcate, their points crossed. ♂ red, ♀ dark and yellowish . . . . . *Loxia*
Bill enormous, nearly = tarsus, *greenish-yellow*. Wings black and white; tail and tibiæ black. (Western.)
Hesperiphona
Bill parrot-like, *whitish*. Head conspicuously crested. ♂ ♀ gray and carmine, face not black. Length 7.50 or more. (S. W. U. S.) . . . . *Pyrrhuloxia*
Bill *reddish*. Head conspicuously crested. ♂ vermilion, face black. ♀ gray and reddish. Length 7.50 or more. (E. and S. U. S.) . . . . *Cardinalis*
Bill with a *ruff*, or pair of *nasal tufts*, of antrorse plumules, at base of upper mandible.
    Length 8.00 or more. ♂ red and gray, ♀ gray and yellowish, uncrested. Bill *turgid*, hooked. (Boreal.)
Pinicola
    — under 8.00. — *Bluish-gray*, below reddish-gray, crown, wings, and tail black. (Alaska.) . . *Pyrrhula*
    — *White*, with black on back, wings, and tail; or washed with clear brown. (Boreal.)
Passerina
    — *Chocolate-brown*, unstreaked, with *rosy edgings*; black or clear *ash* on head. (Western.)
Leucosticte

FRINGILLIDÆ: FINCHES, BUNTINGS, SPARROWS.

— *Streaky;* no yellow; ♂ extensively *red;* ♀ dark and white. Bill *turgid.* (E. and W. U. S.)
Carpodacus
— *Streaky,* with dusky or flaxen-brown and white; crown *crimson.* Bill *acute.* (Boreal.)
Acanthis
— Streaky everywhere; *no* red or pure black, some *yellowish.* Bill *acute.* (*pinus.*) (N. Am.)
Spinus
— much *yellow,* wings and tail *black; no* red. Bill moderate. (N. Am.) . . *Astragalinus*
— *Not* streaky; *red,* black and gold. (*Imported.*) . . . . *Carduelis*

Bill *without* ruff; nostrils exposed.
  Hind claw lengthened, *straightened.* — Bill *moderate;* ♂ with a colored *cervical collar;* oblique white on tail.
  (N. and W. N. A.) . . . . *Calcarius*
        — Bill *turgid; no* cervical collar: transverse white on tail. (Western.)
Rhynchophanes
Hind and fore claws *lengthened;* all much curved; inner reaching at least ½ way to end of middle one —
    — *Spotted and streaked* foxy or slaty sparrows, about 7.00 (or more) long. (N. Am.) . *Passerella*
    — Black, white and chestnut, *in masses.* (A Western species of) . . . . *Pipilo*
Hind and fore claws *not* peculiar.
  Length 4.50 *or less.* — ♂ Black and white, ♀ olivaceous and yellowish. (Texas.) . . . . . *Sporophila*
       ♂ Greenish blackening on head, ♀ greenish. (Florida.) . . . . . *Euetheia*
  Length 7.50 *or more.* — Tail *longer than* wings. Plain brown, etc., or black, white, and chestnut. (U. S.)
Pipilo
    — Tail *shorter than* wings. ♂ breast rose or orange; ♀ sulphur or saffron under wings.
    (U. S.) . . . . *Zamelodia*
  Length *over* 4.50, *under* 7.50.
    Colors *greenish* — with yellow — on edge of wing, and — 2 rufous crown-stripes. (Texas.) *Arremonops*
        — Crown chestnut, breast ashy. (Western.)
Oreospiza
        — on all under parts — no head markings. (♀ of a southern species of)
Cyanospiza
    Colors *not* greenish, and *not* extensively and decidedly spotted or streaked.
      Black, with great *white* wing-patch; longest secondary about = longest primary. (Western.)
Calamospiza
      Blue with *chestnut* on wings, ♂; plain brown, ♀; over 6.00 long. (U. S.) . . . . *Guiraca*
      Blue, with red, purple, gold, white, or not, ♂; brown, with white or not, ♀; under 6.00
      long. (N. Am.) . . . . *Cyanospiza*
      Slate or ashy, red-backed or not, belly and 1-3 tail-feathers *white.* (N. Am.) . . . *Junco*
      Gray, throat and tail *black,* head with 2 white stripes, belly white. (Western.) *Amphispiza*
    Colors *not* greenish, but somewhere or everywhere spotted or streaked.
      Inner secondaries lengthened, about equalling primaries in the closed wing.
        A large *white* wing-patch. (♀ of) . . . . *Calamospiza*
        Upper parts much streaked.
        Bend of wing *chestnut;* outer tail-feather *white; no* yellow anywhere. (N. Am.) *Pocecetes*
        No white or chestnut area on wing, its edge (usually) *yellowish.* . . . *Passerculus*
      Inner secondaries *not* enlarged; wing *decidedly* longer than tail.
        Edge of wing and loral spot *yellow;* breast buff; wing 2.50 or under.
Ammodramus (*Coturniculus*)
        With *yellow* on breast, edge of wing, over eye; *black* throat-patch or stripes. (Eastern.) *Spiza*
        No yellow; head striped with black, white, and *chestnut;* tail *black,* white-tipped. (Western.)
Chondestes
        No yellow; wings *white-barred;* throat black, ♂. (*Imported.*) . . . . *Passer*
      Inner secondaries *not* enlarged; wing not, or not *decidedly,* longer than tail.
        Tail-feathers — very *acute;* bill — very slender. (Eastern, chiefly maritime.) { *Ammodramus*
                — very *stout* (Eastern.) or slender (Interior.) { (*Coturniculus*)
           — *not* acute; tail — *forked.* Length 6.00 or less; *no* yellow on wing. (N. Am.)
Spizella
           — rounded — *black;* edge of wing yellowish. (Western.)
Amphispiza
           — not black. — Streaked below, or crown chestnut.
           (N. Am.) . . . . *Melospiza*
           — not streaked below. (S. and W. U.S.)
           edge of wing — yellow . *Peucæa*
           — not yellow
Hæmophila
           or (N. Am.) large 6.50-7.50
Zonotrichia

*** The commonest "Sparrows" of Eastern U. S., which the student will be most likely to find first, belong to the genera *Passer, Spizella, Melospiza, Zonotrichia, Passerella, Passerculus, Pocecetes, Coturniculus* (these anywhere), *Ammodramus* (marshes only); common but more distinguished Fringillines are *Carpodacus, Astragalinus, Cyanospiza, Spiza,*

Pipilo, and *Cardinalis*. Winter visitors, in flocks, are *Loxia, Pinicola, Passerina, Calcarius, Acanthis, Spinus*, and *Junco*. Genera confined to the West or Southwest are *Hesperiphona, Pyrrhula, Pyrrhuloxia, Leucosticte, Rhynchophanes, Sporophila, Arremonops, Oreospiza, Calamospiza, Hœmophila*, and *Amphispiza*. The genera *Pyrrhula, Sporophila*, and *Euetheia* are properly extralimital. Imported genera are *Passer* and *Carduelis*.

HESPERIPHO'NA. (Gr. ἑσπερίς, *hesperis*, a peculiar feminine of ἑσπέριος, *hesperios*, western, as noun in the plural Ἑσπερίδες, *Hesperides*, the Daughters of Night, who dwelt on the western verge of the world; ἕσπερος or ἑσπέριος, *hesperos* or *hesperios*, as adjective, of time, at evening, at sunset; of place, western, occidental, where the sun goes down, in feminine form ἑσπέρα, *hespera*, Lat. *Vespera*, used as noun, for eve, evening, and for the west, Lat. *occidens;* also Ἑσπερία, *Hesperia*, the west. The forms of the classic word are much confused, and usable both as nouns and adjectives. The second element of the genus name is Gr. φωνή, *phone*, voice, sound.) AMERICAN HAWFINCHES. Bill enormously large, vaulted, nearly as wide as high at base; culmen nearly straight to the decurved end; commissure curved without obvious angulation; gonys very long, and mandibular rami short, not reaching back of base of upper mandible; mandibles of equal thickness, lower not so deep as upper;

FIG. 236. — Evening Grosbeak, reduced. (Sheppard del. Nichols sc.)

lateral outlines of bill converging straight to tip. Nasal fossæ extremely short and broad; nostrils slightly overhung by antrorse plumulæ. Wings long, pointed, folding beyond middle of tail, pointed by first two primaries, the rest rapidly graduated; no peculiar shape of inner primaries or outer secondaries. Tail rather short, emarginate, with long coverts, the under reaching nearly to the forking. Feet small and weak; tarsus shorter than middle toe without claw; lateral toes of about equal lengths, their claws reaching hardly base of middle claw. Coloration black, white, and yellow. Sexes dissimilar. Little different from Old World *Coccothraustes*, excepting in coloration and the simplicity of wing-quills; yet I hardly think it advisable to follow the A. O. U. in reducing this well-marked American group to a subgenus of *Coccothraustes*. We have one strongly marked species, with one subspecies; another good species, *H. abeillæi*, occurs in Mexico and Central America.

H. vesperti'na. (Lat. *vespertina*, of Hesperus. Fig. 236.) EVENING GROSBEAK. Adult ♂: General color sordid yellow, overlaid with a sooty-olive shade, deepest on fore parts, quite black on crown, clearest below behind. Forehead and line over eye, scapulars, and rump, yellow. Wings and tail black; several inner secondaries and inner half of greater coverts white; lining of wings black and yellow. A narrow black line around base of upper mandible; tibiæ black. Bill greenish-yellow; iris brown; feet light brown; claws dark brown. Length 7.50–8.50; wing 4.00–4.50; tail 2.50–3.00; bill 0.75 long, 0.67 deep, 0.60 broad. ♀: Brownish-ash, paler below, whitening on belly; mixed with little, if any, yellowish; white of wings imperfect, or tinged with yellow; primaries, which are quite black in ♂, with large white spaces on inner webs, and sometimes tipped with white; the distinctive head markings of the ♂ wanting. Adults of both sexes differ much in the shade of the yellow and degree of obscuration of the white on the wings; there is also much difference in the extent of yellow and black on the head. Young of both sexes resemble the adult ♀, but the general coloration is duller and more brownish, the under parts are paler and more buffy, and all the special markings less sharply defined; bill plain brownish, or horn-color. In full plumage this is a bird of distinguished appearance, whose very name suggests the far-away land of the dipping sun, and the tuneful romance which the wild bird throws around the fading light of day; clothed in striking color-contrasts of black, white, and gold, he seems to represent the allegory of diurnal transmutation; for his sable pinions close around the brightness of his vesture, as night encompasses the golden hues of sunset, while the clear white space enfolded in these tints foretells the dawn of the morrow. Western U. S., northerly, and adjoining British Provinces, N. to the region of the Saskatchewan, E. regularly to Lake Superior, irregularly to Kansas, Kentucky, Ohio, Ontario, Quebec, New York, Pa., and New England; common in its ordinary range, but somewhat irregularly distributed, especially during its migrations, when it may appear unexpectedly in large roving flocks. In some places it is known as "sugar-bird," from its fondness for the maple (*Acer saccharinum*). A history of the Evening Grosbeak may be read in Bull. Nutt. Club, iv, Apr. 1879, pp. 65-75; and several later articles in the Auk trace its movements and habits in various localities. The nest and eggs remained unknown till those of its western variety were discovered in Yolo Co., Cala., May 10, 1886, as recorded in Bull. Cala. Acad. ii, No. 8, 1887, p. 450. Another nest with eggs, taken June 18, 1896, in El Dorado Co., Cala., is figured in colors in the Nidologist for Sept. 1896. This nest was in a pine tree at a height of 35 feet, in the fork of a limb, substantially built with a foundation of twigs upon which was a neat superstructure of fine rootlets; it contained 4 eggs, averaging 0.92 × 0.64, light bluish-green, spotted and blotched irregularly with dark brown and black.

H. v. monta′na. (Lat., of mountains.) WESTERN EVENING GROSBEAK. As noted in the Key, 2d ed., 1884, p. 343, specimens from the Southern Rocky Mts. were said to have the bill less turgid and the yellow frontlet narrow. This was the alleged character of the present subspecies, to which is now ascribed a range in the U. S., from the Rocky Mts. to the Pacific, and S. in Mexico to Orizaba. The alleged distinction does not hold good; but the ♀ averages browner than that of *vespertina* proper, and is more mixed with yellowish, instead of being plain brownish-ash or gray on most parts. *Hesperiphona vespertina* var. *montana* BD., BREW., and RIDGW., Hist. N. A. B. i, 1874, p. 449; *Coccothraustes vespertina montana* MEARNS, Auk, July, 1890, p. 246; *Coccothraustes (Hesperiphona) vespertinus montanus*, A. O. U. List, 2d ed. 1895, No. 514 *a*. (Not in A. O. U. List, 1st ed. 1886; nor in RIDGW., Man. 1887.)

PINI′COLA. (Lat. *pinus*, a pine; *colere*, to inhabit, cultivate.) PINE BULLFINCHES. Bill short, stout, about as high as broad, sides convex in all directions, culmen convex throughout, tip hooked; commissure gently curved throughout, without decided angulation; gonys relatively long, rami of under mandible short, former nearly straight, latter coming together in a very broad gentle curve; commissural edge inflected. Nostrils small, round, basal, concealed by the ruff of antrorse plumules; nasal fossæ short and broad. Wings of moderate length, tipped by 2d-4th quills, 1st and 5th a little shorter; 2d-5th with outer webs incised; no peculiarity of inner quills. Tail little shorter than wings, emarginate, its short coverts scarcely or not reaching half-way to end. Feet small; tarsus not longer than middle toe without claw, 7-scutellate in front, laminiplantar behind, but the outer of these plates commonly subdivided into 3 or 4 below! Lateral toes short, their claws scarcely surpassing base of middle, outer rather longer than inner; hind toe less in length than inner lateral; its claw shorter, though stouter and more curved than the middle. Sexes unlike; ♂ red, ♀ gray. One species.

P. enuclea′tor canadensis. (Lat. *enucleator*, one who shells out. Fig. 237.) CANADIAN PINE GROSBEAK. PINE BULLFINCH. Adult ♂: Light carmine or rosy-red, feathers of back with dusky centres; lower belly and under tail-coverts gray, and, in general, the red continuous only in highly plumaged specimens. Nasal tufts and lores blackish. Wings blackish; primaries with narrow white or rosy edging, inner secondaries more broadly edged with white, ends of greater and middle coverts white or rosy, forming conspicuous wing-bars. Tail like wings, with narrow edgings like those of primaries. Bill blackish, with or without paler base below; feet blackish. Length about 8.50; wing 4.50 or more; tail 4.00; bill 0.55; tarsus 0.90. The plumage of the ♂ is extremely variable in tint, and some apparently adult individuals of this sex are indistinguishable in colors from the ♀. Adult ♀: Ashy-gray, paler below; feathers of back with darker centres, those of head, rump, and fore parts generally skirted with a saffron or yellowish color, very variable in extent and tint, from dull gamboge-yellow to olive-orange, or rusty-orange, or even reddish; in some specimens crown and rump quite bricky-red. Throat sometimes abruptly paler than surrounding parts. Rather smaller than ♂. Young ♂ resembles ♀. Northern portions of N. Am. to about the limit of trees; in summer, most of British America and N. border of U. S., E. of the Rocky Mts.; in winter, range extended irregularly sometimes to Maryland, Ohio, Illinois, and Kansas. Inhabits chiefly coniferous woods, in flocks when not breeding, feeding upon the fruit of such trees. A fine musician, of amiable disposition and gentle manners, often caged, and in confinement often failing to develop or retain the red color. Nest usually in conifers, composed of a basement of twigs and rootlets, within which is a more compact fabric of finer materials; eggs usually 4, 1.05 × 0.74, greenish-blue, spotted and blotched with dark brown and blackish surface-markings and lilac shell-spots. *P. enucleator* of former eds. of Key, as of A. O. U. Lists, 1886-95. *P. canadensis* CAB., 1851. *P. e. canadensis*, A. O. U. Suppl. List, Auk, Jan. 1899, p. 113, No. 515. I agree with A. O. U. that this is a good subspecies of *P. enucleator*. It is decidedly larger than the European bird: wing sometimes up to 5.00, rarely under 4.50, or tail under 4.00, while these parts in *enucleator* average only 4.25 and 3.65; and the ♀ of *canadensis* fails to show the peculiar olivaceous tint characteristic of ♀ *enucleator*. I have not such a good opinion of the following subspecies, which I spread on my page in deference to the A. O. U.

FIG. 237.—Pine Grosbeak, reduced. (Sheppard del. Nichols sc.)

P. e. monta′na. (Lat., of mountains.) ROCKY MOUNTAIN PINE GROSBEAK. "Similar to *P. e. californica* but decidedly (?) larger and slightly darker, the adult ♂ with the red of a darker, more carmine, hue." Wing 4.71; tail 3.72; culmen 0.63; depth of bill at base 0.47; width 0.39; tarsus 0.92; middle toe without claw 0.66. Rocky Mts., breeding from Montana and Idaho to New Mexico. Included under the stock form in former eds. of the Key. *P. e. montana* RIDGW., Auk, Oct. 1898, p. 319; A. O. U. Suppl. List, Auk, Jan. 1899, p. 113, No. 515 *a*.

P. e. califor′nica. (Lat., Californian.) CALIFORNIAN PINE GROSBEAK. Said to differ from *P. e. canadensis* in "larger, more hooked, and less turgid bill," together with deficiency of dark centres of dorsal and scapular feathers, and to inhabit the Sierras Nevadas of California from 7,000 feet up to timber line. Included under the stock form in former eds. of the Key. *P. e. californica* PRICE, Auk, Apr. 1897, p. 182; A. O. U. Suppl. List, Auk, Jan. 1899, p. 113, No. 515 *b*.

P. e. alascen′sis. (Lat., of Alaska.) ALASKAN PINE GROSBEAK. "Similar to *P. e. canadensis* but decidedly (?) larger, with smaller or shorter bill and paler coloration; both sexes with the gray parts distinctly (?) lighter, more ashy." ♂: Wing 4.61; tail 3.65; culmen 0.57; depth of bill at base 0.48; width 0.40; tarsus 0.90; middle toe without claw 0.60. ♀: Wing 4.57; tail 3.68, etc. Northwestern N. Am., "including wooded portions of Alaska, except Kadiak and the southern coast region;" S. in winter to eastern British Columbia, Montana, etc. Included distinctly in the stock form in former eds. of the Key; I say "Alaska" expressly, though the concordance as published fails to give "C 190 *part*." *P. e. alascensis* RIDGW., Auk, Oct. 1898, p. 319; A. O. U. Suppl. List, Auk, Jan. 1899, p. 114, No. 515 *c*.

P. e. flam′mula. (Lat. *flammula*, a little flame, or other small red thing; dimin. of *flamma*, a flame, blaze, fire.) KADIAK PINE GROSBEAK. "Smaller than *P. e. canadensis*, with proportionally much larger bill and shorter tail." Length given as 8.00–8.50; wing 4.25–4.60; tail 3.60–3.80; culmen 0.60; tarsus 0.90. Kadiak to Sitka, Alaska. *P. flammula* HOMEYER, J. f. O. 1880, p. 156. *P. e. flammula*, A. O. U. Suppl. List, Auk, Jan. 1899, p. 114, No. 515 *d*. *P. e. kodiaca* RIDGW., Man. 1887, p. 388. Included with all the others in former eds. of the Key.

PYR′RHULA. (Lat. *pyrrhula*, a bullfinch.) BULLFINCHES. Generic characters of *Pinicola* as above given; the different shape of the bill and different style of coloration being the principal distinction. Bill as wide at base as long, its under outline twice concave. Colors in masses of black, white or gray, and red.

P. cas′sini. (To John Cassin. Fig. 238.) CASSIN'S BULLFINCH. Adult ♂: Above, clear ashy-gray; below, paler ashy-gray; rump and under wing- and tail-coverts white; wings, tail, crown, chin, and face, black; greater wing-coverts broadly tipped with whitish; bill black; feet dusky. Length 6.50; wing 3.50; tail 3.25. In less perfect plumage, some of the tail-feathers are patched with white, and there may be some white edging of the primaries. The lesser and median wing-coverts are like the back, contrasting with the greater coverts. The ♀ closely resembles the ♂, but has the under parts tinged with cinnamon. Nulato, Alaska, a stray from E. Siberia; the type specimen marked ♂, but having all the characters of a ♀; nearly related to *P. coccinea* of Asia (especially its subspecies *Kamtschatica*), and originally described as a variety of that species; identical with *P. cineracea* CAB., J. f. O. 1872, p. 316, and with *P. cineracea pallida* SEEBOHM, Ibis, 1887, p. 101.

FIG. 238. — Cassin's Bullfinch, reduced. (From Baird.)

PAS′SER. (Lat. *passer*, a sparrow; this very species.) SPARROWS. Form stout and stocky. Bill very stout, shaped somewhat as in *Carpodacus*, but without nasal ruff. Culmen curved; commissure little angulated; gonys convex, ascending; lateral outlines of bill bulging to near end. Wing pointed; 1st, 2d, and 3d primaries nearly equal and longest; 4th little shorter, rest graduated; inner secondaries not elongate. Tail shorter than wings, nearly even. Feet small; tarsus about equal to middle toe and claw; lateral toes of equal lengths, their claws not reaching to base of middle claw. Sexes unlike. ♂ with black and chestnut on head. Middle of back only streaked. Old World: two species naturalized in North America, out of the 26 which compose this genus.

P. domes′ticus. (Lat. *domesticus*, domestic. Figs. 239, 240.) THE SPARROW. PHILIP SPARROW. HOUSE SPARROW. PARASITE. TRAMP. HOODLUM. GAMIN. Adult ♂: Upper parts ashy-gray; middle of back and scapulars boldly streaked with black and bay. A

FIG. 239. — English Sparrow. (L. A. Fuertes.)

dark chestnut or mahogany space behind eye, spreading on side of neck. Lesser wing-coverts deep chestnut; median tipped with white, forming a conspicuous wing-bar, bordering which is a black line. Greater coverts and inner quills with central black field bordered with bay. Tail dusky-gray, unmarked. Lower parts ashy, gray or whitish; chin and throat jet black, spreading on the breast and lores, bordered on side of neck with white. Bill blue-black; feet brown. Wing about 3.00; tail 2.50. Adult ♀: Above, brownish-gray; streaking of back light ochrey-brown and black; wing-edgings light ochrey-brown, the white bar impure. No black, mahogany, or white on head; a pale brown postocular stripe; bill blackish-brown, yellowish at base below. Varies endlessly in purity or dinginess of coloration. Young ♂ at first like ♀. Europe, etc. Repeatedly imported since 1858, and especially in the sixties, during a craze which even affected some ornithologists, making people fancy that a granivorous conirostral species would rid us of insect-pests, this sturdy and invincible little bird has overrun the whole country, and proved a nuisance without a redeeming quality. The original offender in the case is said to have been one Deblois, of Portland, Me., in 1858; but the pernicious activity of Dr. T. M. Brewer affected the city fathers of Boston in 1868–69, and even the Smithsonian Institution at Washington, about the same years. New York had the sparrow-fever in 1860–64, and Philadelphia was not as slow as usual in catching the contagion, in 1869. There is no need to follow the sad record further. Well-informed persons denounced the bird without avail during the years when it might have been abated, but protest has long been futile, for the sparrows have had it all their own way, and can afford to laugh at legislatures, like rats, mice, cockroaches and other parasites of the human race which we must endure. This species, of all birds, naturally attaches itself most closely to man, and easily modifies its habits to suit such artificial surroundings; this ready yielding to conditions of environment, and profiting by them, makes it one of the creatures best fitted to survive in the struggle for existence under whatever conditions man may afford or enforce; hence it wins in every competition with native birds, and in this country has as yet developed no counteractive influences to restore a disturbed balance of forces, nor any check whatever upon its limitless increase. Its habits need not be noted, as they are already better known to every one than those of any native bird whatever, but few realize how many million dollars the bird has already cost us. Nest anywhere about buildings, also in trees, bushes, and vines, built of any rubbish, usually lined with feathers, and making a bulky, unsightly object amidst dirty surroundings; eggs indefinitely numerous, usually 5 to 7, about 0.90 × 0.60, dull whitish thickly marked with dark brown and neutral tints; several broods a year are raised, as the birds breed in and out of season.

FIG. 240. — *P. domesticus* (lowest); *P. montanus;* reduced. (From Brehm.)

P. monta′nus. (Lat. *montanus*, of mountains. Fig. 240.) MOUNTAIN SPARROW. EUROPEAN TREE SPARROW. Somewhat like the last, but smaller and otherwise different. ♂: Crown and nape a peculiar purplish-brown. Lores, chin, and throat black, the throat-patch narrow and short, not spreading on breast, contrasted with ashy-white on side of head and neck; ear-coverts blackish. Back and scapulars streaked with black and bay, the streaking reaching to the purplish nape; rump and tail plain grayish-brown. Wings marked much as in *P. domesticus*, with a black and white bar across tips of median coverts, but also a narrow white bar across tips of greater coverts. Primaries more varied with ochrey-brown on outer webs, forming a basal spot and other edging. Below, ashy-gray, shaded on sides, flanks, and crissum with grayish-brown. Bill blue-black; feet brown. Wing 2.75; tail 2.25. ♀ differs much as before. Europe; naturalized about St. Louis and elsewhere. Nesting and general habits like those of *P. domesticus*; eggs similar, smaller, 0.77 × 0.55.

CARPO'DACUS. (Gr. καρπός, *karpos*, fruit; δάκος, *dakos*, biting.) Purple Bullfinches. Bill smaller and less turgid than in *Pinicola* or *Pyrrhula*, more regularly conic and more acute; sides convex in all directions, but with distinct ridge prolonged in a point on forehead where not concealed by the antiæ, its outline moderately curved; commissure decidedly angulated, about straight before and behind the bend; gonys quite straight. Nasal ruff little developed, barely concealing the slight nasal fossæ, thence falling over sides of bill, but discontinuous across culmen. Wings long and pointed, folding half-way to end of tail or further, pointed by first 3 or 4 quills. Tail much shorter than wings, emarginate to even, with rather narrow feathers; both sets of coverts reaching more than half-way to end. Feet small and weak; tarsus shorter than middle toe; lateral toes subequal, outer rather longer than inner, their claws reaching base of middle claw. Sexes unlike. ♂ extensively red of some shade, ♀ streaky brown and white. Head with erectile feathers, but not fairly crested. A beautiful genus, of 25 or more species of New and Old World.

Fig. 241. — Bill of Purple Finch, nat. size.

Analysis of Species and Subspecies (♂).

Bill conic-acute, with scarcely convex culmen; edgings of wing- and tail-feathers reddish. (Carpodacus proper.)
    Large: length 6.50–7.00; bill at least 0.50 along culmen. Under tail-coverts streaked with dusky centres of the feathers. Crimson crown well distinguished from merely reddish-brown back. (Southwestern U. S.) . *cassini*
    Medium: length 5.75–6.25; bill not 0.50 along culmen. Under tail-coverts scarcely or not streaked. Crimson of crown not well distinguished from that of back. (U. S.) . . . . *purpureus*
    Like the last; coloration darker and more diffuse (Pacific Coast) . . . . *purpureus californicus*
Bill conoid-obtuse, with very convex culmen. Edgings of wing- and tail-feathers whitish. (Subgenus Burrica.)
    Small: length scarcely 6.00; bill about 0.40 along culmen. Front, line over eye, rump and throat red, more or less contrasting with brown or white of other parts.
        Red pretty definitely restricted to the areas said (Southwestern U. S.) . . . . *mexicanus frontalis*
        Red spreading over other parts (Lower California) . . . . *mexicanus ruberrimus*
    Large: length 6.00 or more; bill over 0.40; wing 3.25. (Insular forms.)
        Red of ♂ as usual. Guadalupe Island . . . . *amplus*
                   Santa Barbara Islands . . . . *clementis*
        Red of ♂ replaced by orange. San Benito Island . . . . *macgregori*

C. purpu'reus. (Lat. *purpureus*, purple. Figs. 241, 242.) Purple Finch (better Crimson Finch). Purple Linnet. Red Linnet. Adult ♂: Rose-red, paler below, insensibly whitening on belly and crissum, brightest anteriorly, intensified to crimson on crown, darker and more brownish-red on back, where also streaked with dark brown. Wings and tail dusky, quills edged and coverts tipped with brownish-red. Lores and feathers about base of bill hoary-whitish. Bill and feet brown; under mandible rather paler. Length 6.00–6.25; extent 10.00–10.60; wing 3.00–3.25; tail 2.25–2.50; tarsus 0.62; middle toe and claw 0.87; bill under 0.50. The shade of red is very variable, almost anything but purplish — according to season, and age and vigor of the individual. In high feather, the crown is richer crimson than any other part, but does not form a definite cap. The auriculars are dusky, and there is an appreciably light rosy stripe over them. Younger ♂ ♂ have frequently a bronzy shade. ♀ and young: Olivaceous-brown, more clearly olivaceous on rump, everywhere streaked with dusky. Below, white, marked everywhere except on throat, belly, and crissum with streaks and arrow-heads of dusky olive-brown; the latter pretty evenly distributed on breast, former the same on sides, on sides of neck and throat confluent and gathered into a maxillary series running up to bill, separated by a poorly-defined whitish area from olive-brown auriculars, over which is a whitish postocular streak. Wings and tail as in ♂, but the edgings plain brown. Length 5.70–5.90; extent 9.50–10.00; wing about 3.00. Young ♂ cannot be certainly distinguished from ♀; in general, duller and grayer brown, with less olive shade; the red first shows pale or bronzy in slight touches. Cage-birds sometimes turn yellowish after moulting, as is the case with various other red Finches. U. S. from Atlantic to the Great Plains; N. to Labrador, Hudson's Bay, and the Saskatchewan. Breeds from the Middle States, Minnesota and N. Dakota northward; winters in most of the U. S., particularly the Middle and S. States. An engaging bird, of bright colors, sweet song, and many amiable traits, among them its fondness for the society of man; it comes fearlessly about our houses to build its own, which is generally situated on a horizontal bough or fork, composed of the most miscellaneous materials, almost any vegetable fibre being available for the flat and shallow structure; it is usually lined with hair, and the eggs, to the number of 4 or 5, are pale dull greenish, or almost whitish, sparsely sprinkled and scratched with blackish surface-markings and lilac shell-spots; size about 0.85×0.65; two broods are often reared. When not breeding the birds are generally found in flocks, and it is to be feared they damage in spring the blossoms of fruit-trees.

FIG. 242. — Purple Finch, ♂, reduced. (Sheppard del. Nichols, sc.)

C. p. califor'nicus. (Lat., Californian.) CALIFORNIAN PURPLE FINCH. Like the last; first quill said to be usually shorter than the 4th (not longer as usual in *purpureus*); ♂ with sides and flanks suffused with brownish, the streaks there broad and not sharp; streaks of back indistinct; red of crown and rump dark and dull. ♀ differs correspondingly from that of *purpureus*. Pacific Coast region from British Columbia to southern California. Not in any previous ed. of the Key, in consequence of a consultation held many years ago by Prof. Baird, Mr. Cassin, and myself, in which it was decided against unanimously; and I only admit it now *pro forma*, in my desire to bring about as far as possible nominal conformity of the Key with the A. O. U. Lists. *C. californicus* BD. B. N. A. 1858, p. 413; *C. purpureus californicus* of most authors since 1874; A. O. U. Lists, 1886 and 1895, No. 517 *a*.

C. cas'sini. (To John Cassin.) CASSIN'S PURPLE FINCH. Adult ♂: In highest plumage duller than *C. purpureus*, excepting on crown. Middle of back brown, tinged with red, the feathers dusky-centred, gray-edged; crown crimson, the cap not so extensive as in *purpureus*, and quite well defined, separated by a dusky and gray interval from color of back. Under tail-coverts with dusky shaft-lines, usually wanting in *purpureus*. Larger: length 6.50–7.00; extent 11.00–11.50; wing 3.50; tail 2.50; bill at least 0.50 along culmen, usually more, relatively less turgid than in *purpureus*. Iris brown; feet blackish-brown; bill above dark bluish horn-color, below dusky flesh-tinted. The sexual changes are the same as in the last species; it is not easy to distinguish ♀ and young ♂ from those of *purpureus*, but they are larger, with longer and less tumid bill, and more streaked crissum. Very young birds have an ochraceous or light rufous suffusion, especially noticeable on the under parts; the streaks are more numerous and diffuse. Rocky Mts. of U. S. and westward, especially the Southern Rocky Mt. region, as Utah, Nevada, Arizona, and New Mexico; N. to British Columbia; E. to Wind River

Mountains; S. to table lands of Mexico. Habits the same as those of the common Purple Finch; eggs not fairly distinguishable.

C. mexica′nus fronta′lis. (Lat. *mexicanus*, Mexican; *frontalis*, pertaining to the front.) CRIMSON-FRONTED FINCH. HOUSE FINCH. BURION. Adult ♂ : Grayish-brown above, somewhat varied with darker centres and paler edges of the feathers, and for the most part tinged with red. Below dull white, streaked with dark brown, often tinged with red. Fore part of crown, superciliary line, rump, throat, breast, and sometimes side of head, crimson. Wings and tail dark brown, with narrow pale edgings. Bill dusky-brown above, paler below; feet and eyes brown. Length about 6.00; extent scarcely 10.00; wing 3.00; tail 2.50, scarcely forked; tarsus 0.67; bill 0.40, very turgid, almost as in *Pinicola* or *Pyrrhula*. ♀ : Like ♂, but without any red; upper parts more varied with darker centres and paler edges of the feathers, and entire under parts streaked like belly of ♂. Young ♂ resembles ♀, but at an early age is browner, and apt to have buffy edgings of the wings. Colors of adult ♂ as variable as those of *purpureus* or more so. In winter, the red less intense and more diffuse, and may have a rosy or purplish tint, or be interrupted with grayish edgings of the feathers. Generally in the Colorado Valley, where the typical form is developed, the red is restricted to the parts said, but the constant tendency is to spread; the back and belly have usually in fact a tinge of red, and in some cases the whole head and fore parts are thus encrimsoned. U. S., rather southerly, from the Rocky Mts. to Oregon and California; western Texas, Colorado, Utah, Nevada, Arizona, western Kansas, New Mexico, and northern Mexico; familiar as a Swallow or Chip-bird, nesting in streets and gardens, where its bright colors, hearty song, and sprightly ways make it a welcome visitor. Nesting like that of the Purple Finch in essential particulars; eggs smaller, paler, and of more fugitive bluish tint, with the blackish sprinkling sparser; size 0.68 × 0.60 to 0.75 × 0.54. *C. frontalis* of earlier eds. of the Key, lately ascertained to intergrade with *C. mexicanus* (*Fringilla mexicana* P. L. S. MÜLLER, Syst. Nat. Suppl. 1766, p. 165), and therefore reducible to a subspecies of the latter. See RIDGW. Man. 1887, p. 391; COUES, Key, 4th ed. 1890, p. 899; A. O. U. List, 2d ed. 1895, p. 213, No. 519.

C. m. ruber′rimus. (Lat., superlative degree of *ruber*, red.) RED-BREASTED FINCH. ST. LUCAS HOUSE FINCH. This alleged variety resembles the last; crimson tints more diffuse. Lower California and probably Sonora. This is *C. frontalis rhodocolpus* of the 2d, 3d, and 4th eds. of the Key, p. 348, and A. O. U. List, 1st ed. 1886, p. 257, after *C. rhodocolpus* CAB. Mus. Hein. i, 1851, p. 166; and if the variation be worth any name, I fail to see why this is not available, as it certainly covers the present case, though Dr. Cabanis may not have indicated satisfactorily the geographical distribution. The bird in question has received three different names from Mr. Ridgway; being his *C. frontalis* var. *rhodocolpus* of Am. Journ. Sci. v, Jan. 1873, p. 39; his *C. frontalis ruberrimus*, Man. 1887, p. 391; and his *C. mexicanus ruberrimus*, ibid. p. 594; which latter is the choice of the A. O. U. List, 2d ed. 1895, p. 214, No. 519 *b*.

C. m. clemen′tis. (Lat. *clemens*, gen. *clementis*, adj., clement, mild; proper name of Sanctus Clemens, St. Clement, a person, applied in Spanish form San Clemente to an island.) SAN CLEMENTE HOUSE FINCH. "Intermediate between the form of *frontalis* inhabiting the neighboring mainland of California and *C. Mcgregori*." San Clemente and Santa Barbara Islands. MEARNS, Auk, July, 1898, p. 258; *C. mexicanus clementis*, A. O. U. Suppl. List, Auk, Jan. 1899, p. 114, No. 519 *c*.

C. am′plus. (Lat. *amplus*, ample, large.) GUADALUPE HOUSE FINCH. An insular form, resembling *C. m. frontalis* proper, but with darker tints, and considerably larger; ♂, wing 3.10–3.35; tail 2.60–2.90; bill 0.40–0.45 from nostril, and same in depth; tarsus 0.75-0.85: ♀ somewhat smaller. Guadalupe Island, Lower California. Bull. U. S. Geol. Surv. ii, Apr. 1876, p. 187; Key, 3d ed. 1887, p. 872; A. O. U. Lists, No. 520.

C. mcgreg'ori. (To R. C. McGregor, of Palo Alto, Cal.) MCGREGOR'S HOUSE FINCH. Nearest *C. amplus;* slightly smaller, with more compressed, somewhat grooved bill, and longer tail; red of ♂ replaced by orange. San Benito Island, Lower California. ANTHONY, Auk, Apr. 1897, p. 165, fig. *b;* A. O. U. Suppl. List, Auk, Jan. 1899, p. 114, No. 520. 1.

LOX'IA. (Gr. λοξός, *loxos,* crooked.) CROSSBILLS. *Bill metagnathous;* both mandibles falcate, deflected to opposite sides, their points crossed (unique among birds). Upper mandible stout and broad at base, rapidly narrowing to the elongate, decurved, laterally deflected and overhanging tip, its sides nearly flat, culminal ridge well marked and very convex throughout; its base beset with a ruff of antrorse plumules concealing nostrils and nasal fossæ. Lower mandible with gonys very long, occupying nearly all the exposed part of bill, convex throughout, end of mandible prolonged, curved upward and deflected to one side. Commissural line of either mandible curved in the opposite direction from its fellow. Mouth very narrow anteriorly, ample at base; tongue horny and concave at end; œsophagus with a large special crop, bulging to the right side. Wings long, pointed by tips of first three primaries, rest rapidly graduated. Tail very short, only about ⅔ as long as wing, emarginate and divaricate, covered nearly to the fork by coverts both above and below. Feet small; tarsus shorter than middle toe without claw; covered with 3 or 4 large overlapping plates, and smaller ones above and below; postero-lateral plates much broken up below. Lateral toes of subequal lengths, tips of their claws falling opposite base of middle claw. Hind claw about equal to its digit, longer, stouter, and more curved than middle one. Form stout, thick-set; neck short; head broad and flattened on top. Plumage soft and blended. Sexes dissimilar in color. ♂ red, ♀ brown with olive or yellowish tinge. There are several species of these singular Finches, in which not only the horny envelop of the beak, but the bony framework, and to some extent the ligaments and muscles acting upon it, are unsymmetrical. The conformation is only completed at maturity, for in nestlings the points of the bill are not crossed. The structures concerned in what would appear at first sight to be a deformity constitute a handy tool for cracking nuts of some kinds and shelling out their kernels; it acts like a pair of cutting pliers, — pincers and scissors in one, — and the tongue comes into play at the same time as a scoop to secure the seed or pip thus exposed in a pine-cone or fleshy fruit. Our two species inhabit the northern parts of America, coming southward in flocks in the fall; but they are also resident in northern and mountainous parts of the U. S., where they sometimes breed in winter. They are irregularly migratory according to exigencies of weather and food-supply; are eminently gregarious, and feed principally upon pine seeds, which they skilfully husk out of the cones with their curious bills.

FIG. 243. — White-winged Crossbill, reduced. (After Audubon.)

Analysis of Species and Subspecies.

Wings with two white bars. ♂ rosy-red; ♀ brownish-olive, streaked and spotted with dusky, the rump saffron-yellow . . . . *leucoptera*
Wings without bars. ♂ bricky-red. ♀ as before, without wing-bars.
   Bill small, about ⅔ of an inch long . . . . *curvirostra minor*
   Bill large, ¾–⅞ of an inch long . . . . *stricklandi*

L. leucop'tera. (Gr. λευκός, *leukos*, white; πτερόν, *pteron*, wing. Figs. 243, 244.) WHITE-WINGED CROSSBILL. Adult ♂: Rosy-red, sometimes carmined or even crimsoned, obscured on middle of back, paling on lower belly and crissum, latter whitish with dusky centres of the feathers. Scapulars black, this color sometimes meeting across lower back. Wing- and tail-feathers black, with slight white or rosy edgings; inner secondaries and greater and middle coverts tipped with white, forming two cross-bars, sometimes confluent in one large patch. Rather larger than the next, the bill thinner and more attenuate. ♀ and young: Though the differences are parallel with those of *L. minor*, some peculiarity in tone of color usually serves to distinguish the two species, independently of the white wing-marks,

FIG. 244. — White-winged Crossbill. (L. A. Fuertes.)

which exist in both sexes at all ages. The difference is something like that between the ♀ ♀ of *Piranga rubra* and *P. erythromelas*, in the presence of ochrey or buffy tints, instead of clear olivaceous or yellowish. Upper parts fuscous, closely lined with an ochrey-olive or dingy ochre, the rump bright yellow-ochre. Below, the gray overlaid with ochreous, and further varied with dark gray centres of the feathers, tending to streaks on the flanks. The whole tone of coloration varies interminably; the under parts and rump are sometimes bright tawny yellow, or brownish-orange. Some ♂ ♂ are brilliant carmine, some ♀ ♀ pale orange, almost uniform. North Am., northerly; Alaska; Greenland; casual in Europe. In winter S. to about 38° in U. S., in flocks like the next, not so common. Resident in N. New England, and along whole N. tier of States, probably breeding also in alpine U. S. localities to Pennsylvania and Colorado. Breeds in winter and early spring; nesting like that of the next species; eggs pale blue, dotted chiefly at the larger end with black and lilac; 0.80 × 0.56.

L. curviros'tra mi'nor. (Lat. *curvirostris*, curve-billed; *minor*, lesser, smaller. Figs. 245, 246.) AMERICAN RED CROSSBILL. Adult ♂: Red; wings and tail blackish, without white markings. Middle of back darker, more brownish-red than elsewhere, the feathers with dusky centres. In the highest feather, even, the red is scarcely continuous except on head and rump, where brightest; lower belly and crissum usually gray or pale. Though the shade of red is never rosy or carmine as in the last, it varies interminably. It is usually tile-red or cinnabar, heightening in some cases to vermilion, in others shading to brownish-red, and often mixed not only with gray, but with olivaceous or

FIG. 245. — Common Crossbill, ♂ ♀, reduced. (Sheppard del. Nichols sc.)

saffron-yellowish tints. Orange, chrome, or gamboge ♂ ♂ are sometimes seen, and in captivity the European species of which ours is a variety is well known to lose the red tints; the same is doubtless true under some circumstances of all the members of this genus, in a state of nature. Length about 6.00; wing 3.50; tail 2.25; tarsus 0.65; bill (chord of culmen) 0.67 *or* less, very variable; depth at base 0.35; under mandible usually weaker than upper. ♀ and young: Dull greenish-olive, much mixed with gray or dusky, brighter and more yellowish on head and rump; below, gray, most feathers skirted with dingy yellowish, overcasting most of the plumage. Very young are dusky, streaked with grayish-white, usually no trace of olivaceous; below, gray, streaked with dusky; bill weak. From such state as this the ♂ usually passes through stages resembling the ♀, being found in every possible patchy state of mixed gray, olive and dusky-reddish; sometimes appears to pass directly into the red state, and the same is doubtless the case with other species. N. Am., alpine and northerly; S. in most of the U. S. in winter, sometimes even to South Carolina and Louisiana; resident in Maine, etc., mountains S. to Georgia, and in the Rocky and other mountains of the West; abundant irregularly, in unwary but timid flocks, usually including some individuals of the other species, fluttering and creeping about in the foliage of coniferous trees. Nesting often in winter or early spring when snow still covers the ground; nest in forks or among twigs of a tree, founded on a mass of twigs and bark-strips, the inside felted of finer materials, including small twigs, rootlets, grasses, hair, feathers, etc.; eggs 3–4, 0.75 × 0.57, pale greenish, spotted and dotted about larger end with dark purplish-brown, with lavender shell-markings. *L. c. americana* of former eds. of the Key; name changed because of the prior *Loxia americana* GM. 1788, which is a species of the genus *Sporophila*. Our bird is recognizable as a subspecies distinct from *L. curvirostra* of Europe, and named as *Crucirostra minor* by BREHM, Naum. 1853, p. 93; it has many synonyms, among which is *L. c. bendirei* RIDGW. Pr. Biol. Soc. Wash. ii, Apr. 1884, p. 101, and Man. 1887, p. 392, bestowed upon specimens intermediate between this form and the following:

FIG. 246. — American Red Crossbill. (L. A. Fuertes.)

L. c. strick'landi. (To H. E. Strickland. Fig. 247.) MEXICAN CROSSBILL. Like the last; larger; length about 7.00; wing nearly or quite 4.00; tail 2.50; tarsus 0.70; bill 0.75 or more long, depth at base 0.50; the under mandible especially more robust. Southern Rocky Mts.; westward to the Sierras Nevadas, and southward on the table lands of Mexico to Guatemala. *L. c. mexicana* of former eds. of the Key; *L. c. stricklandi*, Pr. U. S. Nat. Mus. viii, 1885, p. 354; A. O. U. Lists, 1886 and 1895, No. 521 *a*.

LEUCOSTIC'TE. (Gr. λευκός, *leukos*, white; στικτή, *sticte*, varied. Fig. 248.) ROSY FINCHES. Bill small, conic-acute, ruffed at base with antrorse plumules meeting over culmen and concealing short nasal fossæ and small nostrils. Side of under mandible (in typical species) with a sharp ridge running obliquely upward and forward. Culmen

FIG. 247. — Mexican Crossbill. (L. A. Fuertes.)

ridged between two slight depressions parallel with itself, gently convex throughout. No obvious angulation of commissural edge of upper mandible; that of lower with decided bend; gonys straight. Wings long, folding beyond middle of tail, tipped by first 3 primaries, 4th shorter. Tail of moderate length, forked, its feathers rather broad, its coverts reaching about

FRINGILLIDÆ: FINCHES, BUNTINGS, SPARROWS.

½ way to end. Tarsus not shorter than middle toe without claw; lateral toes unequal, inner shorter, its claw not reaching base of middle claw. Hind claw about as long as its digit, more curved and longer than middle claw. Sexes somewhat dissimilar. Coloration peculiar; usually chocolate-brown, enriched with rose or carmine, shaded with silvery-gray or black; one species mostly silvery-gray. The American representative of the Old World genus *Montifringilla*, of which some authorities make it a subgenus. Terrestrial, highly gregarious; nest on ground; eggs immaculate white. Numerous species of this very interesting genus are scarcely stable. I present the forms that are usually recognizable. The nearest American relative is *Acanthis;* the general economy is more that of *Passerina.*

Analysis of Subgenera, Species, and Subspecies.

Under mandible ridged. Nasal tufts white. Body-color chocolate-brown or darker. (*Leucosticte* proper.)
   No ash on head (Colorado and New Mexico) . . . .
                                                                           australis
   Ash on head confined to the top.
      Coloration blackish (Colorado and Utah to Idaho) . . . . *atrata*
      Coloration chocolate (W. America) . . . . *tephrocotis*
   Ash spreading on sides of head.
      Smaller: wing 4.20. (W. America) . . . . *tephrocotis litoralis*
      Larger: wing 4.60. (Alaska) . . . . *griseinucha*
Under mandible smooth. Nasal tufts blackish. (Subgenus *Hypolia*.)
   Dusky-purplish and silvery-gray, with rosy . . . . *arctoa*

L. atra'ta. (Lat. *atrata,* blackened.) RIDGWAY'S ROSY FINCH. BLACK LEUCOSTICTE. Sexes unlike. Adult ♂: Pattern of coloration and distribution of tints as in *tephrocotis* proper (see beyond); nasal tufts white, and occiput ashy, as in that species, but the chocolate-brown of *tephrocotis* replaced by black, deepest anteriorly and on under parts, sooty-brownish on back. Bill black (April) or yellow (September). Size of *tephrocotis.* Adult ♀: Black of ♂ represented by dark slate-gray, more brownish on back, the rosy markings duller and more restricted; size rather less. Rocky Mountain region of the U. S., breeding in Idaho and probably other northern regions, S. in winter to Colorado and Utah.

L. austra'lis. (Lat. *australis,* southern.) ALLEN'S ROSY FINCH. BROWN-CAPPED LEUCOSTICTE. Sexes unlike. ♂, breeding plumage: Rich chocolate or umber-brown; feathers of back with darker shaft-lines and paler edges, those of under parts darker and somewhat purplish-brown. Red parts of the body heightened to intense crimson, extending farther forward than in *tephrocotis,* sometimes skirting all feathers of under parts; especially strong on the wing- and tail-coverts and belly. No pure ash whatever on head; whole pileum black or blackish, purest anteriorly, duller behind. Bill and feet black. Length 6.75; wing 4.00–4.40, averaging in 69 specimens 4.30; tail 2.80–3.35, average 3.10; bill 0.45; tarsus 0.75. When not in highest feather, carmine toned down to more pink or rosy. In winter, bill yellow, changing to black through various cloudings. ♀, in summer: Generally like ♂, having black bill and no ash on head; averages a little smaller, and is much duller colored; brown parts of a grayish cast; rosy reduced or almost extinguished, chiefly traceable on rump and wing-coverts; abdomen scarcely tinted; quills and tail-feathers with whitish instead of rosy edgings. Wing 4.00–4.20, averaging little over 4.00; tail 2.90–3.25, average 3.00. Colorado, breeding up to 12,000 feet; S. in winter to New Mexico.

FIG. 248. — Rosy Finch, reduced. (Sheppard del. Nichols sc.)

L. tephroco'tis. (Gr. τεφρός, *tephros,* gray; οὖς, ὠτός, *ous, otos,* the ear. Fig. 249.) SWAINSON'S ROSY FINCH. GRAY-CROWNED LEUCOSTICTE. Sexes similar. Adult ♂, in breeding plumage or nearly so: Bill and feet black. Frontlet black; rest of pileum hoary-ash, *not* descending below level of eyes and upper border of auriculars (for when ash invades sides of head to any extent, the bird takes the first step toward *litoralis*, in which the head is extensively hooded in ash). General color, sides of head included, chocolate or liver-brown of varying intensity, many feathers skirted with gray or whitish, especially the interscapulars, which also have dusky centres, and inclining to blackish on chin and throat. Hinder parts of body above and below, including tail-coverts, rich rosy or carmine red, this color due to broad edgings of dusky feathers of these parts. Wings and tail blackish; wing-coverts and primaries edged with rosy, showing nearly continuous in the closed wing; edgings of inner secondaries rosy-white or white. Length (average) 6.75; wing 4.00–

FIG. 249. — Swainson's Rosy Finch. (L. A. Fuertes.)

4.45, average 4.25; tail 2.50–3.00, average 2.75; culmen 0.40–0.50, average 0.45; tarsus 0.75–0.85, average 0.80. Adult ♀: Very similar; pattern identical; tone subdued; size a little less; length 6.60; wing 4.10; tail 2.65. ♂ ♀ in winter: Bill yellow; pattern unchanged; coloration less vivid, the brown rather umber than chocolate, the red rather rosy than carmine. Rocky Mt. region, from the Saskatchewan or beyond, through most of W. U. S. in winter; breeding limits unknown, supposed to be Northern Rocky Mts. of U. S. and beyond, known to breed in the Sierras Nevadas of California. This is the central figure in the genus. It runs directly into

L. t. litora′lis. (Lat. *litoralis*, littoral. Fig. 250.) BAIRD'S ROSY FINCH. HEPBURN'S LEUCOSTICTE. Like the last; ash spreading over head more or less, sometimes almost enveloping it like a hood, and even occupying chin in extreme cases. Size of the last.

FIG. 250. — Baird's Rosy Finch. (L. A. Fuertes.)

Northwest coast; in summer, mountains of S. E. Alaska; in winter, Kadiak S. and E. to California, Nevada, Utah, and Colorado; very abundant, in flocks mixed with *tephrocotis* proper.

L. griseinu′cha. (Low Lat. *griseus*, gray, and *nucha*, nape. Figs. 251, 252.) BRANDT'S ROSY FINCH. ALEUTIAN LEUCOSTICTE. Like the littoral variety of *tephrocotis*, in having the ashy extending over sides of head; this color settled in a definite hood, said to never invade chin. The resident form of the N. W. islands, from Kadiak and Unalaska, N. to the Prybilof and Commander Islands. Much larger than the foregoing; length 7.00 or more; wing 4.50 (4.25–4.85); tail 3.50 (3.15–3.90); culmen 0.57; tarsus 0.95. Sexes scarcely distinguishable. Bill black or yellow according to season. Young "uniform brownish-gray, washed with umber; wings and tail dusky-slate, the feathers bordered with paler; the edges of the lesser wing-coverts and remiges

FIG. 251. — Brandt's Rosy Finch. (After Baird.)

very pale pinkish; of the greater wing-coverts and tertials pale dull ochraceous; no black or gray about head; bill horn-color." Nest well made of grasses and mosses, lined with feathers, on ground or among rocks; eggs 3–6, generally 4, pure white, 0.97 × 0.67.

NOTE. *Leucosticte* (*Hypolia*) *arctoa*, the Silver-winged Leucosticte, or Pallas' Rosy Finch, of Siberia, has been admitted to our fauna upon insufficient evidence, and is therefore now withdrawn from the position it has occupied in the 2d, 3d, and 4th eds. of the Key. It may be recognized by the following description: Dusky-purplish; neck above pale yellowish; forehead and nasal feathers blackish; outer webs of quills and wing-coverts, tail-coverts, rump and crissum silvery-gray, rosy-margined. Subgenerically different from any of the foregoing.

FRINGILLIDÆ: FINCHES, BUNTINGS, SPARROWS.

Fig. 252. — Aleutian Leucosticte.

ACAN'THIS. (Gr. ἀκανθίς, *akanthis*, linnet. Fig. 253.) RED-POLL LINNETS. Bill small, short, straight, very acute, more or less compressed, lateral outlines usually a little concave, those of culmen and gonys straight; commissure straight to the slight angulation. Base of bill thickly beset with a ruff of antrorse plumules, concealing small nasal fossæ and round nostrils. Wings longer than tail, pointed by first 3 primaries. Tail rather long for this group, forked. Feet small and weak, but tarsi longer than middle toe without claw; lateral toes of equal lengths, their claw-tips falling beyond base of middle claw. Hind claw much longer, stouter and more curved than the middle, exceeding its digit in length. Size small; plumage streaky with dusky, white, and flaxen colors, crown crimson, face and throat blackish; sexes otherwise dissimilar ; ♂ with rosy or carmine on breast, wanting in ♀. Arboreal, highly boreal, gregarious Finches of circumpolar distribution, breeding in high latitudes and alpine regions, roving south in winter in great flocks. Nest in trees and bushes; eggs colored. The species are much involved; we have five recognizable forms of two distinct species.

FIG. 253. — Details of *Acanthis* (*A. hornemanni*, nat. size). (From Elliot.)

Analysis of Species and Subspecies.

Tarsus as long as middle toe and claw. Heavily streaked below. Rump always fully streaked.
    Smaller: length about 5.50; wing 3.00; bill moderate (N. Am. at large) . . . . *linaria*
    Larger: length about 6.00; wing 3.25; bill large, acute (Canada, etc.) . . . . *linaria holboelli*
                   bill very stout (Greenland) . . . . *linaria rostrata*
Tarsus longer than middle toe and claw. Lightly or scarcely streaked below. Rump of adult ♂ immaculate white to some extent.
    Smaller: length about 5.50; wing 3.00. Bill and feet small (Brit. Am., etc.) . . . . . *hornemanni exilipes*
    Larger: length about 6.00; wing 3.30. Bill and feet large (Greenland) . . . . *hornemanni*

A. lina'ria. (Lat. *linaria*, flaxen; a linnet. Fig. 254.) COMMON RED-POLL. Adult ♂ : Frontlet, lores, and throat-spot sooty-black. Crown crimson. Above, variegated with brownish-yellow and dusky, the feathers having dark centres and flaxen edges. Rump streaked with dusky and white, and tinged with rosy, more or less so according to age and season. Below, white; sides and crissum streaked with dusky; entire fore parts colored with rose-red, more or less rich and extensive according to same circumstances. Wings and tail dusky, the feathers edged with whitish; middle and greater coverts tipped with the same, forming two cross-bars. Bill black or yellow, usually found yellow with dusky tip and edges. Feet blackish. Length 5.50; extent 9.00; wing 3.00; tail 2.40; bill 0.33; tarsus 0.65; middle toe and claw the same. Adult ♀ : Wanting entirely or having but a trace of rosy on rump and under parts. Breast with a dingy yellowish wash, streaked with dusky. Slightly smaller. Young: Like ♀, but ♂ soon showing rosy. Young may usually be distinguished from adult ♀ by a generally buffy suffusion, especially on fore parts; edgings of wing likewise buffy; streaks below less sharply defined; crimson of crown restricted, or of a coppery or bronzy tint. In worn midsummer plumage the bird is very dark colored, almost entirely dusky. This bright little bird inhabits northerly parts of both hemispheres, irregularly south in winter in N. Am. to about 35°; at times abundant, but erratic. Eggs 4-5, very pale bluish, finely speckled all over with reddish-brown, 0.65 × 0.52. Nest in low trees and bushes.

FIG. 254. — Common Red-poll, reduced. (Sheppard del. Nichols sc.)

A. l. hol'boelli. (To C. Holböll, a Danish naturalist.) HOLBÖLL'S RED-POLL. Like the last; larger; length 6.00 or about; wing 3.25; tail 2.45; bill longer and less constricted, with straight lateral outlines and rather curved culmen. Europe, Asia, N. Am., northerly; Canada (Quebec, Ontario) and New England occasionally in winter.

A. l. rostra'ta. (Lat. *rostrata*, beaked.) GREATER RED-POLL. Size of the last; bill very stout. Greenland, S. in winter to New England, New York, and the Great Lake region. I originally described this bird in 1861 upon dark midsummer skins from Greenland. At that time did not know *holboelli*, and was insufficiently informed on seasonal variation in this genus. I do not now see how it differs tangibly from *holboelli*, but others seem to be able to draw a distinction. *Ægiothus rostratus* COUES, Proc. Acad. Philada. 1861, p. 378; *Acanthis linaria rostrata* STEJNEGER, Auk, Apr. 1884, p. 153; RIDGW. Man. 1887, p. 397; A. O. U. Lists, 1886 and 1895, No. 528 *b*.

A. hor'nemanni. (To J. W. Hornemann. Figs. 253, 255.) GREENLAND MEALY RED-POLL. Bill regularly conic, only moderately compressed and acute, as high at base as long, color varying

FIG. 255. — Greenland Red-poll.

with season from black to yellow. Frontlet black, overlaid with hoary. A recognizable light superciliary stripe, reaching to the bill. Crimson cap over nearly all the crown. Upper parts streaked with brownish-black and white, the latter edging and tipping the feathers; this white nearly pure, only slightly flaxen on sides of head and neck. Wings and tail as in other species. Rump and entire under parts from the sooty throat white, free from spots, the rump and breast rosy. Feet large and stout; tarsus rather longer than middle toe and claw. Length 6.00; wing 3.30; tail 2.80; bill 0.34; tarsus 0.65; middle toe and claw 0.58. Sexual and seasonal changes as before; quite dark in midsummer. Greenland, Arctic America, and N. Europe. This large hoary northern form is resident; never known to occur in the U. S.; and most of the continental Red-polls of even Arctic N. Am. belong to the next species.

A. h. exi'lipes. (Lat. *exilis*, exiguous, small; *pes*, foot.) AMERICAN MEALY RED-POLL. Bill small, short, stout at base, regularly conic, little compressed, all its outlines about straight; nasal plumules very heavy, sometimes reaching half-way to tip of bill. Frontlet dusky, but the feathers tipped with hoary; an appreciable light superciliary line; lores and throat-spot dusky. General color of upper parts as in *linaria*, but the dusky streaks are smaller and less distinct, especially on the anterior parts; and the flaxen is very pale, nearly white, disappearing entirely on lower back, leaving a space streaked only with dusky and white. Rump snowy-white, rosy-tinted, immaculate. Wings and tail as in other species; under parts white, the breast with a rosy tint, paler than in *linaria* of same age and season; the sides streaked with dusky, the markings sparser and less definite than in *linaria*; crissum almost immaculate. Feet very small and weak, the toes especially shorter. Length 5.50; extent 9.00; wing 3.00; tail 2.50; tarsus 0.55; middle toe without claw 0.28; middle toe and claw shorter than tarsus; bill 0.32. Seasonal and sexual differences as before. This form inhabits N. Europe, N. Asia, and the whole of boreal N. America, reaching the U. S. regularly along the northern tier of States sometimes in flocks in company with *A. linaria*.

A. brew'steri? (To Wm. Brewster of Cambridge.) BREWSTER'S LINNET. With the general appearance of an immature *A. linaria*, this bird will be recognized by absence of crimson on crown, no black throat-spot, a sulphur-yellowish shade on lower back, and somewhat different proportions. Wing 3.00; tail 2.50; tarsus 0.50. Waltham, Mass., Nov. 1, 1870; one specimen known. *Ægiothus flavirostris*, var. *brewsterii* RIDGW. Am. Nat. vi, July, 1872, p. 433; Hist. N. A. B. i, 1874, p. 501; *Acanthis brewsterii* RIDGW. Man. 1887, p. 398; A. O. U. Hypothetical List, 1895, p. 330, No. 17. See BREWSTER, Bull. Nutt. Orn. Club, vi, 1881, p. 225. Conjectured to be *Acanthis linaria* × *Spinus pinus*.

SPI'NUS. (Gr. σπίνος, *spinos*; Lat. *spinus*, a linnet, siskin, or some other related bird.) LINNETS. SISKINS. Bill exceedingly acute; its lateral outlines concave by compression of sides toward end, culmen and gonys about straight, commissure angulated, cutting edges inflected, no ridges on either mandible. Nasal tufts concealing nostrils in their short fossæ. Wings long, exceeding the short, emarginate tail; point formed by 1-3 or 4 quills, 5 and rest rapidly shorter. Tarsus about as long as middle toe with claw; lateral toes of equal lengths, their claws reaching base of middle claw; hind claw shorter than its digit. Everywhere thickly streaked (*pinus*) or black and yellow (*notatus*). No red. Sexes alike. Habit gregarious. Nest in trees. Eggs speckled.

S. pi'nus. (Lat. *pinus*, a pine. Fig. 256.) PINE LINNET. PINE FINCH. PINE SISKIN. AMERICAN SISKIN. Adult ♂ ♀: Continuously streaked, above with dusky or dark olivaceous-brown and flaxen or whitish, below with dusky and whitish, the whole body usually suffused with yellowish, most evident on rump. Wings dusky, the basal portion of all the quills and their inner webs for some distance sulphury-yellow, usually showing externally as a spot just beyond the coverts, sometimes restricted and hidden. Outer webs of quills also narrowly edged with yellow, separated from the basal yellow patch by a blackish interval. Tail dusky, its basal half yellow, and outer webs edged with yellow. Bill and feet brown. Length about 4.75; extent 8.75; wing 2.75; tail 1.75. Very variable in yellowness of tone, sometimes quite bright, again plain streaky, dusky and whitish or flaxen; but the yellow coloration of the wings and tail is distinctive. Young birds have the markings diffuse, with a general buffy-brownish suffusion. N. Am. at large, breeding northerly and in alpine regions southerly (to Rocky Mts. of New Mexico and Arizona, and Sierra Nevadas of California); N. New England, etc.; in winter through most of U. S. into Lower California and Mexico; abundant. Nest in trees, preferably conifers; a well-concealed, flattish structure, compactly built of small twigs, rootlets, plant fibres, and hair; eggs 3-4, pale greenish, speckled with reddish-brown, and blackish chiefly about the larger end, about 0.70×0.50. Flight undulatory; voice querulous. This bird closely resembles no other of our country, but is the exact representative in America of the European Siskin, Tarin, or Aberdevine, *S. spinus*. (*Chrysomitris pinus* of former eds. of the Key. *Spinus pinus* of A. O. U. Lists.)

FIG. 256. — Pine Finch, reduced. (Sheppard del. Nichols sc.)

S. nota′tus. (Lat. *notatus*, noted.) BLACK-HEADED GOLDFINCH. Adult ♂: Bright yellow, obscured on back; head all around glossy black, extending on fore breast; wings black, with large yellow basal area on all the quills, forming a conspicuous patch; tail black, with basal half or more of all feathers except middle pair yellow. Young: Similar; lacking black on head, and general coloration duller. Length 4.60; wing 2.50-2.70; tail 1.80; bill 0.45, extremely acute. Cent. Am. and Mexico; a straggler in the U. S. (Kentucky, AUDUBON). *Astragalinus notatus* of former eds. of the Key; *Spinus notatus*, A. O. U. Lists, 1886-95, No. [532.]

ASTRAGALI′NUS. (Gr. ἀστραγαλῖνος, *astragalinos*, name of some bird.) AMERICAN GOLDFINCHES. Like *Spinus*. Bill stouter, less acuminate, without extreme lateral compression, culmen rather convex, gonys quite straight; commissure strongly angulated; upper mandible usually showing longitudinal striæ. Nasal ruff evident, though short. Wings and tail as in *Spinus*; feet smaller; toes shorter; lateral digits of unequal lengths; outer claw rather overreaching, inner not reaching, base of middle claw. Coloration massed, not streaky; yellow, olive, black and white, no red. Sexes unlike. Eggs white. The A. O. U., Auk, Jan. 1899, pp. 115, 116, reverted to the nomenclature of this genus, which has stood in the Key since 1884.

FIG. 257. — American Goldfinch, ♂, in summer, reduced. (Sheppard del. Nichols sc.)

Analysis of Species and Subspecies.

♂ yellow (in summer) or flaxen (in winter), with black cap, and black and white wings and tail.
    Eastern . . . . *tristis*
    Western, interior . . . . *t. pallidus*
    Western, Pacific coast . . . . *t.salicamans*
♂ gray, varied with yellow on back and breast, face black, wings black and yellow, tail black and white . *lawrencei*

♂ olive or black above, or mixed with both; yellow below; wings and tail black and white. Western.
Back olive; crown black, not below eyes; large white tail-spots . . . . *psaltria*
Back mixed olive and black; crown black; moderate white tail-spots . . . . *ps. arizonæ*
Back black; crown black to below eyes; small white tail-spots . . . . *ps. mexicanus*

A. tris'tis. (Lat. *tristis*, sad; from its note. Fig. 257.) AMERICAN GOLDFINCH. YELLOW-BIRD. THISTLE-BIRD. " WILD CANARY." ♂, in summer: Rich yellow, changing to whitish on tail-coverts; a black patch on crown; wings black, more or less edged with white; lesser wing-coverts white or yellow; greater coverts tipped with white; tail black, every feather with a white spot; bill and feet flesh-colored. In September, the black cap disappears; the general plumage changes to a pale flaxen-brown above and whitey-brown below, with traces of yellow, especially about head; wings and tail much as in summer; sexes then much alike: this continues until the following April or May. Length 4.80–5.20; extent 8.75–9.25; wing 2.75; tail 2.00; ♀ olivaceous above, including crown; below soiled yellowish; wings and tail dusky, whitish-edged; rather smaller than ♂. Young like winter ♀; when very young, suffused with fulvous, and wings edged with tawny. N. Am., especially Eastern U. S.; an abundant and familiar species, conspicuous by its bright colors, and plaintive lisping notes; in the fall, collects in large flocks, and so remains until the breeding season; irregularly migratory, but winters as far north as New England; feeds especially on seeds of thistle and button-wood; flies in an undulating course. Nest small, compact, built of downy and other soft pliant substances, placed in a crotch of a low tree, bush, or tall weed; eggs 3–6, usually 4 or 5, faintly bluish-white, normally unmarked, 0.65 × 0.50.

FIG. 258. — Lawrence's Goldfinch, reduced. (Altered from Audubon.)

A. t. pal'lidus. (Lat. *pallidus*, pale, pallid.) WESTERN AMERICAN GOLDFINCH. Like the last; paler; the various white markings more extensive; black cap larger. Rocky Mt. plateau district, British Columbia and Manitoba to Mexico; a local race. *Spinus tristis pallidus* MEARNS, Auk, July, 1890, p. 244; see also Auk, July, 1887, p. 198; A. O. U. List, 2d ed. 1895, p. 218, No. 529 a. *Astragalinus t. pallidus* A. O. U. Suppl. List, Auk, Jan. 1899, p. 115, No. 529 a.

A. t. sali'camans. (Lat. *salix*, gen. *salicis*, willow; *amans*, pres. partic. of *amare*, to love.) WILLOW GOLDFINCH. Like *tristis;* darker, with broader wing-markings. Pacific coast form. *Spinus tristis salicamans* GRINNELL, Auk, Oct. 1897, p. 397. *Astragalinus tristis salicamans* A. O. U. Suppl. List, Auk, Jan. 1899, p. 115, No. 529 b.

A. lawren'cei. (To G. N. Lawrence, of New York. Fig. 258.) LAWRENCE'S GOLDFINCH. CALIFORNIA CANARY. ♂, in summer: Gray, more or less tinged with yellowish, whitening on belly and crissum; rump greenish-yellow; a large breast-patch rich yellow; crown, face, and chin black; wings black, variegated with yellow, most of the coverts being of this color, and the same broadly edging the quills; inner secondaries edged with hoary gray; tail black, most of the feathers with large square white spots on inner webs and whitish edging of outer; bill and feet flesh-color more or less obscured. ♀ resembles ♂, but there is no black on head, and the yellow places are not so bright; yellow of back often wanting. ♂ ♀, in winter: yellowish of upper parts changed to olive-gray, but yellow of other parts often as bright as in summer, and black of ♂'s head the same. Young birds like ♀, but may be somewhat streaky. Size of *tristis*, or rather less; an elegant species. California, Arizona, and New

Mexico; N. Lower California; breeds W. of Sierra Nevadas. Nest and eggs similar to those of *tristis;* eggs smaller, 0.60 × 0.45, 3 to 5 in number, pure white or with a creamy tinge.

A. psal′tria. (Gr. ψάλτρια. *psaltria*, a lutist. Fig. 259.) ARKANSAW GOLDFINCH. TARWEED CANARY. Adult ♂ : Upper parts uniform olive-green, without any black; below yellow; crown black, *not* extending below eyes; wings black, most of the quills and greater coverts white-tipped, and primaries white at base; tail black, outermost three pairs of feathers with a long rectangular white spot on inner web. ♀ and young similar, not so bright; no black on head; sometimes, also, no decided white spots on tail. Length 4.25–4.50; wing 2.40; tail 2.00. Plains to the Pacific, U. S., southerly; N. at least to Oregon; S. to Cape St. Lucas. A pretty species, of the same habits as the common Goldfinch; nest and eggs similar, latter rather smaller, 0.60 × 0.45. Southward this form passes directly into

FIG. 259. — Arkansaw Goldfinch, reduced. (After Audubon.)

A. p. arizo′næ. (Lat. of Arizona.) ARIZONA GOLDFINCH. The upper parts mixed olive and black in about equal amounts; in W. Texas, New Mexico, Arizona, portions of Utah and Nevada, and S. California; thus leading directly into

A. p. mexica′nus. (Lat. Mexican. Fig. 260.) MEXICAN GOLDFINCH. The upper parts continuously black, and black of crown extending below eyes, enclosing olive under eyelid. Mexican border and southward. This bird looks quite unlike typical *psaltria*, but the gradation through *arizonæ* is perfect; and *mexicana*, moreover, leads directly into *columbiana*, a Central American form in which the tail-spots are very small or wanting. The females of these several varieties cannot be distinguished with certainty. My original determination of this case may be read in Pr. Philada. Acad. 1866, p. 82.

FIG. 260. — Mexican Goldfinch, reduced. (After Audubon.)

CARDUE′LIS. (Lat. a thistle-bird, from *carduus*, Gr. κάρδος, *kardos*, a thistle.) OLD WORLD GOLDFINCHES. Generic characters of *Spinus*, but bill exceedingly acute, attenuated to a length nearly equalling that of the tarsus, and plumage gaudily variegated with red, yellow, black, and white in both sexes. (Extralimital genus, introduced.)

C. cardue′lis. EUROPEAN GOLDFINCH. Adult ♂ ♀ : Head varied with crimson, black, and whitish; wings and tail varied with rich yellow, black, and white; back brown, whitening on rump and upper tail-coverts; lower parts whitish, shaded with brown on the sides; bill white. Length 5.00–5.50; wing 3.00; tail 2.00; bill nearly or about 0.50. Young birds lack the crimson red, rich yellow, and pure black; but this well-known cage-bird can hardly be mistaken in any plumage. Europe, and portions of Asia and Africa; introduced artificially in the United States, and naturalized to the extent of breeding sometimes, as in New York city and Cambridge, Mass. The nest resembles that of our Goldfinch; eggs different, being marked with reddish-brown spots, chiefly at the larger end, on a pale bluish or greenish ground; 4–6 in number, 0.70 × 0.50.

PASSERI′NA. (Lat. *passerinus*, sparrow-like.) Bill very small and truly conic, well exhibiting "emberizine" or "bunting" characters; *i. e.*, strong angulation of commissure; inflected cutting edges; a palatal knob. Culmen slightly curved; gonys perfectly straight, and very short, less in length than width of bill; lower mandible heavier than upper. A dense nasal ruff. Wings very long and pointed; 1st or 1st and 2d quills longest, rest rapidly graduated. Tail ⅛ shorter than wings, nearly square. Tarsus longer than middle toe without claw; lateral toes of subequal lengths, and much shorter than middle one. Claws slender and compressed, with deep lateral grooves at base; hind claw lengthened and less curved than the rest, but not straight. Gullet very distensible. Sexes alike. Colors very different with season; in summer ♂ entirely black and white. Terrestrial, gregarious; nest on the ground; eggs colored. One species of circumpolar distribution, and another peculiar to Arctic America. (*Plectrophanes* of all former eds. of the Key.)

Analysis of Species and Subspecies.

Adults in summer with black on more than middle pair of tail-feathers, and much black on the back and wings.
Smaller; bill about 0.40 along culmen . . . . *nivalis*
Larger; bill about 0.50 along culmen. Alaska only . . . . *nivalis townsendi*
Adult ♂ in summer with black on tail reduced to spots on two middle feathers, and back white; ♀ with only 4 tail-feathers black, and back only streaked with black. Alaska only . . . . *hyperboreus*

P. niva'lis. (Lat. *nivalis*, snowy; *nix, nivis*, snow. Figs. 261, 262.) SNOW BUNTING. SNOW LARK. SNOWBIRD. SNOWFLAKE. WHITEBIRD. ♂, in full dress: Pure white; bill, feet, middle of back, scapulars, primaries except at base, most inner secondaries, bastard quills, and several tail-feathers, black. Length about 7.00; extent 12.50–13.00; wing 4.00–4.25; tail 2.50–2.75. In less perfect summer dress, black of back, inner secondaries and tail-feathers varied with white. ♀, in breeding plumage: The black impure or brownish, and most or all upper parts brownish-black, varied with white. Rather smaller. Dimensions of many specimens of both sexes: Length 6.50–7.00; extent 12.00–13.00; wing 4.00–4.25; tail 2.50–2.75; bill 0.40; tarsus 0.80;

FIG. 261. — Snow Bunting, winter plumage.

middle toe and claw 0.90; hind toe and claw 0.67–0.75; claw alone 0.33–0.44. Adults, in winter, as generally seen in the U. S. (where black-and-white birds are rarely if ever found): Upper parts overcast with rich warm chestnut-brown and grayish-brown, mixed with black of back, and clouding other upper parts which are white in summer, becoming dusky or even blackish on head; this brown also usually forming a patch on ears, a collar on breast, edging of inner wing- and tail-feathers, and a wash on flanks; but specimens vary interminably; other parts white or black as in summer; bill yellowish, usually black-tipped, but drying reddish-brown. Fledglings: Dark ashy-gray above and on fore parts below, this color overlaid with brown, and streaked on back with dusky; below, from breast, white; lateral tail-feathers mostly white; inner secondaries black with brown edging. The Snowflake is a good example of what. I call *aptosochromatism,* or change of color without moult; for the pure black-and-white coloration is acquired in spring by the wearing away of the edges of the brown feathers; this brown being confined to the surface of the plumage, the deeper parts of which are black or white. It is a notable bird, inhabiting the N. hemisphere, breeding in arctic and subarctic regions, whence migrating south in vast flocks with the snow, as if one with those pure crystallizations. Thousands whirl into the U. S. in the fall on the wings of the storm, relieving by their animated presence the desolation of places exposed to the fury of the blast. South regularly only to the Northern States, but often roving flocks reach 35°. Nest on ground in sphagnum and tussocks of arctic regions, of a great quantity of grass and moss, lined profusely with feathers: eggs 4–6, very variable in size and color, about 0.90 × 0.65, white or whitish, speckled, veined, blotched, and marbled with deep browns and neutral tints.

FIG. 262. — Snow Bunting, in summer, reduced. (Sheppard del. Nichols sc.)

P. n. town'sendi. (To C. H. Townsend.) PRIBILOF SNOW BUNTING. TOWNSEND'S SNOWFLAKE. Like the last; averaging larger, with heavier bill. ♂ : Wing 4.50; tail 3.00; bill 0.50; ♀ not quite so large as this, but exceeding average size of *P. nivalis* proper. Pribilof and Aleutian Islands, Alaska; Commander Islands, Kamtschatka. *Plectrophenax nivalis townsendi* RIDGW. Man. 1887, p. 403; A. O. U. List, 2d ed. 1895, p. 220, No. 534 *a; Plectrophanes nivalis townsendi* COUES, Key, 4th ed. 1890, p. 899; *Passerina n. townsendi* A. O. U. Suppl. List, Auk, Jan. 1899, p. 117.

P. hyperbor'eus. (Lat. *hyperboreus,* hyperborean. Fig. 263.) POLAR SNOW BUNTING. MCKAY'S SNOWFLAKE. Adult ♂, in breeding dress: Pure white, except tips of wings, which are black for about 1.50, one or two black touches on inner secondaries, and a subterminal black spot on middle tail-feather; white edging of black part of wings; bill and feet black. In winter: washed with rusty brown on head, nape, back, rump, and across breast; bill yellowish, with dusky tip. The full plumaged ♀ is less extensively white than ♂, having more black on wings and tail, and back also streaked with black; seasonal changes are correspondent. Larger than *P. nivalis:* ♂ averaging over 7.00; extent over 13.00; wing 4.60; tail 3.10; bill 0.45; tarsus 0.90. ♀ less, about as large as ♂ *P. nivalis*. A beautiful Snowflake, apparently quite distinct from the foregoing. Pacific coast of Alaska in winter; known to breed on Hall Island in Bering Sea. *Plectrophenax hyperboreus* RIDGW. Proc. U. S. Nat. Mus. vii, June, 1884, p. 68; Man. 1887, p. 403; A. O. U. List, 2d ed. 1895, p. 221, No. 535; *Plectrophanes hyperboreus* COUES, Key, 3d and 4th eds. 1887 and 1890, p. 873; *Passerina hyperborea* A. O. U. Suppl. List, Auk, Jan. 1899, p. 117.

CALCA'RIUS. (Lat. *calcar,* a spur; in plural *calcaria,* from *calx,* genitive *calcis,* the heel; *i. e.,* the hind claw lengthened and straightened.) LONGSPURS. Characters of *Passerina;* hind claw and its digit more developed, longer than middle; bill relatively and absolutely larger, rather "fringilline" than thoroughly "emberizine," but still with a palatal knob; no decided nasal ruff, but antrorse plumules in nasal fossæ; a little tuft at base of rictus. Wings less acute, the point formed by 1st–3d primaries, 4th abruptly shorter; tail emarginate. Sexes very unlike: ♂ with black hood and chestnut cervical collar. Gregarious, terrestrial; nest on the ground; eggs 3–6, colored. (*Centrophanes* of former eds. of the Key; but *Calcarius* has priority.)

FRINGILLIDÆ: FINCHES, BUNTINGS, SPARROWS. 397

Fig. 263.— McKay's Snowflake.

Analysis of Adult Males.

Whole head and throat black; belly white; bill yellow; feet black	*lapponicus* and *alascensis*
Crown black; whole under parts fawn-colored; feet flesh-colored	*pictus*
Crown black; throat white; belly black or mahogany; feet dark	*ornatus*

C. lappon'icus. (Lat. *lapponicus*, of Lapponia, Lapp-land. Figs. 43, 264.) LAPLAND LONGSPUR. ♂, in full dress (seldom seen in U. S.): Whole head, throat and breast jet-black, bordered with buffy or whitish, which forms a postocular stripe separating black of crown from that of sides of head, sometimes continued to bill. A broad cervical chestnut collar, separated from black cap by whitish or buffy line and nuchal spot. Upper parts brownish-black completely streaked with buff or whitish edges of the feathers; under parts white, the sides streaked with black. Wings dusky, with pale or brownish edgings of the feathers, but no strong markings. Tail like wings, with large oblique white spaces on outer 3 feathers. Bill yellow, black-tipped. Legs and feet black. Length about 6.50; extent 11.25; wing 3.50–3.75; tail 2.50–2.75; tarsus 0.75; middle toe and claw rather more; hind claw about 0.50, slender, sharp, and little curved. Adult ♂, in winter: The black hood

FIG. 264. — Lapland Longspur, in summer, reduced. (Sheppard del. Nichols sc.)

overcast with brown or gray tips of the feathers, or otherwise imperfect. Chestnut collar also overlaid with gray. Edges of secondaries and wing-coverts ruddy-brown; sides of flanks washed with brown. White tail-spots less extensive. Yellow of bill obscured. ♀, in breeding plumage: Upper parts of body, wings and tail, as in ♂. No continuous pure black on sides of head, chin, or throat. Cervical collar indicated, but dull and obscured. Black of crown overlaid with gray; superciliary and postocular stripe buffy; sides of head blackish, overlaid with gray; throat similarly varied, but chin nearly white; on the whole, the pattern of ♂'s black hood clearly indicated, but interrupted and ill-defined. Sides of breast and belly with few small sharp dark streaks, instead of heavy black stripes; other under parts as in ♂. Bill obscure yellowish, dusky-tipped; feet dark brown, not black. Rather smaller. ♂ ♀, young, in winter, as usually seen in U. S., without any continuous black, resemble adult ♀ as to coloration of head and fore parts, and are like winter ♂ in other respects. The cervical collar may be scarcely appreciable, but usually shows a trace at least; sides often quite brown. Fledglings: Continuously streaked on upper and fore parts with blackish and brownish-yellow; wings and tail broadly edged with chestnut; bill dark; feet pale. A species of circumpolar distribution, like *P. nivalis;* breeding range and winter rovings much the same, but less commonly observed in the U. S. South irregularly to the Middle States, Ohio, Kansas, Colorado, etc., casually to South Carolina. Nesting like that of *P. nivalis;* eggs 4–6, 0.80 × 0.62, dark-colored, very heavily mottled and clouded with chocolate-brown, through which the greenish-gray ground scarcely appears.

C. l. alascen'sis. (Lat. of Alaska.) ALASKAN LONGSPUR. Like the last; paler, especially in winter: in summer, upper parts with a ground color of light grayish-brown with little if any rusty tinge, even on wings, and the black streaks narrow. Alaska, including Aleutian and Prybilov Islands, E. to Fort Simpson, S. in winter to Kansas, Colorado, and Nevada. Included under *lapponicus* in former eds. of the Key, and hardly worth recognition by name. RIDGW. Auk, Oct. 1898, p. 320; A. O. U. Suppl. List, Auk, Jan. 1899, p. 117.

C. pic'tus. (Lat. *pictus*, painted.) PAINTED LONGSPUR. SMITH'S LONGSPUR. Adult ♂: Cervical collar and entire under parts rich fawn color; crown and sides of head black, bounded below by a white line, and interrupted by a white superciliary and auricular line and white occipital spot. Upper parts streaked with black and brownish-yellow. Lesser and middle wing-coverts black, tipped with white, forming conspicuous patches. One or two outer tail-feathers mostly white. No white on the rest. Legs pale or flesh-colored. Length 6.50; extent 11.25; wing 3.75; tail 2.50; tarsus 0.75; middle toe and claw, about the same; hind toe and claw, rather less (*ornatus* is much less in all its dimensions). Young, and generally in winter: Bill dusky-brown above and at tip, paler below; feet light brown (drying darker); toes rather darker. Entire under parts rich yellowish-brown, or buffy (in *ornatus* never thus); paler on chin and throat, which, with fore-breast, are obsoletely streaked with dusky; tibiæ white. Tail white only on two or three outer feathers (in *ornatus* all the feathers, excepting sometimes the central pair, are white at base). Upper parts much as in the adult, but distinctive head-markings wanting, or only obscurely indicated. Interior N. Am. from the region of the Yukon, McKenzie, Saskatchewan, and Upper Missouri to Texas and Illinois in winter. It is not found in the Atlantic States, but is common on prairies of both Dakotas, Montana, and southward, associated in the fall with *ornatus*, but breeding ranges farther north. Habits and general aspect of *ornatus*, but easily distinguished by larger size, buffy under parts, black and white wing-patch, and white only on some lateral instead of all the tail-feathers. Nest on ground; like that of other species of this genus, but a less elaborate structure and less warmly lined than that of *C. lapponicus*; eggs 3–6, about 0.82 × 0.60, less heavily colored than those of *lapponicus* usually are, and thus closely resembling those of *ornatus*.

C. orna'tus. (Lat. *ornatus*, adorned. Fig. 265.) CHESTNUT-COLLARED LONGSPUR. BLACK-SHOULDERED LONGSPUR. WHITE-TAILED LONGSPUR. ♂, in full dress: Cervical collar intense chestnut. Crown black; a whitish spot on nape, and broad white superciliary stripe. Auriculars black, mixed with color of throat; throat and most of sides of head below eyes rusty-white, changing to pure white which extends around sides of neck, partly bordering the chestnut collar. Breast and belly lustrous black, often mixed with intense ferruginous or mahogany feathers, sometimes largely overlaid with this rich sienna color. Lining

FIG. 265. — Chestnut-collared Longspur. (L. A. Fuertes.)

of wings pure white. Sides of body, flanks, lower belly, and under tail-coverts white, all but the last usually rusty-tinged. Back, rump, and scapulars brownish-black, varied with grayish-brown edges of the feathers. Wings dark brown without decided markings, though the feathers are pale-edged, excepting jet-black lesser coverts, with or without white tips. Tail like wings, but two or three lateral feathers entirely white, and all the rest basally white in decreasing amount: in flight the "white tail" is very conspicuous. Bill blackish-plumbeous; feet dark. Smaller than the foregoing: Length 5.75–6.00, rarely 6.25; extent 10.25–10.75, rarely 11.00; wing 3.00–3.30; tail 2.00–2.30. ♀, in full dress: Rather smaller; size averaging about the lesser figures just given. Upper parts, wings, and tail as before, but lesser coverts not black; chestnut collar obscured; crown like back, separated from the back-markings by a slight rufous dusky-streaked interval. Sides of head, and throat, whitish, with dusky speckling on cheeks and ears. Under parts dull brown, fading to white on belly and crissum, the feathers sometimes with dusky streaks. Thus an obscure bird: but observe generic characters, and extensively *white tail*. Adult ♂, after fall moult: The full dress is confined to the breeding season; afterward, the colors are much obscured. Cervical collar and black of head and belly veiled by gray ends of the feathers, but visible on raising the plumage. Crown like back, with concealed black; superciliary stripe and other distinctive head-markings obliterated; bill brownish-plumbeous. Changes in ♀ parallel, but there is less to be altered. Young ♂ ♀, before first moult: Whole upper parts blackish-brown, with semicircular gray or whitish markings, and a slightly lighter cervical interval. Throat definitely white. Under parts dull brown, heavily streaked with dusky, especially on breast. Much light brown edging and tipping of quills and wing-coverts. Feet and bill pale. This stage is transitory; with first moult the young acquire characters above described for winter. A beautiful species of interior plains, British America and U. S. and Mexico; breeds in profusion on prairies of Dakotas, Montana, and whole Upper Missouri and Saskatchewan regions, S. to Kansas or farther; has occurred in Massachusetts; rarely W. of the Rocky Mts. Breeds in June and July; nest on ground, sunken flush with surface, of a few grasses and weed-stalks; eggs usually 4, about 0.75 × 0.55, white clouded with purplish shell-markings, gray the prevailing tone, this irregularly dotted and veined with sharp dark-brown surface-marks. Young covered with whitish down. In the breeding season the birds are fond of soaring and singing as they fly, rising to great height and letting themselves down with the wings held like parachutes; they curiously resemble butterflies when so engaged. The white tail shows very conspicuously. Ordinary flight wayward and vacillating; song weak and twittering, but pleasing. The birds flock as soon as young are fairly on wing, and leave the northern prairies in October. They are associated in the breeding season with *R. maccowni*, and joined in October by *C. pictus* and *lapponicus* from the north.

RHYNCHO'PHANES. (Gr. ῥύγχος, *rhugchos*, beak, and φαίνω, *phaino*, I appear; in allusion to the turgid bill.) LONGSPURS. Similar to *Calcarius*, but departing in the direction of *Montifringilla* (an exotic genus). Bill turgid, very stout and large in comparison; culmen rising high on forehead, its outline almost concave. Hind toe and claw less developed. Hind claw not longer than its digit, not notably straightened. Sexes dissimilar. No cervical collar. ♂ with black pectoral crescent and bay bend of wing. Habits of *Calcarius* strictly.

R. maccown'i. (To Capt. J. P. McCown, U. S. A. Fig. 266.) BLACK-BREASTED LONGSPUR. BAY-WINGED LONGSPUR. ♂, in full dress: Upper parts slate-gray, streaked with dusky and grayish or yellowish-brown, es-

FIG. 266. — Black-breasted Longspur, reduced. (Sheppard del. Nichols sc.)

pecially on interscapulars. No cervical collar, but a chestnut patch on wings, formed by median coverts. Crown jet-black, bounded by a white superciliary line; sides of head whitish, but auriculars more or less slaty. Throat white, bounded by firm black maxillary stripes. Breast jet-black, in broad crescentic form, sharply defined against white throat, shading behind into slaty-blackish, becoming more and more mixed with white on belly and sides, till posteriorly the parts are pure white; lining of wings white. All tail-feathers, except middle pair, and bases and tips of intermediate ones, white, ending squarely across both webs. Bill blackish-plumbeous, pale at base below; feet brownish-black. Length about 6.00; extent 11.00–11.50; wing 3.30–3.60; tail 2.25; bill 0.50; tarsus 0.67; middle toe and claw rather less. ♀, in breeding plumage: Upper parts, wings, and tail as in the ♂ — coverts with at least a trace of chestnut, and tail displaying rectangular shape of white area; crown like back instead of black; no black maxillary stripes, and breast-crescent slaty-gray; throat whitish; bill and feet yellowish-brown, more or less obscured. The seasonal changes of plumage, as well as the sexual differences, are parallel with those of *ornatus*; there is the same veiling of black parts by gray, etc. Though so different from *ornatus* in full dress, the bird is very similar in other conditions, age for age, and sex for sex: but larger; no trace of chestnut on nape; trace at least on wing-coverts; peculiar pattern of tail-feathers shown as soon as they sprout, and never lost. Very young birds have curved edgings of feathers of upper parts; under parts quite purely white, with some dusky streaks, and a buff suffusion on breast. Region of the Upper Missouri and its tributaries; N. to the Saskatchewan; casual W. of Rocky Mts.; S. in winter to Arizona, Texas, and Mexico; E. to probably Iowa and Missouri. Breeds in profusion on prairies from Colorado northward, in parts of Dakotas and in Montana associated with *ornatus*; winters from Colorado southward. Its habits and manners are the same as those of *ornatus*. It has the same soaring singing flight, and parachute-like descent, "sliding down on the scale of its own music;" nesting the same; eggs resembling the paler varieties of *ornatus*; 0.80 × 0.60.

POŒ'CETES. (Gr. πόη, *poe*, grass; οἰκέτης, *oiketes*, an inhabitant.) GRASS SPARROWS. Bill moderate, culmen, gonys, and commissure nearly straight. Wings long, longer than tail, tip formed by first 4 quills; inner secondaries somewhat elongate, less so than in *Passerculus*. Tail emarginate, with rather broad firm feathers, not acuminate at ends. Tarsus nearly equalling middle toe with its claw; lateral toes of about equal lengths, their claws scarcely reaching base of middle claw; hind claw as usual, not longer than its digit. Plumage thickly streaked everywhere above, on sides below and across breast; bend of wing chestnut; 1–3 outer tail-feathers white; crown without light median stripe; no trace of yellow anywhere. *Poœcetes* BAIRD, B. N. A. 1858, p. xx, p. xxxix, and p. 927, misspelled *Poocætes* on pp. 439, 447, and 994. The A. O. U. picked out one of the places where the false orthography occurs citing it for this wrong form of the name. The error was promptly detected and corrected by Dr. Sclater and myself, and the proper form of the word occurs in all the editions of the Key, as well as in both editions of my Check List — in fact, in the works of most authors from 1859 to date. I know that Professor Baird felt sore over this solecism as well as that of "Nephocætes," because he told me so. Neither the spirit nor the letter of the A. O. U. code required us to perpetuate such a perpetration, which was not corrected till 1899: see Gill, Auk, Jan. 1899, p. 20; A. O. U. Suppl. List, *ibid.* p. 117.

P. grami'neus. (Lat. *gramineus*, applied to a grass-loving bird; *gramen*, grass. Fig. 267.) GRASS FINCH. BAY-WINGED BUNTING. VESPER SPARROW. Above, grayish-brown, closely and uniformly marked with dusky-centred brown-edged streaks, and further variegated by pale gray edging of the feathers. Crown quite like back, though the marking is in smaller pattern; superciliary line and eye-ring whitish. Under parts dull white, usually noticeably buff-tinged in the streaked areas, thickly streaked across breast and along sides with dusky-centred brown-edged streaks, anteriorly tending to concentrate in lateral chains bounding the white throat; above this chain a maxillary brown stripe; auriculars varied with light and dark brown.

Fig. 267. — Bay-winged Bunting, reduced. (Sheppard del. Nichols sc.)

Quills fuscous, the longer ones with grayish-white edging, the secondaries and greater and median coverts with broad firm brown and white edges and tips; lesser coverts bright chestnut, whence the name "bay-winged." Outer tail-feather largely or wholly white, next pair or two pairs largely white in decreasing amount. Upper mandible brown; lower, and the feet, flesh-colored or yellowish. Length 5.75–6.25; extent 10.00–10.50; wing 2.80–3.25; tail 2.25–2.75. Eastern N. Am. to the Great Plains, N. to the British Provinces adjoining the U. S., breeding throughout its range, but partially migratory, chiefly nesting northward, and wintering southward. A large, stout, full-chested Sparrow of plain appearance, but recognized on sight by bay bend of wing and white lateral tail-feathers, — the latter conspicuous as it flies. Very abundant in fields, along roadsides; terrestrial, gregarious to some extent when not breeding. Nest sunken in the ground, thick-rimmed, well cupped; eggs 4–6, variously colored, as in *P. savanna*, 0.80 × 0.60; two or three broods may be reared. One of the sweetest songsters among the Sparrows.

P. g. confi′nis. (Lat. *confinis*, near.) WESTERN GRASS FINCH. HESPERIAN-BIRD. Like the last; paler and grayer, with narrower streaks; wings and tail averaging longer, and bill somewhat slenderer. The difference in length, when existent, is due to the tail, which averages near the extreme of length of the common form. Habits, nest, and eggs the same. Western U. S. and adjoining British Provinces, S. into Mexico.

P. g. affi′nis. (Lat. *affinis*, allied, affined.) OREGON GRASS FINCH. MILLER'S VESPER SPARROW. Like *P. g. confinis* in respect of slender bill and narrow dorsal streaks; ground color above buffy brown rather than grayish-brown, and the white of the under parts, including crissum and lining of wings, suffused with pinkish-buff. Size of the Eastern bird. Pacific coast region of Oregon and northern California; apparently a slight local race. G. S. MILLER, Jr., Auk, Oct. 1888, p. 404; COUES, Key, 4th ed. 1890, p. 899; A. O. U. List, 2d ed. 1895, p. 223, No. 540 *b*.

PASSER′CULUS. (Lat. *passerculus*, a little sparrow; diminutive of *passer*, a sparrow.) SAVANNA SPARROWS. GROUND SPARROWS. Bill rather slenderly conical, culmen, commissure, and gonys about straight (bill more turgid in *rostratus* and *r. guttatus*). Wing longer than tail, point formed by outer 4 primaries, of nearly equal lengths; inner secondaries more or less enlarged and flowing, reaching nearly or quite to end of primaries in the closed wing. Tail short, nearly even or emarginate, of narrow pointed feathers. Feet slender, pale-colored, usually reaching when outstretched nearly or quite to end of tail; tarsus and middle toe with claw of about equal lengths; lateral toes of equal lengths, their claws underreaching base of middle claw; hind toe rather longer than its claw, which has no special development. Plumage thickly streaked everywhere above, and below on breast and sides; crown with median light line and lateral dark ones; no decided markings on tail-feathers. In most species edge of wing yellow, and traces at least of yellow on head; no red, blue, or greenish. Sexes alike. Embracing small plain streaked ground Sparrows of slender build, mostly with a touch of lemon-yellow on edge of wing, long inner secondaries and pale slender legs; one species abounding in the East, the others all of more special distribution, and with one exception Western. Nest on the ground, in prairie, meadow, or marsh; eggs colored.

NOTE. — The genus *Passerculus* has been reduced to a subgenus of *Ammodramus* by the A. O. U. Committee, but without sufficient reason, as I think. The "Savanna Sparrow group" is a well-marked one, to which as good characters can be assigned as those of most genera of fringillines, and nothing is gained, either scientifically or conventionally, by putting it under *Ammodramus*; for the same diagnosis has to be drawn up, to distinguish this group from its relatives, whether we call it a genus or only a subgenus. Moreover, in puzzling out the species and subspecies of this difficult group, it is practically most convenient, at the outset, to distinguish them collectively from *Ammodramus*. With this explanation, I must decline to follow my respected colleagues in this needless innovation upon long-established usage, and continue to keep the genus *Passerculus*, as in all former eds. of the Key, and as nearly all writers have done since 1858, until recently. As to *Centronyx*, which I admitted to full rank in the orig. ed., 1872, and suppressed entirely in the later eds., 1884-90, I am quite willing to adopt the middle course of the A. O. U. and give it as a subgenus of *Passerculus*. As to *Coturniculus*, it shades directly into *Ammodramus* through certain of its species, which moreover do not very closely agree among themselves, as stated in former eds. of the work; and I am therefore very well satisfied to degrade it to subgeneric rank, following the A. O. U. We thus have, in the case of the four genera in mention, *Passerculus* (with *Centronyx*) on the one hand, and on the other *Ammodramus* (with *Coturniculus*). This seems to me the most judicious stand to take, and it is also the one taken by eminent British authority: see SHARPE, Cat. B. Brit. Mus. xii, 1888, p. 683.

Analysis of Subgenera, Species, and Subspecies.

Tail short, less than thrice as long as tarsus, obviously emarginate and with very narrow, pointed feathers. (*Subgenus* CENTRONYX.)
  Bill typical. Crown with median light stripe. Inner secondaries seldom quite equalling primaries. No decided lemon-yellow on edge of wing. Top of head with two black stripes, and suffused with rich brownish-yellow
  . . . . *bairdi*

Tail longer, thrice as long as tarsus, not obviously emarginate, and with broader, less pointed feathers. (*Subgenus* PASSERCULUS.)
  Bill typical. Crown with median light stripe. Inner secondaries at full length. Edge of wing with lemon-yellow; same shade on head, if any. Upper parts much variegated; under white, with sharp streaking.
    Large, pale; little or no yellowish; length 6.00 or more; wing 3.25. Atlantic coast . . . . . . *princeps*
    Large, dark, with decided yellow; length about 6.00; wing 3.00. Northwest coast . . . . . *sandwichensis*
    Medium, of average coloration; length about 5.50; wing 2.75. N. Am. at large . . . . *s. savanna*
    Medium; pale; size of *savanna* proper. Interior and western . . . . *s. alaudinus*
    Small, dark; yellow very decided. Length about 5.25; wing 2.50. Cala. coast . . . . *s. bryanti*
    Small, very dark; head stripes obscure; under tail-coverts streaked. Length about 5.00; wing 2.60. Cala. coast . . . . *beldingi*
  Bill enlarged, turgid, with convex culmen. Crown-stripe obsolete. No yellow on head or wing.
    Larger: bill 0.50. Length 5.30; wing near 3.00. Pale brownish-gray, with obsolete streaking; the streaks below light brown. Coast of California . . . . *rostratus*
    Smaller: bill 0.33. Length 5.00; wing 2.50. Darker, the streaks below dusky. L. Cala. . . . . *r. guttatus*
  Bill size of that of *rostratus*, but conic, with straight culmen.
    Like *guttatus*, but larger. San Benito Isl. L. Cala. . . . . *sanctorum*

(Subgenus CENTRONYX.)

P. (C.) baird'i. (To Prof. S. F. Baird. Fig. 268.) BAIRD'S SPARROW. PRAIRIE SPARROW. *Adult ♂ ♀, in breeding plumage:* With a general resemblance to the common Savanna Sparrow. Inner secondaries less elongated, rarely equalling primaries in the closed wings. First 4 quills about equal and longest. Hind toe and claw about equalling middle toe and claw, its claw about equalling the digit. Tail shorter than wing, lightly double-rounded (central and outer pair of feathers both a little shorter than intermediate ones). Top of head streaked with black and rich brownish-yellow, or buff, the former predominating laterally, the latter chiefly as a median stripe, but also suffusing nape and sides of head in greater or less degree. Back varied with brownish-black and gray, together with a little bay, the two latter colors forming edgings of interscapulars and scapulars. Rump variegated with gray and chestnut-brown, different in shade from that of back. Under parts dull white, usually with a faint ochrey tinge on breast, but often without; a circlet of small, sharp, sparse, dusky streaks across breast, continuous with others, longer and mostly lighter, along whole sides, and with others, again, extending up sides of neck into small vague maxillary and auricular markings. When the feathers are perfectly arranged these lateral head-markings are seen to be a postocular stripe just over auriculars, a post-auricular spot, a streak starting from angle of mouth, and another heavier one parallel with and below this, running directly into the pectoral ones. Quills without special markings, excepting elongated inner secondaries, which correspond with scapulars. Tail the same, slightly whitish-edged. Upper mandible mostly dark, lower pale. Feet flesh-colored. Length 5.10–5.85, averaging 5.67; extent 8.60–9.85, average 9.50; wing 2.75–3.00; tail 2.00–2.25; culmen about 0.40; tarsus about 0.75; middle toe and claw, and hind toe and claw, each, rather less; ♀ averages rather smaller. *Autumnal plumage:* Soft, with brighter, more suffused colors, in bolder pattern. Whole top and sides of head, as well as nape and part of neck, suffused with rich buff, in many instances as bright a golden-brown as that on head of *Siurus auricapillus*. A paler, rather ochraceous shade of the same also suffusing the whole fore under-parts. Pectoral and lateral dusky streaks, as well as two rows on each side of throat, large, heavy, diffuse. Bay and whitish edgings of secondaries broad and conspicuous, contrasting with black central fields. Whitish edgings of tail-feathers the same; and, in general, the same character is stamped over all the upper plumage. *Newly-fledged young* have each feather of the dorsal plumage conspicuously bordered with white, producing a set of semicircles, much as in *Anthus spraguei*. There is the same general buffy suffusion of head and fore parts as in autumnal adults, but the tint is dull and ochrey. The markings below have a short, broad, guttiform character. When just from the nest, the edging of secondaries and tail-feathers is of a peculiar pinkish-rusty shade. Central Plains; N. to the Saskatchewan; E. to Red River of the North; S. to Nebraska; S. to Texas, New Mexico, Arizona, and Chihuahua in migration; W. to Rocky Mts., casually beyond, in Idaho, Oregon, and Washington. An interesting Sparrow, long almost unknown till I found it breeding in profusion in Dakota, taking 75 specimens in the summer of 1873. In general habits and appearance in life quite like the Savanna Sparrow; mixing freely with these and *Otocorys*, *Anthus spraguei*, and *Calcarius ornatus*. Song peculiar, of two or three tinkling syllables and a trill, like *zip-zip-zip-zr-r-r-r*. Nest on ground, very hard to find, a slight structure of grasses and weed-stalks, about 4 inches across; eggs 5, 0.80 × 0.60, white, irregularly speckled and blotched with pale and dark reddish-browns, laid in June and July.

FIG. 268. — Baird's Savanna Sparrow, reduced. (Sheppard del. Nichols sc.)

(*Subgenus* PASSERCULUS.)

P. prin'ceps. (Lat. *princeps*, chief.) IPSWICH SPARROW. PALLID SPARROW. BARREN GROUND SPARROW. ♂ : General appearance of a large Savanna Sparrow, but with a resemblance to a Bay-winged Bunting. Upper parts grayish-brown, with blackish rufous-edged centres of the feathers; median crown-stripe not strong, and scarcely yellowish; a whitish superciliary stripe, not yellow anteriorly; ear-coverts grayish, with rufous tinge. Scapulars, coverts, and secondaries blackish-brown, broadly edged with rufous, brightest on secondaries; scapulars also edged with white, and both median and greater coverts white-tipped. Tail brownish, tipped and edged with whitish. Whole under parts white, breast and sides of throat and body streaked, the streaks dusky-centred, rufous-edged. Bill dark brown, base of under mandible paler; eyes and feet brown. Length 6.30; extent 11.00; wing 3.25; tail 2.60; bill 0.45; tarsus 0.95; middle toe and claw 1.05; hind toe and claw 0.72. (Foregoing condensed from original description of the type, taken in winter. Following as redescribed by Ridgway:)

Bill of size and shape as in *P. bairdi* exactly; inner secondaries little lengthened. Outstretched feet not reaching to end of tail. In color almost exactly as in *P. rostratus*, but different in markings; above light ashy, the dorsal feathers light sandy-brown centrally, their shafts black. Surface of wings pale sandy-brown, the feathers darker-centred; inner secondaries with whitish outer webs, and conspicuous black central field. Crown becoming darker brown anteriorly, where an indistinct median line of ochrey-white; an indistinct superciliary stripe, and conspicuous maxillary stripe of the same, the latter bordered above by a narrow dusky stripe; lores and cheeks like the superciliary stripe; auriculars like crown. Below, white, slightly ashy on flanks; whole breast and sides of body with narrow streaks of blackish-centred sandy-brown; belly, crissum, and lining of wings immaculate; throat with a few minute specks, but on each side a bridle of suffuse streaks. ♀: Wing 2.90; tail 2.40; culmen 0.50; tarsus 0.85. (Following notes taken by me of a specimen received from Maynard; ♀, Ipswich, Oct. 18, 1872, No. 73,553, Mus. S. I.: About size of largest *P. sandwichensis* from Alaska. No trace of yellow on head or wing. Upper parts even paler and grayer than extreme of *P. alaudinus* from the West — the streaks of upper parts having only shaft-lines of blackish-brown, brown-edged, the edges of the feathers finally gray; nape, rump, and upper tail-coverts gray, scarcely streaked at all. Crown streaked like interscapulars, but in smaller pattern; divided by a median light line. A long whitish (not yellowish) superciliary line; lore gray below this. Inner secondaries and greater coverts blackish, broadly edged on outer webs with bay, fading to whitish at tips; median coverts similar, but more noticeably whitish-tipped; these edgings of wing-feathers making the strongest coloration of all the upper parts. Below, white; throat and middle of belly only immaculate, flanks a little shaded with gray; whole breast, sides of neck and body, and crissum, with brown streaks, pale in comparison with those of *P. savanna*, and rather suffuse. On the sides of head below auriculars the stripes tend to form two chains — a maxillary one and another above it separated by an immaculate interval.) The breeding plumage shows yellow on the superciliary line anteriorly and on bend of the wing. This curious Sparrow was originally discovered on the sand hills of the Massachusetts coast in Dec. 1868, by C. J. Maynard. It was at first mistaken for *P.* (*c.*) *bairdi*, to which it bears no special resemblance (see Am. Nat. 1869, p. 554 and p. 631; 1872, p. 307; Naturalist's Guide, p. 112, with frontispiece plate; Key, orig. ed. 1872, p. 135). The range of the species was long a mystery; it is now known to be a local form, breeding on Sable Island, 86 miles from Nova Scotia, and ranging in migration along the Atlantic coast from Nova Scotia to Georgia. Its peculiar characters have been developed in direct consequence of its **insular environment in the breeding season.** For full history, synonymy, bibliography, description of breeding plumage. see DWIGHT, Mem. Nutt. Orn. Club, No. 2, Aug. 1895, 4to, pp. 56, colored plate. Eggs 4-5, 0.80 × 0.60, colored like those of the common species. (*P. princeps* MAYN. Am. Nat. vi, 1872, p. 637; *Ammodramus princeps* RIDGW. Pr. U. S. Nat Mus. viii, 1885, p. 354; *A.* (*P.*) *princeps*, A. O. U. Lists, 1886 and 1895, No. 541.)

FIG. 269. — Sandwich Sparrow.

P. sandwichen′sis. (Of the Sandwich, one of the Aleutian Islands. Fig. 269.) SANDWICH SPARROW. Similar to the ordinary Savanna Sparrow; averaging in size about the maximum of the latter; length about 6.00; wing 3.00; tail 2.25; culmen 0.45; depth of bill at base 0.25; tarsus, and middle toe and claw, each 0.80. Bill nearly twice as bulky as that of ordinary *savanna*. A firm bright yellow superciliary stripe from nostril to eye, thence fading over auriculars (*i. e.*, *chrysops* PALL.). Under parts precisely as in *savanna;* upper similar, but grayer — less rufous and more gray in edgings of feathers. Such are the peculiarities of a specimen from the very spot whence Latham described his Sandwich Bunting in 1783, basis of *Emberiza sandwichensis* GM. 1788, the same as the Aoonalashka Bunting of PENNANT, 1785, and *Emberiza arctica* LATH. 1790. The differences are appreciable on laying the skin alongside a large varying series of Eastern *savanna;* but it does not follow that all Alaskan and Aleutian Savanna Sparrows are like this. Birds more or less exactly like this are now known to range along the N. W. coast from Oregon to the Aleutian Islands. They are not specifically distinct from the common Savanna Sparrow, but represent a fairly well marked form. It is unfortunate that, as in the case of our Hermit Thrushes, we have to go to this extreme for our name of the stock species, and treat the common Savanna Sparrow as nominally a subspecies of *sandwichensis;* but our rules of nomenclature leave us no alternative. I first reduced the several subspecies to their proper status in the original edition of the Key, 1872, and first adopted the present nomenclature in the second edition, 1884. This is the same course now taken by the A. O. U., excepting that, *Passerculus* being reduced to a subgenus of *Ammodramus*, the present species appears as *Ammodramus sandwichensis* in the Lists, 1886 and 1895, No. 542.

P. s. savan'na. (Spanish *sabana* or *savana*, a meadow. Fig. 270.) COMMON SAVANNA SPARROW. Adult ♂ ♀, in spring: Thickly streaked everywhere above, on sides, and across breast; a superciliary line, and edge of wing, *yellowish;* lesser wing-coverts *not* chestnut; legs flesh-color; bill rather slender and acute; tail nearly even, its outer feathers not white; longest secondary nearly as long as primaries in the closed wing. Above, brownish-gray, streaked with blackish, whitish-gray and pale bay, the streaks largest on interscapulars, smallest on cervix; crown divided by an obscure whitish line; sometimes an obscure yellowish suffusion about head besides the streak over the eye. Below, white, pure or with faint buffy shade, thickly streaked, as just stated, with dusky — the individual spots edged with brown, mostly arrow-shaped, running in chains along sides, and often aggregated in an obscure blotch on breast. Wings dusky; coverts and inner secondaries black-edged and tipped with bright bay; tail-feathers rather narrow and pointed, dusky, not noticeably marked. Extreme dimensions of both sexes: Length 5.20–6.00; extent 8.50–10.00! wing 2.40–3.00; tail 1.75–2.25; tarsus 0.75–0.88; but such figures are rare. Average of both sexes 5.25; extent 8.75; wing 2.60; tail 2.00; tarsus 0.84. ♂ usually 5.30–5.60; extent 9.00–9.50; wing 2.67–2.75; ♀ usually 5.00–5.30; extent 8.75–9.00; wing 2.50–2.67. Ordinarily, bill about 0.40; tarsus, middle toe and claw together 1.50. Fall and winter specimens much more brightly colored than spring and summer ones; the young particularly having much ochrey or buffy suffusion, instead of clean colors, more brown and bay, instead of dusky and gray. It is not easy for an unpractised person to discriminate the small Sparrows, and so variable a one as this offers special difficulty; attention to the points of *form* as well as of color is requisite. North America, eastern; very abundant N. in its breeding range, in fields, on plains, by the wayside, and along the seashore; a thoroughly terrestrial bird, migratory, and in fall somewhat gregarious. Has an agreeable though weak song in spring. Winters at least from Middle

FIG. 270. — Common Savanna Sparrow, reduced. (Sheppard del. Nichols sc.)

States southward, and breeds at least from New England and other northern States to Labrador and the region of Hudson's Bay. Nest sunken in ground, flush with surface, of a few grasses and weed-stalks; eggs 4-6, 0.70 × 0.50, bluish-white, spotted with brown, varying interminably in their motley coloring; often heavily clouded and blotched with dark brown; most like those of *Poœcetes*, but smaller.

P. s. alaudi'nus. (Lat. *alaudinus*, lark-like; no applicability. Fig. 271.) WESTERN SAVANNA SPARROW. So similar to the last as only to be distinguished by rather duller and paler coloration on an average, and weaker bill, about 0.35 long by 0.20 deep at the base. If the "savanna sparrow" be split into several races, this may possibly be allowed with the rest. Western N. Am. from the Great Plains to the Pacific and Arctic Oceans.

P. s. bry'anti. (To W. E. Bryant.) BRYANT'S MARSH SPARROW. A form from the marshes of the California coast from San Francisco southward. Bill as long as that of *savanna*, but slenderer; general coloration darker; under parts more sharply, darkly, and extensively streaked; yellow eyebrow and bend of wing quite as well marked as in *savanna*, and whole head sometimes suffused with yellowish. This general heaviness of coloration contrasts with the paler and grayer *alaudinus* of

FIG. 271. — Western Savanna Sparrow.

the West; but is not very different from some specimens of true *savanna*; the size averages about the minimum of the latter. This is *P. s. anthinus* of earlier eds. of the Key — a name which has proved inapplicable. *P. s. bryanti* RIDGW. Pr. U. S. Nat. Mus. vii, Jan. 1885, p. 517, and *A. s. bryanti* ID. *ibid.* viii, 1885, p. 354; A. O. U. List, 2d ed. 1896, No. 542 c.

P. bel'dingi. (To L. Belding.) BELDING'S MARSH SPARROW. Similar to the last, but apparently specifically different in lack of distinct median and superciliary stripes, decidedly streaked under tail-coverts, and general dark coloration, being more heavily streaked with black above and with dusky below; wings and tail rather shorter; bill comparatively larger. Length about 5.00; wing 2.60; tail 2.00; bill 0.45; tarsus 0.80. Salt marshes of the Pacific coast from Santa Barbara southward to Todos Santos Island, Lower California. Nest of grass, in the grass, usually lined with hairs; eggs 4, pale bluish, irregularly marked with different shades of brown, laid in April. Not in earlier eds. of the Key. *P. beldingi* RIDGW. Pr. U. S. Nat. Mus. vii, Jan. 1885, p. 516; *Ammodramus beldingi* ID. *ibid.* viii, 1885, p. 354; A. O. U. List, 2d ed. 1895, No. 543.

P. rostra'tus. (Lat. *rostratus*, beaked; *rostrum*, beak. LARGE-BILLED SPARROW. BEAKED SPARROW. SAN DIEGO SPARROW. SEASHORE SPARROW. With the form of a Savanna, but bill elongated as in *Ammodramus*, yet very stout and turgid, with decidedly convex culmen 0.50 long. No yellowish over eye or on edge of wing; no evident median stripe on crown. Brownish-gray, obsoletely streaked with dark brown, most noticeable on crown and middle of back; entire under parts dull white, confluently streaked with clear brown everywhere except on throat, middle of belly, and crissum. Wings and tail dusky gray; rectrices with paler edges, primaries with whitish edges, wing-coverts and secondaries broadly edged and tipped with grayish-bay. An obscure whitish superciliary line. Bill light brown, under mandible paler or yellowish; legs pale. Length 5.25; wing 2.50-2.75; tail 2.00. Pacific coast of Southern and Lower California and N. W. parts of Mexico; a curious species, common, maritime, representing, like the two foregoing, the *Ammodrami* in the marshes of the Pacific seashore. *Emberiza rostrata* CASS. 1852; *Ammodramus rostratus* CASS. Ill. 1855, p. 226, pl. 38; *A. (Passerculus) rostratus* A. O. U. Lists, 1886 and 1895, No. 544; *Passerculus rostratus* of all eds. of the Key, 1872-90.

P. r. gutta′tus. (Lat. *guttatus*, spotted; *gutta*, a drop.) ST. LUCAS SPARROW. Bill shaped as in *rostratus*, relatively as stout, but smaller; culmen 0.45; depth at base 0.25. Bird smaller: pattern of coloration the same, but tone darker; streaking of under parts sharper, heavier, and darker. Instead of the light brownish-gray of *rostratus* the upper parts are here dark, almost olivaceous, brown, so that the dark streaking of crown and interscapulars is less noticeable. The same difference characterizes the under parts. Cape St. Lucas, and some other portions of L. Cala. *Passerculus guttatus* LAWR. 1867; COUES, all eds. of the Key, 1872-90; *Ammodramus* (*P.*) *rostratus guttatus* RIDGW. Pr. U. S. Nat. Mus. viii, 1885, p. 355; Man. 1887, p. 410; A. O. U. Lists, 1886 and 1895, No. 544 *a*.

P. sancto′rum. (Lat. genitive plural of *sanctus*, holy, sacred, saintly; as noun, a saint. There are so many places named in Lower California for persons of such description, that I concluded to dedicate this Sparrow impartially to the whole calendar of them.) ALL SAINTS' SPARROW. Like *guttatus*: larger; wing 2.75; bill 0.50, at base 0.30 deep, thus as large as that of *rostratus*, but regularly conic, with straight culmen suddenly deflected at end, and perfectly straight commissure; upper mandible and tip of lower blackish; rest apparently yellowish. Eggs 0.82 × 0.60, flecked and blotched with umber on bluish white ground, as usual in the genus, laid March and later. San Benito Isl., on the Pacific coast of Lower California, lat. 28° 18′ N., long. 115° 35′ W. (See Pr. U. S. Nat. Mus. v, 1882, pub. March 21, 1883, p. 538.) This species, which has stood in the Key since 1884, p. 364, was ignored by the A. O. U. until confirmed by better specimens than my types in Mus. S. I. See COUES, Auk, Jan. 1897, p. 92. *Ammodramus sanctorum*, A. O. U. List, Eighth Suppl. Auk, Jan. 1897, p. 121, No. 544. 1.

AMMO′DRAMUS. (Gr. ἄμμος, *ammos*, sand; δραμεῖν, *dramein*, to run.) GRASSHOPPER SPARROWS and SEASIDE SPARROWS. (Thus including the two genera *Coturniculus* and *Ammodramus* of all previous editions of the Key, the former being now reduced to a subgenus of the latter: see under *Passerculus* for explanation.) Bill in typical *Ammodramus* remarkably slender and lengthened for this family, with culmen decurved toward end, gonys straight, and sometimes an evident lobation of cutting edge of upper mandible; in some species of *Coturniculus*, bill much shorter and stouter. Wings short and rounded, so that the inner secondaries reach nearly to its tip when closed, without special elongation on their part. Tail variable with the species, in most of them shorter than wings, in some about equal, in *C. lecontei* longer than wings; in form rounded or even graduated, with narrow, pointed feathers, quite stiffish and sharp in some species, in others weak and lanceolate, in *C. lecontei* extremely attenuate and acuminate — in fact, the tails of these Sparrows differ more than is usual among species which are allowed to be of the same genus. Feet large and stout, reaching when outstretched nearly or quite to the end of the tail; tarsus about equal to middle toe and claw; lateral toes equal, short, their claws underreaching base of middle claw. *Coturniculus* contains three remarkably distinct North American species (besides several extralimital ones) of queer little "grasshopper" Sparrows of grass, weeds, and reeds, with greatly variegated plumage and conspicuous buffy tints on under parts; they show a greater range of variation in form than our finical modern genera usually allow, and grade through *C. lecontei* into the *Ammodrami*. The latter are the true "seaside" Sparrows, embracing several species of small Sparrows of marshes, especially of the seacoast, but not so exclusively maritime as was long supposed; they are remarkable for slenderness of bill, sharp, narrow tail-feathers, and stout feet fitted for grasping slender, swaying reeds; they have edge of wing yellow, a yellow spot or buff stripe on head, and upper parts olive-gray or quite blackish, streaky. I have several species and subspecies to add to those given in former editions of the Key; they are best analyzed under separate heads of their respective subgenera.

FRINGILLIDÆ: FINCHES, BUNTINGS, SPARROWS.

(*Subgenus* COTURNICULUS.)

Analysis of Species.

Tail shorter than wing, nearly even or a little double-rounded; outstretched feet reaching to or beyond its end. Bill stout, nearly as deep as long, brown. Adults not evidently streaked below. Edge of wing conspicuously yellow
savannarum passerinus and *s. perpallidus*
Tail about equal to wing. Bill stout, nearly as deep as long, brown. Adults with sharp maxillary, pectoral and lateral black streaks. Edge of wing yellow . . . . *henslowi* and *h. occidentalis*
Tail longer than wing, graduated, with very narrow, tapering, pointed feathers. Bill slender, not nearly as deep as long, bluish. Adults with sharp lateral but not maxillary or pectoral black streaks. Edge of wing not yellow
lecontei

A. (C.) **savanna′rum passeri′nus.** (Lat. *savannarum*, of savannas, genitive plural of *savanna*, Spanish, *sabana*, a meadow. Lat. *passerinus*, sparrow-like. Fig. 272.) YELLOW-WINGED SPARROW. QUAIL SPARROW. GRASSHOPPER SPARROW. Adult ♂ ♀: Edge of wing conspicuously yellow; lesser wing-coverts greenish-yellow; a yellow loral spot; short line over eye buffy yellow. Crown with median stripe of pale brownish-yellow. Below, ochraceous or pale buff or tawny, fading to whitish on belly, not evidently streaked, though a few dark touches may appear on sides of breast. Above, singularly variegated with black, gray, yellowish-brown and a peculiar purplish-bay, in short streaks and specks; crown nearly black with sharp median brownish-yellow stripe; middle of back chiefly black with bay and brownish-yellow edgings of the feathers; cervical region and rump chiefly bay and gray. When the feathers are not disturbed, the peculiar pattern of cervical region separates that of crown and back; the markings extend on sides of neck, but sides of head are plain, like under parts. Wing-coverts and inner secondaries variegated in intricate pattern, in general effect like back. Primaries and tail-feathers plain dusky, with narrow light edgings; outer tail-feathers paler, but not white. Feet flesh-colored. Small: Length 4.80–5.25; extent 8.00–8.50; wing 2.25–2.50; tail 2.00 or less, thus shorter than wing, outstretched feet reaching beyond it; rounded or rather double-rounded at end, the feathers narrow and lanceolate; tarsus 0.75. Bill brownish, very stout and full, culmen about 0.40, depth at base 0.30. In autumn, fresh-moulted birds are as usual richer in color, the markings more blended and diffuse; fore parts below and sides rich buffy brown in which vague lighter and darker markings usually appear. Young: Before moult, like the adult above, but with less of the reddish-brown and more of the buff markings; whitish below, with decided dusky maxillary and pectoral streaks, thus resembling *C. henslowi*. Eastern U. S. and southern Canada, W. to the Plains; breeds throughout its range; resident in the Southern States, elsewhere a migrant and summer visitant, extends in winter to some of the West India Islands, Mexico, and even Central America. Abundant in rank herbage of old fields, but less frequently observed than it would be did it not hide so persistently. This little Sparrow has a curious resemblance to a miniature Quail, whence the subgeneric name *Coturniculus*, diminutive of *coturnix*, a quail. It has a peculiar chirring note, like the stridulation of a grasshopper, which made me give the name of "Grasshopper Sparrows" to this subgeneric group. The nest is built on the ground,

FIG. 272. — Yellow-winged Sparrow, reduced. (Sheppard del. Nichols sc.)

of grasses, rather large for the size of the bird, and often somewhat domed or arched over; eggs 4-5, sometimes only 3, 0.72 × 0.60, crystal white, flecked with reddish-brown, in which markings a few neutral tints or blackish specks may also appear. *Coturniculus passerinus* of all previous editions of the Key, as of most American writers; but our Yellow-winged Sparrow proves to be only a subspecies of that which inhabits some of the West Indies, and was originally described from Jamaica as the savanna bird by SLOANE, Nat. Hist. ii, p. 306, pl. 259, whence *Fringilla savannarum* GM. 1788. This record, in connection with reduction of *Coturniculus* to a subgenus of *Ammodramus*, explains the name which I now adopt, following the A. O. U. List, No. 546.

A. (C.) s. perpal'lidus. (Lat. *perpallidus*, very pale.) BLEACHED YELLOW-WINGED SPARROW. WESTERN GRASSHOPPER SPARROW. Very similar to the last; size the same; coloration paler and grayer; less black and more slaty-gray on upper parts; ochrey crown-stripe and edgings of dorsal feathers, as well as under parts generally, paler. Western U. S., Plains to the Pacific, S. to México and Cape St. Lucas. *Coturniculus passerinus perpallidus* COUES, Key, 1872, p. 137, and of all later eds.; *Ammodramus savannarum perpallidus* RIDGW. Pr. U. S. Nat. Mus. viii, 1885, p. 355; Man. 1887, p. 411; *Ammodramus (Coturniculus) savannarum perpallidus* A. O. U. Lists, 1886 and 1895, No. 546 a.

A. (C.) hen'slowi. (To Prof. J. S. Henslow, of England. Fig. 273.) HENSLOW'S GRASSHOPPER SPARROW. Somewhat resembling a *young* Yellow-winged Sparrow. Adult ♂ ♀:

FIG. 273. — Henslow's Grasshopper Sparrow.

Under parts whitish, tinged strongly along whole sides, across breast, and on flanks and crissum, with buff, all these buff parts sharply and distinctly streaked with blackish in fine pattern; pectoral streaks connecting along sides of neck with decided black maxillary stripes. The brownish-yellow shade is very variable in extent and intensity, but it usually leaves only throat and belly decidedly whitish. Ground-color of head and hind neck peculiar pale olive-gray, with decided greenish-yellow tinge; top of head with broad lateral blackish stripes, continued on cervix in much smaller pattern, divided by a greenish-brownish-yellow median stripe. The peculiar color of hind neck extending far around on sides of neck, and sides of head of much the same tint; a blackish post-ocular stripe bounding auriculars above; below and anterior to them a black maxillary stripe starting from angle of mouth; below this usually other maxillary streaks; dark specks often behind auriculars. Dorsal and scapular feathers with broad black central field, then broadly chestnut, then mostly narrowly edged with whitish, these markings in bold pattern, and contrasting with peculiar greenish-gray cervical region with its fine black streaks. Edge of wing yellow. Greater wing-coverts and most secondaries colored to correspond with back, the closed wing showing chiefly chestnut with black field of three innermost secondaries. Tail-feathers extremely narrow and acute, brown, the inner at least with long blackish shaft-stripe, and reddish-brown on inner webs. Bill brownish, usually quite dusky above, pale below; feet pale. Length 5.00; extent 7.50; wing and tail, each, 2.00-2.10; bill from extreme base of culmen 0.45; 0.30 deep at base; tarsus or middle toe and claw 0.65. Young resemble the adults sufficiently to be unmistakable, but are rather dark buff above, with black streaks and spots, and very pale buff below, without black pectoral and most of the maxillary streaks, though there is one stripe starting from the corner of the mouth. The lesser streaking of the under parts is the reverse of the case of the Yellow-winged Sparrow, the young of which are more streaked below than the adults. Eastern U. S., strictly, N. to New England, Mich., Minn., and Ontario, not very commonly; W. to the edge of the Great Plains; winters in the Gulf States. Not abundant on the whole, nor easily observed; song a simple *zip, zip, zip, zirip, zipzirip, zipzirip, zirip*, with head held back, bill up, and tail down. Common about Washington, D. C., where it breeds, in fields and meadows; nest on the ground, in tufts of grass. Eggs 4–5, greenish-white, profusely speckled with reddish, 0.75 × 0.57. *Coturniculus henslowi* of all previous eds. of the Key; *Emberiza henslowii* AUD. 1831; *Ammodromus henslowi* GRAY, 1849; *Ammodramus* (*Coturniculus*) *henslowii* A. O. U. Lists, 1886 and 1895, No. 547.

A. (C.) h. occidenta'lis. (Lat. of the occident, western.) WESTERN HENSLOW'S SPARROW. DAKOTA GRASSHOPPER SPARROW. Similar to *A.* (*C.*) *henslowi*, but general coloration paler (as in the corresponding case of *A.* (*C.*) *savannarum perpallidus*); under parts whiter; back and scapulars with broader black streaking and much less chestnut, the wings and tail grayer. Wing 2.18; tail 1.95; tarsus 0.69; bill from nostril 0.31; its depth at nostril 0.32. Moody Co., S. Dakota, and probably other places along the E. border of the Great Plains (the Nebraska record of *henslowi* probably belongs here). BREWST. Auk, Apr. 1891, p. 145; A. O. U. List, 2d ed. 1895, No. 547 *a*.

A. (C.) lecon'tei. (To Maj. J. Le Conte, of Philadelphia.) LE CONTE'S GRASSHOPPER SPARROW. LE CONTE'S BUNTING. Adult ♂ ♀: Bill smaller and slenderer than in either of the foregoing, dark horn-blue above, paler bluish below; iris black. Tail long, decidedly exceeding wing when full grown, and remarkably graduated; lateral feathers ⅓–½ inch shorter than central pair; all extremely narrow, tapering, and acuminate, even more so than in the Sharp-tailed Finch (*Ammodramus caudacutus*); outstretched feet not reaching to its end. Wings short and much rounded; primaries in closed wing hardly ¼ inch longer than secondaries. Length 4.90–5.10; extent 6.90–7.10; wing 1.90–2.00; tail 2.00–2.25 or a little more; bill 0.40; tarsus 0.67. No trace of yellow on bend of wing, nor any yellow loral spot. No black maxillary or pectoral streaks; markings of under parts confined to sparse, sharp, blackish streaks on sides. General coloration more or less buff, according to age and season. Crown with black lateral stripes, separated by a whitish stripe becoming ochrey on forehead. Sides of head buff, brightest on long broad superciliary line, enclosing slaty-gray auriculars, which are bordered above by a black postocular line, sometimes chiefly appearing as a dark speck behind them. Cervical feathers bay, black-shafted and whitish-edged, forming a distinct interval between markings of back and crown. Dorsal feathers in bold pattern, with black terminal central field, little rufous and much whitish or buffy edging; streaking extending on rump and upper tail-coverts. Wing-coverts and inner secondaries colored boldly to correspond with the back. Under parts buffy-white, sometimes quite whitish, again much more buffy, with season, usually quite buff with only belly whitish. Fresh moulted fall birds are often entirely deep buff below, excepting belly, which is white, in marked contrast. Young: Bill still smaller, reddish-brown instead of bluish; general color buff above, whitish below, more or less buffy on breast and sides; markings of upper parts black, without bay and brown variegation, except on wings and tail, which are nearly as in the adults; sparse black streaks of under parts usually appearing across breast as well as on sides. An interesting, long-lost species, but rediscovered: Yellowstone region (*Audubon*, 1843); Texas (*Lincecum*); N. Dakota, breeding (*Coues*, 1873); Illinois (*Nelson*, 1875); Iowa (*Newton*, 1875); Minnesota (*Tiffany*, 1878); North Carolina (*Brimley*, 1894); South Carolina (*Loomis*, 1881); New York (*Fuertes*, 1897). The normal range of the species may now be given as the Great Plains of the U. S. and adjoining British Provinces, from Assiniboia and Manitoba to Texas, and E. in migration to the

S. Atlantic and Gulf States, including Florida. It breeds in northerly parts of this range, but only in moist or marshy spots; nest bulky, on the ground or in thick grass or a clump of reeds; eggs 3-5, 0.72 × 0.54, white, profusely flecked with brown, sometimes chiefly marked about the larger end with darker brown or blackish spots. *Coturniculus lecontii* of all previous eds. of the Key; *Emberiza leconteii* AUD., 1843; *Ammodromus leconteii* GRAY, 1849; *Ammodramus* (*Coturniculus*) *leconteii* A. O. U. Lists, 1886 and 1895, No. 548. Approaching *Ammodramus caudacutus* in many respects, and inhabiting similar resorts in the interior.

(Subgenus AMMODRAMUS.)

Analysis of Species.

Coloration much variegated; general tone buffy. No bright yellow on lore or edge of wing; long buff superciliary and malar stripes . . . . *caudacutus, nelsoni, n. subvirgatus*
Coloration little variegated; general tone dark. Loral spot and edge of wing bright yellow.
Upper parts olive-gray, obscurely streaked . . . . *maritimus, m. peninsulæ, m. sennetti*
Upper parts quite blackish . . . . *nigrescens*

A. caudacu'tus. (Lat. *cauda*, tail; *acutus*, sharp. Fig. 274.) SHARP-TAILED FINCH. QUAIL-HEAD. Olive-gray, sharply streaked on back with blackish and whitish, less so on rump with blackish alone. Crown darker than nape, with brownish-black streaks, tending to form lateral stripes and obscure olive-gray median line; no yellow loral spot, but long line over eye and sides of head rich buff or orange-brown, enclosing olive-gray auriculars and a dark speck behind them, or dark postocular stripe over them. Olive-gray of cervix extending on sides of neck. Below, white; fore parts and sides tinged with yellowish-brown or buff of variable intensity, breast and sides sharply streaked with dusky. Greater coverts and inner secondaries with blackish field toward their ends, broadly margined with rusty brown and whitish. Tail-feathers brown, with dusky shaft-stripes

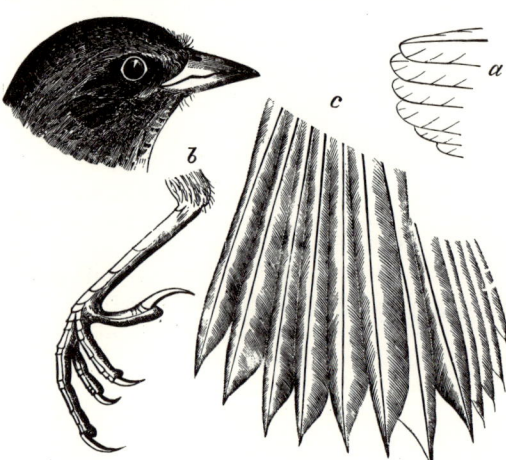

FIG. 274. — Generic details of *Ammodramus* (*A. caudacutus*), nat. size. (Ad. nat. del. E. C.)

and tendency to "water" with crosswise wavy bars. Bill blackish above, pale or not below, feet brown. Coloration in spring and summer clearer and paler, in fall and in young birds more brightly and extensively buff. Rather smaller than *A. maritimus*; bill still slenderer and tail-feathers still narrower and more acute. Length 5.10–5.50; extent 7.50; wing 2.25; tail 2.00; bill 0.45–0.50; tarsus, or middle toe and claw, 0.75. Salt marshes of the Atlantic and Gulf States, N. to Maine, abundant; range similar to that of *A. maritimus*, but on the whole more northerly, especially in the breeding season; nest and eggs similar and scarcely distinguishable; eggs rather smaller, 0.75 × 0.55, and perhaps less boldly marked.

A. nel'soni. (To E. W. Nelson, of Illinois.) NELSON'S SHARP-TAILED FINCH. Similar to the last, but smaller, with bill slenderer and shorter; colors brighter and markings more sharply defined, especially the dead white streaks on the rich brown ground of the back. Fresh marshes of Mississippi Valley; breeds from N. Illinois to Manitoba, winters to Texas; in migrations on the Atlantic coast from New England to South Carolina; has occurred in California (the so-called *A. c. becki* RIDGW. Pr. U. S. Nat. Mus. xiv, 1891, p. 483). *A. c. nelsoni* of 3d–4th eds. of Key. *A. nelsoni* NORTON, Pr. Portland Soc. Nat. Hist. ii, Mar. 15, 1897, p. 102; A. O. U. Suppl. List, Auk, Jan. 1899, p. 118, No. 549. 1 (formerly No. 549 *a*).

A. n. subvirga′tus. (Lat. *sub-*, under, less than, somewhat, and *virgatus*, striped, streaked.) ACADIAN SHARP-TAILED FINCH. Said to be "similar in size and coloring to *A. caudacutus*, but paler and much less conspicuously streaked beneath with pale greenish-gray instead of black or deep brown. Bill averages smaller. Compared with *nelsoni* it is much paler and grayer, generally larger and with a longer bill." Range said to be "coast of southern New Brunswick, Prince Edward Island (and probably Nova Scotia), and southward in migration to South Carolina." *A. caudacutus subvirgatus* J. DWIGHT, Jr. Auk, July, 1887, p. 233; COUES, Key, 4th ed. 1890, p. 900; A. O. U. List, 2d ed. 1895, p. 228, No. 549 *b*. See Auk, Oct. 1896, p. 272, pl. iv. *A. nelsoni subvirgatus* NORTON, Pr. Portland Soc. Nat. Hist. ii, Mar. 15, 1897, p. 102; A. O. U. Suppl. List, Auk, Jan. 1899, p. 118; No. 549. 1 *a*.

A. mari′timus. (Lat. *maritimus*, maritime, coast-wise; *mare*, the sea. Fig. 275.) SEASIDE FINCH. SEASIDE SPARROW. Adult ♂ ♀: Olive-gray, obscurely streaked on back and crown with darker and paler; below, whitish, often washed with brownish, shaded on sides with color of back, and with ill-defined dark streaks on breast and sides; maxillary stripes of the same; wings and tail plain dusky, with slight olivaceous edgings; wing-coverts and inner quills somewhat margined with brown; edge of wing bright yellow; a bright *yellow spot* on lore, and often some vague brownish and dusky markings on side of head; bill plumbeous, or dark horn-blue; feet dark. Length 5.75–6.25; extent 8.50; wing 2.25–2.50; tail about 2.00. Recognizable on sight by bright yellow edge of wing and loral spot, with little varied olive-gray upper parts. Salt marshes of Atlantic coast from southern New England to Florida, abundant; breeds throughout its range, and resident in the south, but screened from casual observation by the nature of its haunts and habits. Nest in a tussock of grass just out of water; eggs 3–5, 0.80 × 0.60, grayish-white, thickly and pretty evenly marked with umber-brown.

FIG. 275. — Seaside Finch, reduced. (Sheppard del. Nichols sc.)

The foregoing description is applicable to all the forms of the stock species, and requires to be particularized if we wish to recognize the several slight local races into which *maritimus* was lately split with the sanction of the A. O. U. For the race to which the name *maritimus* has thus been restricted, observe the following points: Sides of crown olive, with occasionally black shaft-streaks; median bluish-gray line well defined; nape pale greenish-olive; back olive, margined with bluish-gray; breast streaked with bluish-gray, margined with buff; flanks obscurely streaked with bluish-gray and faintly washed with buff. Wing averaging 2.50; tail 2.25.

A. m. macgillivrayi. (To William Macgillivray.) MACGILLIVRAY'S SEASIDE FINCH. Scarcely different from the last. Sides of crown black, margined with brown; median bluish-gray line ill-defined; nape tawny-olive; back black, bordered by greenish-olive and margined with bluish-gray; breast and flanks streaked with dusky, margined with buff. Wing averaging 2.36; tail 2.18. Said to be confined to coasts of South Carolina and Georgia; originally described and figured from Charleston, S. C.; probable type specimen a young bird now No. 2894 U. S. Nat. Mus., examined by me. Included under *maritimus* in former eds. of the Key. *Fringilla macgillivraii* AUD. Orn. Biogr. ii, 1834, p. 285; iv, 1838, p. 394; v, 1839, p. 499; folio pl. ccclv — *in part;* for the account includes the race below given as *fisheri*. *Ammodramus macgillivrayi* AUD. B. Am. iii, 1841, p. 106, 8vo, pl. clxxiii — *in part;* range given from South Carolina to Texas, thus including other races. *A. m. macgillivraii* CHAPM. Auk, Jan. 1899, p. 5 (not of RIDGW. 1896, nor of A. O. U. List, Eighth Suppl. 1897, No. 550 *c*, which is *fisheri*); A. O. U. List, Ninth Suppl. Auk, Jan. 1899, p. 118, No. 550 *d*.

A. m. penin′sulæ. (Lat. *peninsula*, a peninsula, almost an island; *pene*, almost; *insula*, island.) SCOTT'S SEASIDE FINCH. PENINSULAR SEASIDE SPARROW. Crown as in *macgillivrayi;* nape greenish-olive, about as in *maritimus* proper; back dull black, margined with greenish-olive; breast streaked with dusky margined with buff or with bluish-gray; flanks streaked with dusky, margined with grayish or olive-buff. Wing averaging 2.32; tail 2.09. These comparative characters may be thus amplified: Adult ♀: Differing from *maritimus* proper in some points by which it approaches *nigrescens* (see beyond); like the latter in size and proportions, including size and shape of bill. Feathers of upper parts with dull brownish centres, broadly edged with olive and gray. Black streaks of under parts stronger and sharper than in *maritimus*, less so than in *nigrescens*. Throat and belly white; other under parts shaded with brownish-ash, besides the streaks. Young in first plumage: Black prevailing above, the feathers narrowly edged with ochraceous; throat and middle of belly white; sides bright ochraceous, narrowly streaked with black. Wing 2.20; tail 2.00; tarsus 0.83; bill 0.52. Type locality Tarpon Springs, Florida; breeding range not made out; general range given as from South Carolina to Texas, probably by error, in A. O. U. List. Included under *maritimus* in 1st–3d eds. of Key. ALLEN, Auk, July, 1888, pp. 284, 286; COUES, Key, 4th ed. 1890, p. 899; A. O. U. List, 2d ed. 1895, p. 228, No. 550 *a*.

A. m. fish′eri. (To Dr. A. K. Fisher, of the U. S. Biological Survey.) FISHER'S SEASIDE FINCH. LOUISIANA SEASIDE SPARROW. An alleged subspecies, sadly mixed up in synonymy, geography, and diagnosis with both the foregoing races. One account is: Sides of crown deep black, margined with mummy brown, median line ill-defined, bluish-gray; nape mummy brown; back deep black, bordered by mummy brown and margined by bluish-gray; breast and flanks streaked with black, widely margined with pale ochraceous. Wing 2.29; tail 2.12. Another diagnosis is: Upper parts deep black, in fresh plumage the feathers bordered by mummy brown and margined with bluish-gray; breast and flanks streaked with black and more or less heavily washed with pale ochraceous. Type No. 163,722 U. S. Nat. Mus., Grande Isle, Louisiana, June 9, 1886. Range given by its describer as coast of Gulf States; breeding from Grande Isle, Louisiana, westward, probably to N. E. Texas, S. in winter to Corpus Christi, Texas, and Tarpon Springs, Florida. Range restricted by A. O. U. to coast of Louisiana; in migration, coast of Texas. Included under *maritimus* or *peninsulæ* in all former eds. of Key. Said to be *Fringilla macgillivraii* AUD., in part. Said to be *A. m. macgillivrayi* RIDGW. Man. 2d ed. 1896, p. 602. Said to be *A. m. peninsulæ* ALLEN, Auk, July, 1888, p. 284, in part, and CHAPM. Bull. Amer. Mus. iii, 1891, p. 324. Finally, it turned up as *A. m. fisheri* CHAPM. Auk, Jan. 1899, p. 10, pl. i, upper fig.; A. O. U. Suppl. List, Auk, Jan. 1899, p. 118, No. 550 *c*.

A. m. sen′netti. (To George B. Sennett.) SENNETT'S SEASIDE FINCH. TEXAN SEASIDE SPARROW. Closely resembling *maritimus* proper, from which separated geographically by the foregoing races, and otherwise distinguished by the greenishness of the black-centred feathers of upper parts. It thus differs from the typical form in the opposite direction from that taken by *peninsulæ* and *nigrescens*. Adult ♂ ♀: Upper parts lighter than in *maritimus* proper; nape streaked with black. Under parts light gray, white on throat and belly, with distinct narrow blackish streaks on breast and flanks, those of breast bordered with white or pale ochraceous. Young, first plumage : Grayish-brown prevailing above, streaked with black, streaks broadest on middle of back; below pale fulvous, shaded on sides, where also sparsely streaked with black. Confined to coast of Texas, as far as known; resident at Corpus Christi, breeding abundantly in marshes of Nueces Bay. ALLEN, Auk, July, 1888, p. 286; COUES, Key, 4th ed. 1890, p. 899; CHAPM. Bull. Amer. Mus. Nat. Hist. iii, 1891, p. 323 (habits); A. O. U. List, 2d ed. 1895, p. 228, No. 550 b. A. sennetti CHAPM. Auk, Jan. 1899, p. 3, pl. i, lower fig.

A. nigres′cens. (Lat. *nigrescens*, growing black. Fig. 276.) FLORIDA SEASIDE FINCH. DUSKY SEASIDE SPARROW. Like *maritimus;* rather smaller bodied, though members not shorter, and conspicuously different in color, being almost entirely black and white. Upper parts sooty-black, slightly variegated with slate-colored edgings of the feathers, and some pale gray edgings of interscapulars. Below white, heavily streaked

FIG. 276. — Dusky Seaside Sparrow.

with blackish everywhere excepting on throat and middle of belly. A bright yellow loral spot, and bend of wing bright yellow (both very conspicuous in the black plumage). Wing-quills blackish, inner secondaries quite black ; all narrowly edged with brownish. Tail black, with gray edgings of the feathers, these edgings tending to form scallops with the black central field. Bill and feet as in *maritimus*. A curiously localized species, resident in Florida, on the Atlantic side, discovered by C. J. Maynard near Titusville, March, 1891. *A. m. nigrescens* of 2d-4th eds. of Key, now rated as a distinct species. *A. melanoleucus* MAYN. Am. Sportsm. v, 1875, p. 248; Birds of Eastern N. Am. 1881, p. 119, pl. v.

FIG. 277. — Lincoln's Song Sparrow, reduced. (Sheppard del. Nichols sc.)

MELOSPI′ZA. (Gr. μέλος, *melos*, song, melody, and σπίζα, *spiza*, name of some Finch in Aristotle.) SONG SPARROWS. Bill moderate, conic, without special turgidity or compression, outlines of culmen, commissure, gonys, and sides nearly or about straight. Wings short and much rounded, folding little beyond base of tail; 1st primary quite short; point of wing formed by 3d, 4th, and 5th, supported closely by 2d and 6th; inner secondaries not elongated. Tail long, about equalling or rather exceeding wings, much rounded, with firm feathers broad to their rounded ends. Feet moderately stout; tarsus scarcely or not longer than middle toe and claw; lateral toes slightly unequal, outer the longer, its claw scarcely or not reaching base of middle claw. Embracing a large number of middle-sized and large Sparrows, without a trace of yellow anywhere, and of brownish-yellow only in *lincolni;* upper parts, including crown, thickly streaked; under parts white or ashy, thickly streaked across breast and along sides (excepting adult *M. georgiana*). No bright color anywhere, and no colors in masses. The type of the genus is the familiar and beloved Song Sparrow, which the authors of Citizen Bird call "Everybody's Darling"—a bird of constant characters in the East, but which in the West is split into numerous geographical races, some of them looking so different from typical *melodia* that they have been considered as distinct species, and even placed in other genera. This differentiation affects not only color, but size, relative proportion of parts, and particularly shape of bill. Nevertheless, the gradation is complete, and effected by imperceptible degrees. Some Northwestern forms of great size and dark colors are easily discriminated, but there are U. S. birds from Atlantic to Pacific which are not readily told apart. The student should not be discouraged if a subject which has tried the chiefs perplexes him; nor must he expect to find drawn on paper hard and fast lines which do not exist in nature. The curt antithetical expressions used in constructing the analysis of species and varieties necessarily exaggerate the case, and are only true as indicating the typical style of each; plenty of specimens lie "between the lines" as written.

Analysis of Species and leading Subspecies.

Breast streaked, and with a transverse belt of brownish-yellow; tail nearly equal to wings . . . . . . *lincolni*
Breast ashy, unbelted, with few streaks, or none; tail about equal to wings . . . . *georgiana*
Breast white, or brownish-white, with numerous streaks; tail usually longer than the wings, both rounded. Thickly streaked above, on sides, and across breast . . . . *melodia* and its subspecies
    The streaks distinct, decidedly blackish-centred (in breeding plumage).
Tone of upper parts grayish-brown or reddish-gray. Streaked from head to tail. Dorsal streaks black, rufous, and grayish-white. Wing 2.60; tail under 3.00 . . . . *melodia* (Eastern N. A.) and *juddi*
Tone of upper parts gray. Streaks obsolete on rump. Dorsal streaks narrowly blackish and grayish-white, with little rufous. Tail about 3.00. Great Basin and Rocky Mt. regions . . . . . . *fallax* and *montana*
Tone of upper parts ashy-gray. Streaks obsolete on rump. Dorsal streaks broadly black, with little rufous and scarcely any grayish-white. Size of the first. California . . . . *heermanni*
Tone of upper parts olive-gray. Streaks on rump and upper tail-coverts. Dorsal streaks as in the last. Very small. Wing 2.25; tail 2.50. Coast of California . . . . *samuelis*
    The streaks diffuse, not black-centred nor whitish-edged. Bill slender.
Tone of upper parts rufous-brown. Streaks above and below dark rufous. Medium-sized; wing 2.60; tail under 3.00. Pacific coast, U. S. and British Columbia; Idaho . . . . *morphna; merrilli*
Tone of upper parts olive-brown. Streaks sooty. Larger: wing and tail about 3.00. Pacific coast, British Columbia and Alaska . . . . *rufina*
Breast plumbeous, with numerous diffuse streaks.
Tone of upper parts dark cinereous. Streaking reddish-brown. Largest; wing and tail 3.25 or more.
    Kadiak Island . . . . *insignis*
    Aleutian islands at large . . . . *cinerea*

M. lin'colni. (To Thomas Lincoln, who accompanied Audubon to Labrador in 1833. Figs. 277, 278.) LINCOLN'S SONG SPARROW. ♂ ♀: Below, white, with a broad brownish-yellow belt across breast; sides of body and neck, and crissum, washed with the same; extent and intensity of this buff very variable, often leaving only chin, throat, and belly purely white, but a pectoral band is always evident. All the buffy parts sharply and thickly streaked with dusky. Above, grayish-brown, with numerous sharp black-centred, brown-edged streaks. Top of head ashy, with a pair of dark brown black-streaked stripes; *or*, say, top of head brown, streaked with black, and with median and lateral ashy stripes. Below the superciliary ashy stripe is a narrow dark brown one, running from eye over ear; auriculars also bounded below by an indistinct dark brown stripe, below which and behind auriculars the parts are suffused with buff. Wings with much rufous-brown edging of all the quills; inner secondaries and coverts having quite black central fields, with broad bay edging, becoming whitish toward their ends. Tail brown, the feathers with pale edges, and central pair at least with dusky shaft-stripes. Bill blackish, lighter below; feet brownish. Length 5.50–6.00; extent 7.75–8.25; wing and tail, each, about 2.50, latter rather shorter. There is little variation in color, except as above said. Fall specimens are usually most buffy. Very young: Before fall moult, birds of the year are much browner above, with considerable brownish-yellow streaking besides the black markings; top of head quite like back, the ashy stripes not being established; whole under parts brownish-yellow, merely paler on throat and belly, dusky-streaked throughout. N. Am. at large; a peculiar species, not so well known as it might be, less numerous in the Atlantic States than in the interior and West; and keeping very close in shrubbery. Migratory; winters in the South; breeds at least from N. New York and New England to Arctic regions, and in the West S. at least to mountains of Colorado and California; S. in winter to Panama. Nesting like that of the Song Sparrow, and eggs not distinguishable with certainty; they average smaller, about 0.75 × 0.55; the ground color varies from whitish to greenish-white or brownish-white, and the markings are usually coarsely blotched.

FIG. 278. — Lincoln's Song Sparrow.

M. l. stria'ta. (Lat. *striata*, streaked, striped.) FORBUSH'S SONG SPARROW. Similar to *M. lincolni*; superciliary stripe and whole upper parts more strongly olivaceous, with the dark streaks coarser, blacker, and more numerous, especially on pileum, back, and upper tail-coverts. British Columbia. BREWSTER, Auk, April, 1889, p. 89; COUES, Key, 4th ed. 1890, p. 900; A. O. U. List, 2d ed. 1895, p. 243, No. 583 *a*.

M. georgia'na. (Lat. Georgian, *i. e.*, of the State of Georgia, named for King George II. of England, 1683–1760. Figs. 279, 280.) SWAMP SONG SPARROW. ♂ ♀, perfect plumage: Crown bright chestnut, blackening on forehead, the red cap and black vizor as conspicuous as in a Chipping Sparrow; but oftener, crown with obscure median ashy line, and streaked with black. An ashy-gray superciliary line; a dark brown postocular stripe, bordering auriculars; sides of head ashy, with grayish-brown auriculars, dusky speckling on cheeks and lores, and slight dusky maxillary spots or streaks. An ashy cervical collar separating chestnut crown from back, sometimes pure, oftener interrupted with blackish streaks. The general ash of sides of head and neck spreads all over breast and under parts, fading to whitish on throat and belly; sides, flanks, and crissum marked with brown, and obsoletely streaked with darker brown. Back and rump brown, rather darker than sides of body, boldly variegated with black

FIG. 279. — Swamp Song Sparrow. (L. A. Fuertes.)

central streaks of the feathers and their pale brown or grayish edges. Wings so strongly edged with bright bay as to appear almost uniformly brownish-red when closed; but inner secondaries and greater coverts showing some black and whitish besides the bay. Tail likewise strongly edged with bay, and usually showing sharp black shaft-lines. Thus well marked by emphasis of black, bay, and ash. Length 5.40–5.80, usually 5.60; extent 7.50–8.00; wing and tail, each, 2.20–2.40. Varies little except as above noted, and in extent and intensity of the ash on fore and under parts. In birds of the first autumn, the crown may be quite blackish, with little chestnut and an ashy median stripe. Very young birds may be conspicuously streaked below, and a few streaks may persist on sides of breast. Eastern N. Am., W. regularly to the Great Plains, casually Utah, N. to Hudson's Bay and Labrador, but chiefly Eastern U. S. and Canada; breeding from the N. States northward, wintering chiefly in the Southern States. Abundant, but in the breeding season closely confined to watery tracts, and seldom seen by the *profanum vulgus;* a good musician, like all the genus. Nesting and eggs generally like those of the Song Sparrow, the eggs perhaps averaging a little smaller, 0.75 × 0.55, and rather coarsely blotched than finely speckled with the darker colors. *M. palustris* of most authors, and all previous editions of the Key; but there is no doubt that this is *Fringilla georgiana* LATHAM, Ind. Orn. i, 1790, p. 460, as indicated by NUTTALL, Man. 2d ed. 1840, p. 588, and doubtfully by BAIRD, B. N. A. 1858, p. 483.

FIG. 280. — Swamp Song Sparrow, reduced. (Sheppard del. Nichols sc.)

M. melo'dia. (Gr. μελῳδία, *melodia,* Lat. *melodia,* a noun, meaning melody, or a melodious song. The adj. would be *meloda* or *melodicus.* Figs. 281, 282.) SONG SPARROW. SILVER-TONGUE. "EVERYBODY'S DARLING." Below, white, slightly shaded with brownish on flanks and crissum; with numerous black-centred, brown-edged streaks across breast and along sides, usually forming a pectoral blotch and coalescing into maxillary stripes bounding white throat; crown dull bay, with fine black streaks, divided in the middle and bounded on either side by ashy-whitish lines; vague brown or dusky and whitish markings on sides of head; a brown postocular stripe over gray auriculars, and another, not so well defined, from angle of mouth below auriculars; interscapular streaks black, with bay and ashy-white edgings; rump and cervix grayish-brown, with merely a few bay marks; wings with dull bay edgings, coverts and inner quills marked like interscapulars; tail plain brown, with darker shaft-lines, on middle feathers at least, and often with obsolete transverse wavy markings.

FIG. 281. — Song Sparrow. (L. A. Fuertes.)

Very constant in plumage, the chief differences being in sharpness and breadth of markings, due in part to the wear of the feathers. In worn midsummer plumage, the streaking is very sharp, narrow, and black, from wearing of rufous and whitish, especially observable below where the streaks contrast with white, and giving the impression of *heavier* streaking than in fall and winter, when, in fresher feather, the markings are softer and more suffuse. The aggregation of spots into a blotch on middle of breast is usual. Bill dark brown, paler below; feet pale brown. Length 5.90–6.50, usually 6.30; extent 8.25–9.25, usually 8.50–9.00; wing 2.40–2.75, usually about 2.60; tail nearer 3.00. ♀ averaging near the lesser dimensions, but the species remarkably constant in size, form, and coloring. Eastern U. S. and Canada; breeds in nearly all its range, wintering nearly throughout; one of the common winter Sparrows of the Middle States. A very abundant bird everywhere in shrubbery and tangle, garden, orchard, and park, as well as swamp and brake. A hearty, sunny songster, whose quivering pipe is often tuned to the most dreary scenes; the limpid notes being one of the few snatches of bird melody that enlivens winter. Nesting various, usually near the ground in bush or grass tuft, or on the ground: eggs 4–6, 0.75–0.85 × 0.55–0.60, greenish or grayish-white, endlessly varied with browns, from reddish to chocolate as surface-markings, and lavender or purplish shell-markings, either speckled, blotched, or clouded; no general effect describable in few words. Two or three broods may be reared. (*M. fasciata* of 2d–4th eds. of Key and A. O. U. Lists to 1899, after *Fringilla fasciata* GM. 1788, but this specific name is preoccupied by *F. fasciata* MÜLL. 1776 (see Auk, Apr. 1899, p. 183). We may therefore gladly revert to the name *M. melodia* of the orig. ed. of the Key, 1872, after BAIRD, 1858, from *F. melodia* WILS. 1810.)

FIG. 282. — Song Sparrow, reduced. (Sheppard del. Nichols sc.)

M. m. jud'di. (To E. T. Judd, of Cando, N. Dak.) DAKOTA SONG SPARROW. The least departure from *melodia* proper, apparently in the direction of the Oregon Song Sparrow. Ground color of upper parts rather paler than in *melodia*, especially the superciliary streak and sides of neck; interscapulars with broader black centres, narrower reddish-brown portions, and paler gray edgings; markings of under parts restricted and more sharply defined on a clearer white ground. Length 6.75; wing 2.62; tail 2.78; tarsus 0.81; culmen 0.51; depth of bill at base 0.31. North Dakota, breeding about Turtle Mt. in June and July. Eggs 0.75 × 0.60, with alleged tendency to a subpyriform figure unusual in those of *melodia*, but indistinguishable in color. I became familiar with this bird while camping on Turtle Mt. in 1873, without suspecting any difference from the common Song Sparrow of the East; however the A. O. U. Committee admitted it to the List at the Cambridge meeting, Nov. 13, 1896. *M. fasciata juddi* BISHOP, Auk, Apr. 1896, p. 132; A. O. U. List, 8th Suppl. Auk, Jan. 1897, p. 122, No. 581 *j*. (Included under *melodia* or *fasciata* in former eds. of Key.)

M. m. fal'lax. (Lat. *fallax*, fallacious, deceitful: well named.) GRAY SONG SPARROW. DESERT SONG SPARROW. Very similar to both the foregoing; tail rather longer; tone of upper parts paler and grayer; streaks less obviously blackish in centre and with less rufous; obsolete on rump. Southern Rocky Mt. region and portions of the Great Basin, in desert places; type specimen from Pueblo Creek, N. M., on Whipple's route, Jan. 22, 1854; range mainly in New Mexico, Arizona, southern Nevada, southwestern Utah. *Zonotrichia fallax*

BAIRD, 1854; *M. fallax* BAIRD, 1858; *M. m. fallax* COUES, Key, 1872, p. 139; *M. f. fallax* of later eds. of Key, p. 372, and A. O. U. Lists, 1886 and 1895, No. 581 *a*.

M. m. monta'na. (Lat. *montana*, of mountains.) MOUNTAIN SONG SPARROW. Scarcely distinguishable from *fallax*, and the form which most authors have called *fallax*. Upper parts umber-brown with gray margins of the feathers, giving a strong grayish cast; back streaked with blackish-brown; streaks of under parts also of this color. This is the form characteristic of the Great Basin at large. *M. f. montana* HENSH. Auk, July, 1884, p. 224; COUES, Key, 3d ed. 1887, p. 874; A. O. U. Lists, 1886 and 1895, No. 581 *b*. *M. m. montana* OBERH. Auk, Apr. 1899, p. 183. (Included under *fallax* in 1st and 2d eds. of Key.)

M. m. heer'manni. (To Dr. A. L. Heermann, of Philadelphia.) HEERMANN'S SONG SPARROW. Similar to the foregoing, and size of *melodia* proper. Tone of upper parts grayish, the streaks numerous, broad, distinct, with little rufous, mostly lacking pale edgings, and obsolete on rump. Portions of California, and western Nevada; type from Tejon Pass, Cal. *M. heermanni* BD. 1858; *M. m. heermannii* COUES, Key, 1872, p. 139; *M. f. heermanni* of 2d–4th eds. of Key, p. 372, and of A. O. U. Lists, 1886 and 1895, No. 581 *c*.

M. m. samue'lis. (To E. Samuels, of California.) SAMUELS' SONG SPARROW. Similar to *heermanni* in distinctness of the black streaks, which are not obsolete on rump; under tail-coverts also streaked. Bill long, slender, acute; wings very short, much rounded. Size very small. Baird gives length only 5.00; wing 2.20; tail 2.35; Ridgway gives length 4.70–5.75; wing 2.15–2.50; tail 2.00–2.68,—measurements manifestly impossible to a single subspecies so finely drawn as those of *melospiza!* Bill along culmen about 0.50, its depth at base about 0.25; tarsus 0.85. Eggs said to measure 0.74 × 0.58. Coast region of California; type specimens from Petaluma, Cal. *Ammodromus Samuelis* BAIRD, Pr. Bost. Soc. Nat. Hist. June, 1858; B. N. A. 1858, p. 455; later ed. pl. 71, fig. 1. *Melospiza gouldii* BAIRD, B. N. A. 1858, p. 479. *M. m. gouldii* COUES, Key, 1872, p. 139; *M. f. Samuelis*, COUES, Key, 2d–4th eds. p. 372, and of A. O. U. Lists, 1886–95, No. 581 *d*.

M. m. coop'eri. (To Dr. J. G. Cooper.) SAN DIEGO SONG SPARROW. Like *heermani*; slightly smaller; coloration lighter, grayer. Back grayish-olive, broadly streaked with black, the streaks with little or no rusty edging. Wing and tail 2.50 on an average; culmen 0.48; depth of bill at base 0.29; tarsus 0.85. Southern coast region of California, N. to Monterey Bay, S. to San Quentin Bay, Lower California. RIDGW. Auk, Jan 1899, p. 35.

M. m. pusil'lula. (Lat. *pusillula*, very small, dimin. of *pusilla*, small.) SALT MARSH SONG SPARROW. Like *samuelis*; still smaller; coloration less rusty and more olivaceous above; superciliary line and under parts more or less tinged with yellowish. Wing averaging 2.29; tail 2.16: culmen 0.47; depth of bill at base 0.25; tarsus 0.82. Salt marshes of San Francisco Bay, California. RIDGW. Auk, Jan. 1899, p. 35.

M. m. cleonen'sis. (Lat. of Cleone, a town of Mendocino Co., Cal.) MENDOCINO SONG SPARROW. Size of *samuelis*; lighter and more rusty; black marks of back restricted; spots of breast broadly edged with rusty; black markings of sides of head and neck almost entirely replaced by reddish-brown. Wing 2.28–2.38; tail 2.10–2.22; culmen 0.42. Coast of Mendocino Co., California. *M. melodia cleonensis* McGREGOR, Bull. Coop. Club, Sept. 15, 1899, p. 87.

M. m. rivula'ris. (Lat. *rivularis*, of small rivers or creeks, fluviatile; *rivulus*, a rivulet, dimin. of *rivus*, a river.) BROWN'S SONG SPARROW. With this alleged subspecies we pass to some peninsular and insular forms, before resuming the series with forms from northwestern U. S. and northward. Lower California. *M. f. rivularis* BRYANT, Proc. Cala. Acad. 2d ser. i, Sept. 1888, p. 197; COUES, Key, 4th ed. 1890, p. 900; A. O. U. List, 2d ed. 1895, No. 581 *g*. *M. m. rivularis* OBERH. Auk, Apr. 1899, p. 183.

M. m. gramin'ea. (Lat. *graminea*, of grass or herbage.) SANTA BARBARA SONG SPARROW. Described as being of the size of *samuelis*; tail shorter; feet larger; coloration lighter, with an ashy cast; hind neck decidedly ashy; dark markings of back and sides of throat smaller and less blended. Wing given as 2.35; tail 2.25. The alleged characters may be due in part to abrasion of plumage by the coarse grass in which the bird lives. Santa Barbara Isl., breeding, and adjacent coast of California in winter. *M. f. graminea* TOWNSEND, Pr. U. S. Nat. Mus. xiii, 1890, p. 139; A. O. U. List, 2d ed. 1895, p. 242, No. 581 *h*. *M. m. graminea*, OBERH. Auk, Apr. 1899, p. 183.

M. m. clemen'tæ. (Dog-Lat., intended to mean of San Clemente, the island named in Spanish form for St. Clement, bishop of Rome in the 1st century, A. D.; Lat. *clementinus*, from *clementia*, clemency, mildness, from *clemens*, clement, mild; preferable forms of the specific name would be *clementiæ*, *clementina*, or *clementensis*.) SAN CLEMENTE SONG SPARROW. Like the last; larger; bill longer. Wing and tail each 2.50; culmen 0.45; tarsus 0.85. San Clemente and Santa Rosa Islands, California. *M. f. clementæ* TOWNS. Pr. U. S. Nat. Mus. xiii, 1890, p. 139; A. O. U. List, 2d ed. 1895, p. 243, No. 581 *i*.

M. m. morph'na. (Gr. μόρφνος, *morphnos*, Lat. *morphnus*, epithet of an eagle, supposed to mean dark-colored, dusky, swarthy, like Lat. *furvus*.) OREGON SONG SPARROW. RUSTY SONG SPARROW. Decidedly different from any of the foregoing! Streaking diffuse; streaks above and below dark rufous-brown, without black centres or pale edges. Coloration blended; general tone ruddy; under parts extensively shaded with brownish, except on belly. Rather larger than typical *melodia*. Pacific coast, U. S., and British Columbia, breeding northerly, S. in winter to S. California. This well-marked form was first distinguished by Nuttall, Man. 2d ed. 1840, p. 581; he named it *Fringilla guttata*, and compared it with the Fox Sparrow, from its resemblance to *Passerella iliaca;* but the name he bestowed is ruled out by the prior *F. guttata* VIEILL. 1817, an Australian bird. This Song Sparrow was also recognized by Audubon, who wrongly called it *Fringilla cinerea* GM., a name belonging to the distinct species described below. It was not recognized as different from *rufina* by Baird in 1858, the *M. rufina* of this author, B. N. A. p. 480, being a composite. I disengaged the two forms in the orig. ed. of the Key, 1872, p. 139, calling the present one *M. melodia guttata*, and changed the name to *M. fasciata guttata* in the 2d-4th eds. 1884-90, p. 372, after RIDGW. Bull. Nutt. Club, iii, 1878, p. 66. It has since so stood in A. O. U. Lists, No. 581 *e;* but as the name *guttata* has proven untenable, a new one has been proposed by OBERHOLSER, Auk, Apr. 1899, p. 183, and this I now adopt.

M. m. ingersol'li. (To Albert M. Ingersoll, of San Diego, Cal.) TEHAMA SONG SPARROW. Nearest *morphna;* said to be darker, without rusty wash, with under parts more streaky. Types from Sacramento Valley, California. MCGREGOR, Bull. Cooper Club, Mar. 15, 1899, p. 35, and Sept. 15, 1899, p. 88.

M. m. mer'rilli. (To Dr. James C. Merrill, U. S. A.) MERRILL'S SONG SPARROW. Most like *morphna;* bill smaller; ground color of upper parts and sides of head and neck lighter and more ashy, with darker and sharper markings, especially of back; white of under parts clearer and more extensive. Length 6.10; wing 2.63; tail 2.58; tarsus 0.84; bill 0.44, its depth at nostril 0.25. Fort Sherman, Idaho. *M. f. merrilli* BREWST. Auk, Jan. 1896, p. 46; A. O. U. List, 8th Suppl. Auk, Jan. 1897, p. 122, No. 581 *k*.

M. m. rufi'na. (Lat. *rufina*, rufous, reddish.) SOOTY SONG SPARROW (called RUSTY SONG SPARROW in 2d-4th eds. of Key). Quite like *morphna*, of which it is a larger, darker northern form. Tone of upper parts sooty or smoky brown; streaking very dark. Length 6.50 or more; wing and tail about 3.00; tarsus 1.00. Pacific coast region, British Columbia to Sitka; latter the type locality. *Passerella rufina* BP. Consp. Av. i, July 15, 1850, p. 477, as based on *Emberiza rufina* BRANDT, 1836 (Sitka). *M. rufina* BAIRD, 1858, in part (includes *guttata = morphna*). *M. melodia rufina* COUES, Key, orig. ed. 1872, p. 139; *M. fasciata rufina* of Key, 2d-4th eds. p. 372, and of A. O. U. Lists, No. 581 *f*.

M. m. cauri'nas. (Lat. *caurinus*, northwestern; *caurus* or *corus*, the northwest wind.) YAKUTAT SONG SPARROW. Like *rufina;* described as having bill longer, and coloration grayer, the superciliary stripe, most of auriculars, sides of neck, and edges of interscapulars being quite gray, in contrast with the brown markings; streaks below "seal-brown;" ground color of flanks "olive-grayish." Wing 3.00; tail 2.85; culmen 0.56; depth of bill at base 0.30; tarsus 0.95. Range ascribed to Alaska from Cook's Inlet to Cross Sound, and southward in winter. *M. f. caurina* RIDGW. Auk, Jan. 1899, p. 36; *M. m. caurina* OBERH. Auk, Apr. 1899, p. 183.

M. insig'nis. (Lat. *insignis*, signal, notable, well marked, as this species is; *in*, and *signum*, a sign, mark, token. Fig. 283.) BISCHOFF'S SONG SPARROW. KADIAK SONG SPARROW. Specific characters receding from those of *rufina* and approaching those of *cinerea;* coloration most like that of *rufina*, size nearly that of *cinerea*. Wing 3.20; tail 3.10; tarsus 1.00; bill 0.55. Eggs 0.89 × 0.65. An isolated species only known from Kadiak, etc., Alaska. BAIRD, Trans. Chicago Acad. Sci. i, 1869, p. 319, pl. 29, fig. 1. *M. melodia insignis* COUES, Key, orig. ed. 1872, p. 140; in later eds. wrongly combined with *cinerea* under the name of the latter, as it also is in A. O. U. List, 1st ed. 1886; see especially RICHMOND, Auk, Apr. 1895, pp. 144–150, for best account of this species and the next. It is No. 581.1 of A. O. U. List, 2d ed. 1895, p. 243.

FIG. 283.—Kadiak Song Sparrow. (L. A. Fuertes.)

SONG SPARROW (called also "KADIAK SONG SPARROW" in 2d–4th eds. of Key). A distinct species, peculiar in size, shape, and color. Above, brownish-slate color, more rufous on wings; the streaking broad and blended, very dark. Below, plumbeous-whitish, shaded with brown on sides; the streaks broad, diffuse, and dark. Spring and fall plumages differ much, but the bird may always be recognized by its great size and long slender bill. Length about 7.50; wing 3.30; tail 3.50; tarsus 1.10; bill 0.65, its depth at base 0.30. Fort Kenai, Alaska; Aleutian Islands (not Kadiak, however). *Fringilla cinerea* GM. S. N. i, 1788, p. 922; *Melospiza cinerea* FINSCH, Abhandl. Nat. Verein, Bremen, iii, 1872, p. 20. Not in orig. ed. of Key, and *M. cinerea* of Key, 2d–4th eds. p. 372, and of RIDGW. Man. 1887, p. 432, includes both this species and *M. insignis*. It is No. 582 of A. O. U. List, 2d ed. 1895, p. 243, where "Pribilof Islands" are wrongly included in its ascribed habitat.

M. cine'rea. (Lat. *cinerea*, ashy, ash-colored; *cinis*, gen. *cineris*, ashes. Fig. 284.) CINEREOUS SONG SPARROW. ALEUTIAN

FIG. 284.—Aleutian Song Sparrow.

PEUCÆ'A. (Gr. πεύκη, *peuce*, a pine; not well applied except to *P. æstivalis*.) SUMMER

FRINGILLIDÆ: FINCHES, BUNTINGS, SPARROWS. 423

FINCHES. Bill of moderate size, rather elongate-conic, upper mandible declivous toward end, commissure bent. Wings short and much rounded, folding little if any beyond base of tail; inner secondaries not elongated. Tail little or much longer than wing, much rounded; lateral feathers some ½ an inch shorter than middle; of weak, narrowly linear feathers with elliptically rounded ends. Feet small and weak, not reaching when outstretched nearly to end of tail; tarsus about equal to middle toe and claw; lateral toes equal, short, their claws not nearly reaching base of middle claw. Adults scarcely or not streaked below; crown quite like back, streaked with rusty-brown, black, and gray. A superciliary and postocular stripe, but usually none running under auriculars; more or less distinct black maxillary stripes (in *cassini* flanks also striped). Edge of wing yellow. Nest on ground; eggs white. Sexes alike; young different, being more or less streaked below. Aside from this, *seasonal* differences in plumage of adults, due to wear and tear of the feathers, are very great, and in some respects peculiar; they have occasioned much perplexity and confusion in determination of several closely allied species or subspecies.

Analysis of Species (adults).

Edge of wing yellow. Crown not uniform chestnut; no chestnut on lesser wing-coverts. Maxillary stripes slight. Nest on ground; eggs white.
No stripes on the flanks, and no cross-bars on the tail.
Broadly marked above with rufous streaks or blotches on ashy ground, with black centres of streaks on middle of back. Tail-feathers plain, or only with obscure whitish area.
Eastern species, mostly dull whitish on the under parts . . . . *æstivalis* and *æ. bachmani*
Western species, mostly grayish-buff on the under parts.
Southern Arizona and Sonora . . . . *arizonæ*
Texas and Mexico . . . . *mexicana*
Flanks distinctly striped; tail cross-barred.
Marked above with pale brown black-centred streaks, these black centres enlarged transversely at their ends on middle of back. Tail-feathers shafted and barred with blackish, outer broadly edged and tipped with white
cassini

P. æstiva′lis. (Lat. *æstivalis*, like *æstivus*, summery; *æstas*, summer.) FLORIDA SUMMER FINCH. PINE-WOODS SPARROW. Adult ♂ ♀: Upper parts, including crown, continuously streaked with blackish, dull chestnut and ashy-gray; no yellow about head; wing-coverts and inner secondaries marked like back; edge and bend of wing yellow, as in *Coturniculus passerinus*. Below, dull brownish-ash, or brownish-gray, whitening on belly, deepest on sides and across breast, nowhere obviously streaked in adult plumage. Some obscure dusky maxillary streaks, some vague dusky markings on auriculars, a slight ashy superciliary line, and very obscure median ashy line on crown. Bill dark above, pale below; legs very pale; lateral claws falling far short of base of middle claw; hind claw much shorter than its digit; tarsus not longer than middle toe and claw; tail much rounded, with obscure grayish-white area on lateral feathers. *Young* have breast and sides evidently streaked. Length 5.75–6.20, average 5.90; extent 7.60–8.30, average 8.00; wing 2.17–2.55, average 2.40; tail 2.25–2.68, average 2.50. South Atlantic States, strictly, especially Florida and southern Georgia; a bird of pine barrens, common in suitable localities; a fine songster. Nest on ground, of grasses; eggs 4, 0.75 × 0.60, pure white. As the first described species of the genus, this has been used as a standard of comparison; but it is the *most* modified offshoot of a genus which focuses in the Southwest and Mexico.

P. æ. bach′mani. (To the Rev. John Bachman.) BACHMAN'S SUMMER FINCH. OAK-WOODS SPARROW. Adult ♂ ♀: Above, sandy-ferruginous, indistinctly streaked with light ashy-gray; streaks broadest on back and middle line of crown; interscapulars sometimes with narrow black streaks. Wings light ferruginous; greater coverts less reddish and edged with paler; inner secondaries dusky, bordered at ends with pale reddish-ash. Tail plain grayish-brown, with ashy edgings of the feathers. Sides of head, neck, and body and breast quite across, dingy buff-color, deepest on breast, paler on throat and chin; a postocular rusty-brown streak over auriculars; sides of neck streaked with the same; an indistinct dusky streak on side of throat; belly dull white; crissum buff; edge of wing bright yellow; bill pale horn-color, darkest above; feet pale brown; iris brown. Size of *æstivalis;* wing a little longer, 2.35–2.60, average 2.50; tail 2.55–2.80, average 2.70; bill thicker; black streaks of upper parts, instead of being generally distributed, few and confined to the interscapulars; breast and sides more buffy. Thus much like *æstivalis* proper, but quite different from any of the following forms. Southern states at large, from southern Virginia, southern Indiana, and southern Illinois, to Florida and Texas, breeding in most of its range, but migratory to some extent, its range including that of true *æstivalis* only in winter · casually N. to Maryland (Auk, 1897, p. 219). This is the genuine original "Bachman's Finch" of Audubon (type examined: see BREWST. Auk, Jan. 1885, p. 105). When the species was divided into its two subspecies, Mr. Ridgway unluckily named the wrong one; for he identified *P. bachmani* with the dark coast form from Georgia and Florida, which is true *æstivalis*, and accordingly gave a new name to the reddish bird of the interior, calling it *P. illinoensis* in Bull. Nutt. Orn. Club, iv, 1879, p. 219. It consequently stands as *P. æ. illinoensis* in the 2d–4th eds. of the Key, 1884–90; but must be known as *P. æ. bachmani*, as in the A. O. U. Lists, 1886 and 1895, No. 575 *a;* and the required change in the English names of the two forms must also be made.

P. arizo'næ. (Of Arizona.) ARIZONA SUMMER FINCH. With a general likeness to *æstivalis*, in pattern of coloration, streaking of all upper parts, similarity of back to crown, yellow edge of wing, and plain tail-feathers; size same, wing and tail a trifle longer (as in *bachmani*). Colors duller and less variegated; maxillary stripes obscure or obsolete. Upper parts light dull chestnut or reddish-brown, moderately streaked with plumbeous-gray, but reddish the prevailing tone; interscapular feathers, and sometimes those of crown, with blackish centres; a poorly defined light superciliary stripe. Beneath, dull whitish, unstreaked, breast and sides with a decided ochrey-brown tinge. Wings dusky, inner secondaries darker and with more conspicuous rusty-brown edgings than those of longer quills, and also some whitish edging or tipping. Bill blackish above, pale below; legs flesh-color. Young: Above streaked with blackish and yellowish-gray, showing little reddish; under parts more or less streaked with dusky. Southern Arizona and southward in Sonora. (This is in part what I meant by *P.* var. *cassini* of orig. ed. of Key; but true *cassini* is entirely different.) *P. æstivalis arizonæ* RIDGW. Am. Nat. Oct. 1873, p. 615; COUES, Key, 2d–4th eds. 1884–90, p. 374.? *P. arizonæ* RIDGW. Pr. U. S. Nat. Mus. i, Aug. 1878, p. 127; *P. arizonæ* of A. O. U. Lists, 1st and 2d eds. 1886 and 1895, No. 576. The bird is distinct from the foregoing, but I doubt that it is specifically separable from the following:

P. mexica'na. (Lat. Mexican.) MEXICAN SUMMER FINCH. Very similar to the last. Adult ♂ ♀: Upper parts gray suffused with bay, streaked on most of back with bold black bay-edged stripes; crown similar, rather darker in smaller pattern of markings and without lighter median line. Bend of wing yellow; coverts blackish, with broad grayish-bay edgings; flight-feathers dusky, several inner secondaries blackish, with firm light edgings. Tail-feathers dusky, with obsolete scarcely discernible cross-waves, middle pair with paler edges their whole length, lateral ones fading toward their ends. Under parts pale grayish-brown, blanching on throat and abdomen, unstreaked excepting for a slight pair of black maxillary stripes. Bill dark horn-color; feet light brown. Length 6.30; wing 2.65; tail 2.80; tarsus 0.80. (Described from Mexican specimens.) Mexico to the Valley of the Lower Rio Grande in Texas; a late addition to our fauna, not given in the 1st or 2d eds. of the Key. *Coturniculus mexicanus* LAWR. Ann. Lyc. Nat. Hist. N. Y. viii, May, 1867, p. 474, described from Colima, Mex. *Peucæa mexicana* RIDGW. Pr. U. S. Nat. Mus. viii, May, 1885, p. 99; COUES, Key, 3d ed. 1887, p. 874; A. O. U. Lists, 1st and 2d eds. 1886 and 1895, No. 577 (*P. mexicana* of RIDGW. Man. 1887, p. 428, includes both this species and *P. arizonæ*, the latter being abandoned by its author). *P. botterii* SHARPE, Cat. B. Brit. Mus. xii, 1888, p. 711, who considers that *this* is the true *Zonotrichia botterii* of SCL. P. Z. S. 1857, p. 214. Dr. Sharpe indicates by his synonymy that he considers this to be also *P. arizonæ* of RIDGW. Pr. U. S. Nat. Mus. i, Aug. 1878, p. 127. It would appear that Mr. Ridgway has at different times confounded the two supposed species under the one name of *P. arizonæ*. This does not invalidate his original *P. æstivalis arizonæ* of 1873, as above cited, but brings in question his *P. arizonæ* of 1878, which the A. O. U. Committee cites as authority for the name of the foregoing species.

P. cas'sini. (To John Cassin.) CASSIN'S SUMMER FINCH. Belonging to the *æstivalis* group, with yellow edge of wing, and most resembling *arizonæ;* but perfectly distinct. A peculiar character of marking raises groundless suspicion of immaturity. Adult ♂ ♀: Entire upper parts, from bill to tail, alike in pattern of coloration — a peculiarly intimate variegation of ashy-gray, rufous-brown and blackish — the ruddy color occupying most of the feathers, which have a blackish central field and gray edging; the blackish area on each feather, especially of back, rump, and upper tail-coverts, where it is most conspicuous, being hammer-headed, or widened toward end of the feather. Pattern of markings smallest on cervix. No special head-markings, but a tendency toward a lateral browner band on side of crown, and browner postocular stripe, separated by a gray interval. Variegation of upper parts descending on sides of neck; sides of head with vague markings. Innermost secondaries showing quite blackish in general field of upper parts, and edged all around with a firm border of ashy-white or hoary-white. Greater and middle coverts exactly like inner secondaries; primaries similar, but the edging not so clear. Edge of wing clear yellow, and some of the least coverts tinged with this color. Tail curiously particolored; middle pair of feathers light grayish-brown, with a strong dusky shaft-line throwing off numerous dusky cross-bars, so that these feathers seem "watered" with lighter and darker shades. Other tail-feathers, except outermost pair, dusky-brown, with pale grayish-brown terminal spots increasing in size from inner feathers outward. On outermost feather this pale gray space is very large, and rimmed all around with white. An indistinct maxillary stripe on each side of chin. A number of strong well-defined dusky stripes on flanks; otherwise, entire under parts unmarked, and of a dingy whitish color, clearest on belly and throat, more grayish on sides and across breast. Bill brown, pale below; feet pale. Length 6.00–6.25; extent about 8.25; wing 2.50; tail 2.75. Young: Similar, but with a few drop-shaped streaks on jugulum and along sides; feathers of upper parts with a more appreciable terminal border of buff. Texas to California and southern Nevada, N. to Kansas in summer, S. through New Mexico and Arizona into Mexico. Habits, nest, and eggs as in *P. æstivalis* (eggs pure white, 0.75 × 0.55).

HÆMO'PHILA. (Gr. αἷμα, *haima*, blood; φίλος, *philos*, loving: what application?) Related to *Peucæa;* crown chestnut or rufous (in our species); no yellow on edge of wing; eggs not white. This is an extensive and varied genus of chiefly extralimital species, to which our birds of the *ruficeps* group and *carpalis* group prove to be more closely related than they are to the *æstivalis* group, *arizonæ*, and *cassini*. This distinction, first indicated in Key, 2d ed. 1884, p. 374, under head of *ruficeps*, was confirmed by Ridgway, Auk, Jan. 1899, p. 80, and formally adopted in A. O. U. Suppl. List, *ibid.* p. 119, where the name of the genus is misspelled "*Aimophila*," as by Swainson, 1837.

Analysis of Species and Subspecies (adults).

No chestnut on lesser wing-coverts.
    California coast region . . . . *ruficeps*
    Mountains of Lower California . . . . *sororia*
    Southern Arizona, southern New Mexico and southward . . . . *scotti*
    Southwestern Texas and southward . . . . *eremœca*
Chestnut on lesser wing-coverts . . . . *carpalis*

P. ru'ficeps. (Lat. *ruficeps*, red-headed; *rufus*, rufous; *caput*, head.) RUFOUS-CROWNED SUMMER FINCH. Lesser wing-coverts not chestnut as in *carpalis*. Strong maxillary streaks.

Adult ♂ ♀: Crown chestnut, in perfect condition bright and continuous, blackening on forehead, where divided by a short whitish line (whole cap thus as in *Spizella socialis* or *Melospiza georgiana*); crown, however, oftener streaked with olive-ash, especially along a median dividing line, thus assimilating more nearly with colors of other upper parts. An obscure olive-ashy superciliary line, whitening over lores. Back streaked with olive-ash and chestnut-brown, latter sometimes distinct, as bold streaking with ashy edging of the feathers, sometimes spreading almost to extinction of the ashy; brown also varying in shade from a purplish-bay to light rusty-brown, apparently according to wear and tear of plumage. Wings and tail dusky, with varying amount of reddish-brown edgings of the feathers. Under parts dull whitish, strongly shaded with olive-gray or olive-brown, paler on belly, quite whitish on throat, which latter is bounded by strong black maxillary stripes. Size of *P. cassini*, or rather less; length 6.00 or less; wing 2.20–2.40; tail 2.60; tarsus 0.77; bill 0.48, its depth at base 0.22. Young: Crown like back; under parts streaked with dusky, especially the breast. California coast region, from about lat. 40° to Cape St. Lucas; a strongly marked bird, which cannot be mistaken. The eggs are not pure white as in all the foregoing species of the genus, but of a pale bluish or greenish-white ground color, unmarked, somewhat like those of the Indigo Bird or Bluebird; size about 0.77 × 0.58. *Peucæa ruficeps* of all former eds. of Key, "*Aimophila*" *ruficeps* A. O. U. Suppl. List, Auk, Jan. 1899, p. 120.

H. r. soro′ria. (Lat. *sororia*, sisterly, like a sister.) LAGUNA SPARROW. Said to be like *ruficeps* proper, with chestnut of pileum somewhat lighter, supraloral line whiter, and supra-auricular line grayer; to be smaller than *scotti*, with back less ashy, chestnut streaks darker and narrower, and under parts more buffy; and to differ from all our other forms in thicker and relatively shorter bill. Wing 2.20–2.50; tail 2.40–2.58; culmen 0.45; depth of bill at base 0.26; tarsus 0.80. Mountains of Lower California. Included under *Peucæa ruficeps* in all former eds. of Key. RIDGW. Auk, July, 1898, p. 226; A. O. U. Suppl. List, Auk, Jan. 1899, p. 120.

H. r. scot′ti. (To W. E. D. Scott.) SCOTT'S SPARROW. Larger than *ruficeps* proper; length over 6.00, sometimes 6.50; wing 2.50–2.75; tail 2.75–3.00; tarsus 0.80; bill 0.55, its depth at base 0.27. Coloration duller and paler than in *ruficeps*; crown less intensely rufous; upper parts more uniformly brownish, lacking the black shaft-streaks of *eremœca*. S. Arizona and S. New Mexico, S. in Mexico to Puebla. Nest said to be built on ground, and eggs to be 3–4, 0.83 × 0.60, plain white (if this be true, it is a good character, as eggs of *ruficeps* are tinted). *Peucæa ruficeps boucardi, in part,* of 2d–4th eds. of Key; my former description, giving black shaft-streaks, etc., being based upon specimens of *eremœca*, with Arizona habitat assigned. *P. ruficeps boucardi* of A. O. U. Lists, 1886–95, No. 580 *a*, but not the true *boucardi* SCL. *P. r. scottii* SENNETT, Auk, Jan. 1888, p. 42; "*Aimophila*" *r. scottii* A. O. U. Suppl. List, Auk, Jan. 1899, p. 120, No. 580 *a*. *Peucæa homochlamys* SHARPE, Cat. B. Brit. Mus. xii, 1888, p. 713, is this bird, and the name probably has priority, as the Introduction is dated Dec. 10, 1887, and the Preface Jan. 6, 1888.

H. r. eremœ′ca. (Gr. ἔρημος, *eremos*, a desert; οἰκέω, *oikeo*, I inhabit.) DESERT SUMMER FINCH. ROCK SPARROW. Like *scotti* (*homochlamys*), and quite as large; length 6.25; extent 8.60; wing 2.60–2.75; tail 2.75–3.00; tarsus 0.80; bill 0.50. General aspect dull gray; back grayish-ash, the feathers there with brownish centres and black shaft-lines — a good color mark in comparison with *scotti*. Cap mixed rufous and gray, with black frontlet divided by a white median line, as in other members of the *ruficeps* group; ear-coverts conspicuously ashy. Below clear gray, whitening on abdomen, tinged with fulvous on flanks and vent; maxillary stripes indistinct. Southern and middle Texas; S. in Mexico to Orizaba. *Peucæa ruficeps boucardi, in part,* of 2d–4th eds. of Key; for in describing what I thought was *boucardi* I actually had *eremœca* in view, and so could see no difference from the latter! This bird is also *P. r. boucardi* of SENNETT, Auk, Jan. 1888, p. 42, and *P. boucardi* of SHARPE, Cat. B.

Brit. Mus. xii, 1888, p. 714, at least in part (description from Sclater's type of *Zonotrichia boucardi*, synonymy and habitat including *eremœca*). *P. r. eremœca* BROWN, Bull. Nutt. Club, Jan. 1882, p. 26 and p. 38; A. O. U. Lists, 1886-95, No. 580 *b*. "*Aimophila*" *r. eremœca* A. O. U. Suppl. List, Auk, Jan. 1899, p. 120, No. 580 *b*.

H. carpa'lis. (Lat. *carpalis*, relating to *carpus*, wrist-joint.) BAY-WINGED SUMMER FINCH. Adult ♂ ♀ : Lesser wing-coverts chestnut, forming a patch as conspicuous as in *Pocecetes* or *Auriparus*. Strong black maxillary stripes. Whole crown rufous, or dull bay, divided on forehead by a short pale stripe, and bordered with a pale grayish-ash superciliary stripe. Cervix like crown, but mixed with ashy-gray. Middle of back and scapulars grayish-brown, mixed with a little bay, and sharply streaked with blackish; lower back gray, with little or no black or brown. The general effect of the upper parts, crown, and back is like that of *Spizella socialis*. Wings and their greater coverts dusky, with grayish-fulvous edging and tipping; primaries and tail-feathers with whitish edging; one or two outer tail-feathers white-tipped. Under parts white, shaded on breast and sides with ashy; throat pure white, bounded on each side by a sharp black maxillary stripe, above which is another dark line from angle of mouth. Bill apparently reddish flesh color below, dusky above; feet pale brown, toes rather darker. Length about 6.00; extent 8.50; wing 2.25-2.50; tail 2.75, graduated about 0.50; bill 0.40; tarsus 0.67. Less mature: Crown less different from back, being streaked with ashy, blackish, and rufous. Very young: No chestnut on wing-coverts; upper parts, including crown, dull brownish broadly streaked with blackish; under parts streaked with dusky; thus much like the earliest stage of *Spizella socialis*; after this the chestnut bend of the wing is always conspicuous. Arizona and Sonora. A very distinct and curious species, nesting in bushes and laying a plain greenish egg. Eggs 4-5, 0.72 × 0.58, June-September; nest in a fork of bush, deeply cupped, of grasses, rootlets, and hairs. *Peucæa carpalis* COUES, Am. Nat. June, 1873, p. 322, and of 2d-4th eds. of Key, p. 375; A. O. U. Lists, 1886-95, No. 579. "*Aimophila*" *carpalis*, A. O. U. Suppl. List, Auk, Jan. 1899, p. 119.

AMPHISPI'ZA. (Gr. ἀμφί, *amphi*, on both sides; σπίζα, *spiza*, a finch: alluding to the close relation of the genus to those about it.) SAGE SPARROWS. Bill moderate, conical, not peculiar. Wings folding considerably beyond base of tail, without elongated inner secondaries; point of wing formed by 2d-5th quills, 1st between 6th and 7th. Tail nearly equal to wings, of rather broad firm feathers, rounded at ends. Tarsus longer than middle toe and claw; lateral toes of unequal lengths, outer (longer) not reaching base of middle claw. Embracing two Southwestern species, with rounded blackish tail, grayish-brown above, plumbeous-black bill and feet, and few decided streaks, or none. These do not particularly resemble each other, and are very different from the exotic *Poospiza* to which they were formerly referred. I based this genus in 1874 (B. N. W. p. 234) on *A. bilineata*, and also included *A. belli*; since then, several extralimital species have been referred to it, as *A. humeralis*, *A. mystacalis*, and *A. quinquestriata*, which had before been placed in the genera *Hæmophila* and *Zonotrichia*.

Analysis of Species and Subspecies.

Adult with throat black, a long white superciliary stripe, sides not streaked, and no yellow on edge of wing.
    Smaller, darker, with larger white tip of lateral tail-feather. E. Texas and southward . . . . . . *bilineata*
    Larger, lighter, with smaller white tip of lateral tail-feather. W. Texas, westward and southward *b. deserticola*
Adult with throat white, no long white superciliary stripe, sides streaked, and yellow on edge of wing.
    Smaller: wing and tail under 3.00; dorsal streaks obsolete.
        Darker. California . . . . *belli*
        Paler; very small. Lower California . . . . *b. cinerea*
    Larger: wing and tail 3.00 or more; dorsal streaks distinct . . . . *b. nevadensis*

A. bilinea'ta. (Lat. *bilineata*, two-lined; *bis*, twice, *linea*, a line; alluding to the stripes on the head. Fig. 285.) BLACK-THROATED FINCH. BLACK-FACED SAGE SPARROW. The typical form, to which the name is now restricted, averages somewhat smaller, with darker upper parts and more white tipping of lateral tail-feathers, than the next form. Eastern and Central Texas, N. to W. Kansas, S. to San Luis Potosi. *Emberiza bilineata* CASS. Pr. Phila. Acad. v, Oct. 1850, p. 104, pl. 3, and Illust. B. Cal. and Tex. pt. v, 1854, p. 150, pl. xxiii, in part; includes both forms. *Poospiza bilineata* SCL. 1857; BD. 1858; and COUES, Key, 1st ed. 1872, p. 140, in part. *Amphispiza bilineata* COUES, Birds N. W. 1874, p. 234, and Key, 2d-4th eds. 1884-90, p. 258, in part; A. O. U. Lists, 1886-95, No. 573, in part.

FIG. 285. — Black-throated Finch, reduced. (Sheppard del. Nichols sc.)

A. b. desertic′ola. (Lat. *desertus*, deserted, perf. partic. of *deserere*, to desert, forsake; *deserta*, n. pl., deserted places, deserts; *desertum*, a desert; and *colere*, to inhabit.) DESERT BLACK-THROATED FINCH. Adult ♂ ♀: Face, chin, and throat sharply jet-black; a strong white superciliary line, and another bounding black of throat; under eyelid white; auriculars dark slate. No yellow anywhere. Below, pure white; sides, flanks, and crissum shaded with ashy or fulvous-brownish, but no streaks. Above, uniform grayish-brown; clearer ash in high plumage, otherwise browner, generally more ashy anteriorly than behind, and shading insensibly into the black face. Wings dusky; coverts and inner quills edged with color of back. Tail black, with narrow grayish edgings; outer feather sharply edged and tipped with white, and several others similarly tipped; white spot on inner web of outer tail-feather under 0.50, sometimes only 0.10. Bill and feet plumbeous-black. Length 5.00 or more, sometimes 5.50; wing 2.50-2.75; tail nearly same; culmen 0.40; depth of bill at base 0.25; tarsus 0.75. Young: Head-markings obscure; little or no black on throat; a few pectoral streaks. Owing to absence of black on throat, the white maxillary stripe is ill-defined, but the other stripe is conspicuous. Back rather brown than ashy; tail blackish, not pure black. A jaunty little Sparrow, haunting sage-brush and chaparral of the Southwest, from western Texas and New Mexico W. to coast region of California, N. throughout the Great Basin, S. in Chihuahua, Sonora, and Lower California; breeds throughout its U. S. range, migratory from northerly parts. An effective songster, with its sweet simple notes. Nest in bushes slight and frail, close to the ground; eggs 2-5, 0.72 × 0.58, white with a pale greenish or bluish tinge, unmarked; laid in May, June, and later. *A. bilineata*, in part, of 2d-4th eds. of Key, and of A. O. U. Lists, 1886-95. *A. b. deserticola* RIDGW. Auk, July, 1898, p. 229; A. O. U. Suppl. List, Auk, Jan. 1899, p. 119, No. 573 a.

A. bel′li. (To J. G. Bell, of N. Y.) BELL'S FINCH. CALIFORNIA SAGE SPARROW. Adult ♂ ♀: Breast with a black or dusky spot; edge of wing slightly yellowish. Forehead, supraloral spot, and edges of eyelids, inconspicuously white. Below, white, more or less tinged with pale brownish; sides with slight sparse streaks that anteriorly become aggregated into dusky maxillary stripes cutting off from white throat a white stripe that runs from corner of bill; lores and circumocular region dusky. Above, grayish-brown, ashier on head; middle of back with small obscure blackish streaks; wing-coverts and inner quills with much fulvous edging; tail black with slight pale edgings, outer web of outer feather simply whitish. Bill and feet plumbeous-blue. Length ♂ ♀ under 6.00; wing and tail under 3.00 (♂ wing 2.50-2.80; tail 2.60-2.90; in ♀ rather less). Young: Similar; more streaked below, and wings with two grayish-buff bars. Resident in California W. of the Sierra Nevada, N. to 38° at least, S. into Lower California. Breeds nearly or quite throughout its range; nest on ground or very near it, in sage brush, built of bark shreds, grasses, etc. Eggs 3-4, 0.70 × 0.50, pale greenish blue, speckled.

NOTE. — *A. b. clementis* is described as "exactly like *A. belli*, but larger and with relatively larger bill"; but the dimensions assigned do not bear out this statement: Length of skins 5.20-5.70; wing 2.45-2.72; tail 2.30-2.68; culmen 0.38-0.41; depth of bill at base 0.22-0.23; tarsus 0.79-0.85; middle toe 0.49-0.53. San Clemente Island. *A. b. "clementeæ"* RIDGW. Auk, July, 1898, p. 230; not adopted by A. O. U., 1899.

A. b. cine'rea. (Lat. *cinereus*, ashy in color). GRAY SAGE SPARROW. Resembling the next subspecies (*nevadensis*) in lightness of coloration, but even paler and less streaked, lacking dark streaks on back, having those on breast and throat few and small. Very small: length 5.50 or less; wing 2.25; tail 2.15; tarsus 0.75; bill 0.35. Lower California. TOWNS. Pr. U. S. Nat. Mus. xiii. 1890, p. 136; A. O. U. List, 2d ed. 1895, No. 574 *b*.

A. b. nevaden'sis. (Lat. of Nevada; Span. *nevada*, snowy, applied to the Sierra Nevada or main range of mountains of California by Padre Pedro Font in 1775-76.) ARTEMISIA SPARROW. NEVADA SAGE SPARROW. Similar to *A. belli* in coloration. Edge of wing, and sometimes lesser coverts, yellowish. Above, ashy-brown, much as in *deserticola*, clearer ash anteriorly, more brownish behind; also clearer in high plumage, and more overcast with brown in less mature specimens; middle of back and scapulars very notably streaked with fine black lines. Below, white; sides, and sometimes, especially in fall specimens, most under parts shaded with pale fulvous-brown; sides, and sometimes breast, with dusky streaks, which on side of neck tend to run in a chain, partly distinguishing a pure white lateral stripe above them from the general whitish of under parts. Sides of head slaty, becoming dusky on lores; a conspicuous white eye-ring. A short white line above lores, and another on middle of forehead. Wings and tail as in *A. belli;* outer feather edged and tipped with white. Bill dark bluish-plumbeous, under mandible sometimes yellowish. Paler and larger than *belli* proper; wing and tail averaging fully 3.00, if not more; bill 0.35; tarsus 0.75. The strongly marked form of the Great Basin, N. to 40° and beyond, resident breeding throughout its range; abounding in the sage-brush deserts of eastern Oregon, portions of Idaho and Montana, interior California, Wyoming, Nevada, Utah, New Mexico, and Arizona. Nesting as in *belli;* eggs 3-4, 0.80 × 0.60, pale greenish or grayish, profusely speckled with reddish-brown and blackish-brown, with purplish shell-markings.

JUN'CO. (? Lat. *juncus*, a reed.) SNOW SPARROWS. SNOWBIRDS. Bill small, strictly conic. Wings rather long, primaries much surpassing short inner secondaries in the closed wing; usually 2d, 3d, and 4th quills longest, 5th little shorter, then 1st and 6th. Tarsus a little longer than middle toe and claw; lateral toes subequal, their claws about reaching base of middle claw. Tail about as long as wing, slightly emarginate or about even, of rather narrow but firm feathers, rounded oval at ends. A beautiful genus; adults unspotted, unstreaked, the colors massed in large definite areas; belly, crissum, and 2-3 lateral tail-feathers white; bill whitish, or black and yellow. Length 6.00-7.00; wing and tail about 3.00. Sexes subsimilar, but ♂ clearer and purer in coloration; young entirely different, quite streaky. Nest normally on the ground, rarely in a bush; eggs speckled. One common Eastern species; in the West *Junco* is split into numerous forms, which intergrade with one another, and with the Eastern bird; the degree of difference between almost any two of the nearest related ones is about the same. The distinctions between typical styles of each are very nice and easily perceived. The theory of hybridization advanced to account for connecting links simply restates without explaining the case; for interbreeding is just one of the conditions of intergraded species, keeping them from positive distinctness. *Adult male* birds of the several forms afford the following

Analysis of Species or Subspecies.

Two white wing-bars. Ashy, without any reddish tints. Western . . . . *aikeni*
No white wing-bars.
   Bill flesh-color; eyes brown.
      Eastern species. Blackish-ash, with no reddish anywhere . . . . *hiemalis* and *carolinensis*
      Western species.
         Sides pinkish, or of some tint different from that of the breast.

Sooty blackish, with reddish-brown back. Pacific coast region . . . . *oregonus*
    Like the last; coloration less vivid.
        Rocky mountain region at large and eastward . . . . *connectens*
        Sierra Nevada and Coast range of California . . . . *thurberi*
        Bay of Monterey, California . . . . *pinosus*
    Clear ash, with reddish interscapulars and blackish lores. Rocky Mts. at large . . . *annectens*
        Like the last. Arizona and New Mexico to Wyoming . . . . *ridgwayi*
        Like the last. Lower California . . . . *townsendi*
    Sides ashy, like the breast. Clear ash, with reddish interscapulars and blackish lores. Rocky Mts.
        caniceps
Insular species, with very short wings and tail. Guadalupe Isl. . . . . *insularis*
Bill black and yellow; eyes yellow.
    Sides pinkish. Lower California . . . . *bairdi*
    Sides ashy.
        Reddish of back confined to the interscapulars. New Mexico and Arizona . . . . *dorsalis*
        Reddish of back spreading on the wings. Southern Arizona . . . . *palliatus*

J. ai′keni. (To C. E. Aiken, of Colorado.) WHITE-WINGED SNOWBIRD. AIKEN'S JUNCO. Adult: Plain plumbeous-gray, neither blackish on head nor tinged with pinkish anywhere, but uniform on back, head, breast, and sides; belly, crissum, and lateral tail-feathers white, as usual in this genus; wings crossed with two conspicuous white bars formed by tips of greater and median coverts, and sometimes inner secondaries edged with white. Bill nearly or quite as in *hiemalis*. Large, the average being at if not beyond the maximum of *hiemalis*; ♂, length, 6.25 to nearly 7.00; wing 3.20–3.60, averaging about 3.40; tail 3.25 or more; bill over 0.50; tarsus 0.85; ♀ rather smaller. Young of the year after the first moult resemble adults, but differ in having no white wing-bars, or these only indicated by two rows of small white dots, and the gray somewhat overcast with brown. (*J. h. danbyi* of COUES, Nidologist, iii, 1895, p. 14: see COUES, Auk, Jan. 1897, p. 94.) A good species, readily distinguished from *hiemalis* in any plumage; the appearance in life is quite different, as I ascertained during a visit to the Black Hills of S. Dakota and Wyoming in 1895. It breeds abundantly there, but disappears in the fall, retiring S., chiefly in the mountains, to Colorado, where it winters, and also straggles E. to Kansas. The whole geographical range is quite restricted.

FIG. 286. — Eastern Snowbird. (Sheppard del. Nichols sc.)

J. hiema′lis. (Lat. *hiemalis*, wintry; *hiems*, winter. Fig. 286.) EASTERN SNOWBIRD. BLACK SNOWBIRD. SLATE-COLORED JUNCO. Blackish-ash, below abruptly pure white from the breast, the sides shaded with ashy. In the ♀, and most fall and winter specimens, the upper parts have a more grayish, or even a decidedly brownish, cast, and the inner secondaries are edged with pale bay. ♂, in full dress: Slaty-black intense on head; belly and crissum pure white, the line between the two transverse or convex forward; wings and tail blackish, with slightly hoary edging of some feathers; 2–3 lateral tail feathers pure white, wholly or in greatest part. No rusty-brown on back or sides; any shade on sides ashy, not pinkish. Bill pinkish-white, or flesh-color, usually black-tipped. Length 6.00–6.50; extent 9.50–10.00; wing 3.00–3.25; tail rather less. These extremes uncommon; average 6.25—9.75–3.10. ♀, in summer: Slate-color less intense, overlaid with brown (not reddish), sometimes quite brown; edging of inner secondaries rusty-brown;

average less white on tail; rather smaller; average about at the lesser of the above dimensions: sometimes only 5.75—9.25—2.75. ♂ ♀, in winter: Resembling ♀ in summer. Young of the year: General color rather brown than slate, with conspicuous bay edgings of inner secondaries; bill much obscured with dusky. The brown overcast is a general shading, not of particular areas, and not pinkish. Young before first moult: Entirely streaked and spotted, like most very young Sparrows. Upper parts streaked with blackish and rusty-brown; secondaries and wing-coverts conspicuously edged with the latter. Under parts streaked or speckled with dusky and ochrey brown, on all fore parts and sides; belly and crissum soiled whitish. Bill dusky, paler below. Eastern N. Amer., N. W. to Alaska, W. to the Rocky Mts. and sparingly even to Utah, Washington, California, and Arizona; still chiefly Eastern. One of our most abundant and familiar winter birds, in flocks in shrubbery, from October to April. Retires to high latitudes or altitudes to breed. Nests in mountains of the Middle and some of the Northern States, and down to sea level from limits of Canadian fauna in Maine; winters most numerously from Massachusetts southward to the Gulf of Mexico; a cheery bright little bird, coming fearlessly to the threshold and window-sill in bad weather. Its snapping note is better known than is the pleasant song with which it takes leave in spring. Nest on ground; eggs 4–6, white, sprinkled with reddish and darker brown dots, about 0.80×0.60.

J. h. carolinen'sis. (Lat. of Carolina.) CAROLINA SNOWBIRD. BREWSTER'S JUNCO. This is the form which breeds in the southern Alleghany region. I have found it abundant in summer in the mountains of Virginia and North Carolina, where its usual nesting-place is in the cut banks of roadsides, just under the overhanging fringe of weeds and grass. The ascribed characters are not very tangible, but the bird can be distinguished at gunshot range from typical *hiemalis* by one who is familiar with both. I hesitated to accept it in former editions of the Key, but have since seen reason to modify my opinion. See BREWSTER, Auk, Jan. 1886, p. 108; Key, 4th ed. 1890, p. 900; A. O. U. List, 2d ed. 1895, No. 567 *e*.

J. h. connec'tens. (Lat. *connectens*, connecting; *con*, with; *necto*, I join.) HYBRID SNOWBIRD. ROCKY MOUNTAIN JUNCO. Possessing in varying degree characters of *hiemalis* and *oregonus;* rufous back of latter and ashy sides of former, or, oftener, ashy back of former and pink sides of latter; coloration less vivid, with less contrast between the blackish, reddish, and white parts; head and neck with a somewhat mottled appearance; "sides slaty rufous"; wing little over and tail little under 3.00; tarsus 0.73; bill 0.43. This form shades on the one hand into *hiemalis*, on the other into *oregonus*, but more generally resembles the latter. Rocky Mt. region of the U. S. and adjoining British provinces; W. in the Great Basin to California; S. in Arizona, New Mexico, Texas, and adjoining portions of Mexico; straggling E. to Michigan, Illinois, Maryland, and Massachusetts. This form, which I named and characterized in the 2d ed. of the Key, 1884, p. 378, is the one afterward named *J. h. shufeldti* by Mr. H. K. Coale, Auk, Oct. 1887, p. 330; A. O. U. List, 2d ed. 1895, No. 567 *b*. This fact was inadvertently overlooked both by the Committee and by myself in preparing the new List, but the oversight has since been rectified: see COUES, Auk, Jan. 1897, p. 94; A. O. U. Suppl. List, ibid. p. 128, No. 567 *b;* since then my *connectens* has been again renamed *J. montanus* by Mr. Ridgway, Auk, Oct. 1898, p. 321; A. O. U. Suppl. List, Auk, Jan. 1899, p. 119, No. 567. 1, by error.

J. h. ore'gonus. (Lat. of the Oregon River. Fig. 287.) OREGON SNOWBIRD. J. K. TOWNSEND'S JUNCO. Head and neck all around and fore breast sooty-black, ending sharply against white with a rounded outline convex backward; middle of back dull reddish-brown; feathers of wings much edged with the same; below from fore breast abruptly white, tinged on sides with pale reddish-brown — a peculiar "pinkish" shade. Bill white, black-tipped. In ♀ and young the black is obscured by brownish, but the typical form may always be distinguished by an evident contrast in color between interscapulars and head, and fulvous or pinkish wash on sides. The seasonal and sexual changes of plumage are parallel with those of *hiemalis*.

Pacific coast, breeds from British Columbia to Alaska, S. in winter to California and Nevada. Under this form were long included all the black-headed, red-backed, pink-sided Juncos from the Rocky mountain region to the Pacific, but *oregonus* is now restricted to the Pacific coast form, and others have afforded the basis of *connectens* and *thurberi*. This is the bird named *Fringilla oregana* by Townsend in 1837; and this form of the word, assumed by the A. O. U. Committee to be a typographical error, is not necessarily such, for the country used to be called Oregan, Ouragan, etc. However, Townsend changed it to *oregona* in 1839 (Narr., p. 345).

FIG. 287.—Oregon Snowbird.

J. h. thur'beri. (To Eugene Carlton Thurber of California.) SIERRA SNOWBIRD. THURBER'S JUNCO. Like *J. h. oregonus*; sides paler and less extensively pinkish; dorsal patch paler and more sharply defined. Sierra Nevada to southern coast ranges of California. Formerly included under *oregonus*, and named since the last ed. of the Key appeared. ANTHONY, Zoe, i, Oct. 1890, p. 238; A. O. U. List, 2d ed. 1895, No. 567 c.

J. h. pino'sus. (Lat. full of pines, though it appears that the pines were full of the birds, the implication being Point Pinos, a place on Monterey Bay, Cal.) POINT PINOS SNOWBIRD. LOOMIS' JUNCO. "Most nearly like *J. h. thurberi*, but throat, jugulum, and fore breast slate-gray, varying to dark slate-gray, and upper portions of head and neck slate-gray, varying to blackish-slate;" this dark color abruptly defined against the colors of the body; interscapulars and scapulars pale chestnut; rump gray, tinged with chestnut; sides faintly washed with "vinaceous-buff." A local race, breeding down to sea level about the bay of Monterey, Cal., in pine woods. *J. pinosus* LOOMIS, The Auk, Jan. 1893, p. 47; reduced to a subspecies, as *J. h. pinosus* by A. O. U. Committee in The Auk, Jan. 1894, p. 47; A. O. U. List, 2d ed. 1895, No. 567 d.

J. annec'tens. (Lat. *annectens*, annexing; *ad*, to, and *necto*, I join.) PINK-SIDED SNOWBIRD. ANNEX JUNCO. Quite different from any of the foregoing, and resembling *caniceps*. General color clear ashy plumbeous, or leaden gray, that of the breast abruptly defined against the white of the belly; lores distinctly blackish, in contrast with rest of the head; interscapulars and scapulars reddish-brown, or light chestnut rufous, this color spreading more or less over the wing-coverts; sides pinkish, or pale cinnamon fulvous, like a lighter shade of the color of the back, well marked against the white of the belly. Bill in life pinkish white, with more or less dusky tip; iris dark brown. Sexes alike. The general characters are thus those of *caniceps*, from which this species is distinguished by the more abrupt definition of the ashy breast from the white belly, and especially by the pink sides: and by so much it approaches *oregonus*, though it is quite different in most respects. The eggs, 4 or 5 in number, 0.80×0.60, are indistinguishable from those of other species of the genus. This bird, too curtly though not incorrectly described in the 2d, 3d, and 4th eds. of the Key, was originally characterized by BAIRD in Coop. Orn. Cal. i, 1870, p. 564: see also my Birds N. W. 1874, p. 145; it has turned out better than I expected, and may now be given specific rank. It breeds in the mountains of Idaho, Montana, and Wyoming, is especially abundant in winter in Colorado, and extends S. through Arizona and New Mexico, even into northern Mexico. It was among several species of Snowbirds I took at Fort Whipple in Arizona in the winter of 1865–66.

J. ridg'wayi. (To R. Ridgway.) RIDGWAY'S SNOWBIRD. MEARNS' JUNCO. "Above similar to *J. caniceps;* below indistinguishable from *J. annectens.*" The adult ♂ is said to have the outer webs of inner tertiaries tinged with rufous; outer tail-feather white, next white except a dusky line along each edge, third with a long white terminal stripe nearly confined to the inner web. Bill flesh color, slightly tipped with black. Feet and claws light brown. The specific character is given as above by its describer, Dr. E. A. Mearns, U. S. A., in The Auk, Oct. 1890, p. 243; type taken at Fort Whipple, April 22, 1884; range extended in A. O. U. List, 2d ed. 1895, No. 568. 1, to include New Mexico, Colorado, and Wyoming. (Not in former eds. of the Key.) *J. annectens* A. O. U. Suppl. List, Jan. 1897.

J. town'sendi. (To Chas. H. Townsend.) SAN PEDRO SNOWBIRD. C. H. TOWNSEND'S JUNCO. "Similar to *J. annectens,* but differing in smaller size, darker gray of the head, neck, and chest, the back less brown and the sides less extensively pinkish." Bill flesh color, as in all the foregoing and in *caniceps;* iris brown. San Pedro Mts., Lower California, where apparently resident and differentiated as a species. ANTHONY, Proc. Cala. Acad. 2d ser. Oct. 1889, p. 76; A. O. U. List, 2d ed. 1895, No. 571. 1; *J. h. townsendi* of the Key, 4th ed. 1890, p. 900.

J. ca'niceps. (Lat. *caniceps,* gray-headed; *canus,* gray.) GRAY-HEADED SNOWBIRD. WOODHOUSE'S JUNCO. Clear ash, purest on head, paler below, and fading gradually into white on belly; interscapulars abruptly, definitely, chestnut or rusty-brown; lores blackish; bill flesh color; iris brown; no fulvous wash on sides; no chestnut on wings. Rather larger than *hiemalis;* length nearly 7.00; wing over 3.00; tail about 3.00. The sexual and seasonal changes are not so well marked as in the heavily-colored *hiemalis* and *oregonus,* but parallel as far as they go. Very young birds are streaked, like all the rest. Eggs 0.80 × 0.60, white or whitish, specked with reddish-brown, usually minutely and chiefly about the larger end. Rocky Mts. of the U. S., from Wyoming southward to Mexico; Wahsatch and Uintah Mts.

J. phæono'tus dorsa'lis. (Gr. φαιός, *phaios,* of a dun color; νῶτος, *notos,* back. Lat. *dorsalis,* pertaining to the back; *dorsum,* the back.) RED-BACKED SNOWBIRD. HENRY'S JUNCO. Characters in general of *caniceps;* but with the bill black and yellow, as in *palliatus,* and iris yellow. In this case the reddish of the back is confined to the interscapulars, not spreading over the wings, as in *palliatus.* Eggs whitish, with a greenish tinge, immaculate or with only minute reddish-brown sprinkling about the larger end. Mountains of New Mexico and Arizona, and S. into Mexico. This is *J. dorsalis* of HENRY, 1858, long considered a synonym of *caniceps;* but it is one form of a distinct Mexican species, *J. phæonotus* of WAGLER, 1831 (or *Fringilla cinerea* SWAINS. 1827, which name is preoccupied). It is also *J. h. dorsalis* of the 2d–4th eds. of the Key; *J. cinereus dorsalis* A. O. U. List, 1st ed. 1886, and RIDGW. Man. 1887, p. 423; *J. phæonotus dorsalis* A. O. U. List, 2d ed. 1895, No. 570 *a:* see Auk, Oct. 1895, p. 391.

J. p. pallia'tus. (Lat. *palliatus,* palliated, *i. e.,* wearing the *pallium* or mantle, with allusion to the reddish which mantles the back and wings.) CINEREOUS SNOWBIRD. ARIZONA JUNCO. Like the last. Chestnut of back intense, and spreading over wing-coverts and inner secondaries; upper mandible black; lower yellow; iris yellow. Eggs greenish-white, unmarked. Mexico to U. S. border of Arizona. This is the form which most nearly approaches in the U. S. the Mexican stock species *phæonotus* (or *cinereus*), and conducts to the Guatemalan *alticola.* It is *J. h. cinereus* of the 2d–4th eds. of the Key, p. 379; *J. cinereus palliatus* RIDGW. The Auk, Oct. 1885, p. 364, and Man. 1887, p. 424; A. O. U. List, 1st ed. 1886, p. 275; *J. phæonotus palliatus* RIDGW. The Auk, Oct. 1895, p. 391; A. O. U. List, 2d ed. 1895, No. 570.

J. baird'i. (To Prof. S. F. Baird.) BAIRD'S SNOWBIRD. BELDING'S JUNCO. Head and neck ashy-gray, paler on throat, tinged on hind head with brown, the lores distinctly blackish. Back, scapulars and adjoining wing-feathers pale rufous-brown, tinged with olivaceous; rump and upper tail-coverts, with lesser, middle, and outer wing-coverts grayish-olive; inner webs of tertials dusky; primaries gray, edged with paler, outermost with white; outer tail-feather mostly white, two next with white in diminishing amount. Jugulum pale buffy-gray, contrasting with the white of abdomen; sides and flanks cinnamon-buff; crissum dull whitish. Upper mandible dark brown, lower yellow; iris yellow; feet pale brown. Wing 2.80; tail 2.75; bill 0.40; tarsus 0.80. A form lately discovered in the mountains of southern Lower California, resembling a bright-colored ♀ *oregonus*, but presenting the peculiar combination of "pink" sides with yellow eyes and under mandible. BELDING, Pr. U. S. Nat. Mus. vi, Oct. 1883, p. 155; A. O. U. Lists, 1886 and 1895, No. 571; *J. h. bairdi* of 3d and 4th eds. of the Key, p. 875.

J. insula'ris. (Lat. *insularis*, insular; *insula*, an island.) GUADALUPE SNOWBIRD. INSULAR JUNCO. Resembling *annectens*; darker, with somewhat different proportions. Crown and nape dark slate; lower tail-coverts dusky, the feathers edged with whitish; lores blackish. Wings and tail relatively short: wing 2.55-2.85; tail 2.30-2.60; bill 0.37 long, 0.27 deep. (In *annectens*, etc., wing and tail about 3.00.) Added to our Fauna by the inclusion of Guadalupe Island, off Lower California; the characters ascribed are specific, as in the nature of the case intergradation is unlikely to occur. RIDGW. Bull. U. S. Geol. and Geogr. Surv. Terr. ii, No. 2, Apr. 1876, p. 188; Man. 1887, p. 425; COUES, Key, 3d and 4th eds. p. 875; A. O. U. Lists, 1886 and 1895, No. 572.

FIG. 288.—Chippy's head, as large as life. (E. C.)

SPIZEL'LA. (Ital. diminutive form of Lat. *spiza*, from Gr. σπίζα, *spiza*, a finch.) CHIPPING SPARROWS. Embracing small species, 5.00-6.00 long; long, broad-feathered, forked tail about equalling (more or less) rather pointed wings; no yellowish anywhere; no streaks on under parts *when adult;* interscapular region distinctly streaked; rump plain (except *atrigularis*); *young* fully streaked. Point of wing formed by 2d to 4th or 5th quill; 1st usually between 5th and 6th. Bill small, conic. Tarsus little if any longer than middle toe and claw; lateral toes about equal. Tail-feathers widening a little to broadly oval tips. Sexes alike; young somewhat different. Nest usually in bushes; eggs colored. Numerous species, Eastern and Western, inhabiting shrubbery; three of them familiar Eastern birds.

Analysis of Species.

Eastern and Western species with the crown *of the adult* chestnut or bright brown, little or not streaked.
    Bill black and yellow; forehead not black; two distinct white wing-bars; dark spot on breast; large: about 6.00 long . . . . *monticola* and *m. ochracea*
    Bill and forehead black; wing-bars not conspicuous; breast ashy-white, without spot; length under 6.00. Tail decidedly shorter than wing . . . . *socialis* and *s. arizonæ*
    Bill brownish-red; forehead not black; wing-bars indistinct; breast buffy white, without spot. Length under 6.00 . . . . *pusilla* and *p. arenacea*
Southwestern species, with the crown tawny brown, obscurely streaked, rest of head ashy, no dusky postocular streak, and one wing-bar across median coverts; bill reddish-brown . . . . *wortheni*
Western species, with the crown not chestnut, and streaked like the back.
    Crown divided by a median stripe, and its streaks separated from those of the back by an ashy interval. Tail equal to wings . . . . *pallida*
    Crown not evidently divided, and streaked continuously with the back. Tail longer . . . . *breweri*
Southwestern species, with the crown of the adult dark ash. Face and throat black. Bill brownish-red. Tail decidedly longer than wing . . . . *atrigularis*

S. monti'cola. (Lat. *monticola*, inhabiting mountains; *mons, montis*, a mountain; *colo*, I dwell; *incola*, an inhabitant. Fig. 289.) TREE SPARROW. TREE BUNTING. CANADA

Sparrow. Winter Chip-bird. Winter Chippy. Arctic Chipper. Adult ♂ ♀: Bill black above, yellow below; legs brown; toes black. No black on forehead; crown chestnut (in winter specimens the feathers usually skirted with gray), bordered by a grayish-white superciliary and loral line; a postocular chestnut stripe over auriculars, and some vague chestnut marks on cheeks; sides of head and neck otherwise ashy-gray. Below, impurely whitish, tinged with ashy anteriorly, washed with pale brownish posteriorly, middle of breast with an obscure dusky blotch. Middle of back boldly streaked with black, bay, and flaxen; middle and greater wing-coverts black, edged with bay and tipped with white, forming two conspicuous cross-bars; inner secondaries similarly variegated; other quills and tail-feathers plain

Fig. 289. — Tree Sparrow.

dusky, with pale or whitish edges. Remarkably constant in coloration; sexes indistinguishable, and young very similar, the chief variation being in the veiling of the cap with gray. There is a very early streaky stage, however, as in other species. A handsome sparrow, the largest of the genus. Length 5.80–6.20, usually 6.00; extent 8.75–9.75, usually 9.25; wing and tail 2.75–3.10. Eastern N. Am., northerly, W. to the Plains, S. in winter to the Carolinas, Kentucky, Kansas, and corresponding latitudes. Abundant in the U. S. in winter, flocking in shrubbery; breeds N. of the U. S. and E. of the Rocky Mts., even to the Arctic coast. Nest in low bushes or on ground, loosely constructed of bark-strips, weeds, and grasses, warmly lined with feathers. Eggs 4–6 or even 7, 0.75 × 0.55, pale green, minutely and regularly sprinkled with reddish-brown spots.

S. m. ochra′cea. (Lat. *ochracea*, of an ochrey color.) Western Tree Sparrow. Like the last; paler above, with sparser, sharper, and narrower dorsal streaks; sides and throat more ochraceous. Western N. Am., from the Dakotas and Kansas to the Pacific; breeds in Alaska, and extends S. in winter to Texas, New Mexico, and Arizona. Brewst. Bull. Nutt. Ornith. Club, Oct. 1882, p. 228; Coues, Key, 3d and 4th eds. 1887–90, p. 875; Ridgw. Man. 1887, p. 418; A. O. U. Lists, 1st and 2d eds. 1886–95, No. 559 a.

Fig. 290. — Chipping-Sparrow, reduced. (Sheppard del. Nichols sc.)

S. socia′lis. (Lat. *socialis*, given to society, sociable. Figs. 288, 290.) Social Sparrow. Chipping Sparrow. Chip-bird or Chippy. Hair-bird. Adult ♂ ♀: Bill black; feet pale; crown chestnut; extreme forehead black, usually divided by a pale line; a grayish-white superciliary line; below this a blackish stripe through eye and over auriculars; lores dusky. Below, a variable shade of pale ash, nearly uniform and entirely unmarked; back streaked with black, dull bay and grayish-brown; inner secondaries and wing-coverts similarly variegated; tips of greater and median coverts forming whitish bars; rump ashy, with slight blackish streaks or none; primaries and tail-feathers dusky, with paler edges. Smaller than the Tree Sparrow; length 5.00–5.50; extent 8.00–9.00; wing 2.66–2.75; tail less, about 2.50. Sexes alike, but very young birds quite different: crown streaked like back; breast and sides thickly streaked with dusky; bill pale brown; and head lacking definite black. In this stage, which, however, is of brief duration, it resembles some other species, but may be known by a certain ashiness the others lack, and from the small Sparrows that are streaked below when adult, by its generic characters. Eastern N. Am., N. to subarctic regions, W. to Rocky Mts., S. into Mexico; migratory in most regions, but breeding throughout its range; extremely abundant, and the most familiar species about houses, in gardens, and elsewhere, nesting in trees or shrubbery; nest of fine dried grass, lined with hair; eggs 4–5, bluish, speckled sparsely and chiefly about the larger end with blackish-brown, with purplish shell-markings; size about 0.70×0.55. (*S. domestica* of 2d–4th eds. of the Key, after *Passer domesticus* BARTRAM, Trav. 1791, p. 291 — an author to whom North American ornithology owes much, but one whom the A. O. U. Committee decline to recognize on the ground that he was not a strict binomialist.)

S. s. arizo'næ. (Lat. of Arizona.) ARIZONA CHIPPING SPARROW. Like an immature *S. socialis*. Paler than this species, the ashiness in great measure brown; crown grayish-brown streaked with dusky like back, and showing evident traces of rich chestnut, but never becoming wholly chestnut; black frontlet lacking or obscure, and no definite ashy superciliary line, the sides of crown merely lighter brown; bill brown above, pale below. Western N. Am., generally, from the Rocky Mts. to the Pacific; S. in winter in Mexico and Lower California. A curious form, as it were an arrested stage of *socialis*. Some specimens, with least chestnut on head, look remarkably like *breweri*, but this last is evidently smaller, without chestnut on head, and otherwise different.

S. pusil'la. (Lat. *pusillus*, petty, small; *pusus*, a little boy.) FIELD SPARROW. BUSH SPARROW. Adult ♂ ♀: Bill pale reddish; feet very pale; crown dull chestnut; auriculars and postocular stripe the same; no decided black or whitish about head. Below, white, unmarked, but much washed with pale brown on breast and sides; sides of head and neck with some vague brown markings; all the ashy parts of *socialis* replaced by pale brownish. Back bright bay, with black streaks and some pale flaxen edgings; inner secondaries similarly variegated; tips of median and greater coverts forming whitish cross-bars. Size of *socialis*, but more nearly the colors of *monticola*. Length 5.25–5.75; extent 7.75–8.40; wing 2.30–2.50; tail quite as much, or more, thus not shorter than wing, as it is in the last. Young for a short time streaked below, as usual in *Spizella*. Eastern U. S. and Canadian border, strictly; hardly N. throughout New England, W. only to edge of the Plains; migratory; breeds from the Carolinas and corresponding latitudes northward, and winters from the same southward; very abundant in fields, copses, and hedges, in flocks when not breeding. Nest indifferently in low bushes or on ground; eggs 3–5, white or whitish, speckled with rusty-brown, 0.68×0.50. (*S. agrestis* of 2d–4th eds. of the Key, after BARTRAM: see remark under *S. socialis*.)

S. p. arena'cea. (Lat. *arenaceus*, sandy.) WESTERN FIELD SPARROW. Like the last, but with the rufous replaced by brownish-ash; tail somewhat longer. Length about 6.00; wing 2.70; tail 2.80. Western U. S., from Montana and N. Dakota to Texas and Louisiana. This form, described as a migrant or winter resident in southern Texas (CHADBOURNE, Auk, Apr. 1886, p. 248), was recognized by the name of *S. agrestis arenacea* in the First Appendix of the Key, 1887, p. 875, and as *S. pusilla arenacea* by Mr. Ridgway in his Manual, 1887, p. 420, though not admitted to the orig. ed. of the A. O. U. List, 1886; but the Committee, on reconsideration, endorsed its subspecific validity in the A. O. U. List, 2d ed. 1895, p. 233, No. 563 *a*. It seems to me as well entitled to recognition as either of the other subspecies of this genus. It is the form of Field Sparrow which inhabits the Great Plains from Texas to Dakota and Montana. See Auk, Oct. 1897, pp. 345–347, pl. 3.

S. worth'eni. (To C. K. Worthen.) WORTHEN'S SPARROW. Resembling the Field Sparrow, but quite distinct, and in some respects approaching *S. atrigularis*, especially in coloration of upper parts. Much less rufous than *S. pusilla*, with broader black dorsal streaks, no rufous auricular streak nor pectoral spots, a whitish eye-ring, and slender bill. Length about 5.00; wing 2.70; tail about the same; bill 0.35; tarsus 0.70. Western Texas and New Mexico, S. to Puebla, Mex. RIDGW. Pr. U. S. Nat. Mus. vii, Aug. 1884, p. 259; Man. 1887, p. 419; COUES, Key, 3d ed. 1887, p. 875; A. O. U. List, 2d ed. 1895, p. 233, No. 564.

S. pal'lida. (Lat. *pallida*, pale.) CLAY-COLORED SPARROW. Adult ♂ ♀: Crown and back clay-colored or flaxen, distinctly streaked with black, without evident bay; dorsal streaks noticeably separated from those of crown, by an ashier, less streaked, cervical interval; rump brownish-gray. Crown divided by a pale median stripe; a distinct whitish superciliary line; loral and auricular regions decidedly brown, with a dark postocular stripe over auriculars, and another from angle of mouth, bounding the brown area inferiorly; below this a dusky maxillary streak; wing-coverts and inner secondaries variegated like back, being black with broad flaxen-brown edging and whitish tipping. Below, white, soiled with clay-color. Bill dusky above, pale below; feet pale. Small: Length 5.00–5.25, rarely 5.50; extent 7.40–7.75, rarely 8.00; wing and tail, each, about 2.50. Young birds lightly streaked below. Central region of the U. S. into British America, in the Saskatchewan and Red River regions; W. to the Rocky Mts. only in most localities, but reaching Lower California; S. to Texas and thence through much of Mexico; E. to Iowa and Illinois. Abundant; migratory; breeds from Iowa, Kansas, Nebraska and corresponding latitudes northward; nest in bushes close to ground; eggs 3–6, pale green sparsely speckled with rich brown, 0.65 × 0.50.

S. brew'eri. (To Dr. T. M. Brewer, of Boston.) BREWER'S SPARROW. Similar; paler and duller, all the markings indistinct; streaks of crown and back small, numerous, not separated by a cervical interval; no definite markings on sides of head. Upper parts grayish-brown, with marked dorsal area of brighter brown, and continuously streaked from head to tail. Size of the last, but tail relatively longer, equalling wings — about 2.66 long, thus equalling, if it does not somewhat exceed, that of *socialis*, although the latter is a larger bird. Western U. S., especially New Mexico and Arizona, but N. to Montana and even British Columbia; S. to Lower California and some parts of Mexico; accidental in Massachusetts. Breeds throughout its U. S. range; habits those of *pallida*; nest and eggs indistinguishable.

S. atrigula'ris. (Lat. *atrigularis*, black-throated; *ater*, black; *gula*, throat.) BLACK-CHINNED SPARROW. Adult ♂ ♀: Dark ash, fading insensibly into whitish on belly, deepening to black on face and throat; interscapulars bright bay, streaked with black; wing-coverts and inner secondaries variegated with the same colors; tail blackish, with pale edgings; bill coral reddish, as in *S. pusilla;* feet dark brown. A small-bodied species, but full 6.00 long, on account of length of tail (2.75–3.00), which much exceeds wings (2.25–2.50; extent 7.75). The young lack black on face, have crown washed with ashy-brown, middle of back duller chestnut, and bill dusky above; but may be known by length of tail. Mexico, Lower California, Arizona, and New Mexico; N. in California to about 37° in desert regions, S. in Mexico on the tablelands to Puebla. Nest in bushes, eggs 3–5, 0.66 × 0.50, pale greenish-blue, normally unmarked; May, June.

ZONOTRICH'IA. (Gr. ζώνη, *zone*, a girdle, band; τριχιάς, *trichias*, name of a bird. Figs. 291, 292.) CROWN SPARROWS. Embracing our largest and handsomest Sparrows, 6.50 to 7.50 inches long; rounded wings and tail each 3.00 or more; under parts with very few streaks, or none; middle of back streaked; rump plain; wings with two white cross-bars; head of adults with black, and usually with white or yellow also, or both. Bill moderate, conical, culmen and gonys just appreciably curved, commissure very little angulated. Point of wing formed usually by 2d–4th quills, 1st about equal to 5th; folding decidedly beyond inner secondaries, and to near middle of tail. Tail-feathers of moderate width and consistency, rounded

Fig. 291. — Crown Sparrow (white-throated), nat. size. (Ad. nat. del. E. C.)

Fig. 292. — Crown Sparrow (white-crowned), nat. size. (Ad. nat. del. E. C.)

oval at end; tail as a whole rounded. Tarsus about equal to middle toe and claw; lateral toes about equal to each other. Sexes similar; young similar in most respects, but lacking the distinctive markings which adorn the heads of the adults. Nest on the ground or in bushes near it; eggs colored. The Crown Sparrows are peculiar to America, where they are represented by about nine beautiful and perfectly distinct species, four of which (one of them with two subspecies) occur N. of Mexico.

Analysis of Species and Subspecies (adults only).

Crown black and white; no yellow on head; throat ash.
    Lores black. Dorsal streaks purplish-bay; no yellow on wing. Eastern . . . . *leucophrys*
    Lores gray. Dorsal streaks purplish-bay; no yellow on wing. Western . . . . *l. intermedia*
    Lores gray. Dorsal streaks sooty-black; edge of wing yellowish. Pacific coast . . . . *l. gambeli*
Crown black and white; yellow spot before eye; throat white; edge of wing yellow. Eastern . . . . *albicollis*
Crown black, yellow and ash; edge of wing yellow; throat ashy. Pacific coast . . . . *coronata*
Crown, face, and throat black; no yellow on head or wing. Interior regions . . . . *querula*

Z. albicol'lis. (Lat. *albicollis*, white-throated; *albus*, white; *collum*, neck. Fig. 293.) WHITE-THROATED CROWN SPARROW. PEABODY-BIRD. Adult ♂: Crown black, divided by a median white stripe, bounded by a white superciliary line, and *yellow spot* from nostril to eye; below this a black stripe through eye; below this a maxillary black stripe bounding definitely pure white throat, sharply contrasted with dark ash of breast and sides of neck and head. *Edge of wing yellow.* Back continuously streaked with black, chestnut, and fulvous-white; rump ashy, unmarked. Wings much edged with bay; white tips of median and greater coverts forming two conspicuous bars; quills and tail-feathers dusky, with pale edges. Below, white, shaded with ashy-brown on sides, deeper and purer on breast; bill dark; feet pale. ♀, immature birds, and specimens as generally seen in the U. S. in fall and winter, having black of head replaced by brown, the white of throat less conspicuously contrasted with duller ash of surrounding parts, and frequently with obscure dusky streaks on breast and sides; but the species may always be known by the yellow over eye and on edge of wing (these never being imperceptible), coupled with large size and generic characters. Length 6.50–6.90; extent 9.20–9.90; wing 2.75–3.00; tail about the same. A fine Sparrow, abundant throughout eastern N. Am.; W. to Dakota, Montana, and Wyoming, and casually to Utah, California, and Oregon; 9 California specimens have been reported; breeds from N. New England and other Northern States N. to about 65° in the fur countries;

Fig. 293. — White-throated Crown Sparrow, reduced. (Sheppard del. Nichols sc.)

winters from Massachusetts southward. Found in all situations, but especially in shrubbery, generally in flocks, except when breeding; a pleasing if not brilliant songster, with its limpid *pea-peabody, peabody, peabody* in cadence. Nest on the ground, rarely in bushes; eggs 4-6, about 0.90 × 0.66, with the endless diversity of tone and pattern of those of the Song Sparrow, from which only distinguished by greater size.

Z. leuco'phrys. (Gr. λευκός, *leucos*, white; ὀφρύς, *ophrus*, eyebrow. Fig. 294.) WHITE-BROWED CROWN SPARROW. Adult ♂ ♀: Crown pure white, enclosing on either side a broad black stripe that meets its fellow on forehead and descends lores to level of eyes, and bounded by another narrow black stripe that starts behind eye and curves around side of hind head, nearly meeting its fellow on nape; edge of under eyelid white. Or, we may say, crown black, enclosing a median white stripe and two lateral white stripes, all confluent on hind head. No yellow anywhere. General color a fine dark ash, paler below, whitening insensibly on chin and belly, more brownish on rump, changing to dull brownish on flanks and crissum, middle of back streaked with dark purplish-bay and ashy-white. No bright bay, like that of *albicollis*, anywhere, except some edging on wing-coverts and inner secondaries; middle and greater coverts tipped with white, forming two bars. Bill and feet reddish. Length 6.25—7.00; extent 9.20—10.20;

FIG. 294.—White-browed Crown Sparrow, reduced. (Sheppard del. Nichols sc.)

wing and tail 2.90—3.20; usually 6.75—9.50—3.10. Young: Black of head replaced by very rich warm brown, white of head by pale brownish; the general ash has a brownish suffusion, and the back is more like that of *albicollis*, being streaked with dusky and ochrey-brown; but the two species cannot be confounded. Very young: Before the first moult, there are indications of head-markings as last described; but whole upper parts, sides of neck and fore under parts, are streaked with blackish and ochrey-brown or whitish. N. Am., especially eastern and rather northerly; W. in the Rocky Mts., where mixed with *intermedia*; California; Greenland; Cape St. Lucas; S. in winter in Mexico. Not nearly so abundant in the U. S. as *albicollis*, but common in many sections in winter and during migrations. Breeds occasionally in northern New England, and plentifully in Labrador, where it is one of the commonest Sparrows; also, in the Rocky Mountains, and the Sierras Nevadas. Nesting same as that of *albicollis*, and eggs indistinguishable.

Z. l. interme'dia. (Lat. *intermedia*, intermediate, in the middle.) INTERMEDIATE CROWN SPARROW. Exactly like the last, but lores gray or ashy, continuous with white stripe over eye, *i. e.*, black of forehead does not descend to eye. Perhaps averaging a trifle smaller, and duller colored. Some specimens resemble *leucophrys* on one side of head, and *intermedia* on the other. Rocky Mts. to the Pacific, mostly replacing true *leucophrys* from Mexico and Lower California to Alaska; breeds mainly beyond the U. S. (*Z. gambeli* BD. 1858, COUES, 1872, *nec* NUTT.)

Z. l. gam'beli. (To Wm. Gambel, of Phila.) GAMBEL'S CROWN SPARROW. Markings of head much the same as in *intermedia;* body colors entirely different, almost exactly as in *coronata*. Streaking of back sooty-black. Edge of wing yellow, as in *coronata* and *albicollis*. Bill in dried specimens blackish and yellow, not reddish. About *coronata* size. Pacific coast region, from Lower California to British Columbia. (*Z. gambeli* NUTT. 1840, *nec* BAIRD, 1858, COUES, 1872. This is given as a full species in the 2d–4th eds. of the Key, 1884–90; in the A. O. U. List, 1886, No. 556; and RIDGW. Man. 1887, p. 416; but is reduced to a subspecies in the A. O. U. List, 2d ed. 1895, No. 554 b. It certainly seems to me an entirely different bird, and I have seen no intermediates; but such no doubt occur, as otherwise the Committee would not have reversed its former ruling.)

FIG. 295. — Golden-Crowned Sparrow.

Z. corona'ta. (Lat. *coronata*, crowned; *corona*, a crown. Fig. 295.) GOLDEN CROWNED SPARROW. Adult ♂ ♀: Forehead and sides of crown black, enclosing a dull yellow coronal patch anteriorly, an ashy one posteriorly; a yellow spot over eye; lores black. Edge of wing yellow. Above, much like *albicollis*, but with less bay and no whitish; two white wing-bars. Below, including sides of head and neck, ashy, passing insensibly into whitish on belly, and much shaded with brownish on flanks and crissum; thus much like *leucophrys*, but the ashy not so pure; larger than *leucophrys*; length 7.00 or more; wing and tail over 3.00. Young: Black of crown replaced by brown; but always traces of yellow on crown and wings. The yellow eye-spot is small, and not always evident. This large and handsome species inhabits the Pacific coast region, from Alaska S. to Lower California, and has occurred casually in Colorado, Wisconsin, and on Guadaloupe island; breeds in Alaska. Eggs 0.85 × 0.65.

Z. que'rula. (Lat. *querula*, querulous, plaintive; *queror*, I complain, lament. Fig. 296.) HOODED CROWN SPARROW. HARRIS'S SPARROW. Adult ♂, in breeding plumage: Whole crown, face, and throat, jet-black; sides of head pale ash; auriculars darker ash, bounded by a black line starting behind eye and curving around them.

FIG. 296. — Harris's Sparrow.

Under parts nearly pure white, but slightly ashy before and faintly brownish-washed behind; sides with a few dusky streaks; breast with a few black spots continued from the black throat-patch. Back nearly as in *coronata*, streaked with dusky and reddish-brown. Bill coral-red; toes dark; tarsi pale. No yellow anywhere. Very large: Length 7.00–7.75; extent 10.75–11.25; wing 3.25–3.50; tail 3.40–3.60; bill 0.45; tarsus 1.00; middle toe and claw rather less. ♀ similar, but with much less black on head and throat, the hood being restricted or imperfect; but its outline usually traceable. ♂ ♀, in fall: Bill light reddish-brown, usually obscured on ridge and at tip, and paler at base below; feet flesh-colored, obscured on toes; eyes brown. Crown grayish-black, every feather with a distinct, narrow, pale gray edge all around, producing a peculiar effect; this area bounded with a light ochrey-brown superciliary and frontal line. Sides of head like the superciliary, but auricular patch rather darker grayish-brown, and loral region obscurely whitish. Chin pure white, bounded on each side by a sharp maxillary line of blackish, with a rusty-red tinge. On lower throat, a large, diffuse and partially discontinuous blotch of this same blackish-red, cutting off white chin from white of rest of under parts, connecting with maxillary streaks, and stretching along sides of neck and breast in a series of rich dusky-chestnut streaks. On middle of breast the blotch generally runs out into the white in a sharp point, but its size and shape vary interminably. The markings here described are all included in the jet-black hood and breast-plate of the perfect spring dress; and between the two extremes every intermediate condition may be observed at various seasons. The rest of the plumage does not differ very materially from that of the adult ♂ in summer. This is the largest of our Sparrows; a bird of imposing appearance — for a Sparrow! Interior U. S. and British Provinces, especially the valley of the Mississippi, Lower Missouri, and Red River of the North; E. to Minnesota, Missouri, Iowa, and even Illinois; S. to Texas; accidental in Oregon and British Columbia. It is abundant in the line of its migration, as in Kansas, Nebraska, Iowa, Dakota, etc., but its breeding resorts are not well made out — probably Manitoba, Assiniboia, N. to Hudson Bay. I found it in Dakota at 49° coming early in September from the North.

CHONDES'TES. (Gr. χόνδρος, *chondros*, cartilage; also grain, seeds; ἐδεστής, *edestes*, an eater; badly formed.) LARK SPARROWS. Framed for a single species, with long pointed wings, exceeding long rounded tail; point of wing formed by 2d and 3d primaries, but 1st and 4th scarcely shorter; rest rapidly graduated. Tarsus about equal to middle toe and claw; lateral toes short, tips of claws not reaching base of middle claw. Bill swollen-conic, with culmen slightly convex, commissure little angulated. Species large, for a Sparrow, streaked above, white below, head and tail parti-colored.

C. gram'macus. (Gr. γραμμικός, *grammicos*, marked with a γράμμα, *gramma*, a line, word; badly selected to indicate the stripes of the head, and badly spelled. Fig. 297.) LARK SPARROW. LARK FINCH. Adult ♂ ♀ : Head variegated with chestnut, black, and white ; crown chestnut, blackening on forehead, divided by a median stripe, and bounded by superciliary stripes, of white; a black line through eye, and another below eye, enclosing a white streak under eye and chestnut auriculars; next, a sharp black maxillary stripe not quite reaching bill, cutting off a white stripe from white chin and throat. A black blotch on middle of breast. Under parts white, faintly shaded with grayish-brown; upper parts grayish-brown; middle of back with fine black streaks. Tail very long, its central feathers like back, the rest jet-black, broadly tipped with pure white in diminishing amount from the lateral pair inward, and outer web of outer pair entirely white. Length 6.00–6.75 ; wing 3.20–3.50, pointed; tail 3.00 or less, rounded. Very young: Crown, back, and nearly all under parts streaked with dusky; no chestnut on head, nor are the black stripes firm; but with the first moult the peculiar pattern of the head-

FIG. 297. — Lark Sparrow, nat. size. (Ad. nat. del. E. C.)

markings becomes evident, and there is little variation afterward with age, sex, or season. A beautiful species, abundant from the Great Plains to the Great Lake region, in the Mississippi Valley at large, Texas to Ontario, and irregularly or casually farther East, even in various Atlantic localities from Massachusetts to Florida. A sweet songster; breeds throughout its regular range; nest usually on ground, of dried grass; eggs 4-7, white, with straggling zigzag dark lines, as in many *Icteridæ*; size 0.75-0.85 × about 0.65.

C. g. striga'tus. (Lat. *strigatus*, striped, marked with *strigæa*, stripes.) WESTERN LARK SPARROW. Quite like the last; averaging paler or dingier, with duller chestnut on head, and narrower black streaks on back; wings and tail rather longer. Length 6.50-7.25; wing 3.50 or rather more; tail 2.75-3.25. Plains to the Pacific, U. S. and adjoining British Provinces, S. through L. Cala. and Mexico to Guatemala. *C. grammica* of previous eds. of the Key includes this form, which I have hitherto declined to recognize. *C. strigatus* Sw. 1827; *C. grammaca strigata* RIDGW. 1880; *C. grammacus strigatus* A. O. U. Lists, 1886 and 1895, No. 552 a.

FIG. 298. — Bill of Fox Sparrow, nat. size.

PASSEREL'LA. (Ital. diminutive form of Lat. *passer*, a sparrow.) FOX SPARROWS. Remarkable for size of feet and claws: lateral toes elongated to about equal degree; ends of their claws reaching about halfway to end of middle claw; claws all very large; middle toe and claw about as long as tarsus. Wings long and pointed, folding about to middle of tail; point formed by 2d-4th quills, 1st and 5th little shorter. Tail moderate, little rounded or nearly even. Bill strictly conic, with straight outlines and scarcely angulated commissure, very variable in size. Large handsome reddish or slate-colored species, marked below with triangular spots and streaks of the color of the back. Habits terrestrial and somewhat rasorial. Nest indifferently in trees or bushes or on ground; eggs greenish, fully speckled. The species, if more than one, are, like those of *Junco, Melospiza, Peucæa, Pipilo,* etc., still imperfectly differentiated.

Analysis of Subspecies.

Tail decidedly shorter than wing. General coloration foxy or ferruginous. Two whitish wing-bars well marked.
    Eastern (chiefly) . . . . *iliaca*
Tail about equal to wing. General coloration ruddy olive. Wing-bars obsolete. Pacific coast region *i. unalascensis*
Tail little or not shorter than wing. General coloration slaty olive. Markings of upper parts obsolete.
    Bill moderate, 0.30 deep at base. Rocky Mt. region . . . . *i. schistacea*
    Bill immoderate, 0.40 deep at base. Mts. of California . . . . *i. megarhyncha*
    Bill enormous, 0.50 deep at base. Mts. of California . . . . *i. stephensi*

P. Ill'aca. (Lat. *iliaca*, relating to the ilia, or flanks, which are conspicuously marked. Figs. 298, 299.) EASTERN FOX SPARROW. FOXY FINCH. FERRUGINOUS FINCH. FOXTAIL. ♂ ♀: General color above ferruginous or rusty-red, purest and brightest on rump, tail, and wings, on other upper parts appearing in streaks laid on an ashy ground. Below, white, variously but thickly marked except on belly and crissum with rusty-red — the markings anteriorly in the form of diffuse confluent blotches, on breast and sides consisting chiefly of sharp arrow-heads and pointed streaks. Tips of middle and greater wing-coverts forming two whitish bars. Upper mandible dark, lower mostly yellow; feet pale. One of the finest singers of the family; quite unlike any other Eastern Sparrow. A large handsome species: Length 6.50-7.25; extent 10.50-11.50; wing 3.25-3.60, averaging 3.40; tail little or not over 3.00, thus decidedly shorter than wing; bill, along culmen, 0.40; tarsus 0.90; hind claw about 0.35. Sexes alike, and young not particularly different after first moult, though in an early stage much darker; back rufous-brown with darker streaks; no wing-bars; all under parts heavily marked. There is much individual variation in color, independently of age, sex, or season. Eastern N. Am.; W. in the U. S. regularly only to the edge of the Plains, occasionally to Colorado, casually to California; but in Alaska regularly to Bering Sea; N. to the Arctic coast. Breeds throughout British America and in Alaska; not known to do so anywhere in the U. S. Winters from the Middle States southward. Nest on ground or in bushes or trees; eggs 3–5, 0.95 × 0.70, greenish-white, thickly speckled with rusty-brown; general aspect as in *Zonotrichia* and *Melospiza*.

P. i. unalascen′sis. (Of the Island of Unalashka.) TOWNSEND'S FOX SPARROW. ♂ ♀: General color above dark olive-brown, overcast with a reddish-brown tinge, and the streaking obsolete,—thus giving a uniform and continuous ruddy-olive tone, becoming more foxy-red on rump, wings, and tail. Wing-bars obsolete. Beneath, white, thickly marked, excepting on the middle of the belly, with triangular spots of about the same dark color as the back,—aggregated on breast, and entire sides of neck and body almost like back in uniformity of color, still showing ill-defined confluent dark reddish-brown streaks on a more olive-brown ground. Cheeks and auriculars with some whitish speckling. No obvious markings on wings. Bill dusky above, apparently reddish or yellowish below; feet reddish-brown.

FIG. 299. — Fox Sparrow, reduced. (Sheppard del. Nichols sc.)

Size of *iliaca*, but very different in color, and somewhat differently proportioned; wing averaging 3.25, and tail scarcely or not shorter; bill about 0.50; hind claw the same, and as long as its digit. Eggs not distinguishable with certainty from those of *iliaca* in size, form, or color, but tending to be rather distinctly spotted than heavily clouded. A curious form, related to *iliaca* much as *Melospiza rufina* is to the Eastern Song Sparrow. Pacific coast region, from Alaska Peninsula to southern California in winter, breeding north of the United States. (*P. townsendi* AUD. The A. O. U. spells the name *unalaschcensis*, after GM. 1788.) An attempt has been made to split this subspecies into two, distinguishing as *P. i. townsendi* the bird which breeds in the Sitka district of Alaska.

P. i. fuligino′sa. (Lat. *fuliginosa*, fuliginous, sooty.) SOOTY FOX SPARROW. Like Townsend's, but darker and less rufescent; upper parts and sides sooty-brown; upper tail-coverts and tail more rufescent; spots of under parts very dark brown, large and confluent. Coast region of British Columbia, Vancouver's Island, and Washington State, breeding; S. in winter along coast to San Francisco, Cal. RIDGW. Auk, Jan. 1899, p. 36.

P. i. schista′cea. (Lat. *schistacea*, slaty; Gr. σχιστός, *schistos*, fissile or cleft, as slate-stone is; the allusion, however, is to color. Fig. 300.) SLATE-COLORED FOX SPARROW. Adult ♂ ♀: General color above uniform slate with a slight olive tinge, becoming dull foxy-red on wings and tail; streaking of back obsolete, but whitish wing-bars sometimes indicated. Below, white, shaded along sides with color of back, but not so as to obscure the decided markings of the parts; under parts at large spotted and streaked with dusky-brown, usually aggregated into a blotch on breast. This is the connecting link between *iliaca* and *unalascensis*; the upper parts are nearly of the slaty-ash that forms the ground color of *iliaca*, only the foxy streaks of the back are obsolete. The spotting below is correspondingly darker. The form has, however, some peculiarities: tail decidedly longer in comparison with wings. Length

Fig. 300. — Slate-colored Fox Sparrow. (From The Osprey.)

7.00–7.50; wing 3.05–3.45; tail 3.00–3.30; bill 0.45 along culmen, 0.30 deep at base; tarsus 0.90. Eggs 0.85 × 0.65, with the same tendency to distinct spotting seen in those of all the other Western Fox Sparrows. Rocky Mt. region, chiefly, but noted from Kansas to California.

P. i. megarhyn'cha. (Gr. μέγας, *megas*, great; ῥύγχος, *hrugchos*, in Lat. *rhynchus*, beak.) THICK-BILLED FOX SPARROW. Coloration as in *schistacea*. Tail at maximum length, averaging at the extreme of that of *schistacea*; claws and beak highly developed; bill very thick, its depth at base 0.40, rather more than its length from nostril to tip; culmen 0.45; hind claw longer than its digit. A local race, in mountains of California and Nevada.

P. i. ste'phensi. (To F. Stephens.) STEPHENS' FOX SPARROW. Like the last; rather larger; the bill still larger, its average about at the maximum of that of *megarhyncha*, its maximum 0.55 along culmen, 0.50 deep at base. This is simply the extreme differentiation of the foregoing, in the San Jacinto and San Bernardino Mts. of California. *P. i. megarhyncha*, in part, of previous eds. of the Key; of RIDGW. Man. 1887, p. 434; and of the A. O. U. Lists, 1886-95, No. 585 b. *P. i. stephensi* ANTHONY, Auk, Oct. 1895, p. 348; admitted to full communion with other holy subspecies by the A. O. U. Committee on Nov. 13, 1896, at the Union's Congress at Cambridge, Mass. A. O. U. List, Eighth Suppl. Auk, Jan. 1897, p. 122, No. 585 d.

CALAMOSPI'ZA. (Gr. κάλαμος, *kalamos*, Lat. *calamus*, a reed; σπίζα, *spiza*, a finch.) LARK BUNTINGS. Bill large and stout at base, culmen a little curved, commissure well angulated; rictus bristly. Wing long and pointed; tip formed by 1st-4th quills, rest rapidly graduated; inner secondaries enlarged and flowing, one of them about reaching point of wing when closed. Tail shorter than wing, nearly even. Feet stout, adapted to terrestrial habits; middle toe and claw about as long as tarsus; lateral toes nearly equal to each other, scarcely reaching base of middle claw; hind claw about as long as its digit, but not straightened. A well-marked genus, with wing-structure reminding one of *Anthus* or *Alauda*; the turgid strongly-angulated bill resembles that of a Grosbeak. Sexes very dissimilar; ♂ black and white, in masses of color. ♀ brown and white, streaky. Nest on the ground; eggs whole-colored, as in *Spiza*, etc. There is a curious analogy if not affinity of this genus to some of the *Icteridæ*.

C. melano'corys. (Gr. μέλας, gen. μέλανος, *melas, melanos*, black, and κόρυς, *korus*, a lark. Fig. 301.) LARK BUNTING. WHITE-WINGED BLACKBIRD. ♂, in summer. Black, with a large white patch on wings, formed by the median and greater coverts; quills and tail-feathers frequently marked with white; bill dark horn-blue above, paler below; feet brown. Length 6.00-7.00; extent 10.00-11.00; wing 3.25-3.50; tail 2.50-2.75; bill 0.50-0.55; tarsus, or middle toe and claw, 0.90-1.00. Sexes unlike: ♀ more resembling a Sparrow. Above, grayish-brown, streaked with dusky-brown, on the back the edges of the dark streaks often of a purer brown than the general ground-color. Below,

FIG. 301.—Lark Bunting, ♂ ♀, reduced. (Sheppard del. Nichols sc.)

white, shaded on sides with grayish-brown, thickly streaked with blackish-brown everywhere excepting throat and belly, the streaks mostly sharp and distinct, but blended on sides, tending to aggregate on breast, and run forward as a maxillary chain. A poorly-defined light superciliary stripe. Wings dusky, with a large white or whitish speculum, much as in ♂, but not so pure nor so extensive ; inner secondaries edged with brown and white. Tail-feathers, the middle excepted, blackish tipped with white. Young ♂ like ♀, but colors more suffuse and brighter; upper parts pure brown; under parts tinged with fulvous ; wing-markings quite fulvous; under surface of wing quite blackish. In very young birds the markings more motley than streaky; feathers of the upper parts edged with pale buff ; bill brownish, flesh-colored below. ♂ wears the black plumage only during the breeding season, like the Bobolink ; when changing, the characters of the two sexes are confused. The change of the adult ♂ from a winter plumage resembling that of the ♀ to the full breeding dress is accomplished by aptosochromatism — that is, without moulting ; for the black comes to the surface by the wearing away of light tips and edgings of the feathers, as in the Bobolink. In form of bill, this interesting species is closely allied to Grosbeaks ; and this, with the singularly enlarged secondaries, as long as the primaries in the closed wing, renders it unmistakable in any plumage. A prairie bird, abundant on the Great Plains ; N. to 49° at least, in the Missouri and Milk River region, and beyond, in Manitoba and Assiniboia ; W. to the Rocky Mts., and in the winter to Southern and Lower California ; S. in Mexico to Guanajuato; accidental in New York, Massachusetts, and S. Carolina. The male has a habit of soaring and singing on wing like a Lark ; nest on ground, sunken flush with the surface, of grasses ; eggs 4–5, 0.90×0.65, pale bluish ; normally unmarked, occasionally speckled. *C. bicolor* of all previous eds. of the Key, after *Fringilla bicolor* TOWNS. 1837 ; name changed on account of there being already a *Fringilla bicolor* LINN. 1766, which is an entirely different bird, now known as *Euetheia bicolor*. *C. melanocorys* STEJ. Auk, Jan. 1885, p. 49 ; A. O. U. Lists, 1st and 2d eds. 1886-95, No. 605.

SPI'ZA. (Gr. σπίζα, *spiza*, a kind of Finch, probably *F. cœlebs*.) SILK BUNTINGS. Bill much as in *Calamospiza*, but longer for its depth and not so strongly angulated. Wings very long and pointed ; 2d primary usually longest, 1st and 3d little shorter, 4th and rest rapidly graduated ; one inner secondary a little elongated, but not nearly reaching point of wing. Tail short, nearly even, but a little emarginate. Tarsus and middle toe and claw of about equal lengths; lateral toes of nearly equal lengths, not reaching base of middle claw; hind toe with claw as long as middle toe without claw.

FIG. 302. — Black-throated Bunting, reduced. (Sheppard del. Nichols sc.)

S. america'na. (Lat., of America. Fig. 302.) BLACK-THROATED BUNTING. ♂ : Above, grayish-brown ; middle of back streaked with black ; hind neck ashy, becoming on crown yellowish-olive with black touches. A yellow superciliary line, and maxillary touch of the same ; eyelid white ; ear-coverts ashy like cervix ; chin white ; throat with a large jet-black patch. Under parts in general white, shaded with gray on sides, extensively tinged with yellow on breast and belly. Edge of wing yellow ; lesser and middle coverts rich chestnut, other coverts and inner secondaries edged with paler. Bill dark horn-blue ; feet brown. Length 6.00– 7.00; extent 10.50–11.00 ; wing 3.25–3.50, sharp-pointed ; tail 2.30–2.75, emarginate. ♀ : Smaller ; wing under 3.00, etc. ; above, like ♂, but head and neck plainer ; below, less tinged with yellow ; black throat-patch wanting, replaced by sparse sharp maxillary and pectoral streaks ; wing-coverts not chestnut, though so indicated by rufous edgings of individual feathers. Young ♂ : Larger than ♀, but in general similar ; throat-patch indicated by blackish feathers; wing-coverts chestnut. An elegant species, of trim form, tasteful colors, and very smooth plumage, abundant in fertile portions of the Eastern U. S.; N. to Massachusetts; W. to Kansas, Nebraska, Colorado, and S. W. even to Arizona; rather southerly, scarcely reaching the N. border of the U. S. anywhere, except in the region of the Great Lakes, where it extends into southern Ontario; winters wholly extralimital, in Central and even South America; breeds throughout its U. S. range. The local distribution of the birds within their general range is irregular, apparently fortuitous, and seems to have changed of late years, the species being rare E. of the Alleghanies, and absent from many Atlantic localities where it used to be common, as in the District of Columbia, for example. Not a good vocalist; the simple ditty sounds like *chip-chip-chee, chee, chee.* Nest on the ground, or in a low bush; eggs 4-5, normally plain greenish-white, rarely speckled; 0.80 × 0.65.

S. town'sendi. (To J. K. Townsend.) TOWNSEND'S BUNTING. "Upper parts, head and neck all round, sides of body and fore part of breast, slate-blue; back and upper surface of wings tinged with yellowish-brown; interscapulars streaked with black; superciliary and maxillary line, chin and throat and central line of under parts from breast to crissum, white; edge of wing, and gloss on breast and middle of belly, yellow; a black spotted line from lower corner of lower mandible down the side of the throat, connecting with a crescent of streaks in the upper edge of the slate portion of the breast." Chester Co., Pennsylvania, J. K. Townsend, May 11, 1833; one specimen known, a standing puzzle to ornithologists, in the uncertainty whether it is a "good species," or merely an abnormal plumage of the last, or a hybrid, possibly of *S. americana* ♀ × ♂ *Guiraca cærulea.* While it is not improbable that the type came from an egg laid by *S. americana,* even such immediate ancestry would not forbid recognition of "specific characters;" the solitary bird having been killed, it represents a species which died at its birth. The type is extant in the U. S. National Museum. An unfinished sketch of this specimen, different from Audubon's published plate, forms the frontispiece of Audubon and his Journals, Vol. ii. 1897.

ZAMELO'DIA. (Gr. ζά, *za,* much, very; μελῳδία, *melodia,* melody. Fig. 303.) SONG GROSBEAKS. Bill extremely heavy; lower mandible as deep as upper or deeper; commissural angle strong, far in advance of feathered base of bill; rictus overhung with a few long stiff bristles. Wing with outer 4 primaries abruptly longer than 5th. Tail shorter than wing, even or scarcely rounded. Feet short and stout. Embracing two large species, of beautiful and striking colors, the sexes dissimilar. ♂ black and white, with carmine-red or orange-brown; ♀ otherwise, but with lining of wings yellow. Brilliant songsters; nest in trees and bushes; eggs spotted. (*Zamelodia* COUES, 1880, and of 2d-4th eds. of the Key, 1884-90, must stand as against *Habia* REICH. 1850, and of 1st and 2d eds. of the A. O. U. List, 1886-95; for this *Habia* is antedated by *Habia* of Cuvier, 1849, for a genus of South American Tanagers (*Saltator* VIEILL.). Compare STEJ. Auk, Oct. 1884, p. 366, with COUES, Auk, Jan. 1897, p. 39. (The introduction of *Habia* into North-American ornithology for our Song Grosbeaks was rectified by the A. O. U. Committee in 8th Suppl. List, Auk, Jan. 1897, p. 130.)

FIG. 303.—Bill of *Zamelodia ludoviciana,* nat. size. (Ad. nat. del. E. C.)

Analysis of Species.

♂ black and white, with carmine-red on breast and under wings.   ♀ with lining of wings saffron-yellow.   Eastern
ludoviciana

♂ black and white, with orange-brown on breast: ♂ ♀ with lining of wings and belly yellow.   Western
melanocephala

Z. ludovicia′na. (Lat. of Louisiana. Figs. 303, 304, 311.) ROSE-BREASTED SONG GROSBEAK. Adult ♂: Head and neck all around, and most of upper parts, black; rump, upper tail-coverts, and under parts white; breast and under wing-coverts exquisite carmine or rose-red; wings and tail black, variegated with white; bill white; feet grayish-blue; iris brown. ♀ above, streaked with blackish and olive-brown or flaxen-brown, with median white coronal and superciliary line; below, white, more or less tinged with fulvous and streaked with dusky; *under wing-coverts saffron-yellow;* upper coverts and inner quills with a white spot at end; bill brown. Young ♂ at first resembling ♀; but rose color appears with first full feathering of the first autumn. It then resembles the adult winter ♂, but has brown instead of black quills and tail-feathers. At the first spring moult it becomes black, white, and rose as soon as some brownish bordering of the black feathers disappears, apparently by wearing off. Sexes of same size. Length 7.75–8.50; extent 12.00–13.00; wing 3.90–

FIG. 304. — Rose-breasted Grosbeak, reduced. (Sheppard del. Nichols sc.)

4.25; tail 3.25; tarsus 0.90. Eastern U. S. and British Provinces, N. to Labrador and region of the Saskatchewan; W. in U. S. to the Red River Valley, and edge of the Missouri River plains; winters extralimital in Cuba, Central Am., and northern S. Am.; breeds from the Middle States, Kansas, etc., northward, and in mountains S. to the Carolinas. A splendid bird! Few combine such attractions for eye and ear. Nest in bushes and low trees, a thin, flat structure, chiefly composed of rootlets and other slender fibres; eggs 3–5, rarely only 2, 1.00 × 0.75, dull greenish, fully splashed and dotted with various dark browns, laid in June.

Z. melanoce′phala. (Gr. μέλας, μέλανος, *melas, melanos*, black; κεφαλή, *kephale*, head. Fig. 305.) BLACK-HEADED SONG GROSBEAK. Adult ♂: Crown, sides of head, back, wings, and tail black; back usually varied with whitish or cinnamon-brown; wings spotted with white on ends of coverts, and usually also toward ends of quills, and with a large white patch at base of primaries; several lateral tail-feathers with large white spots on inner webs near their ends. Neck all around, rump, and under parts rich orange-brown, changing to bright pure yellow on belly and under wing-coverts; bill and feet dark grayish-blue. Size of the last. The ♀ and young differ much as in the last species, but may be recognized by the *rich sulphur-yellow* under wing-coverts;

FIG. 305. — Black-headed Grosbeak, reduced. (Sheppard del. Nichols sc.)

bill shorter and more tumid, 0.66–0.75 along culmen, 0.60 deep at base. Adult ♀: Under parts like those of ♂, but paler, though belly and lining of wings are as pure yellow. Upper parts dark brown with an olive shade, varied with whitish or brownish-white; head blackish with white or brownish coronal and superciliary stripes. Wings dusky, marked as in ♂, but basal white spot on primaries restricted; tail as in ♂, but the white spots reduced or obsolete. Bill light-colored below. In ♂ the tendency is to perfectly black head, back, tail, and wings, the two former pure and continuous, the two latter boldly spotted with white as described; but such faultless full dress is not often seen. This stylish western representative of the elegant Rose-breast is common in suitable woodland, Plains to the Pacific, U. S., and adjoining British Provinces, wintering in Mexico and portions of Lower California, breeding throughout its U. S. and Brit. Am. range; its habits are similar. Nest in trees of various kinds, often willows, up to 20 feet from ground; a flimsy structure, on a foundation of weed-stalks, openworked with grass and rootlets, 4.50–5.50 in outside diameter, 3.00 inside, cupped 1.00; eggs 2–5, usually 3–4, averaging 1.00 × 0.70, moderately variable in size, fugacious greenish-blue, speckled, spotted, and blotched with reddish and darker brown, with lavender shell spots, mostly laid in May; both sexes incubate. There is a nearer relationship between the Song Grosbeaks and the Evening Grosbeak than would appear from the distance apart of their respective genera in the present book; but I hesitate to remove *Hesperiphona* from the place it has always occupied in the Key.

GUIRA'CA. (*Vox barb.*, Mexican or S. Am. name of some bird. Fig. 306.) BLUE GROSBEAKS. Bill with commissure strongly angulated far beyond base, with deep under mandible and bristly rictus as in *Zamelodia*, but not so swollen, the culmen nearly straight. Wings long and pointed, folding about to middle of tail; tip formed by 2d–4th quills, 1st little shorter, 5th rapidly graduated. Tail shorter than wings, even. Tarsus rather less than middle toe and claw; outer lateral toe slightly longer than inner, but scarcely reaching base of middle claw. One species, large, ♂ blue, ♀ brown.

Fig. 306. — Bill of *Guiraca*, nat. size. (Ad. nat. del. E. C.)

G. cœru'lea. (Lat. *cœrulea*, cerulean. Fig. 307.) BLUE GROSBEAK. Adult ♂: Rich dark blue, nearly uniform, but darker or blackish across middle of back; feathers around base of bill, wings, and tail black; middle and greater wing-coverts tipped with chestnut; bill dark horn-blue, paler below; feet blackish. Length 6.50–7.00; extent 10.50–11.00; wing 3.30–3.60; tail 2.75–3.00; bill 0.60–0.67; tarsus 0.75; middle toe and claw rather more. ♀ smaller, plain warm brown above, paler and rather flaxen-brown below, sometimes whitey-brown on throat and belly, or with slight streaks on belly and crissum; wings and tail fuscous, sometimes slightly bluish-glossed or edged, former with whitey-brown cross-bars; bill and feet brown. Young ♂ at first like ♀; when changing, shows confused brown and blue; afterward, blue interrupted with white below. Eastern U. S., but southerly; rarely N. to Massachusetts, and even Maine; winters wholly extralimital in Cuba and Mexico; breeds throughout its U. S. range. Its limit of northward migration with regularity and in any numbers is about the latitude of Philadelphia, in the Atlantic States,

Fig. 307. — Blue Grosbeak, reduced. (Sheppard del. Nichols sc.)

Illinois, Nebraska, etc., in the interior. Nest in bushes, vines, or other shrubbery, sometimes a low tree, of grasses and rootlets; eggs 3–4, averaging 0.90 × 0.65, palest bluish, normally unspotted; quite like those of the Indigo-bird, but larger.

G. c. eurhyn′cha. (Gr. εὖ, well, as intensive prefix; ῥύγχος, *hrugchos*, beak.) WESTERN BLUE GROSBEAK. Larger; length 7.00 or more; wing nearly 4.00; tail 3.00 or more; bill notably larger. ♂ paler blue, with broader wing-bars, that on the greater coverts paler than the other. ♀ and young ♂ grayish-brown. Western U. S., from Neb., Col., Utah, and Sacramento Valley, Cal. in summer to Mexico; winters to Costa Rica. I first described this form in Am. Nat., 1874, p. 563, from Mexican examples; it was not taken up in the Key till 1890, p. 900, through deference to the A. O. U. Committee, who first recognized it after RIDGW. Man. 1887, p. 446; A. O. U. List, 2d ed. 1895, No. 597 *a*. It was next supposed to be a synonym of *Pitylus lazula* LESS. Rev. Zool. v, 1842, p. 174, and so given as *Guiraca cærulea lazula* in A. O. U. Suppl. List, Auk, Jan. 1899, p. 121; but that remains to be proven.

CYANOSPI′ZA. (Gr. κύανος, *kuanos*, blue; σπίζα, *spiza*, a finch.) PAINTED FINCHES. Bill relatively smaller and weaker than in *Guiraca*, with less conspicuous angulation; culmen regularly a little convex, gonys nearly straight. Outer 4 primaries longest; 1st usually between 4th and 5th, latter much shorter. Tail little shorter than wing, about even or emarginate. Feet moderate; tarsus about equal to middle toe and claw; lateral toes about equal to each other, their claws falling short of base of middle claw. Embracing several elegant Finches of small size; ♂ of very showy hues, especially blue, but also red, purple, yellow, and green, usually in masses; ♀ of simple and tasteful greenish or brownish shades. Nest in bushes and low trees, sometimes close to the ground; eggs oftenest whole-colored, very pale, sometimes spotted. The name of this genus has been changed back from *Passerina* to *Cyanospiza* BAIRD, 1858, as in 1st ed. of Key, 1872, in A. O. U. Suppl. List, Auk, Jan. 1899, p. 121. (*Passerina* of 2d–4th eds. of Key; A. O. U. Lists, 1st and 2d eds. 1886 and 1895.)

Analysis of Species.

♂ rich blue, intense red and golden-green; ♀ greenish and yellow. Southern . . . . *ciris*
♂ purplish-blue, dusky and reddish. ♀ brown. Southwestern . . . . *versicolor* and *v. pulchra*
♂ lazuli-blue and white, the breast brown; ♀ brown and whitish. Western . . . . *amœna*
♂ indigo-blue; ♀ brown. Eastern . . . . *cyanea*

C. ci′ris. (Gr. κείρις, *keiris*, name of a bird into which Scylla, daughter of Nisus, was transformed.) PAINTED FINCH. PAINTED BUNTING. NONPAREIL. POPE. Adult ♂: Crown and hind neck and sides of head and neck rich blue; back and scapulars beautiful golden-green; eyelids and entire under parts intense vermilion-red; rump duller red; wings dusky, glossed with green and reddish; tail dusky reddish or purplish-brown. Bill dark horn-color; feet dark brown. Size of *amœna*; wing 2.70; tail 2.50, a little emarginate. ♀: Above, plain yellowish-green, or light olive, nearly uniform, this color glossing the dusky wings and tail; below, yellowish; bill brownish, pale below: thus quite different from the brown ♀ ♀ of all the following species. Young ♂ at first like ♀, though rather duller, with some buffy and grayish-brown shades; acquiring the red and blue with every possible gradation between the colors of the two sexes. In confinement the ♂ is liable to lose its brilliant colors, the scarlet turning to orange, etc. South Atlantic and Gulf States, abundant; up the coast to Carolina, and in the interior to Illinois and Kansas; winters in Mexico, C. Am., Cuba, etc.; accidental in Mass. An exquisite little creature of matchless hues, well named the "incomparable"; a fair songster, and a favorite cage-bird in Louisiana. Nest in bushes, hedges, and low trees; eggs 0.75 × 0.55, pearly white, speckled with reddish and purplish browns, chiefly about the larger end.

C. versi′color. (Lat. *versicolor*, various in color; *verto*, I turn, *color*, color.) PURPLE PAINTED FINCH. VARIED BUNTING. WESTERN NONPAREIL. PRUSIANO. Adult ♂: Hind head, throat, and fore breast brownish-red or claret-color, the former sometimes scarlet;

hind neck and middle of back similar, but more obscured; fore part of crown purplish-red; rump and upper tail-coverts purplish-blue; below, from breast, and wings and tail, dusky, tinged or glossed with purplish ; concealed white in feathers of side of rump ; lores and circumrostral feathers black. Bill horn-bluish, paler below, stouter than in other species, with very convex culmen and concave cutting edge of upper mandible. Feet dark. The versicoloration is difficult to describe ; the general aspect is that of a purplish-dusky bird, redder or bluer here and there. Size of the others. ♀ plain brown above, whitey-brown below, like *amœna* and *cyanea;* no whitish wing-bars; no black stripe on gonys; concealed white on sides of rump. Eastern Mexico, S. to Guatemala, N. to U. S. border, in the Lower Rio Grande Valley, where common in some localities; accidental in Michigan. Eggs 0.78 × 0.58, plain bluish-white, like those of the Indigo-bird.

C. v. pul'chra. (Lat. *pulcher* (masc.) or *pulchra* (fem.), beautiful.) PENINSULA PAINTED FINCH. BEAUTIFUL BUNTING. Like the last; wings and tail said to be shorter; wing about 2.50; tail 2.20; ♂ said to have the "red on occiput brighter, purple on throat less reddish (never decidedly red?), flanks brighter plum-purple, and rump more purplish-blue or lavender." Lower California and western Mexico. RIDGW. Man. 1887, p. 448; COUES, Key, 4th ed. 1890, p. 900; A. O. U. List, 2d ed. 1895, No. 600 *a.*

C. amœ'na. (Lat. *amœna*, delightful, charming, dressy.) LAZULI PAINTED FINCH. Adult ♂ : Head and neck all around, entire upper parts, and lining of wings, rich azure or lapis-lazuli blue, more or less obscured on middle of back; lores black. Below, from the blue neck, chestnut-brown, changing to white on belly and crissum. A firm white wing-bar across ends of median coverts, and usually another weaker one across tips of greater coverts. Wings and tail dusky, glossed with blue. Bill and feet bluish-black. Length 5.25-5.50; extent 8.00-8.50; wing 2.75-3.00; tail 2.25-2.50; bill 0.37; tarsus 0.65. Adult ♀ : Above, flaxen-brown, nearly uniform, but with slightly darker centres of the feathers, and sometimes a faint bluish gloss. Below, buffy or brownish-white, most colored on breast, palest on throat and belly. Wings and tail fuscous, with faint bluish edgings usually, crossed with two decided brownish-white bars, — the chief distinction from ♀ *cyanea*. Young ♂ : Like ♀ ; when changing, patched with brown and blue ; when very young, ♂ ♀ somewhat streaky, especially on under parts. Replacing *cyanea* from the Plains to the Pacific, U. S. and interior of British Columbia, S. into Mexico ; common in suitable places ; habits, nest, and eggs the same.

C. cya'nea. (Lat. *cyanea*, Gr. κυάνεος, *kuaneos*, dark blue. Fig. 308.) INDIGO PAINTED FINCH. INDIGO-BIRD. Adult ♂ : Indigo-blue, intense and constant on head, glancing greenish with different lights on other parts; wings and tail blackish, glossed with greenish-blue; feathers around base of bill black ; bill dark above, rather paler below, with a curious black stripe along gonys. ♀ : Above, plain warm brown, below whitey-brown, obsoletely streaky on breast and sides; wing-coverts and inner quills pale-edged, but not whitish ; no whitish wing-bars; upper mandible blackish, lower pale, with the black stripe just mentioned, — this is a pretty constant feature, and will distinguish the species from any of our Eastern little brown birds. Young ♂ : Like ♀, but soon shows blue traces, and afterward is blue with white variegation below. Size of the foregoing. Eastern U. S., N. to Maine and some parts of Canada ; W. to Kansas, Indian Terri-

FIG. 308. — Indigo-bird, reduced. (Sheppard del. Nichols sc.)

tory, and Texas; S. in winter to Central America; breeds throughout its N. Am. range. Abundant in fields and open woodland, in summer; a well meaning but rather weak vocalist, whose low rambling strain is delivered as if the little performer were tired or indifferent. Nest in crotch of a bush, large for size of the bird, and not at all artistic; eggs usually 4, rarely 5 averaging 0.72 × 0.52, white, usually with a faint bluish tint, and normally plain, though not seldom a little speckled.

OBS. It is probable that yet another species of this beautiful genus is to be added to our Fauna, as follows: **C. parelli′na.** (Lat. uncertain, perhaps from *paralius*, from Gr. παράλιος, *paralios*, beside the sea, maritime, with reference to the "ultra marine" blue of the ♂; compare *Paralus*, name of a man beautifully painted by Protogenes; the expression *Paralum pictum* occurs in Cicero.) PARALINE PAINTED FINCH. MEXICAN BLUE BUNTING. Adult ♂: Rich dark blue, brightening on the front and sides of the head, the rump, and lesser wing-coverts, into azure blue; lores, chin, tail, and bill black; eyes brown; feet dark. ♀: Brown, paler below, whitening on throat and belly. Young ♂: Like ♀; but in any plumage this species may be recognized by the large turgid bill and much rounded wings with 3d-5th quills longest, 2d about equal to 6th, 1st shortest of all. The species represents a connecting link between *Cyanospiza* and *Guiraca*, and is type of the subgenus *Cyanocompsa* CAB. J. f. O. 1861, p. 4. Very small; length 5.00 or little more; wing 2.70; tail 2.30; culmen 0.40-0.45; gonys up to 0.30. Eastern Mexico, said to extend into Texas in the valley of the Lower Rio Grande. Not noted in previous eds. of the Key. *Cyanoloxia parellina* BP. Consp. Av. i, Aug. 1850, p. 502; *Cyanospiza parellina* BAIRD, B. N. A. 1858, p. 502, Tamaulipas and New Leon; *Passerina parellina* RIDGW. Pr. U. S. Nat. Mus. iii, 1880, p. 182; *P.* (*Cyanocompsa*) *parellina* RIDGW. Man. 1887, p. 446.

SPORO′PHILA. (Gr. σπόρος, *sporos*, seed; φίλος, *philos*, loving.) PYGMY FINCHES. FINCHLETS. Bill like that of a Bullfinch in miniature, short and extremely turgid; swollen in all directions, culmen convex nearly in the sextant of a circle; cutting edge of upper mandible very concave; gonys short, about straight in outline. Wings short and greatly rounded; 2d-4th quills longest, 1st, 5th, and even 6th, little shorter, and secondaries nearly covering primaries in the closed wing. Tail rather shorter than wings, slightly rounded, with abruptly pointed tips of the feathers. Tarsus equal to middle toe and claw, and lateral toes to each other, their claws about reaching base of middle claw. A large Central and South American genus of Pygmy Finches, one of which reaches our border. (Name changed from *Spermophila* of former eds. of the Key because this is preoccupied for the mammalian genus *Spermophilus* of F. CUVIER, 1822, or *Spermophila* RICHARDSON, 1825.)

S. morelet′i sharpe′i. (To Arthur Morelet, a French traveller, shell-collector, and author, and to Dr. R. Bowdler Sharpe, the famous English ornithologist.) SHARPE'S PYGMY FINCH. SHARPE'S FINCHLET. LITTLE SEED-EATER. ♂: Top and sides of head, back of neck, broad band across upper part of breast, middle of back, wings, and tail, black; chin, upper throat, neck nearly all around, rump, and remaining under parts, white, the latter often tinged with pale buff; two wing-bands, and bases of all the quills, also white, that on secondaries hidden by coverts, that on primaries forming an exposed spot; inner secondaries usually edged with white; tail-feathers sometimes with obscurely whitish tip. Bill blue-black; feet dark. ♀ olivaceous-brown above, brownish-yellow or dull buff below; wings with whitish bars, but no white bases of quills; bill brown; feet dark. Length about 4.00; wing 2.00-2.10; tail 1.90; tarsus 0.60. Mexico to Texas, in the Lower Rio Grande Valley. *Spermophila moreleti* of most American writers, and of 1st-3d eds. of the Key, but subspecifically different from the true *S. morelleti* BP. of Guatemala; *Sporophila morelleti sharpei* LAWR. Auk, Jan. 1889, p. 53; COUES, Key, 4th ed. 1890, p. 900; A. O. U. List, 2d ed. 1895, No. 602. (*S. parva* SHARPE. Cat. Brit. Mus. Birds, xii, 1888, p. 124, includes this form, doubtless distinct from *S. parva* LAWR. Ann. N. Y. Acad. ii, 1883, p. 382.)

EUETHEI'A. (Gr. εὐήθεια, *euetheia*, guilelessness, simplicity, innocence.) GRASS QUITS. Bill small, acute, culmen slightly convex, commissure about straight to the angulation at base. Wings short, rounded, 2d–5th primaries subequal and little longer than 1st, 6th, 7th. Tail still shorter, about even. Tarsus if anything shorter than middle toe and claw; lateral toes subequal to each other in length, scarcely reaching base of middle claw. West Indian and tropical American genus of diminutive finches, two of which occur casually in Florida. (*Phonipara* of previous eds. of the Key, and of most writers; but *Phonipara* Bp. Consp. Av. i, p. 494, July 30, 1850, is antedated by *Euetheia* REICH. Syst. Av. pl. lxxix, fig. 13, June 1, 1850, the correct form of which is **Euethi'a** CAB. Mus. Hein. i, 1851, p. 146.)

E. bi'color. (Lat. *bicolor*, of two colors.) BLACK-FACED GRASS QUIT. Adult ♂: Upper parts, including exposed surfaces of wings and tail, dull olivaceous, passing on face, throat, and breast, into sooty-black, fading on other under parts into olive-gray, more or less varied with whitish; wings and tail unmarked; no decided demarcation of colors and no yellow anywhere. Bill blue-black; feet dark brown. ♀ lighter olivaceous, passing to olive-ashy where ♂ is black; bill pale below; feet light brown. Length about 4.00; wing 2.00–2.10; tail 1.75. West Indies and of rare or casual occurrence in southern Florida, where it was taken in 1871 by C. J. Maynard. One of the common House Finches in various West Indian Islands; nest in bushes and shrubbery, large, domed, with lateral entrance; eggs 3–6, 0.65 × 0.50, white, speckled with umber-brown. (*Phonipara zena* of 2d–4th eds. of the Key, and most American writers, after *Fringilla zena* LINN. Syst. Nat. 10th ed. 1758, p. 183, as based on Catesby, 1731, pl. 37 (but not *F. zena* LINN. *ibid.* p. 181, which is the Bahaman Tanager now called *Spinadalis zena*); *F. bicolor* LINN. 12th ed. p. 324; *Phonipara bicolor* BP. Consp. i, 1850, p. 494; *Euethia bicolor* GUNDLACH, J. f. O. xxii, 1874, p. 312; A. O. U. Lists, 1st and 2d eds. 1886 and 1895, No. 603 or [603]]) : see Auk, Jan. 1885, p. 48.

E. cano'ra. (Lat. *canorus* (masc.) or *canora* (fem.), singing, tuneful, melodious; *canor*, song, melody; *cano*, I sing.) MELODIOUS GRASS QUIT. Adult ♂: Upper parts bright olive-green; lower parts gray, whitening on the crissum; most of head black; a black pectoral band, and a broad bar of bright yellow curving upon each side of the head behind the ears to the eyes. ♀ similar, but the black of the ♂ replaced by chestnut-brown, the yellow curve paler or broken. About the size of the last. A Cuban Quit, one specimen of which was taken on Sombrero Key, Florida, April 17, 1888, by M. E. Spencer: see Auk, July, 1888, p. 322. (*Loxia canora* GM. S. N. I. 1788, p. 858, based on the Brown-cheeked Grosbeak of Brown, Ill. 1776, pl. xxiv, fig. 1; *Phonipara canora* BP. 1850; *Euetheia canora* BREWER, 1860; COUES, Key, 4th ed. 1890, p. 900; A. O. U. List, 2d ed. 1895, No. [603. 1]).

PYRRHULO'XIA. (Lat. *pyrrhula* + *loxia*; *pyrrhula*, a bullfinch; *loxia*, a cross-bill. Gr. πυρρός, *purhros*, red; λοξίας, *loxias*, crooked.) BULLFINCH CARDINALS. PYRRHULOXIAS. Bill very short and stout, hooked almost like a Parrot's, its depth at base exceeding its length; under mandible deeper than upper at nostrils; culmen curved almost to the quadrant of a circle; commissure forcibly angulated in advance of nostrils; gonys about straight. Otherwise generally like *Cardinalis*. Colors grayish and red; head crested; sexes unlike. One large species, with two subspecies.

P. sinua'ta. (Lat. *sinuata*, bent, bowed, curved; *sinus*, a bend, bay: alluding to the bill. Fig. 309.) BECKHAM'S CARDINAL. ARIZONA PYRRHULOXIA. Like the common Bullfinch Cardinal or Texan Pyrrhuloxia as below described; said to differ in lighter and browner tone of the gray parts, greater extent of red on tail, little if any blackish suffusion in red of the capistrum of ♂, and lighter red of crest; ♀ less grayish on fore breast and along sides. ♂, wing 3.60–3.90; tail 4.25; depth of bill 0.51. S. W. Texas, S. New Mexico, S. Arizona, and southward. This is the true *P. sinuata*, originally described by Bonaparte from W. Mexico as *Cardinalis sinuatus*, and erroneously renamed *P. sinuata beckhami* by RIDGW. Auk, Oct. 1887, p. 347; COUES, Key, 4th ed. 1890, p. 900; A. O. U. List, 2d ed. 1895, No. 594 *a*:

see PALMER, Nidologist, iii, May, 1896, p. 102, and RIDGW. Auk, Jan. 1897, p. 95. *P. sinuata*, A. O. U. Suppl. List, Auk, Jan. 1897, p. 130, No. 594 (not 594 *a*).

P. s. penin'sulæ. (Lat. of a peninsula, to wit, that of Lower California.) PENINSULA CARDINAL. ST. LUCAS PYRRHULOXIA. Said to be colored like the last, but smaller, with larger bill: wing of ♂ 3.30–3.60; tail 3.80–4.15; depth of bill 0.52–0.55. Lower California. References as above; A. O. U. No. 594 *b*. Both of these forms are included under *P. sinuata* in earlier eds. of the Key.

FIG. 309. — Arizona Pyrrhuloxia.

P. s. texa'na. (Lat. Texan.) COMMON BULLFINCH CARDINAL. TEXAS PYRRHULOXIA. Conspicuously crested, and otherwise like the common Cardinal in form, but bill extremely short and crooked. ♂: Ashy brown, paler or whitish below; crest, face, throat, breast, middle line of belly, wings, and tail, more or less perfectly crimson or carmine red; bill whitish. Length 8.00–8.50; extent 11.00–12.00; wing 3.50–4.00; tail 3.75–4.25. ♀ similar to ♂, more so than ♀ *Cardinalis*: red of crest, wings, and tail much the same; rather brownish-yellow below, usually with traces of red on breast and belly, sometimes without. Young ♂ like ♀. At an early age, both sexes have the bill obscured. In this species the crest is long, but thin, consisting of a few coronal feathers, without general elongation of head-plumage. The shade of red is very variable in equally adult males. In highest feather it is continuous on under parts from bill to tail along median line; but it is often broken into patches on throat, belly, and crissum. The tint is always carmine, not vermilion as usual in the common Cardinal. The intense rose-color is well displayed on spreading the wings. A singular bird, inhabiting Texas near the Mexican border; abundant in the valley of the Lower Rio Grande, sometimes extending thence into Louisiana; S. through much of E. Mexico. The habits, nest, and eggs are substantially the same as those of the common Cardinal: eggs rather smaller, averaging 0.95 × 0.75. (*P. sinuata* of former eds. of Key; A. O. U. Lists, 1886–95, No. 594; *P. s. texana* RIDGW. Auk, Jan. 1897, p. 95; A. O. U. Suppl. List, Auk, Jan. 1897, p. 129, No. 594 *a*.)

CARDINA'LIS. (Lat. *cardinalis*, pertaining to *cardo*, a door-hinge; *cardinal*, that upon which something hinges or depends; hence important, principal, *cardinal* point; *cardinal*, a chief ecclesiastical official, wearing the red hat; hence *cardinal-red*, from which color the bird is named. Fig. 310.) CARDINAL GROSBEAKS. Bill very large and stout, but quite conic; culmen a little convex; gonys about straight; commissure sinuate, not abruptly angulated; lower mandible about as deep as upper; rictus bristled. Wings very short and rounded; usually 4th and 5th quills longest, others rapidly graduated both ways — 5th to 1st, 5th to 9th. Tail longer than wings, rounded, of broad feathers with obliquely oval tips. Tarsus longer than middle toe and claw; lateral toes subequal. Size large. Head crested. Color mostly red, including bill. Sexes

FIG. 310. — Head of Cardinal Grosbeak, nat. size. (Ad. nat. del. E. C.)

FRINGILLIDÆ: FINCHES, BUNTINGS, SPARROWS.

subsimilar. There are several species of this strikingly beautiful genus, as *C. carneus* and *C. phœniceus*, but only one of them, with several of its subspecies, occurs in our Fauna.

C. cardina′lis. (Figs. 310, 311.) CARDINAL GROSBEAK. CARDINAL REDBIRD. CRESTED REDBIRD. VIRGINIA REDBIRD. VIRGINIA NIGHTINGALE. Adult ♂ : Rich red, usually vermilion, sometimes rosy; pure and intense on crest and under parts, darker on back, where obscured with ashy-gray, as it is also on upper surfaces of wings and tail; feathers of wings

FIG. 311. — Cardinal Grosbeak, upper; Rose-breasted Grosbeak, lower; reduced. (From Brehm.)

fuscous on inner webs. A jet-black mask on face, entirely surrounding bill, extending on throat. Bill coral-red; feet brown. Length 8.00–9.00; extent 11.00–12.00; wing 3.50–4.00; tail 4.25–4.75; bill 0.67–0.75; tarsus 0.90–1.00. ♀ rather less: Ashy-brown, paler and somewhat yellowish-brown below, with traces of red; reddening much as in ♂ on crest, wings, and tail. Young ♂: At first like ♀, but soon reddening; at an early age, bill dark. Eastern U. S. southerly, seldom N. to the Connecticut Valley; lower Hudson Valley, and Great Lakes region, and only casually further N.; W. to the Great Plains; resident in the Bermudas; along the Mexican border shading into other varieties. A bird of striking appearance and brilliant vocal powers, resident and abundant from the Middle States southward; inhabits thickets, tangle, and undergrowth of all kinds, whence issue its rich rolling whistling notes, while the performer, brightly clad as he is, often eludes observation by his shyness, vigilance, and activity. The nest, built loosely of bark-strips, twigs, leaves, and grasses, is placed in a bush, vine, or low thick tree; eggs 1.00–1.10 × 0.70–0.80, profusely marked with browns, from reddish to dark chocolate, with neutral tint in the shell, usually in fine dotting or marbling pattern. Two or three broods are reared in the South. Like the Rose-breasted Grosbeak, the Cardinal is a favorite cage-bird. (*C. virginianus* of all former eds. of the Key; but by the canons of the A. O. U., in the formulation of which I took part, I am obliged to use the miserable tautonymy of *Cardinalis cardinalis*, so offensive to literary good taste.)

C. c. florida'nus. FLORIDA CARDINAL. Resident birds of Florida are attempted to be distinguished by somewhat brighter color by RIDGW. Man. 2d ed. 1896, p. 606. The alleged distinction was denied by a majority of the A. O. U. at Cambridge in Nov. 1896, and affirmed by a majority of the same at Washington in December. Hence A. O. U. Suppl. List, Auk, Jan. 1897, p. 122, No. 593 *d*.

C. c. canicau'dus. (Alleged Lat. for *canicaudatus*, having a gray tail; Lat. *canus*, gray, *cauda*, tail.) GRAY-TAILED CARDINAL. ♂ like that of true *cardinalis*, but with a less conspicuous black frontlet, in this respect approaching *superbus*. ♀ grayer than ♀ *cardinalis*, "and with the tail-feathers broadly margined with gray, instead of being narrowly edged with olivaceous brown." This form seems to be of the "new woman" type, the ♀ being more distinguished than the ♂. Vicinity of Corpus Christi, Texas, and southward. *C. cardinalis canicaudus* CHAPM. Bull. Am. Mus. Nat. Hist. N. Y. iii, Aug. 1891, p. 324; A. O. U. List, 2d ed. 1895, No. 593 *c*. Included under *cardinalis* proper in all former eds. of the Key.

C. c. super'bus. (Lat. *superbus*, proud, haughty.) SUPERB CARDINAL. ARIZONA CARDINAL. Like the next form, but larger, and ♀ more richly colored. ♂, wing 4.10; tail 5.00; tarsus 1.05; bill along culmen 0.85, its depth at base 0.70: ♀ smaller. S. Arizona and N. W. Mexico. *C. c. superbus* RIDGW. Auk, Oct. 1885, p. 344; A. O. U. Lists, 1st and 2d eds. 1886 and 1895, No. 593 *a*; *C. v. superbus* COUES, Key, 3d and 4th eds. 1887 and 1890, p. 876.

C. c. ig'neus. (Lat. *igneus*, fiery.) FIERY-RED CARDINAL. ST. LUCAS CARDINAL. Like the typical form; not redder, but if anything lighter red; black mask narrowed on forehead, or so interrupted there that the red reaches bill; crest inclining to light red, more like that of belly than of back. Bill tending to swell, with more decidedly curved culmen. Tail rather longer, on an average. Lower California, common. This form, described in 1859 by Baird as a full species, was reduced to its proper subspecific grade in the Key, orig. ed. 1872; the 2d ed. included *superbus* under the name of *igneus* from the valley of the Colorado and Gila, these two forms being discriminated in the 3d ed. 1887.

PI'PILO. (Lat. *pipilo* or *pipio*, I pip, peep, chirp.) TOWHEE BUNTINGS. Embracing numerous species and subspecies of large *Fringillidæ*, varying much in system of coloration and details of form, and therefore not easy to characterize concisely. Excepting one species, *all are over seven inches long*. Bill moderate in size, conic without extremes of turgidity or compression, but varying much in precise shape with the species. Feet large and strong, fitted for ground work; tarsus about equalling or rather exceeding middle toe and claw; lateral toes subequal, outer usually a little the longer, its claw reaching, in some cases exceeding, base of middle claw; claws all stout and much curved, in some species highly developed. Wings short and greatly rounded; 4th–5th primary longest, whence the quills are rapidly graduated to 1st and 9th; 1st very short. Tail long, exceeding wings, rounded or much graduated, of broad firm feathers with rounded ends. Large species, inhabiting shrubbery, and partly terrestrial. They fall in two subgenera. I. *Black Towhees* or PIPILO proper: of which the only eastern species is a typical example. In this, the sexes are very unlike, but the sexual difference is less in western subspecies of *P. maculatus*: all North American forms are black on head and upper parts, with black, white-marked wings or tail, the back also white-marked or not; belly white, sides chestnut. II. *Brown Towhees* or subgenus KIENERIA: variously brown above, paler, etc., below, the sexes alike. These are confined to the southwest, where they stand in the same relation to *Fringillidæ* that the southwestern forms of *Harporhynchus* bear to *Miminæ*. (On recent rupture of the genus, see COUES, Auk, 1897, p. 221.)

OBS. I. The black series of *Pipilo* offers a case nearly parallel with those of *Melospiza*, *Peucæa*, *Passerella*, and *Junco*, already discussed. There is one eastern form much more distinct from the several western ones than these are from one another. It is uniform black above, seldom with a trace of white spotting on scapulars : ♀ distinctively brown where ♂ is black. The western ones all have spotted scapulars and sometimes also interscapulars; and ♀♀ are blackish, much like ♂♂. (These furthermore shade into the *olivaceous* Mexican stock-form *P. maculatus*.) It might be more consistent to treat all the black Towhees as races of one incompletely differentiated stock; but it is not easy to so far ignore the sexual distinctiveness, nor the fact that though *erythrophthalmus* has occasional spots on the scapulars, its intergradation with the Mexican *maculatus* is not established. II. The Brown Towhees afford one remarkably distinct species, *P. aberti*, to be likened to *Harporhynchus crissalis*; and several subspecies of the Mexican *P. fuscus*, incompletely separated from one another, like some of the forms of *Harporhynchus*.

Analysis of Species and Subspecies.

1. **Black Towhees.** Colors of the male black, white, and chestnut in definite areas.
   No white on the scapulars or wing-coverts. Sexes very unlike.
   Eyes red. Eastern U. S. at large . . . . *erythrophthalmus*
   Eyes white. Florida, resident . . . . *e. alleni*
   Scapulars and wing-coverts with white spots ; sexes more alike. Western.
   Little if any white at bases of primaries ; none on outer web of outer tail-feathers except at end. Pacific Coast region . . . . *maculatus oregonus*
   White on wings and tail as in *erythrophthalmus*, but interscapulars streaked. Western, interior *m. arcticus*
   Like the last ; claws highly developed ; sexes nearly alike. Rocky Mt. region . . . . . *m. megalonyx*
   Like *m. oregonus* in color ; much smaller ; wing about 3.00; tail 3.50. Guadalupe Isl. . . . *consobrinus*
2. **Brown Towhees.** Colors not definitely black, white, and chestnut ; no greenish ; sexes alike. Southwestern.
   Grayish-brown, paler below, without blackish face ; throat and crissum fulvous or rufescent.
   Light ; belly whitening ; crissum yellowish-brown ; necklace of dusky streaks. Texas to Arizona.
   fuscus mesoleucus
   Similar ; more white on throat. L. Cala. . . . . *f. albigula*
   Dark ; belly only paler ; crissum cinnamon-brown ; throat fulvous, speckled. Pacific Coast region.
   f. crissalis
   Like the last ; darker above, grayer below ; smaller. S. and L. Cala. . . . . *f. seniculus*
   Grayish-brown, paler below ; face blackish ; no other decided markings . . . . *aberti*

(*Black Towhees: subgenus* PIPILO.)

P. erythrophthal'mus. (Gr. ἐρυθρός, *eruthros*, red; ὀφθαλμός. *ophthalmos*, eye. Fig. 312.) TOWHEE BUNTING. MARSH ROBIN. GROUND ROBIN. TURKEY SPARROW. BUSH-BIRD. CHEWINK. JOREE-GRASEL. Adult ♂: Glossy black; belly white; sides chestnut; crissum fulvous-brown; primaries and inner secondaries with white touches on outer webs; outer tail-feather with outer web and nearly terminal half of inner web white, next two or three with white spots decreasing in size; bill black; feet pale brown; *iris red* in the adult, ashy or brown in the young. Normally, the black pure and continuous; occasionally, white touches on wing-coverts and scapulars. White on primaries confined to bases of outer 6, and their outer webs at about their middle; on secondaries to outer webs of inner 2 or 3. Black feathers of throat with concealed whitish bases. Length 7.50–8.75 ; extent 10.00–

12.00; wing 3.20-3.90; tail 3.35-4.00; tarsus 1.00-1.12; but these extremes are rare; average length 8.00; extent 11.25; wing 3.75; tail 4.50. ♀: Rich warm brown where the male is black; otherwise similar, but smaller. *Very young* birds are streaked brown and dusky above, below whitish tinged with brown and streaked with dusky; but this plumage is of brief duration; sexual distinctions may be noted in birds just from the nest, and they rapidly become much like the adults. Eastern U. S. and adjoining parts of the British Provinces; N. to Canada, Manitoba, and N Dakota, where meeting *arcticus;* W. to Kansas, and in Missouri River region to about 43°. Northerly perfectly migratory; winters from middle U. S. southward; breeds nearly throughout its range. An abundant and familiar inhabitant of thickets, undergrowth, and briery tracts, spending much of its time on the ground, scratching among fallen leaves. Nest on the ground, bulky, of leaves, grasses, and other fibrous material; eggs 4-5, 0.95 × 0.70, white, thickly speckled with reddish. The curious names "Towhee," "Joree," and "Chewink" are from its cry; "Ground Robin" from its haunts and the chestnut of the sides.

P. e. al'leni. (To J. A. Allen, the eminent naturalist.) WHITE-EYED TOWHEE BUNTING. Similar; smaller; less white on wings and tail; claws longer; *iris yellowish-white.* ♂, extremes: Length 7.25-8.50; extent 9.50-11.55; wing 2.80-3.50; tail 3.25-4.00; tarsus 0.80-1.10; average length 7.90; extent 9.90; wing 3.12; tail 3.50; tail *relatively* longer than in northern specimens, producing less difference in total length than there is in length and extent of wings. White on outer tail-feather about as much as on next feather of *erythrophthalmus.* Florida; resident; a local race.

[**P. macula'tus.** (Lat. *maculatus,* spotted.) OLIVE-BLACK SPOTTED TOWHEE. A Mexican species, with extensively olivaceous coloration and streaked back, into which the following three subspecies are supposed to shade imperceptibly, — *oregonus* being furthest removed and most like *erythrophthalmus, arcticus,* and *megalonyx* successively nearing the Mexican stock-form.]

P. m. orego'nus. (To the Territory of the Oregon.) OREGON TOWHEE. ♂: Very similar to *erythrophthalmus;* quite as black, but not continuously so; chestnut of sides dark; wing-coverts with small rounded, and scapulars with larger oval, white spots on outer webs near end. (Interscapulars sometimes also with white touches?) White marks on primaries and inner secondaries very small or wanting, usually none at bases of the former; white spots on tail-feathers very small; outer web of outer rectrix not white except at end; greatest extent of white on tail 1.00 or less. Excepting these particulars, this form looks more like *erythrophthalmus* than like typical *maculatus,* in which the body colors are *olivaceous.* ♀ dark umber-brown, but not quite blackish. About the same size as *erythrophthalmus;* but averaging rather less; ♂, wing 3.40; tail 3.90; tarsus 1.10; culmen 0.54. Pacific coast region, N. to British Columbia, S. to Southern California, melting eastward into *arcticus,* southeastward into *megalonyx.*

P. m. arc'ticus. (Lat. *arcticus,* arctic.) ARCTIC TOWHEE. Similar to the foregoing; less purely and continuously black, with tendency to olivaceous on back and rump; white spots of wing-coverts larger, those of scapulars still larger and lengthening into streaks; interscapulars also streaked with white; white on quills and tail-feathers at a maximum, as in *erythrophthalmus;* usually, also, concealed white specks in black of throat. ♀ comparatively dark, but not quite blackish. In this form, the white on the wing-quills and tail-feathers, so much reduced in the glossy black *oregonus,* is as extensive as in *erythrophthalmus;* but the wing-coverts, scapulars, and interscapulars are fully marked with white; the black tends to olive, at least on rump, and the ♀ is not fairly brown. The dimensions do not differ appreciably from those of *oregonus.* Central region of N. Am., from limit of *erythrophthalmus* in Kansas, Nebraska, and Dakota, to that of *oregonus* in Oregon, Washington, and British Columbia; N. in summer to the region of the Saskatchewan; S. in winter to Texas; in the S. Rocky Mt. region melting into *megalonyx.*

Fig. 312.—Towhee.

P. m. megalo'nyx. (μεγάλη, *megale*, great; ὄνυξ, *onux*, claw.) SPURRED TOWHEE. The prevailing form in the S. Rocky Mt. region, New Mexico, Arizona, and California. Similar to *arcticus*, but feet larger, with highly-developed claws; hind claw decidedly longer than its digit; lateral claws reaching to or beyond middle of middle claw. In this form at any rate, the ♀ is hardly distinguishable in color from the ♂, being slaty-blackish with an appreciable olivaceous shade, thus exhibiting a decided approach to the typical Mexican stock. The note is entirely different from that of the eastern Towhee, being so exactly like the scolding "mew" of a Catbird, that I have heard persons stoutly contend that there are Catbirds in Arizona. The general habits, nest, and eggs of all these western Towhees are substantially the same as those of the eastern. (*P. m. magnirostris,* BREWST. Auk, Apr. 1891, p. 146, is described as similar to *megalonyx*; but bill much larger, rufous of under parts paler, upper parts browner, and tinged with olive; ♀ decidedly lighter than ♂; bill from nostril 0.42, its depth there 0.40. Laguna, L. Cala. Not admitted in A. O. U. List, 1895.)

P. m. atra'tus. (Lat. *atratus*, blackened; *ater*, black.) SAN DIEGO TOWHEE. Like *megalonyx*; white markings more restricted; ♂ black even on rump; ♀ dark brown, even sooty on throat and breast. Type from Pasadena, Los Angeles Co., Cal.; range ascribed from southern coast region into Lower California. RIDGW. Auk, July, 1899, p. 254.

P. m. clemen'tis. (For etym. see under *Carpodacus clementis*.) SAN CLEMENTE TOWHEE. Insular form of *megalonyx*, scarcely differentiated; slightly larger, and lighter colored; ♂ about like ♀ *megalonyx* in tone of dark parts; call note said to be different. Average dimensions of ♂: Length 8.65; wing 3.45; tail 4.25; tarsus 1.10; hind claw 0.53; culmen 0.55. San Clemente Island. Not in former eds. of Key. *Pipilo* "*clementæ*," by error for *clementis*. GRINNELL, Auk, July, 1897, p. 294. *P. maculatus* "*clementæ*" (*sic!*) A. O. U. Suppl. List, Auk, Jan. 1899, p. 120, No. 588 c.

P. consobri'nus. (Lat. as adj., related, as are those who are the children of brothers or sisters; as noun, *consobrinus*, a first-cousin, ♂, or *consobrina*, the same, ♀; *con*, with, and *sobrinus* for *sororinus*, sisterly, from *soror*, a sister; originally referring only to the children of sisters.) GUADALUPE TOWHEE. An insular form, distinct from any of the foregoing, though belonging to the same group. Coloration most nearly as in *oregonus*, in the reduction, restriction, or extinction of the white markings in the black of the ♂. Head, neck, and back black; white on outer webs of scapulars usually bordered with black; indications of white wing-bars in rows of spots on ends of median and greater coverts; inner secondaries and a middle portion of primaries with narrow and short white edgings; two or three lateral tail-feathers with short white patches. Below, as usual, white with chestnut sides and buff crissum. ♀ similar, but dull brownish-black where the ♂ is black, and smaller white tail spots. Decidedly smaller than the three foregoing; ♂ wing 3.00–3.25; tail 3.50–3.75; ♀ somewhat less. The dimensions are the main characteristic of the species in comparison with *oregonus*. Guadalupe Island, Lower California. *P. maculatus consobrinus,* RIDGW. Bull. U. S. Geol. Surv. Terr. ii. Apr. 1876, p. 189, and as such taken up in the 3d ed. of the Key, 1887, p. 876, after the admission of Lower California and its islands to ornithological union with the North American Fauna; *P. consobrinus* RIDGW. Bull. Nutt. Club, July, 1877, p. 60; Man. 1887, p. 437; A. O. U. Lists, 1st and 2d eds. 1886 and 1895, No. 589.

(Brown Towhees: subgenus KIENERIA.)

[**P. fus'cus.** (Lat. *fuscus*, dark brown.) MEXICAN BROWN TOWHEE. An obscure Mexican stock-form, carelessly described by Swainson, to which the five following N. Am. birds are probably referable as subspecies.]

P. f. mesoleu'cus. (Gr. μέσος, *mesos*, middle; λευκός, *leucos*, white; the middle under parts whiter than in *crissalis*.) BROWN TOWHEE. CAÑON TOWHEE. ♂ ♀: Above, uniform grayish-brown with slight olivaceous shade; crown brown in decided contrast; wings and tail like back, unmarked, or some tail-feathers with rusty tips. Below, a paler shade of color of back, whitening on belly, tinged with fulvous and streaked with dusky on sides of throat and middle of breast, washed with rich rusty-brown on flanks and crissum; belly usually quite white, contrasting with rusty flanks and vent; throat ochrey, usually immaculate and embraced necklace-wise with dusky spots in series on each side, aggregated and blotched on breast. Bill dusky, paler below; feet brown, toes usually darker than tarsus. Sexes indistinguishable. In fresh fall specimens, tawny suffuses nearly all the under parts except middle of belly, and the throat-spots are diffused instead of being in series. In the very early streaked stage, there is no distinction of a brown cap; wing-coverts rusty-edged; whole under parts dusky-streaked. Length 8.00–8.50; wing 3.60–4.00; tail 4.25–4.60; tarsus 1.05; bill 0.60. S. W. U. S., chiefly New Mexico and Arizona; but also W. Texas, S. Colorado, Utah, and Nevada; south to N. Sonora and Chihuahua. Nest in bushes; eggs, as in all the Brown Towhees, speckled and scratched with dark brown and blackish on a pale greenish ground, 1.00 × 0.70. (*P. fuscus* of the Key, orig. ed. 1872.)

P. f. albi′gula. (Lat. *albus*, white; *gula*, throat.) WHITE-THROATED BROWN TOWHEE. ST. LUCAS TOWHEE. Exactly like the last, but white of under parts extending farther up breast; gular spots more restricted, sparser, and better defined. Slightly distinguished; but in good spring specimens rusty is restricted to crissum; ochraceous of the throat less extensive, paler, and mainly confined within the necklace, and the size averages less: wing 3.40–3.70; tail 3.85–4.25. Lower California, N. to about lat. 30°.

P. f. seni′culus. (Lat. *seniculus*, diminutive of *senex*, an old man.) ANTHONY'S TOWHEE. Intermediate between the last and the next. Said to be distinguished from *albigula* by its darker lower parts, more pronounced throat-patch (which is very pale buffy in *albigula*), and chestnut lower tail-coverts; and to differ from *crissalis* in smaller size, less rusty on lower parts, darker upper parts and more grayish lower parts. " Above, clear grayish sepia; pileum indistinctly Vandyke brown; below, smoky grayish with rusty wash on flanks and buffy on lower abdomen; lower tail-coverts chestnut; throat tawny clay-color, about as in *crissalis;* malar region grayish-brown." Size of the foregoing. Southern California and S. in Lower California to lat. 29°. *P. f. senicula*, ANTHONY, Auk, Apr. 1895, p. 111; A. O. U. List, 2d ed. 1895, No. 591 *c;* but whatever may be the question regarding the propriety of recognizing this connecting link by name, there can be none respecting the gender of the word *seniculus*.

P. f. crissa′lis. (Low Lat. *crissalis*, relating to the *crissum*, the under tail-coverts, which are highly colored.) CRISSAL TOWHEE. CALIFORNIA TOWHEE. Similar to *mesoleucus;* crown like back; rather darker above, with an olivaceous tinge, decidedly so below; middle of belly scarcely or not whitening, gula fulvous strong, and, with its dusky streaks, definitely restricted to throat; flanks and crissum chestnut or deep cinnamon-brown. Rather larger. ♂: Length 8.50–9.00; wing 3.75–4.00; tail 4.50–5.00; tarsus 1.12; culmen 0.60; ♀ rather less. Pacific Coast region, N. to Umpqua Valley, Oregon, S. through southern California, abundant. Nest in bushes, probably also on ground; eggs 3–4, 0.95 × 0.72, pale greenish or bluish-white, fully spotted with blackish and neutral tints. This is the dark coast form, bearing the same relation to *mesoleucus* that the coast *Harporhynchus redivivus* bears to the paler *H. lecontei* of the interior. The crown is brownish, but not forming a cap contrasting with back; throat fulvous rather than ochrey; this color of very limited extent, and speckled with dusky throughout; crissum rich rusty. *P. fuscus*, CASS. Ill. 1854, pl. 17; BD. B. N. A. 1858, p. 517; but not true *fuscus* of Sw. Philos. Mag. i, 1827, p. 434; *Fringilla crissalis* VIGORS, Voy. Blossom, 1839, p. 19; *P. fuscus*, var. *crissalis* COUES, Key, orig. ed. 1872, p. 153, and of later writers.

P. f. carolæ. (To Charlotte C. McGregor.) NORTHERN BROWN TOWHEE. Described as grayer and more uniform above, with paler throat and slightly longer tail. Battle Creek, Cal. MCGREGOR, Bull. Cooper Club, i, No. 1, Jan. 1899, p. 11.

P. a'berti. (To Lieut. J. W. Abert.) ABERT'S TOWHEE. GRAY TOWHEE. Somewhat similar to foregoing species of this section of the genus, but entirely distinct; a very large, long-tailed form, with no decided markings anywhere excepting the dark face. Adult ♂ ♀: Above, grayish-brown, with a slight fulvous tinge; wings and tail darker and purer brown; tail-feathers slightly rusty-tipped. Below as above, but paler, by dilution with a peculiar pale pinkish-brown shade (like that on sides of an Oregon Snowbird), particularly on throat; crissum more cinnamon-brown; lores and chin blackish. Bill and feet brown; under mandible paler than upper. Young more rusty. There is much individual variation in shade, but this large dingy whole-colored bird with dark face is always easily recognized. Length 8.50–9.00; wing 3.40–3.90; tail 4.50–5.00; tarsus 1.00–1.10. New Mexico and Arizona, abundant, especially in the valley of the Gila and Colorado, where we find it a wild and shy inhabitant of thickets and chaparral; N. to Colorado, Nevada, and Utah. Nest in bushes, loose and bulky; eggs 3–4, 1.00 × 0.75, bluish-white, sparingly speckled and scrawled with blackish-brown, chiefly about the large end.

OREOSPI'ZA. (Gr. ὄρος, *oros*, gen. ὄρεος, *oreos*; σπίζα, *spiza*, a fringilline bird, perhaps the Chaffinch.) Related to *Pipilo*, especially to the section of that genus which contains greenish species; smaller than any of the foregoing Towhees; best recognized by the pattern of coloration, which is olivaceous, with yellow under the wing, rufous cap, and white throat in ashy surroundings, the latter feature strikingly as in *Zonotrichia albicollis* — indeed it is not easy to see how *Oreospiza* differs in form from *Zonotrichia*. One western species. (*Pipilo*, section III. of Key, 2d–4th eds. 1884–90; *Oreospiza* RIDGW. Man. 2d ed. 1896, p. 605; A. O. U. Suppl. List, Auk, Jan. 1897, p. 129.)

O. chloru'ra. (Gr. χλωρός, *chloros*, green; οὐρα, *oura*, tail.) GREEN-TAILED TOWHEE. BLANDING'S FINCH. Adult ♂ ♀: Above, grayish-green, sometimes quite olive-gray, at others bright olive-green; exposed surfaces of wings and tail with brighter greenish edgings. Edge of wing and under coverts and axillaries bright yellow. Crown rich chestnut; forehead blackish, with a whitish loral spot on each side. Chin and throat pure white, bounded by dusky maxillary stripes, as sharply contrasted with dark surroundings as in the White-throated Sparrow. Whole breast and sides of head, neck, and body fine clear-ash, or slate-gray, obscured on flanks and crissum with brownish, fading to white on belly — completing the resemblance to *Zonotrichia albicollis*. Bill blackish-plumbeous; feet brown, toes darker. Length about 7.00; extent 9.50; wing 2.80–3.20; tail 3.40–3.70; tarsus 0.95. Less mature birds have the chestnut cap veiled by gray tips of the feathers. Young: Crown like back. Upper parts dull brown tinged with greenish in places, streaked throughout with dusky, but wings and tail as in adult; under parts forecasting pattern of adults, but dusky-streaked throughout. This stage is brief; birds resemble adults after first fall moult. Western U. S., especially S. Rocky Mt. region and across the Great Basin to Coast Range of Cal.; N. to Wyoming, Montana, Idaho, and eastern Oregon; S. in Lower California and Mexico; migratory; winters over our border. A sprightly inhabitant of shrubbery; nest in bush or on the ground; eggs 0.90 × 0.68, pale greenish or grayish-white, freckled all over with bright reddish-brown, usually aggregating or wreathing at the larger end. (*Pipilo chlorurus* BAIRD, 1858, and of most later authors, as of all former eds. of the Key. *Oreospiza chlorura* RIDGW. 1896; A. O. U. Suppl. List, Auk, Jan. 1899, p. 121, No. 592. 1.)

ARRE'MONOPS. (Gr. ἀρρήμων, *arhremon*, speechless, silent; ὤψ, *ops*, face, aspect: *i. e.* resembling the S. Am. genus *Arremon*.) Bill not notable in any way. Tarsus exceeding middle toe and claw; lateral toes short; outer a little longer than inner; claw of neither reaching base of middle claw; fore claws all small and weak; hind claw about as long as its digit. Wings very short and much rounded; 4th–7th primaries about equal and longest; 2d as long as 9th; 1st equalling 3d from innermost secondary. Tail about as long as wings, much rounded; outer feathers 0.50 shorter than middle ones; all broad to their rounded ends. Coloration olivaceous with yellow edge of wing and inconspicuous head-stripes. (*Embernagra* of former eds. of the Key, and of A. O. U. Lists as of U. S. writers generally; but our bird proves not to belong to that S. Am. genus. *Arremonops* RIDGW. Man. 1896, 2d ed. p. 434; A. O. U. Suppl. List, Auk, Jan. 1897, p. 129.)

A. rufivirga'ta. (Lat. *rufus*, rufous, *virgata*, striped; *virga*, a rod.) GREEN FINCH. TEXAS SPARROW. Adult ♂ : Above, dull olive-green, brighter on wings and tail. Under parts shading from color of the upper through grayish-olive and olive-gray to sordid whitish, purest on middle of belly. Inner webs of wing-quills fuscous; tail the same, but more glossed with greenish, and sometimes showing traces of crosswise watering with darker waves, as often seen in the Song Sparrow. Whole bend and lining of wing bright clear yellow. Crown like back, with two broad stripes of dull rufous from nostrils to nape; a similar rufous stripe behind eye, sometimes traceable past eye to the lore, then defining a superciliary line of light olive-gray or whitish. A whitish eye-ring. Upper mandible light brown, lower drying yellowish; feet pale. Length 6.25–6.75 (not 5.50, as in Baird); extent 8.50–9.00; wing 2.40–2.75; tail the same; bill 0.50; tarsus 0.90; middle toe and claw 0.75. ♀ does not differ materially, and young lacks the head-stripes. Young, first plumage: Above, mixed brown and olive-tawny; wings brown, edged with olive, the coverts edged and tipped with tawny; breast like back; belly tawny. Texas, in Lower Rio Grande Valley. Inhabits shrubbery, chaparral, and close cover of all kinds, where it is difficult to discover, owing to its quiet ways and greenish tints. Keeps near the ground, but builds a domed nest of twigs and grasses in bushes and low trees; two broods are reared, in May–June and Aug.–Sept. Eggs 2–4, pure white, unmarked, averaging 0.85 × 0.65, but from 0.75–0.90 by 0.60–0.70. (*Embernagra rufovirgata* of 2d–4th eds. of the Key. *Arremonops rufivirgata* A. O. U. Suppl. List, Auk, Jan. 1897, p. 129, No. 586.)

Family ICTERIDÆ: American Starlings; Blackbirds, etc.

Cultrirostral Oscines with 9 primaries, 9 secondaries, 12 rectrices and scutellate tarsi. — A family of moderate extent, confined to America, where it represents the *Sturnidæ*, or Starlings of the Old World, and to some extent the *Ploceidæ* or Weaver-birds and their allies; but the latter family is well distinguished by conirostral bill and 10 primaries. It consists of the Blackbirds and Orioles, among the former being included the Bobolinks, Cowbirds, and Meadow "Larks." The family *Icteridæ* is composed of about 130 species, distributed among over 30 genera or subgenera. The relationships are very close with *Fringillidæ*, on the one hand; on the other, they grade toward Crows (*Corvidæ*). They share with fringilline birds the characters of angulated commissure and only 9 developed primaries, which distinguish them from all other families whatsoever; but the distinctions from *Fringillidæ* are not easily expressed. In fact, I know of no character that will relegate the Bobolink and Cowbird to *Icteridæ* rather

FIG. 313. — A typical *Icterus* (*I. bullocki*). (After Audubon.)

than to *Fringillidæ*, in the current acceptation of these terms; and *Dolichonyx oryzivorus* is curiously similar in some respects to *Ammodramus caudacutus*. In general, however, *Icteridæ* are *cultrirostral* rather than strictly *conirostral* Oscines, having that cutting rather than crushing style of bill seen in perfection in the Crows, toward which some *Icteridæ* approach; being thus distinguished by length, acuteness, and not strictly conical shape of the unnotched, unbristled bill, which has a peculiar extension of culmen on forehead, dividing the prominent antiæ of close-set velvety feathers that reach to or on nasal scale, — a character well exhibited in *Sturnella*, for instance. In length, the bill *usually* equals if it does not exceed the head; the tip is unnotched, rictus unbristled, commissure obtusely but evidently angulated. The bill is shortest and most fringilline in *Dolichonyx* and *Molothrus*; most acute in Orioles (*Icterus*), where it is sometimes actually decurved; most thrush-like in the genus *Scolecophagus*; most crow-like in Grackles (*Quiscalus*). (See any figs. beyond.) In some exotic genera (of the subfamily *Cassicinæ* or Cassiques) the bill acquires enormous dimensions and very peculiar shapes, from expansion of the mesorhinium into a frontal shield; and in these the nostrils open flush with the bill. Excepting in arboreal Orioles and Cassiques, the feet are gressorial, large and strong, fitted for the more or less terrestrial life which most of the species lead, walking on the ground with ease instead of hopping like most *Fringillidæ*. No specialties of wing or tail; former usually pointed, latter rounded, sometimes very large and fan-shaped.

Among our moderate number of species are representatives of four of the five subfamilies into which *Icteridæ* are conveniently and quite naturally divisible. In most genera black is the prevailing color — either uniform and of intense metallic lustre, or contrasted with masses of red or yellow. In *Sturnella* the pattern is "niggled." In nearly all, the sexes are conspicuously dissimilar, the ♀ being smaller and brownish or streaky in the iridescent black species, greenish and yellowish in the brilliantly colored ones. All are migratory in this country. As a rule they are strictly monogamous, and build elaborate nests; but our genera *Molothrus* and *Callothrus* offer the striking exception of polygamy and polyandry among Oscines; for these, like the Old World Cuckoos, do not pair and make no nest. Other details are best given under heads of the four North American subfamilies first established in the 2d ed. of the Key, 1884. The A. O. U. Lists take no note of these; but they are recognized in the British Museum Catalogue of Birds, xi, 1886, by so eminent an authority as Dr. P. L. Sclater, and should not have been ignored by our Committee. These groups, with their component genera, may be analyzed as follows by the salient features more likely to attract the attention of the student than less obvious technical characters: —

Analysis of Subfamilies and Genera.

AGELÆINÆ. *Marsh Blackbirds, etc.* Terrestrial and gregarious. Bill conic-acute, sometimes quite fringilline, shorter, or scarcely longer than head. Feet stout.
    Bobolinks. Sexes unlike in summer only. Black, white, and buff ♂, or yellowish ♂ in winter and ♀; no red.
    Bill fringilline. Tail-feathers very acute. Tarsus shorter than middle toe and claw . . . . . *Dolichonyx*
    Cowbirds. Sexes unlike. Lustrous black ♂, brown ♀; no red or yellow . . . . *Molothrus*
    Cowbirds. Sexes less unlike. ♂ with erectile ruff on neck, sinuated primaries and red eyes . . *Callothrus*
    Marsh Blackbirds. Sexes unlike. Lustrous black ♂, red on wing; streaky ♀; no yellow . . . . *Agelæus*
    Prairie Blackbirds. Sexes unlike. Lustrous black ♂, brown ♀, both with yellow head . . *Xanthocephalus*
STURNELLINÆ. *Meadow Larks.* Terrestrial and imperfectly gregarious. Bill of peculiar shape. Tail very short. Some of the secondaries elongated. Feet large and stout.
    Sexes alike. Motley-colored, extensively yellow below, with black breastplate . . . . *Sturnella*
ICTERINÆ. *Orioles.* Arboreal, non-gressorial, non-gregarious. Bill extremely acute, sometimes decurved. Feet weak. Sexes unlike.
    Black, with yellow or orange or chestnut in masses, in the ♂; ♀ greenish and yellowish . . . . . . *Icterus*
QUISCALINÆ. *Grackles.* Terrestrial and gregarious. Bill elongate, turdine or corvine. Feet stout, gressorial. Color of ♂ entirely iridescent black; ♀ brown or blackish.
    Thrush Blackbirds. Bill shorter than head, turdine; even tail shorter than wings . . . . . . *Scolecophagus*
    Crow Blackbirds. Bill not shorter than head, corvine; graduated tail shorter or not than wings   *Quiscalus*

Subfamily AGELÆINÆ : Marsh Blackbirds.

Gregarious, granivorous species, more or less completely terrestrial, and chiefly palustrine, not ordinarily conspicuous vocalists; building rather rude, not pensile, nests, laying 4-6 spotted or curiously limned eggs. Feet strong, fitted both for walking and for grasping swaying reeds; wings more or less pointed, equalling or exceeding the tail in length; bill conic-acute, shorter or little longer than head, its cutting edges more or less inflected. Five well-marked genera, species of four of which abound in the U. S., on plain and prairie, in marsh and meadow. In the West, they swarm about the settlements, stage and railroad stations, military posts, and other habitable places. We have half of the ten genera which compose this subfamily, the extra-limital ones being *Amblyrhamphus, Leistes, Pseudoleistes, Nesopsar,* and *Curæus.*

DOLICHO′NYX. (Gr. δολιχός, *dolichos*, long; ὄνυξ, *onux*, claw.) BOBOLINKS. Sexes unlike, but only in the breeding season: ♂ black, buff, and white; ♀ brownish and yellowish. Bill short, conic, fringilline, not nearly as long as head. Wings long and pointed, 1st and 2d quills longest, others rapidly graduated. Tail stiffened, with rigid very acute feathers, almost like a Woodpecker's, shorter than wing. Feet stout; tarsus shorter than middle toe and claw; claws very large. One remarkable species noted for the peculiar changes of plumage and the "mad music" of the ♂; abundant in marsh and meadow of eastern U. S.

D. oryzi′vorus. (Gr. ὄρυζα, *oruza*, Lat. *oryza*, rice; *voro*, I devour. Fig. 314.) BOBOLINK. MEADOW-WINK. MAYBIRD. SKUNK BLACKBIRD, Northern States. REED-BIRD, Middle States. RICE-BIRD and BUTTER-BIRD, Southern States. ♂, in breeding plumage: Black; cervix buff; scapulars, rump, and upper tail-coverts ashy-white; interscapulars streaked with black, buff, and ashy; outer quills edged with yellowish; bill blackish-horn; feet brown. The faultless full dress of black, white, and buff is worn only for a brief period; and even in spring and summer, most males are found to have yellowish touches in the black, especially of the under parts. The change occurs in spring by aptosochromatism, without moult; the yellow ends of the feathers are dropped, bringing the black to the surface. A similar whitening of the buff cervix occurs in summer, whence the untenable *D. o. albinucha* RIDGW., based on specimens

FIG. 314. — Bobolink, ♂, reduced. (Sheppard del. Nichols sc.)

I collected in N. Dakota in 1873. The "delirious song," which has stimulated so many poets to the exercise of their versifying craft, is only heard while the males are trooping their way to their breeding-grounds, and before the midsummer change of feather. ♂ in fall, ♀, and young, entirely different in color: Yellowish-brown above, brownish-yellow below; crown and back conspicuously, nape, rump, and sides less broadly, streaked with black; crown with median and lateral light stripes; wings and tail blackish, pale-edged; bill brown, paler below. In this, the ordinary condition, ♂ is best known by superior size. Fall birds are more buffy than spring ♀; ♂ changing shows confused characters of both sexes (see p. 94); but in any plumage the species may be recognized by stiffish, extremely acute tail-feathers, in connection with special dimensions. ♂: Length 7.00-7.50; extent 11.50-12.25; wing 3.50-3.80; tail 2.75-3.00; tarsus 1.00; middle toe and claw 1.25. ♀: Length 6.50-7.00; extent 10.50-11.25; wing 3.25-3.50, etc., averaging 0.50 in length and 1.00 in extent less than ♂. Chiefly eastern U. S. and Canada; N. to 54° in the region of the Saskatchewan, W. not ordinarily beyond the central plains, but occurs in Montana, Idaho, Washington, Utah, Nevada, etc. Winters wholly extralimital; breeds mainly in about the northern half of its N. Am. range, but sometimes S. to the Gulf States. In May, the vivacious, voluble, and eccentric "Bobolinks" pass North, spreading over meadows of the Middle and Northern States from the Atlantic to Kansas and Dakota, perfecting their black dress, and breeding in June and July. After the midsummer change the "Reed-bird" or "Rice-bird" comes back, thronging the marshes in immense flocks with Blackbirds, has simply a clinking note, feeds on wild oats and rice, to which it is highly destructive while the grain is in the milk, becomes extremely fat, and is accounted a great delicacy for the table, as well as a pest in the field. The name "Ortolan," applied by some gunners and restaurateurs to this bird, as well as to the Carolina Rail (*Porzana carolina*), is in either case a strange misnomer, the Ortolan being a fringilline bird of Europe, *Emberiza hortulana* L. (Lat. *hortulanus*, relating to a garden). In the West Indies, where this bird retires in winter, as it does also to Central and South America, it is called "Butter-bird." The names "Bobolink" and "Meadow-wink" are in imitation of its cry; "Skunk Blackbird" notes the resemblance in color to the obnoxious quadruped. The migrations are performed mostly at night, when in May and early September one may hear the mellow metallic "chink" of the invisible passengers. Nest on the ground or close to it, artfully concealed in the grass, composed of weed stems, grasses, and finer materials, 4.00 × 2.50 outside, cupped 2.50 × 1.50 inside; eggs 4–7, usually 5 or 6, from 0.90 × 0.65 to 0.70 × 0.60, averaging 0.82 × 0.63, stone-gray, dotted, mottled, and clouded with dark browns, and lighter neutral tints, usually also marked with some fine blackish scrawls, the whole pattern intricate and very variable.

MOL'OTHRUS. (Gr. μολοθρός, or μολοβρός, vagabond, tramp, parasite.) COWBIRDS. Bill short, stout, conic, and fringilline, about ⅔ as long as head; entirely unnotched and unbristled, with little bend of commissure, the broad culmen running well up on forehead, nostrils well in advance of the feathers. Wings long and pointed; first 3 primaries entering into tip; rest rapidly graduated. Tail shorter than wings, nearly even or a little rounded, tending to divaricate in the middle, the feathers broad and plane to their rounded ends. Feet strong; tarsus not shorter than middle toe. ♂ black and lustrous on the body, brown on the head, without red or yellow; ♀ plain brown. Terrestrial, but not specially palustrine; eminently gregarious and polygamous, or rather communistic, never mating or building nests; thus parasitic, like Old World Cuckoos; no musical ability. There is a single notorious species in the U. S., and another subspecies. Several other species in the warmer parts of America, all of the same irregular and objectionable tendencies, are usually brought under *Molothrus*, but sometimes dissociated in other genera.

FIG. 315. — Cowbird, reduced. (Sheppard del. Nichols sc.)

M. a'ter. (Lat. *ater*, black. Fig. 315.) BROWN - HEADED BLACKBIRD. COW BLACKBIRD. COMMON COWBIRD. COWBUNTING. COW-TROOPIAL. COWPENBIRD. BUFFALO - BIRD. LAZY - BIRD. CLODHOPPER. CUCKOLD. SHINY-EYE. Adult ♂: Lustrous green-black, with steel-blue, purple, and violet iridescence. Head and neck deep wood-brown, with some purplish lustre. Bill and feet black. Length 7.50–8.00; extent 13.50; wing about 4.50, at least over 4.00; tail about 3.25; bill 0.70; tarsus 1.00–1.10. Adult ♀: An obscure-looking bird, dusky grayish-brown, nearly uniform, but paler below than above, where most of the feathers have dusky centres; and most of those of the under parts have dark shaft-lines, giving a somewhat streaky appearance. Some gloss on upper parts, particularly on wings and tail, where a slight greenish lustre is usually evident. Bill blackish-brown, paler below; feet blackish-brown. Smaller than ♂: Length 7.00–7.50; wing about 3.75; tail 2.75. Young ♂ ♀ : Similar to ♀ adult; still duller, and more variegated; upper parts dusky brown, the feathers skirted with gray, producing a set of semicircles on back; below, pale grayish, or even ochrey-brown, everywhere streaked with dusky. Sexual difference in size is soon appreciable, and black of ♂ soon begins to appear in patches. Temperate N. Am., S. in winter through Mexico; migratory, abundant, gregarious, polygynous, polyandrous, parasitic. The singular habits of this bird, shared by others of the genus, form one of the most interesting chapters in ornithology. Like the European Cuckoo, it builds no nest, laying its eggs by stealth in nests of various other birds, especially Warblers, Vireos, and Sparrows; and it appears to constitute, furthermore, a remarkable exception to the rule of conjugal affection and fidelity among birds. A wonderful provision for perpetuation of the species is seen in its instinctive selection of smaller birds as the foster-parents of its offspring ; for the larger egg receives the greater share of warmth during incubation, and the lustier young Cowbird asserts its precedence in the nest; while the foster-birds, however reluctant to incubate the strange egg (their devices to avoid the duty are sometimes astonishing), become assiduous in their care of the foundling, even to the neglect of their own young, which usually perish in consequence. The Cowbird's egg hatches in 10 or 11 days, and thus sooner than that of most birds; this obviously confers additional advantage. The list of birds in whose nests Cowbirds' eggs have been found is now about 100, and includes a large number of Finches, Warblers, Chats, Greenlets, Wrens, Larks, Thrushes, Gnatcatchers, Flycatchers, etc.; there seems to be really little choice. While *small* species are usually victimized, this is not always the case; we have found eggs in nests of the Kingbird, Towhee, Robin, Bobolink, Marsh Blackbird, Brewer's Blackbird, Yellow-headed Blackbird, various Orioles, Red-headed Woodpecker, Yellow-billed Cuckoo, and Carolina Dove. In the West, where Cowbirds swarm about the ranches and settlements, it is the rule, I had almost said, to find their eggs in nests of the prairie *Fringillidæ*, etc. Egg usually single; sometimes 2, 3, and even 4 are found in a nest; they range 0.75–1.00×0.60–0.70, averaging 0.85×0.65, and are white or whitish, fully speckled and dashed with browns and neutral tints, in very variable details of pattern. The number which may be laid by any one ♀ is unknown, supposed to be 8 or more; the laying season is from middle of May to end of July.

M. a. obscu'rus. (Lat. *obscurus*, dark.) DWARF COWBIRD. Similar; smaller; ♂ the size of ♀ *ater;* ♀ under 7.00; wing 3.33; tail 2.33. The difference is strongly marked, and apparently constant. Southwestern U. S., Texas to S. Arizona, and S. into Mexico; the resident form, breeding there, while *ater* passes on, though the two are associated during the migration of the latter. Swarming like *ater;* eggs as in that species, but smaller;·only up to about 0.80×0.60, laid from middle of April to end of July in the nests of such birds as the Common Cowbird usually selects; the ascertained list of species victimized is now 25.

CAL'LOTHRUS. (A word apparently formed to agree in termination with *Molothrus*, from Gr. κάλλος, beauty, + (*Mol-*)*othrus:* CASSIN, Pr. Acad. Nat. Sci. Phila. 1866, p. 18.) BRAZEN COWBIRDS. Generic characters nearly those of *Molothrus;* but feathers of neck of ♂ elongated, forming an erectile ruff, like the pile of velvet, and inner webs of 4 outer primaries sinuated and emarginated in a peculiar manner. Sexual differences in coloration less marked than in *Molothrus.* Eggs whole-colored. Our single species has been given in 2d–4th eds. of Key as *M. æneus;* but the Mexican birds have proven to be of two species, to one of which, from western Mexico, the original *Psarocolius æneus* of Wagler has been restricted, while for the other, *M. robustus* of CABANIS, Mus. Hein. i, 1851, p. 193, has been adopted, this being the bird of eastern Mexico, with which ours is identical. See RIDGW. Man. 1887, p. 589; COUES, Key, 4th ed. 1890, p. 900; A. O. U. List, 2d ed. 1895, p. 203, No. 496.

C. robus'tus. (Lat. *robustus*, stout, strong, robust; *robur*, strength.) BRASS COWBIRD. BRONZED COWBIRD. RED-EYED COWBIRD. Adult ♂: Entire body and head black, splendidly lustrous with bronzy reflections, the tint much like that of the back of *Quiscalus æneus*. This rich brassy-black uniform over the whole bird, there being no distinction of color between head and body, as in *M. ater*. The bronze only on ends of the feathers, the covered parts of which are violet-black, with plain dusky roots. Wings and tail black, with violet, purple, and especially green metallic lustre on upper surfaces. Under wing- and tail-coverts chiefly violaceous-black; purplish and violaceous tints most noticeable on upper coverts of both wings and tail; reflections of quill-feathers themselves chiefly green. Bill ebony-black. Feet black. Iris red. Length 8.00–8.50; extent about 15.00; wing 4.50–4.75; tail 3.25–3.50; tarsus 1.15–1.25; bill 0.90 along culmen, very stout and especially deep at base, much compressed; lateral outlines concave; under outline straight; upper gently convex throughout; tip very acute. ♀ notably smaller: wing scarcely over 4.00; tail about 3.00; culmen scarcely 0.75; tarsus 1.00. Color not brown, as in *M. ater* ♀, but uniformly quite black, with considerable gloss, though nothing like the brassy splendor of ♂. Wings and tail with greenish reflections. Young ♂: Uniform dull black, faintly violaceous on back and rump, greenish on wings and tail. Early spring birds, in imperfect dress, are exactly like adult ♀ in color, but much larger. Central America and Mexico to the Lower Rio Grande of Texas, abounding in some places; a large and very handsome Cowbird, added to our Fauna in 1877. It is a bird of striking aspect, with its bloody eyes and top-heavy attitudes. Polygamous and parasitic like the others, but egg entirely different, being greenish-white, or pale bluish-green, without markings; size 0.85–0.95 × 0.65–0.75; average 0.90 × 0.70. Found in nests of *Icteria, Icterus, Cardinalis, Guiraca, Milvulus, Tyrannus*, etc., the birds victimized thus being much larger than the average of those selected by the common Cowbird.

AGELÆ'US. (Gr. ἀγελαῖος, *agelaios*, gregarious; ἀγέλη, a flock. The A. O. U. continues to misspell the word "*Agelaius*," after Vieillot's original error.) RED-WING MARSH BLACKBIRDS. MAIZE-BIRDS. MAIZERS. Bill about as long as head, stout at base, where deeper than broad, upper and under outlines on an average about straight; commissure variously sinuate or bent; culmen high on forehead, where flattish and broadly parting the feathers; bill rapidly tapering to acute tip. Wings pointed, but 1st primary not longest; usually 2d–4th entering point of wing. Tail even or little rounded, of broad feathers widening a little to very obtuse ends, somewhat divaricate in the middle. Tarsus a little longer than bill. Our three species are very closely related: ♂ uniform lustrous black, with bend of wing red; 8.00–9.00 long; wing 4.50–5.00; tail 3.50–4.00. ♀ everywhere streaked; above blackish-brown with pale streaks, inclining on head to form median and superciliary stripes; below, whitish, with many sharp dusky streaks; sides of head, throat, and bend of wing, tinged with reddish or fulvous; length under 8.00; wing about 4.00; tail 3.00. The young ♂ at first like the ♀, but larger, apt to have a general buffy or fulvous suffusion, with bright bay edgings of feathers of back, wings, and tail, and soon showing black patches. The ♀ ♀ are scarcely distinguishable: the ♂ ♂ may be determined as follows:

Analysis of Species.

♂ Middle wing-coverts buff, bordering the bright red patch . . . . *phœniceus*
♂ Middle wing-coverts buff, but black-tipped, usually leaving red patch without buff border . . . . *gubernator*
♂ Middle wing-coverts white, bordering the dark red patch . . . . *tricolor*

A. phœni'ceus. (Gr. φοινίκεος, *phoinikeos*, Lat. *phœniceus*, red, of a color introduced in Greece by the Phœnicians. Fig. 316.) BLACKBIRD. MARSH BLACKBIRD. SWAMP BLACKBIRD. RED-WINGED BLACKBIRD. RED-AND-BUFF-SHOULDERED MARSH BLACKBIRD. MAIZE-THIEF. HUSSAR. ♂: Lesser wing-coverts scarlet, like arterial blood, broadly bordered by brownish-yellow, or brownish-white, the middle row of coverts being entirely of this color;

sometimes the greater row, likewise, are mostly similar, producing a patch on the wing nearly as large as the red one; occasionally, there are traces of red on the edge of the wing and below; in some specimens the bordering is almost pure white, instead of buff. Extremes: ♂, length 8.25–9.85; extent 13.60–15.30; wing 4.35–5.00; tail 3.12–3.90; bill 0.75–1.00; average: Length 9.00; extent 14.50; wing 4.65; tail 3.60. ♀, length 7.35–8.55; extent 11.85–13.55; wing 3.65–4.25; tail 2.65–3.20; bill 0.70–0.80; average: Length 7.65; extent 12.35; wing 3.85; tail 3.00; bill 0.75. The extremes here given not often seen. Southern-bred birds are much smaller as well as glossier (see varieties given below). Temperate N. Am., N. to lat. 62°, but chiefly E. of the Rocky Mts.; breeding anywhere in its range, wintering from about lat. 35° S. to Central America; accidental in Europe. From its general dispersion in low or wet thickets or fields, swamps, and marshes, the Blackbird collects in August and September in immense flocks, thronging extensive tracts of wild oats and other aquatic plants in marshes and along water courses, also visiting and doing much damage to grain-fields. Thousands are destroyed by boys and pot-hunters, but the hosts scarcely diminish, and every known artifice fails to protect the crops from the invasion of the dusky hordes. At other seasons the "maize-thief" is innocuous, if not positively beneficial, as it destroys its share of injurious insects and seeds of troublesome weeds. Nest usually in reeds or bushes near the ground, or in a tussock of grass, or on the ground; occasionally in small trees, vines, and shrubbery; a bulky structure of coarse fibrous materials, usually strips of bark, rushes, sedges, or marsh grass, lined with finer grasses, sometimes hair, occasionally snake skins; size 4 or 5 inches broad outside, 4 to 6 deep outside; cavity about 3 either way. The breeding season in northerly parts of the U. S. and Brit. Am. is mostly from the middle of May to that of June, and often again in July; in the south it begins a month earlier; incubation about 14 days. Eggs 2–6, usually 3 or 4, ranging from the rare extremes of 0.80 to 1.10 × 0.62 to 0.75, averaging scant 1.00 × 0.70; color pale bluish, bluish-green, or smoky-gray, fantastically dotted, blotched, clouded, and scrawled over with dark or even blackish-brown, and paler or purplish shell-marks; in very rare instances an unmarked egg is laid. The usual note is a guttural *chuck;* in the breeding season the "creaking chorus" makes an indescribable medley.

Fig. 316.—Marsh Blackbird, ♂, reduced. (Sheppard del. Nichols sc.)

A. p. sonorien'sis. (Lat. of Sonora.) SONORAN RED-WING. Like the typical form, but averaging rather smaller in each sex than northern-bred birds. ♂ indistinguishable in plumage from *phœniceus* proper; ♀ lighter colored, with more conspicuous light markings of the upper parts, and white in excess of dusky in the streaking of the under parts; tinge of throat rather pinkish than creamy or buff. Southwestern U. S., from the valley of the Lower Rio Grande to that of the Lower Colorado in southern California; south into Mexico. *A. p. sonoriensis* RIDGW. Man. 1887, p. 370; COUES, Key, 4th ed. 1890, p. 901; A. O. U. List, 2d ed. 1895, No. 498 *a*.

A. p. bryant'i. (To H. Bryant.) BAHAMAN RED-WING. Somewhat smaller than *A. p. sonoriensis*, with relatively larger bill; coloration of ♀ darker, and therefore about as in ♀ of *phœniceus* proper. Length of ♂ 8.00–8.50; wing 4.50; tail 3.50; culmen 1.00–1.05; depth of bill at base 0.40–0.42; ♀: length 6.50–7.00; wing 3.65; tail 2.80; culmen 0.80. Bahamas and southern Florida to Louisiana, S. to Yucatan and Nicaragua. RIDGW. Man. 1887, p. 370; COUES, Key, 4th ed. 1890, p. 901; A. O. U. List, 2d ed. 1895, No. 498 b.

OBS. *A. p. floridanus* MAYN. Birds E. N. Am. pt. xl, 1896, p. 689, is another form, accepted by the A. O. U. in Eighth Suppl. Auk, Jan. 1897, p. 121, No. 498 c; but I find there is nothing in it.

A. guberna'tor califor'nicus. (Lat. *gubernator*, a governor, alluding to the red epaulettes, as if a sign of rank or command.) RED-SHOULDERED MARSH BLACKBIRD. BICOLOR BLACKBIRD. ♂: Lesser wing-coverts scarlet, as before, narrowly or not at all bordered with buff, the next row having black tips for all or most of their exposed portion, so that the brownish-yellow of their bases does not show much, if any. ♀ indistinguishable from ♀ *phœniceus*. Coast region of central and northern California; western Oregon; N. to Cape Disappointment, Washington. Nest and eggs indistinguishable from those of *phœniceus*, and general habits identical. (Given as a subspecies of *phœniceus* in all former eds. of the Key. The further separation of our bird from typical Mexican *gubernator* is made by NELSON, Auk, Jan. 1897, p. 59, on ground of rather smaller size, slenderer bill, and more streaking of upper parts of the ♀. A. O. U. List, Eighth Suppl. Auk, Jan. 1897, p. 128, No. 499.)

A. tri'color. (Lat. *tricolor*, three-colored; red, white, and black.) RED-AND-WHITE-SHOULDERED MARSH BLACKBIRD. TRICOLOR BLACKBIRD. ♂: Lesser wing-coverts dark-red (like venous blood), bordered with pure white. Besides this obvious distinction from *phœniceus*, bill is usually slenderer and tail less rounded; gloss of plumage bluish, not greenish (appreciably so in ♀ as well as in ♂?). ♀ with median wing-coverts white-edged. California and Oregon, especially coastwise, or at any rate W. of the Sierras Nevadas; northern L. Cala.; scarcely migratory. General habits like those of *phœniceus*; nest and eggs indistinguishable; average size of eggs a trifle less, and sets of 3 eggs the rule; first sets are found late in April and early in May, and there is usually a second brood. The congregations of this Blackbird in some favorite breeding-places are enormous, and vast flocks may be seen at other times.

XANTHOCE'PHALUS. (Gr. ξανθός, *xanthos*, yellow; κεφαλή, *kephale*, head.) YELLOW-HEADED BLACKBIRDS. PRAIRIE BLACKBIRDS. General characters of *Agelæus;* claws more developed, lateral reaching much beyond base of middle; feet relatively longer. Tail more nearly even, with narrower feathers. Wings long and pointed; tip formed by outer 3 quills. Colors black, white, and yellow. Eggs spotted, not scrawled.

X. xanthoce'phalus. (Fig. 317.) YELLOW-HEADED BLACKBIRD. ♂: Black, including lores and small space around eye and bill; whole head otherwise, neck, and breast, rich yellow, orange in high feather, the color extending interruptedly to or toward belly; some feathers around vent, and the tibiæ, usually yellow also. A large white patch on wing, formed by primary and many greater secondary coverts, interrupted by black of bastard quills. Bill and feet black. Length 10.00–11.00; extent 16.50–17.50; wing about 5.50; tail 4.50; bill 0.75–1.00; tarsus 1.25. In less perfect dress, the yellow overcast with dusky. Adult ♀: Dark brown, including back of head and neck; line over eye, throat, and breast, dull yellow, with dusky maxillary streaks; usually whitish feathers in the yellow, and sometimes the same in black of breast. No white wing-patch. Bill dark brownish horn-color; feet blackish. Much smaller: Length 8.00–9.50; extent scarcely 14.00; wing under 5.00; tail under 4.00. Nestlings are snuffy-brown; sprouting wing-feathers black, already showing white; feet flesh-color. It is useless to pursue the endless color variations; the species is unmistakable. Western U. S. and British Provinces, N. to lat. 58°; E. regularly to Illinois, Iowa, Wisconsin, etc., casually to Ontario, Quebec, Pennsylvania, New York, New England, District of Columbia, S. Carolina, and Florida, accidentally to Cuba and Greenland; S. into Mexico; migratory, very abundant. Its distribution is general on the prairies, but irregular; it flocks about ranches and settlements, and collects in colonies to breed in marshy spots, sloughs, and coulées, anywhere in its general range; I have myself found it breeding from New Mexico to Manitoba. Nest a light but large thick-brimmed fabric of dried reeds and grasses, slung to growing ones, at no considerable elevation above the water; it is usually built late in May and in June; 5-6 inches in outside diameter, and about as deep; eggs 2-6, usually 3-4, 0.95 to 1.12 long by 0.69 to 0.78 broad, averaging 1.00×0.70; pale grayish-green, spotted as in *Scolecophagus* with reddish and other browns, and neutral tints, but seldom scrawled as in *Agelæus*. A fine large species, conspicuous by its yellow head among the several Blackbirds that troop together in the West. (*Icterus icterocephalus* and *I. xanthocephalus* BP.; *Xanthocephalus icterocephalus* BAIRD, 1858, and of all former eds. of the Key; *Xanthocephalus xanthocephalus* JORDAN,

FIG. 317. — Yellow-headed Blackbird, reduced. (Sheppard del. Nichols sc.)

1884, and A. O. U. Lists, 1886 and 1895, No. 497; *Agelaius longipes* Sw. 1827; *Xanthocephalus longipes* SCL. 1884; *Psarocolius perspicillatus* WAGLER, 1829; *Xanthocephalus perspicillatus* BP. 1850; and with all these names to select from, it is sad to think that the A. O. U. rules require us to perpetuate the tautonym above adopted.)

Subfamily STURNELLINÆ: Meadow Starlings.

If Marsh Blackbirds, Orioles, and Grackles be respectively considered to represent subfamilies of *Icteridæ*, Meadow Starlings seem to be equally entitled to such distinction; and I find that by making *Sturnella* (with *Trupialis*) type of a subfamily, *Agelæinæ* are susceptible of better definition. The characters are included under head of the type genus, as follows:

STURNEL'LA. (Irregular dimin. of Lat. *sturnus*, a starling. Fig. 318.) MEADOW LARKS. (Name "lark" objectionable and misleading, but apparently ineradicable.) A remarkable genus of *Icteridæ*. Bill along culmen longer than head, shorter than tarsus; depth at base about $\frac{1}{3}$ the length; outlines about straight above and below, and along commissure to the strong bend near its base. Culmen flattened throughout, extending broad and far into feathers of forehead; laterally, frontal feathers reaching narrow scaled nostrils. Inner lateral toe rather longer than outer, claw of neither reaching base of middle claw. Hind toe long, with a great claw twice as large as middle one. Feet very large and stout, reaching beyond end of tail when outstretched; eminently fitted for terrestrial locomotion. Wings short and much rounded; little difference in lengths of 1st-5th quills; enlarged inner secondaries nearly covering them in

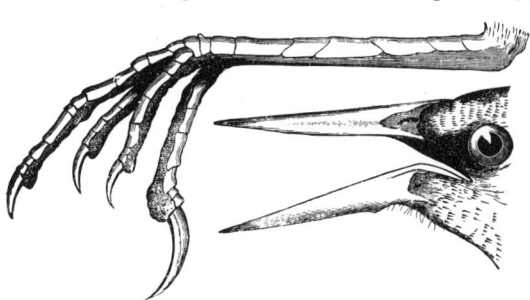

FIG. 318. — Bill and foot of *Sturnella*, nat. size. (Ad. nat. del. E. C.)

closed wing. Tail very short, rounded, of narrow, acute feathers. Feathers of crown stiffish, bristle-tipped. No other genus approaches *Sturnella*, excepting *Trupialis*, which is much the same, with red instead of yellow. Contains several imperfectly differentiated conspecies, 3 of this country.

Analysis of Conspecies.

Common Characters. — Plumage highly variegated; each feather of back blackish, with terminal reddish-brown area, and sharp brownish-yellow borders; neck similar, the pattern smaller; crown streaked with black and brown, and with a pale median and superciliary stripe; a blackish line behind eye; several lateral tail-feathers white, the others, with inner quills and wing-coverts, barred or scalloped with black, and brown or gray. Edge of wing, spot over eye, and under parts generally, bright yellow; sides and crissum flaxen-brown, with numerous sharp blackish streaks; breast with a large black crescent (obscure in young).

Prevailing tone brown above: yellow of chin confined to space between forks of jaw; wings and tail with confluent black bars and gray scallops.
Larger; black less predominant; wing 4.50 or more . . . . *magna*
Smaller; black more predominant; wing 4.50 or less . . . . *m. hoopesi*
Prevailing tone gray above: yellow of chin spreading on cheeks; wings and tail with alternating black and gray bars
neglecta

S. mag'na. (Lat. *magna*, large. Fig. 319.) FIELD LARK. OLD-FIELD LARK. MEADOW LARK. Colors as above described rich and pure, the prevailing aspect brown; black streaks prevailing on crown; yellow of chin usually confined between rami of under mandible; black bars on wings and tail usually confluent along shaft of the feathers, leaving the gray in scallops. Sexes similar: ♀ duller colored, the yellow paler. Young at first have little if any pale yellow, and pectoral crescent indicated by a few streaks. Length of ♂ 10.00–11.00; extent about 17.00; wing 4.50 or more; tail 3.50; bill 1.35; tarsus 1.40. ♀: Length 9.00–9.50; extent about 15.00; wing 4.25; tail 3.00. Varies greatly in size, like *Agelæus*; southern-bred birds much smaller than northern. Eastern U. S. and British Provinces; N. to about 54°; mixing in the Upper Mississippi valley with *neglecta*, and extending to edge of the Plains; everywhere abundant in open country; winters usually from the Middle States southward; imperfectly migratory; partially gregarious when not breeding; strictly terrestrial; an agreeable vocalist. Breeds throughout its range; nest of dried grass, etc., on the ground, usually domed or covered in some way in the grass-clump, occasionally at the end of a long arch-way; the fabric is thick-walled, with comparatively small cavity, measuring usually 6 or 7 inches across outside, and 3 or 4 in depth, with a cavity of only about 3×2 inches. Eggs 3–7, usually 4–6, oftenest 5, crystal white, rarely tinged, speckled with reddish and purplish, in endless variation of size, number, and shade of the markings, but neither veined nor clouded; very variable in size, from 0.85×0.72 to 1.20×0.90, averaging 1.10×0.80. Two or three broods may be reared.

S. m. argu'tula. (Lat. *argutula*, rather noisy, somewhat talkative.) FLORIDA MEADOW LARK. Slightly different from the foregoing; averaging a little smaller, yellower below, browner above. Florida to Louisiana; Mississippi valley to S. E. Ill. and S. W. Ind. BANGS, Proc. N. E. Zoöl. Club, Feb. 28, 1899, p. 20.

S. m. hoopes'i. (To Josiah Hoopes, of Westchester, Pa.) RIO GRANDE MEADOW LARK. Very similar; the browns intense, approaching reddish-brown; black at a maximum; yellow very rich. Size smaller; wing of ♂ about 4.25; bill and feet relatively larger; bill 1.20; tarsus 1.60. Northern Mexico to S. border of Texas, New Mexico, and Arizona; not so well marked as the next. (*S. m. mexicana* of 2d–4th eds. of Key. *S. m. hoopesi* STONE, Pr. Phila. Acad. 1897, p. 149; A. O. U. Suppl. List, Auk, Jan. 1899, p. 113, No. 501 *a*.)

S. neglec'ta. (Lat. *neglecta*, not selected, overlooked; as the variety long was.) WESTERN MEADOW LARK. The colors duller and paler, the prevailing aspect gray; black at a minimum, not prevailing over gray on the crown; yellow of chin usually encroaching on sides of lower jaw; black on wings and tail usually resolved into distinct bars, alternating with gray bars. Western U. S. and British Provinces; N. to Manitoba, Assiniboia, Saskatchewan, Alberta, and British Columbia, chiefly in southern portions of these provinces; E. regularly to the edge of the Great Plains, as in both Dakotas, Nebraska, Kansas, Indian Territory, and Texas, thence less regularly or more sparingly in Minnesota, Iowa, Missouri, Wisconsin, and Illinois; S. in Lower California and some parts of Mexico; in the Upper Mississippi valley at large

FIG. 319. — Meadow Lark.

preserving its own characteristics, though there often associated with *magna* proper. The general habits are the same as those of the eastern bird; but the appearance in life is quite different, and the peculiarities of the song are attested by numberless hearers of this fine melody, from the time when the notes fell on the surprised ears of Audubon, Sprague, Harris, and Bell, in ascending the Missouri together in 1843, to the present day. I am a competent witness to these facts, and also to the fact that I have never seen a specimen that could not be distinguished from *magna*; under which circumstances I do not follow the A. O. U. in reducing *neglecta* to a subspecies of *magna*. The nest and eggs are indistinguishable from those of *magna*, though the average of very extensive series is slightly larger, and the average spotting slightly less. As in the case of the eastern species, the Western Meadow Larks are affectionate and faithful mates; both sexes share the labors of nidification and incubation; the period of the latter is 15 or 16 days; the young leave the nest in two weeks or less, and run about before they can fly, like young quails.

Subfamily ICTERINÆ: American Orioles; Hang-nests.

Non-gregarious, insectivorous, and frugivorous species, strictly arboricole; of brilliant or strikingly contrasted colors, and pleasing song; distinguished as architects, constructing elaborately woven pensile nests. Bill relatively longer, as well as slenderer and more acute than in most *Icteridæ*; feet weaker, non-gressorial, exclusively fitted for perching. Three of our species are migratory birds, abundant in summer; the rest merely reach our southern border from tropical America, where the subfamily focuses. *Icterinæ* number altogether about 40 species, all referable to the genus *Icterus*, with the single exception of *Gymnomystax melanicterus*, a remarkable species with naked circumorbital region, commonly referred to *Agelæinæ*, but by Sclater brought under *Icterinæ*. In their modes of nidification *Icterinæ* agree with *Cassicinæ*; and the extraordinary fabrics constructed by some members of both these subfamilies recall those of the Old World *Ploceidæ* or Weaver-birds. To call our *Icterinæ* " Orioles " is to misapply to them the name which belongs to Old World *Oriolidæ* — an entirely different family; but " Orioles " will they continue to be miscalled, to the end of ornithological time.

IC′TERUS. (Gr. ἴκτερος, *ikteros*, Lat. *icterus*, yellow. Fig. 320.) AMERICAN ORIOLES. TROUPIALS. HANG-NESTS. Our single genus of the subfamily: characters practically the same. Bill averaging as long as head (more or less); very acute, sometimes decurved. Feet fitted for perching, not for walking; tarsus not longer than middle toe and claw. Lateral toes, if not of equal lengths, outer longest (the rule in *Fringillidæ*; in *Icteridæ* the reverse). Wings usually pointed and averaging equal to (longer or shorter than) the rounded or graduated tail. A large and beautiful genus of about 40 species, which vary much in details of form, but are not easily divided otherwise than specifically. The colors are striking: ♂ black with orange or yellow, usually also with white; in one species, black and chest-

Fig. 320. — Bill of an Oriole.

nut. Sexes very unlike in some species, in others quite alike. ♀ ♀ of several species closely resemble one another, though ♂ ♂ are very different. We have two eastern species; one western; three southwestern; and one southern straggler. These seven species represent the three current subgenera of the genus which, as Dr. Sclater observes, " may be used as a makeshift "; for when we come to consider the whole genus, we find the numerous species so variously interrelated that no satisfactory sections can be established. To my eye, *Icterus icterus* looks more different from all the rest than any of these are from one another. I also observe that though Dr. Sclater and the A. O. U. adopt the same three subgenera — *Hyphantes* (or *Yphantes*), *Pendulinus*, and *Icterus* proper — these authorities disagree in the way they respectively allocate the species under two of the three sections. My respect for the A. O. U. and B. O. U. being equal, my patriotism must be allowed weight in a case in which I have no prejudice and no preference. I accordingly follow the American method in the following sorry

Analysis of Subgenera.

Bill stoutly conic, straight; its depth at base equal to half the length of culmen . . . . *Hyphantes*
  (Species *galbula* and *bullocki*; the ♂ black and orange.)
Bill slenderly conic, not quite straight; its depth at base not equal to half the length of culmen . . . *Pendulinus*
  (Species *spurius*, ♂ black and chestnut; and *cucullatus*, ♂ black and orange.)

Bill slenderly conic, straight; its depth at base not equal to half the length of culmen . . . . *Icterus*
(Species *parisorum*, ♂ black and yellow; *auduboni* ♂ ♀ black and yellow; and *icterus* ♂ ♀ black and yellow, with throat-feathers lanceolate and orbits naked.)

⁎ Further refinement of the foregoing would place *auduboni* (with *melanocephalus*) in the subgenus *Ateleopsar* Cass. 1867; and make *parisorum* type of the subgenus *Cassiculoides* Cass. 1867.

Analysis of Species.

The ♂ black and chestnut: *spurius.*
The ♂ black and orange: *galbula, bullocki, cucullatus.*
The ♂ black and clear yellow: *parisorum, auduboni, icterus.*
Feathers of the throat soft and normal.
   ♂ black and chestnut; ♀ olivaceous and yellowish. Length 7.00 or less . . . . *spurius*
   ♂ black and orange, or flame-color.
     Tail rounded, not longer than wings.
        ♂ head and neck all around black; white on wings in bars . . . . *galbula*
        ♂ crown and throat black, sides of head orange. White patch on wings . . . . *bullocki*
     Tail graduated; outer feathers an inch shorter than middle ones; longer than wings.
        ♂ head orange, with black mask; ♀ olivaceous and yellow . . . . *cucullatus*
   ♂ black and pure yellow.
        ♂ head, neck, breast, and back black. Sexes unlike; length about 8.00 . . . . *parisorum*
        ♂ ♀ head, neck, and breast black; body yellow, greenish on back; length about 9.50 . . . . . *auduboni*
   Feathers of throat elongate and lanceolate. Sexes alike. Length about 10.00.
     ♂ ♀ Black and yellow, with white on wings . . . . *icterus*

(*Subgenus* Hyphantes.)

(*Yphantes* Vieillot, 1816, and so misspelled by the A. O. U. Gr. ὑφάντες, *hyphantes*, a weaver.)

1. gal′bula. (Lat. *galgula* or *galbula*, some small yellow bird of the ancients. "Baltimore" is not from the city of that name, but from the title of Sir George Calvert, first baron of Baltimore; the colors of the bird being chosen for his livery, or resembling those of his coat-of-arms. Fig. 321.) Baltimore Oriole. Golden Robin. Fire-bird. Pea-bird. Hammock-bird. Hanging-bird. Hang-nest. Adult ♂: Black and orange. Head and neck all round, and back, black; rump, upper tail-coverts, lesser and under wing-coverts, most tail-feathers, and all under parts from throat fiery orange, of varying intensity according to age and season. Middle tail-feathers black; wings black, the middle and greater coverts, and inner quills, more or less edged and tipped with white, but white on coverts not forming a continuous patch. Bill and feet blue-black, or dark grayish-blue. Length 7.50–8.00; extent 11.50–12.50; wing 3.66; tail 3.00. ♀ smaller, and much paler, the

Fig. 321. — Baltimore Oriole, reduced. (Sheppard del. Nichols sc.)

black obscured by olive, sometimes entirely wanting. Above, mixed dusky and yellowish-olive, somewhat overcast with a gray shade. Below, dull orange, more or less mixed with whitish, and usually with black traces on throat. Tail and its upper coverts dull yellowish, the central feathers usually blackish. Bill and feet lighter plumbeous than in ♂. Young ♂ entirely without black on throat and head, otherwise colored nearly like ♀. Below, dull orange yellow, whitening on throat, shaded with olive on sides. Above, olive, more yellowish on rump and tail, but latter without black; middle of back obscured with dusky centres of the feathers; wings dusky, with two white bars and white edgings of inner quills. In some splendid featherings, particularly from the Mississippi valley, the orange becomes intense flame-color, and there is so much white on the wings as to approach the character of *I. bullocki.* U. S. and adjoining British Provinces; in the interior N. to Saskatchewan and Keewatin, about lat. 55°; W. in the U. S. nearly or quite to the Rocky Mts. of Montana, Wyoming, and Colorado, and in Brit. Am. to Assiniboia; S. in winter through Mexico and Central Am. to Panama; accidental in Cuba and the Shetland Islands; migratory; breeds nearly throughout its N. Am. range, the Gulf coast region probably only excepted. It passes N. in late April and May, reaching our northern districts about the middle of the latter month. This is one of our famous beauties of bird-life, noted alike for its flash of color, its assiduity in singing, and its skill at the loom; its elaborately fabricated and perfectly pensile nests swaying from the tops of our shade-trees, which have one charm added when fired with such brilliancy as the Oriole brings to contrast with verdure. Both sexes work diligently and intelligently at the nest, in the composition of which scarcely anything that can be woven or fitted seems to come amiss, and the materials consequently vary interminably; the shape is pouch-like, with the entrance somewhat contracted; the walls are firm, but thin, so that the cavity is comparatively large. The depth of the nest is commonly 5 or 6 inches, sometimes more, the width less; the situation is generally high in large trees, and out at the end of a branch, where it may be quite inaccessible. Eggs 4–6, oftenest 4 or 5, from 0.85×0.60 to 1.00×0.65, thus rather elongate; ground color a shaded white, irregularly spotted, blotched, clouded, and especially scrawled with blackish-brown and other heavy surface colors, together with subdued shell-markings.

I. bul'locki. (To Wm. Bullock, of London. Fig. 313.) BULLOCK'S ORIOLE. Adult ♂: Black and orange, like the last, but orange invading sides of head and neck and forehead, leaving only a narrow space on throat, lores, and a line through eye, black; a large continuous white patch on wing, formed by middle and greater coverts. Larger than the Baltimore. Length 8.00–8.50; extent 12.50–13.50; wing 4.00; tail 3.40. ♀: Olive-gray, below whitish, all fore parts of body and head tinged with yellow; wings dusky, with two white bars, but tail and its under coverts quite yellowish. ♀ thus very closely resembling ♀ Baltimore, and more detailed description may be desirable. Larger: Length about 8.00; extent 12.00; wing 3.75; tail 3.25. Above olive-gray, becoming quite gray on rump, brightening into olive-yellow on nape, upper tail-coverts, and tail. Forehead, superciliary line, sides of head and neck, and large space on breast, bright yellow; lores and throat white. Other under parts grayish-white, tinged with yellow on under tail-coverts. Edge and lining of wing yellow; middle coverts broadly edged and tipped with white; greater coverts and quills less conspicuously edged. Young ♂ at first like ♀, soon, however, showing black and orange; in one stage with a black throat patch. Western U. S. and adjoining British Provinces of Assiniboia, Alberta, and British Columbia, E. to both Dakotas, Nebraska, Colorado, Kansas, and W. Texas; Lower California, and in winter S. into Mexico; accidental in Maine. It is abundant in woodland, replacing the Baltimore, to which it is so closely allied, and with which it corresponds in habits and manners. The nest and eggs are indistinguishable with any certainty; sets run, however, from 3 to 6. The third species of this section of the genus is the Mexican *I. abeillei*, with a black rump.

(*Subgenus* PENDULINUS.)

I. spu'rius. (Lat. *spurius*, spurious; the species was formerly called "Bastard Baltimore Oriole," whence the undeserved name.) ORCHARD ORIOLE. BASKET-BIRD. Adult ♂: Black and chestnut. Head and neck all around, fore breast and back, black. Rump and upper tail-coverts, lesser and under wing-coverts, and whole under-parts from breast, chestnut or chocolate-brown. Wings and tail black, former except as said, and some white or whitish edging of quills and tipping of greater coverts, latter forming a wing-bar; outer tail-feathers sometimes with a touch of chestnut. Bill and feet blue-black. Length about 7.00; extent about 10.00; wing 3.00–3.25; tail nearly as long, much rounded, its graduation nearly 0.50; bill

0.70 along culmen, very slender and acute, somewhat decurved; tarsus 0.90. Adult ♀: Smaller than ♂. Above, dull yellowish-olive, clearest on head, rump, and tail, obscured on the back. Below, sordid yellowish. Wings plain dusky, glossed with olivaceous, with whitish edging, much as in ♂. An inconspicuous object, but known from other ♀ Orioles by its small size and slender bill, a little curved. Young ♂: First year like ♀, but larger; second year like ♀, but with black mask on face and throat. Afterward showing confused characters of both sexes. Three years required to assume full dress. Eastern U. S., strictly; rarely N. to Maine, and even New Brunswick, but regularly reaching Ontario; W. to the high central plains of the Dakotas, Kansas, Nebraska, Colorado, Indian Territory, and Texas. Breeds throughout its N. Am. range; winters extralimital. Abundant in orchards, parks, streets, skirts of woods, etc., from April to August. The song is loud, clear, and volubly delivered during the whole breeding season. The nest is one of the most perfect examples of a woven fabric, even in a group of birds distinguished as the Orioles are for the dexterity and assiduity they display in their elaborate textile *rostrifactures*. They antedate Howe in the expedient of placing the eye of a needle at its point — that which revolutionized hand-sewing, and made sewing-machines practicable: for their bill works precisely to the same effect. The Orchard Oriole's nest is generally more compact and homogeneous than the Baltimore's, woven chiefly of slender grass-blades which cure in the sun like good hay, long retaining some greenness, which tends to its concealment in the foliage. It is smaller, less deep in proportion, often not strictly pendant from its forked twig, and generally placed lower down in a tree. Both sexes work at its speedy construction, but only the ♀ incubates. Eggs 4–6, oftenest 5, smaller than the Baltimore's, ranging from 0.72×0.56 to 0.85×0.60, averaging about 0.80×0.55, and spotty rather than scrawly, with predominance of the heavy markings over the neutral ones; the markings prevail about the larger end, but the general Icterine tracery is always unmistakable.

I. s. affi'nis ? (Lat. *affinis*, affined, allied.) TEXAS ORCHARD ORIOLE. Smaller: ♂ little over 6.00; wing usually under 3.00. Texas: Southern race, scarcely distinguishable; ignored by the A. O. U.

I. cuculla'tus. (Lat. *cucullatus*, wearing the *cuculla*, a kind of hood or cowl.) HOODED ORIOLE. Adult ♂: Orange and black. General color orange — from rich chrome yellow to flame-color. Middle of back (scapulars and interscapulars) black. A black mask, embracing eyes, narrow frontal line, and patch on chin, cheeks, and throat. Wings black, with white edging of quills and coverts. Tail black, some or all feathers usually with narrow whitish tips. Bill and feet blue-black, former extremely slender and somewhat decurved, 0.80; tarsus 0.90. Length 8.00; extent 10.50; wing 3.30–3.60; tail 3.50–4.25, thus longer than wings, feathers narrow and lanceolate, outermost an inch or so shorter than central pair; such length, narrowness, and extreme graduation of tail being a strong character. Adult ♀: Above, dull grayish-olive; tail and under parts dull yellowish; wings dusky, the quills and coverts edged with dull white. ♀ thus resembles other species, but the long slender graduated tail and attenuated decurved bill are diagnostic. Fairly smaller than ♂. Young ♂: At first like ♀, but bill pale at base below. Various intermediate states during progress to maturity; sometimes the black dorsal band interrupted by yellowish-gray, and the general orange obscured with the same. A frequent condition, when the general plumage is like that of ♀, is to have a black frontlet and gorget, like *I. spurius* under the same circumstances. Texas, chiefly near the Mexican border, and southward to Honduras. Nest woven like that of other Orioles, very substantial and durable though thin-walled, and more like a saucer than a cup; in places where Spanish moss grows, it is usually made of this material, and placed in a truss of the same. Eggs 3–4, sometimes 5, varying from 0.75 to 0.90 long by 0.60 to 0 65 broad, usually quite pointed at both ends; color white or whitish, irregularly spotted and blotched with shades of brown and neutral tints, especially about the larger end, with less scrawling than usual in this genus, and altogether less heavily marked. In the Lower Rio Grande valley this is the commonest Oriole in some places, Apr.-Sept.; nests with full sets of eggs are found from middle of April to first week in July.

I. c. nel′soni. (To E. W Nelson.) ARIZONA HOODED ORIOLE. PALM-LEAF ORIOLE. A paler-colored race, in which the yellow is not supposed to become orange or flame-color, from New Mexico, Arizona, California, and southward to Mazatlan. The distinction is trivial, hardly indicating a geographical race. Pr. U. S. Nat. Mus. viii, Apr. 1885, p. 19; Key, 3d ed. 1887, p. 877; A. O. U. List, 2d ed. 1896, p. 208, No. 505 a.

(*Subgenus* ICTERUS.)

I. pariso′rum. (To the brothers Paris.) BLACK-AND-YELLOW ORIOLE. PARIS' ORIOLE. SCOTT'S ORIOLE. MOUNTAIN ORIOLE. Adult ♂ : Black and clear yellow. Below from breast, rump, and upper tail-coverts, lesser, middle, and under wing-coverts, and basal portions of all the tail-feathers, except central ones, clear yellow; greater wing-coverts tipped, inner quills edged, with white. Head, neck, breast, back, and wings, except as said, black. On the tail, the yellow occupies the basal half of lateral feathers, but only extreme base of central pair. Length 8.00; extent 12.00; wing 4.00; tail 3.40–3.60, moderately rounded, lateral feathers graduated about 0.50; bill 0.90, attenuate and slightly decurved; tarsus 1.00. Young ♂ : Black parts all overcast with grayish-olive skirting of the feathers, giving the prevailing tone on upper parts, but on breast the black showing more clearly; yellow likewise obscured with grayish-olive, especially on rump. Tail greenish-yellow, middle feathers blackening. Wings dusky, all quills and greater and middle coverts broadly edged and tipped with white. Adult ♀ : Dull greenish or grayish olive above, with dusky shaft-streaks on the back; dull yellowish below; greater and median wing-coverts tipped with white, forming two bars; tail like under parts, but darker on middle feathers and toward the ends of the others. Smaller than the ♂ on an average. Western Texas, New Mexico, Arizona, Southern and Lower California, and some portions of Nevada and Utah; S. in Mexico to Puebla and Vera Cruz; migratory, entering the U. S. late in March and early in April, and breeding throughout its U. S. range; a voluble and persistent songster. Nesting essentially the same as that of other Orioles; the purse-like fabric is well woven of grasses and fibres, the latter oftenest of the yuccas, in which the structure is habitually suspended at little elevation; but it is also placed in various other trees or bushes, sometimes in bunches of moss or vines hanging in cactuses, quite near the ground; eggs 2–4, oftenest 3, averaging 0.95×0.66, ranging from 1.05×0.70 to 0.90×0.60, white with a fugacious pale bluish tint, variously blotched and dotted with purplish and blackish-browns, chiefly about the larger end, and with little if any tracery; to be found in May, June, and even July. Best biography in SCOTT, Auk, Jan. 1885, pp. 1–7; and BEND. ii, "1895" (pub. Sept. 1896), pp. 471–474.

I. aud′uboni. (To J. J. Audubon.) BLACK-HEADED ORIOLE. AUDUBON'S ORIOLE. Adult ♂ : Black and clear yellow. Entire body rich gamboge-yellow, without orange or flame tint, but shaded with greenish on back, sides, and upper tail-coverts; under tail-coverts pure yellow, like belly. Middle and lesser wing-coverts and lining of wings pure yellow, former with black bases concealed by yellow tips. Head all around, fore neck and breast, glossy jet-black, without any concealed yellow, except at edges of the black on breast — the black there thus ending ragged, different from the clean-cut border of *cucullatus*. Wings black; outer webs of quills white-edged, especially on inner secondaries and outer primaries toward their end; greater coverts with white spot at end of outer web. Tail black; outer feathers more or less edged and tipped with white. Bill and feet plumbeous-blackish, former paler at base below. Length 9.25–9.75 or more; extent 12.50–13.00 or more; wing averaging 4.00; tail rather more, much graduated, outer feathers 1.00 or more shorter than middle. Bill stout, straight, almost as in

Agelæus; culmen 0.90-1.10, averaging 1.00. Tarsus 1.10; middle toe and claw the same. Adult ♀: Quite like ♂; not smaller, and little different in color, contrary to the rule in this genus and family. Back rather more olivaceous; wings rather more edged with white; outer tail-feather edged and tipped with whitish. Young ♂ ♀: No black or white; plain olive-green above, yellow below, shaded on the sides with olive. This is a large, beautiful Oriole, occurring in the U. S. only, as far as known, in the Lower Rio Grande valley; thence S. in Mexico to Oaxaca; a magnificent songster, and a favorite cage bird. Nest half-pensile, woven of grasses like that of the Orchard Oriole, placed in trees and bushes, oftenest mezquite, at no great elevation; eggs laid Apr.–June, 3 to 5 in number, the set often incomplete from imposition of Red-eyed Cowbird's eggs; they measure from 1.05×0.75 to 0.90×0.70, averaging 1.00×0.72, and are pale bluish or grayish white, dusted with fine brown specks, over which are stains and splashes of dark brown and lilac, with occasionally some of the blackish hieroglyphs usual in this genus. *I. melanocephalus audubonii* of former eds. of the Key, and I do not feel sure of its specific distinction, as its difference from the Mexican *melanocephalus* consists only in the white markings on the wings, extent of greenish on the scapulars, and of yellow on the middle wing-coverts, these parts being nearly or quite black in the stock form; however, I follow the A. O. U. List in now presenting it as a good species.

I. ic'terus. TROUPIAL. Bill elongate, attenuated, acute, straight, or scarcely decurved. Throat feathers lengthened, loosened, and lanceolate. Bare space about eye, and in other respects entirely different from any of the foregoing species. Adult ♂ ♀: Head and neck all round, fore breast, isolated dorsal area, wings, and tail, black, the wings with a white patch on the coverts, and much whitish edging of the secondaries. Rump, upper tail-coverts, lesser wing-coverts, cervical collar, and under parts from the breast, including lining of wings, rich yellow, ordinarily clear and pure, sometimes intensified to orange. Large: length nearly or quite 10.00; wing 4.60; tail less; bill 1.25-1.50; tarsus about the same. A common and well-known species of Tropical America, also introduced in the West Indies, and often seen as a cage-bird, said by Audubon to have occurred at Charleston, S. C. This case is its only claim to a place in our Fauna. (*I. vulgaris* of former eds. of the Key; *I. icterus* A. O. U. Lists, No. [502].)

Subfamily QUISCALINÆ: American Grackles.

Closely resembling *Agelæinæ* both in structure and in habits, these birds are distinguished by length and attenuation of bill, with decidedly curved culmen, especially toward end, more or less sinuate commissure, and strongly inflected tomia. The bill is quite cultrirostral, and typical *Quiscali* have a certain crow-like aspect, but are readily distinguished by several features, besides 9 instead of 10 primaries; one species of *Scolecophagus* so much resembles a Thrush that it was originally classed as a *Turdus*. In *Scolecophagus* the tail is slightly rounded and shorter than wings; in *Quiscalus* the tail is graduated, and nearly equals or exceeds wings. They are not specially palustrine. The feet are large and strong, and the birds spend much time on the ground, where they walk or run instead of advancing by leaps. The *Quiscalinæ* generally build rude, bulky, non-pensile nests, lay spotted, clouded, or streaked eggs, and their best vocal efforts are hardly to be called musical. The ♂ of all our species is lustrous black, with various iridescence, the ♀ merely blackish or brown, and *much* smaller. Individuals of all our species abound, especially in the South and West; only two are common eastern birds. The equivocal extralimital genus *Cassidix*, usually referred to the *Quiscalinæ*, is placed by Sclater in the *Cassicinæ*. *C. oryzivora* is glossy black, with a ruff on the neck of the ♂. Other extralimital forms of this subfamily, according to the same authority, are *Lampropsar tanagrinus*, black, with a frontal hood of erect feathers; *Aphobus chopi* and *Hypopyrrhus pyrohypogaster*, in both of which the feathers of the head are lanceolate; together with *Macragelæus subalaris* and several species of the genus *Dives*. In the cases of the two last-named genera, the relationship of *Dives* appears to be with *Scolecophagus*, and that of *Macragelæus* with *Quiscalus*.

SCOLECO'PHAGUS. (Gr. σκώληξ, gen. σκώληκος, *scolex, scolecos*, a worm : φαγος, *phagos*, eating.) RUSTY GRACKLES. THRUSH BLACKBIRDS. Bill shorter or not longer than head, slender for the subfamily — somewhat like a Robin's, for instance; culmen little convex, if any, except at decurved tip; gonys slightly convex; cutting edges inflexed; commissure little sinuated. Wings pointed, decidedly longer than nearly even tail; point formed by outer 4 primaries. Tail much as in *Agelæus* in size and shape. Tarsus rather longer than middle toe and claw. Lateral toes short, with moderate claws, scarcely or not reaching base of middle claw. Nest in bushes. Eggs spotty, not veiny and streaky.

Analysis of Species.

Smaller : wing under 5.00. Bill slender, thrush-like. ♂ greenish-black, including head. Sexes very unlike : ♀ quite rusty-brown, even with chestnut ; a light line over eye . . . . *carolinus*
Larger : wing 5.00 or more. Bill stouter, more blackbird-like. ♂ greenish-black, head more violet. ♀ subsimilar, sooty-brown ; no pale superciliary stripe . . . . *cyanocephalus*

S. caroli'nus. (Lat. *Carolinus*, of or pertaining to *Carolus*, Charles (whether King Charles II. of England or IX. of France), referring to the present N. or S. Carolina, name of which is found in French as early as 1564; see COUES, Check List, 2d ed. 1882, p. 25.) RUSTY GRACKLE. THRUSH BLACKBIRD. Adult ♂, in summer : One lustrous black with green metallic reflections; head not notably different from other parts in its iridescence. Bill and feet black. Iris creamy or lemon. (Not ordinarily seen in the U. S. in this full dress — usually with some rusty.) Length 9.00–9.50; extent 14.00–15.00; wing under 5.00; tail 4.00 or less; bill 0.80, only about 0.35 deep at base; tarsus 1.20; middle toe and claw less. Adult ♀ in summer : Slaty-blackish, duller below, with greenish reflections chiefly on wings and tail; in winter the upper parts overlaid with rich rusty-brown, and under parts with a paler shade of the same; inner secondaries brown-edged; a whitey-brown streak over eye; iris brown. Moderately smaller than ♂. Young ♂ at first resembles ♀, but is larger, and shows more decidedly lustrous black, especially on wings and tail. As usually found in flocks in the U. S., in fall, winter, and early spring, young and old of both sexes are very rusty, with light line over eye. E. N. Am., N. to Labrador and the Hudson Bay region, thence N. W. to Alaska and Behring Sea; in the U. S. extending W. regularly to the Dakotas, Nebraska, Indian Territory, and Texas, sparingly to the Rocky Mts., in these regions meeting and mixing in the fall with the next species; accidental in Greenland and Lower California. Migratory, abundant; in winter, more or less dispersed in the U. S.; in summer, breeding from N. New England, New York, and Michigan northward, in loose colonies, in swampy tangle. Nest in bushes and low trees, of sticks and grasses mixed with moss and mud, lined with fine grasses and rootlets; it is a firm, durable structure, quite bulky, 7 inches or more across outside, and 5 deep, with a cavity about 3.50 × 2.50. Eggs 4–5, usually 4, from 1.05 × 0.80 down to 0.90 × 0.70, dull greenish, bluish, or grayish white, flecked and mottled with dark browns, but with little or no line-tracery, and thus resembling those of the Yellow-headed rather than of the Red-winged or Crow Blackbird. Incubation is said to last 14 days, and the young to leave the nest in 16 days; the nestling plumage is gray. (*S. ferrugineus* of most writers, as of all former eds. of the Key; but *Turdus carolinus* MÜLL., 1776, is prior to *Oriolus ferrugineus* GM., 1788, and I am glad to be able to do away with a name which does not apply to the adult ♂.)

S. cyanoce'phalus. (Gr. κύανος, *kuanos*, Lat. *cyanus*, blue; κεφαλή, *kephale*, head.) BLUE-HEADED GRACKLE. BREWER'S BLACKBIRD. Similar to the last, but quite a different bird, type of the subgenus *Euphagus* (CASS. 1866). Adult ♂, in summer : Very lustrous green-black, as before, but with purple and violet iridescence, especially on head, where the violet or steel-blue sheen contrasts with the general greenish hue. Bill and feet black. Iris creamy or lemon. Larger: length averaging 10.00 — 9.75-10.25; extent 16.00 or more; wing 5.00-5.25; tail 4.00-4.25; bill 0.80, stout at base, where about 0.40 deep — more like an abbreviated *Quiscalus* bill than a Thrush's; tarsus 1.25–1.30; middle toe and claw 1.10–1.15. Adult ♀, in summer: Blackish, with dull-greenish shade on back, wings, and tail; more slaty-blackish below. Fore parts of body above, head and most under parts overlaid with brownish-gray, lightest on head and throat, never rich rusty-brown. No light superciliary line. Iris brown. There is thus much less sexual difference than in *S. carolinus*. Smaller; size about that of ♂ *carolinus* ; length 9.00–9.50; extent 14.50–15.50; wing 4.50–4.90, etc. Young ♂ resembling ♀ ; soon, however, showing more lustre, overcast with grayish (not rusty) brown, in same style as *carolinus*, but different shade. Western U. S. and British Provinces ; E. regularly to eastern edge of the plains, overlapping the migratory range of *S. carolinus*, occasionally extending to Iowa, Missouri, Wisconsin, and Illinois, casually to Louisiana, and South Carolina; N. to the region of the Saskatchewan; S. in Mexico to Oaxaca; very abundant in most parts of the west, both in prairie and mountain, in large flocks when not mated; then in small colonies. Breeds nearly throughout its range, in suitable places; migratory to and from extremes of its range. Nidification substantially the same as that of the Rusty Grackle, but the nest is sometimes built on the ground; eggs. 4–8 rarely, usually 5 or 6, ranging from 1.10 × 0.80 to 0.80 × 0.60, but such extremes exceptional, the average 1.00 × 0.70 ; pattern of the markings fundamentally, as in *S. carolinus*, but in the endless diversity of coloration some specimens show tracery. The spotting is sometimes so heavy and uniform as to produce a dark brown egg; but the pale greenish or grayish ground-color is usually visible in the profuse blotching and marking with dark browns, reddish browns, and neutral tints.

QUIS′CALUS. Derivation questionable. In New Latin of the Linnæan period and back to Gesner, about 1550, *quiscula* appears as a name of the European Quail ; *quisquila* is said to be a Portuguese name of the same bird; compare Spanish *quisquilla*. Middle Latin *quiscula*, *quisquila*, *quisquilla*, *quaquila*, *quaquara*, and *quaquadra* mean *quail*, which English word is the same, etymologically, as French *caille*, Port. *calha*, Ital. *quaglia*, etc., all being no doubt onomatopoetic. (See Coues, Check List, 2d ed. p. 64; Stej. Auk, Jan. 1885, p. 43.)

CROW BLACKBIRDS. Bill about as long as head, quite cultrate and crow-like, but more attenuate and acute, with deflected cutting edges; upper and under outlines straightish to terminal curve of culmen, but variable ; commissure variously sinuate. Wings relatively shorter and less acute than in *Scolecophagus*, usually pointed by 2d–4th quills, 1st and 5th

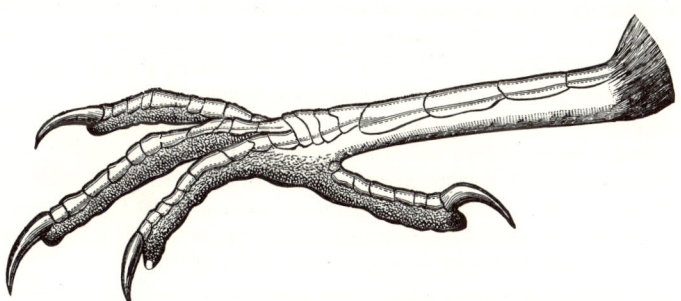

FIG. 322. — Foot of a *Quiscalus* (*Q. macrurus*, nat. size). (From Baird.)

shorter. Tail of varying development with the species; at its greatest, much longer than wings; at its least, decidedly shorter; always graduated, lateral feathers 1–3 inches shorter than middle pair, in life capable of slanting upward on each side, so that the middle feathers make a keel below; whence the name "boat-tail." (Tail usually described as "longer than wings" in *Quiscalus* ; but in most species it is decidedly shorter.) Feet stout; tarsus about equal to middle toe and claw. The ♂ ♂ in species "black," but so magnificently iridescent that little dead black is seen, being brassy, steel-blue, violet, purple, greenish, etc. ♀ subsimilar (in *Quiscalus* proper), or plain brown, and much smaller than the ♂ (in the subgenus *Megaquiscalus*).

Analysis of Subgenera, Species, and Subspecies.

Sexes subsimilar in size and color. (Subgenus *Quiscalus*.)
   Tail decidedly shorter than wings, graduated 1.00–1.50.
      Iridescence various — green, blue, purple, violet. ♂ usually over 12.00 . . . . *purpureus.*
      Iridescence of back brassy; head steel-blue. ♂ usually over 12.00 . . . . *p. æneus*
      Iridescence greenish, neck purple. ♂ usually under 12.00 . . . . *p. aglæus*
Sexes dissimilar; ♀ brown much smaller than the ♂. (Subgenus *Megaquiscalus*.)
   Tail about equal to wings, graduated about 2.50 . . . . *major*
   Tail decidedly longer than wings, graduated 2.50–3.50 . . . . *macrurus*

(*Subgenus* MEGAQUISCALUS.)

Q. macru′rus. (Gr. μακρός, *macros*, long, large; οὐρα, *oura*, tail. Fig. 322.) FAN-TAILED CROW BLACKBIRD. TEXAS GRACKLE. Of largest size, with longest, most keeled and graduated tail. Sexes very unlike. Bill very stout at base, tapering to strongly deflected tip. Adult ♂: Iridescence chiefly purplish and violet, more greenish posteriorly. Length 17.00–20.00, averaging about 18.50; extent 23.00–24.00; wing 7.50–8.00; tail about 9.00, graduated 2.50–3.50; bill 1.75. Adult ♀: Dark brown; paler, grayish, or whitish below. Length 11.50–13.50; extent 18.00–19.00; wing 5.50–6.50; tail about the same; bill 1.30. The species thus presents dimensions *Q. major* has not shown. Lower Rio Grande of Texas and S. through Mexico to Nicaragua, very abundant, swarming in towns, where conspicuous by its curious antics as well as great size and numbers. Breeds in colonies, either in reedy marshes, when the nest is placed in the rushes over water, or anywhere about settlements in trees away from water; sometimes there are many nests in one tree, some at an altitude of 30 or 40 feet. Nests built of any trash, usually with mud. Eggs in April, May, and June, usually 3, often 4, rarely 5, 1.12–1.45 by 0.82–0.90, averaging 1.25 × 0.85; greenish- or purplish-white, clouded oftener over smaller end than at the other, irregularly spotted, veined, and scratched with dark brown, blackish, and neutral tints.

Q. ma′jor. (Lat. *major*, greater (than *Q. purpureus*).) BOAT-TAILED CROW BLACKBIRD. BOAT-TAILED GRACKLE. JACKDAW. Of large size, with long, much keeled and graduated tail. Sexes very unlike. Bill stout at base, tapering to deflected tip. Adult ♂: Iridescence mostly green, becoming purple or violet chiefly on head and neck. Length 15.50–17.00, average 16.50; extent 21.00–23.50, average 22.50; wing and tail, each, 6.25–7.25, average 7.00, latter rather the longer of the two; its graduation about 2.50; bill 1.50; tarsus nearly 2.00; middle toe and claw about the same. Adult ♀: Astonishingly smaller than ♂, lacking any great development of tail, and easily to be mistaken for another species. Length 12.00–13.50, average 13.00; extent 17.25–18.25, average 17.75; wing 5.25–6.00, average 5.67; tail 4.75–5.50, average 5.25. General color plain brown, only darker on wings and tail; below brownish-gray, frequently whitening on throat. S. Atlantic and Gulf States, coastwise, abundant; N. regularly to Virginia and Maryland, casually to New Jersey; breeds throughout its range, which meets that of *Q. macrurus* in Texas, and winters from Virginia southward. This species differs from the common Crow Blackbird in being strictly maritime, with consequent modification in food and habits; it may be seen at times wading in water, and small fish and crustaceans form much of its fare. Nesting and eggs as in *macrurus*; eggs averaging smaller, but not distinguishable with certainty.

(*Subgenus* QUISCALUS.)

Q. quis′cula. (For etymology, see the generic name, which is another form of the same word. Fig. 323.) PURPLE CROW BLACKBIRD. COMMON CROW BLACKBIRD. KEEL-TAILED GRACKLE. PURPLE GRACKLE. RUSTY HINGE. Of medium size, with moderately keeled and graduated tail, shorter than wings. Sexes subsimilar. Bill usually less tapering and deflected at tip, but very variable. Adult ♂: Iridescence very variable with season, age, and sexual vigor, as well as on different parts of the body; but always intense in healthy adults, and at its height during the love-ardor; variously purple, green, blue, violet, and bronzy; not the extensive green of the last species, nor usually the *decided* brassy of the next variety; wings and tail mostly purplish; dark purplish and steel-blue on head, neck, and breast; back more greenish or bronzy. Bill and feet ebony black. Iris straw-yellow. Length 12.00–13.50; extent 17.00–18.50; wing 5.00–6.00, averaging 5.60; tail 4.50–6.00, usually under 5.50; bill 1.25, very variable; tarsus 1.25; graduation of tail 1.00–1.50. Adult ♀: Blackish, quite lustrous; sufficiently similar to ♂; length 11.00–12.00; wing about 5.00; tail about 4.50. Birds of this character, without perfectly brassy back and steel-blue head, are usual in the Atlantic States; abundant and generally distributed, migratory and gregarious, breeding anywhere in their range, but chiefly northerly. Nesting variable, in tree or bush, on bough or in a hollow, at any height; sometimes in an artificial retreat, or a Fish-hawk's nest. Nest bulky, of any trash, usually with mud; eggs of the character and with all the indescribable variability of others of the genus; usually bluish or greenish, with purplish veining and clouding, zigzagged and flourished with dark browns or blackish; averaging about 1.15 × 0.85, but ranging from 1.25 × 0.90 to 1.00 × 0.80 in size; 4–6 in number, rarely 7, oftenest 5. Grackles are absent from their northerly breeding-grounds for only a small part of the year, when they flock southerly, often in immense bands scouring about for food. At times they are very injurious to crops, but this is offset by their destruction of noxious insects. The courtships of the males look very curious to a dispassionate observer, being carried on with the most grotesque actions and ludicrous attitudes, as well as curious vocalization. (*Q. purpureus* of all previous eds. of the Key. *Gracula quiscula* LINN. S. N. 1758, p. 109, and 1766, p. 165, whence necessarily, by our rules, the peculiar literary atrocity of the pseudotautonym *Quiscalus quiscula* of the A. O. U. Lists, No. 511.)

FIG. 323. — Purple Grackle, reduced. (Sheppard del. Nichols sc.)

Q. q. æ'neus. (Lat. *æneus*, brassy.) BRASS CROW BLACKBIRD. WESTERN CROW BLACKBIRD. BRONZED GRACKLE. Birds from the interior of N. Am., especially the Mississippi valley, acquire in full plumage a splendid iridescence of three kinds, in pretty distinct areas. Body uniform shining brassy. Hind neck and breast chiefly steel-blue. Wings and tail chiefly violet and purple. This brilliant coloration is represented by Audubon, folio pl. 7, 8vo, pl. 221. Such birds occur from New England, Nova Scotia, Newfoundland, Labrador, Hudson's Bay, the region of the Saskatchewan and Great Slave Lake, and the Rocky Mts. to Texas and the Gulf States; also passing to some extent into Mexico, and frequently invading those Atlantic States which our Lists reserve as the peculiar demesne of the foregoing species. Nest and eggs indistinguishable from those of *quiscula* proper; general habits the same. (*Q. purpureus æneus* of 2d–4th eds. of the Key.)

Q. q. aglæ'us. (Gr. ἀγλαίος, *aglaios*, splendid.) FLORIDA CROW BLACKBIRD. GREEN GRACKLE. Birds resident in Florida and others of the S. Atlantic and Gulf States are smaller than average *quiscula*, with relatively longer and slenderer bill more decurved at tip; body-lustre chiefly greenish; head and neck chiefly violaceous steel-blue; wings and tail steel-blue, becoming violet on coverts. Averaging an inch less in length than *quiscula*, and other parts in proportion, excepting bill and feet, which are quite as long. The eggs are said to *average* 1.20

× 0.82, thence running up to 1.43 × 0.84, and down to 1.06 × 0.76; and to be only 3–5 in number. (*Q. baritus* BD. 1858, nec auct. *Q. aglæus* BD. 1866. *Q. purpureus aglæus* COUES, 1872, and all other eds. of the Key; *Q. quiscula aglæus*, A. O. U. Lists, No. 511 *a*.)

Family CORVIDÆ: Crows, Jays, Pies, etc.

Cultrirostral Oscines with 10 *primaries.* — A rather large and important family, comprising such familiar birds as Ravens, Crows, Rooks, Jackdaws, Magpies, Jays, Choughs, with their allies, and a few diverging forms not so well known; nearly related to the famous Birds of Paradise (*Paradiseidæ*), to the Old World Orioles (*Oriolidæ*), and to the Old World Starlings (*Sturnidæ*). There are 10 primaries, of which 1st is short, generally about half as long as 2d, and several outer ones are more or less sinuate-attenuate on inner web toward end. The tail has 12 rectrices, as usual among higher birds; it varies much in shape, but is generally rounded — sometimes extremely graduated, as in the Magpie; and is not forked in any of our forms. The tarsus has scutella in front, separated on one or both sides from rest of tarsal envelope by a groove, sometimes naked, sometimes filled in by small scales. The bill is stout, about as long as head or shorter, tapering, rather acute, generally unnotched, with convex culmen; it lacks the commissural angulation of *Fringillidæ* and *Icteridæ*, the deep cleavage of *Hirundinidæ*, the slenderness of *Certhiidæ*, *Sittidæ*, and most small insectivorous birds. The rictus usually has a few stiffish bristles, and there are others about base of bill. The gonys is rather short, *i. e.*, the mandibular rami usually unite in advance of a perpendicular line let down from the nostrils; and these are normally placed high up, near the culmen (they are lower in the Choughs, *Fregilinæ*). An essential character is seen in dense covering of nostrils with large long tufts of close-pressed antrorse bristly feathers (excepting, among our forms, in *Cyanocephalus* and *Psilorhinus*). These last features (in connection with the presence of 10 primaries) distinguish *Corvidæ* from all our other birds excepting *Paridæ;* the mutual resemblance is here so close, that I cannot point out any obvious technical character of external form to distinguish, for example, *Cyanocitta* from *Lophophanes*, or *Perisoreus* from *Parus*. But as already remarked, *size* is here perfectly distinctive, all *Corvidæ* being much larger than any *Paridæ*.

Although technically Oscine, *Corvidæ* are non-melodious; their vocal organs are well developed, but none of them can sing. This shows that musical ability depends upon something more than mere complexity of the syrinx or sound-making apparatus. The voice of the larger corvine birds is hoarse and raucous, that of the smaller garruline ones harsh and strident — hear the ominous croak of the Raven, the cacophony of the Crow's cawing, the shrill scream of the Jay.

Owing to uniformity of color in leading groups of the family, and an apparent plasticity of organization in many forms, the number of species is difficult to determine, and is very variously estimated by different writers. Mr. G. R. Gray admits upward of 200, which he distributes in 50 genera and subgenera; but these figures are certainly excessive. Dr. R. B. Sharpe, in the Brit. Mus. Catalogue of 1877, describes about 160 species or subspecies, arranged in 43 genera, with 4 genera under a subfamily *Fregilinæ*, all the rest under *Corvinæ*. *Corvidæ* have also been divided into 5 subfamilies; 3 of these are small specialized groups confined to the Old World, where they are represented most largely in the Australian and Indian regions; the other two, constituting the great bulk of the family, are more nearly cosmopolitan. These are *Corvinæ* and *Garrulinæ*, or Crows and Jays, readily distinguishable, at least so far as our forms are concerned, by the longer pointed wings and shorter, less rounded tail of the former, as contrasted with the shorter, rounded wings and longer, more rounded or graduated tail of the latter. This is the subdivision of the family which I have kept in all the eds. of the Key, and the one followed by the A. O. U.

Subfamily CORVINÆ : Crows.

Wings long and pointed, much exceeding tail, tip formed by 3d, 4th, and 5th quills; 2d much shorter, 1st only about ½ as long as 3d. Legs stout, fitted for walking as well as perching. As a rule, the plumage is sombre or at least unvariegated, — blue, the characteristic color of Jays, being here rare. Sexes alike, and changes of plumage slight. Crows frequent all situations, and walk firmly and easily on the ground, where Jays hop. They are among the most nearly omnivorous of birds, and as a consequence, in connection with their hardy nature, they are rarely if ever truly migratory. Their nesting is various, according to circumstances, but the fabric is usually rude and bulky; the eggs, of average oscine number, are commonly bluish or greenish, speckled. Although not properly gregarious, as a rule, they often associate in large numbers, drawn together by community of interest. In illustration of this may be instanced the extensive roosting-places in the Atlantic States, comparable to the rookeries of Europe, whither immense troops of Crows resort nightly, often from great distances, recalling the fine line of the poet, —

FIG. 324. — Typical Corvine bill.

"The blackening trains of crows to their repose."

Our 3 genera of *Corvinæ* are readily known by the black color of *Corvus*, the gray, white, and black of *Nucifraga* (*Picicorvus*), and the blue of *Cyanocephalus*. In the latter, as in *Psilorhinus* of *Garrulinæ*, the nostrils are exposed, contrary to the rule in each subfamily.

COR'VUS. (Lat. *corvus*, a crow. Fig. 324.) RAVENS. CROWS. The species throughout uniform lustrous black, including bill and feet; nasal bristles about half as long as bill, which exhibits the typical cultrirostral style. Nostrils large, entirely concealed. Wings much longer than tail, folding about to its end. Several outer primaries sinuate-attenuate on inner webs. Tail rounded, with broad feathers, sinuate-truncate at ends, with mucronate shafts. Feet stout; tarsus more or less nearly equal to middle toe and claw, roughly scutellate in front, laminar behind, with a set of small plates between.

Analysis of Species and Subspecies.

Ravens, with throat-feathers acute, lengthened, disconnected.
    About 24.00 long; wing 16.00-18.00; tail about 10.00. Bases of cervical feathers gray.
        Largest : bill averaging 3.00. Chiefly northern . . . . *corax principalis*
        Not so large; bill not averaging 3.00. Chiefly western . . . . *corax sinuatus*
    About 20.00 long; wing 13.00-14.00; tail 7.50-8.50; concealed *bases* of cervical feathers pure white. Southwestern . . . . *cryptoleucus*
Crows, with throat-feathers oval and blended.
    Length 18-20; wing 12-14; tail 7-8; bill 1¾-2, its height at base ¾; tarsus about equal to middle toe and claw, longer than bill; 1st quill not longer than 10th. Chiefly eastern . . . . *americanus* and *pascuus*
    Small. Length 14-16; wing 10-11; tail 6-7; bill 1¾-2; tarsus rather longer than bill or middle toe and claw; 1st quill longer than 10th. Northwestern . . . . *caurinus*
    Small; 14-16 inches long; wing 10-11; tail 6-7; tarsus shorter than middle toe and claw, longer than bill; 1st quill not longer than 10th. Eastern, chiefly southerly and maritime . . . . *ossifragus*

C. co'rax sinua'tus. (Gr. κόραξ, *korax*, Lat. *corax*, a croaker — the raven. Lat. *sinuatus*, have a sinus, re-entrance, or incision; sinuated, as the inner webs of the outer primaries are. Fig. 325.) AMERICAN RAVEN. Feathers of throat somewhat stiffened, lengthened, pointed, lying loose from one another; those of neck with gray downy bases, as elsewhere on the body. Color entirely lustrous black, with chiefly purplish and violet burnishing. Length about 2 feet — at least over 20 inches; expanse of wings 4 or 4½ feet — much over a yard. Wing about 1½ feet — at least over 15 inches. Tail about 10 inches; its feathers graduated 1.50-2.50

inches. Bill along chord of culmen, and tarsus, about 2.50, the bill ranging up to 3.00. Varies much in size. Greenland, Labrador, and boreal or arctic specimens generally, are of great size, with immense bill averaging 3.00 (in the so-called var. *principalis*). The bill is usually longer and relatively less deep in the American than in the European Raven (*corax* proper); whole bird more sturdy and robust. The usual wing-formula is: primary $4 > 3 = 5 > 2 > 6 > 1 = 8$; but these quills grow and moult so gradually the proportionate lengths differ much in specimens examined. ♀ is indistinguishable from ♂, though averaging smaller. N. Am.; but now rare in the U. S. east of the Mississippi, and altogether wanting in most localities; Labrador, ranging southward, rarely, along the coast and in mountainous regions to the Middle districts, casually even to South Carolina, Georgia, and Alabama; very abundant in the West, where the sable plume and the bleaching skeleton, the ominous croak and the Indian war-whoop, are not entirely things of the past. Wherever in the West the Raven

FIG. 325. — Head of a very large American Raven, nat. size. (Ad. nat. del. E. C.)

abounds, the Crow seems to be supplanted. Nests sometimes in trees, but as a rule on cliffs or in other rocky places, selecting the most inaccessible sites. Eggs 4–8, oftener 5 or 6, about 2.00×1.30 on an average, ranging from 1.60×1.25 to 2.35×1.50, though such extremes of length are rare; the color is pale green, often shaded with drab or olive, and the whole surface is profusely dotted, blotched, and clouded with neutral tints, purplish, and various shades of brown.

Regarding the vexed question of relationship of the American to the European Raven, I have throughout successive eds. of the Key, and in other works, since 1872, contended against specific distinction; and I observe that the two forms are united in one by such high authority as that of Dr. SHARPE, in the British Museum Cat. iii, 1877, p. 14. But we may have gone too far in ignoring some differences, particularly in average size, which appear to exist, and I am now willing to take the safest middle course of recognizing subspecific distinction. Our bird has plenty of names from which to choose. The earliest of these is *C. carnivorus* BAR-TRAM, 1791, against which certain technical objections have been alleged, though it was adopted by BAIRD in 1858, and became current for some years. The next in order of date appears to be *C. sinuatus* WAGLER, Isis, 1829, p. 748, based on Mexican specimens; this I am willing to adopt, in deference to my colleagues of the A. O. U., though my well-known contention has long been in favor of Bartram. Other names are *C. lugubris* of AGASSIZ, 1846, denounced as a nomen nudum, though nobody doubts what he meant by it; and *C. catototl* or *cacalotl* of BONAPARTE, 1838 and 1850, and of BAIRD, 1858 (after WAGLER, Isis, 1831, p. 748). For the case of *C. littoralis* or *principalis*, see next article.

C. c. principa′lis. (Lat. principal, foremost, chief; *princeps*, adj. first in time or order, and as noun a chief, a prince; from *primus*, first, and *capere*, to take, choose. Fig. 326.) NORTHERN RAVEN. Size at a maximum of the dimensions above given, with very large bill

FIG. 326. — Northern Raven.

and stout feet; chord of culmen averaging 3.00, and depth of bill at base 1.00. Individuals answering to such requirements occur chiefly in Greenland, Labrador, and British America at large, but also in northerly parts of the United States, and on the Atlantic coast even to North Carolina. Figure 325, drawn *precisely* of life size, is fully up to average *principalis*; the specimen was taken by me at Fort Randall, South Dakota, Feb. 4, 1873. The large northern bird of Greenland and Labrador was first named *C. c. littoralis* by HOLBÖLL, in Kröyer's Tidsk. iv, 1843, p. 390; but this name is preoccupied in the genus by A. E. BREHM, 1831. It is *C. c. principalis* RIDGW. Man. 1887, p. 361; COUES, Key, 4th ed. 1890, p. 901; A. O. U. List, 2d ed. 1895, p. 200, No. 486 *a*. With this may be compared the Kamtschatkan *C. c. behringianus* of DYBOWSKI, Bull. Soc. Zool. France, 1883, p. 363.

C. cryptoleu′cus. (Gr. κρυπτός, *kruptos*, crypted or hidden; λευκός, *leukos*, white.) WHITE-NECKED RAVEN. Throat-feathers as in *corax*; but bases of feathers of neck snowy-white.

Smaller than the Raven; length 19.00–21.00; wing 13.25–14.25; tail 7.50–8.50; bill along culmen 2.00–2.25, its depth at base about 0.85; tarsus 2.25–2.50; thus this Raven is about as large as a good-sized Crow, and often mistaken for one in those regions where it occurs with the common Raven, the difference between them being obvious in life; the accounts of "Crows" in some regions where *americanus* does *not* occur being based upon the presence of *cryptoleucus*. Southwestern U. S., Llano Estacado, and higher Rio Grande of Texas, Indian Territory, Oklahoma, Kansas, Nebraska, Wyoming, Colorado, Utah, Nevada, New Mexico, Arizona, and some portions of California; S. some little distance in Mexico. Nest in trees and bushes, at no great height, and resembling that of the common Crow. Eggs 3–8, usually 4, 5 or 6, averaging 1.75 × 1.20, ranging from 1.90 × 1.30 to 1.50 × 1.10; ground color greenish or grayish, markings lighter and fewer than is usual in this genus, with a tendency to be lengthwise streaky rather than spotty; some eggs are almost unmarked, but as a rule the brown, purplish, and neutral tints are conspicuous. They are laid late in May, and in June, sometimes in April.

C. america'nus. (Lat. American. Fig. 329.) COMMON AMERICAN CROW. The common Crow is a foot and a half long, or rather more, ranging from 17.00 to 21.00 inches; wing 12.00–14.00; tail 7.00–8.00; bill 1.75–2.00, about 0.75 high at base; tarsus 2.25–2.35, about equal to middle toe and claw, rather exceeding the bill. First primary not longer than 10th. Feathers of throat oval, soft, and blended; no snowy-white under-plumage. The burnishing is chiefly on the wings, tail, and back, the head being nearly dead-black. ♀ is decidedly smaller than ♂, and under-sized cabinet specimens are not seldom labelled "ossifragus." N. Am. at large, chiefly U. S. and easterly, not ordinarily found westward in the interior, where Ravens abound; rare or wanting in the Upper Missouri and Southern Rocky Mt. regions; common, however, in some parts of California and other localities on the Pacific slopes; resident or only irregularly migratory. In settled parts of the country the Crow tends to colonize, and some of its "roosts" are of vast extent. Mine is on the Virginia side of the Potomac, near Washington. Crows are always flying west over the city in the afternoon, and when as a boy I used to see the gray of the morning, Crows were flying the other way. Nest in trees, anywhere in the woods, usually high up and concealed with some art, though so bulky as to measure about 24.00 × 12.00 outside, with a cavity 12.00 × 6.00; built of sticks and trash; eggs 3–8, oftenest 5 or 6, about 1.60 × 1.15, with extremes of 1.85 × 1.20 to 1.45 × 1.00, like the Raven's in color and markings, and equally variable. The Crow lays betimes, the season for eggs being from February in the Southern states, March and April in the Middle, and early May in the Northern; incubation occupies about 17 days; the young remain in the nest for about three weeks; there is only one brood annually. In its relations to man the Crow is rather beneficial than injurious on the whole, the damage it unquestionably does under some circumstances being more than offset by its habitual destruction of noxious insects; it should therefore be protected not persecuted. But such is its sagacity that it manages to hold its own, unterrified by scarecrows, undismayed by man's many devices for its destruction, and quite regardless of legislatures which declare it to be an outlaw. (*C. frugivorus* BARTRAM, 1791, of 2d–4th eds. of the Key. *C. americanus* AUD. 1834; Key, orig. ed. 1872, p. 162; A. O. U. List, No. 488. *C. a. hesperis* RIDGW. Man. 1887, p. 362, based on Pacific slope specimens, is ignored by the A. O. U.)

C. a. pas'cuus. (Lat. *pascuus*, relating to meadows; *pascuum*, a pasture. Name intended to connote the same as *floridanus*, with allusion to the Spanish name of the country, said to have been called Pascua Florida or Pascua de Flores by Ponce de Leon, because he discovered it on Paschal or Easter day of 1512. Cf. Lat. *pascualis, paschalis*, paschal, relating to *pascha*, feast of the Passover.) FLORIDA CROW. Represents the greater relative size of bill and feet shown by many resident birds of Florida and corresponding latitudes. Average size somewhat less, not over 20.00; wing 11.50–12.50; tail under 8.00; bill 2.00 or rather more along culmen, its depth at base 0.75–0.85; tarsus 2.45. Eggs 3–5, indistinguishable from those of the common Crow, laid in Feb. and Mar. (*C. f. floridanus* of 2d–4th eds. of Key. *C. a. pascuus* COUES, Auk, Jan. 1899, p. 84. A. O. U. Suppl. List, *ibid.* p. 112, No. 488 *a*.)

C. cauri'nus. (Lat. *caurus*, the N. W. wind, whence *caurinus*, northwestern. Fig. 327.) NORTHWESTERN FISH CROW. Small: about the size of the common Fish Crow, but feet

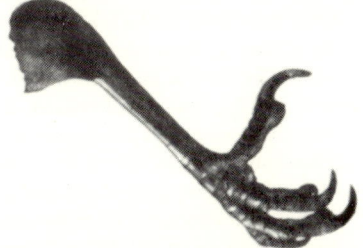

FIG. 327. — Northwestern Fish Crow. (L. A. Fuertes.)   FIG. 328. — Corvus Americanus.

more as in *americanus;* tarsus not shorter than middle toe and claw, though rather less than bill; 1st primary longer than 10th. Length 14.00–16.00; wing 10.50; tail 6.50; bill 1.75–2.00 along culmen, 0.70 deep at base; tarsus averaging under 2.00. N. Pacific coast, from N. California and Oregon to S. Alaska; maritime; piscivorous; voice said to be different from that of *americanus*. The species seems to be well established; it is smaller than the common

FIG. 329. — American Crow.

Crow, with decidedly shorter tarsus, the extreme length of which does not quite equal the least length in *C. americanus*. It abounds from the mouth of the Columbia N. to Sitka, and occupies the same position on the Pacific that *C. ossifragus* has on the Atlantic coast. Eggs usually 4-5, indistinguishable from those of the common Crow, only averaging a trifle smaller, laid in April, May, and June.

C. ossi'fragus. (Lat. *ossifragus*, as adj. ossifragous, bone-breaking; as noun also *ossifraga*, the ossifrage, osprey, or sea-eagle, *i. e.*, the Fish Hawk now called *Pandion haliaëtus*; *os*, gen. *ossis*, a bone; *frangere*, to break.) SOUTHEASTERN FISH CROW. Small. Length 14.00–16.00; wing 10.00–11.00; tail 6.00–7.00; bill 1.50; tarsus 1.60; middle toe and claw 1.75. First primary not longer than 10th; a bare space about gape? South Atlantic and Gulf States, Louisiana to southern New England, rare or casual beyond Long Island, in summer only in the lower Hudson and Connecticut valleys, resident from New Jersey southward. Common; maritime, piscivorous. A different bird from any of the foregoing, as it presents some tangible distinctions, although constantly associated with *C. americanus*. It is decidedly smaller, with maxima not reaching minima of the common species; the voice is different, and the habits are not the same. Nest and eggs not to be distinguished with certainty from those of the common Crow, though averaging smaller. Eggs usually 4 or 5, averaging 1.45 × 1.05, laid from Feb. through May. (*C. maritimus* BARTRAM, 1791, and of COUES, Key, 2d–4th eds. 1884–90, p. 417. *C. ossifragus* WILS. 1812, of orig. ed. of the Key, 1872, p. 163, and of the A. O. U. Lists, No. 490.)

PICICOR'VUS. (Compounded of *picus*, a woodpecker, or *pica*, a magpie, and *corvus*, a crow. Fig. 330.) AMERICAN NUTCRACKERS. General characters of the European *Nucifraga*. Bill slenderer, more acute, with more regularly curved culmen and commissure, and straight ascending gonys; as a whole somewhat decurved. Nostrils circular, concealed by a full tuft of plumules. Wings long and pointed, folding to end of tail; 5th quill longest; 4th, 3d, 6th little less; 2d much shorter, 1st not half as long as 5th. Tail little over half as long as wing, little rounded. Tarsus shorter than middle toe and claw; the envelop divided into small plates on sides behind toward the bottom. Claws very large, strong, acute, and much curved, especially that of the hind toe; the lateral reaching beyond base of the middle claw. Coloration peculiar; gray, with black-and-white wings and tail. Habits much the same as those of *N. caryocatactes;* alpine and sub-boreal, pinicoline, and pinivorous. One species, confined to W. Am., differing from *Nucifraga* chiefly in the pattern of coloration. *Picicorvus* BP. 1850, of all previous eds. of the Key, and of the A. O. U. List, 1st ed. 1886, p. 246; later reduced to a subgenus of *Nucifraga* — a needless procedure.

FIG. 330. — Head of *Picicorvus*, nat. size. (Ad. nat. del. E. C.)

P. columbia'nus. (Of the Columbia River. Fig. 331.) CLARK'S CROW. CLARK'S NUTCRACKER. Adult ♂ ♀: Gray, often bleaching on head; wings glossy black, most of the secondaries broadly tipped with white; tail white, including under coverts; central feathers and usually part of the next pair, together with upper coverts, black. Bill and feet black. Iris brown. Length about 12.50; extent 22.00; wing 7.00–8.00; tail 4.00–5.00; tarsus 1.35; bill averaging 1.67; feet from 1.25–1.75. Sexes alike in color, but ♀ smaller than ♂. Young similar, but browner ash. There is great difference in the shade in adults, the plumage when fresh being more glaucous ash, wearing browner, and also bleaching in patches, especially on head. Coniferous belt of the West, N. to northern Alaska, within the Arctic circle, S. to Mexico and Lower California, W. to the Coast Ranges, E. regularly to the eastern spurs and foothills of the Rocky Mts., as the Black Hills of S. Dakota, and casually to Kansas, Nebraska, Missouri, and Arkansas; the only American representative of the European Nutcracker, *N. caryocatactes*; abundant, imperfectly gregarious. A remarkable bird, wild, restless, and noisy, sometimes congregating by thousands in the pineries of the West, roving in search of food. Breeds in pines, usually on a horizontal bough at no great elevation, in alpine and northerly localities; the comparatively few nests thus far known have been taken in Colorado and Oregon, containing eggs in April and May; nest of sticks as a basis, on which bark-strips, grasses, and other fibrous substances are well matted together. Eggs 2–3, 1.35 × 0.90, light grayish-green, speckled and blotched with brown and lilac,

FIG. 331. — Clark's Crow, reduced. (Sheppard del. Nichols sc.)

especially about the larger end, but often quite evenly over the whole surface; the general effect is of a lighter colored egg than usual in this family. I have observed this bird in many parts of the West, from Arizona to the Black Hills of S. Dakota, the National Yellowstone Park in Wyoming, the Bitter Root valley of Montana, and the Salmon River region of Idaho, and always found it a striking object, with something in its flight and other actions to remind one of a Woodpecker. I have more than once known it to be mistaken for the Rocky Mountain Jay, *Perisoreus canadensis capitalis*, whence probably the reason why it shares with the latter the names of "Moose-bird," "Meat-bird," and "Camp-robber," which are stated to be applied to it by Major Bendire in his biography of the bird. (Life Histories, ii, 1896, p. 418.) The species was discovered by Capt. Wm. Clark near the site of Salmon City, Idaho, Aug. 22, 1805: see my History of the Lewis and Clark Expedition, ii, 1893, p. 530.

CYANOCEPH'ALUS. (Gr. κύανος, *kuanos*, blue; κεφαλή, *kephale*, head.) BLUE CROWS. Bill of peculiar shape, with nearly straight culmen mounting on forehead, thus somewhat as in *Sturnella*, between prominent and somewhat antrorse antiæ, which, however, do not hide nostrils; slender, tapering, acute, not notched; gonys straightish, scarcely ascending. Nostrils small, oval, entirely exposed. Tail nearly square, much shorter than wings. Wings long, pointed, folding nearly to end of tail; 4th primary longest, 3d and 5th scarcely shorter; 2d

FIG. 332. — Blue crow, nat. size.; culmen too convex. (Ad. nat. del. E. C.)

shorter, 1st shorter still. Feet stout, indicating somewhat terrestrial habits; tarsus longer than middle toe without claw, the envelop subdivided behind towards the bottom. Claws all large, strong, and much curved. Color bluish, nearly uniform; sexes alike. One species. (*Gymnokitta* and *Gymnocitta* of former eds. of the Key; but *Cyanocephalus* BP. 1842, antedates *Gymnokitta* MAXIM. 1850 (as given in BP. Consp. 1850, p. 382), the latter being proposed as a substitute for *Gymnorhinus* MAXIM. 1841, which is preoccupied by *Gymnorhina* GRAY, 1840, in another connection.)

C. cyanoce'phalus. (For etym. see the generic name. Fig. 332.) BLUE CROW. MAXIMILIAN'S JAY. CASSIN'S JAY. PIÑON JAY. PIÑONERO. ♂ : Dull blue, very variable in intensity, nearly uniform, but brightest on head, fading on belly; throat with whitish streaks; wings dusky on inner webs. Bill and feet black. Iris brown. Length 11.00–12.00; extent 16.50–19.00; wing 5.50–6.00; tail about 4.50; bill 1.33, but from 1.25–1.50; ♀ smaller, duller. Young grayish-blue, paler below. Rocky Mt. region to the Pacific coast ranges; much the same elevated distribution as the last, in the region of conifers, but rather more southerly; N. only to British Columbia; S. to Lower California, western Texas, and northern Mexico; E. casually to Kansas and Nebraska; decidedly gregarious, and very abundant in some places, especially where the nut-pine (*Pinus edulis*) flourishes. A remarkable bird, combining the form of a Crow with the color and habits of a Jay, and a peculiarly shaped bill. It roves about in noisy restless flocks, sometimes of thousands, in search of food, which is pine seeds, especially piñones, juniper berries, acorns, maize, etc. Breeds in colonies of 10–150 pairs; nest in piñon pines and other evergreens, compact but bulky, measuring about 10.00 × 7.00 outside, with a cavity of 4.00 × 3.00, built of twigs, and fibrous bark-strips, grasses, and rootlets well worked together; eggs 3–5, oftenest 4, 1.05 to 1.20 × 0.87, greenish- or bluish-white, profusely spotted with brown and purplish in small and nearly uniform pattern over the whole surface; mostly laid in April and May; young flocking by July.

Subfamily GARRULINÆ : Jays and Pies.

Wings much shorter than or about equalling tail, both rounded; tip of wing formed by 4th–7th quills. Feet, as well as bill, usually weaker than in true Crows, and the birds are more strictly arboricole, usually advancing by leaps when on the ground, to which they do not habitually resort. In striking contrast to most *Corvinæ*, Jays are usually birds of bright and varied colors, among which blue is most prominent; and the head is frequently crested. The sexes are nearly alike, and the changes of plumage do not appear to be as great as is usual among highly-colored birds, although some differences are frequently observable. Our well-known Blue Jay is a familiar illustration of the habits and traits of the species in general. They are found in most parts of the world, and reach their highest development in the warmer portions of America. With one boreal exception (*Perisoreus*), the genera of the Old and New World are entirely different.

It is proper to observe that while American *Corvinæ* and *Garrulinæ*, upon which the foregoing paragraphs are mainly drawn up, are readily distinguishable, the characters given may require modification in their application to the whole family, the different divisions of which appear to intergrade closely. Our 6 genera are easily discriminated.

Analysis of Genera.

Nostrils large, naked.
  Not crested. General color brown . . . . *Psilorhinus*
Nostrils moderate, covered by feathers.
  First primary attenuated, falcate: tail exceedingly long, graduated.
    Not crested. Colors black, white, and iridescent . . . . *Pica*
  First primary not attenuated. Tail moderate.
    Crested. Blue: wings and tail barred with black . . . . *Cyanocitta*
    Not crested. Blue: wings and tail unbarred . . . . *Aphelocoma*
      Green and yellow, with blue and black on head . . . . *Xanthura*
      Gray, with slaty wings and tail . . . . *Perisoreus*

PSILORHI'NUS. (Gr. ψιλός, *psilos*, smooth, bare, bald; ῥίς, ῥινός, *hris, hrinos*, nose.) BROWN JAYS. SMOKY PIES. Nostrils exposed, large, rounded. Bill stout, with very convex culmen, curved from the base. Wing and tail of about equal lengths, both rounded. Of large size, and smoky-brown color; not crested.

P. mo'rio. (Lat. *morio*, a fool; Gr. μωρός, *moros*, foolish, silly.) BROWN JAY. Smoky-brown, darker on head, fading on belly; wings and tail with bluish gloss. Bill and feet black, sometimes yellow. Length about 16.00; wing and tail about 8.00; graduation of latter about 2.00; bill 1.25. Rio Grande Valley and southward; not yet actually taken over our border, and not given in the A. O. U. Lists.

PI'CA. (Lat. *pica*, a pie.) MAGPIES. Tail extremely long, when fully developed forming more than $\frac{1}{2}$ the total length, graduated for about $\frac{1}{2}$ its own length; the feathers with rounded ends, the middle pair at least tapering, and specially lengthened beyond the rest. Bill of ordinary corvine shape; nostrils concealed by long nasal tufts. Wings short and rounded, with very short, narrow, falcate first primary. Feet stout; tarsus little longer than middle toe and claw. Head not crested. A naked space about eye. Plumage black, iridescent, with masses of white; bill black or yellow. Sexes alike. Habits arboreal and somewhat terrestrial, — very irregular, in fact, a Magpie's general character being none of the best, though the generic characters are excellent.

P. pi'ca hudson'ica. (Of Hudson's Bay. Fig. 333.) AMERICAN MAGPIE. BLACK-BILLED MAGPIE. Lustrous black, with green, purple, violet, and even golden iridescence, especially on tail and wings. Below, from breast to crissum, a scapular patch, and a great part of inner webs of primary quills, white; some whitish touches on throat; lower back showing gray, owing to mixture of white with black; bill and feet black; eyes blackish. Length 15 or 20 inches, according to development of tail, which is a foot or less long, extremely graduated; extent about 2 feet; wing about 8.00; outer primary short, slender, and falcate; bill 1.25; tarsus 1.67; middle toe and claw 1.50. ♀ rather smaller than ♂, but alike in color. Western N. Am. from the Great Plains to the Pacific, except most parts of California, common; N. to the Yukon valley; occasionally in the upper Mississippi and Great Lake region even to Ontario. The American Magpie is extremely similar to the notorious bird of Europe, and attempts to establish specific characters have failed. It is a rather larger and "better" bird, though quite as much of a rascal.

FIG. 333. — Magpie, reduced. (From Dixon.)

The nest is usually placed in thickets or shrubbery, more rarely high in trees, and is as big as a bushel, bristling with a *chevaux-de-frise* of sticks outside, with a lateral covered way leading to the nest proper inside, which is built of finer materials and is of ordinary dimensions, with a cavity about 6.00 in diameter by 4 deep. Eggs 6–9, even 10, usually 7; commonly 1.20 to

1.40 long by 0.85 to 1.00 broad, averaging 1.30 × 0.90, pale drab, sometimes with a greenish tint, heavily dotted, dashed, and blotched with purplish and various brown shades, which usually cover the whole surface and sometimes hide the ground color: laid April–June or even July, in different latitudes.

P. nut′talli. (To Thos. Nuttall.) YELLOW-BILLED MAGPIE. Bill and bare space about eye yellow. Smaller. Otherwise like the last, of which it is a perpetuated accident! The European Magpie sometimes shows the same thing, and in some other species, like *Psilorhinus morio*, the bill is indifferently black or yellow. California, common W. of the Sierras Nevadas. General habits, nest, and eggs the same as those of the other species.

CYANOCIT′TA. (Gr. κύανος, *kuanos*, blue; κίττα, *kitta*, a jay.) CRESTED BLUE JAYS. Conspicuously crested; wings and tail blue, black-barred; bill and feet black. Length 11.00–12.00; wing or tail 5.00–6.00. Nostrils large, subcircular, but concealed. Wing and tail of equal lengths, both rounded. Hind claw large, equalling or exceeding its digit in length. There are two subgenera and species of this beautiful genus, one light blue and white, short-crested, eastern, standing quite alone, with its subspecies *florincola*; the other dusky-bodied, long-crested, western, running into three subspecies.

Analysis of Subgenera, Species, and Subspecies.

Purplish-blue, whitening below, with a black collar. (CYANOCITTA proper.)
    The ordinary form of Eastern N. Am. . . . . *cristata*
    The smaller form of Florida . . . . *c. florincola*
Sooty-brownish or -blackish, bluing on body behind, wings and tail; both the latter black-barred. (Subgenus STELLEROCITTA.)
    Sooty-blackish; little if any blue on forehead; none about eye; wing-coverts unbarred . . . . . . *stelleri*
    Sooty-blackish; but blue on forehead and above eye; wing-coverts unbarred . . . . *s. annectens*
    Sooty-brownish, blue on forehead; little if any blue about eye; wing-coverts unbarred . . . . *s. frontalis*
    Sooty-brownish, the crest quite black. Bluish-white streaks on forehead and about eye; wing-coverts black-barred . . . . *s. macrolopha*

(*Subgenus* CYANOCITTA.)

C. crista′ta. (Lat. *cristata*, crested. Fig. 334.) BLUE JAY. ♂: Purplish-blue, below pale purplish-gray, whitening on throat, belly, and crissum. A black collar across lower throat and up sides of neck and head behind crest; a black frontlet bordered with whitish. Wings and tail pure rich blue, with black bars; greater coverts, secondaries, and tail-feathers, except central ones, broadly tipped with pure white; tail much rounded, graduated over an inch. Length 11.00–12.00; extent 16.00–17.50; wing and tail, each, 5.00–6.00; bill 1.25; tarsus 1.35. ♀ similar, not so richly blue; smaller. There is much difference in size between northern and southern bred birds, as in *Agelæus*. Eastern N. Am., especially U. S., but N. to Hudson's Bay; W. to the central plains; a very abundant resident or half-migratory bird, breeding throughout its range;

FIG. 334. — Blue Jay, reduced. (Sheppard del. Nichols sc.)

a well-known character! Nest in trees and bushes, or any odd nook, large and substantial, 7 or 8 inches across outside × 4 or 5 deep, cupped 3 or 4 × 2.50, with twigs outside, inside of mixed materials; eggs in April, May, and June according to locality, 3-6 in number, usually 4 or 5, 1.00–1.20 × 0.80–0.85, drab-colored varying from greenish to buff, irregularly but generally fully spotted and blotched with the usual brown surface spots and purplish shell-markings. The Jay is one of our handsomest birds, of the worst possible reputation; it probably destroys more nests, eggs, and young of other birds than any Shrike or Hawk.

C. c. florin'cola. (Lat. *flos*, gen. *floris*, a flower; *incola*, an inhabitant; with implied allusion to Florida as the "Land of Flowers," though the country was so named in 1512 by Ponce de Leon because he discovered it on Easter Day, Spanish *Pascua florida* or *Pascua de flores*.) FLORIDA BLUE JAY. Like the last; smaller, with relatively larger bill, shorter crest, and less white on wings and tail. Length 10.00–11.50; wing and tail about 5.00, rather less than more; white on outer tail-feather under 1.00 in extent. Florida, resident; a local race maintaining its subspecific character along the Gulf coast to Texas. COUES, Key, 2d–4th eds. 1884–90, p. 421, in text; RIDGW. Man. 1887, p. 353; A. O. U. Lists, 1886–95, No. 477 *a*.

(*Subgenus* STELLEROCITTA.)

C. stel'leri. (To G. W. Steller. Fig. 335.) STELLER'S JAY. MOUNTAIN JAY. PINE JAY. ♂ ♀: Whole head, neck, and back sooty blackish, little if any lighter on throat, and with little if any blue on forehead or about eyes; this sooty color passing insensibly on rump and breast into dull blue. Wings and tail richer blue, crossed with numerous black bars, not on secondary coverts. Bill and feet black. Young more fuliginous; wing-bars faint if not wanting. Size of the Eastern Blue Jay, or rather larger. Pacific coast region, from portions of California through Oregon, Washington, and British Columbia to Cook Inlet, Alaska, especially in pine belts, as of the Coast and Cascade ranges; but E. to the Rocky Mts., where inosculating with *macrolopha*. This is the typical form, with little or no blue, no whitish on head, and unbarred wing-coverts; running through *annectens*, *frontalis*, and *macrolopha* into some very different Mexican forms. Habits, nest, and eggs as described under *macrolopha*. (*C. s. litoralis* MAYNARD, Orn. and Oöl. Apr. 1889, p. 59, Vancouver Island, is rejected as untenable by the A. O. U. Committee: see Auk, Jan. 1890, p. 65 and p. 91.)

FIG. 335. — Steller's Jay.

C. s. annec'tens. (Lat. *annectens*, annexing.) BLACK-HEADED JAY. This name has been given to specimens directly connecting *stelleri* and *macrolopha*. General tone of the former; quite blackish, short-crested, with plain wing-coverts; but blue frontal streaks and whitish eye-patch of the latter. N. Rocky Mts., U. S., W. to eastern Oregon and Washington, S. to the Wahsatch range, N. to B. Col.; originally named by Baird in 1874, but disallowed by the A. O. U. Committee for some years before it was recognized as entitled to the place it had continuously occupied in the 2d and 3d eds. of the Key.

C. s. fronta′lis. (Lat. *frontalis*, pertaining to *frons*, the forehead.) BLUE-FRONTED JAY. SIERRA JAY. An offset from *stelleri;* sooty color rather brownish than blackish; blue of different shade on body from the deep indigo on wings and tail; whole crest glossed with bluish, and conspicuous blue streaks on forehead; no whitish eye-patches; wing-coverts obsoletely or not barred. This form is best developed in the Sierras Nevadas of California, whence it extends less typically in all directions, shading directly into the several other subspecies in different regions.

C. s. macro′lopha. (Gr. μακρός, *makros*, long ; λόφος, *lophos*, crest. Fig. 336.) LONG-CRESTED JAY. Better marked than the connecting links; were these not forthcoming, it would rank as a good species. ♂ ♀ : Upper parts sooty umber-brown, with a faint blue tinge, blackening on head and neck all around in decided contrast, passing on rump and upper tail-coverts into beautiful light cobalt-blue, on fore breast into the same blue which occupies all the rest of the under parts. Crest black, but faced on forehead with bluish-white, which, when the feathers are not disturbed, runs in two

FIG. 336. — Long-crested Jay, nat. size. (Ad. nat. del. E. C.)

parallel lines from nostrils upward — these colored tips of the feathers of firmer texture than their basal portions. One or both eyelids patched with white. Chin abruptly whitish, streaky. Exposed surfaces of wings rich indigo-blue, most intense on inner secondaries, which, with greater coverts, are regularly and firmly barred across both webs with black; outer webs of primaries lighter blue, more like that of rump or under parts. Upper surface of tail rich indigo, like the secondaries, and similarly black-barred; these bands most distinct towards the ends and on outer webs of the feathers; tail viewed from below appearing mostly blackish. Iris dark. Bill and feet black. Length 12.00–13.00; extent 17.00–19.00; wing 5.50–6.50; tail the same; bill 1.12; tarsus 1.50; middle toe and claw 1.33. Sexes quite alike, but ♀ at the lesser dimensions given. Crest longer than in northern *stelleri*, sometimes 3.00. Young: Much more sooty; below entirely fuliginous, with the future blue indicated by an ashy or grayish shade. Wings and tail nearly as bright blue as in the adult, but black bars faint or wanting. Crest shorter, not quite black, not faced with blue, and no white about eyes. Rocky Mt. region, U. S., especially southerly; N. to Wyoming, where grading into *annectens;* W. to Utah, where melting into *frontalis;* S. into Mexico, where intergrading with the bluer *diademata*, which latter in its turn is directly connected with the quite blue *coronata*. The Long-crested Jay is a common resident of the pine belt, displaying in marked degree the notorious attributes of its genus, or genius. Nest in trees and bushes, usually concealed with art, though bulky; eggs 3–6, usually 4 or 5, from 1.10 to 1.30 × 0.85 to 0.95, averaging 1.20 × 0.87, pale bluish-green, profusely spotted and blotched with dark olive-brown and lighter brown surface markings, with the usual neutral-tint shell-spots, commonly called "lavender" or "lilac"; the pigmentation being pretty evenly distributed, with little tendency to aggregation about the large end of the egg. They are mostly laid in May, but the season runs from April to June.

APHELO′COMA. (Gr. ἀφελής, *apheles*, smooth, sleek; κόμη, *kome*, hair: alluding to the lack of crest.) CRESTLESS BLUE JAYS. Generally as in *Cyanocitta*. Head uncrested. Tail longer or shorter than wings, instead of about equal, graduated (in some extralimital forms about equal to wing and even). Tarsus rather longer than middle toe and claw. Wings and tail blue, without black bars, and blue the chief body-color; whitish underneath, with (usually) or without a gray patch on the back. All southern and western. Several species abound in those portions of the U. S. and in Mexico, where they are as characteristic of thickets of scrub-oak and other low deciduous trees as the western forms of *Cyanocitta* are of the pineries. The nest is placed in such trees and bushes, and is rather a saucer than a cup, being a less substantial structure than usual in this group; the eggs are particolored in *Aphelocoma*, but whole-colored in the subgenus *Sieberocitta*, which I now base upon our representative of Sieber's Jay.

Analysis of Species, Subgenera, and Subspecies.

Tail longer than wings, graduated. Blue above, with gray dorsal area; belly white or whitish; usually a superciliary stripe, and streaks on the throat. Eggs spotted. (APHELOCOMA proper.)
  Crissum blue or bluish, more or less contrasted with white or whitish belly.
    Continental species.
      Forehead hoary white; superciliary stripe ill-defined; dorsal area well-defined; crissum blue, contrasting with grayish belly . . . . *cyanea*
      Forehead blue; superciliary stripe distinct; dorsal area ill-defined, spreading and bluish; crissum bluish, but not well contrasted with dingy whitish belly. Southern Rocky Mts. . . . . *woodhousei*
    Insular species, resembling the last. Santa Cruz Isl. . . . . *insularis*
  Crissum white or whitish, like the belly.
    Sides of head not decidedly blue, but rather blackish.
      Forehead blue; superciliary stripe distinct; dorsal area well defined.
        Larger, medium colored. California, Oregon, Nevada . . . . *californica*
        Smaller, lighter colored. Lower California . . . . *c. hypoleuca*
        Larger, darker colored. San Pedro Mts., Lower California . . . . *c. obscura*
    Sides of head decidedly blue, like the crown.
      Forehead blue; superciliary stripe indistinct or obsolete. Mexico and W Texas . . . . *cyanotis*
Tail shorter than wings, rounded. Blue above, without definite dorsal area; no superciliary stripe or streaks on the throat. Eggs plain. (Subgenus SIEBEROCITTA.) . . . . *sieberi arizonæ*

A. cya′nea. (Gr. κυάνεος, *kuaneos*, Lat. *cyaneus*, blue.) FLORIDA JAY. SCRUB JAY. Adult ♂ ♀: Blue; back with a small well-defined gray patch not invading scapulars; belly and sides pale grayish; under tail-coverts and tibiæ blue in marked contrast; much hoary whitish on forehead and sides of crown, but no sharp white superciliary stripe; chin, throat, and middle of breast vague streaky whitish and bluish; ear-coverts dusky; the blue that seems to encircle head and neck well defined against the gray of back and breast. Bill comparatively short, very stout at base. Length 11.00–12.50, average 11.75; extent 13.50–15.00, average 14.50; wing 4.00–4.75, average 4.40; tail 4.50–5.50, average 5.00, always longer than wing; bill about 1.00. Florida (and Gulf States?), abundant. Very local, and not authentic as occurring outside of Florida. Usual habits of Jays. Nest a flat structure, in the scrubs, of twigs lined with fibres. Eggs 3–5, bluish-green, sparingly speckled, chiefly at larger end, with brown, 1.05 × 0.80 on an average, but ranging from 1.00 to 1.20 in length, laid mostly in April and May. (*A. floridana* of former eds. of Key, after BARTRAM, 1791. *Garrulus cyaneus* VIEILL. 1817. *Aphelocoma cyanea* COUES, Auk, Jan. 1899, p. 84; A. O. U. Suppl. List, *ibid*. p. 112.)

A. woodhou'sei. (To S. W. Woodhouse.) WOODHOUSE'S JAY. Dorsal patch dark, glossed with blue, shading into the blue of surrounding parts; under parts rather darker than in *C. cyanea*, somewhat bluish-gray; the under tail-coverts bluish but not contrasted; on breast the blue and gray shading into each other, gular and pectoral streaks whitish and well defined, superciliary line definite white, but no hoary on forehead; bill slenderer. Adult ♂ ♀: General color blue, rich and pure on wings, tail, rump, crown, back and sides of neck, and on breast surrounding the streaky white area. Middle of back and scapulars dark gray much tinged with blue, shading insensibly into surrounding blue. Upper and under tail-coverts blue. Under parts from breast gray, with blue tinge (in *californica* nearly white). Chin, throat, and breast with a series of whitish blue-edged streaks, enclosed in surrounding blue. Lores, orbits, and auriculars dusky. A series of sharp white streaks over and behind eye. Wings and tail blue; the inner webs of most quills, and tail viewed from below, dusky. The inner secondaries and tail-feathers, closely examined, show obsolete barring, like that which becomes pronounced in *Cyanocitta*, but the traces are faint, and the feathers may be properly called plain. Iris brown; bill and feet black. Length of ♂, about 12.00; extent 16.50; wing 5.00; tail 6.00; bill 1.12; tarsus 1.50; middle toe and claw 1.33. ♀ smaller: average 11.25; extent 15.50, etc. Young: Wings and tail as in adult; upper parts mostly gray; under parts grayish-white, with little or no blue on breast; pectoral streaks undefined, as are those over eye. Rocky Mt. region, from S. E. Oregon, Idaho, Montana, Wyoming, S. through Colorado, parts of Utah and Nevada, S. E. California, Arizona, New Mexico, W. Texas, into N. Mexico. In regions where Woodhouse's and Long-crested Jays occur together, the latter lives chiefly in pines, the former in scrub-oak and other thickets, like its Florida relative. Nest in such situations, rather frail and flattish in comparison with those built by Jays of the genus *Cyanocitta*, made of twigs as a basement, with the inner structure of rootlets, hairs, etc. Eggs laid mostly in April and May, but from late in March to early in June; 3-6 in number, oftenest 4 or 5, pale greenish, rather sparingly flecked all over the surface with rusty brown and duller shell-markings; 1.10 × 0.80 on an average, with a variation from 1.00 to 1.15 in length.

A. insula'ris. (Lat. of an island, insular.) SANTA CRUZ JAY. Above, dark azure blue, including exposed surface of wing- and tail-feathers, this color deepest on crown, and extending on sides of head and well down on neck and breast; back dark sepia brown. A white superciliary line; a black loral and auricular spot. Feathers of throat and breast ashy-white edged with blue; crissum blue; other under parts dull white. Wing 5.35; tail 6.25; tarsus 1.80; bill 1.25. Santa Cruz Island, one of the Santa Barbara group, off the coast of California. The relationships of this species are rather with *woodhousei* than with *californica*, as it has the bluish under tail-coverts of the former; but its insulation keeps it apart from both, and it may be allowed to stand. Nesting as usual in the genus: eggs 2-3, averaging 1.18 × 0.88; markings rather light brown, lavender, and grayish. HENSHAW, Auk, Oct. 1886, p. 452; A. O. U. List, No. 481. 1. *A. floridana insularis* of the Key, 3d and 4th eds. 1887 and 1890, p. 878 and p. 901.

A. cyano'tis. (Gr. κύανος, *kuanos*, a dark blue substance, and as adj. blue, like Lat. *cyaneus*; and *-otis*, combining form of Gr. οὖς, gen. ὠτός, the ear.) BLUE-EARED JAY. Closely resembling Woodhouse's and the California Jay, especially the latter, having the belly and crissum white; but sides of head bright blue, like the crown, and superciliary stripe obsolete; interscapular patch dark gray tinged with blue. Size of the others. A Mexican species to which specimens taken in July, 1890, in western Texas near the border have been referred. See Auk, Oct. 1894, p. 327, and Apr. 1895, p. 165; A. O. U. List, 2d ed. 1895, No. 480. 1. This bird is new to the Key: orig. descr. in RIDGW. Man. 1887, p. 357.

A. califor'nica. (Of California.) CALIFORNIA JAY. Dorsal patch light and distinct, as in *cyanea*, but under parts, including tail-coverts and tibiæ, nearly white; gular streaks very large, aggregated, and white, causing throat to be nearly uniform; a white superciliary line, as in *woodhousei*, but no hoary on forehead; bill slender. Thus it is seen that each of the three forms presents a varying emphasis of common characters. Adult ♂ ♀ : General color blue. Scapulars and interscapulars gray, with little if any tinge of blue; rump and upper tail-coverts bluish-gray, usually mixed with some white. Forehead and nasal tufts blue like crown; a sharp white superciliary stripe over and behind eye; lores, eyelids, and auriculars blackish. Under parts from breast soiled white, with little or no tinge of blue except on crissum; breast appearing as if blue, overlaid with broad white stripes, which become continuous on throat and chin; the breast is really white, in streaks edged with blue, and with a surrounding of blue in which the streaks are as if framed. Iris brown; bill and feet black. Length 12.00 or less; wing 5.00; tail 5.50; bill 1.00; tarsus 1.50; middle toe and claw 1.25. In comparison with *woodhousei*, differences are seen in the well-defined gray dorsal patch; nearly white underparts without decidedly blue crissum; broader and more continuously white gular streaks. The general habits, nest, and eggs are the same. This species is common in the State for which it is named, and there generally distributed, on both sides of the main mountain range; but it extends S. into Lower California, N. through Oregon to the Columbia and thus to Washington, in the Pacific coast region, and E. into some parts of Nevada.

A. c. hypoleu'ca. (Gr. ὑπό, *hupo*, under, below, and λευκός, *leucos*, white.) XANTUS' JAY. Said to be smaller than the last (though the dimensions as alleged do not bear this out), with larger bill and feet, paler blue back and whiter under parts. Lower California. RIDGW. Man. 1887, p. 356; COUES, Key, 4th ed. 1890, p. 901; A. O. U. List, 2d ed. 1895, No. 481 *a*.

A. c. obscu'ra. (Lat. *obscura*, fem. of *obscurus*, dark, obscure: applicable in a double sense to the alleged distinctness of this local race.) SAN PEDRO JAY. BELDING'S JAY. Differing from *A. californica* in much darker colors and weaker feet. San Pedro range, L. Cala., up to 10,000 feet. ANTHONY, Pr. Cal. Acad. Sci. 2d ser. ii, Oct. 1889, p. 75; A. O. U. List, 2d ed. 1895, No. 481 *b*. *A. floridana obscura* COUES, Key, 4th ed. 1890, p. 901.

(Subgenus SIEBEROCITTA.)

A. sieb'eri arizonæ. (To Sieber. Of Arizona.) ARIZONA JAY. Belonging to a different section of the genus, distinguished by having tail rather shorter than longer than wings, upper parts uniform blue, no throat-streaks, and eggs plain. Adult ♂ ♀ : Above, light blue, purer on head, wings, and tail than on back, where rather dull. Beneath, sordid bluish-gray, bluest on breast, paler on throat, whitening on belly, flanks, and crissum. Lores blackish; orbits and auriculars dark. No superciliary stripe, nor decided streaks on throat or breast. Bill normally black, sometimes irregularly patched with whitish. Feet black. Length about 13.00; wing 6.25-6.75; tail 6.00-6.50, rounded, the lateral feathers graduated about 0.50; bill 1.25, 0.40 deep at base; tarsus 1.67; middle toe and claw 1.33. Young: Little if any blue excepting on wings and tail, being dull gray above; below, much like the adult. Bill flesh-colored on most of under mandible. Arizona and New Mexico, N. to about 35°, S. into Sonora and Chihuahua. This Jay abounds in the foothills of the mountains of southern Arizona and southwestern New Mexico, where it goes in troops. The nest is built in scrub-oaks at no great height, rather flattish, sometimes quite flimsy, with small sticks and twigs as a basis, upon which are woven rootlets and horse hairs; some nests measure 10.00 across outside, and 4.00 deep, with a shallow cup 4.50 × 2.00. In some cases additional "cock-nests" are built, but never used for eggs, as is also the case with various other birds, — the Long-billed Marsh Wren, for example. Eggs 3-7, usually 4 or 5, averaging 1.18 × 0.87, but ranging in length from 1.05 to 1.35, laid in April and May. They are remarkable in this genus, indeed in the family, for being whole-colored, of the peculiar light greenish-blue tint commonly called "robin-blue," entirely free from spots. The synonymy of this bird became much involved while authors were groping their way to its identification. Waiving any question of *Garrulus sordidus* SWAINS. Phil. Mag. i, 1827, p. 437, Zool. Ill. 2d ser. pl. 86, it is now regarded as the northern form of

Pica sieberii WAGL. Syst. Av. 1827, Pica No. 23, and its synonymy is as follows: *Cyanocitta sordida* BD. B. N. A. 1858, p. 587, and Mex. B. Surv. ii, p. 21, pl. 22, fig. 1 ; COOPER, Orn. Cal. i, 1870, p. 305; COUES, Proc. Acad. Nat. Sci. Phila. 1866, p. 92 (p. 56 of reprint) ; *Sieber's Jay, Aphelocoma sordida* COUES, Key, 1st ed. 1872, p. 166; *Cyanocitta ultramarina* var. *arizonæ* RIDGW. Bull. Essex Inst. v, Dec. 1873, p. 199; BD. BREW. and RIDGW. B. N. A. 1874, ii, p. 292; Subsp. *a, Aphelocoma* [*sordida*] *arizonæ* SHARPE, Cat. B. Brit. Mus. iii, 1877; p. 117 ; *Aphelocoma ultramarina arizonæ* COUES, Key, 2d–4th eds. 1884–90, p. 424, where the reference of this bird to *ultramarinus* of BONAPARTE, 1825, is criticised; and finally *Aphelocoma sieberii arizonæ* RIDGW. Proc. U. S. Nat. Mus. viii, 1885, p. 355; Man. 1887, p. 357 ; A. O. U. Lists, 1st and 2d eds. 1886–95, No. 482.

XANTHU'RA. (Gr. ξανθός, *xanthos*, yellow; οὐρα, *oura*, tail.) GREEN JAYS. No crest. Wings short, much rounded, with lengthened inner secondaries folding nearly over primaries. Tail longer than wings, graduated. Bill short and deep, with culmen curved from the base. Colors green and yellow, with black and blue on head. Several tropical species of these luxurious Jays, one reaching our border. (Name originally and now usually in the bad form of *Xanthoura*.)

X. luxuo'sa. (Lat. *luxuosa*, luxurious.) GREEN JAY. RIO GRANDE JAY. Adult ♂ : Back and exposed surface of wings yellowish-green; inner webs of most quills blackish edged with clear yellow; their shafts black above, yellow or whitish below ; lining of wings clear yellow. Four middle tail-feathers greenish-blue, at base little different from back, bluing toward ends; these feathers, seen from below, quite black ; other tail-feathers all clear rich yellow, including their shafts. Under parts from breast light greenish-yellow, yielding to pure yellow on middle of belly. Top of head and nasal plumules beautiful rich blue, yielding on forehead to hoary-white. Sides of head to above eyes, and whole chin, throat, and fore breast jet black, enclosing a large triangular patch of blue on side of lower jaw, and blue touches on eyelids. Bill and feet black. Length 11.25–12.00 ; extent 14.50–15.50 ; wing 4.50–5.00; tail 5.25– 5.75; tarsus 1.50; middle toe and claw 1.25; bill 1.00, very stout. ♀ near the lesser dimensions given. This gay and gaudy bird is abundant in some localities in the Lower Rio Grande valley as high up as Rio Grande city, and extends thence S. in Mexico to Puebla and Vera Cruz. As in the case of the Blue Jay its truly elegant attire hides a heart full of mischief and malice; it is an equally merciless despoiler of other birds' nests, eggs, and young. Nest in bushes and small trees, bulky, of twigs, oftenest thorny, with finer lining of rootlets, etc. ; eggs 3–5, usually 4, 1.10 × 0.80, on an average ranging from 1.00 × 0.75 to 1.20 × 0.85, the ground color varying from greenish-drab to whitish, profusely and evenly marked as usual with browns and neutral tints; they are laid in April and May.

PERISO'REUS. (Gr. περισωρεύω, *perisoreuo*, I heap up; probably in allusion to the hoarding or thievish propensities of Jays.) GRAY JAYS. Not crested. Plumage soft, full and lax, grayish or sooty. Bill very short, not deep but wide at base; culmen little curved ; gonys ascending. Wings and tail of approximately equal lengths ; latter graduated. A circumpolar and boreal or alpine genus, type *P. infaustus* of Europe; with two species in America, one of them with three subspecies.

Analysis of Species and Subspecies.

Back without distinct whitish shaft-lines ; belly gray, darker than throat.
    Dark hood moderate ; white forehead extensive ; back grayish . . . . *canadensis*
    Dark hood extensive ; whitish forehead restricted ; back grayish. Labrador coast . . . . *c. nigricapillus*
    Dark hood extensive ; smoky forehead restricted ; back brownish. Alaska coast . . . . . . *c. fumifrons*
    Dark hood restricted ; white forehead extensive ; back clear ash. Rocky Mts. . . . . *c. capitalis*
Back with distinct whitish shaft-lines ; belly white like throat.
    Dark hood extensive ; smoky forehead restricted ; back brownish. N. Pacific coast . . . . *obscurus*

P. canaden'sis. (Of Canada. Fig. 337.) CANADA JAY. WHISKEY JOHN. WHISKEY JACK. MOOSE-BIRD. GREASE BIRD. MEAT HAWK. CARRION BIRD. CAMP ROBBER.

CORVIDÆ — GARRULINÆ: JAYS AND PIES. 501

Adult ♂ ♀: Gray, paler below than above, whitening on head, neck, and breast; a dark cap on crown, hind head and nape, separated by a gray cervical collar from the ashy-plumbeous back; wings and tail dark plumbeous, the feathers obscurely tipped with whitish. Bill and feet black. The dark hood not quite black, not extending over auriculars, and scarcely including the eyes; the forehead quite white, reaching to the eyes; the throat and sides of neck decidedly lighter than the belly and flanks; the back without any distinct whitish shaft-lines. Young: Much darker, sooty slate color, with black face, and an obscure whitish maxillary streak; the bleaching progresses indefinitely with age. Length 11.00–12.00; extent about 16.00; wing 5.25–5.75; tail rather more, graduated; tarsus 1.33; bill 1.00 or less, shaped like a Titmouse's. Subarctic Am. to the limits of trees, S. into the N. States, N. W. to Alaska; common in some

FIG. 337. — Canada Jay, reduced. (Sheppard del. Nichols sc.)

parts of the White and Green Mts. of N. H. and Vermont, the Adirondacks of N. Y., Michigan, Minnesota, etc.; breeds in Maine and northward; resident, and seldom seen south of its breeding range. The "Wisskachon" (whence "Whiskey John" and then "Whiskey Jack") is noted for the familiarity and impudence with which it hangs about the hunter's camp to steal provisions, for consorting with moose, and for nesting in late winter or early spring, Feb.–March. Young birds may be found flying early in April in the U. S., though eggs may be taken in May in arctic regions. Nest usually on the bough of a spruce or other conifer, a large substantial structure, of twigs, grasses, mosses, plant down, and feathers, measuring 7 or 8 inches across outside by 4 deep, with a cavity about 3.00 × 2.50; eggs 3–4, rarely 5, 1.15 × 0.85, yellowish-gray to pale green, finely dotted and blotched with brown and slate, or lavender, especially about the larger end; others more uniformly and largely blotched; variation wide, as in other Jays, both in size and coloration, the range in length from 1.05 to 1.20.

P. c. nigricapil′lus. (Lat. *niger*, black; *capillus*, hair; meaning black-headed.) LABRADOR JAY. Said to differ from true *canadensis* in altogether darker coloration, blacker hood, black auriculars, less extensive white or pale smoky front, and more marked contrast of white and dark areas of head and neck. Coast region of Labrador, N. to Ungava Bay. RIDGW. Pr. U. S. Nat. Mus. v, June, 1882, p. 15; Man. 1887, p. 359; COUES, Key, 3d–4th eds. 1887–90, p. 878; A. O. U. Lists, 1st and 2d eds. 1886 and 1895, No. 484 c.

P. c. fu′mifrons. (Lat. *fumus*, smoke; *frons*, forehead.) ALASKAN JAY. SMUTTY-NOSED JAY. Similar: coloration darker and dingier throughout; white of forehead obscured or obliterated by smoky-gray. Young differing from the adults as before, but of a dingy brownish-slate rather than blackish-slate as in *canadensis* proper and *c. nigricapillus*. Wings and tail averaging a trifle shorter than in the typical form. Alaska, except in the southern coast district; in the interior melting into *canadensis* proper, on the coast the characters best pronounced. Nesting in March and April.

P. c. capita′lis. (Lat. *capitalis*, capital, relating to the head, *caput*.) ROCKY MOUNTAIN JAY. WHITE-HEADED JAY. General color ashy-plumbeous, or leaden-gray, paler below; wings and tail blackish, with a peculiar glaucous shade, as if frosted or silvered over. The body-color giving way on breast and neck to whitish, established as hoary-white on head, isolating the narrow well-defined nuchal band of sooty-gray. No white lines on back; tail-feathers distinctly tipped with whitish, and much edging of the same on wings. The clearer colors generally — back rather bluish-gray than brownish-gray, very white head with narrow nuchal dark band — produce a bird differing so obviously from the ordinary Canada Jay as to look like a different species at gunshot range, as I can affirm from repeated observation of the bird in various mountains of the West. The changes of plumage with age are parallel. Size at a maximum. Length 12.00–13.00; extent 17.00; wing and tail, each, about 6.00; bill 1.00 or more; tarsus 1.40; middle toe and claw 1.00. Rocky Mt. region of the U. S., especially New Mexico and Arizona, Colorado, Wyoming, Idaho, and Montana, N. to British America, northward shading into typical *canadensis*. The high mountains of Colorado furnish the extreme cases. The bird is resident throughout its range, and breeds up to an elevation of at least 10,000 feet; nest and eggs indistinguishable from those of the stock form; eggs laid in March.

P. obscu'rus. (Lat. *obscurus*, obscure, dark.) OBSCURE JAY. OREGON JAY. In general, similar to *P. canadensis*, but apparently distinct. Adult ♂ ♀: Above, rather brown than plumbeous; feathers of back with obvious whitish shaft-stripes; below, white or whitish, nearly uniform over all under parts; dark hood extensive, sooty black, but forehead and nasal plumules white or whitish; wings and tail brownish-gray, with little whitish edging or tipping. Rather small; length 11.00 or less; wing and tail 5.50; bill 0.90; tarsus 1.25. Young differ as before, but are dark dingy brown rather than sooty blackish. Pacific coast region from N. California through Oregon, Washington, and British Columbia; common; resident; habits the same as those of the Canada Jay; nest and eggs indistinguishable; breeding season Mar.–May. *P. c. obscurus* of former eds. of the Key, now rated as a separate species: see SHARPE, Brit. Mus. Cat. iii, 1877, pl. v. for colored figures of this and *P. c. capitalis*.

P. o. gris'eus. (Lat. *griseus*, gray.) GRAY JAY. Described as like *obscurus*, but larger (except feet) and grayer; back "deep mouse gray instead of brown"; quills and tail "between gray (No. 6) and smoke gray, instead of drab gray"; under parts grayish-white instead of brownish-white. California to British Columbia, E. of the Coast and Cascade ranges. RIDGW. Auk, July, 1899, p. 255.

Family STURNIDÆ: Old World Starlings.

A family confined to the Old World: difficult to characterize, owing to the variety of forms it includes. Apparently related to *Icteridæ*, from which distinguished by presence of *ten* primaries, the 1st short or quite spurious; and certainly close to the *Corvidæ*, with which they share 10 primaries and some other characters. There is also evidence of affinity with the *Ploceidæ*. There are about 40 genera and 140 species of *Sturnidæ*, among them several celebrated birds of Europe and Asia, as those called Religious Grackles, Mina-birds, etc.; many are also splendidly iridescent, as in the genus *Lamprocolius* and others. The only genus with which we have here to do is *Sturnus*, belonging to the

Subfamily STURNINÆ: Typical Starlings.

STUR'NUS. (Lat. *sturnus*, a stare or starling.) STARLINGS. Bill shaped somewhat as in *Sturnella* or *Icterus*, but widened and flattened; rather shorter than head; culmen and gonys about straight, both gently rounded in transverse section, and at tip; culmen rising high on forehead, dividing prominent antiæ which extend into well-marked nasal fossæ; a conspicuous nasal scale, overarching nostrils; tomial edges of mandibles dilated, especially those of upper mandible; commissure obtusely angulated; sides of lower mandible extensively denuded and somewhat excavated; feathers filling interramal space; no bristles about bill. Wings long and pointed; 1st primary spurious and very small; 2d and 3d longest, rest rapidly graduated. Tail of 12 feathers, emarginate, little more than half as long as wing. Feet short; tarsus of strictly Oscine podotheca, scutellate and laminiplantar, about as long as middle toe without its claw. Lateral toes of subequal lengths, their claws falling short of base of middle claw; hind claw about as long as its digit. Plumage metallic and iridescent, the feathers all distinctly outlined.

S. vulga′ris. (Lat. *vulgaris*, vulgar, common. Fig. 338.) THE STARLING. Adult: General plumage of metallic lustre, iridescing dark green on most parts, more steel-blue on under parts, and violet or purplish-blue on fore parts; more or less variegated throughout with pale ochraceous or whitish tips of the feathers. Wings and tail fuscous; exposed parts of feathers somewhat frosty or silvery, with velvety-black and pale ochrey marginings, the former within the latter. Bill yellowish; feet reddish. Young and in winter: Plumage more heavily variegated throughout, with larger tawny-brown spots on upper parts, and white ones below; wings and tail strongly edged with brown; bill dark. Length about 8.50; wing 5.00; tail 2.75; bill 1.00; tarsus 1.00; middle toe and claw 1.25. Europe,

FIG. 338. — The Starling. (From Dixon.)

etc., one of the longest and best known of birds. Has straggled to Greenland in one known instance: imported and now naturalized in New York City, where it breeds about buildings in Central Park and elsewhere, like the European Sparrow; eggs 4–7, 1.15 × 0.85, pale greenish-blue, unmarked.

Family ALAUDIDÆ: Larks.

A rather small group, well defined by the character of the feet, in adaptation to terrestrial life. The subcylindrical tarsi are scutellate and blunt behind as in front, with a deep groove along the inner side, and a slight one, or none, on the outer face. That is to say, there is an anomalous structure of the tarsal envelop; the tarsus being covered with two series of scutella, one lapping around in front, the other around behind, the two meeting along a groove on the inner face of the tarsus, which is consequently blunt behind as well as in front. There is a simple suture of the two series of plates on the outer face of the tarsus; the individual plates of each series alternate. Other characters (shared by some *Motacillidæ*) are the very long, straight, hind claw, which equals or exceeds its digit in length; long, pointed wings (with 1st primary apparently wanting in *Otocorys*), and inner secondaries lengthened and flowing. The nostrils are usually concealed by dense tufts of antrorse feathers. The shape of bill is not diagnostic, being sometimes short, stout, and conic, much as in some *Fringillidæ*, while in other cases it is slenderer, and more like that of insectivorous *Passeres*. *Alaudidæ* differ from *Motacillidæ* in having the moult single. The family is composed, nominally, of 100 or more species; with the exception of two genera and several species or subspecies, it is confined to the Old World. Its systematic position is open to question; some place it at the end of the Oscine series, or remove it from *Oscines* altogether, on account of the peculiarities of the podotheca; authors generally place it near *Fringillidæ*, from the resemblance of the bill of some species to that of some Finches, and especially of some Buntings. In former editions of the Key I put *Alaudidæ* next to *Motacillidæ*, with which it has certain relationships. But I have no prejudices in favor of this arrangement, no convictions to be overcome, and no objection now to transfer the family to a place which will put it in line with the A. O. U. List — *i. e.* as nearly as is possible in the nature of a case which traverses the same series of birds in an opposite direction. Our latest monographer, Dr. Sharpe of the British Museum, handles the *Alaudidæ* in the group of *Passeres sturniformes*, in the same volume with *Sturnidæ* and *Ploceidæ*; and the presentation of the family next after *Sturnidæ*, in the present instance, is in practical conformity with such an arrangement, as we have no *Ploceidæ* in America. The fact that *Alaudidæ* appear to have indifferently 9 or 10 primaries may indicate a natural position between the sets of families in which number of primaries is among the diagnostic features. The musical apparatus is certainly well developed, as testified by the eminent vocal powers of the celebrated Skylark of Europe. The unpractised reader must be careful not to confound the Larks proper with certain birds loosely called "Larks:" thus the Tit*larks*, or Pipits, though sharing the lengthened, straightened hind claw and elongated inner secondaries of *Alaudidæ*, belong to an entirely different family, *Motacillidæ*; while the American Field *Lark* is one of the *Icteridæ*, much farther removed.

According to shape of bill, structure of nostrils, and apparent number of primaries, the family has been divided into two subfamilies; *Alaudinæ*, typified by the celebrated Skylark of Europe, and *Calandritinæ*, of which our well-known Horned Lark is a typical representative. But the development of the spurious quill is very variable in the series of genera, and does not seem to be correlated with other structural characters of bill, feet, and wings. It is therefore ineligible as a classificatory character in this family, and the supposed distinction between *Alaudinæ* and *Calandritinæ* fails of effect. I therefore abolish these groups, heretofore presented in the Key, and proceed directly to consider our two genera — one of them, *Alauda*, only represented in our fauna by stragglers or naturalized residents, the other shared by America with other parts of the Northern Hemisphere. These are but a small percentage of the total of about 20 genera of Larks which are recognized by late authorities, and seem to be established. (See SHARPE, Cat. B. Brit. Mus. xiii, 1890, pp. 512–658.)

OTOC'ORYS. (Gr. οὖς, gen. ὠτός, *ous, otos*, the ear, *i. e.* plumicorn; and κόρυς, *korus*, a helmet, also a lark, supposed to be the κορυδός, *korudos*, or crested lark of Europe now called *Galerita cristata.*) HORNED LARKS. Primaries apparently only 9 (no obvious spurious 1st primary.) Point of wing formed by the first 3 developed primaries; inner secondaries elongated. Tail of medium length, nearly even; middle pair of feathers different in shape and color from the rest. Bill compressed-conoid, acute, shorter than head. Nostrils completely concealed by dense tufts of antrorse feathers. Head not crested, but with erectile plumicorns — a peculiar tuft of feathers over each ear, somewhat like the so-called "horns" of some Owls. Feet of ordinary Alaudine characters, as already given. Coloration peculiar in the presence of yellowish tints and strong black bars on the head and breast. Birds of this genus frequent open places, are strictly terrestrial, and never hop when on the ground, like most *Passeres*; they are migratory in most localities, and gregarious when not breeding; nest on the ground, and lay 4–5 speckled eggs; sing sweetly in the spring time. *Eremophila* of previous editions of the Key, and of most authors, after BOIE, Isis, 1828, p. 322, preoccupied in ichthyology by *Eremophilus* HUMB. 1805; *Otocoris* BP. N. Ann. Sci. Nat. Bolog. ii, 1838, p. 407; but it is better to regard this as a typographical error, even if we have to strain a point to do so, than to uncharitably suppose Bonaparte did not know how to spell *Otocorys*; for we cannot imagine that he meant to compound the word with Gr. κόρις, *koris*, a bug. The bird is not an "eared bug," or any sort of an earwig, but a lark! To the single species, with two subspecies, of former editions of the Key, we have now to add eight other races, lately characterized by Henshaw and by Dwight, largely upon geographical considerations. Some of these could not be distinguished if they were not labelled for locality; and in general, only adult males in the breeding season can be characterized at all. The student of these puzzling birds need not therefore despair if many or most of his specimens are equivocal. The peculiarities of the several races are developed only in their respective breeding ranges, and the extensive wanderings of the birds mix them up during the migrations, so that different races may be found together in the same locality, and only one of them be that bred in that locality. Hence the impossibility of identifying specimens which do not typically represent adult males in breeding dress taken within the area of their breeding range. The following attempt to discriminate them is based on DWIGHT, Auk, Apr. 1890, p. 156.

Analysis of Subspecies (♂ in full plumage).

Back grayish or brownish.
  Coloration pale ; nape, rump, and bend of wing pinkish.
    No yellow anywhere ; wing 4.40. Bred in interior of Brit. Am., westerly, and Alaska . . . . *leucolœma*
    Yellow on throat.
      Back dark ; eyebrows white ; wing 4.10. Bred chiefly in the U. S., Dakota to New England . . . . *praticola*
      Back pale ; wing 4.10. Bred chiefly in the U. S., Great Plains and Great Basin . . . . *arenicola*
      Back " very pallid " ; wing 3.75. Bred in Lower California and Sonora . . . . *pallida*
      Back light gray ; wing 3.85. Bred in E. and S. E. Texas . . . . *giraudi*
  Coloration dark ; nape, rump, and bend of wing reddish.
    Browner, less streaked, eyebrows and throat always yellow ; wing 4.30. Bred in Brit. Am., easterly, Greenland,
      Europe . . . . *alpestris (typica !)*
    Darker, more streaked, eyebrows and throat sometimes white.
      Back blacker, nape paler ; wing 4.00. Bred in interior of Oregon, Wash., Brit. Col. . . . . *merrilli*
      Back " yellower, greenish tinged," nape darker ; wing 3.90. Bred on coast of Oregon, Wash., Brit. Col. *strigata*
Back reddish.
  Color of nape in marked contrast with back; wing 3.90. Bred on coast of Cala. and in Mexico . . . . *chrysolœma*
  Color of nape merging in that of back.
    General appearance rich rufous ; wing 3.90. Bred in Sacramento and San Joaquin Valleys, Cala. . . . . *rubea*
    General appearance " pallid and scorched ; " wing 4.05. Bred in W. Tex., New Mex., and S. Ariz. . . . *adusta*

O alpes'tris. (Lat. *alpestris*, alpine. Figs. 339, 340.) HORNED LARK. SHORE LARK. Adult ♂ ♀, in breeding plumage: Upper parts in general pinkish-brown, this pinkish or vinaceous or lilaceous tint brightest on nape, lesser wing-coverts, and tail-coverts ; rest of

FIG. 339. — Shore Lark, much reduced. (From Tenney, after Baird.)
    FIG. 340. — Shore Lark, nat. size. (Ad. nat. del. E. C.)

upper parts duller and more grayish-brown, boldly variegated with dark brown streaks ; middle pair of tail-feathers and several inner secondaries rufous-brown, with darker centres. Under parts, from breast backward, white ; sides strongly washed with color of upper parts, and mottling of same across lower part of breast. A large, distinct, shield-shaped black area on breast. Tail-feathers, except middle pair, black ; outermost edged with whitish. Wing-quills, except innermost, plain fuscous ; outer web of 1st primary whitish. Lesser wing-coverts usually tipped with grayish-white. Top of head like nape ; bar across front of vertex, thence extended along sides of crown, and produced into a tuft or " horn," black ; front and line over eye, also somewhat produced to form part of the tuft, sulphur-yellowish ; a broad bar from nostrils along lores, thence curving below eye and widening as it descends in front of auriculars, black; remainder of sides of head and whole throat sulphury-yellow. Bill plumbeous-blackish, bluish-plumbeous at base below (sometimes there yellowish); feet and claws black; iris brown. Length of ♂ 7.00–7.50; extent 13.00–14.00; wing 4.25–4.50; tail 2.75–3.00; bill, from extreme base of culmen, 0.40–0.50; tarsus 0.88–0.90; middle toe and claw rather less; hind claw about 0.50, usually longer than its digit, but very variable. ♀ duller and smaller than ♂; length 6.75–7.25; extent 12.75–13.25; wing about 4.00, etc. Adult ♂ ♀, in winter: As usually seen in most of the United States in fall, winter, and early spring, differ from the above in more sordid coloration of the upper parts, which may be simply grayish-brown, heavily streaked with dusky, even on the crown, with little or none of the pinkish tints; and in lack or restriction of the black markings of the head and breast, or their being veiled with whitish tips of the individual feathers; nevertheless the sulphury tinge of the white parts about the head is usually very conspicuous. Fledglings have the upper parts dusky, mixed with some yellowish-brown, and sprinkled all over with whitish or light tawny dots, each feather having a terminal speck. Most of the wing- and tail-feathers have rusty, tawny, or whitish edging and tipping. The under parts are white, mottled with the colors of the upper parts along sides and across back; no traces of definite black markings about head and breast, nor any yellow tinge. Bill and feet pale or yellowish. This peculiar speckled stage is of brief duration; with an early autumnal change, a dress little if at all different from that of the adults in winter is acquired. Nesting of this species, or some of its subspecies, begins very early in April, or even in March, sometimes before the snow is gone, and frequently other broods are reared through the summer; nest of grasses, etc., sunken in the ground; eggs 3–5, 0.90–1.05 × 0.60–0.75, usually about 0.95–0.70, very variable in tone, but always profusely and heavily marked with brownish-gray or dark stone-gray upon a grayish or greenish-white ground; in some cases the whole surface nearly uniform. Northern Hemisphere at large; the typical form, identical with *alpestris* of Europe, etc., breeds beyond the U. S. in easterly parts of British America, as the region about Hudson's Bay, and abundantly in Labrador; also Greenland; common in flocks in the E. U. S. in winter S. to the Carolinas and Illinois, or about lat. 35°; replaced in the West by the following varieties:

O. a. pratic′ola. (Lat. an inhabitant of *pratum*, a meadow; *colo*, I inhabit or cultivate. Fig. 341.) PRAIRIE HORNED LARK. First and least departure from *alpestris typica;* nape, rump, and lesser wing-coverts vinaceous, as before, but not so dark, and back flat gray, in contrast; yellow of throat pale, reduced or even wanting; white over eye; smaller; ♂ wing 4.20 or less. Breeding range along the northern tier of States, from the valley of the Red River of the North in E. Dakota, Upper Mississippi Valley, region of the Great Lakes, and adjoining British Provinces, to New York, New Hampshire, Vermont, and Massachusetts; in migration

FIG. 341. — Prairie Horned Lark. (L. A. Fuertes.)

S. to S. Car. and Texas; mixed with *alpestris* proper in winter, and not separated therefrom in former eds. of the Key. *O. a. praticola* HENSH. Auk, July, 1884, p. 264; DWIGHT, Auk, Apr. 1890, p. 144, area marked "3" on the map; A. O. U. Lists, 1886 and 1895, No. 474 *b*.

O. a. leucolæ'ma. (Gr. λευκός, *leukos*, white; λαιμός, *laimos*, throat.) PALLID HORNED LARK. Size of typical *alpestris*; ♂ wing 4.40, etc. General coloration extremely pale — brownish-gray, the peculiar pinkish tint of certain parts sharing the general pallor. Black markings on head and breast much restricted in extent, and white surroundings correspondingly increased — thus, the black post-frontal bar scarcely or not broader than the white of forehead. No yellow about head, excepting perhaps a slight tinge on chin. Changes of plumage parallel with those already given; even nestlings show the same decided pallor. Breeding range in the interior of British America, westerly, and Alaska; in migration scattering over western U. S., mixed with *arenicola* and other varieties. My *leucolæma* was based primarily (Birds N. W. 1874, p. 38) upon types I shot at Fort Randall, S. Dak., February, 1873, and extended to cover pallid birds I shot in N. Dakota in summer and fall of 1873, along the parallel of 49°, and included breeders of *arenicola*; but the name may conveniently be restricted to the present form, as was done by Mr. Henshaw. *O. a. hoyti* BISHOP, Auk, Apr. 1896, p. 130, is considered not sufficiently different from *leucolæma* by the A. O. U. Committee, Auk, Jan. 1897, p. 133.

O. a. arenic'ola. (Lat. *arena* or *harena*, sand, a sandy place, as the *arena* of a Roman amphitheatre was; hence a desert; *colo*, I inhabit or cultivate.) DESERT HORNED LARK. Smaller than *alpestris* proper; size of *praticola*; ♂ wing 4.20 or less. Coloration pallid, as in *leucolæma*, but with decided yellow on throat. Breeding range extensive in western parts of the U. S. on the Great Plains, in the Rocky Mountains and Great Basin, N. through N. Dakota and Montana beyond 49° to the region of the Saskatchewan; in migration S. into Mexico, scattered about and mixed with other varieties. *O. a. leucolæma, in part*, of previous eds. of the Key; *O. a. arenicola* HENSH. Auk, July, 1884, p. 265; DWIGHT, Auk, Apr. 1890, p. 146, area marked "5" on the map; A. O. U. Lists, 1st and 2d eds. 1886-95, No. 474 *c*.

O. a. giraud'i. (To J. P. Giraud, the writer on 16 species of Texas birds, etc.) GIRAUD'S HORNED LARK. TEXAN HORNED LARK. Smaller still than *leucolæma*; ♂ wing under 4.00. Pallid, like *leucolæma* and *arenicola*, the back gray with very indistinct streaks, but the throat bright yellow, and this color usually tingeing the breast also. A very local race, so far as known, confined to E. and S. E. Texas. *Alauda minor* GIRAUD, 1841, type examined; not in any previous ed. of the Key; *O. a. giraudi* HENSH. Auk, July, 1884, p. 266; DWIGHT, Auk, Apr. 1890, area marked "4" on the map; A. O. U. Lists, 1st and 2d eds. 1886-95, No. 474 *d*.

O. a. pal'lida. (Lat. *pallid*, pale, wan.) SONORAN HORNED LARK. Small as *giraudi*; ♂ wing 3.75. Pallid, like the three last, but with yellow on throat; resembling a miniature *arenicola*; back "very pallid," the whitish edging of the feathers evident. Another local race, supposed to be confined in the breeding season to Sonora and Lower California. DWIGHT, Auk, Apr. 1890, p. 154 (from MS. of C. H. TOWNSEND, pub. same year in Proc. U. S. Nat. Mus. xiii, p. 138), area marked "11." This ends the *pale* series, and we revert to *dark* birds like *alpestris*, with the following:

O. a. striga'ta. (Lat. *strigata*, marked with *strigæ*, streaks or stripes.) STREAKED HORNED LARK. Smaller than *alpestris* proper; ♂ wing 4.00 or rather less. Dark and streaky above, with nape, rump, and bend of wing reddish; more or less extensively yellowish below. Coast region of Oregon, Washington, and British Columbia; also, Santa Cruz Islands, off coast of S. Cala. HENSH. Auk, July, 1884, p. 267; DWIGHT, Auk, Apr. 1890, p. 151, area marked "9" on the map; A. O. U. Lists, 1st and 2d eds. 1886-95, No. 474 *g*. "This race has credit for more streaking and more yellow than it deserves. By rumpling the feathers of the back of almost any of the other forms a heavily streaked effect may be obtained, and the extreme yellowness below of the type specimens is not supported by the small series I have before me" (DWIGHT). Nests May-July; eggs 2-4, oftenest 3, 0.83 × 0.60, pale slate-gray, thickly speckled all over with greenish-brown and reddish-brown. Specimens from the Californian Islands have also been called *O. a. insularis* Towns. Pr. U. S. Nat. Mus. xiii, 1890, p. 140.

O. a. mer'rilli. (To Dr. J. C. Merrill, U. S. A.) MERRILL'S HORNED LARK. DUSKY HORNED LARK. Most like the last in size and color; ♂ wing 4.00; "more broadly streaked above and blacker than *strigata*, with less yellow about the head and throat, the nape pinker." Interior of Oregon, Washington, and British Columbia, between the Cascade range and the Rocky Mts., S. in winter to California and Nevada. *O. a. merrilli* DWIGHT, Auk, Apr. 1890, p. 153, area on the map marked "10"; A. O. U. List, 2d ed. 1895, No. 474 *i*. From these two *dark* forms we turn to three *reddish* ones, as follows:

O. a. chrysolæ ma. (Gr. χρύσεος, *chruseos*, golden; λαιμός, *laimos*, throat.) MEXICAN HORNED LARK. Smaller than *alpestris* proper: ♂ wing scarcely or not 4.00; a very small specimen, probably ♀, has the wing only 3.50; in another, marked ♂, it is 3.75. The pinkish tinge intensified into cinnamon-brown, and pervading all the upper parts except middle of back, which is contrasted with nape; yellow of head intensified, but breast white; black markings very heavy, — the black on the crown widened to occupy more than half the cap, reducing the white frontlet to a mere trace. Coast region of California, N. to Nicasio; coast region of Lower California, northerly; and most of Mexico. U. S. specimens which have been referred to *chrysolæma* belong mostly to *adusta* or *rubea*, both of which were included under *chrysolæma* in earlier eds. of the Key. *Alauda rufa* AUD., type examined, agrees with *rubea*, but habitat assigned includes other races; name unavailable also as antedated by *A. rufa* GM., 1788, which is our Titlark. See HENSH. Auk, July, 1884, p. 261; DWIGHT, Auk, Apr. 1890, p. 149, separate areas marked "7"; A. O. U. Lists, No. 474 *e*.

O. a. adus'ta. (Lat. scorched; *adurere*, to burn, parch, scorch.) SCORCHED HORNED LARK. "Similar to *chrysolæma*, but of a uniform scorched pinkish or vinaceous-cinnamon above," without contrast of color between nape and middle of back; lower parts creamy white, reddish-tinged; ♂ wing about 4.00. S. W. U. S. in W. Texas, New Mexico, Arizona, and probably parts of Utah, Nevada, and S. Cala.; S. into Mexico. *E. chrysolæma* of American writers, for the most part, and of previous eds. of the Key, in part. *O. a. adusta* DWIGHT, Auk, Apr. 1890, p. 148, area on map marked "6"; A. O. U. List, 2d ed. 1895, No. 474 *h*.

O. a. ru'bea. (Lat. red.) RUDDY HORNED LARK. "Bright rufous suffusing the whole plumage and merging into the ruddy brown of the back without abrupt change, distinguishes this race from *chrysolæma*;" ♂ wing 3.90. "General color above, deep cinnamon or ferruginous; throat bright yellow; streaks on dorsum nearly obsolete." An extremely local race, of the Sacramento and San Joaquin Valleys in California. HENSH. Auk, July, 1884, p. 267; DWIGHT, Auk, Apr. 1890, p. 150, small area on map marked "8"; A. O. U. List, 2d ed. 1895, No. 474*f*.

ALAU'DA. (Lat. *alauda*, a lark; supposed Celtic *al*, high, and *aud*, song.) SKYLARKS. Primaries 10; spurious 1st primary minute but evident. Head subcrested, but without lateral ear-tufts. Wings long, pointed, the tip formed by first 3 developed primaries; inner secondaries long and flowing. Tail emarginate, little more than half as long as wing. Tarsus equal to middle toe and claw. Lateral toes of unequal lengths. Sexes alike. Nest on the ground. Eggs 3–6, thickly speckled.

A. arven'sis. (Lat. *arvensis*, relating to arable land; *arvum*, a ploughed field.) SKYLARK. Upper parts grayish-brown, the feathers with darker centres; under parts whitish, tinged with buff across breast and along sides, where streaked with dusky; a pale superciliary line; wings with much whitish edging; outer tail-feather mostly white; next one or two with white borders. Length of ♂ 7.50; extent 14.75; wing about 4.00; tail 2.50; bill 0.50; tarsus or middle toe and claw 1.00; hind toe 0.45, its claw up to nearly 1.00. ♀ smaller. Eggs 0.90 × 0.60. This celebrated bird, whose music so often inspires the poet, occurs as a straggler from Europe in Greenland and Bermuda. It has repeatedly been imported and turned out in this country, where it may become naturalized, as it seems to be already established in some localities on Long Island, N. Y. The closely related Kamtschatkan Skylark (*A. blakistoni*) may occur in Alaska.

Suborder PASSERES MESOMYODI, OR CLAMATORES:
Non-melodious or Songless Passeres.

Mesomyodian scutelliplantar Passeres with ten fully developed primaries. — Syrinx with fewer than four distinct pairs of intrinsic muscles, inserted at *middle* of upper bronchial half-rings, representing the mesomyodian type of voice-organ, and constituting an uncomplicated and ineffective musical apparatus. (The word *mesomyodian* is from the Gr. μέσος, *mesos*, middle, and μυώδης, *muodes*, muscular, referring to such insertion of the syringeal muscles into the middle of the upper bronchial cartilages, not at their ends as in *Oscines* or *Acromyodi;* Gr. ἄκρος, *akros*, at the tip or end, apical.) Side and back of tarsus, as well as the front, covered with variously arranged scutella, so that there is no sharp undivided ridge behind (as, *e. g.* in fig. 344, *a*). Ten fully developed primaries, the 1st of which, if not equalling or exceeding the 2d, is at least two-thirds as long. (See p. 246, where the *Oscines* are defined as acromyodian laminiplantar *Passeres* with 9 fully-developed primaries, or 10 and the 1st short or spurious.)

In most *Mesomyodi* or *Clamatores*, the lower end of the trachea itself, aside from its muscles, undergoes no further modification from an ordinary Passerine type; and the birds which are thus not further affected in their windpipes form a group called *Oligomyodæ* (Gr. ὀλίγος, *oligos*, few, as the syringeal muscles are). All our mesomyodian birds are also oligomyodian; there are also several extralimital families, as the *Oxyrhamphidæ*, *Pipridæ*, and *Phytotomidæ* of Neotropical America, and the Old World *Philepittidæ*, *Pittidæ*, *Xenicidæ*, and *Eurylæmidæ*. Again, some mesomyodian birds exhibit a further modification of the same organ, which affects the structure of the lower end of the trachea itself; and such are called *Tracheophonæ*. These tracheophonous birds are all Neotropical; they form the four families, *Dendrocolaptidæ*, *Formicariidæ*, *Conopophagidæ*, and *Pteroptochidæ*.

The essential character of *Passeres Mesomyodi* or *Clamatores*, as distinguished from *Passeres Acromyodi* or *Oscines*, is thus seen to be an anatomical one, consisting in non-development of a singing apparatus; the vocal muscles of the lower larynx (*syrinx*) being small and few, or else forming simply a fleshy mass, not separated into particular muscles; in either case inserted in a special manner into the bronchial half-rings; in the case of oligomyodian forms without further modification of the trachea itself, such as occurs in tracheophonous forms. This character, though subject to some difficulty of determination, corresponds well with the principal external character assignable to the whole suborder, — namely, a certain condition of the tarsal envelop rarely if ever seen in higher *Passeres*. If the leg of a Kingbird, for example, be closely examined, it will be seen covered with a row of scutella forming cylindrical plates continuously enveloping the tarsus like a segmented scroll, and showing on its postero-internal face a deep groove where the edges of the envelop come together; this groove widening into a naked space above, partially filled in behind with a row of small plates. Such a tarsus is called *exaspidean* (Gr. ἐξ, *ex*, outside, and ἀσπις, *aspis*, a shield, scute, plate); it characterizes the whole family *Tyrannidæ*. When the arrangement of the scrolls is reversed, so that they lap round the inner side of the tarsus, it is called *endaspidean* (Gr. ἐνδο-, *endo-*, within, inside); it is shown by the South American *Dendrocolaptidæ*, for example. When the whole back side of the tarsus is broken up into many little close-set scutella, the formation is termed *pycnaspidean* (Gr. πυκνος, *puknos*, close, firm, compact); *Cotingidæ* show this feature. In some rare cases, as *Philepittidæ*, the plantar laminæ are rectangular, in regular series, giving the *taxaspidean* arrangement (Gr. τάξις, *taxis*, a rank or row). As a rare anomaly in this suborder, *Pittidæ* and *Xenicidæ* have ochreate or booted tarsi, as if they belonged to the highest

Oscines, though they are in fact oligomyodian *Clamatores*. With such modifications, and the exceptions noted, this *scutelliplantar* condition marks all *Clamatorial* birds, and is something tangibly different from the typical *Oscine* or laminiplantar character of tarsus, which consists in the presence on the sides of entire corneous laminæ meeting behind in a sharp ridge. And even when, as in cases of the oscine *Otocorys* and *Ampelis*, there is extensive subdivision of laminæ on the sides or behind, the arrangement does not exactly answer to the above description.

The *Clamatores*, especially the *Tracheophones*, represent the lower *Passeres*, approaching the large order *Picariæ* (see beyond) in the steps by which they recede from *Oscines*, yet well separated from Picarian birds. Of the families composing the suborder, as above named, only one occurs in North America, north of Mexico, to any considerable extent, but another (*Cotingidæ*) is represented on our southern border by at least one species ascertained to occur in Arizona, and I describe others beyond.

Analysis of North American Families.

Tarsus exaspidean . . . . TYRANNIDÆ
Tarsus pycnaspidean . . . . COTINGIDÆ

Family TYRANNIDÆ: American Flycatchers.

While having a close general resemblance to some of the foregoing insectivorous and oscine *Passeres*, the North American *Tyrannidæ* will be instantly distinguished by the above-described condition of the tarsus, together with the presence of 10 primaries, whereof the 1st is long or longest; and from birds of the following Picarian order by the Passerine characters of 12 rectrices, greater wing-coverts not more than half as long as secondaries, and hind claw not smaller than middle claw.

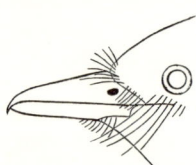

FIG. 342.— Bill of a Flycatcher. (*Tyrannus verticalis*, nat. size.)

This family is peculiar to America; it is one of the most extensive and characteristic groups of its grade in the New World, *Tanagridæ* and *Trochilidæ* alone approaching it in these respects. There are over 400 current species, distributed among about 100 genera and subgenera. Only a small fragment of the family is represented within our limits, giving but a vague idea of the numerous and singularly diversified forms abounding in tropical America. Some of these grade so closely toward other families, that strict definition of *Tyrannidæ* becomes extremely difficult; and I am not prepared to offer a satisfactory diagnosis of the whole group. Our species, however, are closely related to each other, and may readily be defined in a manner answering the requirements of the present volume. With a possible exception, not necessary to insist upon in this connection, they belong to the

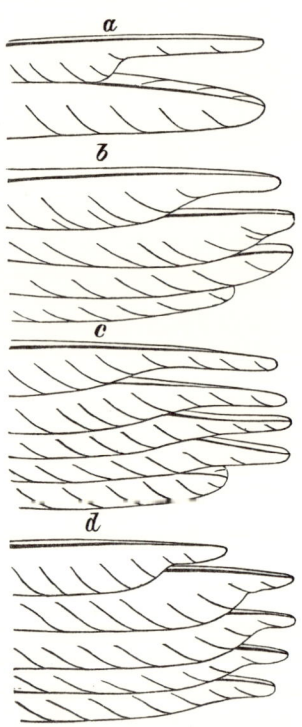

FIG. 343.— Emargination of primaries in *Tyranninæ*. *a. Milvulus forficatus; b. Tyrannus tyrannus; c. Tyrannus verticalis; d. Tyrannus vociferans;* all nat. size. (Ad. nat. del. E. C.)

Subfamily TYRANNINÆ: True Tyrant Flycatchers,

presenting the following characters: Wing of 10 primaries; 1st never spurious nor very short; one or more frequently emarginate or attenuate on inner web near end. Tail of 12 rectrices, usually nearly even, sometimes deeply forficate. Feet small, weak, exclusively fitted for perching; tarsus little if any longer than middle toe and claw; anterior toes, especially the outer, extensively coherent at base. Bill very broad and more or less depressed at base, tapering to a fine point, thus presenting a more or less perfectly triangular outline when viewed from above; tip abruptly deflected and usually plainly notched just behind the bend; culmen smooth and rounded transversely, straight or nearly so lengthwise, except toward end; commissure straight (or slightly curved) except at end; gonys long, flat, not keeled. Nostrils small, circular, strictly basal, overhung but not concealed by bristles. Mouth capacious, its roof somewhat excavated; rictus ample and deeply-cleft; commissural point almost beneath anterior border of eye. Rictus beset with a number of long stiff vibrissæ, sometimes reaching nearly to end of bill; generally shorter, and flaring outward on each side; other bristles or bristle-tipped feathers about base of bill. Bill very light, giving a resonant sound in dried specimens when tapped. On being broken open, the upper mandible will be found extensively hollow. These several peculiarities of the bill (to most of which *Ornithion* offers signal exception) are the most obvious features of the group; and should prevent our small olivaceous Flycatchers from being confounded even by the tyro with insectivorous Oscines, as Warblers and Vireos. (See Figs. 342, 344.)

The structure of the bill is admirably adapted for the capture of winged insects; broad and deeply fissured mandibles form a capacious mouth, while long bristles are of service in entangling the creatures in a trap and restraining their struggles to escape. The shape of the wings and tail confers the power of rapid and varied aërial evolutions necessary for successful pursuit of active flying insects. A little practice in field ornithology will enable one to recognize Flycatchers from their habit of perching in wait for their prey upon some prominent outpost, in a peculiar attitude, with wings and tail drooped and vibrating in readiness for instant action; and of dashing into the air, seizing the passing insect with a quick movement and a click of the bill, and then returning to their stand. Although certain Oscines have somewhat the same habit, these pursue insects from place to place, instead of perching in wait at a particular spot, and their forays are not made with such admirable *élan*. Dependent entirely upon insect food, Flycatchers are necessarily migratory in our latitudes; they appear with great regularity in spring, and depart on the approach of cold weather in fall. They are distributed over temperate North America; many are common birds of the Eastern States.

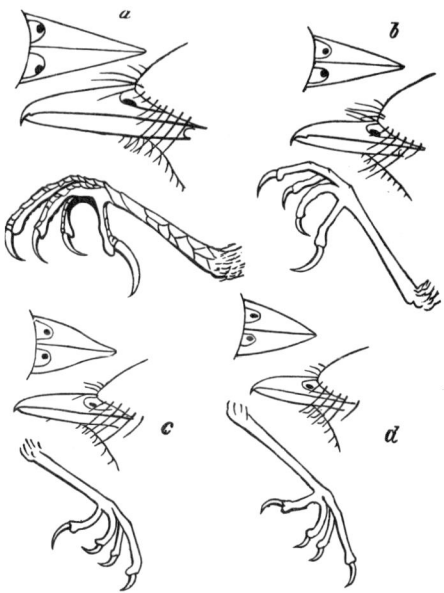

FIG. 344. — Generic details of *Tyranninæ*. a. *Myiarchus;* b. *Sayornis;* c. *Contopus;* d. *Empidonax;* all nat. size. (Ad. nat. del. E. C.)

The voice, susceptible of little modulation, is usually harsh and strident, though some species have no unmusical whistle or twitter. The sexes are not ordinarily distinguishable (remarkable exception in *Pyrocephalus*), and changes of plumage with age and season are not ordinarily great. The modes of nesting are too various to be collectively noted. The larger kinds of Flycatchers are unmistakable, but several of the smaller species, of the genera *Sayornis, Contopus,* and especially *Empidonax,* look much alike, and their discrimination becomes a matter of much tact and diligence.

512   SYSTEMATIC SYNOPSIS. — PASSERES — CLAMATORES.

To the genera of *Tyrannidæ* long known to be North American have been added four from Mexico — the immense-billed *Pitangus;* the short-billed *Myiozetetes* with flaming crown-spot and yellow under parts; the streaky, yellow-bellied, rufous-tailed *Myiodynastes;* and the curious little " beardless " *Ornithion;* while *Mitrephanes* has been merged in *Empidonax*. The 11 genera may be readily discriminated by the following characters:

Analysis of Genera.

Bill flattish, fully bristled and hooked as usual in *Tyrannidæ*.
  One or more outer primaries attenuate at end. A flame or yellow spot on crown.
    Tail deeply forficate, much longer than wings . . . . *Milvulus*
    Tail simple, not longer than wings . . . . *Tyrannus*
  Outer primaries not attenuate. A yellow orange crown-spot. Belly yellow.
    Wings and tail extensively rufous; no streaks except on head . . . . *Pitangus*
    Wings and tail without any rufous; no streaks except on head . . . . *Myiozetetes*
    Tail but not wings extensively rufous. Streaked above and below . . . . *Myiodynastes*
  Outer primaries not attenuate. Tail moderate. No yellow spot on crown.
    Tail chestnut and dusky, in lengthwise pattern. Belly yellow; throat ashy . . . . *Myiarchus*
    Tail without chestnut.
      Tail about equal to or little shorter than wing, slightly or not forked. Bill narrow. Tarsus not shorter or rather longer than middle toe and claw. Coloration black and white, cinnamon-brown, or olivaceous . . . . *Sayornis*
      Tail decidedly shorter than wing, a little forked. Bill broad and flat. Tarsus shorter or not than middle toe and claw. Olivaceous; length 6.25 or more . . . . *Contopus*
      Tail a little shorter than wing, about even. Bill flat. Tarsus not shorter or rather longer than middle toe and claw. Coloration olivaceous and yellowish, but no red, buff or pure brown. Length 6.25 or less — usually under 6.00 . . . . *Empidonax*
      Tail and tarsus as in *Empidonax*. Bill narrow. Hind not longer than lateral toe. Sexes unlike. ♂ full-crested, vermilion and pure brown . . . . *Pyrocephalus*
  Bill compressed, quite parine in appearance, unbristled, unnotched. General color ashy, with yellow lining of wings. Very small: length under 5.00 . . . . *Ornithion*

MIL′VULUS. (Lat. *milvulus*, diminutive of *milvus*, a kite.) SWALLOW-TAILED FLYCATCHERS. Tail in adult deeply forficate, about twice as long as wing. Outer primary or primaries abruptly attenuate, and other characters as in *Tyrannus* proper (beyond). A yellow or flaming crown-spot.

Analysis of Species.

Three or four primaries emarginate. Crown-spot yellow, in black cap . . . . *tyrannus*
One primary emarginate. Crown-spot flaming, in ashy cap . . . . *forficatus*

M. tyran′nus. (Lat. *tyrannus*, a tyrant.) FORK-TAILED FLYCATCHER. Adult ♂ ♀: Outer 3 or 4 primaries emarginate. Crown-patch yellow. Above clear ash; below, white including lining of wings; top and sides of head black; tail black, the outer feather white on outer web for about half its length; wings dusky, unmarked. ♀ duller. Young similar, but primaries not emarginate, nor tail lengthened; no crown-spot; wing- and tail-coverts edged with brown. Wing 4.50; tail up to a foot long, forked 6-8 inches. A beautiful bird of Central and most of S. Am., accidental in the U. S. in four recorded instances (Mississippi, Kentucky, New Jersey, and southern California).

M. forfica′tus. (Lat. *forficatus*, forked like *forfex*, a pair of scissors. Fig. 345.) SWALLOW-TAILED FLYCATCHER. SCISSOR-TAIL. TEXAN BIRD-OF-PARADISE. Adult ♂ ♀: First primary alone emarginate (fig. 343 a). Crown-patch orange or scarlet. General color hoary-ash, paler or white below; sides at insertion of wings and lining of these, scarlet or bloody-red; other parts of body variously tinged with the same, or a paler salmon-red or pink. Wings blackish, with whitish edgings. Tail black, but several of the long feathers extensively white or rosy; these are narrow and linear, sometimes widening somewhat in spoon-shape. Wing 4.50-5.00; extent of wings 14.50-15.50; tail up to a foot long, usually 8.00-10.00 inches, forked 5.00-6.00. ♀ averaging smaller than ♂, with tail commonly less developed. Young:

Similar; primary not abruptly emarginate; tail undeveloped; no crown-spot, and little or no red. Central America, Mexico, and in the U. S., the Lower Mississippi valley, and Texas; usually N. to Indian Territory and Kansas, even Missouri; E. to Louisiana; accidental in Illinois, Florida, Virginia, New Jersey, New England, and Manitoba, at Hudson's Bay, and in the Mackenzie River valley! A most elegant, graceful, and showy bird, abundant in Texas, conspicuous by the display it makes in opening and closing the tail, like scissor-blades; very active, dashing and noisy, like a Kingbird, — all the large Flycatchers sharing this same impetuous, irritable disposition. It makes a very good sort of a "Bird-of-Paradise" to the average apprehension of a Texan. Breeds throughout its regular U. S. range. Nesting like the Kingbird's; eggs 4-6, usually 5, white, boldly blotched with reddish and darker browns on the surface, and lilac shell-spots; size averaging 0.90 × 0.66, but length ranging from 0.80 to 0.95; they are mostly laid in May, but may be taken from April to July.

FIG. 345. — Swallow-tailed Flycatcher. (Sheppard del. Nichols sc.)

TYRAN′NUS. (Lat. *tyrannus*, a tyrant.) KING FLYCATCHERS. Tail moderate in size and shape, rather shorter than wing, even or little rounded, emarginate or lightly forked. Wings long, pointed by 2d–3d quills, 1st and 4th little if any shorter, 5th and rest rapidly graduated. Several outer primaries abruptly emarginate or sinuate-narrowed on inner webs toward end. Bill stout, flattish, fully bristled, notched, and hooked (fig. 342). Feet small and weak; tarsus with scales obviously lapping around. Size large; length 8.00 or more; wing over 4.00. Sexes alike; ♀ sharing the flaming crown-patch. Young lacking crown-spot and attenuation of primaries. Nest bulky, on a bough, compactly woven and felted. Eggs white, boldly marked with oval or tear-shaped spots of reddish-brown, etc. Contains numerous species, 5 of N. Am., which have been divided into several named subgenera, but are closely interrelated through various exotic species. They are the Kingbirds proper.

Analysis of Species.

No olive nor decided yellow; blackish and whitish.
 Only two primaries obviously emarginate. Tail about even, conspicuously white-tipped. Bill small, under 1.00. (*Tyrannus*) . . . . *tyrannus*
 Five or six primaries emarginate. Tail emarginate, merely lighter at end. Bill big, 1.00 long. (*Melittarchus*)
                                    dominicensis
Olivaceous, with pure yellow on belly, ashy on head. Bill moderate. (*Laphyctes*.)
 Tail blackish, merely emarginate; wings dark brown.
  Several outer primaries gradually attenuate for a long distance. Outer web of outer tail-feather white
                                      verticalis
  Several outer primaries abruptly emarginate for a short distance. Outer web of outer tail-feather merely whitish-edged . . . . *vociferans*
 Tail dark brown, like wings, obviously forked.
  Several outer primaries abruptly emarginate for a short distance . . . . *melancholicus couchi*

T. tyran′nus. (Fig. 346.) KINGBIRD. BEE-BIRD. BEE-MARTIN. Adult ♂ ♀: No olive nor decided yellow. Only two outer primaries obviously emarginate (fig. 343, *b*). Tail nearly even — if anything a little rounded. Blackish-ash, still darker or quite black on head, crown with a flaming spot. Below, pure white, the breast shaded with plumbeous. Wings dusky, with much whitish edging. Tail black, broadly and sharply tipped with white, the outer feather sometimes edged with the same. Bill and feet black. Young: Lacking emargination of primaries, and no crown-spot; very young birds show rufous edging of wings and tail. Length about 8.00; extent 14.50; wing 4.50; tail 3.50, even or slightly rounded; bill small, under 1.00. Temperate N. Am., but chiefly U. S. to Rocky Mts.; rare or casual on the Pacific slope; N. in the interior to Saskatchewan and Athabascan regions, about lat. 57°; abundant in summer; migratory mainly in April and September; breeds throughout its range; winters on the southern border and beyond in some of the West Indies, and through Central Am. and S. Am. to Bolivia. This trim and shapely "martinet," in severe black and white but with fiery pompon, is familiar to all, and equally noted for irritability, pugnacity, intrepidity, and its inveterate enmity to Crows, Hawks, and Owls, which it does not hesitate to attack, either in defence of its nest or just to show its spunk; but in its turn it is attacked and sometimes worsted by the Hummingbird. Nest a conspicuous object in the orchard or by the wayside, on the horizontal bough of a tree, large, cupped, compactly woven and matted with fibrous and disintegrated vegetable substances; eggs 3-5, usually 3 or 4, rarely more, 0.85 to 1.05 long, averaging 0.95 × 0.72, white, rosy, or creamy, variously spotted or blotched in bold (often beautiful) pattern with reddish and darker brown surface-spots and lilac shell-markings. They are laid mostly in June, but in different latitudes are found also in May and July. The Kingbird destroys a thousand noxious insects for every bee it eats! (*T. carolinensis* of all previous eds. of the Key.)

Fig. 346. — Kingbird.

T. dominicen'sis. (Of St. Domingo.) GRAY KINGBIRD. Adult ♂ ♀: Five or six outer primaries usually emarginate. Crown-spot as before. Grayish-plumbeous, rather darker on head; auriculars dusky. Below, white, shaded with ashy on breast and sides; under wing- and tail-coverts faintly yellowish; wings and tail dusky, edged with whitish or yellowish; tail-feathers merely indistinctly lighter at extreme tip. Larger than the last: Length about 9.00; wing 4.50; tail nearly 4.00, more or less emarginate; bill 1.00, very turgid. West Indies; Florida regularly, in abundance; N. to S. Carolina rarely, to Massachusetts accidentally; has even been found **astray** in British Columbia! General appearance, habits, and nesting of the Kingbird; eggs indistinguishable, averaging a trifle larger, 1.00×0.73, 3 or 4 in number, sometimes 5, in the U. S. laid in May and June, but much earlier in the West Indies.

T. vertica'lis. (Lat. *verticalis*, relating to the *vertex*, or top of head, which has a flame-patch. Fig. 342.) ARKANSAS TYRANT FLYCATCHER. WESTERN KINGBIRD. Several outer primaries gradually attenuated for a long distance (fig. 343, *c*). Adult ♂ ♀ : Coloration olivaceous and yellow; belly and under wing- and tail-coverts clear yellow; back ashy-olive, changing to clear ash on head, throat, and breast, chin whitening, lores and auriculars dusky; wings dark brown with whitish edging; tail black or blackish; bill and feet black; iris brown. Outer web of outer tail-feather entirely white. Ash of fore parts pale, contrasting with dusky lores and auriculars, fading insensibly into white on chin, and changing gradually to yellow on belly; olive predominating over ashy on back. Length about 9.00; extent about 16.50; wing 5.00; tail 4.00; bill 0.75; tarsus 0.75. Young: Similar; general ash of body dull, with a brownish cast; little or no olivaceous on back; tail not quite black; yellow of under parts pale and sulphury, even whitish; bill light-colored at base below; no color on crown, and primaries scarcely or not attenuate. Very young with rusty edgings, especially on wing- and tail-coverts. Western U. S. and adjoining British Provinces, from the Great Plains to the Pacific, abundant; accidental in Louisiana, Maryland, New Jersey, New York, and Maine; E. regularly to Kansas, Nebraska, Iowa, and Minnesota; N. to British Provinces in the Mouse and Milk River regions and westward; S. in winter in Lower California, Mexico, and Guatemala; migratory; breeds throughout its U. S. and Brit. Am. range, but winters nearly or quite extralimital. General traits those of the King-bird; nest similar, rather larger, with more fluffy and less fibrous material, but very variable in size, shape, make, and position, usually in a tree; eggs 3–5, usually 4, not distinguishable with certainty from those of the common Kingbird, averaging a trifle smaller, under 0.95×0.70.

T. voci'ferans. (Lat. *vociferans*, vociferous, voice-bearing; *vox*, voice, and *fero*, I bear.) CASSIN'S TYRANT FLYCATCHER. Several outer primaries abruptly emarginate for a short distance (fig. 343, *d*). Adult ♂ ♀ : Outer web of outer tail-feather barely or not edged with whitish. General coloration as in *verticalis;* but ash of fore parts dark, little different on lores and auriculars, changing rather abruptly to white on chin and to yellow on belly; ashy predominating over olive on back. The difference is decisive on comparison. The outer primaries are abruptly nicked and narrowed within half an inch of the end. The mere edging of the outer tail-feather with white instead of the whole web being white is also a good character. Changes of plumage the same as in *verticalis;* size the same; bill rather stouter, about 0.85; tarsus slightly longer, on an average. Southwestern U. S., and southward to Guatemala; N. to Wyoming and Idaho, even to Oregon; abundant in the S. Rocky Mt. region, there mostly replacing *verticalis* in the breeding season, and also on the Pacific slopes in California, but rare in the Great Basin; breeds throughout its U. S. range, and resident in some parts. Nesting and eggs like those of the foregoing; nest usually on horizontal bough of a tree at considerable height, bulky, rather flattish, about 8.00 across outside by 3 deep, with a cup 3.50×2.00; eggs 2–5, usually 3 or 4, indistinguishable from those of other Tyrants, averaging 0.95 by 0.70, mostly laid in June, but sometimes earlier.

T. melancho'licus couch'i. (Gr. μελαγχολικός, *melagcholikos*, Lat. *melancholicus*, melancholy, *i. e.*, atrabilious; μέλας, μέλανος, *melas, melanos*, black; χόλος, *cholos*, gall, bile. To Lieut. D. N. Couch.) COUCH'S TYRANT FLYCATCHER. Very similar to the last; primaries abruptly emarginate for a short distance, as in *vociferans*, and outer web of outer tail-feather not white; but tail dark brown, like wings, and obviously forked (about 0.50; in *vociferans* tail quite black, slightly emarginate or nearly even); all its feathers with slight pale edges, and their shafts pale on under surface. Yellow of under parts very bright, reaching high up on breast; throat as well as chin extensively white. Size of the foregoing, and changes of plumage coincident. T. melancholicus is a universally distributed South and Central Am. species, of which this northern subspecies reaches over our Mexican border in the valley of the Lower Rio Grande of Texas, where it is common in some localities, and breeds. Nest in trees at a moderate height, of twigs, Spanish moss, rootlets, etc., outside 6.00 × 2.50, inside 3.00 × 1.50; eggs 3–4, 0.95 × 0.72, indistinguishable from those of the foregoing, laid usually in May in Texas.

PITAN'GUS. (*Vox barb.;* a Mexican or S. Am. name of some bird.) DERBY FLYCATCHERS. Outer primaries not emarginate. An orange crown-patch. Bill as long as head, exceeding tarsus, straight, stout, but narrow, as deep as broad at nostrils, with rigid culmen straight to the hooked end; gonys about straight, ascending, commissure and also lateral outlines perfectly straight. Nostrils rounded, nearer commissure than culmen. Wings rounded, tipped by 3d–5th quills; 2d and 6th about equal and shorter, 1st only about equal to 9th. Tail shorter than wings, nearly even, but somewhat double-rounded. Tarsus about as long as middle toe and claw. Largest-bodied of any N. Am. Flycatcher. Brown above, yellow below, with black, white, and orange head; quills and tail-feathers extensively chestnut, as in *Myiarchus*.

P. derbia'nus. (To Lord Derby, 13th earl of that name, many years president of the Zoölogical Society of London as Lord Stanley, proprietor of the Knowsley Menagerie, died 1850. Fig. 347.) DERBY FLYCATCHER. BULL-HEADED FLYCATCHER. Upper parts light woodbrown, with an olive tinge; wings and tail the same, but the feathers extensively bordered without and within with chestnut, forming a conspicuous continuous area on wing-quills in closed wing, and on most wing- and tail-feathers more extensive than brown portion of inner webs. Below from breast, including lining of wings, clear and continuous lemon-yellow. Chin and throat pure white, widening behind up under ear-coverts. Top and sides of head black, a circle of white from forehead over eyes to nape, the enclosed black enclosing a lemon and orange patch. Or, middle of crown yellow and orange, enclosed and partly concealed in

FIG. 347.—Derby Flycatcher, nat. size. (Ad. nat. del. E. C.)

black, this black enclosed in white, then the long and broad black bar on side of head, separating white of side of crown from that of side of throat. The coronal feathers lengthened and erectile as in a Kingbird, or more so; crown-patch of same character but more extensive. Bill and feet black; iris hazel. Sexes alike. Length of ♂ about 10.50; wing about 5.00; tail about 4.00; bill 1.20; tarsus 1.00. A great Flycatcher of aggressive appearance, long known in Mexico and Central and S. Am., in 1878 ascertained to occur on the Lower Rio Grande in Texas, where it is common in some places, breeds in May and June, and is sometimes called the Bull-headed Flycatcher. Nest in trees at no great height, very large, thick-walled, roofed over with lateral entrance, composed of miscellaneous coarse materials well compacted with the finer lining; size nearly a foot across by half as much in depth, with comparatively small cavity; eggs 4–5, averaging 1.15 × 0.85, creamy white, sparingly speckled and splashed, chiefly about the larger end, with dark brown and neutral tints.

MYIOZETE'TES. (Gr. μυῖα, *muia*, a fly, and ζητητής, a seeker, inquirer.) INQUISITIVE FLYCATCHERS. Bill short, stout, very broad at base, with curved culmen, hooked and notched tip, and heavily bristled rictus; its length from nostril not half the length of tarsus. Primaries not emarginate; 2d–4th longest, 5th shorter, 1st about equal to 6th. Tail shorter than wings, nearly square; feet small; tarsus rather less than middle toe and claw. A widely distributed Neotropical genus, related to *Elainea*, with about 7 species, one of which is believed to reach our border in Texas, on the authority of J. P. Giraud, though nobody has found it there of late years. It has an orange crown-spot, as in *Pitangus, Myiodynastes, Tyrannus, Milvulus*, etc., but its relationships are elsewhere in the family.

M. texen´sis. (Lat. of Texas.) GIRAUD'S FLYCATCHER. TEXAN FLYCATCHER. Crown with concealed orange patch, as in a Kingbird. A conspicuous white superciliary stripe. Under parts, including lining of wings, yellow; but throat definitely white. Above, olivaceous, duller and grayish on head, dusky on lores and auriculars, hoary on forehead. Quills and tail-feathers fuscous, most of them with dull yellowish edging, but no clear rufous or chestnut. Bill and feet black. Young lack the crown-spot, and have some rusty edgings of the feathers, especially of wings and tail. Length 7.00 or less; wing 3.50; tail under 3.00; bill 0.60; tarsus 0.75; middle toe and claw 0.85. Texas, S. through Mexico to Central and S. Am. Nest like that of the Derby Flycatcher, domed; eggs said to be only 2 or 3 in number; they average 0.92×0.66, and are white, sparingly flecked all over with brown and neutral tints. This species was introduced formally in the text of the 2d ed. of the Key, 1884, p. 430; A. O. U. Lists, 1886 and 1895, No. [450.].

MYIODYNAS´TES. (Gr. μυῖα, *muia*, a fly; δυναστής, *dunastes*, a ruler.) STRIPED FLYCATCHERS. Related to *Myiarchus;* tail extensively chestnut, as in that genus, but no chestnut on wings. No primaries emarginate. A yellow crown-spot. Bill shorter than head, as long as tarsus, very turgid, much broader than high at nostrils, lateral outlines slightly convex, culmen nearly straight to little hooked tip; gonys long, ascending; rictus moderately bristled. Wings long and pointed; 3d quill slightly longer than 2d, 4th little shorter, 5th much shorter, 1st between 5th and 6th. Tail shorter than wings, nearly even. Feet very small, relatively as weak as in *Contopus;* tarsus rather shorter than middle toe and claw. Several species of Mexican and tropical American Flycatchers, with crown-spot, rufous tail, and whole plumage *streaked.*

M. luteiven´tris. (Lat. *luteus*, yellow, *ventris*, of *venter*, the belly.) SULPHUR-BELLIED STRIPED FLYCATCHER. Entire upper parts, including the head, streaked; the feathers with broadly dusky centres and olive-brown borders, finally edged slightly with yellowish-brown. A yellow crown-spot, concealed as in the Kingbird. Tail and its upper coverts rich chestnut, all the feathers with blackish shaft-stripes — on the middle feathers about half the width of either web, on the outer narrowed to the shaft itself and a slightly clubbed end; from below, shafts of the feathers white except at ends. Wings blackish: median and greater coverts and inner quills, both externally and internally, conspicuously edged with yellowish-white; some rufous edgings also on lesser coverts. Under parts, including lining of wings, sulphur-yellow, fading to white on the throat; everywhere, excepting on middle of belly and crissum, heavily streaked with blackish, these dark stripes suffused and blended on throat, particularly along its sides. Lores and auriculars dusky; forehead and streak over eye whitish. Bill blackish, pale at base below. Wing 4.40; tail 3.40; bill and tarsus 0.75; middle toe and claw rather more. Central Am. and Mexico to S. Arizona and S. New Mexico, common, breeding in mountainous regions at elevations of 5,000–7,000 feet. It was originally found within our limits, in the Chiricahua Mts. in 1874, and I think that I saw it at Fort Verde, 40 miles E. of Fort Whipple, in May, 1881. Nest in a hole of a tree, preferably a sycamore near running water, 20–50 feet up, mainly composed of walnut leaf-stems without special lining; eggs 2–3, 1.04–0.95×0.77–0.72, creamy white, heavily and profusely spotted and blotched or streaked with light and dark purplish-browns, thus resembling those of *Myiarchus;* laid in July and August.

MYIAR'CHUS. (Gr. μυῖα, *muia*, a fly; ἀρχός, *archos*, a ruler. Fig. 344, *a*.) CRESTED FLYCATCHERS. ASH-THROATED FLYCATCHERS. RUFOUS-TAILED FLYCATCHERS. No colored patch on crown, but head slightly crested by lengthened erectile feathers. Primaries emarginate. Olivaceous; more or less yellow below, throat ash, primaries margined with chestnut, tail-feathers the same or mostly chestnut — such coloration the best mark of the genus. Tail nearly even, if anything rounded, about as long as wings, of broad flat feathers with rounded ends. Wings rounded, tip formed by 2d–4th quills (usually), 5th shorter, 6th and 1st much shorter. Tarsus about as long as middle toe and claw, — if any different, longer. Bill moderate, variable in shape and relative size. Next to the characteristic rufous on wings and tail, *size* is a good clue to this genus among our olivaceous Flycatchers without colored crest; for the *Myiarchi* excepting *M. lawrencei* are much larger than any others excepting *Contopus borealis* and *C. pertinax*. Only one eastern species, but three others in the Southwest, and each of these with a subspecies, requiring nice discrimination. Peculiar, all of them, in nesting in holes, and laying heavily-colored eggs, scratched and snarled, but chiefly scrawled lengthwise, with dark brown, in close and intricate pattern.

Analysis of Species and Subspecies.

Large: Length 8.00 or more. Inner webs of tail-feathers extensively rufous. Bill subconical. (*Subgenus* MYIONAX.)
    Rufous occupying nearly or quite all of the inner webs of several lateral tail-feathers. Eastern N. Am. *crinitus*
    Rufous occupying inner webs of several lateral tail-feathers to nearly equal extent with a fuscous stripe of equal width throughout. S. W. U. S.
        Length 9.00 or less; wing and tail about 4.00; bill about 0.75. Texas . . . . *mexicanus*
        Length over 9.00; wing and tail over 4.00; bill nearly or quite 1.00. Arizona . . . . . . *m. magister*
    Rufous occupying inner webs of several lateral tail-feathers to greater extent than a fuscous stripe which widens at end.
        Outer web of outer tail-feather whitish. Western U. S. . . . . *cinerascens*
        Outer web of outer tail-feather not distinctly whitish. Arizona . . . . *c. nuttingi*
Small: Length about 7.00 or rather less. Inner webs of tail-feathers with little or no rufous. S. W. U. S. Bill flat. (MYIARCHUS proper.)
    Darker and smaller. Texas . . . . *lawrencei*
    Lighter and larger. Arizona . . . . *l. olivascens*

** This genus is unfortunate in having the names of our southwestern species and subspecies much confused and changed about by various writers: see synonymy under head of each, beyond. The several forms now given are the same as those of the 2d–4th eds. of the Key, with different names for *mexicanus* and its subspecies *magister*, and with an additional subspecies for *cinerascens* and *lawrencei*, respectively. My present nomenclature is strictly conformed to the A. O. U. List, with some misgiving in one or two instances.

M. crini'tus. (Lat. *crinitus*, haired, *i. e.* crested; *crinis*, hair. Fig. 348.) GREAT CRESTED FLYCATCHER. Adult ♂ ♀: Decidedly olivaceous above, a little browner on head, where the feathers have dark centres; throat and fore breast pure dark ash; rest of under parts bright yellow, the two colors meeting abruptly; primaries margined on both edges with chestnut; secondaries and coverts edged and tipped with yellowish-white; tail with all the feathers but the central pair chestnut on whole of inner web (excepting perhaps a very narrow space next the shaft); outer web of outer feathers edged with yellowish; middle feathers, outer webs of the rest, and wings except as stated, dusky-brown. The foregoing phrases are intended to be chiefly antithetical to those used in describing *cinerascens*, below. Other diagnostic points are: bill dark but not quite black, pale at base below, stout and comparatively short, hardly or not as long as tarsus, the latter perhaps never 0.90; olive back, ash throat, and yellow belly severally pure in color; all tail-feathers but middle pair so extensively rufous on inner webs that a mere line, if any, of fuscous persists next the shaft (compare *mexicanus* and *m. magister*), and this fuscous line, if any, running of same narrowness to ends of the feathers (compare *cinerascens*); never more than a trace of rufous on outer webs. Very young birds have rufous skirting of many feathers, in addition to the chestnut above described, but this soon disappears. Large: length 8.00-9.00; extent about 13.00; wing and tail about 4.00 (3.80-4.20); bill 0.75-0.80; tarsus 0.70-0.80; middle toe and claw 0.65-0.75; breadth of bill at base 0.33-0.40, or about ½ the length of culmen. Eastern U. S. and adjoining portions of Canada, west to Manitoba, Minnesota, Iowa, Missouri, Kansas, Nebraska, Arkansas, and about half of Texas; S. to Mexico, Central Am., and U. S. of Colombia in winter. Migratory, chiefly in April-May, and Aug.-Sept.; breeds throughout its N. Am. range; winters chiefly extralimital, but a few remain on our extreme southern border. An abundant bird, locally and irregularly distributed in woodland, of loud harsh voice and quarrelsome disposition, noted for its habitual use of cast-off snake-skins in the structure of its nest. Nest in hollows of trees and similar retreats, which are filled with trash of the most miscellaneous description, sometimes accumulated in astonishing bulk; eggs unique (outside this genus) in pattern: ground color buff or rich creamy, heavily overlaid with numberless markings of purplish-chestnut, or purplish-chocolate, and others paler, sharp and scratchy, mostly lengthwise, but especially at the butt tangled up; size about 0.87 × 0.67 on an average, ranging from 0.80 × 0.60 to 0.95 × 0.70; number 4-8, usually 5 or 6; laid in May and June.

Fig. 348. — Great Crested Flycatcher, reduced. (Sheppard del. Nichols sc.)

M. mexica′nus. (Lat. Mexican.) MEXICAN CRESTED FLYCATCHER. On comparing this bird with *crinitus*, it is immediately perceived to be different. The lateral tail-feathers have a stripe of fuscous on inner web adjoining shaft, this stripe equalling or exceeding width of whole outer web of the respective feathers, and being about half-and-half with the rufous, whereas in *crinitus* there is only the narrowest possible dusky stripe on inner web, or none at all. This dusky stripe is of uniform width throughout, not enlarged at the end to occupy most or all of the feather, as is the case with *cinerascens*. Entire upper parts darker than those of *crinitus* — that is, they have a sordid brownish-olive cast, instead of the clearer and purer greenish-olive of *crinitus*; yellow of belly much paler; ash of throat decidedly lighter and clearer, and coming farther down breast, yielding to yellow without intervention of the olivaceous pectoral area which is usually conspicuous in *crinitus*. The general aspect of the under parts is much as in *cinerascens*, both the distribution and shade of the colors being more as witnessed in the latter than as seen in *crinitus*. The light edgings of the wing-feathers are also paler than those of *crinitus*. The bill is black, not dark brown, slenderer than in *crinitus*; nor has it the very constricted shape of that of *cinerascens*. The general body-coloration is almost exactly as in *cinerascens*, from which it is at once distinguished by different shape of bill and different pattern of tail-feathers. Average length 8.75; extent about 12.75; wing 3.60-4.00; tail 3.75; bill 0.75; tarsus 0.85; middle toe and claw 0.75. Lower Rio Grande valley of Texas, and southward to Guatemala. Common, migratory, arriving in Texas early in April, and leaving in Sept., breeding in April and May. Nest and eggs like those of *crinitus*, said to average paler, but not distinguishable; number 4-6, usually 5; size ordinarily 0.88 × 0.69. This bird is now identified with the badly described *Tyrannula mexicana* KAUP, P. Z. S. 1851, p. 51; it is *M. mexicanus* DRESSER, Ibis, 1865, p. 473 (Texas), and LAWR. Ann. Lyc. N. Y. May, 1869, p. 202, but not *M. mexicanus* BD. B. N. A. 1858, p. 179 (which is *M. cinerascens*); A. O. U. Suppl. List, Auk, Jan. 1897, p. 127, No. 453); *M. mexicanus* RIDGW. Pr. U. S. Nat. Mus. ii, p. 14; Man. 1887, p. 333; *M. crinitus* var. *irritabilis* COUES, Pr. Phila. Acad. 1872, p. 65, in part, *nec Tyrannus* *irritabilis* VIEILL.; *M. crinitus* var. *cooperi* BD. BREW. and RIDGW. Hist. N. A. B. ii, 1874, p. 331, in part; *M. mexicanus* var. *cooperi* RIDGW. Pr. Nat. Mus. i, p. 138, in part, *nec M. cooperi* BD.; *M. erythrocercus* BREW. Ibis, 1878, p. 205 (Texas); *M. crinitus erythrocercus* COUES, Bull. U. S. Geol. Surv. iv, 1878, p. 32; v, 1879, p. 402; Key, 2d–4th eds. 1884–95, p. 435 (Texas).

M. m. magis′ter. (Lat. *magister*, a master, magistrate.) LARGE-BILLED CRESTED FLYCATCHER. ARIZONA CRESTED FLYCATCHER. Like the last; differing in greater size, especially of the bill, which runs from 0.80 to 1.10 in length of culmen, equalling or even exceeding the tarsus, which is itself 1.00, and thus fully 0.10 longer than in *mexicanus* proper; wing over 4.00; tail the same; total length 9.00 or more. The coloration of the tail-feathers is as in the stock-species, not as in *crinitus*, of which I formerly regarded both *magister* and *mexicanus* as subspecies. Southern Arizona, S. through western Mexico to Tehuantepec. In our country the bird is characteristic ot the region of the giant cactus, in holes in which, made by the Gila and other Woodpeckers, the nest is placed as a rule. Eggs 3–5, 1.00 × 0.70, like those of *mexicanus* in coloration, laid in May and June. (? *Tyrannula cooperi* KAUP, P. Z. S. 1851, p. 51. *Myiarchus cooperi* BAIRD, B. N. A. 1858, p. 180, and of most authors, wholly or in part. *M. crinitus* var. *cooperi* COUES, Pr. Phila. Acad. 1872, p. 67; Key 2d–4th eds. 1884–90, p. 435 (Arizona); BD. BREW. and RIDGW. Hist. N. A. B. ii, 1874, p. 331, in part (includes *mexicanus* proper). *M. mexicanus magister* RIDGW. Pr. Biol. Soc. Wash. ii, 1884, p. 90; Man. 1887, p. 333; A. O. U. Lists, 1886–95, No. 453 *a*. The name *cooperi* proves to be unavailable for this bird.)

M. cineras′cens. (Lat. *cinerascens*, ashy. Fig. 349.) ASH-THROATED CRESTED FLYCATCHER. Adult ♂ ♀: Rather olivaceous-brown above, quite brown on head; throat very pale ash, sometimes almost whitish, changing gradually to very pale yellow or yellowish-white on rest of under parts. Primaries edged as in *crinitus*, but secondaries and coverts edged with grayish-white. Tail-feathers as in *crinitus*, but rufous of inner webs hardly or not reaching their ends, being cut off from the tip by widening of the fuscous stripe (in young birds, in which the quills and tail-feathers are more extensively rufous-edged, the last distinction does not hold); outer web of outer tail-feather whitish. Size of *crinitus*, or rather less, 8.00–8.50; wing and tail about 4.00; but tarsi longer and bill slenderer; tarsus 0.80–0.90; bill 0.75–0.85, but only 0.27–0.33 broad at base, where only about as wide as high, and obviously narrower than in *crinitus*; though in Cape St. Lucas specimens (*M. pertinax* BD. Pr. Phila. Acad. 1859, p. 303) shaped quite as in *crinitus*, but smaller. Western U. S.; N. to Oregon, Wyoming, Colorado, Utah, and Nevada; S. through Mexico to Guatemala; W. from western Texas through New Mexico, Arizona, Southern and Lower California to the Pacific; said to winter in the Lower Colorado valley, U. S., but ordinarily comes over our border early in March, passes on in that month and April, and lays in May and June; nesting like others of the genus, and eggs indistinguishable, though averaging paler, with finer markings than those of *crinitus*; they number 3–6, usually 4, and measure on an average 0.87 × 0.65. Though so similar to the foregoing, it is a different bird from any of them. (*M. mexicanus* BD. B. N. A. 1858, p. 179, *nec* KAUP, 1851. *Tyrannula cinerascens* LAWR. Ann. Lyc. N. Y. 1851, p. 121. *M. cinerascens* SCL. and SALV. Ibis, 1859, p. 121; COUES, Pr. Phila. Acad. 1872, p. 69; Key, orig. ed. 1872, p. 171. *M. cinerescens* COUES, Key, 2d–4th eds. 1884–90,

FIG. 349. — Ash-throated Flycatcher, reduced. (Sheppard del. Nichols sc.)

p. 436. *M. crinitus* var. *cinerascens* RIDGW. in BD. BREW. and RIDGW. Hist. N. A. B. ii, 1874, p. 332.)

M. c. nut'tingi. (To C. C. Nutting.) NUTTING'S CRESTED FLYCATCHER. Like the last, and especially like its young, which have the tail-feathers more extensively rufous than the adults. Outer web of outer tail-feather not distinctly whitish, and its inner web wholly rufous, or with only a narrow dusky stripe, not widening at the tip. Rather small; wing 3.40–3.70; tail 3.35–3.80; tarsus 0.85; bill from nostril 0.50. Arizona, from the vicinity of Prescott southward through western Mexico to Costa Rica. A nest found in a giant cactus June 12, 1892, contained 4 fresh eggs 0.95 × 0.67, indistinguishable from those of *cinerascens* proper. (*M. nuttingi* RIDGW. Pr. U. S. Nat. Mus. v, 1882, p. 394; Man. 1887, p. 334. *M. cinerascens nuttingi* ALLEN, Bull. Am. Mus. Nat. Hist. iv, Dec. 1892, p. 346; A. O. U. List, 2d ed. 1895, No. 454 *a*. Not in any earlier ed. of the Key.)

M. lawren'cei. (To Geo. N. Lawrence.) LAWRENCE'S CRESTED FLYCATCHER. Similar in color to *M. crinitus*, but *much* smaller, and belonging to a different section of the genus. Bill broad, flat, shaped much as in *Contopus*, about ½ its own length wide at the nostrils. No chestnut on tail-feathers except a narrow bordering on *outer* webs, and, in the young, an inner *margining* also. Wing-coverts and inner secondaries as well as primaries edged with rufous (rarely yellowish on inner secondaries); pileum dark or quite blackish. Very small: length 7.00 or less; wing and tail only 3.00–3.40; bill 0.62–0.70; tarsus 0.75. Lower Rio Grande valley of Texas through eastern Mexico to Guatemala; only included in our fauna on the authority of GIRAUD, 1841. It is a long but not yet well known species.

M. l. olivas'cens. (Lat. *olivascens*, growing olivaceous, somewhat olivaceous.) OLIVACEOUS CRESTED FLYCATCHER. Like the last; lighter colored; crown little darker than back; wing- and tail-feathers usually without rufous edging. Wing 2.90–3.25; tail 3.00–3.25. Western Mexico, S. to Yucatan, N. to Arizona regularly, casually to Colorado (Fort Lyon, May 11, 1883). (*M. lawrencei* BREWST. Bull. Nutt. Club, 1881, p. 252, Santa Rita Mts. Ariz. in May of that year; first record for the species in the U. S. since 1841.) But the bird is now known as a common summer resident of mountains in Arizona and New Mexico, up to about 7,000 feet. Nest in Woodpecker holes and natural cavities in trees, of fur, feathers, and other material; eggs 2–4, with finer markings than usual in this genus, laid in May and June. *M. l. olivascens* RIDGW. Pr. Biol. Soc. Wash. Apr. 1884, p. 91; Man. 1887, p. 335; A. O. U. List, 2d ed. 1895, No. 455 *a*. Not noted in former eds. of the Key.)

SAYOR'NIS. (Name of Thos. Say, with Gr. ὄρνις, *ornis*, a bird.) PEWIT FLYCATCHERS. The 3 following species do not particularly resemble one another; most authors place them in separate genera, and some even under different subfamilies, of *Tyrannidæ*. But the discrepancies of form are not startling, and for the purposes of this work the species may be properly kept together, as they agree in presenting a certain aspect not shown by other N. Am. groups. (Fig. 344, *b*.) They are small — about 7.00 or less in length. Head with a slight crest of erectile feathers. Tarsus rather longer than middle toe and claw (the reverse in *C. borealis*). Bill *narrower* than in other little Flycatchers, with nearly straight lateral outlines, its width at base about ½ length of culmen. Wing pointed by 2d–5th quills, 1st shorter than 6th. Tail about as long as wing, emarginate, with broad feathers tending to divaricate in the middle. One eastern, two western species. Nest affixed to rocks and buildings, with mud; eggs white, normally unmarked, but often sparingly dotted with brownish. (Name spelled *Sayornis* originally by Bonaparte, 1854; A. O. U. Lists, 1886 and 1895; *Sayiornis* COUES, Key, 2d–4th eds. 1884–90. Type of genus *Tyrannula nigricans* SWAINS. 1827.)

Analysis of Species.

Ashy-brown, with cinnamon belly and black tail . . . . *saya*
Blackish, with white belly . . . . *nigricans*
Olivaceous and yellowish. (*Subgenus* or *genus* EMPIDIAS) . . . . *phœbe*

S. say'a. (To Thos. Say.) Say's Pewit Flycatcher. Sayan Phœbe. Adult ♂ ♀ : Grayish-brown, sometimes with faint olivaceous tinge, rather darker on head, where the feathers have dusky centres, paler on throat and breast, then changing to cinnamon-brown on the rest of under parts. Wings dusky, lined with tawny-whitish, edged with whitish on coverts and inner·quills. Tail perfectly black. Bill and feet black. Iris dark brown. Length about 7.50; extent 11.00; wing 3.75–4.35; tail 3.25–3.50; bill 0.50–0.60, narrow and slender for a Flycatcher; tarsus 0.80; middle toe and claw 0.67. Young: More extensively fulvous or paler cinnamon than the adults, this color extending far up the breast, skirting the feathers of back and rump, forming conspicuous cross-bars and edgings on wings, and even tipping tail. But no other bird of our country resembles this one. Western U. S. and British Provinces, N. to Arctic regions in Alaska, E. to Kansas, Iowa, Wisconsin, etc., S. in Lower California and Mexico; accidental in Massachusetts; common in open or rocky country, where seen singly or in pairs; the principal Flycatcher of *unwooded* regions, in weedy, brushy places, displaying the usual activity of its tribe, and uttering a melancholy note of one syllable, or a tremulous twitter. Nests naturally on rocks, but soon adapts itself to buildings like the eastern Pewee. Nest of mud, straw, moss, feathers, etc.; eggs 3–6, usually 4–5, 0.80 × 0.60. (*Sayiornis sayi* of 2d–4th eds. of the Key.)

S. nig'ricans. (Lat. *nigricans*, blackening.) Black Pewit Flycatcher. Black Phœbe. Spider-bird. Adult ♂ ♀ : Sooty-brown or blackish, deepest on head and breast; belly and other under parts pure white, abruptly defined; lining of wings, outer web of outer tail-feathers, and edges of inner secondaries, whitish; bill and feet black; iris red. The coloration is curiously like that of *Junco hiemalis*. Length about 7.00; wing 3.50–3.75; tail 3.25–3.50; bill 0.50 or less, very weak; tarsus 0.67; middle toe and claw 0.60. Southwestern U. S. and southward, but on the Pacific to Oregon, rarely to Washington; S. through Lower California and Mexico to Oaxaca; chiefly in unwooded country, and especially along rocky streams, and in cañons — I have seen it at the *bottom* of the Grand Cañon of the Colorado, some 6,000 feet below the surface of the earth! Breeds throughout its U. S. range, April and later northward; resident southerly. Nest of mud, etc., on rocks and walls; eggs 3–6, usually 4 or 5, averaging 0.75 × 0.56, ranging in length from 0.70 to 0.80.

(Subgenus Empidias.)

S. phœ'be. (Name in form Gr. φοίβη, *Phoibe*, Lat. *Phœbe*, a Titaness, daughter of Uranus and Gæa; also, a title or surname of Diana, as the Moon goddess; but as applied to this bird probably a mere onomatopœia, like "pewit" and "pewee." Fig. 350.) Pewit Flycatcher. Water Pewee. Bridge Phœbe. Phœbe-bird. Adult ♂ ♀ : Dull olivaceous-brown; head much darker fuscous brown, almost blackish, usually in marked contrast with back; below, soiled whitish, or palest possible yellow, particularly on belly; sides, and breast nearly or quite across, shaded with grayish-brown; wings and tail dusky; outer tail-feather, inner secondaries, and usually wing-coverts, edged with whitish; a whitish ring round eye; bill and feet black. Varies greatly in shade; the foregoing is the average spring condition. As summer passes, plumage becomes much duller and darker brown, from wearing of the feathers; then, after moult, fall specimens are much brighter than in spring, the under parts being decidedly yellow, at least on the belly. Very young birds have some feathers skirted with rusty, particularly on edges of wing- and tail-feathers. Sexes alike; ♀ averaging at the lesser dimensions of ♂. The species requires careful discrimination, in the hands of a novice, from any of the little olivaceous species of the next two genera. It is larger; length 6.75–7.25; extent 10.75–11.75; wing 3.00–3.50, usually 3.40; tail about the same, slightly emarginate; bill 0.50 or slightly more, little depressed, not so broad for its length as is usual in *Contopus* and *Empidonax*, its lateral outlines straight; tarsus equalling or slightly exceeding middle toe and claw, these together about 1.33; point of wing formed by 2d–5th quill; 2d shorter than 6th; 3d and

4th generally a little the longest; 1st shorter than 6th. Eastern U. S. and British Provinces, N. to the Fur Countries, W. to the Dakotas, Nebraska, Kansas, Indian Territory and most of Texas, casually to Colorado and British Columbia; S. into Mexico in winter; Cuba; very abundant in open places, fields, along streams, and almost as domestic as the Barn Swallow, or House Wren. One of the very earliest arrivals in spring (whence Wilson's name of *nunciola*, "little messenger"), becoming generally distributed in the U. S. in March, and a late loiterer in fall through September or even October; winters abundantly in the Southern States, and breeds thence northward throughout its range. Its ordinary note is harsh and abrupt, unlike the drawling *pe-a-wee'* of *Contopus virens* — sounding like *pĕ-wĭt' phĕ-bĕ*, whence the name. The typical nest is affixed to the side of a vertical rock over water, often itself moist or dripping, and composed of mud, grass, and especially moss, making a pretty object, lined with hay or feathers. The bird now builds anywhere about houses, bridges, and other buildings; its attachment to particular spots is so strong that it will return year after year, and often persist in nesting under the most discouraging circumstances.

FIG. 350. — Pewit Flycatcher, reduced. (Sheppard del. Nichols sc.)

Eggs 3-8, usually 4-5-6, from 0.80 × 0.60 to 0.67 × 0.55, averaging 0.75 × 0.57; normally pure white, not seldom sparsely dotted with reddish-brown. (*S. fusca* of previous eds. of the Key, as of most late writers, after *Muscicapa fusca* GM. 1788, based on *M. carolinensis fusca* BRISS. 1760, but this is antedated twice, by *M. fusca* MÜLL. 1776, and *M. fusca* BODD. 1783, both of which are different birds. The next name in date, *M. atra* GM. 1788, based on the Dusky Flycatcher of Pennant's Arct. Zoöl. ii, 1785, p. 389, is likewise preoccupied by *M. atra* MÜLL. 1776, a third different bird. The earliest available name is therefore *M. phœbe* LATH. Ind. Orn. ii, 1790, p. 489, based on Pennant as just cited. See Auk, Jan. 1885, p. 51. A. O. U. Lists, 1886 and 1895, No. 456. In the probable event of the removal of this species from the genus *Sayornis*, it will be known as **Empidias phœbe** COUES.)

CON'TOPUS. (Gr. κοντός, *kontos*, an adj., meaning short (not κοντός, *kontos*, noun, a pole or perch) and πούς, *pous*, foot. Fig. 344, *c*.) WOOD PEWEE FLYCATCHERS. Feet extremely small; tarsus shorter or not longer than *bill*, shorter than middle toe and claw (in *Nuttallornis*); tarsus, middle toe, and claw together, barely or not one-third as long as wing; bill flattened, very broad at base; wings pointed, much longer than emarginate tail; proportions of primaries varying with the species. Medium-sized and rather small species, brownish-olivaceous, without any bright colors or very decided markings; coronal feathers lengthened and erectile, but hardly forming a true crest. A small group of woodland species, near *Empidonax*, but characterized, as above described, by the feeble diminutive feet. Nest on boughs; no mud; eggs creamy, spotted. This genus has enjoyed unchallenged the name *Contopus* since 1855; but there is a genus *Contipus*, De Marseul, 1853, in entomology, and if this be held to void *Contopus* in ornithology, our Wood Pewees must be called *Horizopus* OBERH. Auk, Oct. 1899, p. 330.

Analysis of Species.

A conspicuous tuft of white fluffy feathers on the flank, and under parts streaky. (Subgenus NUTTALLORNIS.)
Length 7.00-8.00; tail about 3.00; wing about 4.00, pointed by 2d primary, supported nearly to end by 1st and 3d, 4th much shorter. Tarsus shorter than bill. N. Am. . . . . *borealis*
Less conspicuous white fluffy tuft on flank, or none at all; under parts not streaky. (CONTOPUS proper.)
Large: length about 8.00; tail 3.50 or more; wing about 4.00, pointed by 2d-4th quills, the 1st much shorter.
Tarsus not shorter than middle toe and claw. Western . . . . *pertinax pallidiventris*

Small : length under 7.00. Tarsus, middle toe, and claw, together, hardly or not 1.00 long.
Clearer olivaceous above, paler grayish on breast. Eastern . . . . *virens*
Darker olivaceous above, darker grayish on breast. Western . . . . *richardsoni*

(*Subgenus* NUTTALLORNIS.)

C. (N.) borea'lis. (Lat. *borealis*, northern. Fig. 351.) OLIVE-SIDED FLYCATCHER. NUTTALL'S PEWEE. PEPE-BIRD. Adult ♂ ♀ : Dusky olivaceous-brown, usually darker on crown, where the feathers have blackish centres, and paler on sides below ; chin, throat, belly, crissum, and middle line of breast, *white*, more or less tinged with yellowish and quite streaked ; wings and tail blackish, unmarked, excepting inconspicuous grayish-brown tips of wing-coverts, and some whitish edging on inner quills ; feet and upper mandible black, lower mandible mostly yellowish. The olive-brown below has a peculiar *streaky* appearance hardly seen in other species, and extends almost entirely across breast. This ragged aspect of mixed dusky-olive and whitish, together with the large white fluffy flank-tufts, is diagnostic. *Young* have the feathers, especially of wings and tail, skirted with rufous. Length 7.00–8.00 ; wing 3.87–4.33, averaging 4.00, very long, folding to terminal third of tail, and remarkably pointed ; 2d quill longest, supported nearly to end by 1st and 3d, 4th abruptly shorter ; tail about 3.00, thus about ¾ the wing, emarginate ; tarsus only 0.50, shorter than bill, or than middle toe and claw ; tarsus, middle toe, and claw together only about 1.25 ; bill 0.67–0.75. N. Am. at large, apparently nowhere very abundant, rather common in some New England localities, very rare in the Middle and Southern States, less so in the West. N. in interior British America to lat. 60°, and still farther in Alaska ; accidental in Greenland ; S. through Central America to the U. S. of Colombia ; not known to winter anywhere within our limits. Breeds in most of the Appalachian ranges, more commonly from New England northward, and much farther south in the West ; a common breeder in the mountains of Colorado, New Mexico, Arizona, California, and even Lower California. Generally seen high on some exposed outpost ; note querulous, but loud and harsh. Nest usually high, 30–60 feet, on a horizontal bough of a tree (generally coniferous) rude and flat, of twigs, rootlets, grass, moss ; eggs 2–4, often 3, 0.85 × 0.65, ranging in length from 0.80 to 0.90, buffy or creamy-white, fully spotted, and usually wreathed with lighter and darker reddish-browns, with purplish or lilac shell-markings ; they are laid late, in June and July. A stocky, able-bodied, dark and streaky species, quite unlike any other : type of the subgenus NUTTALLORNIS, RIDGW. Man. 1887, p. 337.

FIG. 351. — Olive-sided Flycatcher.

(*Subgenus* CONTOPUS.)

C. per'tinax pallidiven'tris. (Lat. *pertinax*, pertinacious ; pertaining to *C. borealis* ; *per*, and *tenax*, tenacious. Lat. *pallidus*, pallid, pale ; *venter*, gen. *ventris*, belly.) COUES' FLY-

CATCHER. Adult ♂ ♀ : Somewhat similar to *borealis* ; colors more uniform and more clearly olive; below, dull brownish-olive, lighter on throat, fading insensibly on belly into dingy yellowish-white; lacking the peculiar streaky appearance of *borealis*. Cottony tufts on flanks less conspicuous but observable. Bill longer and comparatively narrower than in *borealis* ; black above, yellow below; feet black. Wing-formula entirely different; 2d, 3d, and 4th quills nearly equal and longest, 1st abruptly 0.50 shorter, about as long as 5th, or between 5th and 6th. Feet small, weak, and properly "contopine," but tarsus if anything longer, not shorter, than middle toe and claw, about equalling bill (reverse proportion of bill, tarsus, and toe obtains in *borealis*). Length of ♂ about 8.00 ; extent 13.00 ; wing 4.00–4.30 ; tail 3.50–3.80 ; bill and tarsus, each, about 0.67 ; middle toe and claw 0.60. ♀ rather less. Young : Lower mandible and mouth orange-yellow ; feathers of wings and tail and their coverts skirted with rusty, and a shade of the same on under parts generally. Midsummer adults wear browner, like the common Wood Pewee ; in fact, the whole coloration of the species is the counterpart of a Wood Pewee's. Northern Mexico, where resident in mountains and on highlands, N. to New Mexico and Arizona, casually to Colorado ; common in mountainous pineries, where it nests in June, both on coniferous and deciduous trees, withdrawing southward in September. I took the first specimen known within our limits at Fort Whipple, Ariz., Aug. 20, 1864. Nest like that of the common Wood Pewee, but larger, 4.00–5.00 in diam. outside by 2.00 deep, cupped 2.00–3.00 by about 1.00, composed mostly of grasses, with some leaves, catkins, mosses, lichens, cobwebs, etc. Eggs 3, about 0.83 × 0.63, creamy buff, spotted with lighter and darker reddish-browns and lilac, the markings sparse and tending to wreathe about large end. *C. pertinax* of all former eds. of the Key. *C. p. pallidiventris* CHAPM. Auk, July, 1897, p. 310; A. O. U. Suppl. List, Auk, Jan. 1899, p. 112.

C. vi'rens. (Lat. *virens*, virent, greenish. Fig. 352.) WOOD PEWEE. Adult ♂ ♀ : Olivaceous-brown, rather darker on head; below, with sides washed with a paler shade of the same, reaching nearly or quite across breast; throat and belly whitish, more or less tinged with dull yellowish; under tail-coverts the same, usually streaked with dusky; tail and wings blackish, former unmarked, inner wing-quills edged, and greater and middle coverts tipped, with whitish; feet and upper mandible black; under mandible usually yellow, sometimes dusky; iris brown. Spring specimens purer olivaceous; early fall birds brighter yellow below; in summer, before the worn feathers are renewed, the plumage is quite brown and dingy whitish. Very young birds have the wing-bars and edging of quills tinged with rusty; feathers of upper parts skirted, and lower plumage tinged, with the same; but in any plumage the species may be known from all birds of the following genus, by these dimensions : Length 6.00–6.50; extent 10.00–11.00; wing 3.25–3.50; tail 2.75–3.00 ; tarsus, middle toe, and claw together hardly 1.00, or evidently less ; tarsus alone about 0.50, not longer than *bill*. Bill very flat, its breadth at base more than ½ its length ; lateral outline bulging. Wings very long and pointed ; 2d quill longest, 3d little if any less, 4th shorter, 1st between 4th and 5th. Tail but little (about 0.50) shorter than wing, emarginate. Eastern N. Am., in woodland ; extremely abundant in most U. S. localities, May–Sept., entering U. S. from the South usually in March, reaching its limit of dispersion in adjoining Canadian localities from New Brunswick to Manitoba by the end of April or early in May. Possibly winters along the southern border, but extends at that season through E. Mexico and Central Am. to equatorial regions. West only to the high central plains, as of the Dakotas, Nebraska, Indian Territory, and Texas. In the breeding season the peculiarly plaintive, drawling note

FIG. 352. — Wood Pewee, reduced. (Sheppard del. Nichols sc.)

may be heard in almost any piece of woods, while the dolorous little bird is at his post, perched on some exposed twig near his nest, and continually raiding after insects, which he captures with a quick twist in the air and a click of the bill, regaining his perch adroitly, and standing erect with hanging tail and wings, saying $p\bar{e}'$-$\check{a}$-wee' — Ah! poor me! Nest a very pretty structure, saddled on a horizontal bough, at little or great height, flat and thin-bottomed, with thick walls and well-turned brim, of fine fibres stuck over with lichens, the whole looking much like a natural excrescence of the tree, or, if in a pine, a lichen-bunch. Eggs 2-4, oftenest 3, but rarely 4, creamy-white, marked with reddish-brown and lilac in various pattern, usually wreathing and blending about the larger end, sparser elsewhere; size about 0.75×0.55, with the usual range of variation both in length and breadth, some being only 0.65×0.50.

C. rich'ardsoni. (To Sir John Richardson.) WESTERN WOOD PEWEE. Similar; darker, more fuscous-olive above; shading of sides reaching almost uninterruptedly across breast; belly rather whitish than yellowish; outer primary usually not obviously white-edged; bill below oftener dusky than yellow, sometimes quite black. I fail to appreciate any reliable differences in size or shape, though some have been alleged; the wing and tail average a trifle longer. It is impracticable to pronounce upon a Pewee, in the closet, without knowing the locality; but those familiar with both eastern and western Pewees in the field will agree with me that they are not the same bird. Note not exactly like that of *virens*, being abrupt and emphatic, rather than drawling and listless. The eggs are indistinguishable, but the nesting is somewhat different; the fabric is often or usually placed in the forking of small horizontal branches, contrary to its saddling on a larger bough by *virens*; conformably with which practice, the shape is usually deeper for its breadth, more like a cup than a saucer, measuring about 2.50×2.00 outside, with a cavity 2.00×1.00 or more. Eggs 2-4, usually 3, averaging 0.70×0.55, with a range of variation in length of at least 0.10. They are laid mostly in June, in any region, but may also be taken fresh during the first half of July. The range of this species extends from the eastern slopes and foothills of the Rocky Mts. to the Pacific, N. to Saskatchewan, Alaska, and British Columbia, S. in winter to equatorial America; breeds throughout its N. Am. range, but winters extralimital; migrates mainly in April, May, and September. The range is for the most part separated from that of *virens* by the treeless plains, but the two species are found together in the valley of the Red River of the North, in Manitoba; "Labrador" (*Audubon*). (*Tyrannula richardsonii* Sw. Fn. Bor.-Am. ii, 1831, p. 146? *Contopus richardsonii* BD. B. N. Am. 1858, p. 189; *C. virens richardsoni* COUES, Key, orig. ed. 1872, p. 174, and of later eds. 1884-90, p. 440. *Muscicapa phœbe* AUD. B. Am. 8vo ed. i, 1840, p. 219, pl. 61; NUTT. Man. i, 2d ed. 1840, p. 319. See COUES, B. N. W. 1874, p. 247.)

C. r. penin'sulæ. (Lat. *peninsula*, that which is almost an island.) LARGE-BILLED WESTERN WOOD PEWEE. BREWSTER'S PEWEE. Like the last; smaller, but with bill absolutely as well as relatively longer and broader; upper parts slightly grayer; yellowish of throat and belly clearer and more extensive; pectoral band narrower and grayer; light edgings of the wings broader and clearer. Wing 3.30; tail 2.38; tarsus 0.52; bill from nostril, 0.42; width there 0.31. Sierra de la Laguna, Lower California. BREWSTER, Auk, Apr. 1891, p. 144; A. O. U. List, 2d ed. 1895, No. 462 a.

EMPIDO'NAX. (Gr. ἐμπίς, gen. ἐμπίδος, *empis, empidos*, a gnat; ἄναξ, *anax*, king. Fig. 344, *d*.) THE LITTLE OLIVACEOUS FLYCATCHERS. Small olivaceous species, 5.00-6.00 (rarely 6.25) long; wing 3.12 *or less;* tail 2.75 *or less;* whole foot at least $\frac{1}{3}$ as long as wing; tarsus more or less obviously longer than middle toe and claw, much longer than bill; 2d, 3d, and 4th quills entering into point of wing, 1st shorter or not obviously longer than 5th; tail not over $\frac{1}{2}$ an inch shorter than wings. As in allied genera, several outer primaries are slightly emarginate on inner web, but this character is obscure, often inappreciable, and may be disregarded. The coronal feathers are lengthened and erectile, but scarcely form a true crest. There are never any more conspicuous color-marks than in *Sayornis phœbe* or *Contopus virens*.

TYRANNIDÆ — TYRANNINÆ: TYRANT FLYCATCHERS. 527

The bill varies with the species in size and shape, from almost as broad and flat as in a Wood Pewee in *E. virescens* (formerly called *acadicus*) to the narrower shape of a Pewit in *E. wrighti;* but it is always much shorter than tarsus. The sexes are alike in this genus, as usual in the family; the young similar, usually rather more yellowish or buffy. The nest is placed in trees or bushes; the eggs are white, spotted or not in different species, thus affording good clews in some cases of doubt. It should not be difficult to recognize *Empidonax* as different from *Contopus*, due attention being given to the nice points of diagnosis; but it is not easy to discriminate the numerous species without much tact, care, and patience. The following account, carefully prepared after examination of a great amount of material from all parts of the country, will probably suffice to determine nearly all specimens. How much alike are these interesting little birds may be inferred from the fact that Wilson in 1810 knew but a single species, *virescens*, which he called *muscicapa querula*, to which Audubon added but one, which he named *traillii*, in 1832, until Baird in 1843 showed him two more, *minimus* and *flaviventris*. Yet these four are perfectly distinct birds. Any experienced collector knows these four to be different, not only when he has them in hand, but in life, by their haunts and habits, their notes, nests, and eggs — indeed, the nests and eggs of each of them are readily discriminated. Three of them occur in New England as breeders — *trailli alnorum*, *minimus*, and *flaviventris;* while *virescens* is the common breeder in the Middle States. The case is complicated, however, in the West. Since 1858, when Baird first fixed our species upon anything like a satisfactory footing, few changes of his determinations and characterizations have been established; but several species which were unknown to him have been added to our Fauna, and some changes of nomenclature have been introduced (see especially the cases of the so-called "Acadian" and of eastern and western "Traill's" Flycatchers, as treated by Brewster, Auk, Apr. 1895, pp. 157–163). It is not reasonably possible to analyze all the forms in concise phrase; the student must go at once to the detailed descriptions; but the following may help him somewhat:

A. Species clearly *olivaceous* of some shade above; below *whitish* more or less shaded on breast, or clearly *yellowish*, but adults never *buffy*. (EMPIDONAX proper.)
Eastern.
    Largest: rather over than under 6.00; wing nearly or over 3.00; tarsus 0.67; middle toe and claw 0.50; bill nearly or quite 0.50. Clear light olive-*green* above, below whitish; wing-bars and eye-ring *tawny*. Nest *flat* in fork of a horizontal bough; eggs *speckled*. Hardly N. to New England. *Virescens* (formerly called *acadicus*)
    Medium: rather under 6.00; wing 2.70; tarsus 0.67, but middle toe and claw 0.60; bill hardly 0.50. Olive-*brown* above, below grayish; wing-bars and eye-ring *whitish*. Nest a bulky cup in a *bush;* eggs *speckled*. New England, etc. . . . . *trailli alnorum*
    Small: rather under 5.50; proportions and colors nearly as in *trailli*. Nest a neat cup in upright crotch of a *tree;* eggs *white*. Commonest breeder in S. New England, etc. . . . . *minimus*
    Medium: under parts *thoroughly yellow*. Nest *near ground* in a stump, moss, etc., bulky. Eggs *speckled*. Northern New England, etc. . . . . *flaviventris*
Western.
    The stock-form of *trailli alnorum* as above described. *Eggs speckled.* Mississippi valley and westward
. . . . *trailli* (formerly called *pusillus*)
    The representative of *flaviventris* in the west. Thoroughly *yellow-bellied*. *Eggs speckled* . . . . *difficilis*
    The representative of *difficilis* in Lower California. *Not* thoroughly yellow-bellied . . . . *cineritius*
    The representative of *difficilis*. Santa Barbara Isls. *insulicola*
    Small, and otherwise like *minimus;* dark below, breast not very different from back; bill extremely narrow. Eggs *white*. Western N. Am. at large . . . . *hammondi*
    Large, about the size of *acadicus;* olive-brown above; breast dark; outer tail-feather white on outer web; bill very narrow. Eggs *white*. Western U. S. . . . . *wrighti* (formerly called *obscurus*)
    The representative of *wrighti* in Lower California. Larger, grayer, etc. . . . . *griseus*
B. Species more or less decidedly *buffy*. Exclusively southwestern. (Section MITREPHANES.) *fulvifrons* and *f. pygmæus*

Another analysis may be made which will suit some students, as follows: —

A. EMPIDONAX proper; no buff-bellied species.
Belly decidedly yellow.
    Eastern . . . . *flaviventris*
    Western . . . . *difficilis*
    Lower California . . . . *cineritius*

Belly not decidedly yellow.
  Width of bill at nostrils more than half the length of culmen.
    Largest: length 6.00 or more. Olive-green. Eastern . . . . *virescens*
    Medium: length 6.00 or less. Olive-brown.
      Western . . . . *trailli*
      Eastern . . . . *trailli alnorum*
    Smallest. Length 5.50 or less. Eastern . . . . *minimus*
  Width of bill at nostrils less than half the length of culmen. Western.
    Outer web of outer tail-feather not decidedly whitish . . . . *hammondi*
    Outer web of outer tail-feather decidedly whitish.
      Western U. S. . . . . *wrighti*
      Lower California . . . . *griseus*
B. MITREPHANES. One buff-bellied species and one subspecies.
  Texas . . . . *fulvifrons*
  New Mexico and Arizona . . . . *fulvifrons pygmæus*

    Observe that the eggs are *speckled* only in the "Acadian" and "Traill" and yellow-bellied groups of species — *white* in all the others.

E. vires'cens. (Lat. *virescens*, growing green, greenish.) SMALL GREEN-CRESTED or so-called "ACADIAN" FLYCATCHER. Adult ♂ ♀: Above, olive-*green*, clear, light, continuous and uniform (though the crown may show rather darker, owing to dusky centres of the slightly lengthened, erectile feathers); below, *whitish*, olive-shaded on sides and nearly across breast, yellowish-washed on belly, flanks, crissum, and axillars; wings dusky, inner quills edged, and coverts tipped, with *tawny* yellow; all quills whitish-edged internally; tail dusky, olive-glossed, unmarked; a *tawny* eye-ring; feet and upper mandible brown, under mandible pale. In midsummer, rather darker; in early fall brighter and especially more yellowish below; in the young, wing-markings more fulvous, general plumage slightly buffy-suffused; when very young, said to be mottled transversely with pale ochraceous. Largest: 5.75–6.25 — rather over than under 6.00; extent rather over than under 9.50; wing 2.75–3.00 (even 3.12); tail 2.50–2.75; bill nearly or quite 0.50, about 0.25 wide at nostrils, broad and flat, like a Pewee's; tarsus 0.66; middle toe and claw 0.50; point of wing reaching nearly *an inch* beyond secondaries; 2d, 3d, and 4th quills nearly equal and *much* ($\frac{1}{4}$ inch or more) longer than 1st and 5th, which about equal each other; 1st *much* longer than 6th. The ♀ near the lesser of all the dimensions given. Eastern U. S., southerly, *scarcely known in New England*, where it is rare or casual as far N. as Massachusetts; N. in the interior to southern New York, southern Michigan, Wisconsin, Minnesota, rarely Manitoba; W. to the limit of trees in Nebraska, Kansas, Indian Territory, and Texas; S. in winter through Mexico to Ecuador; Cuba; an abundant bird of woodland in our middle districts in summer; migratory; breeds throughout its N. Am. range, mostly in June, but in May in the South; winters extralimital. This is the *Empidonax* which most resembles *Contopus virens*, and is readily recognized by the points of size and shape, without regarding coloration; it has a harsh abrupt note of two syllables. Nest in *trees*, low or at no great elevation, semi-pensile in horizontal fork of a slender bough, thin and open-worked, shallow, flat, saucer-shaped, hardly 3.00 in diameter outside, and 2.00 or less deep, with a cavity about 2.00 × 1.00. Eggs 2–4, mostly 3, creamy or pale buff, boldly spotted with reddish and darker browns, especially about the larger end, like a Wood Pewee's; size from 0.78 × 0.56 down to 0.67 × 0.50, averaging about 0.73 × 0.53. (*Muscicapa subviridis* BARTRAM, 1791; *Empidonax* **subviridis** COUES, 1882. *Muscicapa querula* WILS. Am. Orn. 1810, ii, 77, pl. 13, f. 3, nec VIEILL. 1807; *Platyrhynchos virescens* VIEILL. Nouv. Dict. 1818, p. 22; *Empidonax virescens* BREWST. Auk, Apr. 1895, p. 157; A. O. U. List, 2d ed. 1895, No. 465. *Muscicapa acadica* AUD. B. Am. 8vo ed. 1840, i, 221, pl. 62, nec GM. 1788, LATH. 1790; *Empidonax acadicus* BD. B. N. A. 1858, p. 197, and of all previous eds. of the Key, as of most writers since 1858. The long-established name *acadicus*, geographically false for a bird which never reaches Acadia (Nova Scotia), can fortunately be done away with by rules in favor of the entirely appropriate designation *virescens*, if we ignore the earlier name *subviridis* of Bartram, on the ground that it is unaccompanied by a description, though it certainly belongs to this species.)

E. trail'li alno'rum. (To T. S. Traill, of Edinburgh. Lat. *alnorum*, gen. pl. of *alnus*, the alder.) TRAILL'S FLYCATCHER OF THE EAST. ALDER FLYCATCHER. "KEWINK." Adult ♂ ♀: Above, olive-*brown*, lighter and duller brownish posteriorly, darker on head, owing to obviously dusky centres of coronal feathers; below, nearly as in *virescens*, but *darker*, the olive-gray shading quite across breast; wing-markings *grayish-white* with slight yellowish or tawny shade; under mandible pale; upper mandible and feet black. Averaging smaller than *virescens*; length 5.50–6.00; extent under 9.50, usually 8.75–9.00; wing 2.66–2.75, more rounded than in *virescens*, its tip only reaching about ⅔ of an inch beyond secondaries, formed by 2d, 3d, and 4th quills, as before, but 5th not so much shorter (hardly or not ¼ of an inch), 1st ranging between 5th and 6th; tail 2.50, not emarginate, but even or slightly rounded; tarsus 0.66, as in *virescens*, but middle toe and claw 0.60, the feet thus differently proportioned, owing to length of toes; bill not so broad and flat as in *virescens*. Eastern N. Am. to the Plains, common; an entirely different bird from *virescens*, but difficult if not impossible to distinguish from the following western stock-form; almost the same in *color* as *minimus*, but larger, and otherwise perfectly distinct. The Alder Flycatcher, commonly called "Traill's" (though Audubon distinctly says of his *traillii*, "Arkansas to the Columbia"), ranges much farther N. than the foregoing, breeding from the mountains of New York and probably of other Middle States, in much of New England, and most of the Canadian Provinces to 63° or farther; its western limits cannot be given with precision, because this form shades into *trailli* proper in the Mississippi valley; S. in winter to Central America; winters extralimital; migrates chiefly in May and Sept., and breeds in the last half of June and first half of July. This is a bird of thickets and shrubbery rather than of woodland, especially common in low wet places among the alders, willows, and other bushes in which its nest is placed, as a rule in an upright crotch of two or more twigs. It is thick-walled, deeply cupped, more or less compact, sometimes quite slovenly, like an Indigo-bird's, and in any case quite different from the frail flat saucer of *virescens*; it measures about 3.00 across outside by 2.50 high, with a cavity nearly as deep as broad; the materials are miscellaneous, as various grasses, bark strips, weed fibres, plant down, hairs, etc. Eggs 3 or 4, sometimes only 2, indistinguishable from those of *virescens*, quite different from those of *minimus*; ground white, whitish, or buffy, well speckled and blotched with the usual browns, the markings tending to aggregate at or wreathe about the larger end, and occasional specimens being nearly immaculate; average size 0.73 × 0.53, with extremes of 0.78 × 0.55, and 0.68 × 0.50. "Song" notes a harsh *k'wink* or *kewee'* and a soft *ke-wing'*. (*Empidonax traillii* BD. B. N. A. 1858, p. 193; COUES, Key, orig. ed. 1872, p. 175; 2d–4th eds. 1884–90, p. 441, and of most writers, but not the true *Muscicapa traillii* AUD. 1832, which is the western form. *E. traillii alnorum* BREWST. Auk, Apr. 1895, p. 161; A. O. U. List, 2d ed. 1895, No. 466 *a*. *E. pusillus* var. *trailli* RIDGW. in BD. BREW. and RIDGW. Hist. N. A. B. ii, 1874, p. 369. *E. pusillus traillii* A. O. U. 1st ed. 1886, No. 466 *a*; RIDGW. Man. 1887, p. 343; BENDIRE, Life Hist. ii, dated 1895, pub. Sept. 1896, p. 310.)

E. trail'li (proper). TRAILL'S FLYCATCHER OF THE WEST. LITTLE WESTERN FLYCATCHER. The stock-form or species of the foregoing particularly described subspecies. May usually be recognized by its duller or more fuscous coloration, the quite lively olivaceous and yellowish shades of *alnorum* being subdued or overcast; wing-bars duller and less conspicuous; bill larger; tarsi longer, the feet being nearly as in *virescens*. Replaces *alnorum* in western N. Am. from the Plains to the Pacific; but specimens absolutely like *alnorum* are found in the West even to British Columbia, and others like *trailli* proper E. to Michigan, Illinois, Indiana, Ohio, etc., showing that in the Mississippi valley at large no line can be drawn between the two forms. The present species is the usual "Little Flycatcher" of western U. S. and adjoining British Provinces, S. in winter to Central America; abundant, migratory, generally distributed in suitable places; habits, appearance in life, and eggs the counterparts of those of *alnorum* as above described. (The original *Tyrannula pusilla* SWAINS. F. B. A. ii, 1831, p. 144, pl. 46; AUD. B. Am. 8vo ed. ii, 1840, p. 236, pl. 66, is uncertain, and just as likely to have been *minimus* as *trailli;* and the case is further complicated by *Platyrhynchus pusillus* SWAINS. Phil. Mag. i, 1827, p. 366, described from Mexico. I therefore continue to pass over the name, which, if belonging here, antedates *trailli;* and I also now drop *pusillus* BAIRD, 1858, as untenable by our rules, though it certainly belongs here; taking *trailli* for the western stockform, and *alnorum* BREWST. for the eastern subspecies. *Muscicapa traillii* AUD. Orn. Biogr. i, 1832, p. 236, Arkansas River. *Empidonax traillii* BREWST. Auk, Apr. 1895, p. 159, not of BD. 1858, nor of authors referring to the eastern bird; A. O. U. List, 2d ed. 1895, No. 466. ? *E. pusillus* CAB. J. f. O. 1855, p. 480, uncertain, same as *pusillus* SWAINS. *E. pusillus* BAIRD, B. N. A. 1858, p. 194; COUES, Key, 2d–4th eds. 1884–90, p. 442; RIDGW. Man. 1887, p. 343. *E. traillii* var. *pusillus* COUES, Key, orig. ed. 1872, p. 175. *E. pusillus* var. *pusillus* RIDGW. in BD. BREW. and RIDGW. Hist. N. A. B. ii, 1874, p. 365.)

E. min'imus. (Lat. *minimus*, smallest.) LEAST FLYCATCHER. "CHEBEC." Adult ♂ ♀: Colors almost exactly as in *trailli;* usually, however, olive-*gray* rather than olive-brown; wing-markings, eye-ring, and loral feathers plain grayish-*white*, and rather more conspicuous than in *trailli*, especially the wing-bars; whole anterior parts often with a slight *ashy* cast; under mandible ordinarily dusky; feet perfectly black. It is a smaller bird than *trailli*, and not so stoutly built; the wing-tip projects only about 0.50 beyond secondaries; 5th quill but very little shorter than 4th, 1st apt to be nearer 6th than 5th; tail slightly emarginate, not even or slightly rounded; feet differently proportioned, being much as in *virescens;* bill obviously under 0.50. Length 5.00–5.50; extent about 8.00; wing 2.60 *or less;* tail about 2.25. A series of ♂ ♂, measured fresh, runs 5.20–5.50 long, by 7.60–8.30 in extent; several ♀ ♀ are 4.80–5.10 long, by 7.40–7.90 in extent. Although a large ♂ may grade up to ♀ *trailli* in size, and there is no *obviously* different coloration, it is a totally different bird. Eastern N. Am. to the Plains, less commonly to the Rocky Mts., casually to Idaho; very abundant in the U. S. during the migrations in April, May, and again in Sept., in orchards, coppices, hedgerows, and the skirts of woods rather than in heavy forests; ranges N. to about lat. 63°, in the region of Great Slave Lake and the Mackenzie River, but farther East goes little N. of the U. S.; winters wholly extralimital, as far S. as Panama. This is the commonest breeder of its genus in New England, especially Massachusetts, and common thence to the Red River of the North; it also breeds freely in our Middle districts, and sparingly in the Alleghanies, even S. to North Carolina; mostly in June. The nidification resembles that of *trailli* most nearly, in that the nest is as a rule placed in an *upright* crotch. It is small, neat, compact-walled, deeply-cupped, in size about 3.00 × 2.50 outside, and 2.00 × 1.50 inside, thus somewhat like the Goldfinch's structure; it is built of the most miscellaneous materials, exceptionally varies in position to a horizontal bough (like that of *E. virescens* or *Contopus virens*), and is placed in a tree or sapling, 10–20 feet from the ground. Eggs 3–4, oftenest 4, rarely 5 or 6, *white*, normally unmarked, rarely speckled, 0.60–0.67 long, averaging 0.65 × 0.50. Note a sharp *che-beć'*, or *se-wick'*, quickly uttered.

E. flaviven'tris. (Lat. *flavus*, yellow; *ventris*, of the belly.) YELLOW-BELLIED FLYCATCHER. Adult ♂ ♀: Above, olive-*green, clear,* continuous and uniform as in *virescens*, or even brighter; below, not merely yellow*ish*, as in the foregoing, but emphatically *yellow,* bright and pure on belly, shaded on sides and anteriorly with a paler tint of color of back; eye-ring and wing-markings *yellow;* under mandible yellow; feet black. In respect of color, this species differs materially from all the foregoing; none of them, even at their autumnal yellow*est*, quite match it. Size of *trailli*, or rather less; feet proportioned as in *virescens;* bill nearly as in *minimus,* but rather larger; 1st quill usually equal to 6th. Eastern U. S. and British Provinces, N. regularly to Labrador and Hudson's Bay, casually to Greenland, W. only to the eastern edge of the Plains; migratory; winters extralimital, as far S. as Panama, common, in woodland, swamps, and shrubbery. Breeds from the mountains of the Middle States and at any altitude from the northern tier of States, northward; probably also in the Alleghanies S. to the Carolinas. There has been much misunderstanding about the nest and eggs of this bird; the latter are described by Brewer and by Coues (1874) as white. Nest in swamps, *close to ground*, in a stump, log, moss, or among roots of an upturned tree, thick and bulky, deeply-cupped, composed chiefly of mosses and rootlets; eggs 4, sometimes 5, about 0.67 × 0.51, white, *spotted* with rusty brown in fine pattern and mostly about the larger end; laid in June. Thus the nidification is as distinctive as the coloration of this species, in comparison with its eastern congeners. Note a low soft *pe-a'*, slowly delivered; but this species, like others of the genus, has in the breeding season a certain twittering, which may be called by courtesy warbling, quite different from the ordinary call-notes or cries of agitation. This bird is described by NUTTALL, Man. ii, 1834, p. 568, but not named, and not noted in his 2d ed. 1840: see COUES, Auk, Apr. 1897, p. 218.

E. diffi'cilis. (Lat. *difficilis, dis-facilis*, difficult, un-doable; very appropriate!) WESTERN YELLOW-BELLIED FLYCATCHER. BAIRD'S FLYCATCHER. Very closely resembling *flaviventris* in its yellowness, but coloration dingy, instead of pure olivaceous and yellow, the latter dulled with an ochrey or buffy shade, especially on lining of wings; tail said to be longer, but no tangible difference in dimensions from *flaviventris*. Western N. Am., Rocky Mts. to the Pacific, N. in summer to Alaska, S. in winter to Costa Rica; abundant. Nest quite like that of *flaviventris*, and eggs indistinguishable; but the position of the nest extremely variable, on the ground, in trees or bushes, even in odd nooks about buildings; eggs 3-4, rarely 5, laid in May and June. Some individuals of this species winter over our southern border. (*E. difficilis* BD. 1858, A. O. U. Lists, No. 464; *E. flaviventris* var. *difficilis* COUES, Key, orig. ed. 1872, p. 176; *E. f. difficilis?* COUES, Key, 2d-4th eds. 1884-90, p. 442.)

E. cinerit'ius. (Lat. *cineritius*, cinereous, ashy in color.) ST. LUCAS FLYCATCHER. Most like *E. difficilis*: general coloration much duller; upper parts scarcely tinged with greenish; no decided yellowish below, except on jugulum and abdomen; wing-bands brownish-white. Sexes similar. Wing 2.65; tail 2.40; tarsus 0.68. Lower California. BREWSTER, Auk, Jan. 1888, p. 90; COUES, Key, 4th ed. 1890, p. 901; A. O. U. List, 2d ed. 1895, No. 464. 1.

E. insulic'ola. (Lat. *insula*, an island; *colere*, to cultivate, or *incola*, an inhabitant.) ISLAND FLYCATCHER. Like *difficilis*; darker and more brownish above; paler below, breast scarcely washed with ochraceous-brown. Also closely resembling *cineritius*; darker, less ashy and somewhat more olivaceous above, and more continuously yellowish below. Santa Barbara Islands: locality the best diagnostic! OBERHOLSER, Auk, July, 1897, p. 300. For breeding, see Auk, Oct. 1897, p. 405; eggs 2-3, dead white, speckled about large end with reddish; July.

E. ham'mondi. (To Dr. W. A. Hammond, U. S. A.) HAMMOND'S FLYCATCHER. DIRTY LITTLE FLYCATCHER. Adult ♂ ♀: Above, olive-*gray*, decidedly grayer or even ashy on the fore parts; whole throat and *breast* almost continuously *olive-gray*, but little paler than back; belly alone more or less decidedly yellowish; wing-markings and eye-ring dull soiled whitish; bill *very small*, and extremely *narrow*, being hardly or not 0.20 wide at nostrils; this distinguishes the bird from all but *minimus* and *wrighti*; under mandible usually blackish; tail usually decidedly *forked*, more so than in other species (though in all of them it varies from slightly rounded to slightly emarginate); outer tail-feather usually whitish-*edged* externally (a character often shown by *trailli* and *minimus*), but *not* decidedly *white*. About the size of *minimus*; ♂, length under 6.00; wing 2.75; tail 2.40, both thus relatively longer; ♀ a little smaller than ♂, as usual in the genus. Plains to the Pacific, U. S. and Brit. Am., N. to Saskatchewan, Alberta, the N. W. Territory and Alaska, S. to L. Cala. and southern Mexico in winter; migrates in May and Sept., and breeds mainly in June. This is the western representative of *minimus*, but is tangibly distinct; the general tone of coloration is *heavy*, fall specimens in particular giving somewhat the effect of a dirty *flaviventris;* the tiny bill is a good mark. Nesting substantially like *minimus;* eggs normally *white, unmarked*, rarely speckled a little, 3-4 in number, 0.65 × 0.50. Note "a soft *pit.*"

E. wright'i. (To C. Wright.) WRIGHT'S FLYCATCHER. GRAY LITTLE FLYCATCHER. Adult ♂ ♀ : Colors not very tangibly different from those of *trailli* or *minimus*, but *outer web* of outer tail-feather abruptly *white* in decided contrast. General tone quite *gray;* gray below quite across breast, giving the effect there of *Contopus richardsoni;* under mandible obscured; eye-ring and wing-edgings quite whitish. General dimensions approaching those of *virescens*, owing to length of wings and tail. Length doubtless up to 6.00, and extent to 9.50; wing 2.66-3.00; tail 2.50-2.75; tarsi 0.70-0.75; bill 0.50 or more, *extremely narrow* (much as in *Sayornis phœbe*), its width at nostrils only about ½ its length. The bird looks singularly like the western *Contopus*, though of course immediately seen to be *Empidonax*. Western U. S., N. to British Columbia, Rocky Mts. to the Pacific, S. in winter through most of Mexico; especially a bird of the mountains, where found up to 10,000 feet or more, common in woodland, groves, and thickets. To complete the analogies between the eastern and western *Empidonaces*, this may be considered to represent *virescens*. Nesting, however, substantially as in *minimus :* a neat, compact, deep-cupped nest in crotch of a bush or sapling, often deeper than broad, and commonly lined with feathers or hair; eggs 3-4, *white, unmarked*, large, up to 0.75 × 0.58, and averaging 0.68 × 0.52. Note "a weird *sweer*," "a soft liquid *whit.*" (This is *E. obscurus* BAIRD, B. N. A. 1858, p. 200, and of 1st-3d eds. of the Key, but questionably *Tyrannula obscura* Sw. Phil. Mag. i, 1827, p. 367; it is *E. wrightii* of BAIRD, *l. c.*, in text, the name preferably adopted in view of the uncertainty of Swainson's bird : see BREWSTER, Auk, Apr. 1889, p. 89, and COUES, Key, 4th ed. 1890, p. 901 ; A. O. U. List, 2d ed. 1895, No. 469.

E. gris'eus. (Lat. *griseus*, grisly, gray.) GRAY FLYCATCHER. Nearest *E. wrighti;* larger and much grayer; bill longer, flesh-colored on basal half of lower mandible, in contrast with its blackish terminal portion. ♂, wing 2.68; tail 2.45; tarsus 0.72; bill 0.62. Sonora, L. and S. California, and southern Arizona. BREWSTER, Auk, Apr. 1889, p. 87 ; COUES, Key, 4th ed. 1890, p. 901 ; A. O. U. List, 2d ed. 1895, No. 469. 1.)

(*Subgenus* MITREPHANES.)

E. (M.) ful'vifrons. (Lat. *fulvus*, fulvous; *frons*, forehead.) FULVOUS FLYCATCHER. LITTLE BUFF FLYCATCHER. Quite different from any of the foregoing, and type of a genus *Mitrephorus* SCLATER, 1859, or *Mitrephanes* COUES, 1882. Coronal feathers and rictal bristles longer than is usual in *Empidonax*, and general cast of plumage buffy or fulvous rather than olivaceous. Above, umber brown; below, buff, paler or whitish on the belly and under tail-coverts. Length about 5.25; wing 2.70; tail 2.40; bill 0.50; tarsus 0.60. Eastern Mexico to the Rio Grande of Texas. *Mitrephanes fulvifrons* of the Key, 3d ed. 1887, p. 879; *Empidonax fulvifrons*, A. O. U. List, 1886, p. 236; RIDGW. Man. 1887, p. 344; A. O. U. List, 2d ed. 1895, p. 189 No. [470.]

E. f. pygmæ'us. (Lat. *pygmæus*, pigmy, dwarf.) LITTLE BUFF-BREASTED FLY-CATCHER. Adult ♂ ♀ : Above, dull grayish-brown tinged with olive, particularly on back; below, pale fulvous, strongest across breast, whitening on belly; no fulvous on forehead; sides of head light brownish-olive ; wings and tail dusky, outer web of outer tail-feathers, edges of inner primaries except at base, and tips of wing-coverts, whitish ; iris brown; bill yellow below, black above; feet black. Length 4.75; extent 7.33; wing 2.20; tail 2.00; tarsus 0.55; middle toe and claw 0.45; bill 0.40. New Mexico, Arizona, and southward; apparently not common, and not yet well known. I discovered it at Fort Whipple, Ariz., May 9, 1865; it has been seen in the same territory in Sept.; and fledglings were observed at Inscription Rock, N. M., July 24. Nests in mountainous regions up to 9,000 feet, in June and July. The nest is saddled on a limb, 20-50 feet from the ground, resembling that of the Blue-gray

Gnatcatcher, being small, neat, and compact, of leaves, straws, rootlets, and other fibres; eggs 3-4, pale buff or dull whitish, immaculate. (*Empidonax pygmæus* COUES, Ibis, 1865, p. 537; *Mitrephorus pallescens* COUES, Proc. Phila. Acad. 1866, p. 63; ELLIOT, B. N. A. pl. 19; *M. fulvifrons* var. *pallescens* COUES, Key, 1st ed. 1872, p. 176; *Mitrephanes fulvifrons pallescens* COUES, Key, 2d-4th eds. 1884-90, p. 443; *Empidonax fulvifrons pygmæus* RIDGW. Pr. U. S. Nat. Mus. viii, 1885, p. 356; Man. 1887, p. 345; A. O. U. Lists, 1st and 2d eds. 1886-95, No. 470 *a*. My original specimens, affording the descriptions quoted, and the first known to have been taken in the United States, do not appear to be specifically distinct from *Muscicapa fulvifrons* of GIRAUD (B. of Tex. 1841, pl. 2, f. 2); they are clean spring birds, and the species is more fulvous in fall plumage.)

PYROCE′PHALUS. (Gr. πῦρ, gen. πυρός, *pur, puros,* fire; κεφαλή, *kephale,* head.) FIRE-CROWNED FLYCATCHERS. Sexes very dissimilar: head of ♂ with a full globular crest (fig. 353), and all under parts (usually) scarlet-red; other parts deep brown; ♀ brown and whitish. Bill slender, narrow at base, much as in *Sayornis*. Wings moderate, pointed; 2d-4th quills longest, 1st between 5th and 6th. Tail nearly even, shorter than wings, of broad feathers. Tarsus scarcely longer than middle toe and claw. A tropical genus of several species, one of which reaches our border.

P. rubi′neus mexica′nus. (Lat. *rubineus,* ruby-red.) VERMILION FLYCATCHER. Adult ♂: Pure dark brown, including stripe along side of head; wings and tail blackish with slight pale edgings; full globular crest, and all under parts scarlet or vermilion; bill and feet black. ♀: Dull brown, including the little-crested crown; below, white, tinged with red, reddish or orange in some places; breast and sides with slight dusky streaks. Immature ♂ shows gradation between characters of both sexes; at first there is no red whatever, the bird otherwise resembling ♀, but pale yellowish where she is reddish; upper parts gray; all the feathers may be skirted with whitish, especially on the wing-coverts and inner secondaries; tail quite blackish; under parts more purely white than in ♀, and rather speckled than streaked with gray. But reddish soon replaces the yellow of the crissum and axillars. Adult ♂ ♂ are subject to much variation; the red is sometimes rather orange. Length about 6.00; wing 3.25; tail 2.50; bill 0.45; tarsus 0.55; middle toe and claw 0.50.

FIG. 353.—Head of Vermilion Flycatcher.

Valleys of the Rio Grande and Colorado, and N. to the borders of Utah, S. in Lower California and Mexico to Guatemala; common in Arizona on the Gila; a very showy little bird, of the usual flycatcher habits. Some individuals winter over our border, but most enter there in March and depart in October. Breeds from late April to early July, and may raise two broods. Nest in trees or bushes at very variable height, set in a horizontal fork, flat, frail, flimsy, of twigs, plant fibres and down, cobwebs, feathers, fur, hair, etc. Eggs 3, sometimes only 2, 0.70 × 0.52, pale buff or creamy, boldly spotted and blotched with various dark brown and neutral tints, the markings tending to aggregate at or wreathe around the large end.

ORNITH′ION. (Gr. ὀρνίθιον, *ornithion,* dimin. of ὄρνις, *ornis,* a bird.) BEARDLESS FLYCATCHERS. General aspect of *Empidonax,* but remarkably distinguished by *parine* shape of bill and almost entire absence of rictal bristles so conspicuous in most genera of *Tyrannidæ,* though a few slight ones may be seen on close inspection. Bill much shorter than head, stout, *compressed,* not depressed as usual in *Tyrannidæ,* with high-ridged arched culmen and scarcely overhanging tip; commissure gently decurved; gonys about straight. Head a little crested, as in *Empidonax, Contopus,* etc. Wings of moderate length, much rounded; 2d to 5th primaries subequal and longest, 6th shorter, 1st about equal to 7th. Tail a little shorter than wings, even or scarcely rounded. Tarsus long, exceeding middle toe and claw; lateral toes subequal, their claws about reaching base of middle claw; hind claw shorter than its digit. Of diminutive size, and dull plain colors, as in the small olivaceous Flycatchers generally; but for the bill, the species might be mistaken for an *Empidonax*.

O. imber'be. (Lat. *imberbis*, beardless; *in*, not, and *barba*, a beard.) TEXAS BEARDLESS FLYCATCHER. Adult ♂ ♀: Above, dull olive-gray, a little darker (browner) on the lengthened erectile feathers of crown, a little brighter (greener) on rump and upper tail-coverts. Below, pale dull gray, sometimes almost grayish-white anteriorly, clearing on belly and under tail-coverts to pale yellowish. Wings and tail fuscous, with pale gray or whitish edgings of middle and greater coverts and most of the quills of the wings, as in an *Empidonax*. Bill dark brown above, pale below. Worn specimens are quite brownish above, and whitish below, with little edging of the wings and tail. Young and fresh fall specimens are more clearly olivaceous above and yellowish below, shaded with gray across the breast; young with wing-bars tinged with buff or tawny — all quite as usual in *Empidonax*. Very small: Length 4.50; wing 2.10; tail 1.80; bill scarcely 0.30, its depth at nostrils 0.11–0.13; tarsus 0.55; whole foot scarcely 1.00. A curious little Flycatcher of Mexico and Central Am., discovered in the Lower Rio Grande valley at Lomita, Texas, by G. B. Sennett, Apr. 24, 1879. Nest and eggs unknown.

O. i. ridg'wayi. (To R. Ridgway.) RIDGWAY'S FLYCATCHER. ARIZONA BEARDLESS FLYCATCHER. Like the last; bill more robust; coloration darker and ashier; pale ash below, with scarcely any yellowish. ♂, length 4.60; extent 7.20; wing 2.25; tail 2.00; tarsus 0.55; culmen about 0.40; depth of bill 0.15. ♀ somewhat smaller. Southern Arizona and southward to Puebla and Jalisco, Mexico. Discovered at Tucson, Ariz., by F. Stephens, Apr. 20, 1881; young just from the nest May 28; but nest and eggs still unknown. This bird is said to have a sort of "song," besides the usual call note. (Omitted from 2d–4th eds. of the Key. BREWST. Bull. Nutt. Orn. Club, Oct. 1882, p. 208; RIDGW. Man. 1887, p. 346; A. O. U. Lists, 1886 and 1895, No. 472 *a*.)

Family COTINGIDÆ: Cotingas.

An extensive family of tropical and subtropical American clamatorial passerine birds, lately formally added to the North American fauna, in which the *Clamatores* had long been supposed to be represented only by *Tyrannidæ*. *Cotingidæ*, though related to *Tyrannidæ*, may be distinguished by the pycnaspidean instead of exaspidean tarsi (see p. 509), and so far as the two following genera are concerned at least, by the extensive cohesion of inner and middle toes, and especially by shortness of the 2d primary in ♂, together with slight hooking of bill.

"The Cotingidæ are one of the great fruit-eating families of tropical America, and amongst the passerine birds addicted to this kind of diet are the most numerous and important after the *Tanagridæ*. In plumage, structure, and size they are much varied. Nothing can be more brilliant in colour than the typical Cotingas and some allied forms, while the *Lipaugi* and others are of uniformly dull plumage in both sexes. As regards structure, the second aborted primary of the *Tityrinæ*, the feet and crest of *Rupicola*, and the wattles of *Chasmorhynchus* and *Cephalopterus* show such extraordinary excesses of development as are almost unequalled in the Passerine series. . . . Like the *Tyrannidæ* the *Cotingidæ* are dentirostral Oligophones, and have ten well-formed primaries instead of nine, or nine and a shortened outer primary, as is the case with the dentirostral Oscines. They number about 110 species." (SCLATER.) This authority divides the family into six subfamilies: *Tityrinæ*, *Lipauginæ*, *Attilinæ*, *Rupicolinæ*, *Cotinginæ*, and *Gymnoderinæ*. The following genera belong to the

Subfamily TITYRINÆ: Tityrines.

Characterized by the abnormal shortness of the second primary, typical pycnaspidean tarsi, and usually stout, Shrike-like bill. The plumage is not brilliant, and the females differ from the males decidedly.

PLATYPSAR'IS. (Gr. πλατύς, *platus*, broad; ψάρ, *psar*, a starling. BONAPARTE, 1854; SCLATER, P. Z. S. 1857, p. 72. Type *Pachyrhamphus latirostris* BONAP.) BECARDS. Nostrils hidden by bristly feathers; hook of bill very slight, and bill not much flattened; rictal bristles long; head somewhat crested; tail rounded; tarsus with large scutella on the inner side. Sexes dissimilar. Two species occur on and near the Mexican border of the U. S.

P. aglai′æ. (Gr. 'Αγλαία, *Aglaïa*, one of the Three Graces.) ROSE-THROATED BECARD. Adult ♂: Above, slate-gray; crown glossy black; below, ashy-gray, with a rosy patch on throat. ♀ above dark rusty brown, becoming slaty on crown. Length 6.60; wing 3.50; tail 2.75; bill 0.65. Eastern Mexico, north to valley of the Rio Grande. (*Pachyrhynchus aglaiæ* LAFR. Rev. Zoöl. 1839, p. 98; *Pachyrhamphus aglaiæ* BAIRD, Birds North America, 1858, p. 164, and Mex. Bound. Survey, 1859, ii, pt. ii, pl. ix, fig. 1; *Platypsaris aglaiæ* SUMICHRAST, Mem. Bost. Soc. Nat. Hist. i, 1869, p. 558; RIDGW. Man. 1887, p. 324; COUES, Key, 4th ed. 1890, p. 902; not admitted in the A. O. U. List, not having as yet been actually taken over our border.)

P. albiven'tris. (Lat. *albus*, white; *venter*, the belly.) WHITE-BELLIED BECARD. XANTUS' BECARD. Adult ♂: Resembling the preceding, but lighter and more ashy-gray above; crown slaty; under parts pale grayish, whitening on belly. ♀ correspondingly paler than that of *P. aglaiæ*. Western Mexico, north into southern Arizona; Huachuca mountains. (*Hadrostomus albiventris* LAWRENCE, Ann. Lyc. Nat. Hist. viii, 1867, p. 475; *Platypsaris albiventris* RIDGW. Man. 1887, p. 325; COUES, Key, 4th ed. 1890, p. 902; A. O. U. List, 2d ed. 1895, p. 179, No. 441. 1.)

PACHYRHAM'PHUS. (Gr. παχύς, *pachus*, thick; ράμφος, *hramphos*, beak. G. R. GRAY, List. Gen. B. 1838, p. 41.) BILLED BECARDS. Resembling the preceding; bill more flattened, with shorter rictal bristles; tail graduated about ½ an inch; tarsus naked on inner side. Sexes very unlike. One species found near the Mexican border of the U. S.

P. ma'jor. (Lat. *major*, greater.) GREATER BECARD. Adult ♂: Above, ashy-gray, becoming glossy black on the back and crown, and white on scapulars; below, pale ash, whitening on throat, belly, and crissum; wings black, with white edging or tipping of coverts and some inner secondaries; tail black, with white tips of the feathers. ♀ mostly chestnut brown, paler below, black on crown and ends of tail-feathers. Length 6.50; wing 3.25; tail 2.65; bill 0.60. Eastern Mexico, north to valley of the Lower Rio Grande; introduced to our fauna by Baird in 1858 under the name of *Bathmidurus major*, and figured in Report of the Mexican Boundary Survey, 1859, pl. ix, fig. 2, but like *Platypsaris aglaiæ* lost sight of for some years, and not yet recognized in the A. O. U. List: see SCL. P. Z. S. 1857, p. 78; RIDGW. Man. 1887, p. 326; COUES, Key, 4th ed. 1890, p. 902.

NATURAL SCIENCES IN AMERICA

An Arno Press Collection

Allen, J[oel] A[saph]. **The American Bisons,** Living and Extinct. 1876

Allen, Joel Asaph. **History of the North American Pinnipeds:** A Monograph of the Walruses, Sea-Lions, Sea-Bears and Seals of North America. 1880

American Natural History Studies: The Bairdian Period. 1974

American Ornithological Bibliography. 1974

Anker, Jean. **Bird Books and Bird Art.** 1938

Audubon, John James and John Bachman. **The Quadrupeds of North America.** Three vols. 1854

Baird, Spencer F[ullerton]. **Mammals of North America.** 1859

Baird, S[pencer] F[ullerton], T[homas] M. Brewer and R[obert] Ridgway. **A History of North American Birds:** Land Birds. Three vols., 1874

Baird, Spencer F[ullerton], John Cassin and George N. Lawrence. **The Birds of North America.** 1860. Two vols. in one.

Baird, S[pencer] F[ullerton], T[homas] M. Brewer, and R[obert] Ridgway. **The Water Birds of North America.** 1884. Two vols. in one.

Barton, Benjamin Smith. **Notes on the Animals of North America.** Edited, with an Introduction by Keir B. Sterling. 1792

Bendire, Charles [Emil]. **Life Histories of North American Birds** With Special Reference to Their Breeding Habits and Eggs. 1892/1895. Two vols. in one.

Bonaparte, Charles Lucian [Jules Laurent]. **American Ornithology:** Or The Natural History of Birds Inhabiting the United States, Not Given by Wilson. 1825/1828/1833. Four vols. in one.

Cameron, Jenks. **The Bureau of Biological Survey:** Its History, Activities, and Organization. 1929

Caton, John Dean. **The Antelope and Deer of America:** A Comprehensive Scientific Treatise Upon the Natural History, Including the Characteristics, Habits, Affinities, and Capacity for Domestication of the Antilocapra and Cervidae of North America. 1877

Contributions to American Systematics. 1974

Contributions to the Bibliographical Literature of American Mammals. 1974

Contributions to the History of American Natural History. 1974

Contributions to the History of American Ornithology. 1974

Cooper, J[ames] G[raham]. **Ornithology. Volume I, Land Birds.** 1870

Cope, E[dward] D[rinker]. **The Origin of the Fittest:** Essays on Evolution and **The Primary Factors of Organic Evolution.** 1887/1896. Two vols. in one.

Coues, Elliott. **Birds of the Colorado Valley.** 1878

Coues, Elliott. **Birds of the Northwest.** 1874

Coues, Elliott. **Key To North American Birds.** Two vols. 1903

Early Nineteenth-Century Studies and Surveys. 1974

Emmons, Ebenezer. **American Geology:** Containing a Statement of the Principles of the Science. 1855. Two vols. in one.

Fauna Americana. 1825-1826

Fisher, A[lbert] K[enrick]. **The Hawks and Owls of the United States in Their Relation to Agriculture.** 1893

Godman, John D. **American Natural History:** Part I — Mastology and **Rambles of a Naturalist.** 1826-28/1833. Three vols. in one.

Gregory, William King. **Evolution Emerging:** A Survey of Changing Patterns from Primeval Life to Man. Two vols. 1951

Hay, Oliver Perry. **Bibliography and Catalogue of the Fossil Vertebrata of North America.** 1902

Heilprin, Angelo. **The Geographical and Geological Distribution of Animals.** 1887

Hitchcock, Edward. **A Report on the Sandstone of the Connecticut Valley,** Especially Its Fossil Footmarks. 1858

Hubbs, Carl L., editor. **Zoogeography.** 1958

[Kessel, Edward L., editor]. **A Century of Progress in the Natural Sciences: 1853-1953.** 1955

Leidy, Joseph. **The Extinct Mammalian Fauna of Dakota and Nebraska,** Including an Account of Some Allied Forms from Other Localities, Together with a Synopsis of the Mammalian Remains of North America. 1869

Lyon, Marcus Ward, Jr. **Mammals of Indiana.** 1936

Matthew, W[illiam] D[iller]. **Climate and Evolution.** 1915

Mayr, Ernst, editor. **The Species Problem.** 1957

Mearns, Edgar Alexander. **Mammals of the Mexican Boundary of the United States.** Part I: Families Didelphiidae to Muridae. 1907

Merriam, Clinton Hart. **The Mammals of the Adirondack Region,** Northeastern New York. 1884

Nuttall, Thomas. **A Manual of the Ornithology of the United States and of Canada.** Two vols. 1832-1834

Nuttall Ornithological Club. **Bulletin of the Nuttall Ornithological Club:** A Quarterly Journal of Ornithology. 1876-1883. Eight vols. in three.

[Pennant, Thomas]. **Arctic Zoology.** 1784-1787. Two vols. in one.

Richardson, John. **Fauna Boreali-Americana;** Or the Zoology of the Northern Parts of British America, Containing Descriptions of the Objects of Natural History Collected on the Late Northern Land Expeditions Under Command of Captain Sir John Franklin, R. N. Part I: Quadrupeds. 1829

Richardson, John and William Swainson. **Fauna Boreali-Americana:** Or the Zoology of the Northern Parts of British America, Containing Descriptions of the Objects of Natural History Collected by the Late Northern Land Expeditions Under Command of Captain Sir John Franklin, R. N. Part II: The Birds. 1831

Ridgway, Robert. **Ornithology.** 1877

Selected Works By Eighteenth-Century Naturalists and Travellers. 1974

Selected Works in Nineteenth-Century North American Paleontology. 1974

Selected Works of Clinton Hart Merriam. 1974

Selected Works of Joel Asaph Allen. 1974

Selections From the Literature of American Biogeography. 1974

Seton, Ernest Thompson. **Life-Histories of Northern Animals: An Account of the Mammals of Manitoba.** Two vols. 1909

Sterling, Keir Brooks. **Last of the Naturalists:** The Career of C. Hart Merriam. 1974

Vieillot, L. P. **Histoire Naturelle Des Oiseaux de L'Amerique Septentrionale,** Contenant Un Grand Nombre D'Especes Decrites ou Figurees Pour La Premiere Fois. 1807. Two vols. in one.

Wilson, Scott B., assisted by A. H. Evans. **Aves Hawaiienses:** The Birds of the Sandwich Islands. 1890-99

Wood, Casey A., editor. **An Introduction to the Literature of Vertebrate Zoology.** 1931

Zimmer, John Todd. **Catalogue of the Edward E. Ayer Ornithological Library.** 1926